Methods in Enzymology

Volume 172
BIOMEMBRANES
Part S
Transport: Membrane Isolation and Characterization

METHODS IN ENZYMOLOGY

EDITORS-IN-CHIEF

John N. Abelson   Melvin I. Simon

DIVISION OF BIOLOGY
CALIFORNIA INSTITUTE OF TECHNOLOGY
PASADENA, CALIFORNIA

FOUNDING EDITORS

Sidney P. Colowick and Nathan O. Kaplan

Methods in Enzymology

Volume 172

Biomembranes

Part S
Transport: Membrane Isolation and Characterization

EDITED BY

Sidney Fleischer
Becca Fleischer

DEPARTMENT OF MOLECULAR BIOLOGY
VANDERBILT UNIVERSITY
NASHVILLE, TENNESSEE

Editorial Advisory Board

Yasuo Kagawa
Ronald Kaback
Martin Klingenberg
Robert L. Post

George Sachs
Antonio Scarpa
Widmar Tanner
Karl Ullrich

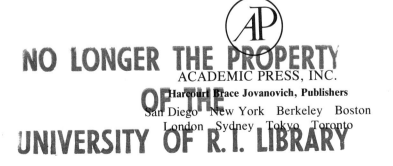

ACADEMIC PRESS, INC.
Harcourt Brace Jovanovich, Publishers
San Diego   New York   Berkeley   Boston
London   Sydney   Tokyo   Toronto

COPYRIGHT © 1989 BY ACADEMIC PRESS, INC.
All Rights Reserved.
No part of this publication may be reproduced or transmitted in any form or by any means, electronic or mechanical, including photocopy, recording, or any information storage and retrieval system, without permission in writing from the publisher.

ACADEMIC PRESS, INC.
San Diego, California 92101

United Kingdom Edition published by
ACADEMIC PRESS LIMITED
24-28 Oval Road, London NW1 7DX

LIBRARY OF CONGRESS CATALOG CARD NUMBER: 54-9110

ISBN 0-12-182073-4 (alk. paper)

PRINTED IN THE UNITED STATES OF AMERICA
89 90 91 92   9 8 7 6 5 4 3 2 1

Table of Contents

CONTRIBUTORS TO VOLUME 172 ix

PREFACE xiii

VOLUMES IN SERIES xv

Section I. Membrane Separation

1. Membrane Isolation: Strategy, Techniques, Markers	E. KINNE-SAFFRAN AND R. K. H. KINNE	3
2. Isolation of Plasma Membranes from Polar Cells and Tissues: Apical/Basolateral Separation, Purity, Function	AUSTIN K. MIRCHEFF	18
3. Transport-Specific Fractionation for Purification of ATP-Dependent Ca^{2+} Pumps	STANLEY M. GOLDIN AND STEVEN C. KING	34
4. Regulation of Membrane Transport by Endocytotic Removal and Exocytotic Insertion of Transporters	QAIS AL-AWQATI	49

Section II. Transport Methods in Cells and Vesicles

5. Proton Electrochemical Potential Gradient in Vesicles, Organelles, and Prokaryotic Cells	HAGAI ROTTENBERG	63
6. Measurement of pH of Intracellular Compartments in Living Cells by Fluorescent Dyes	JANET VAN ADELSBERG, JONATHAN BARASCH, AND QAIS AL-AWQATI	85
7. Use of Carbocyanine Dyes to Assay Membrane Potential of Mouse Ascites Tumor Cells	A. ALAN EDDY	95
8. Optical Measurement of Membrane Potential in Cells, Organelles, and Vesicles	J. C. FREEDMAN AND T. S. NOVAK	102
9. Fluorescence Methods for Continuous Monitoring of Transport in Cells and Vesicles	OFER EIDELMAN AND Z. IOAV CABANTCHIK	122
10. Preparation and Use of Micro- and Macroelectrodes for Measurement of Transmembrane Potentials and Ion Activities	DANIEL AMMANN AND PICO CARONI	136

11. Ion Channel-Mediated Fluxes in Membrane Vesicles: Selective Amplification of Isotope Uptake by Electrical Diffusion Potentials	HAIM GARTY AND STEVEN J. D. KARLISH	155
12. Use of Calcium-Regulated Photoproteins as Intracellular Ca^{2+} Indicators	JOHN R. BLINKS	164
13. Electron Probe X-ray Microanalysis of Ca^{2+}, Mg^{2+}, and Other Ions in Rapidly Frozen Cells	A. V. SOMLYO, H. SHUMAN, AND A. P. SOMLYO	203
14. Measurement of Cytosolic Free Ca^{2+} with Quin2	ROGER TSIEN AND TULLIO POZZAN	230
15. Analyzing Transport Kinetics with Desk-Top Hybrid Computers	HAROLD G. HEMPLING	262
16. Synthesis and Properties of Caged Nucleotides	JEFFERY W. WALKER, GORDON P. REID, AND DAVID R. TRENTHAM	288
17. Measurement of Ion Fluxes in Membrane Vesicles Using Rapid-Reaction Methods	HERBERT S. CHASE, JR., MARK D. GELERNT, AND MARC C. DEBELL	301
18. Tracer Studies with Isolated Membrane Vesicles	ULRICH HOPFER	313
19. Electron Paramagnetic Resonance Methods for Measuring pH Gradients, Transmembrane Potentials, and Membrane Dynamics	DAVID S. CAFISO	331
20. Transport Studies with Renal Proximal Tubular and Small Intestinal Brush Border and Basolateral Membrane Vesicles: Vesicle Heterogeneity, Coexistence of Transport Systems	HEINI MURER, PIOTR GMAJ, BRUNO STIEGER, AND BRUNO HAGENBUCH	346
21. Dynamic Laser Light Scattering to Determine Size Distributions of Vesicles	HORST RUF, YANNIS GEORGALIS, AND ERNST GRELL	364

Section III. Membrane Analysis and Characterization

22. Sensitive Protein Assay in Presence of High Levels of Lipid	RONALD S. KAPLAN AND PETER L. PEDERSEN	393
23. Orienting Synthetic and Native Biological Membranes for Time-Averaged and Time-Resolved Structure Determinations	L. G. HERBETTE AND J. K. BLASIE	399
24. Radiation Inactivation of Membrane Components and Molecular Mass Determination by Target Analysis	E. S. KEMPNER AND SIDNEY FLEISCHER	410

25. Prediction of Bilayer Spanning Domains of Hydrophobic and Amphipathic Membrane Proteins: Application to Cytochrome b and Colicin Families	J. W. Shiver, A. A. Peterson, W. R. Widger, P. N. Furbacher, and W. A. Cramer	439
26. Order and Viscosity of Membranes: Analysis by Time-Resolved Fluorescence Depolarization	Maarten P. Heyn	462
27. Rotational and Translational Diffusion in Membranes Measured by Fluorescence and Phosphorescence Methods	Thomas M. Jovin and Winchil L. C. Vaz	471
28. Membrane Protein Molecular Weight Determined by Low-Angle Laser Light-Scattering Photometry Coupled with High-Performance Gel Chromatography	Yutaro Hayashi, Hideo Matsui, and Toshio Takagi	514
29. Critical Micellar Concentrations of Detergents	M. Zulauf, U. Fürstenberger, M. Grabo, P. Jäggi, M. Regenass, and J. P. Rosenbusch	528
30. High-Performance Liquid Chromatography of Membrane Lipids: Glycosphingolipids and Phospholipids	R. H. McCluer, M. D. Ullman, and F. B. Jungalwala	538
31. Circular Dichroism for Determining Secondary Structure and State of Aggregation of Membrane Proteins	Maarten P. Heyn	575
32. Cross-Linking Techniques	H. G. Bäumert and H. Fasold	584
33. Membrane-Impermeant Cross-Linking Reagents	James V. Staros and P. S. R. Anjaneyulu	609
34. Photochemical Labeling of Apolar Phase of Membranes	Josef Brunner	628
35. Electrophoretic Transfer of High-Molecular-Weight Proteins for Immunostaining	Kuan Wang, Bradford O. Fanger, Cheryl A. Guyer, and James V. Staros	687
36. Size and Shape of Membrane Protein–Detergent Complexes: Hydrodynamic Studies	Steven Clarke and Murray D. Smigel	696

Author Index 711

Subject Index 743

Contributors to Volume 172

Article numbers are in parentheses following the names of contributors.
Affiliations listed are current.

QAIS AL-AWQATI (4, 6), *Departments of Medicine and Physiology, College of Physicians and Surgeons, Columbia University, New York, New York 10032*

DANIEL AMMANN (10), *Department of Organic Chemistry, Swiss Federal Institute of Technology (ETH), CH-8092 Zurich, Switzerland*

P. S. R. ANJANEYULU (33), *Department of Biochemistry, School of Medicine, Vanderbilt University, Nashville, Tennessee 37232*

JONATHAN BARASCH (6), *Department of Medicine, College of Physicians and Surgeons, Columbia University, New York, New York 10032*

H. G. BÄUMERT (32), *Institut für Biochemie, Johann Wolfgang Goethe-Universität, D-6000 Frankfurt am Main 70, Federal Republic of Germany*

J. K. BLASIE (23), *Department of Chemistry, University of Pennsylvania, Philadelphia, Pennsylvania 19104*

JOHN R. BLINKS (12), *Department of Pharmacology, Mayo Foundation, Rochester, Minnesota 55905*

JOSEF BRUNNER (34), *Laboratorium für Biochemie, Eidgenössische Technische Hochschule Zürich, CH-8092 Zurich, Switzerland*

Z. IOAV CABANTCHIK (9), *Department of Biological Chemistry, The Hebrew University of Jerusalem, Jerusalem 91904, Israel*

DAVID S. CAFISO (19), *Department of Chemistry, University of Virginia, Charlottesville, Virginia 22901*

PICO CARONI (10), *Brain Research Institute, University of Zurich, CH-8029 Zurich, Switzerland*

HERBERT S. CHASE, JR. (17), *Department of Medicine, Columbia University, College of Physicians and Surgeons, New York, New York 10032*

STEVEN CLARKE (36), *Department of Chemistry and Biochemistry, and the Molecular Biology Institute, University of California, Los Angeles, California 90024*

W. A. CRAMER (25), *Department of Biological Sciences, Purdue University, West Lafayette, Indiana 47907*

MARC C. DEBELL (17), *Tufts University, Medford, Massachusetts 02155*

A. ALAN EDDY (7), *Department of Biochemistry and Applied Molecular Biology, University of Manchester Institute of Science and Technology, Manchester M60 1QD, England*

OFER EIDELMAN (9), *Department of Biological Chemistry, Hebrew University of Jerusalem, Jerusalem 91904, Israel*

BRADFORD O. FANGER (35), *Merrell Dow Research Institute, Cincinnati, Ohio 45215*

H. FASOLD (32), *Institut für Biochemie, Johann Wolfgang Goethe-Universität, D-6000 Frankfurt am Main 70, Federal Republic of Germany*

SIDNEY FLEISCHER (24), *Department of Molecular Biology, Vanderbilt University, Nashville, Tennessee 37235*

J. C. FREEDMAN (8), *Department of Physiology, SUNY Health Science Center at Syracuse, Syracuse, New York 13210*

P. N. FURBACHER (25), *Department of Biological Sciences, Purdue University, West Lafayette, Indiana 47907*

U. FÜRSTENBERGER (29), *Biozentrum, University of Basel, CH-4056 Basel, Switzerland*

HAIM GARTY (11), *Department of Membrane Research, The Weizmann Institute of Science, Rehovot 76100, Israel*

MARK D. GELERNT (17), *Department of Medicine, St. Lukes—Roosevelt Hospital, New York, New York 10025*

YANNIS GEORGALIS (21), *Institut für Kristallographie, Fachbereich Chemie, Freie Universität Berlin, D-1000 Berlin 33, Federal Republic of Germany*

PIOTR GMAJ (20), *Department of Physiology, University of Zurich, CH-8057 Zurich, Switzerland*

STANLEY M. GOLDIN (3), *Departments of Biological Chemistry and Molecular Pharmacology, Harvard Medical School, Boston, Massachusetts 02115*

M. GRABO (29), *Hoffmann-La Roche, CH-4002 Basel, Switzerland*

ERNST GRELL (21), *Max-Planck-Institut für Biophysik, D-6000 Frankfurt, Federal Republic of Germany*

CHERYL A. GUYER (35), *Department of Pharmacology, School of Medicine, Vanderbilt University, Nashville, Tennessee 37232*

BRUNO HAGENBUCH (20), *Department of Clinical Pharmacology, University Hospital of Zurich, CH-8032 Zurich, Switzerland*

YUTARO HAYASHI (28), *Department of Biochemistry, Kyorin University School of Medicine, Mitaka, Tokyo 181, Japan*

HAROLD G. HEMPLING (15), *Department of Physiology, Medical University of South Carolina, Charleston, South Carolina 29425*

L. G. HERBETTE (23), *Department of Radiology, University of Connecticut Health Center, Farmington, Connecticut 06032*

MAARTEN P. HEYN (26, 31), *Department of Physics, Freie Universität Berlin, D-1000 Berlin 38, Federal Republic of Germany*

ULRICH HOPFER (18), *Departments of Physiology and Biophysics, Case Western Reserve University, Cleveland, Ohio 44106*

P. JÄGGI (29), *Biozentrum, University of Basel, CH-4056 Basel, Switzerland*

THOMAS M. JOVIN (27), *Molecular Biology Department, Max-Planck-Institute for Biophysical Chemistry, D-3400 Göttingen, Federal Republic of Germany*

F. B. JUNGALWALA (30), *Department of Biochemistry, Eunice Kennedy Shriver Center, Waltham, Massachusetts 02254*

RONALD S. KAPLAN (22), *Department of Pharmacology, College of Medicine, University of South Alabama, Mobile, Alabama 36688*

STEVEN J. D. KARLISH (11), *Department of Biochemistry, The Weizmann Institute of Science, Rehovot 76100, Israel*

E. S. KEMPNER (24), *Laboratory of Physical Biology, National Institute of Arthritis and Musculoskeletal and Skin Diseases, National Institutes of Health, Bethesda, Maryland 20892*

STEVEN C. KING (3), *Department of Biological Chemistry, and Molecular Pharmacology, Harvard Medical School, Boston, Massachusetts 02115*

R. K. H. KINNE (1), *Max-Planck-Institut für Systemphysiologie, 4600 Dortmund, Federal Republic of Germany*

E. KINNE-SAFFRAN (1), *Max-Planck-Institut für Systemphysiologie, 4600 Dortmund, Federal Republic of Germany*

HIDEO MATSUI (28), *Department of Biochemistry, Kyorin University School of Medicine, Mitaka, Tokyo 181, Japan*

R. H. MCCLUER (30), *Department of Biochemistry, Eunice Kennedy Shriver Center, Waltham, Massachusetts 02254*

AUSTIN K. MIRCHEFF (2), *Department of Physiology and Biophysics, University of Southern California, School of Medicine, Los Angeles, California 90033*

HEINI MURER (20), *Department of Physiology, University of Zurich, CH-8057 Zurich, Switzerland*

T. S. NOVAK (8), *Department of Physiology, SUNY Health Science Center at Syracuse, Syracuse, New York 13210*

PETER L. PEDERSEN (22), *Department of Biological Chemistry, The Johns Hopkins University School of Medicine, Baltimore, Maryland 21205*

A. A. PETERSON (25), *Diagnostics R & D Quidel, San Diego, California 92121*

TULLIO POZZAN (14), *Institute of General Pathology, University of Padua, 35100 Padua, Italy*

M. REGENASS (29), *Biozentrum, Friedrich Meischer-Institut, CH-4002 Basel, Switzerland*

GORDON P. REID (16), *Physical Biochemistry Division, National Institute for Medical Research, London NW7 1AA, England*

J. P. ROSENBUSCH (29), *Biozentrum, University of Basel, CH-4056 Basel, Switzerland*

HAGAI ROTTENBERG (5), *Department of Pathology, Hahnemann University, Philadelphia, Pennsylvania 19102*

HORST RUF (21), *Max-Planck-Institut für Biophysik, D-6000 Frankfurt, Federal Republic of Germany*

J. W. SHIVER (25), *Experimental Immunology Branch, National Cancer Institute, National Institutes of Health, Bethesda, Maryland 20892*

H. SHUMAN (13), *Pennsylvania Muscle Institute, University of Pennsylvania School of Medicine, Philadelphia, Pennsylvania 19104*

MURRAY D. SMIGEL (36), *Convex Computer Corporation, Richardson, Texas 75083*

A. P. SOMLYO (13), *Pennsylvania Muscle Institute, University of Pennsylvania School of Medicine, Philadelphia, Pennsylvania 19104*

A. V. SOMLYO (13), *Pennsylvania Muscle Institute, University of Pennsylvania School of Medicine, Philadelphia, Pennsylvania 19104*

JAMES V. STAROS (33, 35), *Department of Biochemistry, School of Medicine, Vanderbilt University, Nashville, Tennessee 37232*

BRUNO STIEGER (20), *Department of Clinical Pharmacology, University Hospital of Zurich, CH-8032 Zurich, Switzerland*

TOSHIO TAKAGI (28), *Institute for Protein Research, Osaka University, Suita, Osaka 565, Japan*

DAVID R. TRENTHAM (16), *Physical Biochemistry Division, National Institute for Medical Research, London NW7 1AA, England*

ROGER TSIEN (14), *Department of Physiology-Anatomy, University of California, Berkeley, California 94720*

M. D. ULLMAN (30), *Research Service, ENRM Veterans Hospital, Bedford, Massachusetts 01730*

JANET VAN ADELSBERG (6), *Department of Medicine, Columbia University, College of Physicians and Surgeons, New York, New York 10032*

WINCHIL L. C. VAZ (27), *Molecular Biology Department, Max-Planck-Institute for Biophysical Chemistry, D-3400 Göttingen, Federal Republic of Germany*

JEFFERY W. WALKER (16), *Department of Physiology, University of Wisconsin, Madison, Wisconsin 53706*

KUAN WANG (35), *Clayton Foundation Biochemical Institute, Department of Chemistry, University of Texas, Austin, Texas 78712*

W. R. WIDGER (25), *Department of Biochemical and Biophysical Sciences, University of Houston, Houston, Texas 77004*

M. ZULAUF (29), *Hoffmann-La Roche, CH-4002 Basel, Switzerland*

Preface

Biological transport is part of the Biomembranes series of *Methods in Enzymology*. It is a continuation of methodology concerned with membrane function. This is a particularly good time to cover the topic of biological membrane transport because there is now a strong conceptual basis for its understanding. The field of transport has been subdivided into five topics.

1. Transport in Bacteria, Mitochondria, and Chloroplasts
2. ATP-Driven Pumps and Related Transport
3. General Methodology of Cellular and Subcellular Transport
4. Cellular and Subcellular Transport: Eukaryotic (Nonepithelial) Cells
5. Cellular and Subcellular Transport: Epithelial Cells

Topic 1, covered in Volumes 125 and 126, initiated the series. Topic 2 is covered in Volumes 156 and 157. Topic 3 is covered in Volumes 171 and 172. The remaining two topics will appear in subsequent volumes of the Biomembranes series.

Topic 3 provides general methodology and concepts. Its coverage includes theory and general methodology for the study of biomembranes. It is divided into two parts: Volume 171 (Part R) which covers Transport Theory: Cells and Model Membranes and this volume (Part S) which covers Transport: Membrane Isolation and Characterization.

We are fortunate to have the good counsel of our Advisory Board. Their input insures the quality of these volumes. The same Advisory Board has served for the complete transport series. Valuable input on the outlines of the five topics was also provided by Qais Al-Awqati, Ernesto Carafoli, Halvor Christensen, Isadore Edelman, Joseph Hoffman, Phil Knauf, and Hermann Passow.

The names of our advisory board members were inadvertently omitted in Volumes 125 and 126. When we noted the omission, it was too late to rectify the problem. For volumes 125 and 126, we are also pleased to acknowledge the advice of Angelo Azzi, Youssef Hatefi, Dieter Oesterhelt, and Peter Pedersen.

The enthusiasm and cooperation of the participants have enriched and made these volumes possible. The friendly cooperation of the staff of Academic Press is gratefully acknowledged.

These volumes are dedicated to Professor Sidney Colowick, a dear friend and colleague, who died in 1985. We shall miss his wise counsel, encouragement, and friendship.

SIDNEY FLEISCHER
BECCA FLEISCHER

METHODS IN ENZYMOLOGY

VOLUME I. Preparation and Assay of Enzymes
Edited by SIDNEY P. COLOWICK AND NATHAN O. KAPLAN

VOLUME II. Preparation and Assay of Enzymes
Edited by SIDNEY P. COLOWICK AND NATHAN O. KAPLAN

VOLUME III. Preparation and Assay of Substrates
Edited by SIDNEY P. COLOWICK AND NATHAN O. KAPLAN

VOLUME IV. Special Techniques for the Enzymologist
Edited by SIDNEY P. COLOWICK AND NATHAN O. KAPLAN

VOLUME V. Preparation and Assay of Enzymes
Edited by SIDNEY P. COLOWICK AND NATHAN O. KAPLAN

VOLUME VI. Preparation and Assay of Enzymes (*Continued*)
Preparation and Assay of Substrates
Special Techniques
Edited by SIDNEY P. COLOWICK AND NATHAN O. KAPLAN

VOLUME VII. Cumulative Subject Index
Edited by SIDNEY P. COLOWICK AND NATHAN O. KAPLAN

VOLUME VIII. Complex Carbohydrates
Edited by ELIZABETH F. NEUFELD AND VICTOR GINSBURG

VOLUME IX. Carbohydrate Metabolism
Edited by WILLIS A. WOOD

VOLUME X. Oxidation and Phosphorylation
Edited by RONALD W. ESTABROOK AND MAYNARD E. PULLMAN

VOLUME XI. Enzyme Structure
Edited by C. H. W. HIRS

VOLUME XII. Nucleic Acids (Parts A and B)
Edited by LAWRENCE GROSSMAN AND KIVIE MOLDAVE

VOLUME XIII. Citric Acid Cycle
Edited by J. M. LOWENSTEIN

VOLUME XIV. Lipids
Edited by J. M. LOWENSTEIN

VOLUME XV. Steroids and Terpenoids
Edited by RAYMOND B. CLAYTON

VOLUME XVI. Fast Reactions
Edited by KENNETH KUSTIN

VOLUME XVII. Metabolism of Amino Acids and Amines (Parts A and B)
Edited by HERBERT TABOR AND CELIA WHITE TABOR

VOLUME XVIII. Vitamins and Coenzymes (Parts A, B, and C)
Edited by DONALD B. MCCORMICK AND LEMUEL D. WRIGHT

VOLUME XIX. Proteolytic Enzymes
Edited by GERTRUDE E. PERLMANN AND LASZLO LORAND

VOLUME XX. Nucleic Acids and Protein Synthesis (Part C)
Edited by KIVIE MOLDAVE AND LAWRENCE GROSSMAN

VOLUME XXI. Nucleic Acids (Part D)
Edited by LAWRENCE GROSSMAN AND KIVIE MOLDAVE

VOLUME XXII. Enzyme Purification and Related Techniques
Edited by WILLIAM B. JAKOBY

VOLUME XXIII. Photosynthesis (Part A)
Edited by ANTHONY SAN PIETRO

VOLUME XXIV. Photosynthesis and Nitrogen Fixation (Part B)
Edited by ANTHONY SAN PIETRO

VOLUME XXV. Enzyme Structure (Part B)
Edited by C. H. W. HIRS AND SERGE N. TIMASHEFF

VOLUME XXVI. Enzyme Structure (Part C)
Edited by C. H. W. HIRS AND SERGE N. TIMASHEFF

VOLUME XXVII. Enzyme Structure (Part D)
Edited by C. H. W. HIRS AND SERGE N. TIMASHEFF

VOLUME XXVIII. Complex Carbohydrates (Part B)
Edited by VICTOR GINSBURG

VOLUME XXIX. Nucleic Acids and Protein Synthesis (Part E)
Edited by LAWRENCE GROSSMAN AND KIVIE MOLDAVE

VOLUME XXX. Nucleic Acids and Protein Synthesis (Part F)
Edited by KIVIE MOLDAVE AND LAWRENCE GROSSMAN

VOLUME XXXI. Biomembranes (Part A)
Edited by SIDNEY FLEISCHER AND LESTER PACKER

VOLUME XXXII. Biomembranes (Part B)
Edited by SIDNEY FLEISCHER AND LESTER PACKER

VOLUME XXXIII. Cumulative Subject Index Volumes I–XXX
Edited by MARTHA G. DENNIS AND EDWARD A. DENNIS

VOLUME XXXIV. Affinity Techniques (Enzyme Purification: Part B)
Edited by WILLIAM B. JAKOBY AND MEIR WILCHEK

VOLUME XXXV. Lipids (Part B)
Edited by JOHN M. LOWENSTEIN

VOLUME XXXVI. Hormone Action (Part A: Steroid Hormones)
Edited by BERT W. O'MALLEY AND JOEL G. HARDMAN

VOLUME XXXVII. Hormone Action (Part B: Peptide Hormones)
Edited by BERT W. O'MALLEY AND JOEL G. HARDMAN

VOLUME XXXVIII. Hormone Action (Part C: Cyclic Nucleotides)
Edited by JOEL G. HARDMAN AND BERT W. O'MALLEY

VOLUME XXXIX. Hormone Action (Part D: Isolated Cells, Tissues, and Organ Systems)
Edited by JOEL G. HARDMAN AND BERT W. O'MALLEY

VOLUME XL. Hormone Action (Part E: Nuclear Structure and Function)
Edited by BERT W. O'MALLEY AND JOEL G. HARDMAN

VOLUME XLI. Carbohydrate Metabolism (Part B)
Edited by W. A. WOOD

VOLUME XLII. Carbohydrate Metabolism (Part C)
Edited by W. A. WOOD

VOLUME XLIII. Antibiotics
Edited by JOHN H. HASH

VOLUME XLIV. Immobilized Enzymes
Edited by KLAUS MOSBACH

VOLUME XLV. Proteolytic Enzymes (Part B)
Edited by LASZLO LORAND

VOLUME XLVI. Affinity Labeling
Edited by WILLIAM B. JAKOBY AND MEIR WILCHEK

VOLUME XLVII. Enzyme Structure (Part E)
Edited by C. H. W. HIRS AND SERGE N. TIMASHEFF

VOLUME XLVIII. Enzyme Structure (Part F)
Edited by C. H. W. HIRS AND SERGE N. TIMASHEFF

VOLUME XLIX. Enzyme Structure (Part G)
Edited by C. H. W. HIRS AND SERGE N. TIMASHEFF

VOLUME L. Complex Carbohydrates (Part C)
Edited by VICTOR GINSBURG

VOLUME LI. Purine and Pyrimidine Nucleotide Metabolism
Edited by PATRICIA A. HOFFEE AND MARY ELLEN JONES

VOLUME LII. Biomembranes (Part C: Biological Oxidations)
Edited by SIDNEY FLEISCHER AND LESTER PACKER

VOLUME LIII. Biomembranes (Part D: Biological Oxidations)
Edited by SIDNEY FLEISCHER AND LESTER PACKER

VOLUME LIV. Biomembranes (Part E: Biological Oxidations)
Edited by SIDNEY FLEISCHER AND LESTER PACKER

VOLUME LV. Biomembranes (Part F: Bioenergetics)
Edited by SIDNEY FLEISCHER AND LESTER PACKER

VOLUME LVI. Biomembranes (Part G: Bioenergetics)
Edited by SIDNEY FLEISCHER AND LESTER PACKER

VOLUME LVII. Bioluminescence and Chemiluminescence
Edited by MARLENE A. DELUCA

VOLUME LVIII. Cell Culture
Edited by WILLIAM B. JAKOBY AND IRA PASTAN

VOLUME LIX. Nucleic Acids and Protein Synthesis (Part G)
Edited by KIVIE MOLDAVE AND LAWRENCE GROSSMAN

VOLUME LX. Nucleic Acids and Protein Synthesis (Part H)
Edited by KIVIE MOLDAVE AND LAWRENCE GROSSMAN

VOLUME 61. Enzyme Structure (Part H)
Edited by C. H. W. HIRS AND SERGE N. TIMASHEFF

VOLUME 62. Vitamins and Coenzymes (Part D)
Edited by DONALD B. MCCORMICK AND LEMUEL D. WRIGHT

VOLUME 63. Enzyme Kinetics and Mechanism (Part A: Initial Rate and Inhibitor Methods)
Edited by DANIEL L. PURICH

VOLUME 64. Enzyme Kinetics and Mechanism (Part B: Isotopic Probes and Complex Enzyme Systems)
Edited by DANIEL L. PURICH

VOLUME 65. Nucleic Acids (Part I)
Edited by LAWRENCE GROSSMAN AND KIVIE MOLDAVE

VOLUME 66. Vitamins and Coenzymes (Part E)
Edited by DONALD B. MCCORMICK AND LEMUEL D. WRIGHT

VOLUME 67. Vitamins and Coenzymes (Part F)
Edited by DONALD B. MCCORMICK AND LEMUEL D. WRIGHT

VOLUME 68. Recombinant DNA
Edited by RAY WU

VOLUME 69. Photosynthesis and Nitrogen Fixation (Part C)
Edited by ANTHONY SAN PIETRO

VOLUME 70. Immunochemical Techniques (Part A)
Edited by HELEN VAN VUNAKIS AND JOHN J. LANGONE

VOLUME 71. Lipids (Part C)
Edited by JOHN M. LOWENSTEIN

VOLUME 72. Lipids (Part D)
Edited by JOHN M. LOWENSTEIN

VOLUME 73. Immunochemical Techniques (Part B)
Edited by JOHN J. LANGONE AND HELEN VAN VUNAKIS

VOLUME 74. Immunochemical Techniques (Part C)
Edited by JOHN J. LANGONE AND HELEN VAN VUNAKIS

VOLUME 75. Cumulative Subject Index Volumes XXXI, XXXII, XXXIV–LX
Edited by EDWARD A. DENNIS AND MARTHA G. DENNIS

VOLUME 76. Hemoglobins
Edited by ERALDO ANTONINI, LUIGI ROSSI-BERNARDI, AND EMILIA CHIANCONE

VOLUME 77. Detoxication and Drug Metabolism
Edited by WILLIAM B. JAKOBY

VOLUME 78. Interferons (Part A)
Edited by SIDNEY PESTKA

VOLUME 79. Interferons (Part B)
Edited by SIDNEY PESTKA

VOLUME 80. Proteolytic Enzymes (Part C)
Edited by LASZLO LORAND

VOLUME 81. Biomembranes (Part H: Visual Pigments and Purple Membranes, I)
Edited by LESTER PACKER

VOLUME 82. Structural and Contractile Proteins (Part A: Extracellular Matrix)
Edited by LEON W. CUNNINGHAM AND DIXIE W. FREDERIKSEN

VOLUME 83. Complex Carbohydrates (Part D)
Edited by VICTOR GINSBURG

VOLUME 84. Immunochemical Techniques (Part D: Selected Immunoassays)
Edited by JOHN J. LANGONE AND HELEN VAN VUNAKIS

VOLUME 85. Structural and Contractile Proteins (Part B: The Contractile Apparatus and the Cytoskeleton)
Edited by DIXIE W. FREDERIKSEN AND LEON W. CUNNINGHAM

VOLUME 86. Prostaglandins and Arachidonate Metabolites
Edited by WILLIAM E. M. LANDS AND WILLIAM L. SMITH

VOLUME 87. Enzyme Kinetics and Mechanism (Part C: Intermediates, Stereochemistry, and Rate Studies)
Edited by DANIEL L. PURICH

VOLUME 88. Biomembranes (Part I: Visual Pigments and Purple Membranes, II)
Edited by LESTER PACKER

VOLUME 89. Carbohydrate Metabolism (Part D)
Edited by WILLIS A. WOOD

VOLUME 90. Carbohydrate Metabolism (Part E)
Edited by WILLIS A. WOOD

VOLUME 91. Enzyme Structure (Part I)
Edited by C. H. W. HIRS AND SERGE N. TIMASHEFF

VOLUME 92. Immunochemical Techniques (Part E: Monoclonal Antibodies and General Immunoassay Methods)
Edited by JOHN J. LANGONE AND HELEN VAN VUNAKIS

VOLUME 93. Immunochemical Techniques (Part F: Conventional Antibodies, Fc Receptors, and Cytotoxicity)
Edited by JOHN J. LANGONE AND HELEN VAN VUNAKIS

VOLUME 94. Polyamines
Edited by HERBERT TABOR AND CELIA WHITE TABOR

VOLUME 95. Cumulative Subject Index Volumes 61–74, 76–80
Edited by EDWARD A. DENNIS AND MARTHA G. DENNIS

VOLUME 96. Biomembranes [Part J: Membrane Biogenesis: Assembly and Targeting (General Methods; Eukaryotes)]
Edited by SIDNEY FLEISCHER AND BECCA FLEISCHER

VOLUME 97. Biomembranes [Part K: Membrane Biogenesis: Assembly and Targeting (Prokaryotes, Mitochondria, and Chloroplasts)]
Edited by SIDNEY FLEISCHER AND BECCA FLEISCHER

VOLUME 98. Biomembranes (Part L: Membrane Biogenesis: Processing and Recycling)
Edited by SIDNEY FLEISCHER AND BECCA FLEISCHER

VOLUME 99. Hormone Action (Part F: Protein Kinases)
Edited by JACKIE D. CORBIN AND JOEL G. HARDMAN

VOLUME 100. Recombinant DNA (Part B)
Edited by RAY WU, LAWRENCE GROSSMAN, AND KIVIE MOLDAVE

VOLUME 101. Recombinant DNA (Part C)
Edited by RAY WU, LAWRENCE GROSSMAN, AND KIVIE MOLDAVE

VOLUME 102. Hormone Action (Part G: Calmodulin and Calcium-Binding Proteins)
Edited by ANTHONY R. MEANS AND BERT W. O'MALLEY

VOLUME 103. Hormone Action (Part H: Neuroendocrine Peptides)
Edited by P. MICHAEL CONN

VOLUME 104. Enzyme Purification and Related Techniques (Part C)
Edited by WILLIAM B. JAKOBY

VOLUME 105. Oxygen Radicals in Biological Systems
Edited by LESTER PACKER

VOLUME 106. Posttranslational Modifications (Part A)
Edited by FINN WOLD AND KIVIE MOLDAVE

VOLUME 107. Posttranslational Modifications (Part B)
Edited by FINN WOLD AND KIVIE MOLDAVE

VOLUME 108. Immunochemical Techniques (Part G: Separation and Characterization of Lymphoid Cells)
Edited by GIOVANNI DI SABATO, JOHN J. LANGONE, AND HELEN VAN VUNAKIS

VOLUME 109. Hormone Action (Part I: Peptide Hormones)
Edited by LUTZ BIRNBAUMER AND BERT W. O'MALLEY

VOLUME 110. Steroids and Isoprenoids (Part A)
Edited by JOHN H. LAW AND HANS C. RILLING

VOLUME 111. Steroids and Isoprenoids (Part B)
Edited by JOHN H. LAW AND HANS C. RILLING

VOLUME 112. Drug and Enzyme Targeting (Part A)
Edited by KENNETH J. WIDDER AND RALPH GREEN

VOLUME 113. Glutamate, Glutamine, Glutathione, and Related Compounds
Edited by ALTON MEISTER

VOLUME 114. Diffraction Methods for Biological Macromolecules (Part A)
Edited by HAROLD W. WYCKOFF, C. H. W. HIRS, AND SERGE N. TIMASHEFF

VOLUME 115. Diffraction Methods for Biological Macromolecules (Part B)
Edited by HAROLD W. WYCKOFF, C. H. W. HIRS, AND SERGE N. TIMASHEFF

VOLUME 116. Immunochemical Techniques (Part H: Effectors and Mediators of Lymphoid Cell Functions)
Edited by GIOVANNI DI SABATO, JOHN J. LANGONE, AND HELEN VAN VUNAKIS

VOLUME 117. Enzyme Structure (Part J)
Edited by C. H. W. HIRS AND SERGE N. TIMASHEFF

VOLUME 118. Plant Molecular Biology
Edited by ARTHUR WEISSBACH AND HERBERT WEISSBACH

VOLUME 119. Interferons (Part C)
Edited by SIDNEY PESTKA

VOLUME 120. Cumulative Subject Index Volumes 81–94, 96–101

VOLUME 121. Immunochemical Techniques (Part I: Hybridoma Technology and Monoclonal Antibodies)
Edited by JOHN J. LANGONE AND HELEN VAN VUNAKIS

VOLUME 122. Vitamins and Coenzymes (Part G)
Edited by FRANK CHYTIL AND DONALD B. MCCORMICK

VOLUME 123. Vitamins and Coenzymes (Part H)
Edited by FRANK CHYTIL AND DONALD B. MCCORMICK

VOLUME 124. Hormone Action (Part J: Neuroendocrine Peptides)
Edited by P. MICHAEL CONN

VOLUME 125. Biomembranes (Part M: Transport in Bacteria, Mitochondria, and Chloroplasts: General Approaches and Transport Systems)
Edited by SIDNEY FLEISCHER AND BECCA FLEISCHER

VOLUME 126. Biomembranes (Part N: Transport in Bacteria, Mitochondria, and Chloroplasts: Protonmotive Force)
Edited by SIDNEY FLEISCHER AND BECCA FLEISCHER

VOLUME 127. Biomembranes (Part O: Protons and Water: Structure and Translocation)
Edited by LESTER PACKER

VOLUME 128. Plasma Lipoproteins (Part A: Preparation, Structure, and Molecular Biology)
Edited by JERE P. SEGREST AND JOHN J. ALBERS

VOLUME 129. Plasma Lipoproteins (Part B: Characterization, Cell Biology, and Metabolism)
Edited by JOHN J. ALBERS AND JERE P. SEGREST

VOLUME 130. Enzyme Structure (Part K)
Edited by C. H. W. HIRS AND SERGE N. TIMASHEFF

VOLUME 131. Enzyme Structure (Part L)
Edited by C. H. W. HIRS AND SERGE N. TIMASHEFF

VOLUME 132. Immunochemical Techniques (Part J: Phagocytosis and Cell-Mediated Cytotoxicity)
Edited by GIOVANNI DI SABATO AND JOHANNES EVERSE

VOLUME 133. Bioluminescence and Chemiluminescence (Part B)
Edited by MARLENE DELUCA AND WILLIAM D. MCELROY

VOLUME 134. Structural and Contractile Proteins (Part C: The Contractile Apparatus and the Cytoskeleton)
Edited by RICHARD B. VALLEE

VOLUME 135. Immobilized Enzymes and Cells (Part B)
Edited by KLAUS MOSBACH

VOLUME 136. Immobilized Enzymes and Cells (Part C)
Edited by KLAUS MOSBACH

VOLUME 137. Immobilized Enzymes and Cells (Part D)
Edited by KLAUS MOSBACH

VOLUME 138. Complex Carbohydrates (Part E)
Edited by VICTOR GINSBURG

VOLUME 139. Cellular Regulators (Part A: Calcium- and Calmodulin-Binding Proteins)
Edited by ANTHONY R. MEANS AND P. MICHAEL CONN

VOLUME 140. Cumulative Subject Index Volumes 102–119, 121–134

VOLUME 141. Cellular Regulators (Part B: Calcium and Lipids)
Edited by P. MICHAEL CONN AND ANTHONY R. MEANS

VOLUME 142. Metabolism of Aromatic Amino Acids and Amines
Edited by SEYMOUR KAUFMAN

VOLUME 143. Sulfur and Sulfur Amino Acids
Edited by WILLIAM B. JAKOBY AND OWEN GRIFFITH

VOLUME 144. Structural and Contractile Proteins (Part D: Extracellular Matrix)
Edited by LEON W. CUNNINGHAM

VOLUME 145. Structural and Contractile Proteins (Part E: Extracellular Matrix)
Edited by LEON W. CUNNINGHAM

VOLUME 146. Peptide Growth Factors (Part A)
Edited by DAVID BARNES AND DAVID A. SIRBASKU

VOLUME 147. Peptide Growth Factors (Part B)
Edited by DAVID BARNES AND DAVID A. SIRBASKU

VOLUME 148. Plant Cell Membranes
Edited by LESTER PACKER AND ROLAND DOUCE

VOLUME 149. Drug and Enzyme Targeting (Part B)
Edited by RALPH GREEN AND KENNETH J. WIDDER

VOLUME 150. Immunochemical Techniques (Part K: *In Vitro* Models of B and T Cell Functions and Lymphoid Cell Receptors)
Edited by GIOVANNI DI SABATO

VOLUME 151. Molecular Genetics of Mammalian Cells
Edited by MICHAEL M. GOTTESMAN

VOLUME 152. Guide to Molecular Cloning Techniques
Edited by SHELBY L. BERGER AND ALAN R. KIMMEL

VOLUME 153. Recombinant DNA (Part D)
Edited by RAY WU AND LAWRENCE GROSSMAN

VOLUME 154. Recombinant DNA (Part E)
Edited by RAY WU AND LAWRENCE GROSSMAN

VOLUME 155. Recombinant DNA (Part F)
Edited by RAY WU

VOLUME 156. Biomembranes (Part P: ATP-Driven Pumps and Related Transport: The Na,K-Pump)
Edited by SIDNEY FLEISCHER AND BECCA FLEISCHER

VOLUME 157. Biomembranes (Part Q: ATP-Driven Pumps and Related Transport: Calcium, Proton, and Potassium Pumps)
Edited by SIDNEY FLEISCHER AND BECCA FLEISCHER

VOLUME 158. Metalloproteins (Part A)
Edited by JAMES F. RIORDAN AND BERT L. VALLEE

VOLUME 159. Initiation and Termination of Cyclic Nucleotide Action
Edited by JACKIE D. CORBIN AND ROGER A. JOHNSON

VOLUME 160. Biomass (Part A: Cellulose and Hemicellulose)
Edited by WILLIS A. WOOD AND SCOTT T. KELLOGG

VOLUME 161. Biomass (Part B: Lignin, Pectin, and Chitin)
Edited by WILLIS A. WOOD AND SCOTT T. KELLOGG

VOLUME 162. Immunochemical Techniques (Part L: Chemotaxis and Inflammation)
Edited by GIOVANNI DI SABATO

VOLUME 163. Immunochemical Techniques (Part M: Chemotaxis and Inflammation)
Edited by GIOVANNI DI SABATO

VOLUME 164. Ribosomes
Edited by HARRY F. NOLLER, JR., AND KIVIE MOLDAVE

VOLUME 165. Microbial Toxins: Tools for Enzymology
Edited by SIDNEY HARSHMAN

VOLUME 166. Branched-Chain Amino Acids
Edited by ROBERT HARRIS AND JOHN R. SOKATCH

VOLUME 167. Cyanobacteria
Edited by LESTER PACKER AND ALEXANDER N. GLAZER

VOLUME 168. Hormone Action (Part K: Neuroendocrine Peptides)
Edited by P. MICHAEL CONN

VOLUME 169. Platelets: Receptors, Adhesion, Secretion (Part A)
Edited by JACEK HAWIGER

VOLUME 170. Nucleosomes (in preparation)
Edited by PAUL M. WASSARMAN AND ROGER D. KORNBERG

VOLUME 171. Biomembranes (Part R: Transport Theory: Cells and Model Membranes)
Edited by SIDNEY FLEISCHER AND BECCA FLEISCHER

VOLUME 172. Biomembranes (Part S: Transport: Membrane Isolation and Characterization)
Edited by SIDNEY FLEISCHER AND BECCA FLEISCHER

VOLUME 173. Biomembranes [Part T: Cellular and Subcellular Transport: Eukaryotic (Nonepithelial) Cells] (in preparation)
Edited by SIDNEY FLEISCHER AND BECCA FLEISCHER

VOLUME 174. Biomembranes [Part U: Cellular and Subcellular Transport: Eukaryotic (Nonepithelial) Cells] (in preparation)
Edited by SIDNEY FLEISCHER AND BECCA FLEISCHER

VOLUME 175. Cumulative Subject Index Volumes 135–139, 141–167 (in preparation)

Volume 176. Nuclear Magnetic Resonance (Part A: Spectral Techniques and Dynamics) (in preparation)
Edited by NORMAN J. OPPENHEIMER AND THOMAS L. JAMES

VOLUME 177. Nuclear Magnetic Resonance (Part B: Structure and Mechanism) (in preparation)
Edited by NORMAN J. OPPENHEIMER AND THOMAS L. JAMES

VOLUME 178. Antibodies, Antigens, and Molecular Mimicry (in preparation)
Edited by JOHN J. LANGONE

Section I

Membrane Separation

[1] Membrane Isolation: Strategy, Techniques, Markers

By E. KINNE-SAFFRAN and R. K. H. KINNE

Introduction

The purpose of this chapter is to provide some general guidelines for the separation of plasma membranes as a functional entity. Any membrane isolation involves three successive steps: tissue homogenization, fractionation, and analysis.[1] Before addressing the technical aspects of these three steps, we would like to alert the reader to the fact that, as in other fields, in science a precise and unambiguous vocabulary is necessary to avoid misinterpretation of procedures and/or results. Homogenization in the technical sense is defined as the reduction of a sample to evenly distributed particles of uniform size. In the context of isolating cell membranes *homogenization* can be defined as disruption of the cell envelope under conditions where the integrity of cellular organelles such as nuclei, mitochondria, and lysosomes is maintained. Thus, purposely, an incomplete homogenization as compared to the technical term is desired. The structural and biochemical integrity of the different cellular components is the prerequisite for a successful subsequent fractionation, isolation, and meaningful final analysis of homogeneous membrane preparations.

Usually, preparations of cellular organelles are not homogeneous; this means they always contain to a certain extent other cellular organelles. For example, for a preparation enriched in brush border membranes the term brush border membrane *fraction* should be used instead of brush border membranes, since so far no complete separation of the brush border membranes from other cellular membranes and/or organelles has been achieved. The latter organelles can transfer their particular properties to the membrane fraction, which, mistakingly, are then regarded as a function of the brush border membrane proper. Furthermore, the term fraction takes into account potential mechanical, physical, and chemical events that might take place during isolation of the membrane and might have altered the properties of the isolated membrane as compared to the membrane of the intact cell.[1]

In the context of membrane isolation, probably the most ill-considered term is the term *microsomes* or *microsomal fraction*. It was originally proposed as a synonym for small granules and had a purely operational

---

[1] C. de Duve, *J. Theor. Biol.* **6**, 33 (1964).

significance. It is now clear that microsomal fractions contain a variety of membranous elements derived from several if not from all cellular structures. Therefore, all so-called microsomal fractions should be analyzed for the relative contribution of the cell organelles present and should be named after the main constituent, i.e., rough endoplasmic reticulum fraction or pinocytotic vesicle fraction, etc.

With regard to the analysis of an isolated plasma membrane fraction, the most common way is to determine the activity of so-called *marker enzymes,* i.e., enzymes which belong to a single cellular component in the living cell (postulate of "single location" by de Duve[2]). However, in isolating membranes, too much emphasis has often been placed on the attempt to achieve *enrichment factors* as high as possible, overlooking the fact that the enrichment factor in itself is limited by the amount of membrane to be isolated that is present in the starting homogenate. Thus, when a certain cellular component comprises 5% of the total protein content of a given cell, the maximum enrichment factor for the marker enzyme of this component cannot exceed 20. In reality, however, the enrichment factor depends also on the degree of homogeneity of the starting material. If the starting material is less homogeneous and only 50% of the cells contain the cellular component to be isolated, a maximum enrichment factor of 40 can be obtained. This higher value does not, however, imply a higher degree of purity compared to a fraction prepared from a homogeneous starting material, where *all* cells contain the component to be isolated and for which a maximum enrichment factor of 20 can be achieved. Both membrane fractions would, however, show the same specific activity of marker enzymes and, thus, would have the same degree of purity. Therefore, in comparing degrees of purification, the *specific activity of marker enzymes* in the final membrane fractions together with the *enrichment factors* and the *enzyme recoveries* should be considered.

Homogenization

Homogenization of a given tissue should dissociate the cell surface, i.e., the plasma membrane, from intra- and extracellular components. In polarized cells, where the plasma membrane is composed of several domains,[3] the homogenization should also ensure a complete physical separation of these domains in order to make it possible to isolate homogeneous membrane fractions of defined origin. At the same time, the

---

[2] C. de Duve, *Symp. Soc. Exp. Biol.* **10**, 50 (1957).
[3] K. Simons and S. D. Fuller, *Annu. Rev. Cell Biol.* **1**, 243 (1985).

integrity of the other cellular organelles should be maintained in order to reduce contamination of the plasma membrane fraction by other membranes derived from these organelles. In this context, it should be noted that homogenization procedures, including the ratio of tissue to homogenization medium, are usually developed for a particular tissue or a particular cell preparation. When other starting materials are used, the homogenization method has to be adapted to the specific intrinsic properties of the new material. One of the many possible reasons for this is, for example, a different amount of connective tissue in different organs, which alters the ease with which tissues can be homogenized.

Selection of Starting Material

Before applying one of the homogenization procedures as described below, it has to be considered whether inhomogeneity of the starting material could impair the validity of conclusions drawn from the results of experiments performed with the isolated plasma membrane fraction. The kidney is a typical example for these considerations. Brush border membrane fractions isolated from thin slices of the cortex or the medulla will necessarily differ in their properties, because they are derived from different segments of the proximal tubule. It has been shown, for example, that the stoichiometry of the Na^+–D-glucose cotransporter for sodium is higher in the membranes derived from the medulla than in the membranes derived from thin cortical slices.[4,5] In this instance heterogeneity was overcome by simple physical preselection. For the isolation of luminal membranes of other parts of the nephron, as well as for the isolation of basolateral membranes of the kidney in general, inhomogeneity remains a problem. Therefore, the isolation of homogeneous cell populations might have to precede a successful isolation of a defined membrane fraction. Another possibility for obtaining cells of homogeneous origin, state of growth, and metabolism can be provided by cell culture combined with cloning techniques.

Methods of Cell Disruption

Disruption by Shearing Forces. Currently, the most commonly used homogenization method is disruption by shearing forces. One type of homogenizer consists of a cylindrical glass homogenization vessel, in which either a rod-shaped Teflon-coated pestle is inserted (Potter-Elvehjem) or a glass pestle with a ball-shaped tip (Dounce). The glass–Teflon

---

[4] R. J. Turner and A. Moran, *J. Membr. Biol.* **70,** 37 (1982).
[5] R. J. Turner and A. Moran, *Am. J. Physiol.* **241,** F406 (1982).

homogenizer is the most versatile, because different clearances are available (from 0.045 to 0.115 mm). Furthermore, the speed of rotation of the pestle can be modified, ranging from modest homogenization by hand to motor-driven homogenization using a load-independent drive. The variability of the number of strokes and of the speed of the vertical movement of the pestle broadens the versatility (see, for example, Refs. 6 and 7). In the glass–glass homogenizer the clearance between the pestle and the inner wall of the homogenization vessel is smaller (between 0.05 and 0.07 mm for the loose-fitting pestle and between 0.01 and 0.03 mm for the tight-fitting pestle) and the shearing forces are, due to the ball-shaped tip of the pestle, concentrated in a ringlike contact zone. Thus, the Teflon–glass homogenizer will be advantageous for the homogenization of tissue fragments with lower content of connective tissue, whereas the glass–glass homogenizer will be more appropriate for tissue with higher content of connective tissue,[8] for single cell layers, and for single cells.[9]

Another homogenization method based mainly on shearing forces is the use of metal blades rotating at variable speed, i.e., Waring blender[10,11] and Polytron.[12,13]

Disruption by Cavitation. Another way to disrupt cells is by intracytoplasmic cavitation of nitrogen gas.[14] This method allows quantitative disruption of cells in an inert atmosphere without danger of local heating. Cell disruption due to shearing effects and breakage of intracellular organelles is minimal if the pressure and time of equilibration are properly adjusted. This method is mainly used for single cells in suspension or for cells grown in culture.[15]

Each homogenization procedure yields a broad spectrum of plasma membranes differing in size and shape. The size and shape of the predominantly formed plasma membrane fragments is strongly affected by the homogenization technique used for cellular disruption. Accordingly, the

---

[6] V. Scalera, Y.-K. Huang, B. Hildmann, and H. Murer, *Membr. Biochem.* **4**, 49 (1981).
[7] P. Mitchell and J. Moyle, *Biochem. J.* **104**, 588 (1967).
[8] I. L. Schwartz, L. J. Shlatz, E. Kinne-Saffran, and R. Kinne, *Proc. Natl. Acad. Sci. U.S.A.* **71**, 2595 (1974).
[9] W. A. Brodsky, Z. I. Cabantchik, N. Davidson, G. Ehrenspeck, E. Kinne-Saffran, and R. Kinne, *Biochim. Biophys. Acta* **556**, 490 (1979).
[10] A. G. Booth and A. J. Kenny, *Biochem. J.* **142**, 575 (1974).
[11] V. Scalera, D. Storelli, C. Storelli-Joss, W. Haase, and H. Murer, *Biochem. J.* **186**, 177 (1980).
[12] J. Biber, B. Stieger, W. Haase, and H. Murer, *Biochim. Biophys. Acta* **647**, 169 (1981).
[13] J. Flöge, H. Stolte, and R. Kinne, *J. Comp. Physiol. B* **154**, 355 (1984).
[14] D. F. H. Wallach and V. B. Kamat, *Proc. Natl. Acad. Sci. U.S.A.* **52**, 721 (1964).
[15] J. E. Lever, *J. Biol. Chem.* **257**, 8680 (1982).

method of homogenization determines the behavior of the membranes during the subsequent fractionation procedure. Thus, homogenization is one of the most important steps in plasma membrane isolation and should be controlled very carefully in order to get reproducible results.

Homogenization Medium

A broad variety of homogenization media have been used for the preparation of plasma membranes from different starting materials and for different purposes. As a rule, the following specifics in choosing an appropriate homogenization medium should be considered: buffer capacity and pH, ionic strength, osmolytes, and membrane protective agents.

First, attention should be paid to the buffering capacity and the pH of the homogenization medium. The buffering capacity has to be high enough to maintain a constant pH throughout the preparation. Especially during the disruption of the cells, the buffering capacity of the homogenization medium has to be higher than the buffering capacity of proteins and other solutes of a given starting material. Also, acidic equivalents generated frequently in dissected tissue have to be compensated for. Furthermore, the pH of the medium should be identical with the pH to which the membranes were exposed in the intact cell. The most commonly used pH values lie between 7.0 and 8.2; more acidic values should be avoided to prevent activation of lysosomal enzymes, which will disintegrate plasma membranes.

In choosing the ionic strength of a homogenization medium the following limitations have to be considered: Since the plasma membranes and other intracellular organelles carry surface charges at physiological pH,[16] undesired cross-aggregations can occur at too low an ionic strength. At high ionic strength membrane components, which are only loosely associated with the plasma membrane, can be extracted. High ionic strength of the homogenization medium also excludes the use of differential precipitation techniques such as the use of calcium or magnesium for the isolation of luminal plasma membranes.[10]

In the classical homogenization scheme of Hogeboom *et al.*,[17] sucrose was used for the first time to mimic the cytoplasmic tonicity to which intracellular organelles are exposed in order to avoid swelling and even possible disruption. Even today, sucrose, ranging from 250 to 330 mM, is still the most commonly used substance for this purpose, although manni-

---

[16] K. Hannig, *in* "Techniques of Biochemical and Biophysical Morphology" (D. Glick and R. M. Rosenbaum, eds.), Vol. 1, pp. 191–232. Wiley (Interscience), New York, 1972.

[17] G. H. Hogeboom, W. C. Schneider, and G. E. Palade, *J. Biol. Chem.* **172**, 619 (1948).

tol or sorbitol has been introduced. The use of mannitol facilitates the sedimentation of plasma membranes in the subsequent steps of differential centrifugation, because of its lower viscosity at equal osmolality.

Osmolytes cannot be exchanged indiscriminately since the apparent density of a closed membrane vesicle, which includes sucrose-containing homogenization medium, may differ from that of a membrane vesicle, which is filled with a mannitol-containing medium. This is due to the fact that the apparent buoyant density of a vesicle is determined both by the density of the plasma membrane itself and by its content.

After dissociation from the cell, plasma membranes are prone to destruction by external agents or integral membrane systems. In some instances it might therefore be necessary to add reducing agents, such as dithiothreitol, to prevent oxidation of functionally important SH groups. Furthermore, release of proteases from lysosomes can occur. Their attack on membrane proteins can be prevented by addition of protease inhibitors to the homogenization medium, e.g., phenylmethylsulfonyl fluoride (PMSF). Within the membranes themselves phospholipases and proteases can be activated. Metal ions, such as calcium, that activate certain phospholipases should be avoided.[18]

In general, the action of any destructive agent is strongly temperature dependent. It is, therefore, of extreme importance to keep the temperature of the samples between 0 and 4° beginning with the dissection of the tissue and throughout the entire procedure of homogenization (where, in the absence of appropriate cooling devices, temperatures of up to 15° can easily be reached locally) and fractionation.

Control of Homogenization

In developing a successful method for applying an established process to isolate plasma membranes from a given starting material, it is crucial to control the effectiveness of the homogenization procedure. This can be achieved by either optical or biochemical methods. In the phase-contrast microscope the number of intact cells can be compared to the number of single nuclei. The extent of cell rupture can also be judged biochemically from the release of the cytosolic enzyme lactate dehydrogenase into a particle-free supernatant (1 hr, 100,000 g). If, in addition to cell rupture, undesired rupture of intracellular organelles has also occurred, this supernatant will also contain enzymes that are normally particle bound, such as DNA for nuclei, β-glucuronidase for lysosomes, or succinate dehydrogenase for mitochondria.

---

[18] I. Sabolić and G. Burckhardt, *Biochim. Biophys. Acta* **772**, 140 (1982).

Fractionation

Techniques for the isolation of plasma membranes take advantage of differences in intrinsic properties between plasma membranes and intracellular organelles, differences such as size, shape, density, surface charge, and surface components. If these intrinsic differences are not sufficiently large enough for a separation, a selective modification of the plasma membrane may be necessary in order to exaggerate the intrinsic differences, for example, through density perturbation.[19]

In the following sections, isolation methods will be discussed separately. Usually, however, a combination of these methods is necessary in order to obtain plasma membranes of high purity and derived exclusively from a specific domain of the cell envelope.

Isolation Techniques Based on Size, Shape, and Density

Differential Centrifugation. Size, shape, and density are the principal properties which determine the behavior of a particle during differential centrifugation. In preparing plasma membranes differential centrifugation is commonly used as an initial step to obtain crude fractions enriched in plasma membranes. In contrast to nuclei and mitochondria with a predictable behavior during differential centrifugation, plasma membranes can be basically found in every sediment. Depending on the homogenization procedure, plasma membranes will sediment either predominantly as large fragments with nuclei or as smaller fragments with mitochondria, or will be found as very small vesicles in the microsomal fraction. Thus, interdependence of homogenization procedure and differential centrifugation has to be considered and the two have to be coordinated in such a way that enrichment of plasma membranes in a certain sediment is achieved.

Density Gradient Centrifugation. In considering density gradient centrifugation the following parameters are important: in an idealized situation the equation for the sedimentation of a sphere in a centrifugal field is

$$v = \frac{d^2(\rho_p - \rho_l)}{18\eta} g$$

where v, sedimentation rate; d, diameter of the particle (hydrodynamically equivalent sphere); ρ_p, particle density; ρ_l, liquid density; η, viscosity of the medium; and g, centrifugal force.

From this equation the following relationships can be observed:

[19] M. Inoue, R. Kinne, T. Tran, L. Biempica, and I. M. Arias, *J. Biol. Chem.* **258**, 5183 (1983).

(1) The sedimentation rate of a particle is proportional to its size and (2) proportional to the difference between the density of the particle and the surrounding medium. (3) The sedimentation rate is zero when the density of the particle is equal to the density of the surrounding medium. (4) The sedimentation rate decreases as the viscosity of the medium increases and (5) the sedimentation rate increases as the centrifugal force increases.

Of these parameters, density and viscosity of the media chosen for density gradient centrifugation are relatively well defined and reproducible physicochemical properties. They can be measured either directly or from the informational material supplied by the manufacturers (metrizamide, Nyegaard & Co., Oslo, Sweden; Percoll, Pharmacia, Uppsala, Sweden). The variable entities are the size and the apparent density of the membrane vesicles. The size is initially determined by the physical forces exerted during the disruption of the cells, but it can also be changed subsequently by osmotic forces generated by the density gradient media. Depending on the membrane permeability for osmolytes, swelling or shrinkage will occur which, in turn, alters the size.

The effective buoyant density of the membrane vesicles during density centrifugation is determined both by the intrinsic density of the membrane and by the density of the intravesicular content. For a given membrane the intrinsic density is relatively well defined by the lipid/protein ratio. Large differences exist between the intrinsic densities of the membranes derived from various cellular organelles including the plasma membranes, ranging from 1.07 to 1.27. These differences are, however, very small between different domains of the plasma membranes of a polarized cell. The intravesicular content is determined initially by the homogenization medium used, but can subsequently change, depending on the permeability of the membranes.

In isolating plasma membranes the most commonly used media for density gradient centrifugation are sucrose,[20] Percoll,[11,21] and metrizamide.[22,23] For experimental purposes the differences in osmotic activity, molecular size, and the feasibility of generating shallow gradients are important. According to molecular weight, sucrose has the highest osmolarity at a given density. The permeability of membranes decreases in the sequence sucrose, metrizamide, Percoll. Furthermore, Percoll forms

---

[20] H. J. Rodriguez and I. S. Edelman, *J. Membr. Biol.* **45,** 215 (1979).

[21] R. D. Mamelok, D. R. Macrae, L. Z. Benet, and S. B. Prusiner, *J. Neurochem.* **37,** 768 (1981).

[22] K. N. Dzhandzhugazyan and P. L. Jørgensen, *Biochim. Biophys. Acta* **817,** 165 (1985).

[23] C. Burnham, S. J. D. Karlish, and P. L. Jørgensen, *Biochim. Biophys. Acta* **821,** 461 (1985).

transiently very shallow gradients. Such differences can be used advantageously in separating different domains of plasma membranes that differ only slightly in their densities.

Basically, two different types of density gradient centrifugation can be performed: rate zonal centrifugation and isopycnic centrifugation. In rate zonal centrifugation the density range of the gradient medium is chosen in such a way that the density of the particles to be separated is greater than the density of the medium at all points during the separation. In this type of separation the difference in size of the particles is the determining parameter. At a given membrane density, larger particles move faster through the gradient than smaller particles. Besides the size, the final position of the plasma membranes in the gradient is also a function of the total centrifugal force applied and the length of centrifugation time. This means the run has to be terminated before the separated zones reach the bottom of the tube. In isopycnic centrifugation the density range of the gradient medium encompasses all densities of all particles. Each particle will sediment to an equilibrium position in the gradient where the gradient medium density is equal to the apparent density of the particle. Thus, in this type of separation the particles are separated irrespective of their size and solely on the basis of their different apparent densities. This fact, in turn, will influence the choice of the gradient medium, since the gradient medium can change the apparent density of plasma membrane vesicles, as discussed above.

Isolation Techniques Based on Surface Properties

During recent years, difficulties have often been encountered in separating particular domains of plasma membranes from each other and from other cellular organelles based on the difference in size, shape, or density. In these instances, differences in surface properties such as surface charge or membrane surface hydrophobicity have often been successfully employed. Surface charge is the main determinant for the separation by differential precipitation and free flow electrophoresis; hydrophobicity is the main parameter used in partitioning techniques.

Differential Precipitation. This method has been especially advantageous for the isolation of luminal plasma membranes lining the inner cavities of the body such as the intestine, the renal tubule, and others.[10,13,19,24] These membranes are covered by a glycocalyx, which contains more negatively charged sialic acid than the membranes of other cellular components, including the basolateral parts of the cellular enve-

---

[24] P. A. King, R. Kinne, and L. Goldstein, *J. Comp. Physiol. B* **155,** 185 (1985).

lope. Addition of calcium or magnesium in millimolar concentrations to a low-ionic-strength tissue homogenate leads, through *intermembranous* cross-linking, to aggregation of those membranes with a lower sialic acid content. At the same time, the concentration of the bivalent cations and the density of surface charges at the luminal membrane is high enough to produce *intramembranous* saturation of contiguous anionic sites. Thus, intermembranous cross-linking is prevented and the luminal membranes remain nonaggregated.[10]

Although this technique has worked well in a variety of tissues, the use of high concentrations of calcium can also be problematic. Calcium is known to activate membrane-bound phospholipases and interacts directly with membrane phospholipids or transport proteins; thus, changes in membrane permeability can occur.[18,25] In addition, bivalent cations trapped in the intravesicular space of the isolated membrane vesicles can affect studies of functional properties, e.g., transport properties.[26] In using differential precipitation the concentration as well as the kind of bivalent cations have to be adapted to the special properties of the tissue, since the density of surface charges of the luminal membranes and of other cellular organelles can differ from one tissue to another. This is particularly critical for the behavior of lysosomes, which tend to contaminate these luminal membrane fractions. Generally, either magnesium or calcium is used in a range between 10 and 30 mmol/liter.[10,13,19,24]

Free Flow Electrophoresis. In free flow electrophoresis[16] even small differences in surface charge can be exploited for the separation of membrane particles. Free flow electrophoresis is a continuous, preparative separation process in which a mixture of particles is introduced as a fine jet into a separation buffer moving across the field lines of an electric field. The particles are deflected from the flow direction of the medium because of surface charge, flow velocity of the medium, ionic composition of the medium, and strength of the electric field. Appropriate adjustment of the various parameters can lead to a successful separation of particles with very small differences in their electrophoretic mobility. The method is used most effectively when the plasma membrane domain in question has been prepurified by differential or density gradient centrifugation.[27,28]

Phase Partition. Phase partition represents a method whereby membrane particles are separated by partition between two immiscible but

---

[25] M. Chase and Q. Al-Awqati, *J. Gen. Physiol.* **81,** 643 (1983).
[26] S. M. Grassl, E. Heinz, and R. Kinne, *Biochim. Biophys. Acta* **736,** 178 (1983).
[27] H.-G. Heidrich, R. Kinne, E. Kinne-Saffran, and K. Hannig, *J. Cell Biol.* **54,** 232 (1972).
[28] F. Bode, K. Baumann, and R. Kinne, *Biochim. Biophys. Acta* **433,** 294 (1976).

aqueous phases.[29-31] The aqueous phases generally used are polyethylene glycol and dextran solutions. The polymers are sometimes modified to carry additional negative or positive charges. The main property of the membrane that determines its behavior during phase partition is the hydrophobicity of the membrane surface relative to the polymers used. The partition is also influenced to some extent by the surface charge of the membrane. Although, theoretically, this method has great potential due to the many variables that can be adapted to the specific isolation problem, the application of the method to membrane isolation has been hampered by the fact that plasma membranes usually assemble preferentially at the interphase of the two immiscible polymers. For effective purification, therefore, repeated phase partition by countercurrent distribution is necessary.[31]

Isolation Techniques Based on Specific Membrane Components

In order to utilize specific membrane components for the isolation of membranes, these components must be accessible to the medium as surface components. Glycoproteins, which commonly are asymmetrically distributed on the outer surface of cells are an example. These surface components can be cell specific, specific for a certain membrane domain, or specific for a certain orientation of the vesiculated plasma membranes. Their specificity is based on the existence of unique antigenic epitopes or the presence of unique carbohydrates. Accordingly, antibodies and lectins have been employed in isolating plasma membranes. To date, the use of antibodies to purify intact membranes is still limited, since separation of the antigen–antibody complex requires conditions which usually impair the function of the membranes. However, antibodies have been successfully used to remove unwanted contaminants or to assign specific functions to a specific membrane domain by inducing affinity density perturbation.[19]

Theoretically, the reversal of the interaction between the lectin and the carbohydrate components of the membranes should be possible by adding the appropriate amount of carbohydrate known to be bound specifically by the lectin. In reality, however, lectins have so far been employed to sort out right-side-out vesicles, on which the carbohydrates are exposed to the incubation medium from a mixed membrane vesicle popula-

[29] P. A. Albertsson, B. Andersson, C. Larsson, and H. E. Akerlund, *Methods Biochem. Anal.* **28,** 115 (1983).
[30] H. Glossmann and H. Grips, *Naunyn-Schmiedeberg's Arch. Pharmacol.* **282,** 439 (1974).
[31] A. K. Mircheff, H. E. Ives, V. J. Yee, and D. G. Warnock, *Am. J. Physiol.* **246,** F853 (1984).

tion, the final product being a homogeneous inside-out vesicle preparation.[32] For this purpose, either multivalent lectins can be used, which induce direct agglutination of nonsealed membranes and right-side-out vesicles, or nonagglutinating lectins can be immobilized, for example, to Sepharose 4B, to adsorb the right-side-out vesicles. Nonagglutinating lectins can also be used in combination with antilectin antibodies, where the antibodies achieve immunoprecipitation of the lectins bound to the membranes. One of the reasons for the problems encountered in dissociating membrane–lectin complexes used to recover absorbed or aggregated right-side-out membrane vesicles might be the multiplicity of binding sites on the membranes for lectins. This leads to the interaction of more than one sugar residue with one lectin molecule, thus increasing the effective affinity of the lectin far beyond the affinity measured with free sugars in solution.

Assessment of Purification

At the end of each isolation procedure, the purity of the plasma membrane fraction obtained has to be assessed. The presence of markers specific for the membrane to be isolated and of markers specific for other cellular membranes which could contaminate the fraction has to be determined. Ideally, the former should be enriched compared to the initial homogenate and the latter should not be present at all. As already pointed out, the enrichment factor is naturally limited by the relative abundance of the particular membrane in the cell. The absence of markers for the contaminating membranes has to be interpreted with caution. Dissociation and inactivation of markers can occur during homogenization and/or isolation, which then leads to the impression of a lesser contamination than actually exists. Therefore, whenever possible, the amount of marker recovered in all fractions obtained during the isolation has to be determined and compared to that present in the starting material.

Some markers that have been found to be useful for the characterization of isolated plasma membrane fractions include marker enzymes, marker functions, and marker antibodies. These markers are routinely used and are discussed below. Not included in this discussion is the morphological examination of the membrane fractions, although such techniques as negative staining can sometimes provide fast and definitive identification using electron microscopy. Thin sectioning and freeze-fracture[33] will also give insight into the structure and composition of the

[32] J. G. Lindsay, G. P. Reid, and M. P. D'Souza, *Biochim. Biophys. Acta* **640,** 791 (1981).
[33] W. Haase, A. Schaefer, H. Murer, and R. Kinne, *Biochem. J.* **172,** 57 (1978).

membrane fractions. These are not routine procedures, however, since they involve highly sophisticated technology and require very critical assessment of the data.

Membrane Markers

A useful membrane marker should fulfill the following criteria: (1) it should have a so-called single location, i.e., it should belong only to a single intracellular component in the living cell[2,34]; (2) its properties should be highly specific; and (3) fast, reproducible, quantitative assays should be available. Therefore, the measurement of enzymatic markers has found the broadest application, but transport functions and receptor functions of membranes have also been employed. Increasingly, monoclonal antibodies against specific membrane components have been considered. It is advisable to determine, if possible, several markers simultaneously in characterizing isolated plasma membrane fractions.

Enzyme Activities. For a number of years it was assumed that there are some enzymes which can be used ubiquitously as marker enzymes for all plasma membranes, e.g., the 5'-nucleotidase. Recently, however, it has become clear that plasma membranes from different species, organs, cells, or cell surface domains can differ markedly in their enzyme activities and that some of the so-called plasma membrane enzymes are also present in other cellular organelles and thus do not fulfill the criterion of single location. In polarized cells aminopeptidases, especially γ-glutamyltransferase and, where present, disaccharidases, seem to be reliable and useful markers for apical (luminal) plasma membranes in a variety of tissues; alkaline phosphatase can also be used. For basolateral (contraluminal) plasma membranes Na^+,K^+-ATPase and hormone-sensitive adenylate cyclase can usually be employed.[3,35] Some of these enzymes can also be used for the characterization of membrane fractions isolated from nonpolarized cells. It should be stressed, however, that the location of the plasma membrane marker always has to be verified by histochemical and immunohistochemical techniques in the intact cell in instances where a new system is being explored. (See, for example, the luminal location of Na^+,K^+-ATPase in the choroid plexus.[36-38] In using enzyme activities as markers, false results can be obtained when a large percentage of the membranes in the fraction are sealed vesicles. If the substrate,

[34] C. de Duve, *J. Cell Biol.* **50**, 20d (1971).
[35] R. Kinne and G. Sachs, *in* "Physiology of Membrane Disorders" (T. E. Andreoli, J. F. Hoffmann, D. D. Fanestil, and S. G. Shultz, eds.), pp. 83–92. Plenum, New York, 1986.
[36] P. M. Quinton, E. M. Wright, and J. M. Tormey, *J. Cell Biol.* **58**, 724 (1973).
[37] T. Masuzawa, T. Saito, and F. Sato, *Brain Res.* **222**, 309 (1981).
[38] E. M. Wright, *Brain Res.* **44**, 207 (1972).

activators, or inhibitors cannot penetrate the membrane, a large number of enzyme molecules remain cryptic and show no activity. Therefore, in almost all fractions a pretreatment such as freeze-thawing[39] or detergent solubilization[40] is necessary. Without inactivating the enzyme, this pretreatment transforms the membrane vesicles into membrane sheets where the enzyme molecules are freely accessible to substrates, activators, and inhibitors. This point is critical for all membrane-bound enzymes which are inserted asymmetrically into the membranes, but it also holds for glycolytic enzymes which might have been trapped inside plasma membrane vesicles.[41]

Functional Markers. Functional markers are specific transport or receptor functions of plasma membranes. They can be used advantageously in those instances where the plasma membrane to be isolated cannot readily be identified by marker enzymes either because no marker enzyme is known for the particular membrane or because the exact location of a possible marker enzyme is not known. For example, the furosemide-sensitive NaCl/KCl cotransporter[42,43] serves as a signal for the purification of luminal membranes from the thick ascending limb of Henle's loop.[44]

The use of transport functions as markers is less reliable than the use of membrane-bound enzymes because of questions pertaining to location and quantitation. Only the sodium–D-glucose cotransporter[3,13,24,45–47] seems to be always located in the luminal membrane of epithelial cells. Other sodium cotransport systems, such as sodium–amino acid cotransporters and the NaCl/KCl cotransporter, have been found in either the luminal or contraluminal membrane depending on the functional requirements of the polarized cell.[48–51] Transport activity is only exhibited by that

[39] O. J. Møller, *Exp. Cell Res.* **68,** 347 (1971).
[40] P. L. Jørgensen and J. C. Skou, *Biochim. Biophys. Acta* **233,** 366 (1971).
[41] D. F. Keljo, A. Kleinzeller, H. Murer, and R. Kinne, *Biochim. Biophys. Acta* **508,** 500 (1978).
[42] J. A. Hannafin and R. Kinne, *J. Comp. Physiol. B* **155,** 415 (1985).
[43] B. König, S. Ricapito, and R. Kinne, *Pfluegers Arch.* **399,** 173 (1983).
[44] R. Kinne, E. Kinne-Saffran, B. Schölermann, and H. Schütz, *Pfluegers Arch.* **407,** 5168 (1986).
[45] U. Hopfer, K. Nelson, J. Perotto, and K. J. Isselbacher, *J. Biol. Chem.* **248,** 25 (1973).
[46] H. Murer, U. Hopfer, E. Kinne-Saffran, and R. Kinne, *Biochim. Biophys. Acta* **345,** 170 (1974).
[47] R. Kinne, H. Murer, E. Kinne-Saffran, M. Thees, and G. Sachs, *J. Membr. Biol.* **21,** 375 (1975).
[48] J. Evers, H. Murer, and R. Kinne, *Biochim. Biophys. Acta* **426,** 598 (1976).
[49] M. Inoue, R. Kinne, T. Tran, and I. M. Arias, *Hepatology* **2,** 572 (1982).
[50] F. H. Epstein and P. Silva, *Ann. N.Y. Acad. Sci.* **456,** 187 (1985).
[51] R. Kinne, J. A. Hannafin, and B. König, *Ann. N.Y. Acad. Sci.* **456,** 198 (1985).

portion of the plasma membranes which is vesiculated, and nonvesiculated membranes show no transport activity. In addition, the transport activity also depends on the orientation of the membrane vesicles. Therefore, assessment of purity and enrichment based on transport studies has to be regarded cautiously.

Specific receptor functions of plasma membranes are either related to target sites of hormonal regulation or to transport systems for which highly specific inhibitors are known. Examples for the former are insulin-binding sites on liver sinusoidal plasma membranes[52]; examples for the latter are phlorhizin-binding sites on renal proximal tubule luminal plasma membranes.[53]

Antigenic Markers. Monoclonal antibodies which exclusively recognize components on particular plasma membrane domains or on an intracellular organelle have become increasingly available.[54-56] Their single location and specificity fulfill the criteria for useful markers for plasma membranes. However, the techniques for fast quantitation of antigens in fractions obtained during membrane isolation have to be improved.

Concluding Remarks

We have presented a brief review using feasibility, reproducibility, applicability, and limitation of approaches for membrane isolation as the main guidelines. In general, one can state that the more detailed knowledge contributed by cell biological techniques on the heterogeneity of plasma membranes and cellular membranes has been matched by an improved resolution of the separation techniques. Membrane isolation has become a routine in many laboratories and, as with all routine methods, the basic ideas behind the development of the method as well as the basic assumptions and possible pitfalls tend to be forgotten. If this chapter can reestablish a critical approach to membrane isolation, it will have fulfilled its purpose.

[52] K. J. Chang, V. Bennette, and P. Quatrecasas, *J. Biol. Chem.* **250**, 488 (1975).
[53] W. Frasch, P. P. Frohnert, F. Bode, K. Baumann, and R. Kinne, *Pfluegers Arch.* **320**, 265 (1970).
[54] T. A. Spring, *J. Biol. Chem.* **256**, 3833 (1981).
[55] R. J. Turner, J. Thompson, S. Sariban-Sohraby, and J. S. Handler, *J. Cell Biol.* **101**, 2173 (1985).
[56] J. S. Rodman, L. Seidman, and M. G. Farquhar, *J. Cell Biol.* **102**, 77 (1986).

[2] Isolation of Plasma Membranes from Polar Cells and Tissues: Apical/Basolateral Separation, Purity, Function

By AUSTIN K. MIRCHEFF

A number of methods have been described for the isolation of apical and basolateral plasma membrane vesicles of epithelial cells from the renal proximal tubule and the small intestinal villus. These procedures have made it possible to describe kinetic and molecular details of processes which are localized at the topologically and functionally distinct surfaces of the epithelial barriers in these tissues. The success of such studies makes it likely that transport mechanisms in other epithelia could also be studied effectively with plasma membrane vesicles. Therefore, it has been of interest to design a general methodology for epithelial subcellular fractionation which could be applied to other epithelia with only minor modifications. The theoretical background for this general approach has already been summarized,[1] and the present chapter provides practical details of a procedure which has been adapted for use with lacrimal glands,[2,3] parotid glands,[4] and MDCK cells[5] in addition to the more frequently studied intestinal[6] and proximal tubular epithelia.[7] The overall strategy is to use physical separation techniques as analytical tools which permit the delineation of a variety of subcellular membrane populations. An important virtue of this comprehensive, empirical approach is that it accounts for the possibility that enzyme activities traditionally used as plasma membrane markers may undergo turnover and, therefore, may not be exclusively localized to the plasma membranes. By the same token, an empirical approach makes it possible to isolate and characterize membrane populations which lack unique enzymatic markers.

[1] A. K. Mircheff, *Am. J. Physiol.* **244,** G347 (1983).
[2] A. K. Mircheff, C. C. Lu, and C. N. Conteas, *Am. J. Physiol.* **245,** G661 (1983).
[3] A. K. Mircheff and C. C. Lu, *Am. J. Physiol.* **247,** G651 (1985).
[4] C. N. Conteas, A. A. McDonough, T. R. Kozlowski, C. B. Hensley, and A. K. Mircheff, *Am. J. Physiol.* **250,** C430 (1986).
[5] A. A. McDonough and A. K. Mircheff, *Fed. Proc., Fed. Am. Soc. Exp. Biol.* **42,** 1934 (1983).
[6] A. K. Mircheff, D. J. Ahnen, A. Islam, N. A. Santiago, and G. M. Gray, *J. Membr. Biol.* **83,** 95 (1985).
[7] A. K. Mircheff, H. E. Ives, V. J. Yee, and D. G. Warnock, *Am. J. Physiol.* **246,** F853 (1984).

Isolation of Subcellular Membranes

Homogenization and Differential Sedimentation

The basic isolation buffer consists of 5% (w/v) sorbitol, 0.5 mM NaEDTA, and 5 mM histidine–imidazole, pH 7.5. Density gradient media differ from this only in their sorbitol concentrations. A histidine–imidazole stock is prepared by mixing solutions of 25 mM histidine-HCl and 25 mM imidazole to obtain a final pH of 7.5; it is diluted 5-fold during preparation of the isolation buffer and density gradient media. It is also useful to prepare a stock solution of 100 mM NaEDTA with pH adjusted to 7.5 with NaOH. These solutions can be stored for several months at 4°.

Tissue homogenization can be accomplished with a variety of methods, including the Dounce apparatus,[8] the nitrogen cavitation bomb,[9] and the Tekmar Tissumiser. Several recent applications involving the Tissumiser have been described[3,6] in which the instrument was run for a prolonged time, i.e., 10–20 min, at a speed low enough to prevent cavitation and frothing. This approach had several useful results. It provides large yields of plasma membrane fragments in the low-speed supernatant fractions which are then subjected to density gradient centrifugation; the resulting plasma membrane populations consist of vesicles with a sufficient degree of sealing that they can be used in studies of electrolyte and amino acid transport; finally, the mitochondria remain intact, so that they can be separated from plasma membrane and other microsomal samples by procedures exploiting the large difference in sedimentation coefficients. A disadvantage of this homogenization method is that the lysosomes and secretory granules are frequently disrupted.

Isolated cells and intestinal mucosal scrapings can be suspended in isolation buffer for homogenization without further processing; solid tissues, such as exocrine glands and kidney cortex, should first be decapsulated and minced, e.g., with razor blades. The volume of isolation medium used for the initial homogenization and for subsequent washing steps is constrained by several factors, including the capacity of the rotor to be used for density gradient centrifugation and the sample-loading strategy to be employed.

Connective tissue, unbroken cells, and intact nuclei and secretory granules are removed from the initial homogenate by centrifugation at 2,000 g for 10 min. The supernatant is saved at 4° for subsequent density gradient centrifugation. The pellet is resuspended in isolation buffer, ho-

---

[8] A. K. Mircheff and E. M. Wright, *J. Membr. Biol.* **28**, 309 (1976).
[9] A. K. Mircheff, S. D. Hanna, M. W. Walling, and E. M. Wright, *Prep. Biochem.* **9**, 133 (1979).

mogenized in the Tissumiser, and again centrifuged at 2,000 g for 10 min. It is usually possible to obtain more than 75% of the plasma membranes in the supernatants combined from two to four cycles of homogenization and low-speed centrifugation. To construct a balance sheet of the distributions of biochemical markers among the fractions arising in the course of the isolation procedure, the final pellet should be resuspended to a known volume in isolation buffer.

Density Gradient Centrifugation

Zonal Rotors. Zonal rotors offer several practical advantages to swinging bucket rotors for density gradient centrifugation. Because zonal rotors have large capacities, they can accommodate large amounts of tissue in relatively large sample volumes. For example, one of the smaller zonal rotors, the Beckman Z-60, which has a 330 ml total capacity, has been used to separate more than 500 mg of membrane protein loaded in sample volumes of 50–75 ml. Since the rotor can accommodate such a large sample volume, the low-speed supernatants pooled from several cycles of homogenization, centrifugation, and resuspension could be analyzed by density gradient centrifugation without first having been concentrated by high-speed sedimentation. This strategy saves 60 to 90 min of preparation time. The Beckman Z-60 zonal rotor has the additional advantage that its maximum rated speed of 60,000 rpm generates 4-fold greater centrifugal forces than can be attained with swinging bucket rotors of comparable capacity, permitting shorter run times.

Most zonal rotors are designed to be loaded and unloaded while they are spinning at 2000 or more rpm. Samples and density gradient media are pumped into the rotor in the order of increasing density. The first increment of fluid entering the rotor accumulates as a cylindrical annulus held against the rotor edge by the centrifugal field. The lowest-density fluid is continually displaced toward the center by the incoming fluid until it reaches the rotor's central core. At the end of the run, the rotor contents are displaced by a high-density solution. A peristaltic pump can be used for both loading and unloading.

Standard two-chambered gradient markers can be obtained from several suppliers or can be constructed in institutional machine shops. The solution with the composition desired for the high-density end of the gradient is placed in the reservoir chamber, and the solution representing the low-density end of the gradient is placed in the mixing chamber. The outflow valve of the mixing chamber is connected, via the peristaltic pump, to the loading port of the rotor's rotating seal assembly. In principle, linear gradients would be obtained if the fluid levels in the reservoir and mixing chambers were allowed to fall at the same rate. However, since the vortex produced by vigorous mixing distorts the column height in the mixing gradient, it is not usually possible to obtain highly reproducible linear gradients with this simple apparatus. One way to avoid this problem is to place a piston against the surface of the fluid in the mixing chamber before stirring is begun and to allow the volume entering the rotor to equal the original volume of the reservoir chamber. The result will be a hyperbolic gradient of density versus volume. The density will initially increase linearly with effluent volume; it will asymptotically approach the density of the solution in the reservoir chamber. Increasing the volume of fluid in the mixing chamber will expand the linear region of the gradient. A somewhat more complex alternative utilizes parallel pistons, driven by an electric motor to displace fluid from the reservoir and mixing chambers at the same rate. The result in this case will be a highly reproducible linear gradient.

It is useful to have two branches to the tubing line between the gradient maker and the rotating seal, one proximal and one distal to the peristaltic pump. The proximal branch is used for delivering the sample and solutions of constant density. The distal branch is used for diverting air bubbles before they enter the rotor. Hemostats can be used to direct the flow of fluid through one or another of the tubing lines. In the simplest density gradient separations, the rotor is loaded with (1) 5–10 ml of a low-density solution, e.g., isolation buffer, which serves as an overlay; (2) the sample, e.g., the pooled low-speed supernatants; (3) a hyperbolic density gradient; and (4) a 20–30 ml cushion of a high-density solution, e.g., 80% sorbitol, which prevents the membranes from sedimenting against the edge of the rotor.

The basic gradient design can be varied to optimize desired separations. For example, two gradient makers can be used sequentially to obtain a continuous gradient comprising two steep regions and two shallow regions. The volumes of the density gradient solutions can then be adjusted empirically so that the regions of slow density change are exploited to enhance the separations between populations of similar densities. A frequently useful modification of this theme is to load the sample in at a density approximating the limiting concentration of the first hyperbolic gradient. In this case, the sample should be mixed with an appropriate volume of a very-high-density sorbitol solution. Our practice has been to use a stock solution of 87.4% (w/v) sorbitol. Adding 1.35 g of this solution per gram of sample in 5% sorbitol isolation buffer results in a final sorbitol concentration of approximately 55% (w/v). To most conveniently prepare 500 ml of the 87.4% sorbitol solution mix 437 g sorbitol, 100 ml histidine–imidazole buffer, 2.5 ml 100 mM NaEDTA, and water to a final total weight of 639.3 g. Virtually all the sorbitol will dissolve if the mixture is allowed to sit overnight at room temperature. Mixing with a magnetic stirrer will dissolve the remainder.

Once the rotor is completely loaded, the rotating seal assembly is removed and replaced by an air-tight cap. After a sufficient vacuum has been attained the rotor is accelerated to its maximum rated speed. At the end of the run, the rotor is decelerated back to the loading speed. The rotating seal is installed, and a very-high-density solution is pumped into the rotor to displace the gradient contents through the center line of the rotating seal. An 80% (w/v) sucrose solution is more dense than 80% sorbitol and can be used for this purpose. The gradient fractions can be collected directly into ultracentrifuge tubes.

Swinging Bucket Rotors. The sample-loading strategies and density gradient designs discussed above can be adapted for use with swinging bucket rotors, most conveniently so if a peristaltic pump is being used. For example, hyperbolic gradients with the same shape as described for the zonal rotor can be obtained by delivering the fluid through rigid tubing which is held at the bottom of the centrifuge tube, then carefully withdrawn when formation of the gradient is complete. Such gradients can be prepared in advance, and if the sample is to be loaded at any point other than the top of the gradient, it can be delivered by reinserting the delivery tube to the appropriate level. (Note that, in order to avoid air bubbles which will distort the gradient, it will be necessary to prime the delivery tube with sample before reimmersing it into the gradient.)

Several commercially available systems have been designed for fractionating density gradients from swinging bucket rotors. A simple alternative, which is feasible if there is no pellet beneath the density gradient, is to use a peristaltic pump with the flexible pump tubing connected to rigid tubing long enough to reach the bottom of the centrifuge tube. The rigid tubing is slowly passed through the gradient until it reaches the bottom of the centrifuge tube. The pump is then used to fractionate the gradient from bottom to top.

Separation of Membranes from Gradient Media. Because the membranes have sedimented to regions of the gradient corresponding to their equilibrium densities, the fractions will have to be diluted before the membranes can be pelleted. The choice of dilution solution depends on the subsequent goals of the experiment. The fractions must be mixed thoroughly after dilution, since a layer of unmixed high-density solution will prevent membranes from sedimenting completely to the bottom of the centrifuge tube. Vortex mixing will not be adequate if centrifuge tubes have been filled completely; such fractions can be thoroughly mixed with a paddle attached to a stirring motor. The centrifugal force sufficient to sediment the membranes will depend on the extent to which the fractions have been diluted. Centrifugation at 30,000 to 50,000 rpm for 1 to 2 hr is frequently sufficient after a 1:1 dilution. If a balance sheet of marker distributions and recoveries is being kept, the supernatants should be included in the analysis, especially if the membranes had not been concentrated and separated from soluble components by an ultracentrifugation step prior to density gradient centrifugation in a zonal rotor. It usually suffices to pool all the supernatants in a graduated cylinder, mix by inversion, measure the total volume, and save a 1- to 3-ml aliquot for analysis. The membrane pellets can be resuspended in any medium required for the subsequent goals of the experiment. If the fractions are to be analyzed with additional separation by either phase partitioning or density perturbation, they should be resuspended in isolation buffer. If they are to be used for transport studies, the resuspension volume should be kept small, and if both marker determinations and transport studies are to be done, aliquots of the concentrated membrane fractions can be diluted into isolation buffer. After small aliquots have been set aside for marker determinations, the fractions may be quickly frozen in liquid nitrogen and stored for further use at $-70°$.

Separation of Microsomes and Mitochondria by Differential Rate Sedimentation. Occasionally, the mitochondria have a density distribution overlapping that of a plasma membrane population of interest, e.g., the brush border membranes from intestinal epithelial cells.[6] Because the plasma membrane vesicles have sedimentation coefficients in the microsomal range, the two populations can be separated by differential centrifugation techniques. A number of such techniques are available; the technique used in analytical fractionation of intestinal and exorbital gland epithelial cells will be described here.

Density gradient fractions containing the mitochondria and plasma membranes are pooled to a volume of up to 6 ml and mildly homogenized. The sample is layered over a column of 17.5% sorbitol in an ultracentrifuge tube suitable for use in the Beckman SW28 or comparable swinging bucket rotor. Centrifugation for 25 min at 15,000 rpm will sediment the mitochondria to the bottom of the tube while only slightly displacing the membrane vesicles from their initial position. The sorbitol column and sample layer are then decanted and diluted, and the membranes are harvested by ultracentrifugation.

Subfractionation by Phase Partitioning

Third-Dimension Separations. The centrifugation procedures described above separate membranes on the basis of two independent physical properties, or dimensions, i.e., sedimentation coefficient and equilibrium density. More frequently than not, one or more intracellular membrane populations will be identical to the plasma membranes with respect to both these properties. It is also possible that the apical and basolateral membrane populations will have overlapping sedimentation and density distributions. In these situations it will be necessary to employ separation procedures which detect other membrane properties. Even when a plasma membrane preparation obtained by a combination of centrifugation procedures is thought to be pure, use of a third, independent procedure provides a useful test of the sample's homogeneity. The final result can be considered a three-dimensional analysis of the membrane populations present in the initial homogenate.

The procedures currently being used for third-dimension membrane separations include electrophoresis, which operates on the basis of membrane surface charge; density perturbation with digitonin, which operates on the basis of membrane cholesterol content; and phase partitioning, which is essentially a liquid chromatographic procedure that can be made to detect a variety of membrane surface properties. Phase partitioning was adopted for a series of studies involving comprehensive analytical fractionation of epithelial tissues because it promised to be highly versatile and because the literature on phase partitioning contained examples of a number of striking separations with nonepithelial membrane preparations. It has recently been applied to four different epithelial tissues: exorbital lacrimal gland,[2,3] parotid gland,[4] small intestine,[6] and kidney cortex.[7]

Countercurrent Distribution. Aqueous two-phase systems result when solutions of two different polymers, each of which may have a high water solubility, are mixed together at appropriate final polymer concentrations. Membranes will partition into the two phases and the interface at relative concentrations which are determined by factors that are still poorly understood but which can be empirically manipulated to optimize many separations. If two membrane populations are characterized by very large partitioning differences, they can be separated with only a few steps involving manually mixing and separating the upper and lower phases. Membrane populations which have similar, but quantitatively different, partition coefficients can be separated by countercurrent distribution.[10]

In countercurrent distribution with an aqueous two-phase system, the more-dense phase is held stationary, either by gravity or by a centrifugal field. The less-dense phase is caused to move through the stationary

---

[10] P.-A. Albertsson, B. Andersson, C. Larsson, and H.-E. Akerlund, *Methods Biochem. Anal.* **28,** 115 (1982).

phase by a continuous pump-driven flow in toroidal coil centrifuges[11–14] and by a series of discrete transfers in thin-layer devices.[15] Membranes which have the highest partition coefficients, i.e., which most strongly prefer the upper phase, migrate the most rapidly. They elute earliest from the toroidal coil device, and they migrate to the highest numbered fractions (farthest to the right) in the thin-layer apparatus. The thin-layer device designed by Albertsson has been employed for the epithelial cell fractionation studies noted above, and detailed instructions for its use are presented in this chapter.

The thin-layer apparatus used in these studies was purchased from the Central Workshop of the Department of Biochemistry 1, Chemical Center, University of Lund, Lund, Sweden. A similar apparatus is available from the University of Sheffield, Sheffield, England. The essential feature of the thin-layer device is a pair of circular plexiglass plates. Each plate contains at its periphery a symmetrical array of shallow, radially oriented cavities. When the plates are properly positioned, the cavities align to form chambers. The lower plate is secured to a horizontal table by a set of pins. The upper plate is secured to a motor which is programmed to rotate the upper plate with respect to the lower plate. In some models proper alignment is assured with a solenoid-driven positioning mechanism. The entire assembly is attached to a second motor which generates an orbital shaking motion.

Small holes in the upper plate provide access to each chamber; during operation these holes are covered by a plexiglass ring. In principle, enough of the lower phase is added to each chamber to fill the cavity contributed by the lower plate. Subsequently added aliquots of the upper phase will be contained by the cavities contributed by the upper plate. For automatic operation, the instrument is programmed to perform three functions: shaking, which mixes the phases; resting, which permits the phases to separate; and transfer, which transports the upper plate the distance of one cavity.

Preparation of Phase Systems. Membrane-partitioning properties are sensitive to essentially every variable which influences the equilibrium composition of the upper and lower phases, including polymer concentration, pH, buffer, ionic strength, and temperature. The phase systems used in epithelial cell fractionation studies have consisted of 5% dextran T-500

[11] Y. Ito, *J. Chromatogr.* **192,** 75 (1980).
[12] Y. Ito, *Anal. Biochem.* **102,** 150 (1980).
[13] I. A. Sutherland, D. Heywood-Waddington, and T. J. Peters, *J. Liq. Chromatogr.* **7,** 363 (1984).
[14] D. Heywood-Waddington, I. A. Sutherland, W. J. Morris, and T. J. Peters, *Biochem. J.* **21,** 751 (1984).
[15] P.-A. Albertsson, *Anal. Biochem.* **11,** 121 (1965).

and 3.5% PEG (Carbowax 8000, or Carbowax 6000 in Union Carbide's previous nomenclature). They contained 5% sorbitol, 8.3 mM imidazole, and 10 μM NaEDTA, and pH was adjusted with HCl. The phase systems are prepared from three separate stock solutions, one containing 12.5% dextran and 6.25% sorbitol, one containing 8.75% PEG and 6.25% sorbitol, and one containing 1 M imidazole and 1.2 mM NaEDTA. The PEG and imidazole–EDTA solutions are easily prepared. The following steps will facilitate preparation of the dextran solution: Sorbitol is dissolved in about 75% of the total required water. Dextran is weighed into a large beaker; the capacity should be 3 to 4 times greater than the final desired volume of solution in order to accommodate the dry dextran powder. The sorbitol solution is then quickly added and immediately swirled by hand. Water is added to bring the mixture to the desired final weight, and a large magnetic stir bar is inserted. The beaker is sealed with Parafilm or plastic film (which can be held in place with rubber bands). Dissolution of the dextran globules should be complete within 2 hr. The polymer solutions can be prepared in quantities sufficient for several countercurrent distribution runs, i.e., 500–1000 g. If so, they should be divided into 100 g aliquots and stored at −20°. Since the polymer solutions are viscous, a magnetic stirrer will be required for thorough remixing after the solutions have been thawed.

Two hundred and forty grams of phase system is sufficient for a 120-step separation in the Lund thin-layer apparatus. Final dextran and PEG concentrations of 5 and 3.5% are obtained by mixing, at room temperature, 96 g of the dextran solution, 96 g of the PEG solution, and 2 ml of the imidazole–EDTA solution. The pH of the phase system is adjusted with 1 N HCl, which should be added dropwise during continuous stirring. In initial experiments titration of the phase system should be monitored with a pH electrode. Once the volume of HCl giving the optimal pH has been determined, it will no longer be necessary to monitor the titration; however, the HCl should still be added dropwise with continuous stirring. After the titration is complete, the stir bar is removed and rinsed; one must be conservative at this step so as not to exceed the total final weight of 240 g. When the phase system has been brought to its final composition, it is thoroughly swirled by hand and transferred to a separatory funnel which has been prechilled to 4°. The separatory funnel is covered with Parafilm and mixed again by gentle inversion. The phases should be allowed to equilibrate in the cold room for 16–18 hr.

The dextran-rich phase will tend to adhere to the wall of the separatory funnel during separation of the upper and lower phases. The consequences of this phenomenon are most acute as the interface nears the bottom of the funnel, i.e., when the linear displacement of the interface is rapid with respect to the volume flow rate. It can be countered by using a variable-speed peristaltic pump to control the flow rate. A pump rate of 2 ml/min is typically used to withdraw the first 75 ml of the lower phase. The rate is decreased to 1 ml/min and finally to 0.25 ml/min. The stopcock of the separatory funnel is closed once the interface has passed completely through; the peristaltic pump should be turned off at this point, since care must be taken to avoid contaminating the lower phase with interfacial material. The upper phase is collected by pouring through the top of the separatory funnel.

Loading the Thin-Layer Apparatus. The volume loaded into each chamber of the thin-layer device depends on the volume of the chambers and on whether the interface is to be retained with the lower phase (i.e., when the separations of interest are between the upper phase and the interphase + lower phase) or to be transported with the upper phase (i.e., when the separations are between the upper phase + interface and the lower phase). A constant volume of lower phase is loaded into each chamber. The sample will be loaded into the first 10% of the total number of chambers to be used for the separation. The remainder of the chambers are loaded with upper phase containing no sample.

The membrane preparation to be analyzed is concentrated by ultracentrifugation; after the supernatant has been decanted, the walls of the centrifuge tube are thoroughly wiped with an absorbent tissue. The membranes are resuspended in an aliquot of the upper phase. This resuspension should be carried out at the same temperature as the separation itself. When a motor-driven device, e.g., a Tissumiser, is used, the tube containing the sample should be immersed in a water bath that has equilibrated in the cold room.

After sample loading, the shaking motor is operated manually; the apparatus is switched to automatic operation at the end of this shaking interval. It appears that most countercurrent distribution analyses have been performed under dynamic, rather than equilibrium, conditions. This implies that the observed migration of membranes will be a function of the size of droplets formed during the shaking step and the time allowed for phase separation during the resting step. Therefore, it will be necessary to rigorously control the shaking speed and rest time. In epithelial cell fractionation studies, the rest time was 11 min, the shaking time was 1.5 min, and the shaker setting was 6.3.

It will be necessary to perform pilot studies to design an optimal phase composition for each new preparation to be analyzed by phase partitioning. Two particularly useful parameters to vary are the polymer concentration and the pH. Increasing the PEG concentration tends to shift membranes from the upper phase to the lower phase + interface, while increasing the pH has the opposite effect. It is possible to test several phase systems simultaneously in thin-layer devices having plates containing a sufficiently large number of chambers. For example, a 120-chamber system can be used to perform three simultaneous 20-step analyses. In this case the plates are divided into three sectors of 40 chambers of each. Sector A comprises chambers 1–40, sector B, chambers 41–80, and sector C, chambers 81–120. Each sector is used for a separate phase system. Chambers 1–40 are loaded with the lower phase from the phase system to be tested in sector A. Chambers 1–20 and 23–40 are loaded with the upper phase from the same phase system. Chambers 21 and 22 are loaded with sample resuspended in an aliquot of the upper phase from this phase system. The remaining sectors are loaded in the same manner, and after manual mixing, the apparatus is allowed to perform 18 cycles of phase separation, transfer, and mixing. At the end of this sequence, position 22 of the upper plate will be over position 40 of the lower plate; position 62 of the upper plate will be over position 80 of the lower plate; and position 102 of the upper plate will be over position 120 of the lower plate. The contents of chambers 21–40, 61–80, and 101–120 (numbered with respect to the lower plate) are then collected for analysis. It is also possible to use a similar strategy to analyze simultaneously several different samples with the same phase system. For example, if three samples were to be analyzed in a 120-chamber apparatus, the samples would be resuspended in upper phase as usual; they would be loaded into chambers 1–4, 41–44, and 81–84; and the apparatus would be programmed to perform 36 steps.

Fraction Recovery and Membrane Concentration. The thin-layer countercurrent distribution apparatus is provided with a fraction collector ring. This is designed to hold plastic culture tubes at the openings in the upper plate. At the end of the countercurrent distribution run, sufficient buffer is added to each chamber to dilute the polymers to concentrations below the critical point for phase separation. The cover ring is replaced, and the shaker is turned on for 5 min. After this time, the cover plate is replaced by the fraction collector ring with tubes in place and labeled. The holders and pins securing the plates to the apparatus are released, and the entire three-piece assembly (lower plate, upper plate, and fraction collector ring) is carefully lifted free and inverted. Some shaking will be necessary to overcome surface tension and allow the chamber contents to flow into the fraction collector tubes.

The individual fractions can be analyzed directly for most commonly used membrane markers. If a single sample has been analyzed with a relatively large number of transfers (40 or more), it usually suffices to analyze every other fraction. Many experimental goals, e.g., transport studies or determination of low-specific-activity markers, will require that the membranes be concentrated prior to analysis. In this case the fractions can be pooled into groups of 5 or 10, diluted with buffer, and centrifuged at high speed, i.e., at least 30,000 rpm for 60 min. It has frequently been the case that the membranes precipitate to a cloud which remains suspended at the bottom of the tube; this probably reflects the presence of a high-density cushion formed by sedimentation of dextran. The membranes will form a tight pellet if they are diluted and centrifuged a second time. In this procedure, roughly 90% of the initial supernatant is withdrawn; care should be taken so that the membrane cloud is not disturbed. The membranes are resuspended in the residual supernatant fluid, then diluted with a buffer appropriate for the subsequent experimental goals.

Analysis

The physical separation procedures outlined above are potentially capable of resolving a large number of physically distinct membrane populations. The populations, once resolved, must be delineated through the use of various membrane marker activities. Some of the most frequently used markers and their major subcellular loci are presented in Table I.[16-24] Detailed analytical fractionation studies have shown that many of these markers have diverse subcellular distributions, so one should always recognize that each marker may be associated in measurable quantities with a multiplicity of membrane populations.

In attempting to use biochemical markers to delineate isolated membrane populations, one will initially view the distribution patterns in a statistical sense. For example, a peak on a gradient after equilibrium density gradient centrifugation reveals the modal density of a population of membranes bearing that marker. The appearance of more than one distinct peak indicates that the marker is associated with several populations distinguished by their modal densities. A broad, asymmetrical peak might reflect either a single population with a skewed density distribution

---

[16] H. Murer, E. Amman, and U. Hopfer, *Biochim. Biophys. Acta* **433,** 509 (1976).
[17] R. J. Turner and A. Moran, *Am. J. Physiol.* **242,** F406 (1982).
[18] L. Naftalin, M. Sexton, J. Whitaker, and D. Tracey, *Clin. Chim. Acta* **26,** 293 (1969).
[19] A. K. Rao, F. Garner, and J. Mendocino, *Biochemistry* **15,** 5001 (1976).
[20] R. H. Michell, M. J. Karnovsky, and M. L. Karnovsky, *Biochem. J.* **116,** 207 (1970).
[21] A. J. Barnett and M. F. Heath, in "Lysosomes: A Laboratory Handbook" (J. T. Dingle, ed.), p. 118. Elsevier/North-Holland, Amsterdam, 1975.
[22] A. K. Mircheff, G. Sachs, S. D. Hanna, C. S. Labiner, E. Rabon, A. P. Douglas, M. W. Walling, and E. M. Wright, *J. Membr. Biol.* **50,** 543 (1979).
[23] R. J. Pennington, *Biochem. J.* **80,** 649 (1961).
[24] Bio-Rad Laboratories, "Bio-Rad Instructional Manual #78-0791." Bio-Rad Laboratories, Richmond, California.

TABLE I
TYPICAL MEMBRANE MARKERS

Marker	Principle location	Notes	Ref.
Na^+,K^+-ATPase (K^+-pNPPase)	Basolateral membranes	The reaction may be quenched with 1 N NaOH containing 0.1% Triton and 50 mM NaEDTA	16
Sucrase and maltase	Brush border membranes	Sucrase is found only in small intestine; maltase is found in small intestine and proximal tubule	17
Alkaline phosphatase	Apical membranes		8
γ-Glutamyltransferase	Apical membranes		18
Galactosyltransferase	Trans Golgi elements		19
Acid phosphatase	Various intracellular membranes	Modified by addition of EDTA and use of p-nitrophenyl phosphate as substrate as described by Mircheff and Wright[18]	20
N-Acetyl-β-D-glucosaminidase	Lysosomes		21
NADPH–cytochrome-c reductase	Endoplasmic reticulum	The linear phase of the reaction can be extended by increasing the NADPH and cytochrome c concentrations to 2 and 0.9 mg/ml, respectively, and by adding 5 mM sodium azide	22
Succinate dehydrogenase	Mitochondria		23
Protein		This procedure is preferable to other standard protein determinations because it is less subject to interference by sorbitol, dextran, and PEG	24

or the summation of several incompletely resolved membrane populations. It is usually possible to distinguish between these alternative explanations by pooling gradient fractions containing distinct features of the peak, then analyzing each pooled fraction by the same third-dimension separation procedure. (Since these pooled fractions represent spans of the density gradient, they have been referred to as density windows.) Both density perturbation with digitonin[25] and countercurrent distribution[3,4,6]

[25] A. K. Mircheff, C. N. Conteas, C. C. Lu, N. A. Santiago, G. M. Gray, and L. G. Lipson, Am. J. Physiol. **245**, G133 (1983).

have been used for this explicit purpose; countercurrent distribution appears to give both better resolution and better preservation of membrane permeability barriers essential for transport studies. If the peaks occurring after countercurrent distribution of adjacent density windows are located in different fractions, one would conclude that the two windows contain distinct membrane populations which differ on the basis of their partitioning properties. If the positions of the peaks indicate that the samples have similar partitioning properties, it will be necessary to compare the marker specific activities in the modal partitioning fractions from each density window. If the specific activities vary substantially, the samples can be assigned, at least provisionally, to separate populations. Further work might ultimately show that these populations are, in fact, subpopulations of the fragments generated by disruption of a single organelle, but it is also possible that the physical differences between such subpopulations might reflect functional differences.

A second step in the analysis is to use biochemical markers as the basis for identifying the isolated membrane populations. This step should take advantage of available biophysical and cytochemical data on the subcellular distributions of biochemcial markers. In most cases, the basolateral plasma membrane population can be identified as the major locus of high-specific-activity Na^+,K^+-ATPase. Alkaline phosphatase and γ-glutamyltransferase are probably the most generally useful apical membrane markers. In exocrine glands, where the epithelial cells have a pyramidal shape with relatively few microvilli, the apical membrane surface will be smaller than in the columnar absorptive cells of the proximal tubule and small intestine. Therefore, the isolated apical membranes may not be the major locus of the conventional apical membrane markers, but they may be the population at which these markers have their highest specific activities.[3,4]

Calculation of Enrichment Factors. It is imperative to keep a balance sheet of the marker contents of the various fractions during design of a new membrane isolation procedure. This will make it possible to calculate overall recoveries and to determine whether any markers have been activated or inhibited. For example, if a plasma membrane marker is activated during the procedure, it will lead to an overestimate of the marker purification factor. On the other hand, if an endoplasmic reticulum marker is inhibited, it might suggest that the plasma membranes have been more highly purified than is in fact the case. Recovery data make it possible to correct for these phenomena.

It frequently happens that a marker is partially inactivated during each step of a three-dimensional isolation procedure. One can still estimate the overall enrichment of the membranes with which the marker is associated by calculating an incremental enrichment factor for each separation step; the cumulative enrichment factor is then the product of the individual incremental enrichment factors. One obtains the incremental enrichment factors by expressing the marker activity and protein content of each fraction as percentages of the total activity and protein recovered in all the fractions generated by that separation step. The ratios of these two quantities represent the incremental enrichment factors.

The final marker cumulative enrichment factor in a membrane population isolated through a series of three different separation steps serves as a measure of the relative specific activity of the marker in the population. This factor will still underestimate the extent to which the membrane population has been purified if the marker is associated to any great degree with other membrane populations. An approximate final membrane purification factor can be calculated by dividing the cumulative enrichment factor by an estimate of the fraction of the total recovered marker activity which can be assigned to the population of interest. For example, the cumulative enrichment factor for alkaline phosphatase was 33 in a population provisionally identified as apical membranes from exorbital gland acinar cells. Since this population appeared to account for only 10% of the total recovered alkaline phosphatase, it was estimated that the membrane population had been enriched 330-fold with respect to the initial homogenate.[3]

Function

One of the most frequent goals of a plasma membrane isolation procedure is to obtain a simplified preparation for characterizing epithelial transport mechanisms. Membrane preparations obtained through the isolation procedures presented above have already been found to retain a number of transport activities, including a sodium-stimulated and sodium-independent amino acid transport, ATP-dependent calcium transport, and Na^+/H^+ antiport. Recovery of transport activity is generally good after differential and density gradient centrifugation but has been quite variable after countercurrent distribution.

Detailed procedures for radiotracer and electrophysiological measurements of transport activities in epithelial membrane vesicles will be given under Topic 5 of this series. This section will focus on a strategy for using amino acid transport activities to substantiate the provisional identifications of plasma membrane populations that have been made on the basis of enzymatic marker activities. The premise of this strategy is that amino acid transport systems are differentially distributed between the apical and basolateral surfaces of epithelial cells. The sodium-independent neutral, L-α-amino acid transporter, System L, and the classical sodium-dependent transporters, Systems A and ASC, are concentrated in the basolateral membranes.[2,26] Two different sodium-dependent transporters for neutral, L-α-amino acids appear to be concentrated in the apical membranes.[27] The apical membrane systems, which can be referred to as system 1 and system 2, share the characteristic that they have a high affinity for phenylalanine but do not interact with the test amino acid, methylaminoisobutyric acid. This characteristic distinguishes them from the classical, basolateral membrane systems, since System A interacts with both methylaminoisobutyric acid and phenylalanine, and System ASC interacts with neither phenylalanine nor methylaminoisobutyric acid. Therefore, it is possible to delineate the contributions of the three different kinds of systems by determining the effects of phenylalanine and methylaminoisobutyric acid on the rate of sodium-dependent uptake of L-alanine.[2]

Delineation of Parallel Amino Acid Transport Pathways

Preparation of Membranes. Membrane samples obtained after density gradient centrifugation or countercurrent distribution as described above are diluted with a loading solution containing 30 mM KCl, 340 mM sorbitol, and 5 mM Tris–HEPES buffer, pH 7.6, concentrated by centrifugation at forces of at least 100,000 g for 60 min. The resulting pellets are resuspended in loading solution, ideally at protein concentrations of roughly 4 mg/ml. Samples may be quickly frozen and stored at $-70°$ as discussed above.

Reaction Media. The complete transport reactions contain 10 μM L-[^{14}C]alanine (1.7 μCi/ml), 340 mM sorbitol, 5 mM Tris–HEPES, pH 7.6, and either 50 mM NaCl or 50 mM KCl. When the competing amino acids, phenylalanine and methylaminoisobutyric acid, are to be added, their final concentrations are 62.5 mM, and the sorbitol concentration is reduced accordingly. It is convenient to prepare stock solutions containing the alanine, NaCl or KCl, and competing amino acids at 1.33-fold higher concentrations, so that the desired final concentrations result when 75 μl is mixed with the 25-μl membrane sample to initiate the transport reactions.

Procedure. Detailed procedures will be given under Topic 5. Briefly, transport reactions are initiated by adding an aliquot of membrane sample to the reaction medium and rapidly mixing. They are quenched (at a time

---

[26] A. K. Mircheff, C. H. van Os, and E. M. Wright, *J. Membr. Biol.* **52**, 83 (1980).
[27] A. K. Mircheff, I. Kippen, B. Hirayama, and E. M. Wright, *J. Membr. Biol.* **64**, 113 (1982).

point within the linear phase of the transport reaction) by addition of a 10-fold excess of an ice-cold stop solution which resembles the reaction medium but contains no alanine. Membrane vesicles containing transported alanine are collected by ultrafiltration on nitrocellulose filters, which are rinsed with additional stop solution before being counted in a liquid scintillation spectrometer.

Analysis. Sodium-dependent alanine fluxes are calculated as the difference in the rates of uptake in the presence and absence of external sodium. The degree of inhibition by methylaminoisobutyric acid represents the component of the alanine influx mediated by System A. The flux remaining in the presence of phenylalanine represents the component mediated by System ASC. Finally, the difference between inhibition by phenylalanine and inhibition by methylaminoisobutyric acid represents the collective contributions of the apical membrane systems.

[3] Transport-Specific Fractionation for Purification of ATP-Dependent Ca^{2+} Pumps

By STANLEY M. GOLDIN and STEVEN C. KING

Introduction

We have developed an approach, transport-specific fractionation, as a means for the identification of membrane transport proteins and as a purification procedure. Like affinity chromatography, the method depends on a specific biological property of the protein one wishes to isolate; however, in this case we employ the transport activity itself of the protein (rather than a ligand-binding property as would be used in affinity chromatography), as a physical tool for its purification. Previous articles have presented details for transport-specific fractionation of the human erythrocyte sugar transport system,[1,2] and application of the method to the detergent-solubilized saxitoxin receptor/action potential Na^+ channel has been presented elsewhere.[3,4] Another chapter in this series[5] discussed

---

[1] S. M. Goldin and V. Rhoden, *J. Biol. Chem.* **253**, 2575 (1978).
[2] S. M. Goldin, *in* "Red Cell Membranes—A Methodological Approach," p. 301. Academic Press, London, 1982.
[3] S. M. Goldin, V. Rhoden, and E. J. Hess, *Proc. Natl. Acad. Sci. U.S.A.* **77**, 6884 (1980).
[4] S. M. Goldin, E. G. Moczydlowski, and D. M. Papazian, *Annu. Rev. Neurosci.* **6**, 419 (1983).
[5] S. M. Goldin, M. Forgac, and G. Chin, this series, Vol. 156, p. 127.

the use of transport specific fractionation to separate inside-out from right-side-out vesicles containing the purified, reconstituted Na^+,K^+-ATPase.

The most substantial success of this method in terms of degree of purification (up to ~100-fold) has been its application by several laboratories to the purification of Ca^{2+} transport proteins, notable ATP-dependent Ca^{2+} pumps[6-10] and the Na^+/Ca^{2+} exchanger.[11,12] This article details the methodology for purification of reconstituted ATP-dependent Ca^{2+} pumps from mammalian brain synaptosomes.

Strategy

Reconstitution of Transport System of Interest into Vesicles Before Purification

A membrane fraction containing the transport system is solubilized with cholate in the presence of a large excess of added lipid. The supernatant fraction contains the transport activity as well as other membrane proteins. Vesicles containing the transport activity are formed by the hollow fiber dialysis technique.[5,13] Previous studies on the reconstitution of purified Na^+,K^+-ATPase by cholate dialysis[13,14] indicate that one can obtain the apparently random distribution of functional single molecules of the pump among the resulting small, dimensionally homogeneous (400–600 Å diameter) unilamellar vesicles. As is the case for synaptosomal Ca^{2+} pumps, when the starting material employed for reconstitution is a heterogeneous membrane fraction rather than a purified membrane protein (Fig. 1a), the solubilized proteins are also randomly distributed among the vesicles upon reconstitution; at a sufficiently high lipid/protein ratio, each vesicle will again contain one or only a few protein molecules, and just a fraction of the vesicles will contain the membrane transport protein of interest (Fig. 1b).

[6] D. M. Papazian, H. Rahamimoff, and S. M. Goldin, *Proc. Natl. Acad. Sci. U.S.A.* **76,** 3708 (1979).
[7] S. Y. Chan, E. J. Hess, H. Rahamimoff, and S. M. Goldin, *J. Neurosci.* **4,** 1468 (1984).
[8] D. M. Papazian, H. Rahamimoff, and S. M. Goldin, *J. Neurosci.* **4,** 1933 (1984).
[9] S. M. Goldin, S. Y. Chan, D. M. Papazian, and H. Rahamimoff, *Cold Spring Harbor Symp. Quant. Biol.* **48,** 287 (1983).
[10] S. H. Lin, *J. Biol. Chem.* **260,** 7850 (1985).
[11] S. Luciane, *Biochim. Biophys. Acta* **772,** 127 (1984).
[12] A. Barzilai, R. Spanier, and H. Rahamimoff, *Proc. Natl. Acad. Sci. U.S.A.* **81,** 6521 (1984).
[13] S. M. Goldin, *J. Biol. Chem.* **252,** 5630 (1977).
[14] G. Chin and M. Forgac, *J. Biol. Chem.* **259,** 5255 (1984).

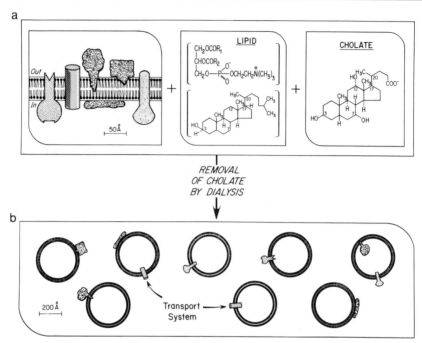

FIG. 1. Schematic representation of the process of reconstitution of membrane proteins into artificial lipid vesicles by the "cholate dialysis" technique. (a) The membranes containing the transport protein of interest are solubilized with cholate in the presence of a large excess of exogenous lipid. Phosphatidylcholine and cholesterol are depicted here, although a variety of other lipid components have been used as well. (b) After removing the cholate by dialysis, small unilamellar vesicles are formed which contain only one or a few protein molecules per vesicle.

Isolation by Use of Specific Permeability Properties of the Transport System

Vesicles containing the transport system are fractionated from the rest of the vesicle population by density gradient centrifugation. This is done by specifically altering intravesicular density according to the transport capabilities of a given vesicle. Because functional reconstitution of transport is required for this approach to work, one knows that a transport activity is being purified, not merely a binding protein or an enzyme that might be missing crucial transport components.

The strategy for purification of reconstituted Ca^{2+} pumps[6-10] and the Na^+/Ca^{2+} exchanger[11,12] takes advantage of the fact that intravesicular formation of a Ca^{2+} precipitate increases the density of only those vesicles exhibiting active uptake of Ca^{2+}, enabling their separation from the rest of the vesicle population on isopycnic density gradients. For purification of reconstituted synaptosomal Ca^{2+} pumps, a two-stage procedure is employed.

(1) Reconstituted vesicles with oxalate trapped inside are incubated first with Mg^{2+} and Ca^{2+} but without ATP, and subjected to the first round of density gradient centrifugation. The purpose of the first gradient is to remove nonspecifically leaky vesicles and other material that sediment to a position below that of the sealed lipid vesicles, since these would interfere with the subsequent purification.

(2) The sealed vesicle peak from the first gradient is then collected and incubated as before but in the *presence* of ATP, and resedimented on a second gradient (Fig. 2). A calcium oxalate precipitate forms inside only those vesicles containing ATP-dependent Ca^{2+} uptake activity, rendering them more dense than the vast majority of other sealed vesicles.

Since the starting material is reconstituted so that there is, on average, one protein molecule per vesicle, substantial purification can result. When applied to membrane proteins reconstituted from rat brain synaptosomes, transport specific fractionation results in up to ~100-fold purification of the ATP-dependent Ca^{2+} uptake activity. Three components of the original synaptosomal fraction, of molecular weights ~230,000, 140,000, and 94,000, are purified[6] (C230, C140, and C94, respectively). Subsequent studies revealed that these components contained two distinct Ca^{2+} pumps: one of them, C140, is a calmodulin-stimulated Ca^{2+} pump that can be separately purified by combining transport specific fractionation with calmodulin affinity chromatography.[7] The use of a vesicle-enriched subfraction of osmotically lysed bovine brain synaptosomes results in copurification of C94 and C230 in the absence of significant amounts of C140.[8] Immunological evidence indicates that C94 and C230 are structurally related, and are antigenically distinct from the Ca^{2+} pumps both of sarcoplasmic reticulum and of erythrocytes of the same species.[8]

The procedural details for transport-specific fractionation of the synaptosomal Ca^{2+} pumps are provided below. The method can be employed on both an analytical scale (yielding a few micrograms of substantially purified protein per run) and on a larger, preparative scale (yielding ~100 μg or more per run). With some modification, the method should be adaptable for purification of other Ca^{2+} transport proteins.

Analytical-Scale Transport-Specific Fractionation

Synaptosomal vesicles are prepared from rat brain[6] or bovine brain[7] as previously described. This membrane fraction, derived from osmotically lysed synaptosomes, is stored as a concentrated suspension (~10 mg/ml

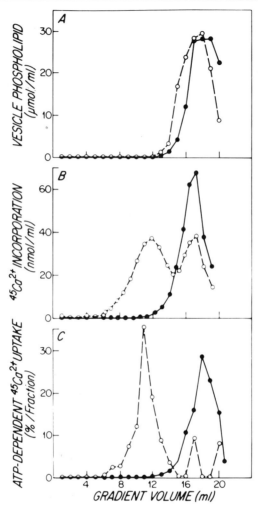

FIG. 2. Results of the two-stage preparative-scale purification of the reconstituted Ca^{2+} transport system from bovine synaptosomal vesicles, as compared with the results of a control preparation. Before being placed on the second gradient, the liposome peak obtained from the first gradient was preincubated with Mg^{2+} and Ca^{2+} in the presence (○) or absence (●) of MgATP. The distribution of $^{45}Ca^{2+}$ within vesicles (B) and the ATP-dependent Ca^{2+} uptake activity, assayed for 2 min at 24°, were determined in fractions of this second gradient. When the liposomes were preincubated in the absence of MgATP prior to being placed on the second gradient, both the $^{45}Ca^{2+}$ distribution (B) and the ATP-dependent Ca^{2+} uptake activity (C) coincided with the major vesicle phospholipid peak (A). From Ref. 7.

protein) at $-70°$ in aliquots of ~ 1 mg protein. With regard to its use as the starting material for reconstitution, the fraction is stable for several months under these conditions. Repetitive freezing and thawing substantially reduces the recovery of reconstitutable Ca^{2+} pump activity.

Crude soybean phospholipid (asolectin) is obtained from Associated Concentrates, Woodside, NY. This solid granular material is stored desiccated below $-20°$. The desiccator should be warmed to room temperature before the moisture barrier is broken and the sample is weighed. The acetone washing procedure previously employed for pretreatment of asolectin[15] has been found to be unnecessary in this case, and in fact $^{45}Ca^{2+}$ transport activity appears to be more stable when the reconstituted vesicles are prepared from unwashed asolectin. Stock solutions of the lipid may be prepared in bulk in the presence of 5 mM 2-mercaptoethanol (as an antioxidant) and may be stored frozen at $-70°$. Because the efficiency of the reconstitution procedure is critically dependent on the ratio of cholate to phospholipid and the asolectin is hygroscopic, it is important to employ lipid phosphate analysis rather than gravimetric procedures to determine accurately lipid concentration: the solid asolectin is uniformly dispersed into the 5 mM 2-mercaptoethanol by homogenization; lipid phosphate is determined by total phosphate analysis as previously described[13] and, using 750 g/mol as the average molecular weight of phospholipid, the concentration of phospholipid in the stock solution is adjusted to 80 mg/ml.

Cholic acid (Sigma Chemical Co.) is decolorized with activated charcoal, recrystallized from 95% ethanol, and titrated into the sodium form as detailed in an earlier chapter in this series.[5] Alternatively, dissolve 20 g of cholic acid in 100 ml of hot 70% ethanol and recrystallize slowly at room temperature and then at 4° to increase the yield. Repeat the crystallization procedure to complete decolorization.

Two critical variables to consider when solubilizing the Ca^{2+}-ATPase for use in transport specific fractionation are the phospholipid to protein ratio (about 80:1 by weight) and the cholate to phospholipid ratio (optimally 1–1.2:1 by weight). To prepare the solubilization buffer, the phospholipid is suspended at 20 mg/ml in a solution whose final concentrations are 20 mg/ml cholate, 5 mM 2-mercaptoethanol, 0.2 mM EDTA, and 0.4 M potassium phosphate, pH 7.5. The synaptosomal membranes are added as a concentrated suspension to 3 ml of this solution, to a final concentration of 0.25 mg/ml protein. After incubation for 2 min at 22°, the mixture is chilled for 5 min in an ice-water bath and dialyzed in a hollow fiber apparatus at 4°. The loading of the hollow fiber dialysis unit and the rate of

---

[15] Y. Kagawa and E. Racker, *J. Biol. Chem.* **246**, 5477 (1971).

dialysis are identical to the methods presented in a previous chapter in this series.[5] The material is dialyzed initially against 1 liter of "high-oxalate buffer" (300 mM oxalic acid, 5 mM 2-mercaptoethanol, pH 7.8 with KOH), over a 12-hr period. The resulting vesicles are then dialyzed against "low-oxalate buffer" (5 mM oxalic acid, 700 mM glycerol, 25 mM KCl, 50 mM Tris, 25 mM ammonium acetate, pH 7.5 with KOH), for 2 hr, and then the vesicles (~2 ml) are removed from the hollow fiber unit.

Linear density gradients are formed in tubes for a Beckman VTI 50 rotor, from 18 ml of 5 mM oxalic acid, 700 mM sucrose, 25 mM KCl, 50 mM Tris, 25 mM ammonium acetate, 5 mM 2-mercaptoethanol, pH 7.5 with KOH (heavy phase) and 16 ml of low-oxalate buffer (light phase). The gradient is layered over a 2-ml cushion of 50% sucrose. Vesicles are incubated for 20 min at 23° after MgCl$_2$ and CaCl$_2$ are added from 100-fold concentrated stock solutions to final concentrations of 5.5 mM and 0.1 mM, respectively (the bivalent cation solutions are added while the vesicles are rapidly vortexed, to minimize vesicle aggregation). The vesicles are then rechilled for ~5 min, and the 2 ml of vesicles is layered on top of the gradient, and overlayered with about 5 ml of gradient light phase.

The gradient tube and a second "blank" counterbalancing tube are heat sealed and subjected to centrifugation for 1.5 hr, 50,000 rpm at 2°. Because the gradient is relatively shallow, it is important to accelerate the rotor as slowly as possible during the gradient reorientation which occurs at a speed below 3000 rpm. Therefore, we have always used the minimum level of acceleration in a Beckman L5-50 centrifuge at speeds below 3000 rpm, and maximum acceleration above this speed. Alternatively, a slow acceleration mode may be used when the centrifuge is so equipped (as is the case for the Beckman L8 series). At the end of a run, the brake is turned off at 3000 rpm and the rotor is allowed to coast to a stop.

After centrifugation, 1-ml fractions are collected by piercing the bottom of the tube. The main lipid vesicle peak consists of four to six easily visible turbid fractions. The top three-fourths of these peak fractions are pooled and reincubated for 20 min, 23° with 0.1 mM CaCl$_2$/5.5 mM MgCl$_2$/2 mM MgATP; the sample is rechilled and then diluted with an equal volume of chilled low-oxalate buffer, overlayered on a second identical gradient, and subjected to centrifugation as before. Fractions (1 ml) of the second gradient are collected and assayed for ATPase activity and/or ATP-dependent Ca^{2+} uptake activity. Once the investigator establishes that the procedure works properly, the easiest way to monitor the purification is to assay ^{45}Ca^{2+} incorporated during the second preincubation by the procedure described below in the section "Preparative-Scale Transport-Specific Fractionation."

Assaying Calcium Transport

We use Dowex 50 minicolumns to separate intravesicular $^{45}Ca^{2+}$ from $^{45}Ca^{2+}$ in the medium. The distinct advantage of ion-exchange chromatography is its high sensitivity and its excellent reproducibility.[16] Below, we will discuss the practical aspects of chromatographing reconstituted vesicles on Dowex 50 minicolumns, and we will introduce a modification of the basic method which will allow (1) the use of chelating agents to control the free Ca^{2+} concentration, and (2) the rapid quenching of the $^{45}Ca^{2+}$ uptake reaction prior to applying samples to the column.

Preparation of the Resin

Before use, Dowex 50-X2-200 (Sigma Chemical Co.) must be converted from the hydrogen form to the Tris form (Dowex purchased in the Tris form and used without pretreatment has not been satisfactory). Add 250 g of the resin to 1.0 liter of 1 M Tris base. Mix for at least 15 hr on a magnetic stir plate (do not be concerned about grinding the resin). Pour off the Tris, add distilled water, and continue to stir. Pour off the water and add fresh distilled water several times per day over a 3-day period. Finally, pour off the water such that the volume above the packed resin is about one-half the total resin volume, creating a convenient slurry for pouring columns.

Pouring the Minicolumns

The minicolumns are $5\frac{3}{4}$ inch Pasteur pipets plugged with polyester wool (not glass wool) which can be obtained from local aquarium supply stores. The tiny amount of wool gently stuffed into each column weighs about 1 mg. Place the plugged columns in a support, and completely fill each column with the Dowex slurry while it is constantly stirred on a magnetic stir plate. To avoid trapping air, squirt the resin directly to the bottom of the column rather than applying it to the side. Continue to fill each column with the slurry such that, after settling, the resin level will be even with the indentation in the barrel of the Pasteur pipet.

Equilibrating the Columns

For assaying vesicles produced by transport-specific fractionation, the columns should be washed three times with 2 ml of column buffer (700

---

[16] O. D. Gasko, A. F. Knowles, H. G. Shertzer, E. M. Suolinna, and E. Racker, *Anal. Biochem.* **72**, 57 (1976).

mM glycerol, 10 mM Tris, pH 7.5 with HCl), and then once with 1% bovine serum albumin (Sigma Chemical Co., "fraction V") in the same buffer. Convenient buffer reservoirs can be constructed by fitting together Tygon tubing of two sizes: (i) ID 3/8 inch, OD 9/16 inch and (ii) ID 1/4 inch, OD 3/8 inch. These buffer reservoirs also serve as syringe adapters used during sample elution (below). After treating the columns with BSA, wrap the tops with Parafilm and place them upright in a sealed container. Such columns may be stored for at least a month at 4°. During an experiment, the columns may be kept cold by thrusting them directly into crushed ice. (Note: A repeating syringe speeds the washing procedure; equilibrating the resin in bulk prior to packing the columns does not appear to be a useful shortcut.)

Positive-Pressure Syringes

The samples are eluted under positive pressure applied to the columns with a 20-ml disposable syringe fitted with a tubing connector (similar to Fisher catalog # 15-315-6C) via a small length of Tygon tubing. This device is inserted into the buffer reservoir and pressure is applied such that 2 ml of column buffer is collected into a scintillation vial within about 90 sec. It is desirable that the syringe maintain positive pressure on its own so that the hands are free to perform other assays simultaneously; this capability is enhanced by removing the plunger lubricant (wipe with paper towels and acetone).

The Assay Conditions

We perform assays in the presence or absence of ATP using a total reaction volume of 0.06 ml. Reactions are conveniently carried out in 0.5-ml conical microfuge tubes. The column buffer contains 700 mM glycerol and 50 mM Tris, titrated to pH 7.5 with HCl. Two 6-fold concentrated substrate solutions are prepared in the same buffer, with 36 mM $MgCl_2$ and 0.6 mM $^{45}CaCl_2$ with or without 12 mM MgATP, respectively. A 0.03-ml sample from the peak fraction of the second density gradient in transport-specific fractionation containing 0.5 μCi of $^{45}Ca^{2+}$ incubated for 2 min at 30° will accumulate ~70 pmol of $^{45}Ca^{2+}$ with ATP and 7 pmol without ATP. The background without vesicles is usually about 0.35 pmol. Standard errors of the mean are usually about 3%.

To carry out the analysis, aliquots of the reaction volume (typically 50 μl) are layered directly into the top of the Dowex and then eluted through the resin under pressure (allowing the column to run dry spoils the assay because vesicles become trapped in the air spaces). Then, 1 ml of ice-cold column buffer is drawn into a pipetter and about 0.2 ml is gently deposited on the resin. This is again forced to the level of the resin; the remaining 0.8 ml is added to the buffer reservoir together with another 1.0 ml of the column buffer. Under these conditions the vesicles elute from the column while the extravesicular $^{45}Ca^{2+}$ is retained. Collect the entire volume directly into a 4-ml glass scintillation vial; the sample can be counted in the gel phase by bringing the vial to near capacity with Hydrofluor (National Diagnostics). With practice, two assays with a 2-min incubation can be performed almost continuously every 3 min.

The Use of Chelating Agents

The small amount of oxalate in the low-oxalate buffer is insufficient to cause $^{45}Ca^{2+}$ to pass through the column as a calcium oxalate complex. On the other hand, stronger chelating agents as are commonly found in buffers designed to control the free concentrations of Ca^{2+} and Mg^{2+} competitively compromise the ability of the Dowex columns to bind $^{45}Ca^{2+}$. The following modification to the above protocol circumvents this chelation problem and therefore allows one to use EDTA or EGTA to control the $[Ca^{2+}]$ in the transport assay.

La^{3+} is similar to Ca^{2+} in terms of ionic radius, but because of its high charge density it usually has very high affinity for Ca^{2+} binding sites.[17] Thus, addition of La^{3+} before the ion-exchange step has two beneficial effects: (1) it rapidly quenches Ca^{2+}-dependent phosphorylation of the calcium pump[18] and (2) by completely displacing $^{45}Ca^{2+}$ from the EDTA or EGTA, the isotope becomes available for binding to the Dowex column. Therefore, after a suitable incubation period, we rapidly stop the $^{45}Ca^{2+}$ uptake reaction by adding an equal volume of ice-cold quenching solution (5 mM $LaCl_3$, 3 mM $MgCl_2$ in 700 mM glycerol, 40 mM HEPES, pH 7.5 with KOH). The mixture is vigorously vortexed and then eluted through Dowex columns in the usual fashion.

An important factor which limits the general usefulness of the La^{3+} quenching technique is the tendency of this ion to cause severe aggregation of membranes, particularly in turbid vesicle suspensions. Such aggregates are tolerated to only a limited extent during chromatography on the Dowex minicolumns. Therefore, the La^{3+} quench method is most suitable for use with the purified material from transport-specific fractionation, while the standard method is more suitable for assaying across the entire density gradient. La^{3+} also precipitates ATP, but this does not interfere with chromatography.

---

[17] D. M. Bers and G. A. Langer, *Am. J. Physiol.* **237,** H332 (1979).
[18] M. Chiesi and G. Inesi, *J. Biol. Chem.* **254,** 10370 (1979).

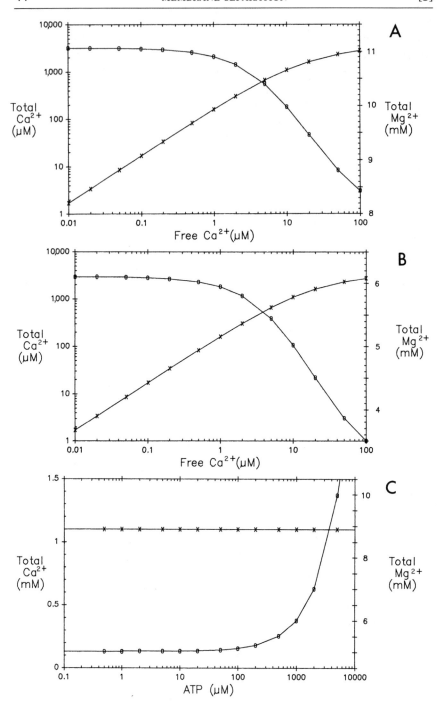

Figure 3[19-21] shows the total reagent concentrations required in order to approximately achieve the indicated free divalent cation concentrations under various conditions. Typically, a $^{45}Ca^{2+}$ uptake assay is carried out for 2 min at 30° using 0.03 ml of the purified Ca^{2+} pump (the fractions pooled from the second sucrose gradient in the preparative-scale Transport-Specific Fractionation, as described below). When the $^{45}Ca^{2+}$ (80 cpm/pmol) and Mg^{2+} are held at free concentrations of 10 μM and 2 mM, respectively, about 125 pmol is trapped in the vesicles with 1 mM ATP and about 12 pmol without ATP. The background without vesicles is about 3.5 pmol. Standard errors of the mean are about 5%. Initial rates of $^{45}Ca^{2+}$ transport can be measured for at least 4 min under these conditions, and the ATP-dependent component is about 70 nmol/min/mg protein.

Preparative-Scale Transport-Specific Fractionation

The analytical-scale reconstitution procedure can be scaled up. A large volume of cholate-solubilized synaptosomal membrane/soybean phospholipid mixture is dialyzed in a 90- or 135-ml C-DAK artificial kidney unit manufactured by Cordis Dow (C-DAK 90 sce and C-DAK 135

[19] E. Graf and J. T. Penniston, *J. Biol. Chem.* **256**, 1587 (1981).
[20] R. M. C. Dawson, D. C. Elliot, W. H. Elliot, and K. M. Jones (eds.), "Data for Biochemical Research," 2nd Ed. Clarendon, Oxford, England, 1969.
[21] A. Young and T. R. Sweet, *Anal. Chem.* **27**, 418 (1955).

FIG. 3. Control of the free divalent cation concentration in calcium transport assays can be achieved as shown. These plots were generated by a computer program based on published association constants[19,20] between chelators and Ca^{2+}, Mg^{2+}, and H^+. The calculations account simultaneously for equilibria between the ATP, EDTA, oxalate, Mg^{2+}, Ca^{2+}, and H^+. The calculations include contributions from the singly protonated forms of ATP and EDTA, but the diprotonated species as well as the protonated forms of oxalate were taken to be inconsequential. (A) Solutions at pH 7.50 containing 5 mM ATP, 3 mM EDTA, and 2.5 mM oxalic acid can be adjusted to approximately the desired free Ca^{2+} concentration and 2 mM free Mg^{2+} by preparing the solutions to contain the indicated total concentrations of Ca^{2+} ($\times$) and Mg^{2+} ($\bigcirc$). (B) Solutions at pH 7.50 containing 3 mM EDTA and 2.5 mM oxalic acid can be adjusted as in A by preparing the solutions to contain the indicated total concentrations of Ca^{2+} ($\times$) and Mg^{2+} ($\bigcirc$). (C) Solutions at pH 7.50 containing 3 mM EDTA, 2.5 mM oxalate, and the desired concentration of ATP can be adjusted approximately to 2 mM free Mg^{2+} and 0.01 mM free Ca^{2+} by preparing the solutions to contain the indicated total concentrations of Ca^{2+} ($\times$) and Mg^{2+} ($\bigcirc$). In order to prepare the above solutions with precision, the following information is useful. A standard solution of $CaCl_2$ can be purchased (Corning) and used both as a source of Ca^{2+} and to determine the concentration of EDTA solutions by spectrophotometric end-point titration using 0.025 mM Eriochrome Black T at pH 11.5 as an indicator.[21] $MgCl_2$ can be purchased as a 4.9 M stock solution (Sigma) and the Mg^{2+} concentration in working solutions can be checked by titration against an EDTA standard.

sce, respectively). Although the procedure is analogous to the analytical preparation described above, there are significant differences which are detailed below.

Solubilization Procedure

The following example illustrates the procedure which would be used to carry out reconstitution with a 90-ml artificial kidney when the batch of asolectin contains 59.4% phospholipid by mass. In a 125-ml Erlenmeyer flask, mix together:
 1.6 M potassium phosphate (pH 7.4), 22.5 ml
 Cholate buffer (below), 14.4 ml
 2-Mercaptoethanol, 0.0315 ml
 Asolectin (59.4% phospholipid), 2.7 g

Bring the asolectin into solution through a combination of magnetic stirring and bath sonication under nitrogen. Finally, bring the volume up to 90 ml with distilled water and add 20 mg of synaptosomal vesicle membrane protein. Solubilize with stirring at room temperature for 10 min and then chill the solution in an ice bath.

The cholate buffer is prepared so that the final concentrations are as follows:
 Recrystallized cholic acid, 95 mg/ml
 NaCl, 100 mM
 Tris (base), 20 mM
 EDTA, 0.1 mM

Dissolve all of the reagents except cholic acid in 0.18 M NaOH. Then add the cholic acid stepwise, allowing time for it to dissolve between additions. Finally, use NaOH to adjust the pH to 7.4. It is convenient to prepare large quantities of both this solution and the solubilization mixture in advance and store them frozen (at $-70°$ if possible). The membrane protein can be added directly to a freshly thawed solubilization mixture.

Loading and Care of the Artificial Kidney

It is important to achieve complete filling of the dialysis fibers. The following method will minimize the tendency for microscopic air bubbles to become lodged in the fibers, which decreases both the overall capacity of the unit as well as its efficiency. The outer chamber of the vertically oriented artificial kidney is filled with high-oxalate buffer (same as in the analytical method except supplemented with 0.5 mM HEPES to stabilize the pH at 7.8). Use a peristaltic pump driving $\frac{3}{32}$ inch ID × $\frac{5}{32}$ inch OD

Tygon tubing at a flow rate of ~2 ml/min to push the solubilization mix up through the hollow fibers from the bottom. Long before the fibers are completely filled, the solubilization mix will begin to emerge from the top. Recycle this material into a reservoir (e.g., a 50-ml conical culture tube) via a piece of Tygon tubing. Every 5 to 10 min, attach a 50-ml disposable syringe to the top of the artificial kidney; gently apply enough negative pressure to dislodge bubbles trapped inside the fibers. Return the solubilization mixture drawn into the syringe to the reservoir. Continue the recirculation and negative pressure routine until the entire solubilization mixture is inside the artificial kidney. Usually, loading can be completed within 45 to 90 min.

A new artificial kidney can always be loaded to full capacity. With repeated use, some of the fibers will become permanently blocked, resulting in reduced capacity. This tendency can be retarded by circulating a lipid-free solubilization mix through the fibers for several hours after every few reconstitutions. The artificial kidney unit should always be washed with low-oxalate buffer immediately after each use. Never attempt to unclog the fibers by application of positive pressure within the fibers. Similarly, the fibers should be treated like a living cell in the sense that both tend to burst when the external solution is hypotonic.

First-Stage Dialysis

The dialysis unit should be configured for countercurrent exchange as described elsewhere in this series.[5] The flow rate through the internal volume of the fibers is maintained at 1.5 ml/min. Dialyze rapidly (~15 ml/min) in the first hour against about 1 liter of high-oxalate buffer. It is useful to notice that, during this period, the solution visible through the Tygon tubing will become opalescent (like milk in water), which is a consequence of the lipid vesicle formation process as the detergent is removed. Continue to dialyze more slowly overnight (15 hr) against 2.5 to 3 liters of high-oxalate buffer (~3 ml/min).

Second-Stage Dialysis

Stop the peristaltic pump that recycles the vesicles through the hollow fibers, and pour out the high-oxalate buffer from the outer chamber of the artificial kidney. Replace it with low-oxalate buffer, and rapidly dialyze against 2 to 3 liters of low-oxalate buffer over the course of 2.5 to 3 hr. After dialysis, recover the vesicles by using air from a 50-ml syringe to push the vesicles into a graduated cylinder. In terms of volume, up to 80% recovery is typical. The material appears quite turbid.

Two-Stage Density Gradient Centrifugation

The density gradients are formed from 16 ml of heavy phase and 14 ml of light phase. A 2-ml 65% sucrose cushion is added via a Pasteur pipet. Up to 4 ml can be removed from the top of each density gradient to accommodate additional sample volume. In this way, material recovered from a 135-ml artificial kidney can usually be centrifuged in a single Beckman VTI50 rotor with satisfactory results. The benefits of greater sample volume must, however, be balanced against the decreased resolution which results from gradient overloading.

Fractions from the first set of density gradients are collected based on their color. It is necessary to collect fractions (~1 ml) from each gradient only from the region containing the (golden or straw colored) main lipid peak. Discard the milky, white material formed on either side of this peak. When in doubt about whether to save or discard a fraction, favor the material of lesser density. It is convenient to clamp the vertical rotor tube to a ring stand, cut off the heat seal, pierce the bottom with a 16-gauge needle, and then regulate the flow by using the index finger to partially cover the hole at the top. Place the relevant fractions on ice. After incubating the pooled lipid peak with Ca^{2+} and ATP as for the analytical-scale procedure, dilute the mixture with one-half volume of low-oxalate buffer, load the material onto gradients in the VTI50 tubes, and centrifuge as before.

We use a four-channel peristaltic pump (Buchler) to gently deposit samples onto four density gradients simultaneously. Four lengths of Tygon tubing are fitted over the ends of four 100-μl Clay Adams micropipets which are each inserted through a rubber stopper. The pipets are inserted into the centrifuge tube to the level of the sucrose. The position of each tip is maintained by placing the rubber stoppers inside the holes of a dismantled centrifuge tube rack which is suspended on a ring stand just above the gradient tubes. The gradient tubes can be raised and lowered as needed by means of a "labjack." The pump speed is adjusted so as to minimally disturb the gradient.

The first 5 ml from the bottom of the second series of sucrose density gradients can be discarded. Collect 1-ml fractions for the rest of the gradient through the major lipid peak. The vesicles containing the calcium pump are most accurately located by using $^{45}Ca^{2+}$ transport assays (described above). It is worth noting that occasionally the separation between the ATP-dependent and the ATP-independent $^{45}Ca^{2+}$ uptake activities is less distinct than is shown in Fig. 2C, particularly if the sucrose gradents are overloaded. The following guidelines may improve the outcome.

In our hands, the fractions containing the purified, reconstituted Ca^{2+} pump, which are pooled, are always crystal-clear and nonturbid. Above this gradient region, there is a slightly turbid region in the gradient (visualized under good lighting against a dark background) that has a markedly higher level of ATP-independent $^{45}Ca^{2+}$ uptake activity, and this should be discarded. A reasonable strategy is to perform $^{45}Ca^{2+}$ transport assays on a single density gradient, selecting only those fractions which contain the ATP-dependent $^{45}Ca^{2+}$ uptake activity *and* which are also free of the ATP-independent activity. Interpret the position of these fractions in terms of the turbidity of adjacent fractions, and identify the analogous fractions in the other gradients which were not actually assayed for $^{45}Ca^{2+}$ transport.

Another way to easily identify the region of the second gradient containing the purified, reconstituted Ca^{2+} pump activity is to incubate a portion of the pooled lipid vesicle peak from the *first* gradient for 20 min with 0.1 mM $^{45}CaCl_2$ (New England Nuclear) of specific activity of 10^5 cpm/nmol, 7.5 mM $MgCl_2$, and 2 mM Na_2ATP. This mixture is then diluted and placed on the second gradient. Twenty microliters of each 1-ml fraction from the second gradient is counted on a liquid scintillation counter. The peak of $^{45}CaCl_2$ which migrates to greater density than the main lipid vesicle peak (Fig. 2B) contains the purified, reconstituted Ca^{2+} pump. This method, of course, contaminates the sample with $^{45}Ca^{2+}$.

The final step, once the appropriate fractions from each gradient have been pooled, is to stabilize the Ca^{2+} pump with glycerol. Add 1 g of glycerol per 8 ml of the pooled preparation. Usually, the preparation may be stored on ice for up to 2 weeks without loss of activity.

[4] Regulation of Membrane Transport by Endocytotic Removal and Exocytotic Insertion of Transporters

By QAIS AL-AWQATI

Although the processes that regulate transport across membranes are numerous, they fall into two fundamental classes. In one, the transport protein is modified in such a way that its turnover rate is changed. There exist many mechanisms by which such a change may be effected, the best studied of which is phosphorylation of the transport protein. In the other mechanism, the number of transporters is changed without any change in the kinetic properties of individual proteins. This latter process was thought to be due largely to changes in the rate of synthesis or degradation of the protein, hence the effect was considered to be too slow to participate in rapid regulation of transport. However, recent evidence from a number of systems has established that a change in the number of transporters can occur very rapidly in response to a variety of stimuli. The mechanisms by which these changes occur are well-known cellular processes, those of regulated exocytosis, except that in this case it is the membrane that is the object of regulation rather than the contents of the vesicles.

This process has been invoked in the regulation of transport of water in response to vasopressin in epithelia,[1,2] the stimulation of glucose uptake in fat cells by insulin,[3,4] the stimulation of gastric H^+ transport by histamine,[5] the stimulation of Na^+ transport in urinary bladder by hydrostatic pressure,[6] and the stimulation of H^+ transport in urinary epithelia by CO_2.[7-9] The methods which we used for demonstration of the latter process will be described in detail below. To demonstrate exocytotic insertion or endocytotic removal of transporters, investigators have used diverse methods. I will discuss first the criteria that are needed to demonstrate this process in any system and evaluate the various methods that have been used. As will become apparent, most assays are flawed largely because of the lack of specificity. However, the use of multiple lines of evidence has allowed reasonable conclusions to be drawn regarding the presence or absence of such a regulatory process.

The transport protein should be located in a pool of vesicles that exist underneath the membrane of interest. The best method is that using immunoelectron microscopy where the location of the protein and its carrier vesicle can be unequivocally determined. This is usually a difficult task since most transport proteins are characterized more by their function than by their structure. An alternative, though less satisfactory approach, is to isolate the vesicles and demonstrate the presence of the relevant transport function in them. This is frequently feasible using a variety of

[1] S. K. Masur, E. Holtzman, and R. Walter, *J. Cell Biol.* **52,** 211 (1972).
[2] J. B. Wade, *Curr. Top. Membr. Transp.* **13,** 123 (1980).
[3] T. Kono, F. W. Robinson, T. L. Bevins, and O. Ezaki, *J. Biol. Chem.* **257,** 10942 (1982).
[4] E. Karniele, M. J. Zarnowski, P. J. Hissin, I. A. Simpson, L. B. Salans, and S. W. Cushman, *J. Biol. Chem.* **256,** 4772 (1982).
[5] T. M. Forte, T. E. Machen, and J. G. Forte, *Gastroenterology* **73,** 941 (1977).
[6] S. A. Lewis and J. L. de Moura, *Nature (London)* **297,** 685 (1985).
[7] S. Gluck, C. Cannon, and Q. Al-Awqati, *Proc. Natl. Acad. Sci. U.S.A.* **79,** 4327 (1982).
[8] D. Stetson and P. R. Steinmetz, *Am. J. Physiol.* **245,** C113 (1983).
[9] K. M. Madsen and C. C. Tisher, *Am. J. Physiol.* **245,** F670 (1983).

subcellular fractionation schemes. Once the vesicles are isolated in a pure form, it is usually straightforward to develop a specific assay that will measure the transport activity in question. Recent studies have uncovered a wealth of complexity in intracellular compartmentation, especially as it relates to the traffic of molecules from the cell surface to various vesicular compartments. Considering that subcellular fractionation is at present an imperfect art form, it would be very difficult to guarantee that the vesicles isolated actually are the ones that participate in fusion.

Ideally, one would like to develop an assay which measures the transport activity of the vesicles in living cells. In one instance, the case of the H^+-ATPase, it was possible to locate the protein by virtue of its function since the pH of these vesicles can be measured.[7] We used a fluorescent weak base, acridine orange, whose emission spectrum was sensitive to the concentration. When it accumulated in acid compartments, it changed color from green to orange.

Addition of the stimulus should cause fusion of the vesicles and its removal should result in endocytosis of the protein into these "reserve" vesicles. A number of assays for exocytosis exist which depend on secretion of the contents of the vesicles into the outside medium. In many of the newly described processes under discussion, the vesicles are probably empty. We developed an alternative method recently, in which we were able to incorporate a fluorescent macromolecule into the relevant vesicular compartment and test the effect of the stimulus on the secretion of this dye.[7] Since in many of the processes discussed, there is both endocytic retrieval and exocytic insertion of the transport protein, this method should be generally applicable. However, one should keep in mind the problem that the internalized molecule may rapidly redistribute into another compartment.

Ultrastructural morphometry was recently used to analyze the volume fraction of a cell occupied by vesicles before and after application of the stimulus.[2,5,8,9] The major problem here is that it is not clear that the vesicles counted are the ones that contain the transporter of interest. For instance, vasopressin causes the exocytosis of two types of vesicles, granules,[1] and tubulovesicles whose membranes contain "aggregates."[2] Which one of them contains the water channels remains debatable. The time course of the appearance of the aggregates on the apical membrane correlates well with that of the increase in water permeability; however, since the granule membrane does not contain any observable marker, it is not possible to ascertain the time course of its fusion and relate it to that of the water permeability. At any rate, the temporal resolution of this method is quite poor unless one uses the "freeze-slam" method of Heuser and Reese.[10] However, this method forms an excellent adjunct to other methods since it will confirm that the size of the "reserve" pool of vesicles changes in response to the stimulus.

An alternative approach, that of subcellular fractionation, has been used in the case of the glucose transporter in fat cells. Using cytochalasin binding as a measure of the number of glucose carriers, Kono et al.[3] and Karniele et al.[4] found that this marker was present in a fraction on density gradients that did not comigrate with plasma membrane markers. On addition of insulin, the cytochalasin-binding sites moved to a fraction that had the density of plasma membranes. Again, much depends here on the specificity of the markers used and the purity of the fractions tested.

Another class of assays measures the area of the membrane before and after addition of the stimulus. This method measures net changes, i.e., the result of both endocytosis and exocytosis. Morphometric analysis can be used, but an electrophysiological method, that of measuring the capacitance of the membrane of interest, yields probably the best results. The capacitance of a membrane is a function of its area and its dielectric constant. Since the latter is largely unaffected by changes in lipid composition, it follows that most changes in capacitance are due to changes in area. The major problem of this method is that it is in many instances "model dependent" i.e., it depends largely on the choice of appropriate equivalent circuit parameters, many of which are not unambiguously known. Recent studies by Fernandez et al.[11] have shown the utility of this method in individual cells by the use of the patch-clamp method in the cell-attached recording mode. The potential of this method is that it allows measurement of exocytosis of a single vesicle. In epithelia this method has been used to estimate the apical membrane area. Obviously, the number of assumptions is much larger here, and care should be exercised in choosing the appropriate equivalent circuit parameters, and measuring them by the appropriate intracellular recording methods.[6,12,13] The change in area is the net result of exocytosis and endocytosis; hence it is a complex parameter at best. Further, this method is relatively nonspecific since it monitors the addition of all membranes rather than the addition of the membranes that contain the transporters. The great advantage of this method is its excellent time resolution.

Exocytosis should insert functional transporters into the membrane. Demonstration of vesicle fusion in response to a stimulus should be sup-

[10] J. E. Heuser and T. S. Reese, *J. Cell Biol.* **88,** 564 (1981).
[11] J. M. Fernandez, E. Neher, and B. D. Gomperts, *Nature (London)* **312,** 453 (1984).
[12] T. E. Dixon, C. Clausen, and D. Coachman, *Kidney Int.* **27,** 280 (Abstr.) (1985).
[13] L. Palmer and N. Speez, *Am. J. Physiol.* **246,** F501 (1984).

plemented by measurement of transport to show that changes in transport correlate well with fusion events. The relationship between the two events can be solidified by studying the effects of various maneuvers on the simultaneously measured rates of exocytosis and transport. An additional series of experiments using inhibitors of exocytosis or endocytosis might also be useful. Unfortunately, many of the inhibitors have diverse effects which might complicate the interpretation of these findings. Hence, studies which rely only on the use of colchicine, cytochalasins, or other drugs are not very convincing and frequently confusing.

In an interesting variation on this approach, Garty and Edelman[14] used trypsin to digest the amiloride-sensitive sodium channels already resident in the apical membrane of the toad bladder. Addition of vasopressin (but not aldosterone or metabolic substrates) stimulated the appearance of channels that were not exposed to the trypsin.

Many of the methods described above have been used to measure the time course of fusion and transport with variable success. Electrophysiological methods have the best time resolution and if the process studied results in some change in membrane currents, e.g., by insertion of ion channels or pumps, then the use of the capacitance method would be ideal to evaluate the time course of this process. Electrophysiological methods have the potential of measuring phenomena with millisecond resolution. Other methods are also available, including the use of density gradient centrifugation after the reaction is stopped at different times following the addition or removal of the stimulus. The time resolution of this method is of the order of minutes.

Endocytosis should remove the transporters from the membrane. Many receptors and other membrane proteins recycle by constant removal and insertion into the plasma membrane. To arrive at a definitive demonstration of removal of transport proteins it will be necessary to label the plasma membrane with a high-affinity ligand, e.g., an antibody, in the absence of endocytosis and then to induce endocytosis to allow internalization of the transporter (and its ligand). Of course, one needs to develop an assay which distinguishes surface-attached from internalized ligand. A useful method is to quantitate surface binding at 4°, a temperature that inhibits endocytosis, and then raise the temperature to 37° and study the internalization of the ligand. Using this method, it should be possible to follow the time course of internalization and to relate it to the transport activity of the whole cell. Other methods which measure pinocytosis by the use of fluid-phase markers suffer from the problems of lack of specificity. Since most cells have endocytosis, it will be difficult to test

---

[14] H. Garty and I. S. Edelman, *J. Gen. Physiol.* **81**, 785 (1983).

whether the endocytosed membrane removed from the surface contains the transporter of interest.

It would also be interesting to attempt to vary the rate of endocytosis and test its effect on the rate of transport. However, there are no known stimuli of specific (receptor-mediated) endocytosis. Phorbol esters, were recently found to stimulate pinocytosis.[15] No drugs are known that specifically inhibit endocytosis; however, low temperature and metabolic inhibitors do so "nonspecifically."

Regulation of H^+ Transport in Urinary Epithelia by CO_2-Induced Exocytotic Insertion of H^+-ATPases[16]

Hydrogen ions are secreted by the renal tubular cell in at least two segments, the proximal and the collecting tubule. In amphibia and reptiles, the urinary bladder serves the same function as the collecting tubule. In the proximal tubule, protons are secreted by two processes, a $NA^+ : H^+$ exchanger and a H^+-ATPase located in the luminal membrane, while in the collecting tubule and urinary bladder, only the H^+-ATPase is responsible for urinary acidification. In the proximal tubules, all the cells participate in proton transport, whereas in the collecting tubule and bladder, the mitochondria-rich (intercalated) cell, which constitutes a minority of the cell population, is the cell that transports protons. We recently found that CO_2 stimulates H^+ secretion by initially acidifying the cell, which leads to increases in cell calcium.[17] It is this increase in cell calcium which leads to exocytotic insertion of H^+-ATPases into the luminal membrane with consequent increase in the rate of transepithelial H^+ transport.

H^+-secreting cells contain a reserve of vesicles underneath the apical membrane whose contents are acidified by proton pumps.[7] Ultrastructural studies show that the H^+-secreting cell contains a large number of subapical vesicles that exchange material with the luminal medium. Using this feature, we loaded these vesicles with a pH-sensitive macromolecule, fluorescein coupled to a large-molecular-weight dextran. The pH in the environment of this probe was then measured using excitation-ratio fluorometry. Here, patches of cytoplasm containing a few vesicles were centered on a microscope fluorometer and the cell was alternately excited with 490 and 460 nm light. The emission intensity was measured at 530

---

[15] J. A. Swanson, B. D. Yirinec, and S. C. Silverstein, *J. Cell Biol.* **100**, 851 (1985).
[16] G. J. Schwartz and Q. Al-Awqati, *Annu. Rev. Physiol.* **48**, 153 (1986).
[17] C. Cannon, J. van Adelsberg, S. Kelly, and Q. Al-Awqati, *Nature (London)* **314**, 443 (1985).

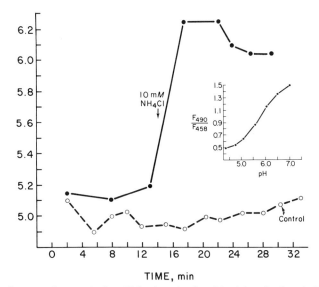

FIG. 1. Apparent intravesicular pH in the mitochondria-rich cell of turtle bladder. The bladder was exposed to fluorescent dextran from the luminal side and then washed and mounted on the stage of a fluorescence microscope. A patch of cytoplasm containing a few vesicles was focused in the photometer aperture and was excited alternately with monochromatic light at 460 and 490 nm using interference filters. The emission was measured at 520 nm. The inset contains a calibration curve performed in buffered 100 mM KCl media in a fluorometer.

nm. The ratio is a function of pH (Fig. 1, inset). As seen in Fig. 1, the apparent pH of the vesicles was near 5.0 and increased on addition of a permeant weak base. The proton ionophore, FCCP, increased the pH, suggesting that the low pH was not due to a Donnan potential. Metabolic inhibitors also increased the pH, indicating that acidification of the vesicle contents was due to a proton pump. A quantitative assessment of the actual intravesicular pH requires an *in situ* calibration curve. This can be achieved by perfusing the cell with highly buffered solutions at the desired pH values and which contain proton ionophores. This was used by Horowitz *et al.* and the apparent pK of the dye in the vesicle was found to be the same as in salt solutions.[18] The situation is not the same for cytoplasmic dyes; we recently found that the pK of cytosolic fluorescein was higher than that in salt solutions by 0.3 pH units.[19]

[18] M. A. Horowitz and F. R. Maxfield, *J. Cell Biol.* **99**, 1936 (1984).
[19] J. van Adelsberg and Q. Al-Awqati, *J. Cell Biol.* **102**, 1639 (1986).

Exoctyosis of Fluorescent Dextran[7,17]

The vesicles of the acid-secreting cells of the turtle bladder do not contain any known material; hence to measure exocytosis we had to incorporate into them a substance that could be easily measured. We used the impermeant molecule, 70,000 M_r dextran coupled to fluorescein isothiocyanate (fluorescent dextran), which was rapidly endocytosed into the revelant vesicles (as judged by the decrease in pH), and developed an assay for release of the dye from intracellular vesicles into the luminal solution when the bladder was stimulated with CO_2. The bladders were mounted in Ussing chambers having an exposed surface area of 8 cm^2. The volume of the luminal side was 3 ml and its lateral wall contained a magnet-driven disk that allowed vigorous stirring. The bladders were perfused on both sides by continuously circulating freshly bubbled buffer through the chamber with a peristaltic pump. The transepithelial potential difference was set to zero with an automatic voltage clamp and the short-circuit current was recorded continuously. The short-circuit current represents net transepithelial ion transport and, in the turtle bladder, is due to Na absorption and H secretion.

The bladder was initially perfused with Ringer's solution equilibrated with 5% CO_2 in air with fluorescent dextran added to the luminal solution at 100 mg/ml and ouabain added to the serosal solution at 1 mM to inhibit transepithelial Na^+ transport. Following the addition of ouabain, the short-circuit current reversed in polarity and reached a steady level. The short-circuit current under these conditions has been shown to be due solely to H^+ transport. After the initial steady H^+ current was attained, the perfusing solutions were changed to CO_2-free Ringer's solution containing fluorescent dextran at 100 mg/ml in the luminal side and 1 mM ouabain in the serosal side. During the next 45 min the H^+ current fell, as expected, and reached a new steady level. At this time the luminal solution was changed to CO_2-free Ringer's solution without fluorescent dextran. Fresh CO_2-free Ringer's solution was pumped continuously through the luminal side and the effluent was passed through a flow cell in a fluorometer (ratio fluorometer; Farrand Optical). The fluorescence of the effluent (470-nm excitation, 520-nm emission) was continuously recorded. After 20–30 min the fluorescence of the effluent fell to a very low level with either a steady or slightly decreasing value. At this time the perfusion solution was changed to 5% CO_2–Ringer's solution, which produced a rise in the H^+ current and release of fluorescent dextran into the luminal solution, detected as a small broad peak in the fluorescence tracing (Fig. 2). The recording was continued until the H^+ current reached a new steady level. Measured amounts of Ringer's solution containing fluores-

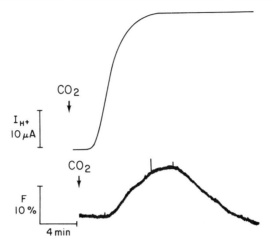

FIG. 2. A turtle bladder was loaded with fluorescent dextran, mounted in an Ussing chamber, and its short-circuit current measured in the presence of ouabain. The luminal side was continuously perfused through a flow-through cell in a fluorometer. On increasing the serosal pCO_2 (at arrow), there was an increase in the H^+ current (I_H) simultaneously with a burst of secretion of fluorescent dextran. The time course of the latter is dominated by that of the perfusion system.

cent dextran were then injected into the luminal chamber to obtain a calibration curve for each experiment. Plotting the area of these calibrating injections against the volume injected gave the standard curve that was used to estimate the exocytosed volume. The areas of the peaks were measured by cutting out and weighing tracings of them.

Single-Cell Assay for Exocytosis[20]

The assay described above used a whole epithelium and measured the signal from 10^5-10^6 cells. When using the isolated perfused tubule which contained 100 intercalated cells/mm it was necessary to develop a more sensitive assay. The single-cell assay described below should be useful for a variety of cells, provided one can load the compartment of interest with the appropriate fluorescent dyes. In brief, the assay measures the total fluorescence of a cell before and after exposure to a stimulus. It is assumed that release of the fluorescent dye from the cell can only occur by exocytosis, hence only fluorescent macromolecules will yield unequivocal results. This implies that the compartment of interest can be loaded

[20] G. J. Schwartz and Q. Al-Awqati, *J. Clin. Invest.* **75**, 1638 (1985).

from the outside. It is further assumed that the stimulus itself has no effect on the intrinsic fluorescence of the dye.

To show that CO_2 causes exocytosis in the cortical collecting tubule, we isolated these segments and perfused both luminal and basolateral surfaces with a solution that contained NaCl, 120 mM; $NaHCO_3$, 25 mM; K_2HPO_4, 2.5 mM; $MgSO_4$, 1.2 mM; $CaCl_2$, 2.0 mM; sodium lactate, 4.0 mM; trisodium citrate, 1.0 mM; L-alanine, 6.0 mM; and glucose, 5.5 mM; pH 7.4 after gassing with 95% O_2/5% CO_2 at room temperature. At 37°, the pH of this solution was 7.62, the total CO_2 was 26 mm, and the calculated pCO_2 was 25 mm Hg. The luminal perfusate contained fluorescein coupled to 70,000 MW dextran at a concentration of 1–5 mg/ml. Then we increased the pCO_2 of perfusate and bath isohydrically to 114 mm Hg. The cell was centered in the photometer aperture and the fluorescent intensity at 520 nm was measured while exciting at 460 nm. A reduction in fluorescence would imply loss of fluorescent dextran, presumably by exocytosis. We showed that colchicine inhibited this decline in fluorescence intensity, further supporting the idea that it was due to exocytotic release of the dye. The time course of change in fluorescence intensity was measured in a single cell before and during exposure to CO_2. Since CO_2 might acidify the vesicle contents, leading to a reduction in the fluorescence intensity of fluorescein, we excited the cell with 460 nm light rather than 490 nm. Although this wavelength (460 nm) is not the isosbestic point of fluorescein, it shows a smaller change in intensity with a change of pH making it useful for the measurement of total dye content. (An alternative approach would be to use a fluorescent dye whose emission intensity is independent of pH, e.g., Lucifer Yellow.) To test for spontaneous declines in the fluorescence signal (due to bleaching of the dye), we performed time control experiments in which the fluorescence intensity was measured at the same ambient pCO_2 for periods of time comparable in length to those of the CO_2-induced exocytosis experiments. To guard against bleaching of the dye, the cell was exposed to the exciting light only during the measurement and the intensity of the light could be reduced using neutral density filters.

In principle at least, it should be possible to develop a single-cell assay for exocytosis using a permeant fluorescent probe. This would be very useful if one wanted to correlate electrophysiological recordings with exocytosis. Most secretory granules contain a proton-translocating ATPase that is capable of generating a pH difference or a potential across the granule membrane. One can use acridine orange, a fluorescent weak base whose emission spectrum is a function of its concentration. At low concentration it fluoresces in the green and at higher concentration there is a red shift in the spectrum. In secretory cells the pH of the granules is usually acid, hence these organelles will accumulate acridine orange which will fluoresce in the yellow or orange range while the rest of the cytoplasm will appear green. If the cell is placed in the aperture of a photometer and red fluorescence measured, then exocytosis should result in a reduction in the fluorescence, provided the medium is perfused continuously and replaced with fluorophore-free media. Unfortunately, acridine orange bleaches quite readily, hence the exciting light intensity should be very low. Also, I fear that the constant perfusion will make the electrophysiological recording difficult. Similar results could be obtained using a permeant voltage-sensitive anionic dye such as one of the oxonols. Again, the major problem is constant perfusion with fluorophore-free medium. Note that without this constant perfusion, the dye, released from one granule, will simply redistribute into another.

Section II

Transport Methods in Cells and Vesicles

[5] Proton Electrochemical Potential Gradient in Vesicles, Organelles, and Prokaryotic Cells

By HAGAI ROTTENBERG

The measurement of the proton electrochemical potential gradient, $\Delta\bar{\mu}_H$, is central to the analysis of energy conversion processes in mitochondria, chloroplasts, storage organelles, bacterial cells, algae, yeast, and other systems in which proton pumps play a primary role in energy conversion processes. The methods for the determination of $\Delta\bar{\mu}_H$ in cells, organelles, and vesicles have been described in some detail in Vol. 55 of this series). The purpose of this review is to update the information of the previous contributions.[1-5] Several other methodological reviews appeared more recently.[6-9] We shall discuss here only simple, two-compartment systems, such as vesicles, organelles, and prokaryotic cells. Similar measurement in more complex cells and tissues is outside the scope of this review.

The proton electrochemical potential difference between the inner volume and the suspending medium ($\Delta\bar{\mu}_H$) is composed of two terms, the transmembrane electrical potential difference ($\Delta\psi$) and the transmembrane pH difference (ΔpH), thus

$$\Delta\bar{\mu}_H = F\Delta\psi - 2.3RT\,\Delta\text{pH}\ (\text{kcal/mol}) \qquad (1)$$

It is customary to express $\Delta\bar{\mu}_H$ in electrical potential units (mV); hence

$$\Delta\bar{\mu}_H\ (\text{mV}) = \Delta\bar{\mu}_H/F = \Delta\psi - (2.3RT/F)\,\Delta\text{pH} \qquad (2)$$

where R is the gas constants, T the absolute temperature (°k), F the Faraday constant.

In practice, the determination of $\Delta\bar{\mu}_H$ requires separate but simultaneous determinations of $\Delta\psi$ and ΔpH.

---

[1] H. Rottenberg, this series, Vol. 55 [64].
[2] C. L. Bashford and J. C. Smith, this series, Vol. 55 [65].
[3] V. P. Skulachev, this series, Vol. 55 [66].
[4] S. Ramos, S. Schuldiner, and H. R. Kaback, this series, Vol. 55 [74].
[5] A. S. Waggoner, this series, Vol. 55 [75].
[6] C. L. Bashford, *Biosci. Rep.* **1**, 183 (1981).
[7] G. F. Azzone, D. Pietrobon, and M. Zorati, *Curr. Top. Bioenerg.* **13**, 2 (1984).
[8] J. B. Jackson and D. B. Nicholls, this series, Vol. 127 [41].
[9] A. Roos and W. F. Boron, *Physiol. Rev.* **61**, 296 (1981).

Measurement of Membrane Potential

The most direct measurement of the transmembrane electrical potential difference is obtained from the use of microelectrodes. The potential between two reference electrodes located at the bulk solutions bathing the two sides of the membrane is, by definition, the membrane potential. However, the application of microelectrodes to organelles, small cells, and vesicles is severely limited due to size limitation. Thus, the methods used in these systems are still mostly indirect.

Equilibrium Distribution of Permeable Ions

The most widely used techniques for estimation of $\Delta\psi$ are based on a very simple and fundamental thermodynamic principle. For a vesicular system at equilibrium, the electrochemical potential gradient of permeable ions vanishes, i.e., at equilibrium

$$\Delta\bar{\mu}_i = 2.3RT \log(a_{in}/a_{out}) + zF \Delta\psi = 0 \qquad (3)$$

therefore

$$\Delta\psi = -2.3(RT/zF) \log(a_{in}/a_{out}) \qquad (4)$$

where a is the ion activity and z is the ion electrical charge. For monovalent ions at room temperature $2.3RT/zF = 59$ mV. To estimate $\Delta\psi$ using Eq. (4), it is necessary to determine the activities of an equilibrated ion on both sides of the membrane. It is difficult to determine ion activities inside vesicles when the activity in the surrounding medium is higher. Therefore, when the membrane potential is negative, it is necessary to choose as the test ion a cation (which accumulates in the vesicle). Similarly, an anion should be used when the membrane potential is positive. The potential-induced accumulation of the proper test ion compensates for the much smaller internal volume of cells and vesicles in the suspension. The available methods for $\Delta\psi$ determination differ from each other by the choice of the test ion and the technique which is used to determine its concentration gradient. In fact, it is now clear that, with few exceptions, most of the potential sensitive optical probes depend on the same fundamental mechanism, namely, the establishment of electrochemical equilibrium distribution of a test ion.

Distribution of Lipophilic Ions. In biological membranes, most ions permeate through specific transport systems. The mechanism of these transport systems often involves coupling between the movement of ions, either as cotransport with other ions or as exchange for other ions. Also, there are cases of coupling of the ion transport to enzymatic reactions (pumps). Obviously, in these cases the electrochemical potential gradient of the test ion at steady state does not necessarily vanish, and hence the apparent equilibrium distribution is not equal to the membrane potential. To apply Eq. (4) correctly, the rate of independent electrogenic transport of the test ion must be much faster than other transport processes.

A transport mechanism which often ensures a true equilibration involves a partition of the ion into the membrane lipid phase, diffusion in the membrane, and dissociation. Fast equilibration of ions across the membrane by this mechanism requires that the ion be lipophilic, i.e., the ion must have a high partition coefficient between water and lipids. In addition, a high diffusion coefficient in the membrane is necessary. Organic ions with a diffuse charge distribution are often sufficiently lipophilic to provide rapid equilibration by this mechanism. This is particularly true in biological membranes (in contrast to artificial phospholipid membranes) since the increase of the membrane dielectric constant by membrane proteins and the presence of mobile charged species reduce the potential barrier for ion diffusion across the membrane.[10]

Several synthetic lipophilic ions in which the charge is shielded by a hydrophobic shell are highly permeable.[11] The rate of membrane diffusion of lipophilic ions may be increased by addition of trace amounts of a lipophilic ion of the opposite charge.[11] Another strategy to ensure true electrochemical equilibration of ions is to load the membrane with an ionophore (i.e., a lipophilic ligand of a specific ion), which results in fast diffusion of the lipophilic ion complex and net transport of the test ion across the membrane.[12]

The classical example of this method is the use of valinomycin to increase the electrogenic transport of K^+ or Rb^+. When K^+–valinomycin transport rate far exceeds other transport pathways (with excess valinomycin), potassium may reach electrochemical equilibrium even when other K^+ transport pathways exist which are not electrogenic. The advantage of the K^+ (or Rb^+)–valinomycin method is that the content of the test ion (i.e., the cation–valinomycin complex) in the membrane is usually negligible in comparison with the content of free ion in the bulk phases. Hence, there is no need to correct for the amount of bound ions when measuring the ion distribution, and the membrane potential can be calculated accurately from measuring the concentration of Rb^+ (or K^+) in the medium and the vesicles.

The main drawback of this method is that the potassium concentration

---

[10] T. P. Dilger, S. G. A. McLaughlin, T. S. McIntosh, and S. A. Simons, *Science* **206**, 1196 (1974).

[11] L. L. Grinius, A. A. Jasaitis, T. P. Kadziauskas, E. A. Liberman, V. P. Skulachev, V. P. Topali, L. M. Tsotine, and M. A. Vladimirov, *Biochim. Biophys. Acta* **216**, 1 (1970).

[12] P. W. Reed, this series, Vol. 55 [50].

in the medium must be kept to a very low level, preferably below 50 μM, in order to reduce the magnitude of the K^+ current at steady state which affects the measured potential. This may be too restrictive for many physiological experiments. Another disadvantage of this method is that it is necessary to determine accurately the internal volume of the vesicles. These measurements are often tedious and not very accurate. (For a description of internal volume determinations, and the measurement of $\Delta\psi$ by the Rb^+-valinomycin method, see Rottenberg, this series, Vol. 55 [64].)

Because of the above-mentioned disadvantages, many investigators have preferred the use of synthetic lipophilic ions, particularly triphenylmethylphosphonium ($TPMP^+$) and tetraphenylphosphonium (TPP^+) for the measurement of $\Delta\psi$. However, these ions are potent inhibitors of membrane enzymes and transport systems which limit the use of these probes in physiological and biochemical experiments. This problem can be partially averted by using a very low concentration of the probe ion, preferably less than 3 μM. In addition, a considerable fraction of these ions is bound to the membrane. Hence, when estimating the internal concentration from the measured uptake of the ion, it is necessary to correct for the amount of bound ions.

A reliable correction procedure must depend on an adequate physical model. Several attempts have been made to describe lipophilic cation binding and distribution.[13-16] The most successful model assumes independent binding sites on the internal and external membrane surfaces which are equilibrated with their respective bathing solutions.[17] This model was applied for the estimation of membrane potential from the electron paramagnetic resonance (EPR) spectra of lipophilic cations[17] and from the distribution of radioactively labeled lipophilic cations.[18] One of the unexpected advantages of the use of lipophilic ions that bind extensively to the membrane (e.g., TPP^+) is that it is not necessary to determine the exact value of the internal volume because most of the accumulated ion is bound internally. Since the binding is proportional to the internal *concentration* and not the internal free ion *content*, the volume (which determines the content) is less important.

[13] A. Zaritsky, M. Kihara, and R. M. McNab, *J. Membr. Biol.* **63**, 215 (1981).
[14] T. S. Lolkema, A. Abbing, K. S. Hellingwerf, and W. N. Koning, *Eur. J. Biochem.* **130**, 287 (1983).
[15] R. S. Richie, *Prog. Biophys. Mol. Biol.* **43**, 1 (1987).
[16] M. Demura, N. Kamo, and Y. Kobatake, *Biochim. Biophys. Acta* **812**, 377 (1985).
[17] D. S. Cafiso and W. L. Hubbel, *Biochemistry* **17**, 187 (1978).
[18] H. Rottenberg, *J. Membr. Biol.* **81**, 127 (1984).

Methods of Determination of Lipophilic Ion Distribution. There are two basic methods for the determination of these gradients. In the first method, the net uptake of the test ion by the vesicles is determined from the disappearance of the ion from the suspending medium. In the second method, the vesicles are separated from the suspending medium and the ion content of both the vesicles and the medium is determined.

The advantage of the net uptake method is that it does not disturb the equilibrium. With a separation there is often a loss of ions from the vesicles during the separation. Also, provided that the test ion equilibrates quickly and that a fast detection of the external concentration is possible (see below), the net ion uptake method can be used for kinetics studies. The disadvantage of the net uptake method is that, since the medium volume is much larger than the vesicle volume, the change in the concentration of the test ion externally is very small compared to the change internally. Hence, the sensitivity of this method is quite limited and the method is reliable only when there are relatively large potentials.

In contrast, the separation method is very sensitive and can reliably detect potential gradients of a few millivolts. However, care must be taken that no significant amount of ion is lost during separation, and it is often difficult to find a reliable separation method, particularly for vesicles of small size.

SELECTIVE ION ELECTRODES. In this method, a selective electrode is inserted into the vesicle suspension and the net uptake is determined from the change in external concentration as detected by the electrode. (For the use of selective electrodes, see this series, Vol. 56 [30].) To calculate the concentration gradient, the amount of ion taken by the vesicle is divided by the total internal volume of the vesicles in the suspension, which must be determined separately. Also, depending on the test ion, it is usually necessary to correct for binding, as described below. There are many commercial selective ion electrodes of good quality. In addition, it is possible to construct such an electrode for any desirable test ion from either planar phospholipid membranes[3] or ion-exchange membranes.[19]

FLOW DIALYSIS. Another variant of this method is the use of flow dialysis (cf. Ref. 4). Here, the concentration of the test ion in the suspension is estimated from the rate of diffusion of the test ion across a dialysis membrane since, to a first approximation, the rate is proportional to the concentration. The rate of diffusion is estimated from collecting fractions of a solution flowing past a dialysis membrane which contain the vesicle suspension. Usually, in this method, the ion is radiolabeled and the con-

---

[19] N. Kamo, M. Muratsugu, R. Hongoh, and Y. Kobatake, *J. Membr. Biol.* **49**, 105 (1979).

tent of the dialyzate is estimated by scintillation counting. An example of the measurement of $\Delta\psi$ and ΔpH in rat liver mitochondria by flow dialysis is shown in Fig. 1. In principle, other detection methods, i.e., optical absorbance, fluorescence, etc. can be used with the appropriate probe. With most probes the response is slow (when compared to ion-selective

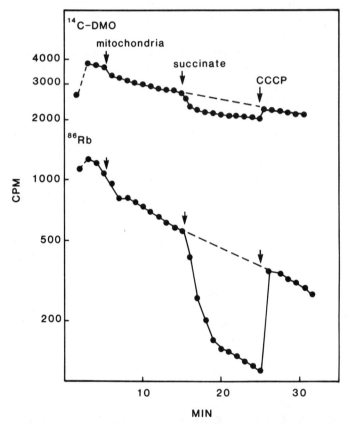

FIG. 1. The determination of membrane potential and ΔpH in rat liver mitochondria from the distribution of ^{86}Rb(+valinomycin) and [^{14}C]DMO (5,5-dimethyloxazolidine-2,4-dione) using flow dialysis.[21] The upper chamber contained 300 μl of 0.2 M sucrose/50 mM NaCl/10 mM PIPES/4 mM MgCl$_2$/1 μM rotenone/1 mM Na$_2$HPO$_4$ (pH 7.0). Mitochondria when added was 1.5 mg protein, 10 mM succinate, and 1 μM carbonyl cyanide m-chlorophenylhydrazone (CCCP). For $\Delta\psi$ determination, 0.1 μM valinomycin and 3 μM ^{86}Rb were added. $\Delta\psi$ is calculated from the amount of Rb$^+$ taken by the mitochondria. In this example, 350 of 500 cpm were taken by energized mitochondria at steady state. With 1 μl/mg protein of internal volume, the volume ratio is 300/1.5 = 200 and the Rb$^+$ concentration gradient is (350/150)200 = 467. The calculated $\Delta\psi = -59 \log 467 = -157$ mV.

electrodes, see Fig. 1), hence the method is primarily used to estimate steady-state potentials.

VESICLE SEPARATION TECHNIQUE. In these methods the test ion is usually radioactively labeled, although optical absorption or fluorescence detection may be used with the appropriate ion. The separation can be accomplished by fast centrifugation or filtration. Centrifugation and filtration through a layer of silicone oil is effective for large organelles and cells. With vesicles that can be sedimated in a few minutes, either by a conventional microfuge (Eppendorf or Beckman) or by airfuge (Beckman), we find that centrifugation (without filtration through silicone oil) is quite adequate. The filtration through silicone oil may help to reduce the loss of ions from the pellet; however, the passage of the vesicles through the oil causes osmotic imbalance and often results in shrinking, which may lead to artificially increased internal concentrations and overestimation of the potential.

When using the centrifugation method without silicone oil, the pellet must be separated from the medium immediately after centrifugation. With the microfuge we use narrow 400-μl tubes, in which a very small area of the pellet surface is in contact with the supernatant. The inclusion of a radiolabel marker for external volume (i.e., sucrose or other sugar) is often necessary (see Ref. 1 for more details). As an example for the use of lipophilic cations for measurement of $\Delta\psi$, the centrifugation method with mitochondria is described below.

METHOD OF DETERMINATION OF $\Delta\psi$ IN MITOCHONDRIA BY LIPOPHILIC CATION DISTRIBUTION.[18] Two 1.5-ml samples are prepared for the parallel measurement of (1) the lipophilic cation distribution of ^{14}C- or ^{3}H-labeled TPP$^+$ or TPMP$^+$ and (2) the mitochondrial internal volume and the pellet trapped volume with ^{3}H$_2$O and ^{14}C-labeled sucrose. The incubation medium composition depends on the particular experiment. For example, for measurement of $\Delta\psi$ at state 4, we use 0.2 M sucrose, 50 mM KCl, 10 mM HEPES (pH 7.4), 2 mM MgCl$_2$, 1 μM rotenone, and 2–3 mg mitochondrial protein/ml. The medium is saturated with 100% O$_2$ prior to the addition of the mitochondria. Sample (1) contains 0.2 μCi ^{14}C-labeled or 1 μCi ^{3}H-labeled TPP$^+$ or TPMP$^+$ and sample (2) contains 1 μCi ^{3}H$_2$O. After addition of mitochondria, the samples are vortexed and allowed to incubate for 3 min. Then succinate is added to both samples (10 mM) and 0.2 μCi [^{14}C]sucrose is added to sample (2).

The samples are vortexed, incubated for a further 2 min, distributed among 400-μl microfuge tubes, and immediately sedimented at high speed (2–3 min). Immediately after centrifugation, samples of 50 μl are withdrawn from each of the supernatants and mixed with 1 ml of 1% sodium dodecyl sulfate (SDS), and the pellets are cut quantitatively with a razor

TABLE I
BINDING CONSTANTS FOR LIPOPHILIC CATIONS
IN MITOCHONDRIA[a]

Cation	K_i	K_o (μl/mg protein)
TPMP$^+$	2.4	5.0
TPP$^+$	7.9	14.3

[a] From Ref. 18.

through the transparent tubes. There is usually a meniscus at the pellet center that contains a drop of the supernatant. This drop is removed carefully by touching it with the tip of an absorbant tissue paper. The pellet, (still in the cut bottom of the tube) is dropped into 1 ml of 1% SDS and immediately vortexed vigorously. It is important to dislodge the pellet from the tube by the vortexing to accelerate the dissolution of the pellet. The samples in the 1% SDS are allowed to stand for a few hours for complete dissolution. The process can be accelerated by frequent vortexing. It is important to verify, before sampling for counting, that all the pellets have dissolved.

Samples of 0.75 ml are taken from each SDS solution and mixed with high-quality scintillation liquid. The samples are counted with a dual-channel counter for ^{3}H and ^{14}C counts. It is best to set the channel gates manually so that there is less than 1% spill over from ^{3}H to ^{14}C and less than 20% spill over from ^{14}C to ^{3}H. Standard samples containing only ^{3}H and ^{14}C are prepared in exactly the same way (i.e., mitochondria suspension, 1% SDS, etc.) and used to calculate counting efficiency and spillover factors from one channel to the other.

To calculate $\Delta\psi$ from these measurements, it is necessary to know the internal and external binding constant of the test ion in mitochondria. The determination of this constant is described elsewhere.[18] The constants for TPP$^+$ and TPMP$^+$ binding in rat liver mitochondria are given in Table I. However, since these constants depend on pH, salt composition, and possibly other factors, it may be necessary to determine these for particular incubation conditions.

CALCULATION OF INTERNAL VOLUME. The counts of ^{3}H and ^{14}C in the supernatant and pellet fractions of sample (2) are corrected for efficiency and spill over. The internal volume (V_i) of the mitochondrial pellet is calculated using the following equation:

$$\text{Internal volume } (\mu\text{l/mg}) = V_i = 50(^3H_p/^3H_s - {^{14}C_p}/{^{14}C_s})/m \quad (5)$$

where the subscripts p and s stand for the pellet and supernant fractions and m is the pellet protein content (mg protein). The external pellet volume is calculated as follows:

$$V_e = (50^3 H_p/^3H_s m) - V_i \qquad (6)$$

CALCULATION OF $\Delta\psi$. First the pellet/supernatant accumulation ration (R_c) (per mg protein) of the lipophilic cation from sample (1) is calculated as follows:

$$R_c = (C_p/8C_s m - V_e/400) \qquad (7)$$

The factors 8, 50, and 400 are based on the assumption that 400 μl was placed in each microtube and 50-μl supernatant samples were taken for counting. If other volumes are used, the factors should be corrected accordingly. The second term in Eq. (7) corrects for lipophilic cation in the external trapped volume. From the value of R_c and using the binding constant K_i and K_o (Table I), we calculate $\Delta\psi$ as follows:

$$\Delta\psi = -59 \log[(R_c V_o - K_o)/(V_i + K_i)] \qquad (8)$$

where V_o is the normalized total volume of the suspension (i.e., 400 μl/mg protein), R_c is the accumulation ratio from Eq. (7), K_o and K_i are the external and internal binding constants (Table I), and V_i is the internal volume from Eq. (5).

Determination of $\Delta\psi$ from Binding of Charged Lipophilic EPR Probes.[17] This method is, in principle, identical to the method described above, except for the detection method. Since both the amounts of bound probes and free probes can be estimated from the EPR spectra without separation of vesicles, some of the advantages of the electrode and the centrifugation techniques can be combined. However, because of the difficulty in estimating small amounts of bound probes in the presence of high amounts of free probes and vice versa, it is necessary to adjust the probe/membrane ratio carefully for each value of $\Delta\psi$. Also, the weakness of the EPR signal requires relatively high concentrations of both membranes and probe and the probes are susceptible to reduction by membrane enzymes. A detailed description of the method is found elsewhere.[17]

Equation (8) should also be used to correct estimates of $\Delta\psi$ obtained from the distribution of lipophilic ions, as measured by specific electrode, flow dialysis, and filtration. In these methods the quantities that are measured are the amount of the ion taken by the vesicles and the amount of the ion that remains in the suspending medium. Hence

$$R_c = (\text{nmoles in/nmoles out})/\text{mg protein} \qquad (9)$$

To estimate $\Delta\psi$, this value is used in Eq. (8), where V_o is the total suspension volume divided by the vesicle content, i.e., $V_o = V/$mg protein.

Optical Probes

Many probes for the determination of membrane potential from changes in optical signals (absorption or fluorescence) have been introduced in recent years.[2,5,6,20] Several of these probes are potentially very useful since they often offer both high sensitivity to small changes in membrane potential and fast response. However, there are two general drawbacks in working with these probes, both of which result form the fact that the precise mechanism of the response of these probes is not always known. First, in general, it is not possible to calculate the magnitude of $\Delta\psi$ directly from the response since the mechanism of the response is unknown. The solution for this problem is to calibrate the signal, either by imposing known diffusion potentials or by calibrating against the ion distribution method. This procedure is not always successful since results often depend on the particular conditions of the calibration procedure and are not always applicable to other conditions.

Second, there are factors other than $\Delta\psi$ that also affect the probe signal. Often the probe responds to changes in surface charge, surface pH, solvent effects, quenchers, etc. Since it is seldom possible to correct properly for all of these effects, there is often uncertainty even with regard to qualitative $\Delta\psi$ estimates with these probes. While the precise mechanism which is responsible for the optical response is not known, the underlying mechanism of the "slow" dyes, which are used in bioenergetics, is believed to be the same as that of the lipophilic ion response described above.[20] With few exceptions (e.g., the "515" shift of chlorophylls and carotenes in photosynthetic membrane[8]), most optical probes used in bioenergetics are permeable ions. Hence, the response is the result of the potential driven redistribution of the probe between the internal and external solution and the inner and outer membrane surface.

In principle, the distribution of the probe between the membrane surfaces and compartments should be related to $\Delta\psi$ by the same relationships derived for lipophilic ions.[17,18] The added complication, which usually precludes a fully quantitative description, is the unknown relationship between the probe concentration and the optical signal in each compartment. The "fast" probes, mostly used in neurophysiology, are impermeable and their response depends on a different mechanism.[20]

---

[20] A. S. Waggoner *in* "The Enzymes of Biological Membranes" (A. N. Martonesi, ed.) Vol. 1, p. 313. Plenum, New York, 1985.

In some cases, the probe fluorescence characteristics are such that the free probe is not fluorescent and only the bound probes contribute to the fluorescence. In such cases the potential could be calculated from the fluorescence by equations similar to those used for the determination of $\Delta\psi$ from the signal of EPR probes.[17] For example, anilinonaphthalene sulfonate (ANS) would have been a good candidate for such a method, if it were not for the fact that a considerable fraction of the bound probe is not fluorescent.[21]

More commonly, the probe fluorescence or absorption is modulated when the probe aggregates either in the compartment of high concentration or on the membrane surface where it is concentrated. However, without exact knowledge of the aggregation equilibrium constants and the fluorescence characteristic of the various species, it is hard to give a satisfactory, fully quantitative description of such processes. Moreover, the extent of aggregation will depend sharply on dye/membrane ratios, requiring a precise calibration for each dye/membrane ratio. Thus, at present, practical application of these probes largely depends on proper calibration procedures.

Calibration by the Imposition of Known Diffusion Potentials. In this method, a series of measurements in which known diffusion potentials are artificially imposed is used to construct a calibration curve. The most common procedure is to use saturating amounts of valinomycin to make the membrane fully permeable to potassium. The gradient is imposed by using various concentrations of a potassium salt in the suspending medium. If one knows the initial internal potassium concentration and provided that valinomycin-induced potassium permeability is far greater than the permeability of all other ions in the suspension, the potential can be calculated from the Nernst equation.

$$\Delta\psi = -59 \log[K_{in}]/[K_{out}] \tag{10}$$

While this calibration procedure is widely used, it is not always adequate. First, in most biological membranes (unlike phospholipid vesicles), the membrane is relatively permeable to several other ions, such as Cl^-, H^+, Ca^{2+}, and Na^+. While it is sometimes possible to eliminate some of these by choosing a proper medium and using chelators (e.g., EDTA), it is often impossible to eliminate all relevant permeable ions. For instance, in mitochondria, even at neutral pH the permeability of H^+ is sufficient to collapse the valinomycin-induced potassium potential within a few seconds. Figure 2 shows the effect of valinomycin-induced potassium diffusion potential on the fluorescence of the cationic carbocyanine dye diS-

---

[21] D. E. Robertson and H. Rottenberg, *J. Biol. Chem.* **258**, 11039 (1983).

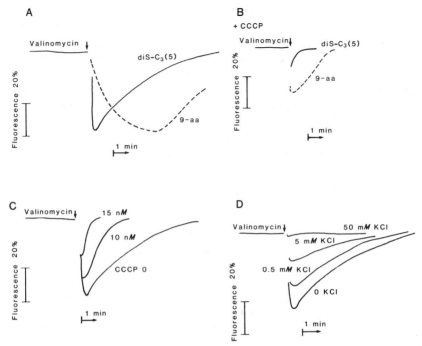

FIG. 2. Effect of valinomycin, KCl, and CCCP on the fluorescence of diS-C_3(5) and 9-aminoacridine in rat liver mitochondria. Mitochondria (1 mg/ml) are incubated in 0.2 M sucrose/10 mM MgCl$_2$/10 mM HEPES/1 mM EGTA/1 μM rotenone. Additions: 2 μM diS-C_3(5) (excitation = 585, emission = >640) or 2 μM 9-aminoacridine (excitation = 400, emission = 407–530), 0.1 μM valinomycin, and CCCP and KCl, as indicated. Fluorescence was measured in an Eppendorf fluorimeter.

C_3(5) (3,3-diethylthiocarbocyanine). In the absence of external potassium (Fig. 2A) the potassium concentration gradient is ~500 and the addition of valinomycin induced 40% quenching of the fluorescence. However, the diffusion potential decays fairly quickly, and the fluorescence returns to its original value after 4–5 min. This decay is due to the uptake of H$^+$ in exchange for K$^+$, as demonstrated by the reduction of the matrix pH, as followed with the ΔpH indicator 9-aminoacridine. Indeed, if the membrane is made more permeable to protons by the addition of CCCP, both the formation of ΔpH and the decay of the K$^+$ diffusion potential are greatly accelerated (Fig. 2B and C). These results indicate that the H$^+$ permeability in mitochondria is sufficiently large to affect the diffusion potential induced by valinomycin. Figure 2D shows that increasing the external potassium concentration also reduced the magnitude of the diffusion potential as expected. A proper calculation of the diffusion potential

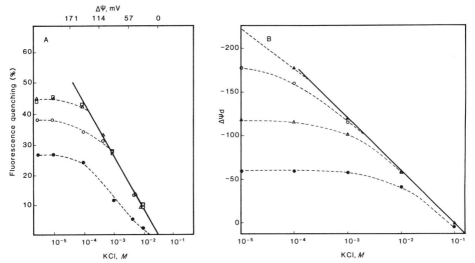

FIG. 3. Effect of valinomycin and KCl on diS-C$_3$(5) fluorescence and the calculated diffusion potentials. Conditions as in Fig. 2. Potential was calculated from Eq. (12). Concentrations of valinomycin: ●, 10^{-9} M; ○, 10^{-8} M; △, 10^{-7} M; □, 10^{-6} M.

requires consideration of all permeable species, as expressed in the Goldman equation:

$$\Delta\psi = -59 \log\left[\left(\sum P^+C_{in}^+ + \sum P^-C_{out}^-\right)\Big/\left(\sum P^+C_{out}^+ + \sum P^-C_{in}^-\right)\right] \quad (11)$$

In mitochondria, adequate description of the potential is obtained (in the presence of EGTA) by considering only potassium and proton permeabilities

$$\Delta\psi = -59 \log[(P_k[K_{in}^+] + P_H[H_{in}^+])/P_k[K_{out}^+] + P_H[H_{out}^+])] \quad (12)$$

Such calibration of the fluoresence quenching of diS-C$_3$(5) in rat liver mitochondria is shown in Fig. 3. Figure 3A shows the effect of valinomycin and potassium on the dye fluorescence. The diffusion potential calculated from Eq. (12) is shown in Fig. 3B. Only at high concentration of both valinomycin and potassium is the potential described by the Nernst equation. Moreover, at high potential, saturation of the quenching response is observed.

In calibrating the membrane potential (as in measuring the potential), it is important to remember that most probes respond also to the surface charge.[21,22] Hence a proper calibration requires changing the potassium

---

[22] K. Masamoto, K. Matsura, S. Itoh, and M. Nishimura, *Biochim. Biophys. Acta* **638**, 108 (1981).

concentration while keeping the ionic strength high and constant. (This can be accomplished by including a high concentration of Mg^{2+} in the medium.)

Another problem with this calibration procedure is that the potentials obtained in the experimental system are often considerably higher than the actual potential that can be reliably obtained with this calibration. This means that the experimental points would fall outside the calibration curve. Since the calibration curve is often not linear and tends to saturate at high potentials (Fig. 3), it is not possible to estimate correctly these potentials from such calibration curves.

Calibration against Ion Distribution Method. Optical probe signals have been calibrated against microelectrode measurements and against the distribution of radiolabeled ions. In general, this is a reliable method of calibration. One selects general conditions that are expected to yield potential values over a wide range and measures the potential by a reliable method in parallel with the optical signal. Besides the obvious precaution that the method of calibration itself must be as reliable as exists, it is important to remember that this calibration is only valid for the conditions in which the calibration curve was constructed. Often, changing the condition (i.e., salt, pH, temperature, substrates, dye/membrane ratio) may yield entirely different calibration curves. Here, we review briefly the most widely used optical probes in bioenergetics.

ANS—This is the oldest and a widely used probe of "energization" in bioenergetics.[23] It is now clear that the energy-induced response is a reflection of the membrane potential.[21,24] Because the probe is negatively charged, it accumulates in vesicles of positive potential (e.g., submitochondrial particles, chromatophores), which enhances the internal binding and fluorescence. The enhancement is a sensitive measure of the potential in this system. In systems of negative potential, such as mitochondria, the generation of potential causes efflux of dye, which leads to dissociating of internal bound dye and quenching. This response is not as sensitive since most of the dye is bound externally to begin with. The dye also responds to both surface charge and pH, and caution must be exercised to eliminate interference from these side effects.[21] The fluorescence has been calibrated by external diffusion potential and by Rb^+ distribution (Fig. 4). As the figure shows, there is a considerable difference between the calibration curve of $\Delta\psi$ generated by the substrates ATP and succinate.[21]

CARBOCYANINE DYES. Many different positively charged dyes of this class have been used for monitoring changes in $\Delta\psi$.[6,20] The choice is

[23] A. Azzi, *Biochem. Biophys. Res. Commun.* **37**, 254 (1969).
[24] A. A. Jasaitis, L. Van Chu, and V. P. Skulachev, *FEBS Lett.* **31**, 241 (1973).

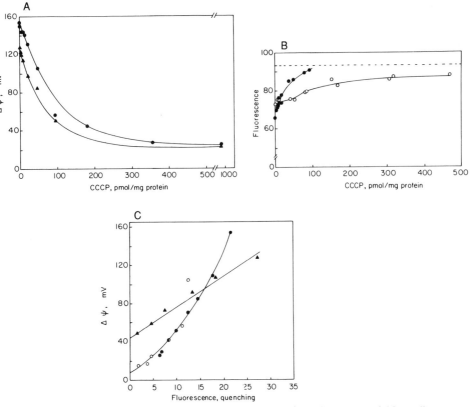

FIG. 4. Calibration of ANS fluorescence as a measure of membrane potential in rat liver mitochondria. Membrane potential was generated by either 3 mM ATP (▲), 10 mM succinate (●), or potassium diffusion potential (○). Medium was 0.25 M sucrose/10 mM Tris-Cl (pH 7.2)/1 mM MgCl$_2$/2 μM rotenone/ANS, when added, 8.3 μM (excitation = 408, emission = 430–470). Potential was estimated from ^{86}Rb distribution in the presence of 0.1 μM valinomycin. The potential was modulated by CCCP (▲, ●) or KCl (○).

dictated somewhat by their absorption and emission characteristics. The most commonly used dye in systems with negative potentials is the positively charged diS-C$_3$(5). Usually, the dye is calibrated with potassium–valinomycin diffusion potential, using the Nernst equation and assuming linear response versus the entire range. This is inadequate, as discussed above (see Fig. 3). A more adequate calibration is obtained by the use of Rb$^+$ distribution. Here, too, the response is not linear with succinate as a substrate but is fairly linear with ATP (cf. Fig. 5). This dye was also shown to respond to surface charge and pH and the same precautions must be taken, as discussed above.

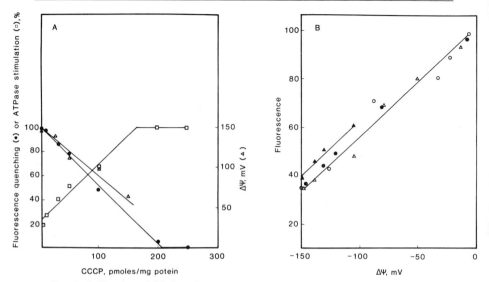

FIG. 5. Calibration of diS-C_3(5) fluorescence as a measure of $\Delta\psi$ in rat liver mitochondria. $\Delta\psi$ was generated by ATP (3 mM) and modulated by either CCCP (●), Ca^{2+} (○), K^+ + valinomycin (△), or the ATP/ADPXP_i ratio (▲). Fluorescence measured as in Fig. 2.

OXONOL IV. This negatively charge dye accumulates in positively charged vesicles where it binds to the internal membrane surface, resulting in shifts in its absorbance and fluorescence.[2,25] Thus, its response is in essence identical to that of other lipophilic ions. It is widely used in photosynthetic membrane vesicles and submitochondrial particles where the membrane potential is positive.[26] Its use in SMP is described in detail elsewhere in this series.[2]

SAFRANINE. This positively charged dye accumulates in negatively charged vesicles and organelles where it aggregates, leading to shifts in its absorption spectra.[27] While it has been used fairly widely, the high dye concentrations required for accurate measurement may produce various artifacts.[28]

[25] H. S. Apell and B. Bersch, *Biochim. Biophys. Acta* **909,** 480 (1987).
[26] K. Krab, E. J. Hotting, H. S. Van Walraven, M. S. G. Scholts, and R. Kraaynhof, *Bioelect. Bioeng.* **16,** 55 (1986).
[27] K. F. O. Akerman and M. F. K. Wikstrom, *FEBS Lett.* **68,** 191 (1976).
[28] V. G. R. Valle, L. Pereiza-da-Silva, and A. E. Vercesi, *Biochem. Biophys. Res. Commun.* **135,** 189 (1986).

Various other dyes have been suggested as good indicators of $\Delta\psi$ (e.g., berberines,[29] rhodamine 123[30]). At present, there are insufficient studies with these dyes to assess their suitability for these measurements.

Measurement of ΔpH

There are two basic approaches to the measurement of ΔpH in suspension of small cells, organelles, and other membrane vesicles. One approach is based on the equilibrium distribution of weak acids or weak bases and, in practice, is very similar to the techniques for $\Delta\psi$ measurement that were described in the preceding section. The other approach is based on direct measurement of the internal pH by the use of internal pH indicators (the pH of external suspension could be easily measured with a pH electrode or external pH indicator).

Distribution of Weak Acids and Bases

These techniques, which have been described in great detail previously, are still widely used in a large variety of systems.[1,9] The basic principle of these techniques is to measure the equilibrium distribution of either a weak acid anion or a weak base cation which is dictated by the pH gradient. Many (but not all) weak acids and bases are highly permeable in their neutral form but impermeable in their charged form. Thus, at equilibrium, the concentration of the neutral form must be identical on both sides of the membrane. Since the concentration of the charged species is determined by the pH at each compartment according to the relationship

$$A^- = K_a AH/H^+ \tag{13}$$

it follows that $H_{in}^+/H_{out}^+ = A_{in}^-/A_{out}^-$ or

$$\Delta pH = \log(A_{in}^-/A_{out}^-) \tag{14}$$

Similarly, for weak bases,

$$\Delta pH = -\log(AH_{in}^+/AH_{out}^+) \tag{15}$$

From these considerations it follows that, to obtain accumulation of the test species in the vesicles, we must use a weak acid if the internal pH is higher than the external pH, and a weak base if the internal pH is lower than the external pH. A useful probe should meet the following criteria: (1) high permeability of the neutral form, (2) low permeability of the

[29] V. Mikes and V. Dadak, *Biochim. Biophys. Acta* **723**, 231 (1983).
[30] R. K. Emaus, R. Grunwald, and S. T. Lemasters, *Biochim. Biophys. Acta* **850**, 436 (1986).

charged form, (3) minimal binding to membranes and proteins, (4) no alternative transport pathways, (5) no participation in metabolism or effects on metabolism.

In general, the smallest carboxylic acids and primary organic amines meet these criteria. Larger molecules are usually more lipophilic, and hence there is an increased permeability of the charged species and increased binding to membranes and proteins. The most widely used acid is still DMO although acetic acid, propionic acid, benzoic acid, salicylic acid, and other small organic acids are frequently used. The most widely used amine is methylamine, while larger amines (e.g., ethylamine, hexylamine), which are more lipophilic, are used less frequently.

In choosing a weak acid or base for ΔpH determination, particular attention must be paid to the pK in relation to expected pH values internally and externally. Unlike pH indicators (see below), the use of weak acids and bases is simpler if the pK is out of the range of the measured pH values. If the pK is within the range of the measured pH, a correction must be made for the concentration of the undissociated acid (since the estimation of the concentration with radiolabeled compounds does not distinguish between the ionized and nonionized species). In this case, the total concentration of both ionized and neutral species is related to the internal pH according to the following relationships:

$$\text{pH}_{in} \text{ (acids)} = \log[A_{in}/A_{out}(10^{pK} + 10^{pH_o} - 10^{pK}] \tag{16}$$

$$\text{pH}_{in} \text{ (bases)} = \log[A_{in}/A_{out}(10^{-pK} + 10^{-pH_o}) - 10^{-pK}] \tag{17}$$

When the pK of an acid is more than 1.5 units below pH$_o$ (or for base 1.5 unit above pH$_o$), this correction becomes negligible and Eqs. (14) or (15) may be used.

Since the calculated pH$_{in}$ according to Eqs. (16) or (17) is very sensitive to the pK value, an estimation of the pK under the incubation condition is often necessary. Moreover, often because of different salt composition inside and outside the vesicle, the actual pK might be different on the two sides of the membrane, introducing error. It should also be noticed that, in many cases, the pK is strongly temperature-dependent, and unless the pK is known for the experimental temperature, another uncertainty is introduced. Although binding is usually not a serious problem with the most commonly used ΔpH probes, it is occasionally necessary, particularly in the case of amines, to correct for binding.[31] This can be done in a similar manner to the procedure described for $\Delta\psi$ probes. The binding constant (K_b) in the absence of pH gradient is determined first (for simplicity one can assume equal binding on both sides of the membrane).

---

[31] E. A. Berry and P. C. Hinkle, *J. Biol. Chem.* **258**, 1474 (1983).

Then a corrected concentration ratio is calculated from the following relationship:

$$A_{in}/A_{out} = (R_c V_o - K_b)/(V_i + K_b) \qquad (18)$$

where R_c is the ratio of the total amount of amine or acid in the vesicle to the amount in the suspension, and V_o and V_i are the suspension volume and the vesicle volume, respectively. In effect, this correction amounts to the determination of an apparent internal volume $(V_i + K_b)$.[31,32] This may be accomplished by a determination of R_c as a function of pH and extrapolation to $\Delta pH = 0$. Since $\Delta pH \simeq \log R_c + \log[V_o/(V_i + K_b)]$, an extrapolation of the plot $\log R_c$ versus ΔpH to $\Delta pH = 0$ yields the value of $\log[V_o/(V_i + K_b)]$.

The various methods used for the determination of the acid or amine concentration gradient are similar to those described above for determination of $\Delta \psi$. Radiolabeled amine or acids can be used following a procedure identical to the procedure described for the lipophilic cations distribution using separation by centrifugation. Alternatively, flow dialysis can be used to measure the distribution for radiolabeled acids or amines (see Fig. 1). Specific electrodes are available for some amines (e.g., NH_4^+) and may be constructed for other acids or amines.

Another common method for the estimation of amine distribution is the use of fluorescent amines such as 9-aminoacridine, 9-amino-6-chloro-2-methoxyacridine (ACMA), acridine orange, and similar amines.[33-36] As is the case with other amines, these amines are concentrated inside acidic vesicles. The fluorescence of the dye concentrated inside the vesicles is quenched; hence, when ΔpH is generated, the uptake of the dye is associated with quenching of the suspension fluorescence. If we assume that the amine is distributed according to the pH gradient and that all the internal amine is quenched, then ΔpH can be calculated according to the following expression[33]:

$$\Delta pH = \log[Q/(1 - Q)] + \log(V_{out}/V_{in}) \qquad (19)$$

where Q is the fraction of total amine fluorescence that was quenched, and V_{out} and V_{in} are the external and internal volumes, respectively.

A serious problem with determination of ΔpH by fluorescent amines is the fact that these dyes bind to the membrane. In fact, the quenching appears to be largely the result of internal binding.[32] This binding could be

---

[32] R. Casadio and B. A. Melandri, *Arch. Biochem. Biophys.* **238**, 219 (1985).
[33] S. Schuldiner, H. Rottenberg, and M. Avron, *Eur. J. Biochem.* **25**, 64 (1972).
[34] H. Rottenberg and C. P. Lee, *Biochemistry* **14**, 2675 (1975).
[35] D. W. Deamer, R. C. Prince, and R. A. Crofts, *Biochim. Biophys. Acta* **274**, 323 (1972).
[36] G. Salama, R. G. Johnson, and A. Scarpa, *J. Gen. Physiol.* **75**, 109 (1980).

corrected by the methods described above. However, because not all the internal free dye is quenched, and the binding appears to saturate at high ΔpH, the overestimated ΔpH (due to binding) is often compensated by underestimation (due to incomplete quenching). Hence, even though in some cases the assumptions behind the calculation [Eq. (19)] are not justified, the various errors tend to cancel each other, giving results that are comparable to other methods.[33-35] For that reason, and despite repeated warning of the perils of the use of these dyes, they are still widely used successfully. Nevertheless, the prudent investigator should calibrate the response against another method of ΔpH or artificially imposed ΔpH. A similar weak-acid fluorescent probe, dansylglycine, which accumulates in vesicles with pH higher than the medium, was introduced recently.[37]

Another method to estimate amine distribution for ΔpH calculation is the use of amines labeled with spin probes.[38,39] The signal in the external medium may be quenched by spin broadening reagents and hence the amount of internal and external amine can be evaluated. These probes have not been used extensively because of various problems such as probe reduction catalyzed by many biological membranes, binding, and permeation of the broadening reagents.

Internal pH Indicator

Optical Probes. Considerable progress has been made in recent years in the use of absorbance and/or florescence pH indicators to monitor internal pH. The earlier attempts were largely unsuccessful because of the use of membrane-permeable (hence lipophilic) indicators which were distributed between the internal and external volume, and also were bound to the membrane surface. Because the distribution itself may depend on ΔpH and/or $\Delta\psi$, it is very difficult, except in special cases (cf. Ref. 40), to interpret changes in optical signal in such systems.

It is now apparent that the only useful internal pH indicators for ΔpH determination are hydrophilic indicators that are located exclusively in the internal water volume. Various strategies have been developed to ensure a successful application. In general, an indicator is chosen which is highly charged, large, and extremely hydrophilic. This ensures that the indicator is impermeable and hence cannot diffuse out of the vesicles and also prevents any binding to the membrane surface. In this case, it is

[37] J. Bramhall, *Biochemistry* **25**, 3458 (1986).
[38] B. A. Melandri, R. J. Melhorn, and L. Packer, *Arch. Biochem. Biophys.* **235**, 97 (1984).
[39] R. S. Melhorn, L. Packer, R. Macy, A. T. Balaban, and I. Dragutan, this series, Vol. 127 [55].
[40] W. Junge, G. Schonknect, and V. Forster, *Biochim. Biophys. Acta* **852**, 93 (1986).

TABLE II
FLUORESCENT INTERNAL pH INDICATORS

Indicator	Loading	pK	Fluorescence	Ref.
Fluorescein	Fluorescein diacetate[a]	6.3	495 → 520	41
Carboxyfluorescein	Carboxyfluorescein diacetate[a]	6.4	495 → 520	42, 43
BCECF [biscarboxyethyl-5(6)-carboxyfluorescein]	BCECF-acetoxymethyl[a]	7.0	500 → 530	44–46
Quene-2	Tetraacetoxymethylquene-1[a]	7.3	390 → 530	47
Pyranine	Trapping[b]	7.2	460 → 510	48
FITC-dextran	Trapping[b]	6.3	480 → 530	1, 49

[a] Uptake of ester follows by hydrolysis by internal esterases.
[b] Trapping by sonication of vesicles or inclusion in lipid mixture before forming liposomes.

necessary to choose indicators with a pK within the range of expected pH$_{in}$.

The problem that remains, of course, is to get the indicator into the vesicles. In vesicles that can be sonicated, either during preparation or prior to the experiments, a brief sonication in a medium that contains the indicator is often sufficient for introducing the indicator in the vesicles. The external indicator can be washed away or dialyzed. This strategy is not applicable to intact cells or delicate organelles (e.g., mitochondria). In these cases it is often possible to introduce the indicator into the cell as a neutral compound in which the carboxylic groups have been esterified. Inside the cell, nonspecific esterases will hydrolyze the esters, producing the charged, impermeable, indicator which is now trapped inside the cells. Table II[41–49] lists some of the most widely used internal pH indicators together with relevant information and references.

[41] J. Slavik, *FEBS Lett.* **140,** 22 (1982).
[42] S. A. Thomas, R. N. Buchsbaum, A. Zimniak, and E. Racker, *Biochemistry* **18,** 2210 (1979).
[43] E. Shechter, L. Letellier, and R. E. Simons, *FEBS Lett.* **139,** 121 (1982).
[44] T. S. Rink, R. Y. Tsien, and T. Pozzan, *J. Cell Biol.* **95,** 189 (1982).
[45] A. M. Paradiso, R. Y. Tsien, T. R. Demarest, and T. E. Machen, *Am. J. Physiol.* **253,** 830 (1987).
[46] M. H. Davis, R. A. Altshuld, D. W. Jung, and G. P. Brierley, *Biochem. Biophys. Res. Commun.* **149,** 40 (1987).
[47] T. Rogers, T. R. Hesketh, G. A. Smith, and J. C. Metcalfe, *J. Biol. Chem.* **258,** 5994 (1983).
[48] M. Seigneuret and J. L. Rigaud, *FEBS Lett.* **188,** 101 (1985).
[49] J. M. Heiple and D. L. Taylor, *in* "Intracellular pH" (R. Nuccitelli and D. W. Deamer, eds.). Liss, New York, 1982.

NMR pH Indicators. A very useful recent addition to the arsenal of methods using internal pH indicators has come from the biological application of NMR.[50] The chemical shifts of several nuclei are sufficiently sensitive to the ionization state of the molecular species to allow accurate pH determination using high-resolution NMR. The first, and still the most widely used indicator, is the ^{31}P chemical shift of inorganic phosphate, or more specifically, of the $H_2PO_4^-/HPO_4^-$ mixture.[51] The chemical shifts of these two species are separated by about 2.0 ppm at high field. A mixture (which depends on the pH, with a pK of about 6.8) would give intermediate chemical shift values which allow accurate estimation of the pH. Since most cells and organelles have a relatively high inorganic phosphate content, this is a truly natural indicator. Moreover, external phosphate would appear as a separate signal provided ΔpH is sufficiently large. In bacterial cells[52] and isolated mitochondria,[53] the estimated ΔpH agree well with measurements with DMO. Some of the problems associated with the use of this method are sensitivity of the pK to the medium salt composition, limited pH range, broadening of the signal, and overlap with chemical shifts of various organic phosphates.

Other phosphate esters with appropriate pK may serve as alternative pH indicators when present at sufficiently high concentrations.[54] Cells may also be loaded with various other phosphates with a pK more suitable for the desired measurement provided that they do not interfere with metabolism.[55] Other nuclei, particularly ^{19}F, are similarly sensitive to pH. While there are no naturally occurring fluorinated compounds in cells which may serve as internal pH indicators, several fluorinated amino acids and amines have been used as pH indicators with promising results.[56,57] These compounds may also be incorporated into the cell as esters where esterases release the trapped indicators.

---

[50] M. S. Avison, H. P. Hetherington, and R. G. Shulman, *Annu. Rev. Biophys. Biophys. Chem.* **15**, 377 (1986).
[51] R. B. Moon and T. H. Richards, *J. Biol. Chem.* **248**, 7276 (1973).
[52] G. Navon, S. Ogawa, R. G. Shulman, and T. Yamane, *Proc. Natl. Acad. Sci. U.S.A.* **74**, 87 (1977).
[53] S. Ogawa, H. Rottenberg, T. R. Brown, R. G. Shulman, C. L. Castillo, and P. Glynn, *Proc. Natl. Acad. Sci. U.S.A.* **75**, 1796 (1978).
[54] S. M. Cohen, S. Ogawa, H. Rottenberg, P. Glynn, T. Yamane, T. R. Brown, R. G. Shulman, and J. R. Williamson, *Nature (London)* **273**, 554 (1978).
[55] R. S. Labotka and R. A. Kleps, *Biochemistry* **22**, 6084 (1983).
[56] J. C. Taylor and C. Deutsch, *Biophys. J.* **43**, 261 (1983).
[57] C. J. Deutsch and J. S. Taylor, *Ann. N.Y. Acad. Sci.* **508**, 33 (1987).

[6] Measurement of pH of Intracellular Compartments in Living Cells by Fluorescent Dyes

By JANET VAN ADELSBERG, JONATHAN BARASCH, and QAIS AL-AWQATI

A number of subcellular organelles, for example, lysosomes, secretory vesicles, and endosomes, generate and maintain a transmembrane pH gradient by proton-translocating ATPases.[1] The acid compartments serve varied functions which include routing and uncoupling of receptor–ligand complexes, providing appropriate pH for enzyme activity, and producing an electrochemical gradient for solute transport, such as amino acid uptake into yeast vacuoles and amine uptake into platelets and into neurotransmitter storage granules. There are two general techniques for measuring pH in cell compartments that are too small to impale with microelectrodes. In one, subcellular particles can be isolated by centrifugation and pH gradients demonstrated by the accumulation of weak bases. Such methods have been discussed in previous volumes of this series.[2,3] In this chapter we will present *in situ* methods that are useful for measuring cytosolic and organellar pH in living cells using fluorescent probes that specifically label these compartments.

Measurement of Cytosolic pH

Measurement of pH in living cells has recently been achieved by loading cells with ester derivatives of fluorescein, a method first described by Thomas *et al.*[4] The ester is permeant and nonfluorescent, and the dye is trapped in the cytosol following its conversion to the charged pH-sensitive fluorophore by cytoplasmic esterases. This technique can be used to measure the cell pH of suspended or single cells. Although one can estimate cell pH from absorbance as well as from fluorescence measurements, fluorescence emission is a more sensitive method than is absorbance, allowing the use of lower concentrations of the fluorophore. Low probe concentration prevents significant changes in intracellular buffering power.

[1] Q. Al-Awqati, *Annu. Rev. Cell Biol.* **2**, 179 (1986).
[2] A. Lowe and Q. Al-Awqati, this series, Vol. 157, p. 611.
[3] H. Rottenberg, this series, Vol. 55, p. 547.
[4] J. A. Thomas, R. S. Buchsbaum, A. Zimniak, and E. Racker, *Biochemistry* **18**, 2210 (1979).

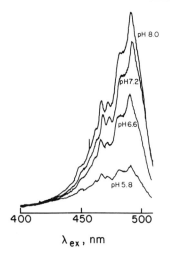

FIG. 1. Excitation spectra of 5,6-carboxyfluorescein in 110 mM KCl buffers of different pH. The peak emission wavelength was 520 nm; this peak does not vary with pH. Slit widths of 2.5–5 nm were used for excitation and emission.

The excitation spectrum of fluorescein is pH-sensitive. Reducing the pH quenches the dye and changes its spectral characteristics (Fig. 1). The pH can be determined by calculating the ratio of the intensity of fluorescence emission at two excitation wavelengths, one pH-sensitive and one relatively pH-insensitive (Fig. 2). These wavelengths vary with the fluorescein derivatives employed. The fluorescein derivatives commonly used to measure intracellular pH are the ester derivatives of 6- or 5,6-

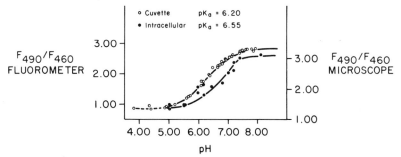

FIG. 2. Calibration curve of the ratio of emission intensity at two excitation wavelengths (F_{490}/F_{460}) versus pH for 5,6-carboxyfluorescein. The curve labeled "cuvette" was constructed from data obtained from dye alone in high-KCl buffers of varying pH. The curve labeled "intracellular" was constructed from data obtained in 5,6-carboxyfluorescein-loaded epithelia mounted on an epifluorescence microscope in high-KCl buffers with nigericin and valinomycin. From Ref. 6.

carboxyfluorescein.[5,6] These compounds have the disadvantage of having relatively low pK_a, about 6.2, and are therefore not useful for measuring pH greater than 7.3. However, in some systems, the pK_a of intracellular dye appears to be increased,[6] which extends the useful pH range of the dye (Fig. 2). The other disadvantage of these compounds is that they are relatively permeant and may leak out of cells at an unacceptably rapid rate. A newer fluorescein derivative which may avoid both these problems is 2',7'-bis(carboxyethyl)-5(6)-carboxyfluorescein (BCECF), a tetracarboxylate derivative of fluorescein with a pK_a of 6.98 which has been used successfully to measure cell pH.[7-9]

A potential difficulty with all esterified dyes is the release of acetate or formaldehyde into the cytosol upon hydrolysis. We have not seen toxic effects; however, with this potential problem in mind, we use the lowest possible concentration of dye for labeling.

Measurement of Cell pH in Cell Suspensions

Although the methods described below were devised for turtle bladder epithelial cells, they should apply to mammalian cells using isotonic buffers and 37° incubations.

To label cells with fluorescent dyes, 4 to 5 × 10^6 isolated turtle bladder epithelial cells are incubated in turtle Ringer's solution (110 mM NaCl, 3.5 mM KCl, 0.3 mM NaH$_2$PO$_4$, 1.65 mM Na$_2$HPO$_4$, 1 mM MgSO$_4$, 1 mM CaCl$_2$) containing 12 μM 5,6-dicarboxyfluorescein diacetate, the ester derivative of 5,6-carboxyfluorescein, for 60 min. After loading, the cells are washed three times and resuspended to final concentration of 1 × 10^6 and filtered to remove clumped cells. For mammalian cells, any balanced salt solution can be used. The use of phenol red should be avoided since it increases background fluorescence. Also, serum contains esterases which will cleave the esterified dyes to the charged nonpermeant species and therefore should not be used. Turtle bladder cells do not cleave the ester in the presence of 1 mM EGTA. Stock solutions of the esters were made in dimethyl sulfoxide (DMSO) and stored desiccated at 4°. Final DMSO concentration in the incubation media never exceeds 0.1%.

The acid-secreting cells of turtle bladder, comprising 15% of the total

---

[5] C. Cannon, J. van Adelsberg, S. Kelly, and Q. Al-Awqati, *Nature (London)* **314**, 443 (1985).

[6] J. van Adelsberg and Q. Al-Awqati, *J. Cell Biol.* **102**, 1638 (1986).

[7] T. J. Rink, R. Y. Tsien, and T. Pozzan, *J. Cell Biol.* **95**, 189 (1982).

[8] W. H. Moolenar, R. Y. Tsien, P. T. van der Saag, and S. W. de Laat, *Nature (London)* **304**, 645 (1983).

[9] A. M. Paradiso, R. Y. Tsien, and T. E. Machen, *Proc. Natl. Acad. Sci. U.S.A.* **81**, 7436 (1984).

cells, contain high concentrations of carbonate dehydratase (carbonic anhydrate) which, incidentally, is also a potent esterase that cleaves these esterified dyes.[10] These cells concentrate esterified dyes preferentially compared to other cells in the epithelium and therefore are responsible for most of the fluorescence emission measured. It is probable that all cells that are enriched in carbonate dehydratase (or other esterases) will concentrate 5,6-carboxyfluorescein or other esterified dyes. Of course, the dye concentration in each cell type depends not only on the rate of ester cleavage but also on the rate of efflux of the dye from the cell. Since fluorescein and other dyes are frequently organic anions, there may be transport processes in these cells that will facilitate their diffusion from the cytoplasm. Obtaining an adequate dye concentration is the major problem of this method and is dependent on the cell type.

To measure intracellular pH, 1.5-ml aliquots containing 5×10^5 cells suspended in turtle Ringer's solution are placed in quartz cuvettes with a magnetic stir bar. We have observed that increasing the concentration of cells above 10^6/ml tends to decrease the signal, presumably because of light scattering. The cells are alternately excited with 460 and 490 nm light and the emission measured at 520 nm in a Farrand spectrofluorometer with 2.5-nm slits for excitation and emission. As the esterified dye has different fluorescence properties from the hydrolyzed compound, excitation and emission spectra are performed on the initial sample in order to ensure that complete hydrolysis of 5,6-carboxyfluorescein diacetate has occurred.

Cell pH is estimated from the ratio of emission intensity at the two exciting wavelengths, F_{490}/F_{460}, using a calibration curve, shown in Fig. 2. Calibration curves have been obtained not only in cuvettes (using the free acid) but also in isolated dye-loaded cells and whole epithelia, both of which are incubated for 5 min in sodium-free buffers containing 90 mM KCl, 2 mM KHPO$_4$, 1 mM MgCl$_2$, 1 mM CaCl$_2$, 20 μM valinomycin, 30 μM nigericin, and 20 mM acetate, MES, MOPS, or HEPES at the desired pH. These solutions were designed to equilibrate intra- and extracellular pH in the absence of a membrane potential. The 5-min incubation in valinomycin/nigericin is intended to avoid the optical effects of changing cell volume on measurement of fluorescence emission. F_{490}/F_{460} can be calculated from measurements made using suspended cells stirred in a fluorometer as well as on single cells mounted on an epifluorescence microscope, as described below. The apparent pK_a of 5,6,-carboxyfluorescein was increased by 0.35 pH units to 6.55 when measured intracellularly. Neither valinomycin, nigericin, or the buffers used above are

[10] D. A. Hopkinson, *Ann. Hum. Genet.* **38**, 155 (1974).

fluorescent at these wavelengths, nor do they quench the fluorescence of 5,6-carboxyfluorescein.

Measurement of Cell pH in Single Identified Cells

The turtle urinary bladder contains two surface cell types, the granular cell responsible for sodium transport and the mitochondria-rich or carbonate dehydratase-rich cell responsible for proton and HCO_3 transport. The latter cell comprises only about 15% of the population. To measure cell pH in identified cells, whole epithelia mounted on rings are incubated on the luminal side only with 10 μM 5,6-carboxyfluorescein diacetate in turtle Ringer's solution for 10 min. The epithelium is then mounted mucosal side down on the stage of an inverted epifluorescence microscope (Zeiss IM 35) using a 63× fluorite objective and a mercury arc lamp. The microscope is equipped with a Farrand microspectrofluorometer (Farrand Optical Co., Valhalla, NY). The serosal side of the epithelium is continuously perfused with the desired media. Because the mitochondria-rich cells are also enriched in carbonate dehydratase, they can be readily identified by the accumulation of free carboxyfluorescein. Further, because the acid-secreting cells can endocytose luminal macromolecules, one can expose the cells to rhodamine-labeled albumin and then stain them with carboxyfluorescein. The identified cell is centered in the photometer aperture and alternately excited with 460 and 490 nm light using interference filters (Ditric Optics, Inc.). The emission intensity is measured at 520 nm. The intensity of the exciting light can be reduced by using neutral density filters placed in series with the exciting filters since fluorescein compounds are readily bleached. Intracellular calibration was performed by incubating the epithelium in the calibrating solution with nigericin and valinomycin.

This method has been used successfully in other cell types. We used it in the carbonate dehydratase-rich cells of isolated perfused rabbit renal tubules,[11] and in a number of cultured cells grown on glass disks. The advantages of this method are that constant perfusion washes away any leaked dye and that one can measure cell pH in an identified cell.

Measurement of pH of Intracellular Compartments
  Using Impermeant Dyes

Endosomal and lysosomal pH can be measured directly by allowing cells to take up fluorescein isothiocyanate linked to high-molecular-weight dextran, MW 60,000–70,000 (FITC-dextran), or to albumin (FITC-

---

[11] G. Schwartz, J. Barasch, and Q. Al-Awqati, *Nature* (*London*) **318**, 368 (1985).

albumin), a method first described by Ohkuma and Poole.[12] The fluorescence excitation spectrum of FITC-dextran, as of any fluorescein derivative, varies with pH. The method of measuring organelle pH is the same as that described above for measurement of cell pH in identified cells. To construct a calibration curve the intensity of fluorescence emission of FITC-dextran in a balanced salt solution is measured at various pH values with a microfluorometer mounted on an inverted epifluorescence microscope using two excitation wavelengths, 460 and 490 nm. These wavelengths are selected by inserting narrow band-pass filters between the mercury arc lamp and the microscope stage. As described for carboxyfluorescein, the ratio F_{490}/F_{460}, is proportional to pH. To measure intravesicular pH, fibroblasts (or the cells of interest) grown on coverslips are incubated for 20 min at 37° in 0.1 mg/ml FITC-dextran in a balanced salt solution without phenol red. Commercial FITC-dextran is dialyzed against H_2O to remove free fluorescein and lyophilized. The time required for sufficient endocytosis of the fluorescent macromolecule varies with temperature and cell type and must be determined experimentally. After incubation, the coverslips are washed extensively with buffer solution at 20° and mounted on the microscope. Measurements of F_{490}/F_{460} are performed with the aperture of the fluorometer limited to one or a few endocytic vesicle(s) and the pH of the vesicles calculated by comparing the measured ratio to the calibration curve (Fig. 3, 10 μg/ml).

Recently, Horowitz and Maxfield have described a method for calibrating fluorescence measurements with the dye accumulated in lysosomes.[13] The cells are allowed to take up FITC-dextran, washed free of extracellular fluorophore, and fixed in formalin. The fixed cells are incubated in buffers of various pH in the presence of 30 μM monensin in order to equilibrate intracellular, intravesicular, and extracellular pH. F_{490}/F_{460} is measured in these preparations to construct a standard curve.

Fluorescence emission intensity of FITC-dextran or FITC-albumin varies with concentration; there is significant quenching at concentrations greater than 2 mg/ml. In addition, as seen in Fig. 3, the pH sensitivity of the F_{490}/F_{460} is blunted or abolished at high concentrations. It is likely that the dye forms oligomers at high concentration; consequently the excitation spectra reflect the behavior of an ensemble of different molecular species, each with its own characteristic pH sensitivity.

Another problem is that the signal may be too weak for the measurement of pH when the aperture of the microfluorometer is limited to a small field. The signal can be increased by enlarging that aperture in order to

[12] S. Ohkuma and B. Poole, *Proc. Natl. Acad. Sci. U.S.A.* **75,** 3327 (1978).
[13] M. A. Horowitz and F. R. Maxfield, *J. Cell Biol.* **99,** 1936 (1984).

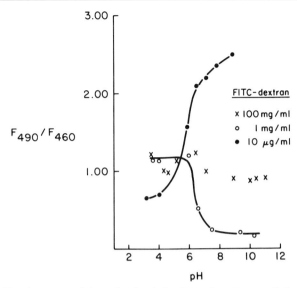

FIG. 3. Calibration curve of the ratio of emission intensity at two excitation wavelengths (F_{490}/F_{460}) versus pH for FITC-dextran at three concentrations in a balanced salt solution. The wavelength of maximum emission is shifted to 565 nm at 100 mg/ml. Fluorescence emission was measured at 520 nm for all three samples.

include greater numbers of vesicles; however, this risks missing the presence of two populations of endocytic vesicles having different pH. Recently, Maxfield et al. developed a method using video intensification and image analysis for the measurement of the pH of individual fluorescent vesicles in a single cell.[13] This would be the method of choice if pH heterogeneity of the vesicles is suspected.

The advantages of this technique for the measurement of intravesicular pH are that the technique is relatively rapid, not toxic, and subject to few artifacts. The disadvantage is that only endocytic vesicles and their derivatives can be labeled by endocytosis. In addition, pH measurements may become less accurate over time since the intravesicular (or intralysosomal) concentration of fluorophore may rise due to continuing endocytosis.

Measurement of pH Gradients in Cells

Acridines are permeant fluorescent weak bases that accumulate in acid intracellular compartments. At high concentration, some of these agents, especially acridine orange, undergo a red shift in their emission

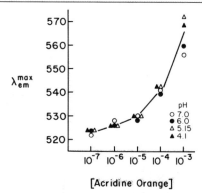

FIG. 4. Correlation of the wavelength of maximal fluorescence emission with concentration of acridine orange at different pH, showing that emission wavelength is concentration- and not pH-dependent. From Ref. 14.

spectrum (Fig. 4).[14] This is due to the formation of dimers and higher order multimers with consequent shifts in spectral properties of the stacks of dye molecules. Other commonly used acridines, such as 9-aminoacridine, have a much lower tendency toward stacking. Acridines are well-known nucleic acid-binding reagents and hence they always stain the nuclei of cells.

The cells are stained by incubation for less than 1 min in 1 to 10 μM acridine orange dissolved in balanced salt solution without phenol red (whose composition in mM is, 137 NaCl, 5.4 KCl, 0.44 KH_2PO_4, 0.34 Na_2HPO_4, 1 $MgSO_4$, 1.26 $CaCl_2$, 5.5 dextrose). Since the dye partitions among the various compartments of the cell and also among cells, the concentration of acridine orange necessary for optimal staining will vary with the number of cells being stained. We had found that when cells die they frequently start staining a diffuse red-orange color. Optimal cellular staining is obtained when a green cytoplasmic background is seen with discrete orange (i.e., acid) vesicles. Nuclei stain a bright green due to the binding of acridine orange to nucleic acids. The coverslip is mounted on an inverted epifluorescence microscope in a chamber which allows application of solutions to the cells. The cells are observed using a 400 to 490 nm band-pass filter for excitation, although any of the excitation filter combinations conventionally used to visualize fluorescein fluorescence can be used for acridine orange.

To ensure that the orange color is due to intravesicular acidity rather than to binding, the effect of addition of weak bases (e.g., 20 mM NH_4Cl)

[14] S. Gluck, C. Cannon, and Q. Al-Awqati, *Proc. Natl. Acad. Sci. U.S.A.* **79**, 4327 (1982).

or proton ionophores (e.g., 1 μM monensin or nigericin) is tested. These agents should abolish the pH gradient and hence the orange color if the vesicles are acid.

Since acridine orange is permeant, it can be used to estimate relative pH in cellular compartments inaccessible to other pH-sensitive dyes. This could be done using the calibration curve given in Fig. 4, and by assuming that the cytoplasmic pH is of the order of 7.0–7.4. An alternative method involves lowering the extracellular pH, keeping the bulk acridine orange concentration constant. When the external pH is lowered to the level of the vesicle or granule pH, there should be no difference in the concentration of acridine orange between them, hence the color of the granule should be the same as that of the outside. We emphasize that these measurements should not be interpreted too quantitatively due to the many problems with binding of this dye to intracellular components.

Fluorescence-activated Cell Sorter

If a weak base is localized predominantly to a single cellular compartment, then a technique that can screen a population of cells becomes useful. For example, cells can be labeled with acridine orange and the accumulation of acridine orange in acid compartments can be measured spectrophotometrically. If the acridine orange accumulation can be assigned to one particular compartment by independent methods, for example, electron microscopy (see Anderson and Pathak[15]), changes in the acridine orange signal are attributable to changes in the transmembrane pH gradient of that organelle.

Measurement of the acridine orange fluorescence emission of a large number of cells is facilitated by a fluorescence-activated cell sorter. This instrument is capable of analyzing the orange and red fluorescence emission intensities of thousands of cells per minute. Cells are stained with acridine orange and excited with a 5-W argon gas laser at 488 nm, 300–400 mW. A cell population is selected for analysis by determining the forward, narrow-angle light scatter of the sample. These data can be displayed as a histogram of cell number versus scatter intensity. Since forward light scatter is proportional to particle volume, one can analyze the fluorescence of particles of a certain size. In this way, the contribution of debris, bacteria, yeast, and red blood cells can be eliminated from analysis. Alternatively, a specific population of cells can be selected by analyzing both forward and right-angle light scatter. The latter is sensitive to internal cell structure, and a frequency distribution of 90° versus forward light scatter may reveal select populations of cells.

[15] R. G. W. Anderson and R. K. Pathak, *Cell* **40,** 635 (1985).

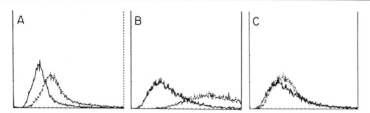

FIG. 5. Fluorescence-activated cell sorter analysis of parafollicular cells in 5 μM acridine orange. The intensity of orange-red fluorescence is plotted on the x axis and number of cells is shown on the y axis. (A) The effect of 2.5 μM valinomycin, (B) the effect of 10 μM nigericin for 10 min, and (C) the effect of 10 μM nigericin for 30 min.

Relative fluorescence of acridine orange-loaded cells is analyzed with standard collection and focusing lenses; primary blocking filters, a 520 nm long-pass dichroic and a 530 nm long-pass absorbance are followed by a beam splitter which directs total fluorescence to two photomultiplier tubes (PMT). PMT 1 receives fluorescence emission after passing a 580 nm long-pass and a 580 nm dichroic filter. This PMT measures the intensity of orange fluorescence. PMT 2 receives fluorescent light through a 650/50 bandpass filter and measures intensity of red fluorescence. Fluorescence is converted to voltage pulses that are proportional to fluorescence intensity. These pulses are quantitated and displayed on a linear scale of cell number versus fluorescence intensity. Analysis of acridine orange fluorescence with a cell sorter allows quantitation of the range of pH gradients formed in a cell population. However, its best use is in demonstrating changes in acidity of an organelle induced by certain treatments. For instance, we recently used this method to demonstrate that the serotonin granules of thyroid parafollicular cells become more acid in response to secretagogues.[16]

The parafollicular granules, like many other secretory granules, contain an electrogenic proton-translocating ATPase.[1] The membrane conductance of these granules is apparently low, hence the ATPase generates a large membrane potential sufficiently rapidly to prevent the formation of significant pH gradients. Only when the membrane potential is shunted by valinomycin can one demonstrate development of a large pH gradient in isolated granules. Similarly, we find that addition of valinomycin increases the mean fluorescence intensity of cells exhibiting the orange red fluorescence (Fig. 5A). Cautious interpretation of whole-cell acridine orange fluorescence is essential since nongranular compartments are also responsive to perturbations that might affect the pH gradient. For exam-

[16] J. M. Barasch, M. D. Gershon, E. A. Nunez, and H. Tamir, Soc. J. Neurosci. **7**, 4017 (1987).

ple, the addition of nigericin (a $K^+:H^+$ exchanger) to acridine orange-loaded cells initially increases the mean orange-red fluorescence due to the outward flux of K linked to the inward proton flux (Fig. 5B). Prolonged incubation in nigericin abolishes this enhanced orange-red fluorescence, and is likely due to equilibration of K^+ and proton gradients across the cell membrane (Fig. 5C).

[7] Use of Carbocyanine Dyes to Assay Membrane Potential of Mouse Ascites Tumor Cells

By A. ALAN EDDY

Introduction

As discussed elsewhere in this series,[1] certain carbocyanine dyes, such as 3,3'-dipropyloxadicarbocyanine and 3,3'-dipropylthiadicarbocyanine, belong to the general class of lipophilic cations which move electrophoretically across natural or artificial lipid membranes in response to the prevailing membrane potential. Accumulation of the dye in the compartment bounded by the membrane results in characteristic changes in the absorption spectrum of the dye as well as in its fluorescence emission spectrum which can readily be detected at appropriate wavelengths in the visible region. Assays are conducted in such a way that the relevant changes in absorbance or fluorescence intensity provide a measure of the fraction of the dye which has accumulated in the membrane compartment.[2] It is convenient to regard the dye monomer as being accumulated, through the bulk phase of the membrane, in accordance with the Nernst equation describing the equilibrium of a monovalent cation between two compartments differing in electrical potential. However, the accumulation of dye into liposomes, or into living cells exhibiting a negative membrane potential ($\Delta\psi$) of up to 60 mV, is greatly amplified by subsequent events. These include (1) the tendency of the dye to bind to the plasma membrane itself, to other cellular membranes, and possibly to other cellular constituents; (2) accumulation of the dye in other cellular compartments such as the mitochondria. A small change in $\Delta\psi$ at the plasma membrane may therefore lead to a relatively large fraction of the dye moving into or out of the cells. In a typical assay this process should

---

[1] L. Cohen, this series.
[2] A. S. Waggoner, *Annu. Rev. Biophys. Bioeng.* **8**, 47 (1979).

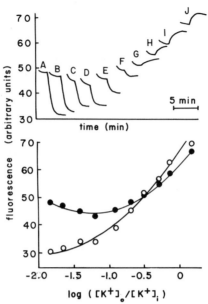

FIG. 1. Observations leading to an estimate of the null point at which the ratio $[K^+]_o/[K^+]_i$ is such that the addition of valinomycin causes no change in fluorescence intensity. The ratio $[K^+]_o/[K]_i$ was systematically varied by changing $[K]_o$ from about 1 mM in A to about 150 mM in J in the steps indicated in the text. Thus the 10 traces (A to J) each represent the effect of adding valinomycin to a single-cell suspension. The abrupt change in fluorescence intensity in each instance was effected by the addition of valinomycin. The lower part of the diagram illustrates the magnitude of the stepwise change in fluorescence as a function of log $[K^+_o/[K^+]_i$; ●, valinomycin absent; ○, valinomycin present. Note membrane hyperpolarization corresponds to a diminished fluorescence intensity.

be completed in 1–2 min (compare Fig. 1). Assessment of the actual plasma membrane potential requires that the mitochondrial membrane potential remains constant or, in the recommended procedure, that it is abolished. Furthermore, the fact that the overall distribution of the dye between the cell and its environment depends on an ill-defined intracellular binding process, in which aggregation of the dye may also be involved, means that care must be taken to ensure that apparent changes in $\Delta\psi$ are not in fact due to changes in binding of the dye at intracellular sites. In particular, changes in intracellular pH in human erythrocytes are known to lead to alterations in dye binding.[3] The magnitude of this problem in other systems, such as mouse ascites tumor cells, is unknown.

[3] T. J. Rink and S. B. Hladky, in "Red Cell Membranes: A Methodological Approach" (J. C. Ellory and J. D. Young, eds.), pp. 321–334. Academic Press, New York, 1982.

General Procedures

(1) A mixture of oligomycin, antimycin, and 2,4-dinitrophenol is added to the mouse ascites tumor cells in amounts that inhibit both oxidative phosphorylation and the generation of a mitochondrial membrane potential at the expense of cytosolic ATP.[4] The subsequent addition of valinomycin to this system would result in insignificant changes in the latter potential. The plasma membrane potential, on the other hand, will approach the equilibrium potential for K^+ ions.

(2) The tumor cells are suspended at 37° for 10–30 min in various Ringer's solutions containing the mitochondrial inhibitors, the object being to obtain a number of preparations exhibiting membrane potentials in the range from -20 to -100 mV. To achieve this, glucose is added to one series of cell suspensions to maintain energy metabolism. Other series contain glucose and selected concentrations of ouabain or amino acids to lower $\Delta\psi$, while a further series becomes hyperpolarized as a result of initial treatments that elevate the cellular Na^+ contents. Each of these series comprises several cell suspensions containing a range of extracellular concentrations of K^+ ions.

(3) A portion of each cell suspension is transferred to the cuvette of a fluorimeter and valinomycin is added (Fig. 1). Comparison of the behavior of a given series of cell suspensions, differing only in the extracellular concentration of K^+ ($[K^+]_o$), permits the identification of the null value of $[K^+]_o$ at which valinomycin causes no fluorescence change. It is assumed that, in these circumstances, the membrane potential is in equilibrium with the distribution of K^+ across the plasma membrane. Thus,

$$\Delta\psi = RT/F \ \ln([K^+]_o/[K^+]_i)$$

Another similar series of cell suspensions, each of which has been partially depolarized or hyperpolarized, is treated in a similar fashion to obtain a different null value of $[K^+]_o$ and a different value of $\Delta\psi$.

(4) A calibration curve is constructed relating the fluorescence intensity observed at the null point to the corresponding value of $\Delta\psi$. The fluorescence intensity and $\Delta\psi$ were found to be approximately linearly related in the range of -20 to -100 mV, irrespective of the method used to change $\Delta\psi$.[4] This shows that, under the conditions of the assay, the fluorescence intensity can be used as a measure of $\Delta\psi$ without actually titrating the cell suspension with valinomycin over a range of concentrations of K^+. Such titrations only lead to the value of $\Delta\psi$ under the ionic conditions prevailing at the null point, which may well differ from those in the original cell suspension.

[4] R. D. Philo and A. A. Eddy, *Biochem. J.* **174**, 801 (1978).

Limitations

(1) The above procedures permit $\Delta\psi$ to be estimated only in the presence of the inhibitors of mitochondrial energy metabolism. Provided glycolysis is maintained, the tumor cells would therefore pump ions and other solutes across the plasma membrane. In this connection the amount of dye used in the assay must itself be compatible with these physiological processes. For instance, in one such study where the magnitude of the sodium electrochemical gradient across the plasma membrane was compared with the prevailing methionine gradient, the recommended dose of carbocyanine dye was shown not to interfere with amino acid accumulation nor to affect the retention of cellular K^+ ions.[4]

(2) In the erythrocyte, carbocyanine dyes may block the Ca^{2+}-activated K^+ channel[5]; evidence for similar behavior in mouse ascites tumor cells is lacking. Indeed, an increase in potassium permeability has been claimed to occur[6] and might, in principle, result from an increase in cytosolic Ca^{2+} when respiration is inhibited by the dye.

(3) It has been argued that because valinomycin rapidly depletes the cellular content of ATP maintained by endogenous energy metabolism, the sodium pump would be inhibited and might give rise to an artifactual null point during a titration in the presence of the carbocyanine dye. At this null point valinomycin would cause a diminution in electrogenic ion pumping which would be exactly balanced by an accelerated efflux of K^+ through the ionophore.[6,7] If such behavior does occur, it can only do so when energy metabolism depends exclusively on oxidative phosphorylation. The presence of glucose in the recommended assay obviates this particular difficulty. It is of some interest that a study, based on the carbocyanine dye technique, of the membrane potential of mouse ascites tumor cells respiring lactate was vitiated by a similar problem.[8]

(4) It has been suggested that the oligomycin used to inhibit mitochondrial functions might inhibit the plasma membrane ATPase.[9] However, the amounts used in the recommended assay procedure are chosen so as to inhibit the mitochondrial ATPase selectively, very much larger amounts being needed to inhibit the sodium pump. Likewise, the recommended concentration of 2,4-dinitrophenol appears not to affect $\Delta\psi$ at the plasma membrane of mouse ascites tumor cells.[4]

[5] T. J. B. Simons, *J. Physiol. (London)* **288**, 481 (1979).
[6] T. C. Smith, *J. Cell. Physiol.* **112**, 302 (1982).
[7] R. M. Johnstone, P. C. Laris, and A. A. Eddy, *J. Cell. Physiol.* **112**, 298 (1982).
[8] E. Johnson and A. A. Eddy, *Biochem. J.* **226**, 773 (1985).
[9] C. L. Bashford and C. A. Pasternak, *J. Membr. Biol.* **79**, 275 (1984).

(5) Great care is needed in interpreting the effects of selected drugs, ionophores, or proteins such as albumin on the magnitude of $\Delta\psi$ as assayed by the carbocyanine dye technique. The possibility that these may physically associate with the dye or alter its fluorescence or absorbance characteristics should be studied in appropriate controls.

(6) Studies with the anionic oxonol dyes have led to estimates of $\Delta\psi$ in fair agreement with those observed using a carbocyanine dye.[10] An advantage of the dyes of the former class is that they appear to be excluded from the mitochondria. However, further work appears to be needed to define the optimal conditions needed to avoid artifactual interactions between ionophores such as valinomycin and the oxonol series. It has been claimed that such interactions can be insignificant when oxonol absorbance is studied at appropriate wavelengths.[10] One rapid procedure advocated for titrating the membrane potential with K^+ ions in the presence of valinomycin is to add successive relatively small volumes of concentrated KCl solution to the cell suspension which has been hyperpolarized in the presence of valinomycin.[10,11] This procedure involves simultaneous changes in tonicity and extracellular $[K^+]$, as compared with the more conventional use of isotonic solutions containing graded concentrations of K^+. The relative merits of the two methods have yet to be assessed in relation to the possibility that dye binding may be affected by large changes in ionic strength. Errors due to the latter cause may only be significant when $\Delta\psi$ itself is small.

Materials and Methods

Chemicals

3,3'-Dipropyloxadicarbocyanine iodide was prepared as described by Sims et al.,[12] the initial crystallization being performed from an ethanolic solution.[4] Both it and 3,3'-dipropylthiadicarbocyanine are available from commercial sources (e.g., Molecular Probes Inc., 24750 Lawrence Rd., Junction City, OR 97448). Solutions of either dye in ethanol or in dimethyl sulfoxide are stored at 0° in the dark (1 mg of dye/ml), where they appear to be stable for at least a month.

---

[10] C. L. Bashford, G. M. Alder, M. A. Gray, K. J. Micklem, C. C. Taylor, P. J. Turek, and C. A. Pasternak, *J. Cell. Physiol.* **123**, 326 (1985).
[11] T. J. Rink, C. Montecucco, T. R. Hesketh, and R. Y. Tsien, *Biochim. Biophys. Acta* **595**, 15 (1980).
[12] P. J. Sims, A. S. Waggoner, C. H. Wang, and J. F. Hoffman, *Biochemistry* **13**, 3315 (1974).

Ringer's Solutions

The standard Ringer's solution contains 155 mM Na$^+$, 8 mM K$^+$, 131 mM Cl$^-$, 16 mM orthophosphate, and 1.2 mM MgSO$_4$, at pH 7.4. The concentration of K$^+$ is changed by keeping [Na$^+$] + [K$^+$] constant at 163 mM and replacing the one ion by the other. Such solutions are prepared daily by mixing 0.77 M solutions of NaCl, KCl, or MgSO$_4$ with 0.1 M Na$_2$HPO$_4$ at pH 7.4 and distilled water. These concentrated stock solutions are stored at 0° for up to 1 month.

Cell Preparations

After collection and washing, the tumor cells (1 vol) are incubated for 30 min at 37° in the standard Ringer's solution (25 vol). It is often useful to filter the warm cell suspension through a nylon coffee strainer at that stage. The cells are collected in the bottom of a graduated centrifugation tube at about 1500 g for 1 min. The volume of packed cells is noted (1–2 ml). Loading the tumor cells with Na$^+$ ions is subsequently achieved by incubating them (1 vol) for 30 min at 37° in a Ringer's solution (30 vol) from which K$^+$ ions are omitted.

Fluorimetry

Fluorescence is excited at 580 nm and recorded at 620 nm in the presence of 3,3'-dipropyloxadicarbocyanine iodide. The corresponding wavelengths for 3,3'-dipropylthiadicarbocyanine iodide are 630 and 680 nm, respectively. The cell suspension (4 ml) containing the dye (4 μg) is maintained at 37° in the magnetically stirred glass cuvette of the spectrofluorimeter whose output is connected to a pen recorder. Suitable commercial instruments are available, such as the Perkin-Elmer luminescence spectrometer (model LS-5). Attention must be paid to both the choice of nominal slit widths for governing light input and output and to maintaining a rate of stirring of the thermostatted solution in the cuvette which disperses the tumor cells without damaging them. In some instruments (that cited above), it may be necessary to screen that part of the cuvette in which a stirrer "flea" might otherwise enter the incoming light beam. Careful initial washing and filtration of the mouse ascites tumor cells usually obviates problems due to scattering of light by clumped tumor cells.

To facilitate comparison of the observations made on different days, the fluorescence intensity of a standard solution of the dye is assayed daily.[4] Deterioration of the light source in the fluorimeter, any change either in the geometry of the cuvette or in the cleanliness and opacity of its surfaces can be expected to lead to anomalous results.

Null Point Experiments

The packed cells (1 vol) are suspended in the standard Ringer's solution (5 vol) and portions (1 ml) are stored on ice. (1) Each 1-ml sample (about 26 mg dry weight of tumor cells) is added to 80 ml of Ringer solution containing a selected concentration of K^+. A typical series would contain 1 mM, 2 mM, 3 mM, 5 mM, 8 mM, 15 mM, 25 mM, 50 mM, 75 mM, or 140 mM of K^+, with $[Na^+] + [K^+]$ constant at 163 mM.

(2) The Ringer's solution also contains 60 μM 2,4-dinitrophenol, 6 ng of oligomycin/ml, and 1 μg of antimycin/ml to inhibit mitochondrial energy metabolism. (3) Optional further components are glucose (2 mg/ml), 1 mM or 2.5 mM glycine, 0.5 mM ouabain. In general, the presence of glucose increases the membrane potential whereas the amino acid or ouabain tends to depolarize the cell membrane. Changing of the cellular concentration of Na^+ provides a further means of manipulating $\Delta \psi$. In the presence of glucose an elevated concentration of Na^+ leads to acceleration of the sodium pump and membrane hyperpolarization.

(4) A given cell suspension, at one selected concentration of K^+, is then incubated at 37° for an appropriate interval (10–30 min). The dye (80 μg) dissolved in 40 μl of ethanol is added, the solution mixed, and 4 ml of the suspension placed in the cuvette. About 4.5 min later valinomycin (10 μg) is added to the cuvette. At the same time, the cells in the remainder of the cell suspension are collected by centrifugation and the cellular contents of water and K^+ determined by conventional procedures (see Refs. 4 and 9). When this whole procedure is repeated with the cellular preparation containing the other selected concentrations of K^+, the observations are combined as shown in Fig. 1. A null point is represented by the value of the ratio $[K^+]_o/[K^+]_i$ at which the addition of valinomycin causes no change in fluorescence intensity. The Nernst potential corresponding to that ratio, on the one hand, and the corresponding fluorescence intensity, on the other hand, provided one pair of observations defining the relationship between $\Delta \psi$ and fluorescence intensity. Further such pairs of observations are obtained at other values of $\Delta \psi$, which is systematically varied as explained above. These are used to construct a calibration curve, allowing $\Delta \psi$ to be estimated directly from the fluorescence intensity.[4,13]

---

[13] E. K. Hoffman and I. H. Lambert, *J. Physiol. (London)* **338**, 613 (1983).

[8] Optical Measurement of Membrane Potential in Cells, Organelles, and Vesicles

By J. C. FREEDMAN and T. S. NOVAK

Optical methods for measuring and monitoring the membrane potential (E_m) in cell, organelle, and vesicle suspensions have been widely used during the past decade to study many fundamental problems in cell physiology. The principles of this new electrophysiological technique, along with examples of applications, have been described in general review articles,[1] in previous methods papers,[2,3] and in recent reviews focused on specific systems including red blood cells and Ehrlich ascites tumor cells,[4] neutrophils,[5] lymphocytes,[6] renal membrane vesicles,[7] muscle,[8] and nerve axons, ganglia, and secretory tissue.[9] A partial summary of cellular electrophysiological problems which are being studied by means of optical potentiometric indicators is given in Table I. This article describes current methodology, gives illustrative specific protocols for the use of slow and fast dyes in red blood cell suspensions, and discusses various methods proposed to relate optical potentiometric signals to millivolt values.

---

[1] L. B. Cohen and B. M. Salzberg, *Rev. Physiol. Biochem. Pharmacol.* **83**, 35 (1978); J. C. Freedman and P. C. Laris, *Int. Rev. Cytol., Suppl.* **12**, 177 (1981); *in* "Spectroscopic Membrane Probes" (L. M. Loew, ed.), CRC Press, Boca Raton, FL, 1988; *in* "Cell Physiology of Blood" (R. B. Gunn and J. C. Parker, eds.), Rockefeller University Press, New York, 1988.

[2] A. S. Waggoner, this series, Vol. 55, p. 689; J. C. Freedman and J. F. Hoffman, *in* "Frontiers of Biological Energetics" (P. L. Dutton, J. Leigh, and A. Scarpa, eds.), Vol. 2, p. 1323. Academic Press, New York, 1978; L. B. Cohen and J. F. Hoffman, *Tech. Cell. Physiol.* **P118**, 1 (1982).

[3] C. L. Bashford and J. C. Smith, this series, Vol. 55, p. 569.

[4] P. C. Laris and J. F. Hoffman, *in* "Optical Methods in Cell Physiology" (P. De Weer and B. M. Salzberg, eds.), p. 199. Wiley, New York, 1986.

[5] B. Seligmann and J. I. Gallin, *Adv. Exp. Med. Biol.* **141**, 335 (1982).

[6] T. J. Rink and C. Deutsch, *Cell Calcium* **4**, 463 (1983).

[7] E. M. Wright, *Am. J. Physiol.* **246**, F363 (1984).

[8] Baylor, S. M., *in* "Handbook of Physiology: Section #10: Skeletal Muscle" (L. D. Peachey, R. H. Adrian, and S. R. Geiger, eds.), p. 355. Am. Physiol. Soc., Bethesda, Maryland, 1983.

[9] B. M. Salzberg, *in* "Current Methods in Cellular Neurobiology" (J. L. Basker, ed.), p. 139. Wiley, New York, 1983; L. B. Cohen, D. Landowne, L. M. Loew, and B. M. Salzberg, *Curr. Top. Membr. Transp.* **22**, 423 (1984); A. Grinvald, *Annu. Rev. Neurosci.* **8**, 263 (1985).

TABLE I
SOME CELLULAR ELECTROPHYSIOLOGICAL PROBLEMS AMENABLE TO STUDY WITH
OPTICAL POTENTIOMETRIC INDICATORS

Property	Study
Gibbs–Donnan equilibrium	Effects of impermeant electrolytes and of intracellular charge
Resting potential	Magnitude; ionic determinants; changes during maturation
Diffusion potentials	Magnitude; effects of ionophores; effects of drugs; measurement of individual ionic conductances; voltage dependence of ionic conductances
Membrane resistance	Magnitude; characteristic current–voltage curve
Electrogenicity	
Ionic pumps	Voltage dependence of overall reaction and partial reactions of Na^+,K^+-ATPase, Ca^{2+}-ATPase, and H^+-ATPase; effect of pump activation and inhibition on E_m
Coupled transport	Magnitude of Na^+ and H^+ electrochemical gradients coupled to amino acid and sugar transport; effects of solute gradients on E_m and of E_m on solute gradients
Cell volume regulation	Effect of E_m on coupled and conductive fluxes
Bioenergetics	
Mitochondrial membrane potential	Magnitude; kinetics of development, effects of e^- donors, inhibitors, and uncouplers; other mitochondrial transport systems including ATP/ADP exchange and Ca^{2+} transport
Bacterial membrane potential	Effects of metabolic substrates, uncouplers, and antibiotics
Photopotentials	Activation of E_m by light in purple membranes containing bacteriorhodopsin, in bacterial chromatophores, and in chloroplasts
Circadian rhythms	Oscillations of E_m
Stimulus–response coupling	Receptor-mediated changes in E_m; effects of intracellular Ca^{2+}
Red blood cells	Ca^{2+}-activated K^+ conductance
Neutrophils	Activation by chemoattractants and by secretagogues including concanavalin A, phorbol myristate acetate (PMA), and N-formylmethionyl-leucylphenylalanine (FMLP)
Macrophages	Activation of phagocytosis by PMA and by zymosan particles
Mast cells	Secretion of histamine
Platelets	Secretion of serotonin induced by ADP and by thrombin
Lymphocytes	Activation of mitogenesis

(*continued*)

TABLE I (continued)

Property	Study
Synaptosomes	Excitation–secretion coupling and effects of pharmacological reagents
Contraction and motility	Sperm motility; bacterial and protozoan motility and chemotaxis; excitation–contraction coupling in intact skeletal muscle, in single fibers, and in skinned fibers; E_m of isolated sarcoplasmic reticulum vesicles
Action potentials	Detection in squid giant axon and in multiple neurons of simple ganglia; propagation in simple ganglia; propagation in heart
Behavior	Reflexes in invertebrate organisms

Choice of Dye

The principal dyes used in recent electrophysiological studies of cells, organelles, and vesicles are from the cyanine, oxonol, safranine, and styryl classes. Structures of some frequently used dyes are depicted in Fig. 1, with chemical names given in Table II; structural analogs and many other dyes also give optical potentiometric signals. Experience indicates that different dyes are optimal for different cell types, and that, even for the same type of cell, different dyes may be optimal for different problems. Some pertinent considerations in selecting a dye are as follows: (1) the sensitivity of dye response in relation to the range of magnitudes of the expected changes in E_m, (2) the required time resolution, and (3) the selectivity of the optical response to E_m and the desired lack of response to other experimental parameters and variables. Ideally, the dye should be nontoxic to cell viability, should not interact with other reagents utilized in the experiments, should neither inhibit nor stimulate the transport systems or other phenomena under investigation, and should not otherwise interfere with normal cell function. The available data suggest that some dyes alter the rate or characteristics of certain passive ion fluxes[10–12] or even the resting potential itself.[13,14] When using optical indicators, it is

[10] P. J. Sims, A. S. Waggoner, C. H. Wang, and J. F. Hoffman, *Biochemistry* **13**, 3315 (1974).
[11] T. J. B. Simons, *Nature (London)* **264**, 467 (1976); T. J. B. Simons, *J. Physiol. (London)* **288**, 481 (1979).
[12] J. C. Freedman and T. S. Novak, *J. Membr. Biol.* **72**, 59 (1983).
[13] T. J. Rink, C. Montecucco, T. R. Hesketh, and R. Y. Tsien, *Biochim. Biophys. Acta* **595**, 15 (1980).
[14] T. C. Smith and S. C. Robinson, *Biochem. Biophys. Res. Commun.* **95**, 722 (1980).

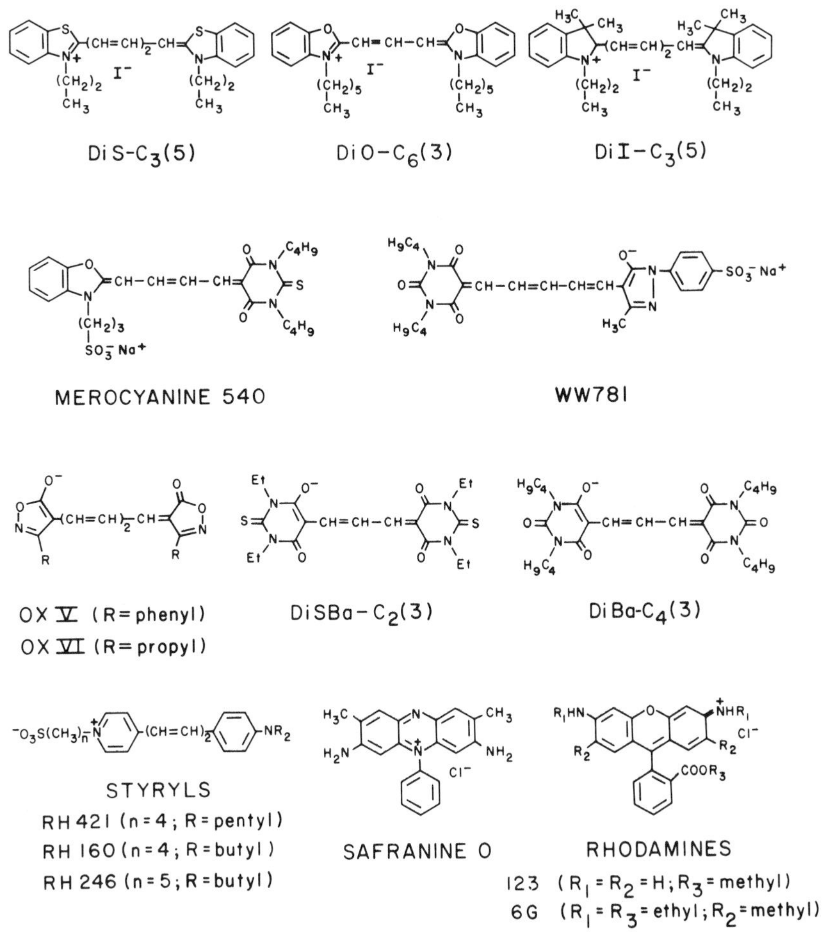

FIG. 1. Structures of optical potentiometric indicators of membrane potential. Cyanine dyes are abbreviated diY-$C_n(m)$ where Y indicates the heterocyclic nucleus, n is the number of carbon atoms in the alkyl chains, and m is the number of carbon atoms in the bridge between the two nuclei. WW indicates that the dye was synthesized by A. S. Waggoner and C.-H. Wang. RH indicates dye synthesis by R. Hildesheim. OX indicates an oxonol dye. See Table II for chemical names.

essential to perform ample controls and to evaluate whether membrane perturbations are quantitatively significant and relevant to each specific biological and electrophysiological question under consideration. The question of the influence of mitochondrial potentials on the measurement of plasma membrane potentials by permeant cyanine dyes is controversial. When mitochondrial respiration is inhibited by diS-C_3(5) (see below),

TABLE II
ABBREVIATIONS AND CHEMICAL NAMES OF SOME OPTICAL
POTENTIOMETRIC INDICATORS

Abbreviation	Chemical nomenclature
Cyanines	
$diS-C_3(5)$	3,3'-Dipropylthiodicarbocyanine
$dI-C_3(5)$	1,1'-Dipropyl-3,3,3',3'-tetramethylindodicarbocyanine
$diI-C_1(5)$	1,3,3,1',3',3'-Hexamethylindodicarbocyanine
$diO-C_6(3)$	3,3'-Dihexyloxacarbocyanine
$diO-C_5(3)$	3,3'-Dipentyloxacarbocyanine
Oxonols	
WW781	Bis[3-methyl-1-p-sulfophenyl-5-pyrazolone-(4)]pentamethine oxonol
OX-V	Bis[3-phenyl-5-oxoisoxazol-4-yl]pentamethine oxonol
OX-VI	Bis[3-propyl-5-oxoisoxazol-4-yl]pentamethine oxonol
$diSBa-C_2(3)$	Bisoxonol, bis[1,3-diethylthiobarbiturate]trimethine oxonol
$diBa-C_4(3)$	Bis[1,3-dibutylbarbituric acid]trimethine oxonol
Styryls	
di4-ANEPPS	N-[4-Sulfobutyl]-4-[4-(p-dibutylaminonaphthyl)butadienyl]pyridinium
RH421	N-[4-Sulfobutyl]-4-[4-(p-dipentylaminophenyl)butadienyl]pyridinium
RH160	N-[4-Sulfobutyl]-4-[4-(p-dibutylaminophenyl)butadienyl]pyridinium
RH246	N-[4-Sulfopentyl]-4-[4-(p-dibutylaminophenyl)butadienyl]pyridinium

it is reasonable to suppose that mitochondrial potentials are abolished. Some investigators have added mitochondrial inhibitors and uncouplers in order to separate the two potentials,[15-18] but with this approach controls are needed to evaluate possible direct effects of the reagents on dye fluorescence or on the plasma membrane potential. Others have circumvented this issue by preparing cytoplasts devoid of organelles[19] or by using anionic oxonols.[20] In any new investigation, it may be necessary to compromise among various desirable criteria while defining the advantages and limitations of specific dyes, and then to continue the study of E_m as better dyes become available.

[15] R. D. Philo and A. A. Eddy, *Biochem. Soc. Trans.* **3**, 904 (1975); H. M. Korchak, A. M. Rich, C. Wilkenfeld, L. E. Rutherford, and G. Weissman, *Biochem. Biophys. Res. Commun.* **108**, 1495 (1982).
[16] R. Levenson, I. G. Macara, R. L. Smith, L. Cantley, and D. Housman, *Cell* **28**, 855 (1982).
[17] S. M. Felber and M. D. Brand, *Biochem. J.* **204**, 577 (1982); S. M. Felber and M. D. Brand, *Biochem. J.* **210**, 885 (1983).
[18] H. A. Wilson, B. E. Seligmann, and T. M. Chused, *J. Cell. Physiol.* **125**, 61 (1985).
[19] G. V. Henius and P. C. Laris, *Biochem. Biophys. Res. Commun.* **91**, 1430 (1979); H. M. Korchak, D. Roos, K. N. Giedd, E. M. Wynkoop, K. Vienne, L. E. Rutherford, J. P. Buyon, A. M. Rich, and G. Weissman, *Proc. Natl. Acad. Sci. U.S.A.* **80**, 4968 (1983).
[20] H. A. Wilson and T. M. Chused, *J. Cell. Physiol.* **125**, 72 (1985).

Optical potentiometric signals have been classified as either "slow" or "fast" depending on the speed of the mechanism of response. Slowly responding permeant dyes undergo a bulk redistribution in accordance with an altered E_m,[10] and typically respond on a time scale of seconds. Rapidly responding impermeant dyes undergo a voltage-dependent association with the membrane, or reorientation within the membrane, and respond on a time scale of milli- or microseconds. In general, the slow redistribution mechanism results in larger changes in fluorescence than fast membrane mechanisms but may be linear over a shorter range of E_m. A brief description of several dyes useful for cell suspensions is given next.

DiS-C_3(5), a dithiodicarbocyanine, emerged as the dye of choice for cell suspensions after an initial survey of cyanine dyes with human red blood cells.[10,21] Upon membrane hyperpolarization, this blue dye with a positive delocalized charge redistributes from the medium into the cells; the fluorescence of cell-associated dye is quenched due to dimerization and binding to cell constituents.[22,23] Under optimal conditions with red cells the changes in suspension fluorescence amount to $1.7\% \Delta F$/mV, with a linear response from about -40 to $+50$ mV.[22,23] After adding valinomycin at low K_o, dye fluorescence decreases (Fig. 2) with a half-time of 3.5 sec. Disadvantages of diS-C_3(5) with red cells include a sensitivity to changes in internal pH[22,23] and inhibition of Ca^{2+}-induced K^+ conductance.[11] DiS-C_3(5) also inhibits oxidative phosphorylation at site I of the respiratory chain in isolated mitochondria.[24] Inhibition of respiration and ATP depletion have been reported in studies of Ehrlich ascites tumor cells[25,26] and lymphocytes.[27] Cyanine dyes also inhibit lymphocyte capping[27] and mitosis and growth of fertilized sea urchin eggs.[28] If mitochondrial respiration is inhibited such that ATP is depleted, indirect effects on active transport, cell electrolyte contents, electrogenic potentials, and other energy-requiring cell functions are expected. However, with

---

[21] J. F. Hoffman and P. C. Laris, *J. Physiol. (London)* **239**, 519 (1974).
[22] J. C. Freedman and J. F. Hoffman, *J. Gen. Physiol.* **74**, 187 (1979).
[23] S. B. Hladky and T. J. Rink, *J. Physiol. (London)* **263**, 287 (1976); R. Y. Tsien and S. B. Hladky, *J. Membr. Biol.* **38**, 73 (1978).
[24] A. Waggoner, *J. Membr. Biol.* **27**, 317 (1976); K. W. Kinnally and H. Tedeschi, *Biochim. Biophys. Acta* **503**, 380 (1978); P. H. Howard and S. B. Wilson, *Biochem. J.* **180**, 669 (1979); T. E. Conover and R. F. Schneider, *J. Biol. Chem.* **256**, 402 (1981).
[25] P. C. Laris, M. Bootman, H. A. Pershadsingh, and R. M. Johnstone, *Biochim. Biophys. Acta* **512**, 397 (1978).
[26] E. Okimasu, J. Akiyama, N. Shiraishi, and K. Utsumi, *Physiol. Chem. Phys.* **11**, 425 (1979); T. C. Smith, J. T. Herlihy, and S. C. Robinson, *J. Biol. Chem.* **256**, 1108 (1981).
[27] C. Montecucco, T. Pozzan, and T. Rink, *Biochim. Biophys. Acta* **552**, 552 (1979).
[28] S. Zigman and P. Gilman, Jr., *Science* **208**, 188 (1980).

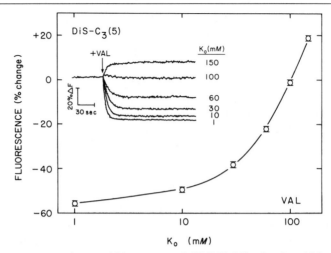

FIG. 2. Percentage change of fluorescence of diS-C_3(5) following the addition of valinomycin (VAL) to suspensions of human red blood cells in x mM KCl (as indicated), $150 - x$ mM NaCl, and 5 mM HEPES (pH 7.4 at 23°). See text for concentrations of dye, cells, and VAL. The inset shows the initial level of fluorescence (after 3 min for equilibration), and the time course to new steady levels of fluorescence after adding VAL at varied K_o. Downward deflections indicate hyperpolarization while upward deflections indicate depolarization.

Ehrlich ascites tumor cells, depletion of ATP caused by diS-C_3(5) is largely prevented by supplying glucose as glycolytic substrate.[25] Photodynamic effects on bacterial motility have also been noted with diS-C_3(5),[29] but the use of submicromolar dye concentration circumvented this problem.[30] While diS-C_3(5) has been the most commonly used optical potentiometric indicator in cells, organelles, and vesicles, and remains the best characterized dye, its pharmacological effects have shifted attention to the use of oxonol and styryl dyes for certain applications.

DiO-C_6(3), an oxocarbocyanine, was the first dye found to exhibit large changes in fluorescence in the original studies of red cells.[21] At micromolar concentrations this yellow cyanine dye, also having a delocalized positive charge, shows a slow decrease in fluorescence on hyperpolarization, much like diS-C_3(5), but without spectral indications of dimerization. However, at submicromolar dye concentrations, membrane mechanisms apparently dominate because hyperpolarization now results in a rapid increase instead of a slow decrease of fluorescence. The potential-dependent fluorescence of cell-associated dye and the excitation

[29] J. B. Miller and D. E. Koshland, Jr., *Nature (London)* **272**, 83 (1978).
[30] J. B. Miller and D. E. Koshland, Jr., *Proc. Natl. Acad. Sci. U.S.A.* **74**, 4752 (1977).

wavelength at 484 nm, corresponding to the 488 nm argon ion laser line, have made this dye and diO-C_5(3) attractive for studies utilizing flow cytometry. Histograms of fluorescence intensity per cell have been obtained for suspensions of lymphocytes,[18,31] neutrophils,[32] mouse red blood cells infected with malaria parasite,[33] pancreatic islet cells,[34] and cultured cells including murine erythroleukemia (MEL),[16] murine leukemia (L1210),[35] human promyelocytic leukemia (HL-60),[36] and rabbit type II pneumocytes.[36] Illustrative fluorescence histograms determined by flow cytometry before and after hyperpolarization of human red blood cells by valinomycin are shown in Fig. 3.

DiI-C_3(5), an indodicarbocyanine, is worthy of mention because its structure imparts steric restraints to dimerization and aggregation, yet its ΔF is comparable to that of diS-C_3(5).[22] Its mechanism and properties are deserving of further study. An analog, DiI-C_1(5), was recently utilized in studies of acetylcholine receptors[37] and of Na^+,K^+-ATPase[38] reconstituted into lipid vesicles.

WW781, a fast oxonol dye, is thought to be impermeant because of its two separated negative charges, one of which is delocalized. It responds within the mixing time of magnetically stirred suspensions in cuvettes, about 1–2 sec, and calibrates at 0.13%ΔF/mV between 0 and -120 mV with red cells.[12] Known pharmacological effects include stimulation of Ca^{2+}-induced ^{86}Rb efflux and also of valinomycin-induced net KCl efflux

[31] H. M. Shapiro, P. J. Natale, and L. A. Kamentsky, *Proc. Natl. Acad. Sci. U.S.A.* **76**, 5728 (1979); H. M. Shapiro and T. B. Strom, *Proc. Natl. Acad. Sci. U.S.A.* **77**, 4317 (1980); C. Nerl, G. Valet, D. J. Schendel, and R. Wank, *Naturwissenschaften* **69**, 292 (1982); K. S. Rosenthal and H. M. Shapiro, *J. Cell. Physiol.* **117**, 39 (1983); J. C. Cambier and J. G. Monroe, this series, Vol. 103, p. 227; P. E. R. Tatham and P. J. Delves, *Biochem. J.* **221**, 137 (1984).

[32] B. Seligmann, T. M. Chused, and J. I. Gallin, *J. Clin. Invest.* **68**, 1125 (1981); B. Seligmann, T. M. Chused, and J. I. Gallin, *J. Immunol.* **133**, 2641 (1984); M. C. Seeds, J. W. Parce, P. Szejda, and D. A. Bass, *Blood* **65**, 233 (1985); M. P. Fletcher and B. E. Seligmann, *J. Leukocyte Biol.* **37**, 431 (1985).

[33] J. W. Jacobberger, P. K. Horan, and J. D. Hare, *Cytometry* **4**, 228 (1983).

[34] H. Zühlke, B. Hehmke, H.-L. Jenssen, and H. Schulz, *Acta Biol. Med. Germ.* **41**, 1139 (1982).

[35] J. A. Hickman, O. C. Blair, A. L. Stepanowski, and A. C. Sartorelli, *Biochim. Biophys. Acta* **778**, 457 (1984); J.-Y. Charcosset, A. Jacquemin-Sablon, and J.-B. LePecq, *Biochem. Pharmacol.* **33**, 2271 (1984).

[36] R. L. Gallo, R. P. Wersto, R. H. Notter, and J. N. Finkelstein, *Arch. Biochem. Biophys.* **235**, 544 (1984).

[37] H. Lüdi, H. Oetliker, U. Brodbeck, P. Ott, B. Schwendimann, and B. W. Fulpius, *J. Membr. Biol.* **74**, 75 (1983).

[38] H.-J. Apell, M. M. Marcus, B. M. Anner, H. Oetliker, H. Oetliker, and P. Läuger, *J. Membr. Biol.* **85**, 49 (1985).

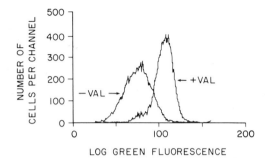

FIG. 3. Fluorescence histograms of human red blood cells determined by flow cytometry. Fresh, washed cells at 0.12% hematocrit in medium containing 1 mM KCl, 149 mM NaCl, and 5 mM HEPES buffer, pH 7.6 at 25°, were stained with 60 nM diO-C_6(3). Then, 10,000 cells were analyzed in a Coulter Epics V flow cytometer. Fluorescence intensity is indicated on the x-axis (log scale) and the number of cells having a given fluorescence is indicated on the y-axis. Hyperpolarization by addition of 1 μM valinomycin (VAL) increases mean cell-associated fluorescence, as indicated by the right-shifted histogram. Data of D. M. Meissner, L. Kozick, and J. C. Freedman.

from human red blood cells. WW781 tracks the propagation of nerve[39] and cardiac[40] action potentials and also gives signals associated with excitation–contraction coupling in skeletal muscle.[41]

Bisoxonol, or diSBa-C_2(3), and diBa-C_4(3) have been used in several laboratories to study lymphocytes,[13,17,20,42] and have also recently been tested with cultured rat mammory tumor cells and mouse embryo L cells.[43]

OX-V and OX-VI are two oxonols originally synthesized at the Johnson Foundation for the study of mitochondrial and chloroplast bioenergetics.[3] Their use has recently been extended to intact cells, including neutrophils,[44] cultured Lettré cells,[45] *E. coli*,[46] and organelles such

[39] R. K. Gupta, B. M. Salzberg, A. Grinvald, L. B. Cohen, K. Kamino, S. Lesher, M. B. Boyle, A. S. Waggoner, and C. H. Wang, *J. Membr. Biol.* **58**, 1 (1981).
[40] S. Dillon and M. Morad, *Science* **214**, 453 (1981); B. C. Hill and K. R. Courtney, *Biophys. J.* **40**, 255 (1982).
[41] S. M. Baylor, W. K. Chandler, and M. W. Marshall, *J. Physiol. (London)* **348**, 209 (1984).
[42] C. Montecucco, T. J. Rink, T. Pozzan, and J. C. Metcalfe, *Biochim. Biophys. Acta* **595**, 65 (1980); R. Y. Tsien, T. Pozzan, and T. J. Rink, *Nature (London)* **295**, 68 (1982); P. E. R. Tatham and P. J. Delves, *Biochem. J.* **221**, 137 (1984).
[43] T. Bräuner, D. F. Hülser, and R. J. Strasser, *Biochim. Biophys. Acta* **771**, 208 (1984).
[44] C. L. Bashford and C. A. Pasternak, *Biochim. Biophys. Acta* **817**, 174 (1985).
[45] C. L. Bashford and C. A. Pasternak, *J. Membr. Biol.* **79**, 275 (1984).
[46] J. P. Armitage and M. C. W. Evans, *FEBS Lett.* **126**, 98 (1981); M. Eisenbach, Y. Margolin, A. Ciobotariu, and H. Rottenberg, *Biophys. J.* **45**, 463 (1984).

as chromaffin granules,[47] pituitary secretory vesicles,[48] and liver lysosomes.[49]

RH-421 is a new styryl dye which gave the largest signal (0.21%ΔF/mV) so far recorded from the soma and 2-μm-wide axonal processes of cultured neuroblastoma cells.[50] An analog, RH160, has been used with bacteriorhodopsin vesicles.[51] Another analog, RH246, was the best of 10 styryl dyes tested with human red blood cells in the authors' laboratory.

Di4-ANEPPS at 0.08%ΔF/mV is the most sensitive of a series of styryl dyes which exhibit voltage-dependent electrochromic shifts on hemispherical lipid bilayers.[52] When tested on squid giant axons, nonelectrochromic mechanisms contribute to the optical responses of styryl dyes.[53]

Safranine O is a permeant dye with a positive delocalized charge which gives absorption shifts on binding and aggregation. This dye was originally used with isolated rat liver mitochondria,[54,55] liposomes,[54] and bacterial membrane vesicles.[56] Its use has been extended to study mitochondrial potentials *in situ* in Ehrlich ascites tumor cells,[57] plant cells,[58] hepatocytes,[59] and perfused rat heart.[60]

Rhodamine dyes have also been used to study membrane potentials. The cationic rhodamines (3B, 6G, and 123) accumulate in the mitochondria of a variety of cell lines.[61] Rhodamines have been used with *Tetrahy-*

---

[47] D. Scherman and J. P. Henry, *Biochim. Biophys. Acta* **599**, 150 (1980); J. Giraudat, M. P. Roisin, and J. P. Henry, *Biochemistry* **19**, 4499 (1980).
[48] Y. P. Loh, W. W. H. Tam, and J. T. Russell, *J. Biol. Chem.* **259**, 8238 (1984); J. T. Russell, *J. Biol. Chem.* **259**, 9496 (1984); J. T. Russell, M. Levine, and D. Njus, *J. Biol. Chem.* **260**, 226 (1985).
[49] Y. Moriyama, T. Takano, and S. Ohkuma, *J. Biochem. (Tokyo)* **95**, 995 (1984).
[50] A. Grinvald, R. Hildesheim, I. C. Farber, and L. Anglister, *Biophys. J.* **39**, 301 (1982).
[51] B. Ehrenberg, Z. Meiri, and L. M. Loew, *Photochem. Photobiol.* **39**, 199 (1984).
[52] L. M. Loew and L. L. Simpson, *Biophys. J.* **34**, 353 (1981); E. Fluhler, V. G. Burnham, and L. M. Loew, *Biochemistry* **24**, 5749 (1985).
[53] L. M. Loew, L. B. Cohen, B. M. Salzberg, A. L. Obaid, and F. Bezanilla, *Biophys. J.* **47**, 71 (1985).
[54] K. E. O. Akerman, *Microsc. Acta* **81**, 147 (1978).
[55] A. Zanotti and G. F. Azzone, *Arch. Biochem. Biophys.* **201**, 255 (1980).
[56] S. Schuldiner and H. R. Kaback, *Biochemistry* **14**, 5451 (1975).
[57] K. E. O. Akerman, *Biochim. Biophys. Acta* **546**, 341 (1979).
[58] S. B. Wilson, *Biochem. J.* **190**, 349 (1980); A. L. Moore and W. D. Bonner, Jr., *Plant Physiol.* **70**, 1271 (1982).
[59] K. E. O. Akerman and J. O. Jarvisalo, *Biochem. J.* **192**, 183 (1980).
[60] R. A. Kauppinen and I. E. Hassinen, *Am. J. Physiol.* **247**, H508 (1984).
[61] L. V. Johnson, M. L. Walsh, and L. B. Chen, *Proc. Natl. Acad. Sci. U.S.A.* **77**, 990 (1980); L. V. Johnson, M. L. Walsh, B. J. Bockus, and L. B. Chen, *J. Cell Biol.* **88**, 526 (1981).

mena,[62] with cyanobacteria,[63] with the protozoan parasite *Toxoplasma*,[64] and also with rat brain synaptosomes.[65] Toxic effects have been detected in cultured carcinoma[66] and myocardial cells.[67]

Most of the dyes mentioned above, as well as numerous structural analogs and dyes of other classes, are available from Molecular Probes, Inc., Eugene, OR, whose catalog contains many pertinent references. Other useful dyes are available from Eastman Kodak Co., Rochester, NY, and from Nippon Kankoh-Shikeso Kenkyusho Co., Ltd., Okayama, Japan. A list of 31 other sources used in an extensive screening of over 1000 dyes with squid giant axon has been published elsewhere in this series.[3]

Optimization of the Signal

Although any fluorimeter can be used to measure the fluorescence of optical potentiometric indicators, a modular design is advantageous because the emission monochromator may be replaced by a cutoff filter to increase the amount of light collected. A frosted quartz plate in series with the cutoff filter reduces the noise envelope of turbid cell suspensions.

Large optical drifts due to settling of nonspherical cells are easily prevented by continuous magnetic stirring. If the available fluorimeter, such as the Aminco-Bowman is not so equipped, a convenient stirring motor is model 2034B 015 SG with 15:1 gear ratio, available from Micro Mo Electronics Inc., St. Petersburg, FL. A 1-cm diameter cylindrical magnet is cemented to the shaft, and the stirring speed may be controlled with an economical power supply such as Model LV3607, made by Electro Motor and Control Corp., Somerville, MA. When the distance between the drive magnet and the magnetic stir bar in the cuvette is kept to within a few millimeters, the mixing time is 1–2 sec. In some instruments where the distances are larger, rare earth alloy magnets have a greater coupling efficiency, resulting in somewhat better stirring. Using a mini-

---

[62] T. Aiuchi, H. Tanabe, K. Kurihara, and Y. Kobatake, *Biochim. Biophys. Acta* **628,** 355 (1980); H. Tanabe, K. Kurihara, and Y. Kobatake, *Biochemistry* **19,** 5339 (1980).

[63] G. V. Murvanidze, I. I. Severina, and V. P. Skulachev, *Dobl. Akad. Nauk. U.S.S.R.* **261,** 215 (1981); I. I. Severina and V. P. Skulachev, *FEBS Lett.* **165,** 67 (1984).

[64] H. Tanabe and K. Murakami, *J. Cell Sci.* **70,** 73 (1984).

[65] T. Aiuchi, T. Daimatsu, K. Nakaya, and Y. Nakamura, *Biochim. Biophys. Acta* **685,** 289 (1982); T. Aiuchi, M. Matsunaga, T. Daimatsu, K. Nakaya, and Y. Nakamura, *Biochim. Biophys. Acta* **771,** 228 (1984).

[66] T. J. Lampidis, S. D. Bernal, I. C. Summerhayes, and L. B. Chen, *Ann. N.Y. Acad. Sci.* **397,** 299 (1982).

[67] T. J. Lampidis, C. Salet, G. Moreno, and L. B. Chen, *Agents Actions* **14,** 751 (1984).

mum volume in the cuvette, e.g., 2.5 ml, also facilitates mixing. Motor-driven stirring paddles for insertion into cuvettes are also available (Instech Laboratories, Horsham, PA).

When diS-C_3(5) is added to buffered saline in the absence of cells, its fluorescence declines by 1–2%/min during 30 min. The rate of fluorescence decline is equivalent in the dark and in the light, and occurs in Pyrex, quartz, polyacrylic, or polystyrene cuvettes. The most likely cause is binding of the dye to the cuvette and stir bar with aggregation and self-quenching of bound dye. It is important to be aware of this phenomenon because addition of some reagents might cause an artifactual increase of fluorescence by displacing dye from the cuvette walls. Significant drifts are encountered with dirty cuvettes. Glass cuvettes should be scrupulously cleaned each day of use with chromsulfuric acid; otherwise, disposable polyacrylic cuvettes are satisfactory. In the presence of a sufficient concentration of cells, drifts due to binding of dye to the cuvette are negligible, and a stable baseline fluorescence is attained. Thus, when diS-C_3(5) is added to human red blood cells in buffered saline, its fluorescence reaches a steady level within 3 min and drifts by less than 0.1%ΔF/min during the subsequent 20 min.

Procedure for a Slow Dye

Typical kinetic traces of the changes in diS-C_3(5) fluorescence when the membrane potential of human red blood cells is altered by valinomycin at varied external [K^+], or K_o, are shown in Fig. 2. These records were obtained by the protocol described below.

Reagents

diS-C_3(5): 0.2 mg/ml in ethanol (546.5 g/mol, Molecular Probes, Inc., Eugene, OR)

Valinomycin: 2.5 mM in ethanol (1111 g/mol, Calbiochem-Behring Corp., La Jolla, CA)

Cell wash medium: 150 mM NaCl, 5 mM HEPES, adjusted to pH 7.4 at 25° with NaOH

Variable K media: x mM KCl, 150 $- x$ mM NaCl, 5 mM HEPES, pH 7.4 at 25°

SLM 8000S fluorimeter settings: Excitation at 616 nm with 16 nm bandpass; emission at 676 nm with 3 mm thick RG665 cutoff filter (Schott Optical Glass Inc., Duryea, PA) with 8 nm bandpass; analog mode, about 800 V at 100 sensitivity, 0.5 sec acquisition time. Cuvettes thermostatted at 23°.

Protocol

1. Blood from healthy human donors is drawn by venipuncture into heparinized tubes (green top Vacutainers, Becton Dickinson, Rutherford, NJ) and immediately centrifuged at 13,800 g for 3 min at 4°. The plasma and buffy coat are aspirated and discarded and the packed cells are then washed three or four times by centrifugation, each time resuspending in about 5 vol of chilled cell wash medium. The cells are then adjusted to 50% hematocrit (HCT) in the cold cell wash medium and kept on ice for use on the same day.

2. To a cuvette containing 2.5 ml of medium is added 20 μl of 50% HCT red cells (final HCT = 0.40%). A Gilson pipet (Rainin Instrument Co., Woburn, MA) or its equivalent is suitable for the cells; syringes with narrow gauge needles (e.g., Hamilton Co., Reno, NV) may cause hemolysis but are useful for adding reagents. Then 8.3 μl of diS-C_3(5) is added to a final concentration of 1.2 μM, and the fluorescence is recorded for 3–5 min until a steady level is reached.

3. An outward KCl diffusion potential is induced by addition of 1.0 μl of valinomycin to a final concentration of 1.0 μM, and the fluorescence is recorded until a new steady level is attained.

With this simple procedure using valinomycin, a decrease in fluorescence at low K_o corresponds to hyperpolarization due to the outward K^+ gradient while an increase in fluorescence at high K_o corresponds to depolarization due to the inward K^+ gradient. Whenever E_m is changed with other ionophores, or by manipulation of internal and external ionic concentrations, diS-C_3(5) fluorescence decreases upon hyperpolarization and increases upon depolarization. However, in any new situation whenever a reagent is added to a cell suspension and a change in fluorescence is detected, many controls are needed before it can be concluded that E_m has changed. For example, chemical interaction between the dye and the added reagent may result in quenching of dye fluorescence. With stimulated neutrophils, oxidative products secreted by the cells quenched diS-C_3(5) fluorescence, and it was necessary to include catalase, cysteamine, or ascorbate to avoid this artifact.[68,69] Changes in dye distribution or binding may also result from changes in internal pH.[22,23] At a low ratio of dye to cell concentration, where dye–membrane interactions dominate the mechanism of response, displacement of dye from the membrane by a

---

[68] J. C. Whitin, C. E. Chapman, E. R. Simons, M. E. Chovaniec, and H. J. Cohen, *J. Biol. Chem.* **255**, 1874 (1980); J. C. Whitin, R. A. Clark, E. R. Simons, and H. J. Cohen, *J. Biol. Chem.* **256**, 8904 (1981); V. Castranova and K. Van Dyke, *Microchem. J.* **29**, 151 (1984).

[69] B. E. Seligmann and J. I. Gallin, *J. Cell. Physiol.* **115**, 105 (1983).

reagent is another possible source of artifactual changes in fluorescence.[70] Another method of monitoring E_m with cyanine dyes is by means of dual-wavelength absorbance signals. This method has been used with diS-C_3(5) in red cells[22] and with diS-C_2(5) in guinea pig cortical synaptosomes.[71] Whatever method is chosen, it is essential to verify that the same pattern of putative changes in E_m is obtained with several different dyes and to correlate the changes in E_m with specific ionic fluxes.

Procedure for a Fast Dye

Representative changes in the fluorescence of WW781 when human red cells are hyperpolarized at low K_o by the addition of valinomycin or of A23187 in the presence of Ca_o at 23° are shown in Fig. 4.[12] To obtain these traces, differential fluorescence was used with the reference cuvette containing sufficient dye in octanol such that its fluorescence approximately matched that of the sample cuvette which contained cells with dye. Records equivalent to those shown in Fig. 4 are obtained by a single-beam fluorescence protocol similar to that described above for diS-C_3(5) except that 62.5 µl of the 50% HCT red cells is added to give a final HCT of 1.2%. Then 6.7 µl of WW781 (1 mg/ml stock in ethanol) is added to a final concentration of 3.2 µM. With the SLM 8000S fluorimeter, excitation is at 615 nm with a 16 nm bandpass and emission is recorded through a 3 mm thick RG645 cutoff filter in series with a frosted quartz diffusion plate. The analog mode is used at about 800 V at sensitivity 10. The initial fluorescence of cells with dye is subtracted by means of the background subtract feature, and the photomultiplier tube voltage is then slightly increased to enable a subsequent decrease in fluorescence to be recorded. Hyperpolarization is induced by 1.0 µM valinomycin, or by 1.0 µM A23187 in the presence of 0.1 µM–1.0 mM Ca_o. Since addition of 1 µl ethanol decreases the fluorescence of WW781 by 0.4%, the volume of ionophore solution added is kept to a minimum, and all results are corrected for solvent controls.

Millimolar concentrations of Ca^{2+} affect the baseline fluorescence of WW781, probably by altering the surface potential and binding of dye to the membrane. This problem precluded use of WW781 in studies of Ca^{2+} transport with sarcoplasmic reticulum vesicles.[72] With red cells, changes

---

[70] A. Zaritsky, M. Kihara, and R. M. MacNab, *J. Membr. Biol.* **63**, 215 (1981).
[71] E. Heinonen, K. E. O. Akerman, K. Kaila, and I. G. Scott, *Biochim. Biophys. Acta* **815**, 203 (1985).
[72] T. J. Beeler, R. H. Farmen, and A. N. Martonosi, *J. Membr. Biol.* **62**, 113 (1981).

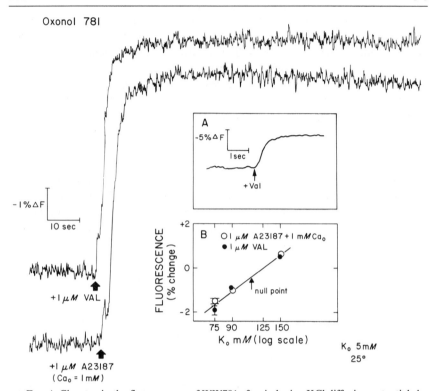

FIG. 4. Changes in the fluorescence of WW781 after inducing KCl diffusion potentials by adding valinomycin (upper trace) or A23187 (lower trace) to suspensions of human red blood cells. See text for protocol. Inset A at higher chart speed shows the response to be complete within 1 sec. Inset B shows identical null points obtained with valinomycin and with Ca plus A23187. Upward deflections indicate a decrease in fluorescence. The arrow designating the null point is displaced to $-0.4\%\Delta F$ to correct for the effect of ethanol. Data reprinted from Freedman and Novak,[12] with permission of the publisher.

in WW781 fluorescence correlated well with the known characteristics of Ca^{2+}-activated K^+ conductance.[12] In the red cell studies, the dye proved usable because K^+ conductance is activated at micromolar concentrations of intracellular Ca^{2+}, a level too low to change the membrane surface potential appreciably.

Screening of Dyes

In order to evaluate a series of new dyes with red cells, each dye is first dissolved in ethanol at 1 mg/ml. The absorption spectra (400 to 800 nm) are determined in ethanol and in isotonic buffered medium (1 mM

KCl, 149 mM NaCl, 5 mM HEPES, pH 7.4 at 23°) at varied [dye], and the absorption maxima, λ_{max}, and corresponding extinction coefficients noted. Next the emission spectrum is determined in buffer and ethanol by exciting the dye at the λ_{max} obtained from the absorption spectrum. The excitation spectrum is then determined at the peak emission wavelength.

With the monochromators set at λ_{max} for excitation and emission, the fluorescence, F, is determined upon sequential addition of cells to dye and then of dye to cells, and plots of F versus [HCT] and F versus [dye] are constructed. Excitation and emission spectra are determined again to optimize the wavelength settings in the presence of cells, and are compared with those in the absence of cells. Finally, the spectra are determined after addition of 1 μM valinomycin to check for any spectral shifts or intensity changes. As an example, the fluorescence excitation and emission spectra of WW781 before and after addition of valinomycin to red cells are shown in Fig. 5. For comparison the spectra in the absence of cells are also shown.[12]

It is next desirable to maximize the percentage change in fluorescence after addition of valinomycin (VAL) at low K_o, or %ΔF_{VAL} = 100 ($F_{initial}$ − F_{final})/$F_{initial}$, by varying dye at constant HCT and by varying HCT at constant [dye]. Illustrative results for WW781 are shown in Fig. 6. By systematically determining F and %ΔF_{VAL} versus HCT and [dye], the signal can be optimized and routinely determined on the plateau part of

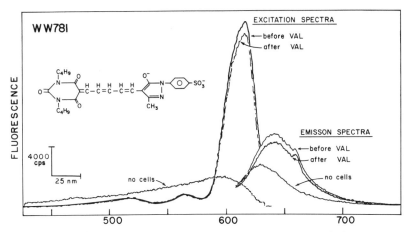

FIG. 5. Excitation and emission spectra (uncorrected) of WW781. The lower traces are for dye in buffer. The upper traces are for dye in a suspension of human red blood cells before and after hyperpolarization with 1 μM valinomycin (VAL). Excitation spectra (λ_{em} = 643 nM) are on the left and emission spectra (λ_{ex} = 593 nM) are on the right. The medium contained 5 mM KCl, 145 mM NaCl, and 5 mM HEPES (pH 7.4 at 37°). See text for protocol. Data reprinted from Freedman and Novak,[12] with permission.

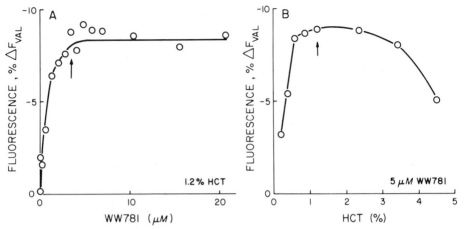

FIG. 6. Optimization of dye, cell, and VAL concentrations. (A) Variation of dye. To a series of cuvettes containing 2.5 ml of 1 mM KCl plus 149 mM NaCl were added 62.5 μl of 50% HCT cells (1.2% HCT final concentration) followed by 0.03 to 21 μM WW781 (final concentrations). After 3.5 min, 0.83 μl of 3 mM VAL was added (1 μM final concentration) and the percent decrease in fluorescence, %ΔF_{VAL}, defined as 100 (final − initial)/initial, was noted. To compute %ΔF_{VAL}, the fluorescence traces before and after VAL were first extrapolated by eye to the time of addition of VAL. (B) Variation of cells. The same protocol as in A was used except that each cuvette contained 5 μM WW781 and the HCT was varied between 0.2 and 4.5%.

the curves, where the results are less subject to variations due to minor differences of HCT and [dye] in a series of cuvettes. With diS-C_3(5), %ΔF_{VAL} is a monotonically decreasing function of HCT and increases slightly with increased [dye].[22] It is important to keep in mind that the kinetics of dye response are also a complex function of [dye] and [HCT], and that the conditions which give the largest %ΔF_{VAL} may not in general be the same as those which give the fastest dye response.[22]

After the signal (%ΔF_{VAL}) is maximized at 1 mM K_o, the experiments are repeated at the null point of 90 mM K_o and at 150 mM K_o to check whether %ΔF_{VAL} is dependent on K_o. If changes in fluorescence occur at the null point, then interactions between the dye and the ionophore are suspected. Such interactions can be detected by mixing the reagents in aqueous solution in the absence of cells and checking for changes in dye fluorescence or for spectral shifts indicative of the formation of complexes. Spectral evidence indicates that in the presence of potassium, valinomycin forms complexes in solution with 1-anilino-8-naphthalene sulfonate (ANS$^-$),[73] tetraphenylboron (TPB$^-$),[74] carbonyl cyanide p-tri-

---

[73] M. B. Feinstein and H. Felsenfeld, *Proc. Natl. Acad. Sci. U.S.A.* **68**, 2037 (1971).
[74] D. G. Davis and D. C. Tosteson, *Biochemistry* **14**, 3962 (1975).

fluoromethoxyphenylhydrazone (FCCP$^-$),[75] and with the anionic dyes merocyanine 540,[76] bisoxonol,[13] and WW781, but not with diS-C$_3$(5).[76] However, with hydrophobic or amphiphilic reagents, a better control is to verify the absence of signals with leaky membranes unable to support a membrane potential.

Calibration of Fluorescence to Millivolts

Several procedures have been devised to calibrate the changes in fluorescence of optical potentiometric indicators to millivolt values. Dyes have been calibrated in cell suspensions by a null point method,[21] by measurement of dye binding,[23] by use of Gibbs–Donnan equilibrium potentials,[22] by calculations according to the Nernst equation or the constant field equation,[21] by a ΔpH method,[12,77] and by simultaneous measurement of the distribution of radioactively labeled permeant ions such as [^{3}H]triphenylmethylphosphonium (TPMP).[30]

In the null point method, valinomycin is added at varied K$_o$ and the K$_o$ is noted where no change in fluorescence occurs, at which point (the null point) E_m equals the K equilibrium potential, or $E_m = E_K = -(2.3\ RT/\mathscr{F})$ log K$_c$/K$_o$, where $2.3RT/\mathscr{F}$ is 58 mV at 23° and 61 mV at 37°. This method is useful when E_m has changed from the resting potential to some new equilibrium value, whereupon VAL can be used to ascertain the extent of the change. The null point method only permits estimation of diffusion potentials away from E_K, such as with VAL at a K$_o$ different from the null point K$_o$, in circumstances where it is known that ionic fluxes other than K$^+$ do not also contribute to E_m.

Calibration by measurement of dye binding is based on the original observation that hyperpolarization is accompanied by a bulk accumulation of cationic dye.[10] The binding method[23] is theoretically elegant but requires measurement of cell-associated dye at varied dye concentration and at varied K$_o$ with VAL. The tendency of diS-C$_3$(5) and many other hydrophobic and amphiphilic dyes to bind to glass, Parafilm, and even Teflon introduces uncertainties when routine chemical procedures are employed.

Gibbs–Donnan equilibrium potentials in theory can serve as a useful primary standard for optical potentiometric indicators. In this method[22] a series of cell suspensions is prepared, each with a known E_m, and the equilibrium fluorescence is determined. E_m is either calculated from basic principles[78] or determined from the Nernst equation for a permeant ion at

---

[75] T. A. O'Brien, D. Nieva-Gomez, and R. B. Gennis, *J. Biol. Chem.* **253**, 1749 (1978).
[76] E. Lavie and M. Sonenberg, *FEBS Lett.* **111**, 281 (1980).
[77] E. M. Bifano, T. S. Novak, and J. C. Freedman, *J. Membr. Biol.* **82**, 1 (1984).
[78] J. C. Freedman and J. F. Hoffman, *J. Gen. Physiol.* **74**, 157 (1979).

equilibrium, such as chloride for red cells. In one version of this method, pH_o is varied, thus changing pH_c. The resultant change in the charge on internal proteins resets the Gibbs–Donnan equilibrium. Three disadvantages of this method are as follows: (1) the range of E_m attained is limited to about 15 mV between pH 6.5 and 8.0, (2) changes in internal pH independently affect diS-C_3(5) fluorescence, and (3) it is desirable to avoid large excursions in external pH so as to minimize possible effects on membrane transport systems. Another version of this method in red cells is to substitute impermeant anions for Cl_o, with simultaneous changes in pH_o to maintain constant pH_c.[21,22] This alternative method sets the Gibbs–Donnan equilibrium over a range of E_m from -5 to $+50$ mV.[22] The equilibrium method does not permit the calibration of hyperpolarizing (negative) potentials in intact red cells. While this method is readily applied to slow redistribution dyes with high sensitivity to E_m (1–2%ΔF/mV), it is not useful with the less sensitive fast dyes (0.1–0.2%ΔF/mV) because the reproducibility of equilibrium fluorescence in a series of cuvettes containing cells is insufficient to enable precise results. With fast dyes, it is preferable to calibrate changes in fluorescence in each cuvette.

A widely used method of dye calibration is to assume that $E_m = E_K$ in the presence of valinomycin at varied K_o. This assumption is valid at the null point, but only in the absence of significant electrogenic potentials.[79] Another limitation is that when K_o is varied away from the null point, other ionic fluxes may contribute to E_m as expressed in the constant field equation as applied to valinomycin-treated red cells, $E_m = -RT/\mathscr{F} \ln[(P_{K \cdot VAL}K_c + P_{Cl}Cl_o)/(P_{K \cdot VAL}K_o + P_{Cl}Cl_c)]$. At low K_o, the Cl term becomes significant and imparts a systematic error if dye calibration is based solely on E_K. Specific ionic permeabilities and their possible voltage and concentration dependencies are not sufficiently characterized in most cells to apply the constant field equation for the purpose of dye calibration. With lipid vesicles reconstituted with a single membrane transport protein, and which contain K and an impermeant anion,[38] the assumption that $E_m = E_K$ seems quite valid for quantitative analysis. But for cells with internal organelles, the free cytoplasmic K^+ concentration and activity are not even known with certainty from bulk chemical measurements to enable precise computation of E_K. The assumption that the constant field equation accurately describes the relation between E_m and ionic concentrations would have to be verified for a given cell type before this equation could serve as the basis for dye calibration.

Recently, a new method of dye calibration for red cells was devised which relies on measurements of changes in the external pH, or ΔpH_o, of

---

[79] L. Simchowitz, I. Spilberg, and P. DeWeer, *J. Gen. Physiol.* **79**, 453 (1982).

unbuffered suspensions.[12,77] In this method, a proton ionophore such as FCCP is added to the suspension after valinomycin to facilitate equilibration of protons with E_m. When the cell concentration is sufficiently large and the internal cytoplasm sufficiently well buffered to maintain constant pH_c, then $\Delta E_m = -(RT/\mathscr{F})\, \Delta pH_o$.[80] Again, there is the disadvantage of large changes in pH_o during the measurements but at least the method applies to both hyperpolarizing and depolarizing ranges of E_m. One precaution with the ΔpH_o method is that coupling of the proton gradient to other ion gradients would impart a systematic error whenever such coupled transport systems are present and significantly active.[81] Moreover, the proton ionophore is yet another reagent in the membrane in addition to dye and valinomycin, and controls must establish that the proton ionophore does not affect the transport systems being studied or the dye fluorescence itself. Other pH calibration methods have been used with mitochondria[54] and chromaffin granules.[82]

Yet another approach to dye calibration is to use radioactively labeled lipid-permeable cations such as [^{3}H]triphenylmethylphosphonium (TPMP$^+$).[83] TPMP has been used in red cells with anionic tetraphenylboron (TPB$^-$) in order to hasten equilibration,[84] as well as in many other systems. [^{3}H]Tetraphenylphosphonium (TPP$^+$) has also been used in red cells.[85] The problem of binding of TPMP to cellular constituents can be circumvented by measuring the total amount of cell-associated probe at varied external [TPMP], and at varied K_o with VAL, and computing ΔE_m as was done with the dye binding method.[23] Several studies have included comparisons of changes in dye fluorescence with changes in TPMP distribution. Good correlations have been reported with bacteria,[30] mitochondria,[55] neutrophils,[69] lymphocytes,[17] and platelets.[86] However, in rat white adipocytes, changes in TPMP distribution induced by hormones were due primarily to mitochondrial depolarization.[87] The rationale for using TPMP distribution for dye calibration in red cells is that once the dyes are cali-

---

[80] R. I. Macey, J. S. Adorante, and F. W. Orme, *Biochim. Biophys. Acta* **512**, 284 (1978).
[81] J. O. Wieth, J. Brahm, and J. Funder, *Ann. N.Y. Acad. Sci.* **341**, 394 (1980).
[82] G. Salama, R. G. Johnson, and A. Scarpa, *J. Gen. Physiol.* **75**, 109 (1980).
[83] L. E. Bakeeva, L. L. Grinius, A. A. Jasaitis, V. V. Kuliene, D. O. Levitsky, E. A. Liberman, I. I. Severina, and V. P. Skulachev, *Biochim. Biophys. Acta* **216**, 13 (1970).
[84] C. J. Deutsch, A. Holian, S. K. Holian, R. P. Daniele, and D. F. Wilson, *J. Cell. Physiol.* **99**, 79 (1979); K. Cheng, H. C. Haspel, M. L. Vallano, B. Osotimehin, and M. Sonenberg, *J. Membr. Biol.* **56**, 191 (1980).
[85] C. A. Dise and D. B. P. Goodman, *J. Biol. Chem.* **260**, 2869 (1985).
[86] L. T. Friedhoff and M. Sonenberg, *Blood* **61**, 180 (1983).
[87] M. L. Vallano and M. Sonenberg, *J. Membr. Biol.* **65**, 57 (1982); M. L. Vallano, M. Y. Lee, and M. Sonenberg, *Am. J. Physiol.* **245**, E266 (1983).

brated, then the optical method provides a continuous monitor of E_m with far better time resolution than the measurement of tracer distribution.

While much progress has been made, the problem of rigorously quantitating the optical signals, and deriving unambiguous and accurate millivolt values, is a complex issue not yet completely solved, even for the relatively simple case of valinomycin-treated human red blood cells. The problems of quantitation are even more severe for cells with internal organelles where cytoplasmic ion activities and dye distributions are subject to some uncertainty. Consequently, at the present time optical potentiometric indicators are best regarded as qualitative, or in certain restricted applications, semiquantitative indicators of changes in E_m, with many controls needed for unambiguous interpretation of the results. The widespread application of this technique to so many fundamental problems in cellular electrophysiology attests to the desirability of continued efforts to find more sensitive and less toxic dyes, acting by well-characterized mechanisms, and able to be rigorously and accurately calibrated for quantitative as well as qualitative measurements of membrane potential in cells, organelles, and vesicles.

Acknowledgment

We gratefully acknowledge the support of a grant from the National Institutes of Health (GM28839). We also thank L. B. Cohen for comments on the manuscript.

[9] Fluorescence Methods for Continuous Monitoring of Transport in Cells and Vesicles

By OFER EIDELMAN and Z. IOAV CABANTCHIK

Introduction

Transport mechanisms in biological systems are studied by following the movement of physiologically relevant substrates (or substrate analogs) from one compartment to another. This flux can be traced by a variety of physical or chemical techniques.[1,2] The most widely used approach for tracing fluxes across membrane-sealed compartments is fol-

[1] G. Gardos, J. F. Hoffman, and H. Passow, *in* "Laboratory Techniques in Membrane Biophysics: An Introductory Course" (H. Passow and R. Stampfli, eds.), p. 9. Springer-Verlag, Berlin, Federal Republic of Germany, 1969.
[2] Y. Eilam and W. D. Stein, *Methods Membr. Biol.* **2,** 283 (1974).

lowing the movement of a radiolabeled substrate by separation of the compartments at various time points, sampling of their contents, and measurement of the label in one of the compartments. This method is, however, limited by the space available to the substrate, its specific activity and total radioactivity, the efficiency and speed of the separation relative to the actual transport rates, and the capacity of the biological system to retain its structural integrity during separation. Since this method is discrete in nature, the amount of information it can yield is fragmentary and often not precise enough for a thorough kinetic evaluation due to sampling and counting errors. Moreover, since its application is limited to homogeneous systems, the information gathered from heterogeneous biological systems is only average in nature.

Fluorescence has been considered to be a suitable alternative to the classical radiotracer method for various reasons[3]: (1) it has the requisite sensitivity and temporal resolution for following movement of fluorescent substrates between relatively small compartments; (2) it allows selection of a discrete area of a given compartment (cell, tissue, or medium) by using physical (e.g., optical focusing)[4,5] or chemical (e.g., impermeant light quencher[6,7]) means; and (3) as a result of (1) and (2), it permits continuous monitoring of transport by fluorescence (CMTF) without requiring periodic separation of compartments as an obligatory step of the transport protocol. A further bonus of the method is its modest demand for biological material (cells, reconstituted proteoliposomes, etc.), both because of the high sensitivity of fluorescence monitoring and because the same biological sample is used for the entire flux measurement.

General Principles

Monitoring of transport by fluorescence is based on the possibility of having compartment-dependent variations in fluorescence properties. These variations can be utilized to follow substrate movement across the compartment membrane by the ensuing change in fluorescence signal. Several factors can cause compartment-dependent variations in fluorescence properties: (1) absorption of the exciting or the emitted light by a substance confined to the internal compartment (e.g., pigment-containing cells or organelles such as hemoglobin in erythrocytes or chlorophyll in chloroplasts); (2) complexation between the fluorophore and a suitable

[3] O. Eidelman and Z. I. Cabantchik, *Alfred Benzon Symp.* **14**, 531 (1980).
[4] R. Peters, *J. Biol. Chem.* **258**, 11427 (1983).
[5] J. K. Foskett, *Am. J. Physiol.* **249**, C56 (1985).
[6] A. Darmon, O. Eidelman, and Z. I. Cabantchik, *Anal. Biochem.* **119**, 313 (1982).
[7] O. Eidelman and Z. I. Cabantchik, *Anal. Biochem.* **106**, 335 (1980).

substance (e.g., antifluorophore antibody, fluorescent chelator or a small collision quencher); (3) another approach is to monitor the fluorescence emitted out of a single compartment (i.e., a single cell or organelle) by focusing of a light detector which is mounted on a fluorescence microscope.

In this work we shall refer to the CMTF method as applied for the anion transport system of red blood cells,[6,7] although the principles are applicable to most other systems (and combinations of substances). The method is portrayed schematically in Figs. 1 and 2. The first step is loading the substrate into the internal compartment. This is most conveniently achieved by prolonged incubation or by lysis and resealing.[6,7] Second, the external substrate is removed by gel filtration, ion-exchange sieving, or centrifugation (in conditions of minimal anion translocation). The cells are then swiftly injected into a well-stirred cuvette placed in a thermostatted spectrofluorometer. The efflux of substrate from the cells is signaled by the time-dependent increase in fluorescence. The fluorescence intensity data are being recorded "on-line" or stored in an appropriate device (storage oscilloscope, microprocessor). For red blood cells

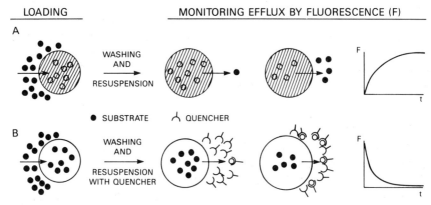

FIG. 1. Schematic representation of the CMTF method as applied to red blood cells (A) and to resealed red blood cell membranes or vesicles (B). The fluorescent substrate (full circles) is loaded either into red cells (A) or into other cells, ghosts, or vesicles (B) by incubation. In (A) the fluorescence of the intracellular substrate is quenched (empty circles), whereas in (B) it is not. After reaching transmembrane equilibrium, the extracellular substrate is removed either by centrifugation in the cold or sequestration by an appropriate resin, and the cells or vesicles are jetted into a flux solution, which for B contains the impermeant fluorophore quencher. Efflux of substrate from red cells is manifested by an increase in the fluorescence signal (F) with time (t), whereas for other cells or vesicles the fluorescence of probe which has exited the cells is quenched so that F decreases with t. This method is applicable to cells or organelles in suspension as well as cells grown on coverslips and mounted on a special holder in the cuvette.

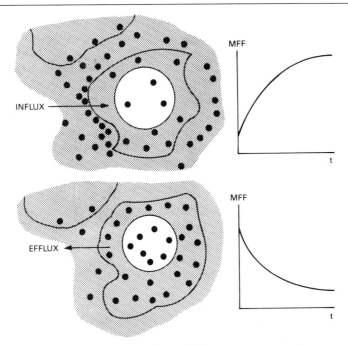

FIG. 2. Schematic representation of the CMTF method as applied to single cells or organelles (nonpigmented). The selective monitoring of a particular area in a field of several cells, a single cell, or an organelle is accomplished by physical devices. The microscopic field fluorescence (MFF) reflects the fluorescence in the selected area of the cells, with minimal signal contributed by fluorophore (full circles) outside the field. The method is applicable both in the influx[4] (A) and efflux[5] (B) mode. The fluorescence profiles are usually of an exponential nature (see further in the text).

(Fig. 1A), the fluorescence signal increases as the probe molecules leave the opaque volume of the cell.[7] For hemoglobin-free systems (Fig. 1B), the substrate-loaded cells or organelles are swiftly jetted into a solution containing antifluorophore antibodies. The decay of the fluorescence signal provides a measure for the egress of probe via the transport system.[6]

Permeant Substrate Analogs and Impermeant Quencher Couples

A substrate, specific for a given transport system (the permeant), is designed or selected, together with a suitable impermeant substance, such that one of them is fluorescent or potentially fluorescent and that interactions between the two will lead to a swift and sizeable change in the fluorescence signal properties. Transport is followed either as an increase or a decrease in fluorescence, depending on the characteristics of the permeant–impermeant couple. The design of N-(2-aminoethyl sulfonate)–7-nitrobenz-2-oxa-3-diazole (NBD–taurine) as a fluorescent substrate of the anion transport system[8] was based on structure–activity relationships of probes for the transporter,[9] the spectral properties of the NBD group (i.e., the relatively large spacing between excitation and emission and their overlapping with hemoglobin absorption[7]), and the possibility of raising a specific and efficient quenching antibody against the NBD hapten group,[6] which is similar in structure to the classic immunogenic dinitrophenyl group.[10] After thoroughly characterizing the specificity of the substrate for the anion transport system,[8] detailed kinetic analyses were performed on the intact system,[11,12] and the functional reconstitution of the anion-exchange protein was quantitatively assessed.[13] Other analogs of NBD–taurine which are suitable for monitoring anion transport are NBD–aminomethane sulfonate (NBD–AMS) and NBD–aminopropane sulfonate (NBD–APS). The first is approximately 20-fold faster and the second about 20-fold slower than NBD–taurine in permeating red cell membranes (Eidelman and Cabantchik, unpublished observations).

A mirror image approach to the CMTF method as described above is based on the use of a chemical quencher as the permeant (Tl^+ as mimic for K^+) and the fluorescent probe aminonaphthalenetrisulfonic acid (ANTS) as the impermeant substrate couple.[14]

Another widely used application of fluorescent couples is the use of quin 2 for measuring intracellular Ca^{2+} levels and fluxes.[15] Similarly, calcein leakage from unilamellar vesicles during fusion was assessed by using extravesicular Co^{2+} as a complexation quencher of calcein fluorescence,[16] and leakage from vesicles was checked with the fluorophore ANTS and the quencher p-xylene bis(pyridinium)bromide (DPX).[17]

A different kind of interaction between transport and a fluorescent dye is given by the use of the potential sensitive dye diS-C_3(5) to study ion permeability and ion gradient dissipation.[18] The movement of ions across

[8] O. Eidelman, M. Zangvil, M. Razin, H. Ginsburg, and Z. I. Cabantchik, *Biochem. J.* **195**, 503 (1981).
[9] M. Barzilay, S. Ship, and Z. I. Cabantchik, *Membr. Biochem.* **2**, 227 (1979).
[10] D. Lancet and I. Pecht, *Biochemistry* **16**, 5150 (1977).
[11] O. Eidelman and Z. I. Cabantchik, *J. Membr. Biol.* **71**, 141 (1983).
[12] O. Eidelman and Z. I. Cabantchik, *J. Membr. Biol.* **71**, 149 (1983).
[13] A. Darmon and Z. I. Cabantchik, *Biochim. Biophys. Acta* **727**, 77 (1983).
[14] H. P. H. Moore and M. A. Raftery, *Proc. Natl. Acad. Sci. U.S.A.* **77**, 4509 (1980).
[15] T. R. Hesketh, G. A. Smith, J. P. Moore, M. V. Taylor, and J. C. Metcalfe, *J. Biol. Chem.* **258**, 4867 (1983).
[16] N. Oku, D. A. Kendall, and R. C. MacDonald, *Biochim. Biophys. Acta* **691**, 332 (1983).
[17] H. Ellens, J. Bentz, and F. Szoka, *Biochemistry* **23**, 1532 (1984).
[18] L. M. Loew, I. Rosenberg, M. Bridge, and C. Gitler, *Biochemistry* **22**, 844 (1983).

the membrane changes the membrane potential, which in turn affects the aggregation of internal dye and its fluorescence. Fluorescence quenching by self-complexation was used for qualitative assessment of leakage of carboxyfluorescein.[19,20]

Experimental

Source of Material

NBD–taurine can be synthesized as described elsewhere[8] or procured from a commercial firm (Molecular Probes, Inc.). Anti-NBD antibodies are prepared using standard immunological procedures such as that described below.

The antigens are prepared by reacting NBD–Cl (Molecular Probes) with either albumin (bovine serum)[6] or hemocyanin (keyhole limpet) at (10 mg/ml) and a 50:1 molar ratio in phosphate-buffered saline (PBS), pH 8, for 4 hr at room temperature and subsequently dialyzing until unreacted material and reaction by-products are completely removed. The final product gives 8 mg/ml protein and probe : protein ratio of either a 10:1 (NBD–albumin) or 15:1 (hemocyanin) as determined by NBD absorption[8] ($\varepsilon_M^1\text{cm} = 23{,}200$ at 478 nm, pH 8) and Lowry-SDS[21] for protein assay (corrected for NBD absorption).

The antigen (1 mg/ml) was injected to female young (1.5–2 kg) rabbits as follows: (1) subcutaneous injection of 1 ml of a 2 mg/ml protein suspension diluted 1:1 with Freund's adjuvant (Difco) and homogenized in a syringe taped on a vortex mixer; (2) injection of the same after 6 weeks (booster); and (3) first bleeding from the ear (20–30 ml blood) at week 14. Booster injections and blood collection were repeated every 8 weeks. Blood was coagulated in the cold (4°) and the serum collected and centrifuged to remove debris and then precipitated with $(NH_4)_2SO_4$ (33% saturation) for 5 to 10 min at room temperature followed by centrifugation (5 min at 1600 rpm). The precipitate was resuspended in PBS to the original serum volume, giving a cloudy suspension, which clarified after three to four reprecipitations with $(NH_4)_2SO_4$ (33% saturation). The latter was resuspended to the original serum volume, dialyzed overnight at 5° against at least 500 volumes of PBS, divided into aliquots, and kept frozen at −70°. Refreezing of thawed samples was avoided in order to minimize antibody inactivation. An aliquot of 10–40 µl was usually sufficient for

---

[19] R. Blumenthal, J. N. Weinstein, S. O. Sharrow, and P. Henkart, *Proc. Natl. Acad. Sci. U.S.A.* **74**, 5603 (1977).
[20] J. N. Weinstein, S. Yoshikami, P. Henkart, and R. Blumenthal, *Science* **195**, 489 (1977).
[21] S. V. Chi and L. Smith, *Anal. Biochem.* **13**, 414 (1975).

complete (>90%) quenching of a 1.5-ml solution containing 150 pmol NBD–taurine (100 nM). No further purifications of antisera were found necessary for the present purposes.

Transport Assay: Human Red Blood Cells

The number of cells required for a single CMTF assay using a regular spectrofluorometer (e.g., Perkin-Elmer MPF-4) is between 1×10^6 and 5×10^6 (~1 μl whole blood). Blood is washed off in an isotonic NaCl solution (145 mM), buffered with sodium phosphate, HEPES, Tris, or other buffers (at 5–20 mM), pH 7.4, 25°, using repeated centrifugation and aspiration of supernatant and buffy coat.

Loading of NBD–Taurine. Cells at 10 to 20% hematocrit in any of the above buffers are incubated with 1 mM NBD–taurine for 1 hr at 37°. After equilibration with the probe, the cells are transferred to 5°. The cells can be kept for up to 2 days in these conditions with no observable changes in their transport properties.

Removal of External Probe. One of the following two methods can be used for this purpose. (1) An aliquot (50 μl) of the loaded cell suspension is diluted in 1 ml isotonic sulfate buffer (5°), placed in a plastic 1.5-ml conical tube, and centrifuged for 4 sec in a microcentrifuge (12,000 g, 5°). The supernatant is discarded, and the cells washed twice again with the same medium and finally resuspended in 100 μl sulfate-buffered solution (5°). These suspensions can be kept at 0–5° for at least 10 hr with insignificant leakage of fluorescent substrate. (2) For faster transport conditions, a 5–10 μl suspension of NBD–taurine-containing cells is loaded on a 0.5-ml Dowex AG 1-X8 (25–50 mesh, sulfate or citrate form) minicolumn precooled to 5°, followed by a quick rinse with 100–200 μl sulfate medium (5°).

Flux Measurement. An aliquot of the NDB–taurine-loaded cells free of extracellular probe (0.5–5 $\times 10^6$ cells) is jetted into a cuvette of a spectrofluorometer containing 0.2–2 ml buffer preequilibrated at the desired temperature. Continuous stirring is recommended when fluxes are followed for more than 20 min. For most cases, however, periodic mixing of cells with a pipette is adequate. The value of the fluorescence signal F_∞, attained when all the probe has exited the cells, is needed for quantitative data analysis. In most cases, F_∞ can be obtained by permeabilizing the cells with detergent [e.g., Nonidet P-40 (BGH) at 0.1 mg/ml final concentration; $C_{12}E_9$ (Calbiochem) at 0.5 mg/ml final concentration]. With Nonidet P-40, care has to be exercised to avoid clouding of the detergent at temperatures above 30° (this occurs generally 1 to 2 min after addition of the detergent). Since the amount of detergent to be added may vary with cell type and cell number, it is recommended that the amount be determined in preliminary studies.

Transport Assay: Pigment-Free Cells, Ghosts, or Vesicles

Loading of NBD–Taurine. The amount of biological material and the concentration of fluorophore required for a transport assay are in the same range as those of erythrocytes, although the loading times for other cells may vary considerably. For ghosts, resealing in the presence of probe[6] circumvents subsequent loading. Similarly, for vesicles, freezing and thawing commonly used to seal membranous systems can also facilitate entrapment of probe.

Removal of External Probe. Initial removal of external probe after loading can be done by centrifugation. Since the cells, or vesicles, are still in the loading solution, there are no limitations on time of centrifugation. However, there might be time limitations on the second separation step, that is, the time of centrifugation has to be much shorter than the half-time for probe efflux in the washing solution. The latter can be increased by lowering the temperature, and/or by addition of transport inhibitors, or removing a required factor for the flux. For cells and resealed red cell ghosts, quick centrifugation (Eppendorf, 1 sec to 5 min) in the cold (5°) in isotonic sulfate medium can serve the purpose. However, for vesicles as well as ghosts, chromatography through a Dowex AG 1-X8 column (100–200 mesh), as shown before for erythrocytes, is faster and more convenient. For fast transporting systems (e.g., reconstituted band 3 vesicles), separation can be avoided by quenching all external probe (after centrifugation and resuspension) with excess anti-NBD antibodies.

Determination of the Requisite Amount of External Quencher. In the anti-fluorophore CMTF assay, efflux of fluorophore has to be performed in a medium containing excess antifluorophore antibody, the purported impermeant fluorescence quencher. To assess the required amount of antibody, an aliquot of probe-loaded and washed cells or vesicles is added to a cuvette containing buffered medium (PBS or equivalent) and sufficient detergent to permeabilize the membranes. Then, small aliquots of the antibody solution are added until about 90% of the fluorescence is quenched. The time required for the antibody to fully elicit its effect at 37° is about 1 to 2 sec. Approximately 40 μl of antisera (diluted to the original serum level) is required to fully quench 150 pmol of NBD-containing molecules.

Flux Measurement. Flux measurement is essentially as described above except that the membrane material and antibody are added in commensurate amounts.

Setting of the Spectrofluorometer

The selection of excitation and emission wavelengths and slits constitutes a trade-off between optimizing sensitivity to emitted light detection and minimizing background signal (due to light scattering, etc.). These settings, therefore, may differ from optima found with clear probe solutions. For NBD–taurine in solution (pH 7.2–7.4), the optimal wavelength settings are 473 nm excitation and 540 nm emission, although for cells the corresponding settings are 468 and 550 nm.[7] In the case of antifluorophore antisera present in the flux media, high background (cloudiness) is often observed unless the antisera are $(NH_4)_2SO_4$ precipitated at least four to five times.[6] Further antibody purification by DEAE-cellulose or affinity chromatography is possible, although background and quenching efficiency improvements are dismal.

Data Acquisition and Kinetic Analysis

The analysis of CMTF kinetic data is based on regular transport (enzyme) kinetic analysis. These procedures can be utilized to their full extent because the flux record is continuous, and, depending on available equipment, good temporal resolution from the fractional second to the hour range can readily be attained.

CMTF is commonly used in a mode whereby the fluorescent substrate is present in trace concentrations. In these conditions, the initial concentration of probe, S_0, is much smaller than the apparent Michaelis constant for the probe K_m. Thus, the probe does not interfere with the normal performance of the transport function (with the primary substrate) but only reports about the kinetic status of the system, i.e., the transport rate of the probe is proportional to the concentration of free transport sites at the *cis* side of probe addition.[11] A more versatile and informative mode of the CMTF method is utilizing substrate–analog concentrations higher than the apparent K_m ($S_0 \gg K_m$). In this mode, a single fluorescence trace provides all of the information required to calculate the apparent K_m and V_{max} for transport of the probe.

Data Acquisition

The raw data from a CMTF experiment are obtained as a F_t versus t trace which is either recorded on-line as an analog signal on a strip-chart recorder or storage oscilloscope, or in a digital form on a microprocessor attached to a spectrofluorometer or an equivalent data acquisition system.

The initial fluorescence immediately after addition of cells or vesicles, F_0, is usually obtained by extrapolation to $t = 0$ from the recorded fluorescence trace. This signal is composed of the background signal F_b coming from light scattering, some remaining external probe after the wash, and from instrumental noise and stray light. For the particular case of pigment-free cells or vesicles used in conjunction with antifluorophore antibodies, F_0 also can be obtained by addition of excess nonfluorescent hapten to restore the original fluorescence, such as dinitrophenyltaurine (Sigma) added to NBD–taurine–anti-NBD complexes formed in the assay mixture at the end of the flux.[6] The F_b normally is assumed to be constant when no major volume changes of the cell or vesicle compartment occur during the flux monitoring time, otherwise, the data should be corrected by the results of the blank (i.e., without fluorophore) experiment.

The final fluorescence (F_∞), that is, when all the fluorescent substrate has exited from cells or vesicles, can be obtained either by waiting the required time after appropriate stirring of the suspension or by adding a small aliquot of detergent (NP-40, $C_{12}E_9$, or equivalent).

Calibration of Fluorescence Signal and Determination of the Concentration Range of the Fluorescent Substrate

The relationship between the concentration of fluorescent substrate–analog and the associated fluorescence signal has to be measured (in either compartment) before flux measurement can be quantitatively analyzed. In the external medium, the calibration is usually done with dye concentrations in the micromolar range. Deviation from linearity occurs mainly because of inner filter effects. Care has therefore to be exercised to limit the total amount of dye in a flux assay so that the final concentration is considerably lower than the plateau level, preferably in the linear range. For the internal compartment, samples are prepared by loading the cells with different concentrations of probe. The calibration is done by measuring the initial signal either in the presence of flux inhibitors, or by extrapolation of the signal intensity to $t = 0$.

In the following analysis, we shall assume that either the fluorescence signal is linear with substrate concentration or that the original data record has been corrected by appropriate calibration methods.

Data Analysis

The fluorescence intensity originating from probe present either in the internal (i) or the external (e) compartment (e.g., red cells) is given by

$$F_t = F_b + \alpha_i N_i(t) + \alpha_e N_e(t) \tag{1}$$

where N_i and N_e are the amounts of probe in the respective compartments and α_i and α_e are the fluorescence signals monitored by the detector from a unit amount of probe present in the given compartment (internal or external). In the efflux mode at $t = 0$, all the probe is in the inner compartment, and at $t = \infty$ all probe is in the external one, so that

$$F_0 = F_b + \alpha_i N_T \tag{2}$$

and

$$F_\infty = F_b + \alpha_e N_T \tag{3}$$

where $N_T = N_e + N_i$ is the total amount of probe, which can be measured experimentally by calibrating the fluorescence.

Combining Eqs. (1) through (3), we obtain

$$N_i(t)/N_T = (F_\infty - F_t)/(F_\infty - F_0) \tag{4a}$$
$$N_e(t)/N_T = (F_t - F_0)/(F_\infty - F_0) \tag{4b}$$

Therefore, the probe concentration in the cellular or vesicular volume is given by

$$\begin{aligned} S_i(t) &= N_i/V_i = (N_T/V_i)[(F_\infty - F_t)/(F_\infty - F_0)] \\ &= S_0[(F_\infty - F_t)/(F_\infty - F_0)] \end{aligned} \tag{5}$$

where S_0 is the initial internal concentration of probe and V_i is the effective volume of cells or vesicles present in the assay system. It should be stressed that S_0 inside the cells is not necessarily equal to that concentration used in the loading medium, since factors such as Donnan effects (or membrane potentials) have a marked effect on the transmembrane distribution of charged species.

Trace Mode: ($S_0 \ll K_m$)

In this mode the efflux of probe (from either system) follows a single exponential (decay for pigment-free system and raise for pigmented cells) with a characteristic rate constant, k. This constant can be evaluated by a variety of methods (Fig. 3).

1. Obtaining the tangent to the curve immediately after flux initiation and dividing it by the total fluorescence:

$$k = \text{initial slope}/(F_\infty - F_0) \tag{6a}$$

Although simple, the method is subjective and often unsuitable for either fast or very slow transport profiles.

2. Converting the fluorescence data into a logarithmic form:

$$\log(F_\infty - F_t) \tag{6b}$$

which, when plotted against t, yields the slope k.

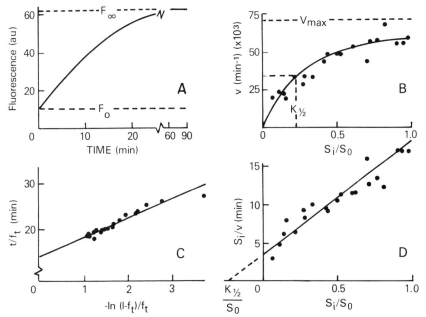

FIG. 3. Methods for determining the kinetic parameters of fluorescent substrate efflux by CMTF. Adapted from Ref. 7 with permission. (A) The efflux profile of NBD–taurine from human red cells. (B) Plot of the velocity of NBD–taurine fluorescence change ($v = \Delta F/\Delta t$) versus S_i/S_0 as computed from the data shown in A. The continuous curve is the nonlinear least squares fit of the data according to the Michaelis–Menten equation. (C) Plot of the data shown in A according to the integrated Michaelis equation [Eq. (8)]. The solid line is the linear least squares fit of the data. (D) Reciprocal Hanes–Woolf plot of the data shown in B.

3. The fluorescence data can be fitted either directly to an exponential function and analyzed by a nonlinear least squares regression program or, alternatively, it can be linearized as in (2) and analyzed by a linear least squares regression.

4. Calculating the instantaneous rate constant k_t for each point along the whole fluorescence trace[11]:

$$k_t = -[1/(F_\infty - F_t)](d/dt)F_t \qquad (6c)$$

For a simple exponential, k_t should be constant throughout the entire profile (and equal to k), although as the plateau region is approached the noise increases.

In the trace mode, the calculated k values are useful for assessing and comparing the effects of external conditions such as temperature, pH, substrate composition, inhibitors, etc.[8] They can also be used to accurately examine the transport system in nonequilibrium conditions, that is, when k changes with time as a result of changes in physical or chemical factors which affect the system. For such a case, calculation of the instantaneous rate parameter [Eq. (6c)] was found to be very useful.[11]

Michaelian Mode: $(S_0 \gg K_m)$

The unidirectional flux of fluorescent substrates can be used to determine the kinetic constants of the system, (K_m and V_{max}) when the concentration of probe is larger than K_m. In this mode, the experimental data are fitted to the integrated form of the Michaelis–Menten equation, using time as the dependent variable:

$$t = (N_T/V_{max})f_t - (K_m N_T/V_{max} S_0)[\ln(1 - f_t)] \tag{7a}$$

where

$$f_t = (F_t - F_0)/(F_\infty - F_0) = 1 - S_i(t)/S_0 \tag{7b}$$

and analyzed by nonlinear least squares regression.

Analoguously, one can linearize the equation and obtain K_m (apparent K_m) and V_{max} by linear least square analysis of

$$t/f_t \text{ versus } \ln(1 - f_t)/f_t$$

based on

$$t/f_t = (N_T/V_{max}) - (K_m N_T/V_{max} S_0)[\ln(1 - f_t)/f_t] \tag{8}$$

The N_T/V_{max} is given by the y-intercept and S_0/K_m by the x-intercept.

Applications

The system to which the CMTF method has been most extensively applied is that of human red blood cells anion exchange[11,12] and the reconstituted band 3 protein, using NBD–taurine as the fluorescent substrate and anti-NBD antibodies as quenchers where applicable. Murihead *et al.*[22] have studied anion transport heterogeneity in KS62 erythroleukemia cells with NBD–taurine and flow cytometry. A successful application of the above to the study of transport in single cells (and single resealed ghosts) was recently presented.[23] The method is based on fluorescence microphotolysis of single cells,[24] a technique which was also employed to

[22] K. A. Muirhead, R. C. Steinfeld, M. C. Seversky, and P. A. Knauf, *Cytometry* **5**, 268 (1984).
[23] R. Peters and H. Passow, *Biochim. Biophys. Acta* **777**, 334 (1984).
[24] R. Peters, *Eur. Biophys. J.* **11**, 43 (1984).

measure fluxes in primary hepatocytes,[25] isolated liver cell nuclei,[4] and nuclear membrane ghosts.[26] Using a similar approach on intact *Necturus* gallbladder, Foskett[5] demonstrated the activation of an anion exchanger following regulatory volume increase of epithelial cells after being subjected to osmotic shrinkage.

The fluorescent quencher Tl^+ has been used as the permeant analog of K^+ in conjunction with aminonaphthalenetrisulfonic acid as the impermeant fluorophore.[14] With this couple and stopped-flow fluorescence assay, Moore and Raftery succeeded in following cation transport across membrane vesicles containing the acetylcholine receptor in the physiologically relevant scale of a few milliseconds.[14]

At present, the possibility of using CMTF with other transport systems is entirely dependent on the design of fluorescent substrates for those systems. Particularly important and useful will be the fluorescent analogs of amino acids, sugars, nucleosides, and nucleotides. The fluorescent probe formycin triphosphate has been quoted to be transported by the ADP–ATP exchange carrier of mitochrondia,[27] although little kinetic information has been presented.[28] The availability of antisera with quenching activity against fluorophores such as NBD, fluorescein, lucifer yellow, and others should certainly be taken into consideration in the chemical design of potential fluoropermeants.

In summary, the CMTF method, with its high temporal resolution, high signal sensitivity, and modest demand for biological material, provides the transport kineticist and biochemist with a most versatile experimental tool for extracting a wealth of kinetic data and for thoroughly analyzing transport mechanisms.

Acknowledgments

We wish to thank Dr. Harvey Pollard and Dr. Robert Blumenthal for their support during the writing of this work and for critical reading of the manuscript. Dr. Anne Walter is gratefully acknowledged for carefully reading the manuscript and many helpful suggestions. Part of this work was supported by NIH grants AI20342 and HL40158 to ZIC.

---

[25] R. Peters, *EMBO J.* **3**, 1831 (1984).
[26] I. Lang and R. Peters, in "Information and Energy Transduction in Biological Membranes" (E. G. Helmreich, ed.), p. 377. Liss, New York, 1984.
[27] E. Schlimme, K. S. Boos, and E. J. deGroot, *Biochemistry* **19**, 5569 (1980).
[28] G. Brandolin, Y. Dupont, and P. V. Vignais, *Biochemistry* **21**, 6348 (1982).

[10] Preparation and Use of Micro- and Macroelectrodes for Measurement of Transmembrane Potentials and Ion Activities

By DANIEL AMMANN and PICO CARONI

Introduction

Today, microelectrodes are a ubiquitous tool in electrophysiology for potentiometric studies of transmembrane potentials and intracellular ion activities of single cells. Comprehensive descriptions of reference microelectrodes,[1] ion-selective microelectrodes,[2,3] and principles of the potentiometry with electrodes[4] have been given in the literature.

Since the introduction of small reference microelectrodes (Ling-Gerard microelectrodes, PD microelectrodes, conventional microelectrodes), pipet preparation, electrode filling, observation of cell contamination, and theoretical understanding of interfering potentials (liquid-junction potentials, tip potentials) have been continually improved. Transmembrane potentials of most cells and, under favorable conditions, of cell organelles can be easily and continuously recorded. Reference microelectrodes can also be used for electrical stimulations (current injection), ionophoresis (application of chemicals), and voltage-clamp experiments.

Liquid membrane microelectrodes based on classical ion-exchanger or highly selective carrier molecules have almost completely replaced the glass membrane and the solid-state microelectrodes. The potentiometry with these ion-selective microelectrodes is the only intracellular approach that gives simultaneous insight into both ion activities and electrical parameters of the cell. Further typical features of ion-selective microelectrodes are quantitative measurements of ion activities or free ion concentrations, extremely local detection (detection volume of about 1 fl), simultaneous measurement of the activities of different ions with multibarrelled microelectrodes, the measurement of ion activity transients

---

[1] R. D. Purves, "Microelectrode Methods for Intracellular Recording and Ionophoresis." Academic Press, London, 1981.

[2] R. C. Thomas, "Ion-Sensitive Intracellular Microelectrodes: How to Make and Use Them." Academic Press, London, 1978.

[3] D. Ammann, "Ion-Selective Microelectrodes—Principles, Design and Applications." Springer-Verlag, Berlin, Federal Republic of Germany, 1986.

[4] W. E. Morf, "The Principles of Ion-Selective Electrodes and of Membrane Transport." Académiai Kiadó, Budapest, 1981/Elsevier, Amsterdam, 1981.

(above the millisecond-range), cell membrane surface recordings, measurements in subcellular organelles, depth profile recordings in tissues, and *in vitro* as well as *in vivo* studies (for a discussion, see Ref. 3).

Macroelectrodes, and in particular the multiparameter-vessel minielectrode version, are a powerful tool for the study of ion transport and membrane potential phenomena in subcellular and reconstituted systems. As biochemists and cell biologists are increasingly interested in the reconstitution of isolated biological processes, ion transport measurements in small reaction vessels become an essential component of such studies. Salient features of the minielectrode setup include continuous recording of single ionic activities in the extravesicular space, 90% response times between 0.1 and 1 sec, possibility of rapid manipulations of the reaction medium during the course of an experiment, and simultaneous measurement of up to three independent ionic parameters.

Preparation of Microelectrodes

The procedure for the preparation of small-tipped glass capillaries (micropipettes) is the same for reference microelectrodes and for ion-selective liquid membrane microelectrodes. The filling of the micropipettes with either an aqueous reference electrolyte or an organic ion-selective membrane is, however, different.

Glass Tubings

Micropipettes are fabricated from glass tubings with diameters in the range of about 1 to 3 mm. Different types of glass (soda lime glass, borosilicates, aluminosilicates, quartz glass, etc.) are available in many different shapes (Fig. 1). Reference microelectrodes always are prepared from single-barrelled tubings. Double- and multibarrelled tubings are suited for the fabrication of combined ion-selective microelectrodes which allow the simultaneous measurement of the transmembrane potential and of one or several intracellular ion activities (Fig. 2). In attempting to select glass types for micropipettes, specific resistance, softening point, and water resistivity of the glass are of particular importance. Ion-exchange selectivity properties, electrical shunt pathways through the glass wall at the tip of the microelectrode, and the ease of fabrication of very small micropipettes heavily depend on these glass properties. The borosilicate Pyrex followed by aluminosilicates are especially well-suited glass types. They are sold as capillaries exhibiting many different shapes and configurations (e.g., from Clark Electromedical Instruments, Pangbourne, England; A-M Systems, Everett, WA; WP Instruments,

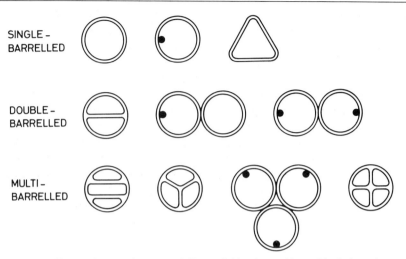

FIG. 1. Different shapes of commercially available glass tubings. Black dots: internal glass filaments.

New Haven, CT). There is no agreement about the importance of cleaning the glass tubings before pulling micropipettes. Usually, the purification involves several steps: the glass tubings are treated for about 24 hr in a diluted (1:1) nitric acid solution. Afterward, the capillaries are washed six times with distilled water and three times with pure acetone. The purified tubings are dried for about 24 hr at 150° in an oven and are then stored in a desiccator. With respect to the reactivity of the glass surface during the silanization step (see below), a further acid treatment of the pulled micropipette seems to be necessary since, after pulling, 99.98% of the surface near the tip is freshly exposed glass.[5]

Pulling of Micropipettes

For the fabrication of micropipettes with tip diameters larger than about 0.5 μm, any commercially available (e.g., David Kopf Instruments, Tujunga, CA; Narashige Scientific Instruments Lab, Tokyo, Japan) or home-built puller[6] is adequate. Using of such instruments implies the heating of a fixed tubing by a Ni/Cr, Ta, or Pt/Ir coil and the simultaneous pulling by electromagnetic forces into two micropipettes. The parameters of the puller (i.e., temperature, strength of pulling) have to be chosen for each type of glass and often have to be readjusted for a new batch of glass tubings. Furthermore, the optimal choice of parameters may vary consid-

[5] F. Deyhimi and J. A. Coles, *Helv. Chim. Acta* **65,** 1752 (1982).
[6] K. T. Brown and D. G. Flaming, *Neuroscience* **2,** 813 (1977).

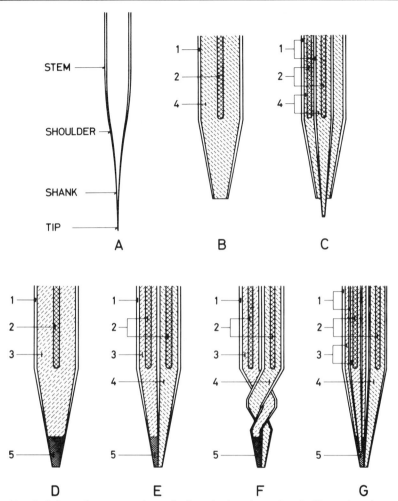

FIG. 2. Schematic cross-sections of selected microelectrodes. 1, Glass micropipette; 2, chlorinated silver wire; 3, internal filling solution; 4, reference electrolyte; 5, ion-selective membrane solution; A, customary nomenclature of parts of micropipettes; B, single-barrelled reference microelectrode; C, coaxial reference microelectrode; D, single-barrelled liquid membrane microelectrode; E, double-barrelled liquid membrane microelectrode (theta-like); F, double-barrelled liquid membrane microelectrode (fused and twisted tubings); G, coaxial double-barrelled liquid membrane microelectrode (theta-like).

erably with the puller construction. For very demanding needs (i.e., tip diameters well below 0.5 μm, multibarrelled micropipettes) more sophisticated pullers such as two-stage airjet instruments have to be employed. Adequate cooling of the heating coil, optimal loop geometry of the coil, and a two-stage pulling process allow the fabrication of very small micro-

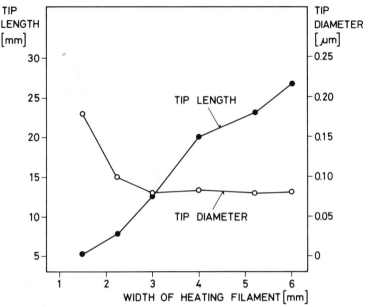

FIG. 3. Influence of heating filament width (trough-type, airjet puller, constant airflow) on the tip length and tip diameter of micropipette. Redrawn from Flaming and Brown.[7]

pipettes with various lengths of the shank[7] (Fig. 3). To date, the smallest tip diameters of single-barrelled and double-barrelled (theta-like) micropipettes are 0.045 and 0.055 μm, respectively.[6]

Breaking and Bevelling of Micropipettes

It is often advantageous to work with microelectrodes of known tip diameter. Within a limited accuracy, defined tip diameters can be produced by mechanically breaking very small tips. It is easy to contact a microelectrode tip with the polished end of a plexiglass rod while observing the microelectrode tip and its mirror image on the plexiglass surface under a light microscope. When the tip and the rod are in direct contact, a slight tap on the table near the microscope causes the tip to break in steps of about 0.5 μm.

Bevelling is another means by which well-defined tip diameters can be obtained. Other major purposes of bevelling are the reduction of electrical electrode resistance, the improvement of the ability to puncture cells, and

[7] D. G. Flaming and K. T. Brown, *J. Neurosci. Methods* **6**, 91 (1982).

the facilitation of current or drug injection into cells. During the process of bevelling, the circular opened tip of the micropipette is transformed into a tip having a hypodermic needlelike shape. Bevelling is performed on rotating wobble-free surfaces containing grinding materials such as alumina, diamond dust, or ceric oxide (e.g., from David Kopf Instruments, Tujunga, CA). The optimal angle of the bevelled part is between 20° and 30°. Reference microelectrodes are bevelled when filled with reference electrolyte and the microelectrode resistance is then usually continuously monitored. The resistances may then be used as a measure of the tip diameter. Bevelling of ion-selective microelectrodes in the presence of the membrane solution is difficult. Therefore, liquid membrane microelectrodes are usually bevelled before the filling procedure.

Tip Observation

Tip diameters of microelectrodes can be estimated or exactly determined by light microscopy (resolution of about ±0.2 μm), electron microscopy (typical resolution of about 2 nm), and measurements of electrical electrode resistances. Thus, the light microscope does not allow an exact measurement of small microelectrode tips. Unfortunately, electron microscopic observations are usually accompanied by destruction of the microelectrode. When micropipettes are prepared under well-controlled and constant conditions, however, highly reproducible tip diameters can be produced [e.g., 0.382 ± 0.003 μm ($n = 20$)[8]].

Filling of Reference Microelectrodes

Several filling methods have been described. Thirteen techniques have been summarized and advantages and disadvantages have been discussed.[1] Today, the so-called glass fiber technique is most convenient. Tubings are available that contain a glass fiber fused to the inner glass wall (see Fig. 1). The fiber causes the reference electrolyte to advance to the tip by capillarity while the air is simultaneously moving in the reverse direction. This way, small reference microelectrodes can be easily filled immediately before use.

Filling of Ion-Selective Liquid Membrane Microelectrodes

Ion-Selective Membrane Solutions. Classical ion-exchanger solutions for the measurement of K^+ and Cl^- are known as Corning 477317 and Corning 477315 products, respectively. The Cl^- exchanger has been im-

---

[8] J. R. Lopez, L. Alamo, C. Caputo, R. DiPolo, and J. Vergara, *Biophys. J.* **43**, 1 (1983).

proved[9] and is available as the Corning 477913 ion exchanger. These membrane solutions are sold by Corning Glass [Medfield, MA; see also the products IE-190 (K^+) and IE-170 (Cl^-), from WP Instruments, New Haven, CT].

A complete assortment of carrier-based membrane solutions for the measurement of H^+, Li^+, Na^+, K^+, Mg^{2+}, and Ca^{2+} is available at Fluka AG, (Fluka, CH-9470 Buchs, Switzerland). The mixtures contain all components and are ready for use. The pure components (carriers, membrane solvents, additives) are available too, and can be mixed to give a membrane solution. However, the complete membrane solutions (termed "cocktail" by Fluka), have to be recommended since they correspond to the current state of membrane technology.

Silanization of Micropipettes

Glass surfaces exhibiting a high density of free hydroxyl groups are quite hydrophilic and therefore strongly repel organic membrane solutions. In order to avoid an easy replacement of the membrane solution by the aqueous outside solution, the glass of microelectrodes is made hydrophobic by reaction with organic silicon compounds. Reactive chlorosilanes or silazanes bind covalently to the free hydroxyl groups, and the organic substituents of the bound silicon compound render the surface lipophilic. A series of silanization techniques, including treatments with liquids, with polymeric silicon oils, and with vapor phases has been proposed. The most sophisticated method is based on a vapor phase treatment and is the result of a detailed study on the influence of the type of silanization reagent and of the experimental conditions on the silanization of micropipettes.[5,10] An arrangement for the silanization of single- and double-barrelled micropipettes and the optimal conditions for the treatment of Pyrex glass have been given (Fig. 4).[5,10] The procedure starts with an acid pretreatment of the freshly drawn micropipettes (5 min, 60% HNO_3). Optimal vapor reaction is then achieved when dimethylaminotrimethylsilane is applied to the acid-treated and dried micropipettes for 5 min at 250°. Advantages of the use of the aminosilane are its relatively high vapor pressure, relative inertness toward formation of polymers when in contact with water, and the formation of the noncorrosive by-product dimethylamine. The procedure is, however, relatively complex and time-consuming since only one micropipette can be silanized at a time. A simpler, presumably less effective but sufficient alternative proce-

[9] C. M. Baumgarten, *Am. J. Physiol.* **241**, C258 (1981).
[10] J.-L. Munoz, F. Deyhimi, and J. A. Coles, *J. Neurosci. Methods* **8**, 231 (1983).

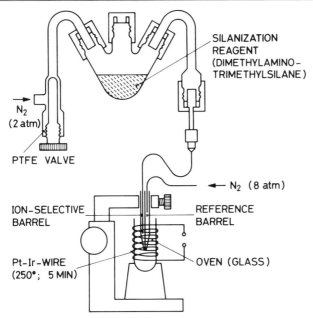

FIG. 4. Apparatus for the selective and efficient vapor silanization of the ion-selective barrel of a double-barrelled microelectrode. A stream of nitrogen carries the vapor of the silanization reagent through the ion-selective barrel while the reference barrel is protected by a separate stream of nitrogen. The microelectrode is placed in a small oven to adjust the optimal temperature. Redrawn from Deyhimi and Coles.[5]

dure for vapor-phase silanization can be employed.[11] A batch of micropipettes is mounted on a glass plate in an oven and covered with an upturned glass beaker (for a schematic illustration showing only one microelectrode, see Fig. 5). After predrying the glass (30 min, 150°) a small amount (e.g., 10 μl if the glass plate and the glass beaker have already been well silanized) of pure dimethylaminotrimethylsilane (Fluka AG) is injected with a syringe into the volume of the beaker. The silicon compound immediately evaporates and is allowed to react under optimal conditions (30 min, 200°). The method is particularly suited for single-barrelled microelectrodes. When using double-barrelled micropipettes both barrels will become hydrophobic (but see, e.g., Ref. 12).

[11] R. Y. Tsien and T. J. Rink, *Biochim. Biophys. Acta* **599**, 623 (1980).
[12] E. Frömter, M. Simon, and B. Gebler, in "Progress in Enzyme and Ion-Selective Electrodes" (D. W. Lübbers, H. Acker, R. P. Buck, G. Eisenman, M. Kessler, and W. Simon, eds.), p. 35. Springer-Verlag, Berlin, Federal Republic of Germany, 1981.

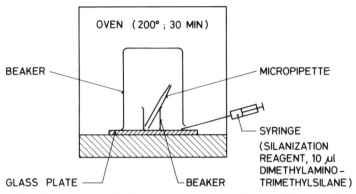

FIG. 5. Simple arrangement for the silanization of micropipettes with organosilanes in the gaseous phase (not drawn to scale, only one micropipette is shown).

Filling of Ion-Selective Liquid Membrane Microelectrodes

The ease of filling a micropipette with an organic membrane solution depends on many factors, including the tip diameter, silanization quality of the glass surface, viscosity of the membrane solution, and the filling procedure itself. A large number of different filling methods have been proposed and cannot be summarized here. The various techniques can be subdivided into back-filling (filling through the stem) and front-filling techniques. Filling through the stem is becoming more and more popular because of the increasing use of very small micropipettes and because of facilitated back-filling due to improved silanization procedures. With the help of injection needles or plastic capillaires,[13] a small drop of membrane solution is inserted through the stem into the shank of the microelectrode. The membrane solution then usually spontaneously flows into the tip (otherwise vacuum or pressure can be applied). Normally, filling heights of several millimeters are obtained. This facilitates, first, proper placement of an internal filling solution-filled injection needle in contact with the surface of the membrane solution and, second, filling of the stem with electrolyte solution without formation of air bubbles.

In order to exploit the beneficial effect of polyvinyl chloride (PVC)-containing membrane solutions, it is necessary to front-fill the micropipette by immersing the tip into the membrane solution. Because of the high viscosity, the PVC membrane solutions must be diluted with tetrahydrofuran.[14,15]

[13] F. Lanter, D. Erne, D. Ammann, and W. Simon, *Anal. Chem.* **52**, 2400 (1980).
[14] R. Y. Tsien and T. J. Rink, *J. Neurosci. Methods* **4**, 73 (1981).
[15] F. Lanter, R. A. Steiner, D. Ammann, and W. Simon, *Anal. Chim. Acta* **135**, 51 (1982).

Use of Microelectrodes

Measuring Arrangement

Block diagrams and equivalent circuits of microelectrode arrangements have been extensively discussed in textbooks[16] and in monographs.[1,2,17,18] Because of the extremely high electrical membrane resistances R_M of microelectrodes (10^9–10^{12} Ω), reliable potentiometric measurements can be performed only if the input impedance R_i of the measuring instruments is very high (usually $>10^{13}$ Ω) and if the input bias current I_i is very low (usually <100 fA). For a microelectrode with a resistance R_M of 10^{10} Ω and for a required resolution of ± 0.1 mV at a measuring range of ± 100 mV, the minimal required value of R_i would amount to 10^{13} Ω. The leakage current I_i then has to be smaller than 10 fA [Eq. (1)].

$$I_i R_M \ll \text{emf} \tag{1}$$

The availability of high-performance operational amplifiers has, however, made the design of appropriate measuring equipments easy. Commercial instruments are available (e.g., WP Instruments, New Haven, CT) or excellent operational amplifiers (e.g., AD 515 KH, Analog Devices, Norwood, MA) can be used in home-built equipment. However, the high-impedance elements of the circuitry must be placed in a Faraday cage to reduce and eliminate interferences from electromagnetic fields.

Calibration and Evaluation

For intracellular recording of the membrane potential, the cell has to be impaled with a reference microelectrode and the external bath (perfusion) solution must contain a conventional reference electrode. The difference of the potential differences between these two electrodes before and after cell puncture corresponds to the membrane potential of the cell. In order to make ion activity studies within a single cell, an intracellular ion-selective microelectrode is required in addition (Fig. 6). In order to evaluate intracellular ion activities, the ion-selective microelectrode cell assembly has to be calibrated. Ideally, calibration solutions should resem-

---

[16] P. Horowitz and W. Hill, "The Art of Electronics." Cambridge Univ. Press, Cambridge, England, 1980.
[17] L. A. Geddes, "Electrodes and the Measurement of Bioelectric Events." Wiley (Interscience), New York, 1972.
[18] E. Neher, "Elektronische Messtechnik in der Physiologie." Springer-Verlag, Berlin, Federal Republic of Germany, 1974.

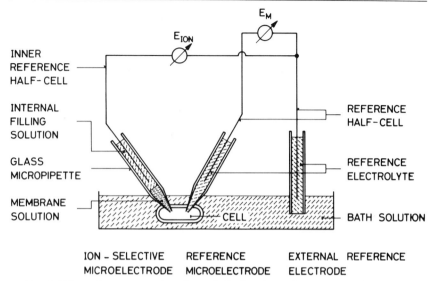

Fig. 6. Schematic representation of a typical microelectrode arrangement for membrane potential and intracellular ion activity measurements.

ble the sample solution (e.g., cytosol) as closely as possible, i.e., the most important interfering ions are considered at their typical physiological concentrations. By evaluating the slope (s) of the electrode function in such calibration solutions and by using one of these solutions for a one-point calibration (E_{cal}, corresponding to the concentration $c_{i,cal}$ of the ion to be measured), the measured electromotive force in the sample (E_S) can be converted to an intracellular sample concentration $c_{S,intra}$ by

$$c_{S,intra} = c_{i,cal}[10(E_S - E_{cal} - E_M)/s] \qquad (2)$$

As shown in Eq. (2), E_S has to be corrected for the cell membrane potential E_M. Usually, the uncertainty in E_M is the dominant source of contributing errors. Calibration for ion activities is simpler since no constant ionic strength is required. The evaluation of the sample activity is the same as described in Eq. (2), i.e., the concentrations have to be replaced by activities. Several activity standards are recommended by the National Bureau of Standards. Activity standards for intracellular Ca^{2+} measurements at submicromolar levels are not available because of the necessity of using Ca^{2+} buffers.

Liquid junction potential, tip potentials, and leakage of reference electrolyte may cause problems in using reference microelectrodes. Even more difficult to evaluate are errors due to cell damage and suspension

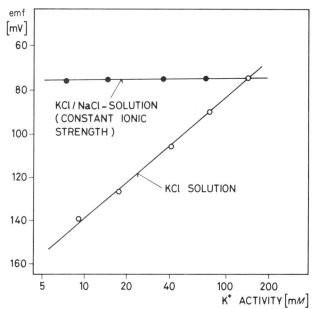

FIG 7. Use of ion-exchanger liquid membrane microelectrodes as reference microelectrodes. The electromotive force (emf) is measured with a cell assembly consisting of a classical 3 M KCl reference microelectrode and a liquid membrane microelectrode of 2% potassium tetrakis(p-chlorophenyl)borate in n-octanol ($R_M \sim 10^{10}$ to 10^{11} Ω). Redrawn from Thomas and Cohen.[21]

effects (for a discussion, see Refs. 3 and 19). It seems that the use of 0.5 M KCl as the reference electrolyte solution is a good compromise to keep all these effects at a tolerable level: changes in the liquid junction potential are still acceptable, cell contamination is clearly reduced when compared with a 3 M KCl solution,[20] and tip potentials should be small. The latter can be further minimized by an appropriate choice of the type of glass, by pretreatment of the micropipette, and by optimization of the filling procedure.[3] Attempts have been made to overcome the difficulties inherent to conventional reference microelectrodes by the use of liquid membrane electrodes as the reference system.[21] The microelectrodes exhibit almost the same selectivity for Na$^+$ and K$^+$ (Fig. 7). Thus, if the sum of the Na$^+$ and K$^+$ activities is constant and exactly the same extra- and intracellularly, the liquid membrane should be useful for measurements of the cell transmembrane potential. The microelectrode would offer clear advan-

[19] I. Tasaki and I. Singer, *Ann. N.Y. Acad. Sci.* **148,** 36 (1968).
[20] M. Fromm and S. G. Schultz, *J. Membr. Biol.* **62,** 239 (1981).
[21] R. C. Thomas and C. J. Cohen, *Pfluegers Arch.* **390,** 96 (1981).

tages: absence of liquid junction potentials and tip potentials, and no cell contamination due to the outflow of reference electrolytes. The relatively rare use (for examples, see Ref. 3) of this alternative reference microelectrode system is probably due to the following disadvantages: strict demands are imposed on the extra- and intracellular Na^+ and K^+ activities, the K_{NaK}^{Pot}-value of the microelectrode has to be exactly 1, other ions should not interfere [Mg^{2+}, Ca^{2+}, organic cations (see Ref. 2)], a considerable leakage of 1-octanol has to be expected, and finally, the microelectrode has a relatively high membrane resistance and is therefore unsuitable for the recording of fast transients.

When using ion-selective liquid membrane microelectrodes, interference by physiologically relevant ions has to be estimated. The degree of interference can be evaluated if the concentrations of the ions to be considered and the selectivity factors of the microelectrodes are known.[22] The performances of the individual electrodes have been described in the original contributions and have been summarized recently.[3] The quality of the cell puncture and the absence of cell damage have to be judged using certain impalement criteria such as instantaneous potential changes upon penetration, low emf drift during intracellular recordings, and reproducible calibrations before and after the impalement. Cell puncture may be improved by the use of small, bevelled micropipettes and by piezoelectric microdrives. Elaborate tests for valid impalements have been proposed[23] (Fig. 8). Thus, for instance, a partial and transient depolarization of the membrane potential is first induced by increasing the extracellular K^+ activity. Both, the intracellular reference microelectrode and the ion-selective microelectrode detect the change in membrane potential. The differential signal, however, remains constant, indicating that the intracellular ion activities have not been changed. In a second testing step, the extracellular activity of Ca^{2+}, the ion to be measured intracellularly, is increased. Stability of the intracellular recording of Ca^{2+} then indicates whether or not influx of Ca^{2+} has occurred.

Figure 9 shows an example of the measurement of a resting intracellular ion activity using a single-barrelled reference microelectrode and a single-barrelled ion-selective microelectrode.[13] After calibration of the cell assembly, a sheep cardiac Purkinje fiber is impaled at (a) with the reference microelectrode and at (b) with a Mg^{2+}-selective microelectrode. E_M is the membrane potential measured with the reference microelectrode

---

[22] P. C. Meier, F. Lanter, D. Ammann, R. A. Steiner, and W. Simon, *Pfluegers Arch.* **393**, 23 (1982).

[23] E. Marban, T. J. Rink, R. W. Tsien, and R. Y. Tsien, *Nature (London)* **286**, 845 (1980).

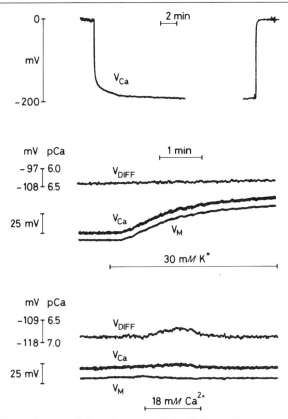

FIG. 8. Evidence for the validity of an intracellular Ca^{2+} activity measurement in a heart muscle cell. The upper trace shows impalement of the cell and withdrawal of the microelectrode after 3 hr of continuous measurement in the single cell. Redrawn from Marban et al.[23]

(3 M KCl) versus a bath reference electrode in the extracellular Tyrode's solution. $E_{Mg} + E_M$ is the potential between the Mg^{2+}-selective electrode and the reference electrode in the Tyrode's solution. The withdrawal of the microelectrode is indicated by (d) and (e). In the range given by (c), the membrane potential E_M is subtracted from the sum $E_{Mg} + E_M$ to get E_{Mg}, which is used for the evaluation of the intracellular Mg^{2+} activity on the basis of the preceding calibration (right-hand ordinate). Obviously, the range between (b) and (d) allows for the execution of a number of experiments while transients of membrane potential and intracellular ion activity are monitored continuously (about 300 examples of such applications have been summarized recently[3]).

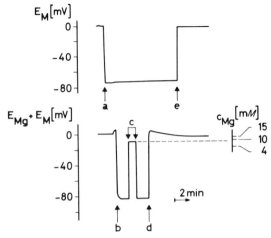

FIG. 9. Example of resting intracellular ion activities measurement. (See text for explanation.)

Preparation of Macroelectrodes

Classical macroelectrodes have diameters of about 10 mm and can therefore exclusively be used for experimental setups with volumes of at least 5 ml. The making and use of these electrodes have been described exhaustively.[24] Although classical macroelectrodes are still used with success, we will concentrate in the following on the making and use of so-called minielectrodes because of their decisive advantageous properties. Thus, small reaction volume (about 0.7 ml), short response times, optimal stirring properties, and the possibility of independent simultaneous recording of up to three ion activities in the reaction vessel make the minielectrode setup into a powerful tool for studying ion-mediated phenomena in in vitro systems. When ion transport is to be recorded in reconstituted systems, the minielectrode setup becomes the instrument of choice where volume limitation is of paramount importance.

Electrode Cell and Membrane Compositions

The Ca^{2+}-selective electrode is used in association with an electrode cell assembly of the following composition: Ag; AgCl, 10 mM CaCl$_2$ | | membrane | | sample solution | | 3 M KCl, AgCl; Ag. The half-cell for the potassium-selective electrode contains 10 mM KCl, that for the

[24] P. C. Meier, D. Ammann, W. E. Morf, and W. Simon, in "Medical and Biological Application of Electrochemical Devices" (J. Koryta, ed.), p. 13. Wiley, New York, 1980.

tetraphenylphosphonium (TPP$^+$)-selective electrode contains 10 mM TPP-Cl instead of 10 mM CaCl$_2$. PVC stock membranes have the following composition: Ca^{2+}-selective membrane: 1.2 mg of neutral carrier ETH 1001, 100 mg o-nitrophenyl-n-octyl ether, 50 mg PVC, 0.5 mg potassium tetrakis(p-chlorophenyl) borate; K$^+$-selective membrane: 2 mg valinomycin, 100 mg sebacic acid bis(2-ethylhexyl) ester, 60 mg PVC; TPP$^+$-selective membrane: 1 mg sodium tetraphenyl borate (NaTPB), 100 mg sebacic acid bis(2-ethylhexyl) ester, 60 mg PVC.

Procedure for Making Electrodes

In some cases, 0.2-mm-thick membranes are commercially available (e.g., Ca^{2+}-selective membranes: W. Möller, Zürich, Switzerland) and can be prepared by dissolving the membrane constituents in tetrahydrofuran followed by evaporation of the solvent within a frame. The PVC is first dissolved into 1.5 ml of distilled tetrahydrofuran, then ion-selective component and membrane solvent are added and the solution is poured into a 20-mm-diameter ring lying on a glass plate. Upon evaporation of the tetrahydrofuran at room temperature, the ~0.2-mm-thick membranes are cut and sealed onto 1- to 3-mm (inner diameter) PVC tubes. Round pieces of membranes with diameters fitting the ones of the PVC tubes are cut out of the parent membrane, which can be stored dry and in the dark for years. Cut membranes are applied onto a glass plate and partially dissolved by the addition of a droplet (2 μl for a 3-mm membrane) of tetrahydrofuran. A PVC tube with a perfectly flat end (razor blade) is then carefully fitted onto the membrane and solvent is evaporated by light aspiration/suction. Finally, overlapping membrane material is removed with a razor blade and degassed reference solution is introduced with a syringe. The minielectrode is completed by plugging the PVC tubing with the ion-selective membrane onto an appropriate connector-containing support.[25] Such a support permits the introduction at the same time of a connected Ag/AgCl wire into the reference solution. The chlorinated Ag wire is prepared as follows: upon cleaning of a 0.5-mm silver wire into concentrated HNO$_3$ and NH$_4$OH, the cleaned wire is connected to the anode of a 1.5 V cell, the cathode of which is connected to a platinum wire. Maximal current is limited to 1 mA while both wires are immersed in 0.1 M HCl for 2 hr.

Reference Electrode

A glass tube (outer diameter: 3 mm) with a ceramic plug at one end and connected to an electrode head at the other end is generally used as

[25] H. Affolter and E. Sigel, *Anal. Biochem.* **97**, 315 (1979).

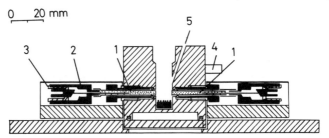

FIG. 10. Minielectrode setup (adapted from Ref. 28). 1, Electrodes [inner reference solution, inner reference half-cell (Ag/AgCl wire), PVC-tube]; 2, electrode head; 3, electric connector; 4, water inlet for temperature control; 5, hole for additions (with microsyringe).

reference electrode. It is positioned analogously to a minielectrode and a chlorinated Ag wire prepared as described above is used as inner reference half-cell. Alternatively, the reference half-cell of a combination pH electrode (Philips, type CA14/02) introduced through the top opening of the vessel (see below) can be used.

Electronics

Electrode potentials are measured by pH meters (e.g., PW 9409/00, Philips) and recorded by two-channel potentiometric recorder (e.g., PE 56, Perkin Elmer). An introduction to the rationale of the detection method is found in Ref. 26. The electrode signals are brought into scale by a bucking voltage device.[27]

Incubation Vessel[25]

Selection of an optimal incubation vessel is an essential factor in the successful application of ion-selective minielectrodes. Such a vessel should allow the simultaneous recording of two to three parameters and should permit the rapid substitution of monitoring electrodes. Essential technical requirements include (1) constant stirring without wobbling to reduce stirring noise, (2) short mixing time, (3) small incubation volume, (4) water jacket for temperature control, (5) optimal electrical shielding, (6) possibility to make additions to the incubation medium, (7) mechanical stability. Such requirements are best satisfied by the setup shown in Fig. 10, which was developed by Affolter and Sigel.[25] The vessel is made of plexiglass, temperature is controlled by an external water bath, and addi-

[26] M. R. Jenkins, *in* "Principles and Techniques of Practical Biochemistry" (B. L. Williams and K. Wilson, eds.), p. 199. Arnold, London, 1975.
[27] V. Madeira, *Biochem. Biophys. Res. Commun.* **64**, 870 (1975).

tions are made either through the top opening or through an appropriate side opening with a microsyringe. Ion-selective electrodes are introduced into the side wall of the vessel through a hollow screw and held in position by an O ring. Optimal stirring is obtained with a round magnetic disk (diameter 10 mm, Radiometer AG, model 912–036). Turbulence is produced by the knobs on top of the stirring disk and the knobs at the bottom of the disk are removed. A close fit of the disk into the vessel is essential to prevent noise-producing lateral movements. The disk is driven by an adjustable magnetic stirrer and mixing times obtained are less than 100 msec. The incubation vessel described in Ref. 25 permits optimal operation with incubation volumes as small as 0.7 ml. Such vessels can be constructed to fit up to three sideways-inserted electrodes plus one electrode inserted from the top.

Use of Macroelectrodes

Calibration, operation, prevention of interferences, and cleaning of classical macroelectrodes and of minielectrodes are essentially identical and will be discussed in the following for the case of the minielectrode setup. Two examples illustrate potential applications.

Calibration and Evaluation

In addition to the presence of interfering ions (whose contributions are expressed by selectivity factors), temperature, stirring, and medium composition all contribute to the potential measured by an ion-selective electrode. It is therefore best to calibrate an electrode under the selected experimental conditions. Simplest and most reliable are relative calibrations during an experiment. Before use, electrodes should be tested for their responsiveness in terms of sensitivity and of response time. Electrodes giving logarithmic activity slopes of less than 90% of the theoretical (Nernst) value should not be used. Typical average response times (time required for 95% of the response to a 10-fold increase in concentration around the central portion of the linear response range of the electrode) for freshly prepared electrodes are K^+ electrode: 0.15 sec (after 6 months, 1 sec); Ca^{2+} electrode: 0.4 sec (after 1 month, 3–5 sec); TPP^+ electrode: 0.4 sec; H^+ electrode (choline-Cl instead of KCl in the reference solution to avoid interferences with K^+ electrodes): 0.15 sec. Average lifetime of the electrodes depends on the frequency of use and varies between about 1 month (Ca^{2+}, TPP^+) and 1 year (K^+).[28]

---

[28] H. Affolter, Ph.D. thesis. Swiss Institute of Technology, Zurich, 1980.

Practical Hints and Troubleshooting

It is essential to be constantly aware of possible interfering factors, as a number of parameters have an effect on the output of ion-selective electrodes. It is best to preincubate solutions at the experimental temperature and to keep volumes of additions during an experiment to a small percentage of the total reaction volume (typically, 0.5–2%). The latter is especially crucial if added substances are dissolved in organic solvents. Every new reagent should be tested for its effect on electrode response by comparing internal calibrations before and after addition of the reagent. If lipid soluble reagents are used, the vessel should be subsequently exposed to excess biological membranes (e.g., liposomes, 1 mg of phospholipid/ml of liposomes suspension) 2–5 times over a period of about 5 min. Such washing has been found to efficiently transfer lipophilic substances from the electrode's liquid membrane into the washing membranes pool. When selecting experimental conditions, it is best to include high concentrations of noninterfering ions in the reaction mixture (e.g., 100 mM choline-Cl; at least 20 mM salt) whenever possible as this minimizes noise and increases stability during measurement. In addition, (1) periodically repeat the chlorination procedure for the Ag wire, (2) check for clogging of the reference electrode's diaphragm, (3) be aware that every liquid membrane electrode will respond to lipophilic ions, e.g., to TPP$^+$, (4) valinomycin and Ca^{2+}-carrier leak out of the membrane according to their partition coefficients between membrane phase and sample; this might interfere with some experiment types, especially when high sensitivities are required.

Example 1: Determination of the H$^+$ and Charge Stoichiometry of Reconstituted Cytochrome Oxidase[29]

The vessel is provided with three electrodes: (1) oxygen electrode to measure enzyme turnover, (2) K$^+$ electrode plus valinomycin to detect charge separation across the liposome membrane, and (3) pH electrode to detect H$^+$ movements across the liposome membrane. Cytochrome-c oxidase vesicles are incubated with 2 μM cytochrome c and 2 μg of valinomycin in a temperature-controlled, air-saturated medium (1.2 ml) containing 100 mM choline-Cl, 0.5 mM KCl, 1 mM choline–HEPES (pH 7.0), 0.1 mM choline–EDTA. The vessel is closed, and calibrations with known amounts of standard HCl and KCl are performed. Potassium is brought to a final concentration of 2 mM and pH is brought to 6.8. After 8 min preincubation, the reaction is started with ascorbate/N,N,N',N'-tetramethylphenylenediamine (TMPD) and changes in electrode potentials are

[29] E. Sigel and E. Carafoli, *Eur. J. Biochem.* **111**, 299 (1980).

continuously recorded over a period of about 30 sec. Initial rates are estimated graphically. Special care is taken to ensure that the pH value and the K^+ concentration of the reductant solution are the same as those of the vesicle suspension. "True" ratios are determined by zero turnover extrapolation of the time dependence of cation to electron ratios.

Example 2: Charge Movements during Na^+/CA^+ Exchange in Heart Sarcolemmal Vesicles[30]

Vesicles (10 μl of a 1 mg of protein/ml suspensions) are preincubated (37°) in either 160 mM NaCl, 20 mM HEPES (pH 7.4) (establishment of a Na^+ gradient) or in 160 mM KCl, 20 mM HEPES (pH 7.4) and then added at time zero to the temperature-equilibrated (37°) vessel solution (0.8 ml) containing 160 mM KCl, 20 mM HEPES (pH 7.4), 10 μM $CaCl_2$. When Ca^{2+} uptake into the vesicles lumen is measured, the vessel is fitted with a Ca^{2+}-selective electrode plus a reference electrode. No uptake is observed in the absence of an outwardly directed Na^+ gradient. To observe charge movements (ratio of 3 or more Na^+ ions per exchanged Ca^{2+}), TPP^+ (10 μM) is included in the reaction vessel, which is fitted with a TPP^+ electrode plus a reference electrode. No TPP^+ uptake is observed in the absence of an outwardly directed Na^+ gradient or in the absence of external Ca^{2+}. In the TPP^+ experiments, vesicles are preincubated in the presence of an amount of TPP^+ selected to give no apparent change in vessel TPP^+ concentration in the absence of net Ca^{2+} uptake. This amount is found empirically. Note that the Ca^{2+} and TPP^+ movements cannot be determined simultaneously in the same vessel since the Ca^{2+}-selective electrode responds to TPP^+.

[30] P. Caroni, L. Reinlib, and E. Carafoli, *Proc. Natl. Acad. Sci. U.S.A.* **77,** 6354 (1980).

[11] Ion Channel-Mediated Fluxes in Membrane Vesicles: Selective Amplification of Isotope Uptake by Electrical Diffusion Potentials

By HAIM GARTY and STEVEN J. D. KARLISH

Introduction

Many different ion-conducting channels have been identified in the cell membrane of higher organisms and their role in inducing transient polarization/depolarization or mediating ion flows is well established.

Functional properties of these proteins are commonly studied in whole cells or phospholipid bilayers using electrophysiological methods.[1-3] However, for isolation and biochemical characterization of channels it is essential to assay the "channel activity," i.e., channel-mediated ion fluxes in defined subcellular systems such as cell membrane vesicles and reconstituted liposomes. But the measurement of isotope fluxes through ion channels in vesicles is complicated by the very high channel "turnover rate" and the small dimensions of membrane vesicles. A typical single-channel conductance is in the order of 1-100 pS.[1-4] Cell membrane vesicles and unilamellar liposomes are usually less than 0.5 μm in diameter. Thus the equilibration time of tracer added to a suspension of vesicles will be in the order of seconds and therefore inaccessible to the manual sampling techniques.[5,6] In addition, the density of ion channels in cell membranes can be as low as 1 μm^{-1}.[3] Therefore, upon isolating plasma membrane and/or forming proteoliposomes with a high lipid/protein ratio, one can expect a heterogeneous population of vesicles where the channel of interest is contained in only a small fraction of the vesicles.

The problem can therefore be summarized as one of having to measure a channel-mediated isotope flux which is (1) very fast, (2) occurs in a small fraction of the vesicles, and (3) must be measured in the presence of many other pathways.

These difficulties may be overcome by imposing an electrical diffusion potential across the vesicle membranes, with the use of ion gradients, and taking advantage of selective permeability characteristics of different vesicle populations to polarize preferentially the vesicles of interest.[7] The result is that isotope is accumulated selectively in the vesicles of interest. Also, due to the large number of isotope molecules taken up, the flux will be prolonged commonly in the time range of minutes. Both of these features, that is selective isotope accumulation and the extended time course, greatly magnify the sensitivity of the flux measurements, and enable one to detect channel-mediated fluxes into minor fractions of heterogeneous vesicle preparations.

Consider measurement of an isotope flux through a cation-specific channel. The diffusion potential can be imposed across the membranes of

---

[1] M. Montal, A. Darszon, and H. G. Schindler, *Rev. Biophys.* **14,** 1 (1981).
[2] B. Sakmann and E. Neher (eds.), "Single-Channel Recordings." Plenum, New York, 1983.
[3] B. Lindemann, *Annu. Rev. Physiol.* **46,** 497 (1984).
[4] R. Latorre and C. J. Miller, *Membr. Biol.* **71,** 11 (1983).
[5] H. Chase and Q. Al-Awqati, *J. Gen. Physiol.* **81,** 643 (1983).
[6] C. Miller, *Annu. Rev. Physiol.* **46,** 549 (1984).
[7] H. Garty, B. Rudy, and S. J. D. Karlish, *J. Biol. Chem.* **258,** 13094 (1983).

those vesicles containing the channel in either of two ways. (A) An outwardly directed gradient of cation which permeates the channel is produced by preparing vesicles to contain that cation and suspending them in a medium containing impermeant ions. Since the membrane potential depends on the relative permeabilities and concentrations of all ions present, only those vesicles which are selectively permeant to the cation of interest, and are impermeant to the other ions present, will be polarized positive outside. (B) Outwardly directed gradients of K^+ or Li^+ are arranged, and the diffusion-potential positive outside is induced with the specific ionophores valinomycin (K^+ gradients) or AS701, a specific Li^+ ionophore[8] (Li^+ gradients). Advantages and disadvantages of procedures A and B are discussed below.

In order to measure the flux, an isotope which permeates the cation channel is added to the exterior, usually in the absence or presence of a selective blocker or activator of the channel. Isotope uptake is detected by taking aliquots of vesicle suspension at time intervals and separating vesicles from their medium (e.g., by filtration, on Dowex columns). Initially, tracer will be accumulated in the control vesicles but will not, or to a lesser extent, in the suspension treated with blocker. During the course of time, the electrical driving force dissipates as a result of progressive collapse of the cation gradients and thus accumulated isotope will flow out of those vesicles into which it had been concentrated. Isotope accumulation will therefore be transient, the height and period of initial accumulation and the rate of the subsequent efflux depending on the relative permeabilities to all the ions present in the system.

Using procedure A above, but not B, transient isotope accumulation can also occur by electroneutral ion exchange via a carrier-mediated countertransport kinetic mechanism. In principle, one can expect the same benefits for assay of a carrier system using countertransport as for the channel mechanism, although this is usually not essential.

Experimental Procedures

Vesicle Preparation

Toad (*Bufo marinus*) bladder membrane vesicles were prepared as described by Garty *et al.*[9] in homogenizing medium that contained 90 mM of either NaCl or KCl, 10 mM EGTA, 5 mM MgCl$_2$, 45 mM sucrose, and 5 mM Tris-Cl (pH 7.8). Crude microsomes from the rabbit renal outer

[8] A. Shanzer, D. Samuel, and R. Korenstein, *J. Am. Chem. Soc.* **105**, 3815 (1983).
[9] H. Garty, E. D. Civan, and M. M. Civan, *J. Membr. Biol.* **87**, 67 (1985).

medulla were prepared as described by Jørgensen,[10] except that the buffer used was 250 mM sucrose, 50 mM KCl, 2 mM MgCl$_2$, 1 mM EGTA, and 10 mM MOPS, and the pH adjusted to 7.2 with Tris.[11]

Transport Assay

The channel-mediated ion flux assay consisted of three stages: (1) establishing an electrical diffusion potential across the vesicle membrane by substituting all external cations by Tris and adding ionophore (if necessary), (2) mixing the vesicles with the radioactive tracer and other reagents such as amiloride and Ba^{2+}, and (3) sampling vesicle aliquots for tracer uptake. Dowex 50W-X8 (50–100 mesh, Tris form) columns were used, both for substituting the external cation by Tris and for separating intra- and extravesicular radioactivity.[12] These columns were poured in Pasteur pipettes plugged with glass wool and prewashed with isotonic sucrose solutions. The columns used to establish the ion gradient were washed with 1.5 ml of sucrose and kept at the assay temperature or at 0°. The columns used to assay the tracer uptake were washed with 1.5 ml of sucrose + 10 mg/ml bovine serum albumin (BSA) (pH 7.5) and kept at 0°. It is advisable to use "crude" BSA for washing the columns, rather than fat-free BSA.

In the first stage, 100–500 μl of the vesicle suspension was eluted through the column with 1–1.5 ml of sucrose. The eluant was mixed with a small volume of buffer and ionophore (if necessary) and immediately distributed among vials containing radioactive tracer and other reagents according to the experimental design in the figure legends.[13] The amount of tracer accumulated in the particles was assayed by taking aliquots from the radioactive suspension and eluting them through the Dowex columns into counting vials with 1.5 ml of an ice-cold sucrose solution.

Amiloride-Blockable Na$^+$ Channels

The apical membrane of several tight epithelia contains Na$^+$-selective amiloride-blockable channels which mediate transepithelial salt reabsorption.[3] In the classical amphibian systems, frog skin, toad bladder etc., the apical membrane conducts Na$^+$ but not K$^+$ and Cl$^-$.[14] The basolateral

[10] P. L. Jørgensen, *Biochim. Biophys. Acta* **356**, 36 (1974).
[11] C. Burnham, S. J. D. Karlish, and P. L. Jørgensen, *Biochim. Biophys. Acta* **821**, 461 (1985).
[12] O. D. Gasko, A. F. Knowles, H. G. Shertzor, E. M. Soulina, and E. Racker, *Anal. Biochem.* **72**, 57 (1976).
[13] Since Dowex substitutes all the cation by protonated Tris and binds many of the common buffers, the eluent pH had to be adjusted.
[14] A. D. C. Macknight, D. R. Dibona, and A. Leaf, *Physiol. Rev.* **60**, 615 (1980).

membranes contain Na^+/K^+ pumps and are permeant to K^+ and Cl^- but not to Na^+.

Amiloride-Blocked Na^+ Uptake into Toad Bladder Membrane Vesicles

Figure 1 shows how the Na^+ channel-mediated $^{22}Na^+$ fluxes can be assayed by the procedures described above using a very crude membrane preparation from toad bladder epithelium. The fraction of vesicles containing the channels in this preparation is estimated to be about 10%. In Fig. 1A $^{22}Na^+$ uptake was measured using procedure A. Cell membrane vesicles containing 90 mM NaCl were prepared as described in the experimental procedure section and Ref. 9. The first step of the assay was to establish the outwardly directed Na^+ gradient by exchanging external Na^+ for Tris ions on a Dowex 50 column. The eluted vesicles were divided,

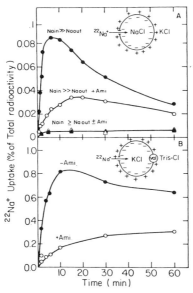

FIG. 1. $^{22}Na^+$ fluxes in toad bladder vesicles. Toad bladder vesicles were prepared to contain 90 mM of either NaCl (A) or KCl (B). Of the vesicle suspensions, 200–400 μl (0.4 mg protein) was eluted through Dowex columns with 1.5 ml 175 mM sucrose, and the pH of the eluent was readjusted to 7.8 with a small volume of Tris buffer. In the KCl-loaded vesicles valinomycin (3 μM) was added. Aliquots were thereupon distributed among vials containing ^{22}NaCl (5 μCi/ml, 0.2–0.4 μM Na^+), 1 μM amiloride (open symbols) or water diluent (closed symbols), and either KCl (11 mM, circles in Fig. 1A) or NaCl (11 mM, triangles in Fig. 1A) or sucrose (175 mM, Fig. 1B). Aliquots, 150 μl (20 μg protein), were removed at different times from the radioactive suspensions, and eluted through Dowex columns into counting vials with 1.5 ml 175 mM sucrose.

and NaCl or KCl (11 mM) was added to either portion. ^{22}Na$^+$ was added to the external medium and the time course of isotope uptake was measured in the absence or presence of amiloride. In the KCl medium, i.e., with a maximal outward Na$^+$ gradient, ^{22}Na$^+$ was taken up linearly for several minutes, uptake passed through a peak at 5 min and the isotope then came out of the vesicles over the course of an hour. The presence of amiloride greatly inhibited the ^{22}Na$^+$ uptake, indicating that this fraction is mediated by the channel. Reduction of the size of the outward Na$^+$ gradient, by an addition of external Na$^+$, greatly reduced ^{22}Na$^+$ uptake and no amiloride sensitivity could be detected.

Comments

1. The difference in results for the KCl or NaCl media demonstrate the dependence of the transient isotope accumulation on the outward Na$^+$ gradient, and the tremendous gain in sensitivity when the maximal Na$^+$ gradient is imposed. This is also the reason why the driving ion gradient was established by passing the vesicles through a column (which removes more than 99.9% of the external Na$^+$ and K$^+$) rather than diluting them into a Na$^+$- or K$^+$-free solution.

2. The presence of KCl in the medium is not coincidental. Since it is known that basolateral membranes are permeant to K$^+$ and Cl$^-$,[14] these vesicles may well be polarized in an opposite sense to apical vesicles, and if anything exclude ^{22}Na$^+$. This reduces the amiloride-insensitive background of ^{22}Na$^+$ uptake. It demonstrates that prior knowledge of permeability characteristics of the different membranes allows one to manipulate the cation compositions so as to polarize vesicle populations of interest selectively.

3. The effectivity of an inhibitor, using procedure A (but not B) to generate the potential, may show the following interesting feature. The inhibitor reduces the channel conductance and thereby also the membrane potential driving tracer uptake. In consequence the inhibitor may appear to be more effective than expected from its binding biotherm and the dose–response curve may not necessarily be hyperbolic. However, computer simulations indicate that these effects are minor (Corcia and Garty, unpublished data).

Figure 1B illustrates the use of procedure B for polarizing the membrane and measuring the channel-mediated ^{22}Na$^+$ flux. In this case, vesicles were prepared to contain 90 mM KCl, external K$^+$ was exchanged for Tris in order to establish an outward K$^+$ gradient and valinomycin was then added to induce the K$^+$ diffusion potential. The curves for ^{22}Na$^+$ uptake are very similar to those in Fig. 1A, demonstrating the Na$^+$ channel is easily assayed using this procedure.

Comment

1. The observation of the valinomycin-induced amiloride-sensitive $^{22}Na^+$ accumulation proves that the $^{22}Na^+$ flux is conductive. The use of ion-specific ionophores to generate the potential is a general method for distinguishing isotope accumulation mediated by conductive or electroneutral exchange pathways, respectively. A disadvantage of this method is that all classes of vesicles may be polarized and hence signal to background isotope fluxes are not expected in general to be as favorable as in method A. However, the high Cl^- permeability of toad bladder basolateral membranes ensures that the membranes are not polarized in the conditions used. Hence in this case the sensitivity of the flux measurement is as good as that using method A.

Kinetic Properties of Amiloride-Blocked $^{22}Na^+$ Uptake

The tracer flux, using our procedure, is a function both of the conductance of the channel to the tracer and the electrical driving force.[7] When inhibitory effects are observed, these can be due either to blocking of the channel or reduction of the driving force (e.g., by permeation of the membrane to other cations and anions). There is no ambiguity when one observes effects of materials, e.g., amiloride, which are known to block the channel from independent electrophysiological observations, but reduction of $^{22}Na^+$ uptake by undefined inhibitors requires one to distinguish between effects on channel conductance and the potential. These possibilities can be distinguished by looking at the rates of both tracer influx and efflux, that is, both rising and falling phases of the full time course of tracer uptake.

Figure 2 shows the time course of the amiloride-sensitive component of $^{22}Na^+$ uptake measured for different compositions of the extravesicular medium. When NaCl-loaded vesicles are suspended in isotonic sucrose/Tris-HCl solution, the channel-mediated flux is relatively large and the internal radioactivity peaks at about $t = 20$ min (upper curve). In the presence of submaximal doses of amiloride (which do not completely block the channel), the initial rate of $^{22}Na^+$ uptake is much lower but the efflux is retarded too. Since amiloride blocked both the rising and falling phases of $^{22}Na^+$ flux there is little doubt that it acts by blocking the channel. The initial rate of tracer uptake could also be reduced by suspending the vesicles in KCl instead of Tris-HCl, but in this case the efflux is accelerated and so the peak of accumulation shifts to shorter times compared to the upper curve. Acceleration of $^{22}Na^+$ efflux cannot be due to blocking a Na^+ channel and the result obviously indicates that the potential is dissipating faster, due to K^+ permeation.

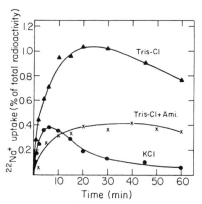

FIG. 2. The time course of the channel-mediated fluxes. $^{22}Na^+$ uptake was measured in vesicles prepared to contain 90 mM NaCl, and the amiloride-blockable fluxes (i.e., the differences between the values obtained in the presence and absence of 200 μM amiloride) are presented. The extravesicular medium was composed of (▲) 4.5% sucrose + 25 mM Tris-Cl, (×) 4.5% sucrose + 25 mM Tris-Cl and 20 μM amiloride, and (●) 4.5% sucrose + 25 mM KCl.

Comments

1. It is possible to distinguish between effects of undefined reagents on channel conductance or the driving potential by looking at the details of both rising and falling phases of the full time course.

2. An alternative and even simpler way of making this distinction is to add the reagent to be tested at the peak of tracer accumulation and observe whether it facilitates or retards trace efflux (see Fig. 3).

Ba^{2+}-Sensitive K^+ Channel in Membranes from Renal Outer Medulla[11,15]

In kidney, transepithelial salt transport across the cells of the thick ascending limbs of the loop of Henle has been studied by electrophysiological techniques using the isolated tubules.[16,17] It is thought to involve the parallel working of a diuretic-sensitive $Na^+/K^+/Cl^-$ cotransporter and a Ba^{2+}-inhibited K^+ channel at the luminal surface and the Na^+/K^+ pump and a net Cl^- conductance at the basolateral surface. Since the red outer medulla is enriched in tubules of the thick ascending limbs, microsomal

[15] C. Burnham, R. Braw, and S. J. D. Karlish, *J. Membr. Biol.* **93**, 177 (1986).
[16] R. Greger and E. Schlatter, *Pfluegers Arch.* **396**, 325 (1983).
[17] S. C. Hebert and T. E. Andreoli, *Am. J. Physiol.* **246**, F745 (1984).

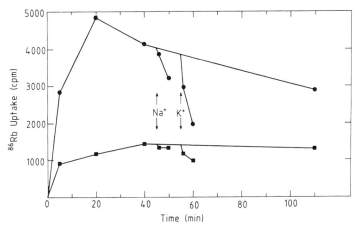

FIG. 3. Ba^{2+}-sensitive $^{86}Rb^{2+}$ uptake in rabbit renal outer medulla microsomes. Two hundred microliters of crude microsomes (6.5 mg protein/ml) was placed on the first Dowex column, and eluted with 750 μl 350 mM sucrose at room temperature. For assay, 250 μl of a reaction mixture was added to the vesicle suspension. The final concentration of components in the reaction mixture was RbCl + $^{86}Rb^{2+}$, 100 μM; MOPS (Tris), 10 mM, pH 7.2; ouabain, 500 μM; vanadate (Tris), 100 μM; $MgCl_2$, 4 mM with (■) or without (●) $BaCl_2$, 10 mM. At the indicated times, 100 μl of the mixture was removed and placed on a second Dowex column, and vesicles eluted with 1.5 ml of ice-cold sucrose solution, 350 mM. Where indicated, 240-μl aliquots of vesicles and reaction medium were added to 60 μl of 250 mM NaCl or KCl solutions (final concentration 50 mM), and then 125-μl aliquots were removed to the Dowex columns at 1 and 5 min.

preparations from this region should be relatively enriched in these transport systems.

Figure 3 shows the time course of $^{86}Rb^+$ uptake into crude outer medulla membrane vesicles loaded with KCl, in the absence or presence of Ba^{2+} ions. In the absence of Ba^{2+} the uptake of the isotope against the opposing K^+ concentration gradient rose to a peak value at about 20 min and then decreased gradually toward the equilibrium value. In the presence of Ba^{2+}, isotope uptake was much reduced and the transient isotope accumulation phenomenon was not observed. These were the features expected for a conductive $^{86}Rb^+$ flux via a Ba^{2+}-inhibited K^+ channel. The maximal level of $^{86}Rb^+$ accumulation was 3- to 4-fold greater than the equilibrium level in the presence of Ba^{2+}, when the opposing chemical gradient K_{in}/Rb_{out} was about 500-fold. Hence specific channel-mediated $^{86}Rb^+$ accumulation was presumably occurring only in a small fraction of the internal vesicle space.

Addition of KCl or NaCl to the microsomes in the absence of Ba^{2+}, after the peak of $^{86}Rb^+$ accumulation, caused a relatively rapid release of the isotope, more so with K^+ than with Na^+. Efflux of the isotope occurs because the electrical potential driving Rb^+ uptake is dissipated and K^+ is more effective than Na^+. The experiment therefore shows that the vesicles of interest are more permeable to K^+ than to Na^+. Other experiments showed that a Ba^{2+}-sensitive Rb^+ uptake could also be detected in Li^+-loaded microsomes suspended in a Li^+-free medium and treated with Li^+ ionophore AS701. Hence the Ba^{2+}-sensitive $^{86}Rb^+$ flux was coupled to the electrical potential, and could not be a K^+/Rb^+ exchange flux by a carrier-type transport system.

Summary and Conclusions

The procedure we have described provides a simple, convenient, and sensitive method to assay conductive ion fluxes in membrane vesicles. It is particularly useful for detecting channels in heterogeneous populations of vesicles. The principal advantages are similar to those of sensitive enzyme assays, namely, screening for existence of channels in different membrane fractions, assaying purified channel proteins, large-scale testing of pharmacological agents, antibodies, etc. and in studies of macroscopic regulatory features, including channel activity or density in different states and interaction with regulatory ligands. In the future one can expect further applications in detecting synthesis of channel proteins, gene expression, etc. The tracer assay does not provide much information on molecular characteristics such as single-channel conductance, voltage sensitivity, and ion specifity. It therefore serves other purposes to those of the modern biophysical methods such as patch-clamp, noise analysis, and study of channels incorporated into bilayers.

[12] Use of Calcium-Regulated Photoproteins as Intracellular Ca^{2+} Indicators[1]

By JOHN R. BLINKS

Photoproteins are self-contained bioluminescent systems in which all of the organic components required for the generation of light are bound

---

[1] In this chapter, frequent reference is made to a chapter in Vol. 57 of this series (J. R. Blinks, P. H. Mattingly, B. R. Jewell, M. van Leeuwen, G. C. Harrer, and D. G. Allen, this series, Vol. 57, p. 292). Readers may wish to have the earlier chapter available for consultation, particularly if they are interested in methods for the isolation, purification, and assay of photoproteins.

together and behave as a single macromolecule (in contrast to "conventional" bioluminescent systems from which discrete luciferins and luciferases can be isolated).[2] This definition says nothing about the biochemical mechanism of luminescence, and the substances that have been termed photoproteins form a functionally diverse group. Here we are concerned only with a subset of that group—those photoproteins in which the rate of the luminescent reaction is influenced by the free calcium ion concentration. These are a group of closely related substances that are responsible for the bioluminescence of a variety of marine organisms (mostly coelenterates). In many of these animals the photoproteins are stored in specialized cells known as photocytes from which light emission takes place, and in which the level of luminescence is controlled by changes in intracellular $[Ca^{2+}]$. In such organisms the photoproteins serve as Ca^{2+}-regulated effector proteins in much the same sense as troponin C or calmodulin, and are referred to as Ca^{2+}-activated photoproteins or Ca^{2+}-regulated photoproteins. In controlling their level of luminescence the photocytes apparently regulate their intracellular $[Ca^{2+}]$ over the same range, and by the same sorts of mechanisms, as many other kinds of cells. A major difference between the photoproteins and other Ca^{2+}-regulated proteins is that the photoproteins perform their physiological function without interacting with any substance other than Ca^{2+}. All of the components required for luminescence (the apoprotein, a low-molecular-weight chromophore, and oxygen) are bound together, and behave as a single 22,000-Da macromolecule. One may, therefore, regard the photoprotein as a conveniently "packaged" system containing all of the ingredients required for the bioluminescent reaction, and in the proper proportions. It is relatively straightforward to isolate and purify the Ca^{2+}-sensitive luminescent system, and to transfer it intact into the cytoplasm of cells not endowed with it as original equipment. This approach has been remarkably successful, and the use of Ca^{2+}-regulated photoproteins as intracellular indicators of ionized calcium is now well established. In the last decade they have been used successfully in more than 50 different types of cells.[3-5]

The easiest photoprotein to obtain from natural sources is aequorin, isolated from jellyfish of the genus *Aequorea*, and the vast majority of the

---

[2] M. J. Cormier, *in* "Bioluminescence in Action" (P. J. Herring, ed.), p. 75. Academic Press, London, 1978.
[3] J. R. Blinks, F. G. Prendergast, and D. G. Allen, *Pharmacol. Rev.* **28**, 1 (1976).
[4] J. R. Blinks, *in* "Techniques in Cellular Physiology" (P. F. Baker, ed.), Vol. P1, Part 2, Chap. P126. Elsevier/North-Holland, Shannon, Ireland, 1982.
[5] J. R. Blinks, W. G. Wier, P. Hess, and F. G. Prendergast, *Prog. Biophys. Mol. Biol.* **40**, 1 (1982).

work done so far with Ca^{2+}-regulated photoproteins has been carried out with that representative of the group. However, at least five other calcium-regulated photoproteins have been described and at least partially characterized. Although it seems clear that these substances are basically very similar, there are significant differences among them, and it may turn out that some of them will be preferable to aequorin for use as Ca^{2+} indicators in certain situations. It is also worth noting that some of the photoproteins, perhaps all, consist of numerous isospecies, and that there are potentially significant differences among the properties of the isospecies of a single photoprotein.[3,6] Although it has not so far been practical to separate the isospecies of aequorin on a preparative scale or to prepare other photoproteins in quantities that would permit their widespread use as biological calcium indicators, it may soon become so. Furthermore, it has been clearly demonstrated that the properties of aequorin can be altered substantially by chemical modification *in vitro*.[7] Thus, while most of what is said about "aequorin" in the body of this chapter refers to the natural mixture of isospecies of that photoprotein, it should be borne in mind that resolved isospecies or chemical modifications of that or other photoproteins may soon be available, and that some of these may be distinctly preferable to native aequorin for certain applications. Photoproteins are not, of course, the only intracellular Ca^{2+} indicators presently available. Four other types of indicators are currently in use: Ca^{2+}-selective microelectrodes, metallochromic dyes, fluorescent indicators, and nuclear magnetic resonance (NMR) indicators. The properties and relative merits of these various indicators have been reviewed in detail in other publications.[8,9,10] Suffice it to say here that none of the methods is ideal, and that for many purposes it may be desirable to use more than one of them. The salient advantage of the photoproteins is that their signals are technically easy to measure and relatively immune to motion artifacts; in this respect they are superior to any other method. In most other important respects, at least one other indicator is as good or better, though no other indicator has the particular balance of properties that the photoproteins have. For example, although the photoproteins permit the detection of fairly low Ca^{2+} concentrations, some of the tetracarboxylate indicators are capable of going even lower, while most of the metallo-

---

[6] O. Shimomura, *Biochem. J.* **234**, 271 (1986).
[7] O. Shimomura and S. Shimomura, *Biochem. J.* **228**, 745 (1985).
[8] P. H. Cobbold and T. J. Rink, *Biochem. J.* **248**, 313 (1987).
[9] J. R. Blinks, *in* "The Heart and Cardiovascular System: Scientific Foundations" (H. A. Fozzard, E. Haber, R. B. Jennings, A. M. Katz, and H. E. Morgan, eds.), p. 671. Raven Press, New York, 1986.
[10] M. V. Thomas, "Techniques in Calcium Research." Academic Press, London, 1982.

chromic dyes are inferior to the photoproteins in this respect. The photoproteins respond moderately quickly to rapid changes of [Ca^{2+}]; in this regard they are not as good as some of the metallochromic dyes, but much better than the Ca^{2+}-selective microelectrodes and the fluorescent indicator quin2. Because of their large molecular size, the photoproteins are probably the most difficult of the Ca^{2+} indicators to introduce into cells: for the same reason, once the photoproteins have been introduced into a cellular compartment, they do not leak out or move to other intracellular compartments, as the tetracarboxylate indicators (particularly fura-2) are prone to do. The photoproteins are difficult enough to work with that their use should not be undertaken casually. However, for certain applications they are clearly still the most satisfactory Ca^{2+} indicators available. This chapter is intended to provide practical details important in their use.

Assay of Photoproteins

General Considerations

A fast and reliable method for measuring the concentration of active photoproteins in solutions is essential in any laboratory where the indicators are used. Repeated measurements of luminescent activity are needed to guide the course of extraction and purification of the protein. It is also important to be able to make frequent checks of the activity of photoprotein solutions as they are used experimentally, and quantitative interpretation of experimental results may depend on knowledge of the exact concentration of active photoprotein in a particular sample. Measurements of protein concentration do not suffice, even when the sample of photoprotein is known to be pure, because they give no indication of the extent to which the protein has been discharged or inactivated since the purification was carried out.

The basic principle in the assay of activity is obvious: one measures the total amount of light emitted when all the photoprotein in a test sample is discharged with Ca^{2+}. Various types of investigator-designed photometers have been described for studies on bioluminescent systems,[11–13] and many of these would be suitable for the assay of photoproteins. In addition, a large number of photometers intended primarily for routine assays

---

[11] G. W. Mitchell and J. W. Hastings, *Anal. Biochem.* **39**, 243 (1971).
[12] J. M. Anderson, G. J. Faini, and J. E. Wampler, this series, Vol. 57, p. 529.
[13] J. E. Wampler and J. C. Gilbert, *in* "Bioluminescence and Chemiluminescence: Instruments and Applications" (K. Van Dyke, ed.), Vol. 1, p. 129. CRC Press, Boca Raton, Florida, 1985.

based on luminescent reactions are now available commercially.[14] While some of these instruments might prove satisfactory for the assay of aequorin, they tend to be rather expensive and to have a lot of "frills" such as digital readouts (which we have tried, and found to be actually less convenient to use than analog meters). Attempts to adapt spectrophotometers or scintillation counters to the purpose have been made, but are not recommended. A good photometer is not difficult to build, and can be constructed for a small fraction of the cost of a spectrophotometer. Essential features are (1) ease and reproducibility of mixing, (2) a wide range of sensitivities, and (3) outputs proportional to both instantaneous light intensity and the total (integrated) light emitted.

For details of the apparatus and procedures used in our laboratory for the assay of aequorin, see our earlier chapter in this series.[1] We use a more elaborate version of the apparatus for the determination of Ca^{2+} concentration–effect curves. It is depicted in Fig. 7,[15] and differs from the assay apparatus principally in that it has provision for temperature control of the reaction chamber. However, it also has a more sensitive photomultiplier, a shutter, and a filter wheel containing three neutral-density absorption filters.

Control of Ca^{2+} Contamination

Although the regeneration of spent aequorin has been accomplished *in vitro*,[16,17] the procedure is not practical for routine use, and for most purposes photoproteins must be regarded as reagents that are consumed in the course of the luminescent reaction. This means that the photoprotein must be rigorously protected from exposure to Ca^{2+} during all stages of its preparation, storage, and handling. During the various steps of extraction and purification this is best accomplished by the presence of a soluble chelator, such as ethylenediaminetetraacetic acid (EDTA). Once the photoprotein has been purified, it usually must be freed of chelators before it can be used experimentally. At this point it becomes critical to take stringent precautions to avoid calcium contamination, since free calcium concentrations as low as 0.1 μM can lead to a fairly rapid loss of luminescent activity, especially when the ionic strength of the solution is

---

[14] K. Van Dyke, in "Bioluminescence and Chemiluminescence: Instruments and Applications" (K. Van Dyke, ed.), Vol. 1, p. 83. CRC Press, Boca Raton, Florida, 1985.

[15] J. R. Blinks, in "Bioluminescence and Chemiluminescence: Instruments and Applications" (K. Van Dyke, ed.), Vol. 2, p. 185. CRC Press, Boca Raton, Florida, 1985.

[16] O. Shimomura and F. H. Johnson, *Nature (London)* **256**, 236 (1975).

[17] D. Prasher, R. O. McCann, and M. J. Cormier, *Biochem. Biophys. Res. Commun.* **126**, 1259 (1985).

low. Calcium leaches out of ordinary glassware no matter how carefully the glass has been cleaned, so it is usually best to use scrupulously cleaned plastic vessels to handle unprotected photoproteins. Water must be distilled at least twice, preferably in quartz, and stored in plastic bottles. Alternatively, ordinary distilled water can be run through a column of chelating resin before it is used. If unprotected photoproteins are to be dissolved in solutions other than distilled water, those solutions must first be passed through columns of a chelating resin such as Chelex 100, because even reagents of the highest grade contain enough calcium to discharge aequorin. Chelex 100 (manufactured and distributed by Bio-Rad Laboratories, Richmond, CA 94804) is similar to Dowex A1 in that it consists of iminodiacetic acid residues covalently linked to a styrene-divinylbenzene copolymer matrix. Methods for the use of Chelex 100 resin were described in detail in Ref. 1.

The affinity of Chelex 100 for Ca^{2+} is not particularly high (binding constant approximately $4.6 \times 10^3 \, M^{-1}$ at pH 7.0),[3] and therefore batch treatment of solutions with Chelex is relatively unsatisfactory as a means of removing contaminating Ca^{2+}. For the same reason, it is important not to stir or repour Chelex columns during use. Chelex that has bound Ca^{2+} must be kept away from the outflow end of the column, and the beads must be rewashed in acid if it is necessary to repour the column. Chelex can be acid-washed and reused more or less indefinitely. EDTA should not be used to remove calcium from Chelex columns because it binds to the Chelex and leaches off slowly thereafter. Obviously, Chelex cannot be used to decalcify solutions that contain Mg^{2+} or other cations with a high affinity for the resin.

Chelating resins consisting of tetracarboxylate chelators bound to polystyrene beads have been prepared recently.[18] The resins prepared to date have had Ca^{2+} affinities much higher than that of Chelex 100, but much lower Ca^{2+}-binding capacities. Because of their low capacity they have not proved as useful as Chelex 100 for the decalcification of large volumes of solutions.

Sources of Photoproteins

One of the major disadvantages presently associated with the use of photoproteins other than aequorin is the difficulty of supply. Aequorin is the only photoprotein so far isolated in quantities sufficient to make its widespread use as a calcium indicator a practical matter. A relatively crude preparation became available commercially in 1975 (through Sigma

---

[18] F. G. Prendergast and A. J. Weaver, personal communication (1987).

Chemical Co., St. Louis, MO 63178). Since 1981 the author's laboratory has made a more highly purified preparation of aequorin available to other investigators at cost (for details, write to the author). Investigators who need other photoproteins or large quantities of aequorin usually collect and purify them themselves. Methods are described in detail in Ref. 1. The purification process described there has worked well, both in our hands and in others,[17,19–21] and has not been changed significantly in the past 10 years. It is not capable, however, of separating the individual isospecies of aequorin, and thus the product is somewhat variable. There are a number of potential improvements with which we are now beginning to experiment in the hope of developing an integrated purification scheme capable of resolving the isospecies of aequorin with acceptable recoveries. The isospecies can be separated by isoelectric focusing and by preparative gel electrophoresis, but both we[5] and Shimomura[6] have despaired of incorporating these steps into a practical preparative scheme. The most obvious potential improvement would be the introduction of high-performance liquid chromatography (HPLC) at some point in the process. Shimomura has recently shown[6] that at least some of the isospecies of aequorin can be separated by HPLC. Chromatofocusing,[22] a relatively new technique that separates proteins by virtue of differences in isoelectric point, might also be expected to be effective in separating the isospecies of aequorin from each other and from other proteins; however, we have not yet had success with the method.

Desalting, Lyophilization, and Storage

As it comes from the purification process, aequorin usually contains about 10 mM EDTA and a substantial concentration of a salt that has been used to elute the protein from an ion-exchange column. These substances must be removed before the photoprotein can be used for most purposes. Furthermore the protein is usually stored for some time, and in some cases shipped before it is used.

Photoproteins in free solution lose their luminescent activity over a period of weeks or months, even in the presence of high concentrations of EDTA. This deterioration can be slowed greatly by (1) freezing the solutions and storing them in an ultracold freezer, (2) lyophilizing the protein,

---

[19] H. Charbonneau, K. A. Walsh, R. O. McCann, F. G. Prendergast, M. J. Cormier, and T. C. Vanaman, *Biochemistry* **24**, 6762 (1985).
[20] B. D. Ray, S. Ho, M. D. Kemple, F. G. Prendergast, and B. D. N. Rao, *Biochemistry* **24**, 4280 (1985).
[21] M. D. Kemple, B. D. Ray, G. K. Jarori, B. D. N. Rao, and F. G. Prendergast, *Biochemistry* **23**, 4383 (1984).
[22] L. A. A. Sluyterman and O. Elgersma, *J. Chromatogr.* **150**, 17 (1978).

or (3) precipitating it with ammonium sulfate. Lyophilized or precipitated photoproteins do deteriorate slowly at room temperature, and should be stored in a freezer, but they are sufficiently stable to be transported without refrigeration.

Photoproteins stored as the ammonium sulfate precipitate must usually be freed of that salt before use. Investigators using photoproteins frequently find it necessary to increase the concentration of a small sample of the protein, or to change the composition of the medium in which it is dissolved. These changes can be readily accomplished by the use of centrifugal microconcentrators such as the Amicon Centricon 10 (Amicon Corporation, Scientific Systems Div., 17 Cherry Hill Drive, Danvers, MA 01923) or the RCF-ConFilt (Bio-Molecular Dynamics, Beaverton, OR 97075). (The latter device allows volumes of solution up to 5 ml to be concentrated down to controlled volumes as small as 25 μl by centrifugation at 1000 g in a device containing ultrafiltration capillaries.) Protein recoveries are excellent, even when very small samples of solution are introduced. Obviously, the concentrated protein can be diluted in a new solution and reconcentrated if necessary. The availability of these devices makes it much less important to be concerned (as we have in the past[1,3]) about developing reliable salt-free preparations of lyophilized aequorin.

Introduction into Cells

One of the more difficult problems associated with the use of photoproteins as intracellular calcium indicators is getting them into the cells of interest. The most obvious, and in many ways the most satisfactory method of doing this is microinjection, but this is technically difficult, time consuming, and incompatible with experimental designs that involve the study of populations of cells in suspension rather than of individual cells or small groups of cells. In the last few years, however, it has become evident that a number of methods for temporarily breaching the integrity of the surface membrane may be used to get photoproteins into cells, apparently without causing irreversible damage.

Microinjection

Some giant cells can be cannulated or fitted with axial perfusion capillaries, and once this has been achieved the introduction of photoproteins is simple enough. Recently developed methods for internally dialyzing cells of ordinary size through patch-clamp electrodes attached with "gigaohm seals" might also be applied,[23] though the electrode would have to be

---

[23] B. Sakmann and E. Neher, *Annu. Rev. Physiol.* **46**, 455 (1984).

removed subsequently or optically shielded to avoid "blinding" the photomultiplier with light from the large amount of photoprotein inside it. Microinjection is usually carried out with pressure, and through conventional glass micropipettes. The methodology of microinjection was reviewed by Meech in 1981[24]; important papers by Kiehart[25] and Cobbold et al.[26] have appeared more recently. Most investigators who have microinjected photoproteins have used gas-filled injection systems rather than liquid-filled ones, probably because it is easier to avoid contaminating the photoprotein with Ca^{2+}. For work with small cells, such as those of cardiac muscle, it is important to use pipettes with the smallest tips possible and therefore to inject under high pressure. The pipettes must be held very firmly to avoid motion when pressure is applied. Pipette holders made for low-pressure work usually grip the pipette with a flexible rubber O-ring or gasket, which may allow too much motion. We use a specially constructed Teflon compression collar that grips the pipette over a length of at least 5 mm. This allows pressures of up to 10 atm to be applied without microscopically detectable motion of the electrode tip (see Figs. 1 and 2).

Because in many cells the only practical way to determine when a cell has been impaled is to monitor the potential sensed by the tip of the pipette, the holder should also have provision for making electrical contact with the solution in the micropipette. The conventional chlorided silver wire will not do because silver rapidly inactivates photoproteins. We use a plain shiny platinum wire, which is of course polarizable, and absolute measurements of resting potentials cannot be made reliably with it, especially over long periods of time.

Another essential piece of equipment for pressure injection is a convenient system to control the pressure applied to the pipette. The operator should be able to turn the pressure on or off at the flick of a switch with decompression of the system when the pressure is turned off, and he should be able to adjust the pressure applied to the pipette while seated at the microscope. Figure 3 is a diagram of a simple and convenient pressure control system employed by Dr. J. E. Brown (illustrated here with his permission). Purely mechanical systems of this sort can be constructed conveniently from miniature components (MTV-3 toggle valves and MAR-1 pressure regulator) available from Clippard Instrument Laboratory, Inc., 7390 Colerain Road, Cincinnati, OH 45239. Electrical control

---

[24] R. W. Meech, in "Techniques in Cellular Physiology" (P. F. Baker, ed.), Vol. P1, Part 1, Chap. P109. Elsevier/North-Holland, Shannon, Ireland, 1981.
[25] D. P. Kiehart, *Methods Cell Biol.* **25**, 13 (1982).
[26] P. H. Cobbold, S. R. Cuthbertson, M. H. Goyns, and V. Rice, *J. Cell Sci.* **61**, 123 (1983).

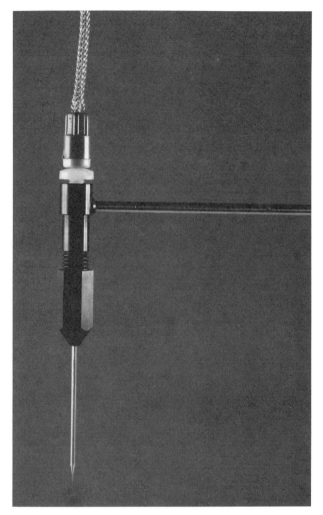

FIG. 1. Pipette holder for microinjection, shown as intact assembly. The holder is clamped by the pressure sidearm at the right.

may be desirable for applying rapid pressure pulses. In this case, it is convenient to use solenoid valves with power requirements low enough so that they can be operated directly from transistorized circuits. Valves of this sort are available from Northeast Fluidics, Inc., Amity Road, Bethany, CT 06525. Convenient systems for applying gas pressure to micropipettes are also available commercially (e.g., from World Precision

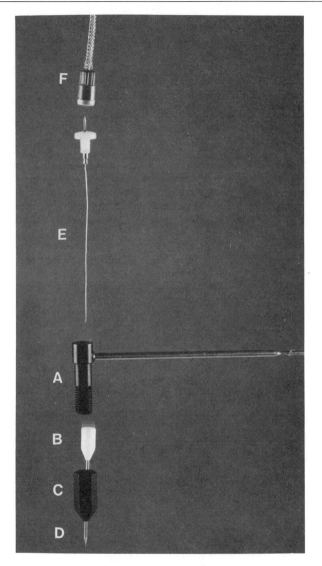

FIG. 2. Pipette holder for microinjection. Component parts of the holder: (A) body with pressure sidearm to which plastic tubing is attached, (B) Teflon packing collar, (C) threaded compression fitting with tapered cavity, (D) pipette, (E) platinum wire attached to pin mounted in nylon screw, (F) electrical connection.

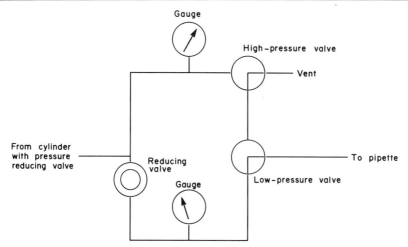

FIG. 3. Pressure system for microinjection. Three-way pressure valves can be either mechanically or electrically operated (see text).

Instruments, Inc., 375 Quinnipiac Ave., New Haven, CT 96513; or General Valve Corporation, P.O. Box 1333, Fairfield, NJ 07006).

The injection pipettes themselves are drawn with a conventional microelectrode puller from borosilicate glass capillary tubing such as Pyrex. (Cobbold has found that aequorin does not survive well in pipettes pulled from soda lime glass.[27]) It is best to fire-polish the ends of the capillary tubing blanks before pulling the electrodes in order to prevent the formation of glass chips and to avoid shaving particles from the Teflon compression collar as the pipette is inserted into its holder. The tubing must, of course, be washed thoroughly before the electrodes are made. We do this by first acid-washing the glass capillaries, then storing them in 10 mM EDTA until they are to be used. Just before the pipettes are pulled, each capillary is rinsed and dried individually by aspirating ultraclean distilled water and then dust-free air through it by means of a fine silicone rubber tube slipped over one end.

Micropipettes can be loaded with photoprotein solution from either the tip or the butt of the pipette. It is neither necessary nor desirable to fill the whole pipette—a small volume in the tip is all that will be injected, and anything more than that will be wasted. If the pipette is to be loaded from the butt, it should be pulled from capillary tubing of the sort that has a fine glass filament fused to the inside (available from Frederick Haer,

[27] P. H. Cobbold, personal communication (1986).

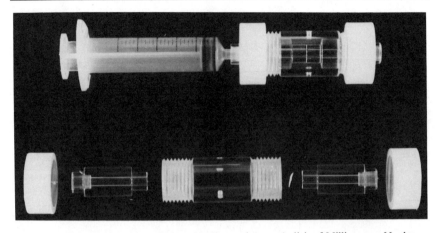

FIG. 4. Filter holder for use with microliter volumes. A disk of Millipore or Nuclepore filter material cut with an ordinary paper punch is soaked in EDTA, and then clamped between the two plastic cylinders. The holes through these cylinders have Luer tapers at their outer ends for the insertion of plastic plugs (not shown) or a syringe. The holes in the middle of the threaded case allow the unit to be washed thoroughly after it is assembled. A drop of aequorin solution is placed on one side of the filter disk and then forced through it by air pressure from the syringe. The filtered solution is removed from the other side with a fine polyethylene tube attached to an Eppendorf pipette.

Inc., P.O. Box 337, Brunswick, ME 04011; World Precision Instruments, Inc., 375 Quinnipiac Ave., New Haven, CT 06515; or Clark Electromedical Instruments, Pangbourne, Reading, UK). The tip of such a pipette will fill by capillarity from a droplet of solution placed anywhere inside the shank. Bubbles in the column of solution usually do not interfere with the function of the pipette because the crevices on either side of the filament fill by capillarity and form conductive channels past the bubbles. The pipette must be kept scrupulously clean and free of dust, and the solution must be filtered before it is loaded into the pipette. Obviously, the filter must remove any particles capable of occluding the tip of the pipette. For this purpose we use a small disk of 0.08 μm Nuclepore filter material in the filter holder illustrated in Fig. 4. The platinum wire contacting the solution must be washed before it is inserted into the pipette. Dipping it successively into a tube of 10 mM EDTA and then into two or three tubes of ultraclean distilled water is a satisfactory method. The wire will usually make satisfactory electrical contact without visibly touching the solution in the electrode tip.

Although relatively few investigators have chosen to load micropipettes from the tip, Cobbold and co-workers have obtained impressive results with this method.[28-32] In their first such studies they loaded their pipettes with a rather elaborate system that permitted them to accelerate filling through the narrow tip by the application of very high pressures to the outside.[26,28] They stated that the use of high pressures eliminated the need to filter the aequorin solution. However, they also reported usually having to break the tip of the micropipette to facilitate filling,[26] and reported that their pipettes had tip bores of about 1 μm.[28] More recently, they have abandoned the pressure system, allowing their pipettes to fill from the tip by capillarity.[27,32] They do not mention having to break the tips of these pipettes, but no information about tip size (or resistance) has been published.

One advantage of filling pipettes from the tip may be that the tips of the pipettes are less likely to plug during injection. Another is that very little photoprotein is wasted. This is particularly important when very high concentrations of aequorin are used, as in the experiments of Cobbold and collaborators. A significant disadvantage of filling pipettes from the tip is that it is very difficult to make electrical contact with the solution in the tip of the pipette. This means that it is not feasible to monitor potential as a means of determining when the pipette has penetrated a cell.

Cobbold and colleagues have used aequorin concentrations as high as 150 mg/ml[8] for their microinjections, and report injecting 0.5–5% of the cell's volume.[8,26] This allowed them to record "resting glows" as well as very bright signals associated with Ca^{2+} transients from individual fibroblasts, hepatocytes, cardiac myocytes, and oocytes. Only a small proportion (3–25%) of the cells survived the larger injections[26,29]; since those that did survive often appeared to function normally for many hours after the injection, it seems likely that the mortality was due to the injection, and not to the presence of aequorin. It seems likely that the high mortality reflects the use of pipettes with rather large tips; another factor contributing to damage during injections may be the speed with which the injection is carried out. (Cobbold uses rather rapid injections.[27])

Pipette tip size inevitably is a compromise between the conflicting goals of facilitating impalement and minimizing cell damage on the one hand and of maintaining pipette patency on the other. The tendency of pipettes to plug tends to increase with the protein concentration of the solution they contain. Without wishing to imply that the combination is

[28] P. H. Cobbold and M. H. Goyns, *Biosci. Rep.* **3**, 79 (1983).
[29] P. H. Cobbold and P. K. Bourne, *Nature (London)* **312**, 444 (1984).
[30] K. S. Cuthbertson and P. H. Cobbold, *Nature (London)* **316**, 541 (1985).
[31] P. H. Cobbold, P. K. Bourne, and S. R. Cuthbertson, *Basic Res. Cardiol. 80* (Suppl. 2), 155 (1985).
[32] N. M. Woods, K. S. Cuthbertson, and P. H. Cobbold, *Nature (London)* **319**, 600 (1986).

necessarily optimal, I offer the information that we are currently using pipettes of tip resistance 10 to 20 MΩ (when filled with 150 mM KCl) in conjunction with a solution of 2 to 10 mg/ml aequorin for work on cardiac muscle. Pipettes of considerably higher resistance have been used,[33] but at a cost of more frequent plugging. If membrane potential is to be monitored as an indicator of penetration, the photoprotein solution must contain a salt to make it conductive. KCl is usually used because K^+ is the most abundant cation in the cytoplasm. The concentration of salt is not crucial—it is probably not important that the injection medium be isotonic with the cytoplasm. Indeed, a case can be made for using the lowest salt concentration compatible with making potential measurements: cells injected with hypotonic solutions would be expected to lose water almost instantly; salts are extruded much more slowly. It follows that if an injection is carried out over a period of seconds to a few minutes, the volume of solution that can be introduced without excessively distending the cell will be inversely related to the osmotic strength of the solution. It is very difficult to estimate or control precisely the amount of solution that has been injected from a pipette during any particular impalement. (Flow rates measured outside the cell cannot be applied to pipettes during injection: flow through the tip of a pipette tends to slow after a cell has been impaled, and frequently stops altogether.) One can only watch for swelling or monitor the membrane potential for depolarization and stop injecting before any major change in cell volume or potential has occurred.

Injection is best carried out with a freshly loaded pipette. When a loaded pipette must be left for an appreciable length of time, it may be wise to leave the tip immersed in distilled water. If the photoprotein in the tip is allowed to dry, the pipette may become permanently plugged; if the tip is left in physiological salt solution, Ca^{2+} will diffuse in and discharge the aequorin in the terminal portion of the pipette. The amount discharged is small, but it is what will next be injected unless the pipette tip is deliberately cleared before the impalement. Pipettes can be cleared and checked for patency by monitoring light output when pressure is applied to the pipette while its tip is immersed in the physiological salt solution bathing the preparation. This maneuver also gives the operator an opportunity to judge how much pressure must be applied to a given pipette to expel aequorin before the pipette is inserted into a cell. When a cell has been impaled with the microelectrode, and a reasonable (and stable) resting membrane potential obtained, pressure is applied to the pipette. The pipette may be less likely to plug if a high pressure is applied in short bursts than if a lower pressure is applied continuously. Injection is discon-

[33] D. G. Allen and J. R. Blinks *Nature (London)* **273**, 509 (1978).

tinued when the recorded membrane potential starts to decline appreciably. (With experience, it is often possible to tell when a pipette has become plugged simply by monitoring the electrical response to the application of pressure.) If multiple cells are to be injected, the pipette is then withdrawn and inserted into another cell. This process is repeated, with occasional checks on the light emission of the preparation in response to stimulation, until an adequate signal intensity is obtained. Pipettes are changed whenever they become plugged or impalement becomes difficult. A dozen or more may be used and discarded in injecting a difficult preparation of cardiac muscle. The number of apparently successful injections required varies greatly with the type of tissue used. It is usually impossible to measure the volume expelled from the pipette into any given cell; it is difficult even to estimate the average amount injected per cell. A rough estimate can be made by measuring the total amount of aequorin in the injected specimen and dividing by the number of attempted injections. The result is inevitably low to the extent that some injections are unsuccessful because of pipette plugging, and others cause such serious damage to the cell that the injected aequorin is discharged by Ca^{2+}.

Alternative Methods

Because of the technical difficulty of microinjection, there have been numerous attempts in recent years to develop alternative methods for introducing photoproteins into cells. Four different methods have been clearly shown to work, and to leave photoprotein-loaded cells that seem to function reasonably normally. All depend on producing reversible damage to the surface membrane while exposing the cells to a solution containing the photoprotein. Not all methods work well with all types of cells.

Obviously, the cells must be incubated in a solution that is essentially Ca^{2+}-free while they are being exposed to the photoprotein, or the photoprotein would be discharged too rapidly to be of any use. The low $[Ca^{2+}]$ in itself may be partly responsible for the increased permeability of the surface membrane achieved by all of the methods, but one of the new techniques apparently depends entirely on reducing the divalent cation concentration. That is the EGTA-loading technique first reported by Sutherland *et al.*,[34] subsequently developed by Morgan and Morgan[35,36] for use in mammalian vascular and cardiac muscle, and then applied to

[34] P. J. Sutherland, D. G. Stephenson, and I. R. Wendt, *Proc. Aust. Physiol. Pharmacol. Soc.* **11**, 160P (1980).
[35] J. P. Morgan and K. G. Morgan, *Pfluegers Arch.* **395**, 75 (1982).
[36] J. P. Morgan, T. T. DeFeo, and K. G. Morgan, *Pfluegers Arch.* **400**, 338 (1984).

suspensions of various kinds of cells by numerous authors.[37–40] In this technique the cells are first made hyperpermeable by incubating them in a solution of very low divalent cation concentration. They are then exposed to the photoprotein, and finally resealed by first increasing the [Mg^{2+}] (to about 10 mM), then gradually increasing the [Ca^{2+}] in the bathing medium to the normal extracellular level. The details of the method are continually being refined; apparently minor variations in technique appear to be important, and optimal conditions appear to be different for different tissues. Those who are considering the use of this method would be well advised to read the latest publications in the field and to consult the authors before trying it themselves.

A second technique, in which the cells are made temporarily leaky by exposing them briefly to hypotonic solutions, was first used by Campbell and Dormer[41] in studies on erythrocyte ghosts, and later refined by Snowdowne and Borle[42] for use in populations of apparently intact renal and hepatic cells. A related method has been described in which cells are first exposed to hypertonic solutions, then returned to solutions of normal or subnormal osmotic strength, all in the presence of the photoprotein.[43] Although in this case the entry of the photoprotein into the cytoplasm was ascribed to osmotic lysis of pinocytotic vesicles,[43] it is not evident to me that the mechanism involved is clearly distinct from that of the hypoosmotic shock technique.

A third technique, recently described by McNeil *et al.*[44,45] depends on the observation that macromolecules gain access to the cytoplasm of cultured cells during almost any maneuver that causes the cells to detach from a substrate to which they have been adhering.[44] Useful amounts of aequorin can be introduced into populations of cultured cells simply by scraping them off the bottom of a culture dish with a rubber "policeman" while they are bathed in an aequorin-containing solution.[45] The method is simple, but is restricted to cultured cells that can be grown as a monolayer.

A fourth technique, which promises to be the simplest of all, involves

[37] R. M. Snider, M. McKinney, C. Forray, and E. Richelson, *Proc. Natl. Acad. Sci. U.S.A.* **81**, 3905 (1984).
[38] P. C. Johnson, J. A. Ware, P. B. Cliveden, M. Smith, A. M. Dvorak, and E. W. Salzman, *J. Biol. Chem.* **260**, 2069 (1985).
[39] M. R. James-Kracke, *Am. J. Physiol.* **251**, C512 (1986).
[40] B. Styrt, P. C. Johnson, and M. S. Klempner, *Tissue Cell* **17**, 793 (1985).
[41] A. K. Campbell and R. L. Dormer, *Biochem. J.* **152**, 255 (1975).
[42] K. W. Snowdowne and A. B. Borle, *Am. J. Physiol.* **247**, C396 (1984).
[43] M. B. Hallett and A. K. Campbell, *Immunology* **50**, 487 (1983).
[44] P. L. McNeil, M. P. McKenna, and D. L. Taylor, *J. Cell Biol.* **101**, 372 (1985).
[45] P. L. McNeil and D. L. Taylor, *Cell Calcium* **6**, 83 (1985).

nothing more than a brief (30 sec) centrifugation (at ~200 g) of suspensions of cells in the presence of a Ca^{2+}-free solution containing aequorin (10–20 μg/ml). The method has been applied successfully to freshly isolated cells as well as to those grown in culture.[46]

In a recent volume of this series, Borle and Snowdowne have summarized in one place their experience with the hypoosmic shock, scrape-loading, and centrifugation methods, and have illustrated some modifications in equipment and technique that they have adopted for use with suspensions of cells.[47] Although I am not yet aware of any examples of its having been used successfully for the purpose, it seems reasonable to suppose that the temporary membrane disruption produced by the brief application of high-voltage electrical fields might well be developed into a technique for introducing photoproteins into populations of dissociated cells (see Refs. 48 and 49). Liposomes containing photoproteins offer another seemingly attractive approach; while some initial results with adipocytes were interpreted as giving encouraging results, the investigators involved later concluded that although some photoprotein might have entered the cells, it probably did not have access to the cytoplasmic space.[50] Several other (unpublished) attempts to introduce photoproteins into cells by the liposome technique have been unsuccessful.

Detection of Light Signals

Because they are consumed in the light-yielding reaction with Ca^{2+}, photoproteins are most useful under conditions in which their reaction rate is low. Fortunately, this is the case within the range of Ca^{2+} concentrations normally encountered in most kinds of cells. Photoproteins are well-suited to, and are frequently used for, the measurement of [Ca^{2+}] in small volumes, such as single cells. Like all other indicators, they should be used in the lowest concentrations feasible because they bind Ca^{2+}, and may possibly have other undesirable effects that are concentration-related. For all of these reasons, it is important to be able to detect and quantify exceedingly low light levels in work with the photoproteins.

---

[46] A. B. Borle, C. C. Freudenrich, and K. W. Snowdowne, *Am. J. Physiol.* **251,** C323 (1986).
[47] A. B. Borle and K. W. Snowdowne, this series, Vol. 124, p. 90.
[48] D. E. Knight, *in* "Techniques in Cellular Physiology" (P. F. Baker, ed.), Vol. P1, Part 1, Chap. P113. Elsevier/North-Holland, Shannon, Ireland, 1981.
[49] D. E. Knight and M. C. Scrutton, *Biochem. J.* **234,** 497 (1986).
[50] A. K. Campbell, M. B. Hallett, R. A. Daw, M. E. T. Ryall, R. C. Hart, and P. J. Herring, *in* "Bioluminescence and Chemiluminescence: Basic Chemistry and Analytical Applications" (M. A. DeLuca and W. D. McElroy, eds.), p. 601. Academic Press, New York, 1981.

Fortunately, photomultipliers and image intensifiers capable of detecting very small numbers of photons are readily available. Photomultipliers are so widely employed that general aspects of their use need not be dealt with here. (For example, brief, readable accounts are available in the catalogs published by Thorn-EMI and by Hamamatsu; a more detailed treatment has been published in book form by RCA.[51]) A few special points specific to their use with photoproteins do deserve mention, however.

For optimum sensitivity, the photocathode should have a high quantum efficiency in the blue (λ_{max} for aequorin = 469 nm), and for low dark current it should have low sensitivity in the red and infrared portions of the spectrum. Bialkali cathodes meet these requirements reasonably well. In setting up a photomultiplier for work on cells injected with photoproteins, one is almost inevitably faced with the need to compromise between sensitivity and convenience. To keep the dark current of the photomultiplier low, it is important not to expose the photocathode to bright light, even when the tube is not under power. This usually means that a shutter is needed. However, for optimum sensitivity the photomultiplier should be mounted as close to the preparation as possible and be separated from it by the minimum possible number of reflecting interfaces. These are clearly conflicting goals. Large photocathodes make it easier to gather light, but also increase the dark current of the photomultiplier. The dark current can be reduced by cooling the photomultiplier, but the improvement in performance is much less in the case of tubes with bialkali photocathodes than with some other types. Overcooling can decrease the quantum efficiency of bialkali photocathodes.

Lenses, fiber-optic probes, and acrylic plastic light guides make it possible to increase considerably the distance between the preparation and the photocathode without intolerable losses of signal intensity. Water-immersion microscope objectives of numerical aperture as high as 1.0 have been used to record light from aequorin-injected cells, and salt-water immersion lenses of NA >1.0 are available (from E. Leitz, Inc.). The light-gathering ability of such lenses is higher than that of glass fiber-optic probes, which usually have a numerical aperature of about 0.5 and accept only about 70% of the incident light within the cone defined by that aperture. Transmission losses in conventional fiber-optic probes are relatively slight, about 0.3%/cm. The losses in microscope lens systems are variable, but may be expected to increase with the number of optical elements in the light path. An advantage of lens systems is that the gathered light can be focused to a small spot. This makes it possible to use

[51] RCA Corporation, "RCA Photomultiplier Manual." RCA Corporation, Harrison, New Jersey, 1970.

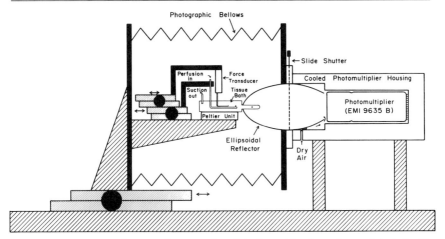

FIG. 5. Apparatus for recording very low light intensities from tissues injected with photoproteins. The heart of the apparatus is the ellipsoidal reflector, which has one point of focus in a glass tube projecting from the tissue bath, and the other at the photocathode of the photomultiplier. The tissue is suspended from two hooks, which can be advanced together into the glass tube by means of a micromanipulator movement. A photographic bellows can be closed to make the apparatus completely light-tight. A slow flow of dry air keeps moisture from accumulating on the photomultiplier or dynode chain under humid conditions or when the photomultiplier is cooled. From Ref. 4, with permission of the publisher.

photomultipliers with small photocathodes and correspondingly low dark currents. An additional advantage of recording light through the microscope is that by visual inspection one can readily define the area from which light is being recorded. Although they do not provide the mechanical flexibility of fiber-optic probes, excellent light guides can be made out of polished rods of acrylic plastic. In our experience their efficiency is considerably higher than that of fiber-optic guides. Mirrors can often be used to increase the fraction of emitted light that reaches the photomultiplier.

For applications in which a high efficiency of light gathering is required, we use various versions of the apparatus illustrated in Fig. 5, which incorporates an ellipsoidal reflector that directs a large fraction of the emitted light to the photocathode and still permits the use of a shutter. The variant of the apparatus pictured here is particularly well suited for use with single skeletal muscle fibers. We have constructed two others, one for use with preparations of intact cardiac muscle[52] (Fig. 6) and one for suspensions of cells, in which the axis of the reflector and photomulti-

---

[52] J. R. Blinks, in "Methods in Studying Cardiac Membranes" (N. S. Dhalla, ed.), Vol. 2, p. 237. CRC Press, Boca Raton, Florida, 1984.

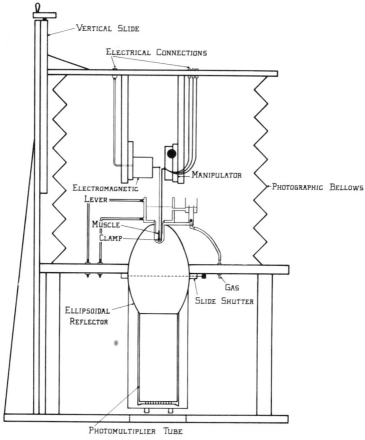

FIG. 6. Experimental setup used in our laboratory for recording light from aequorin-injected heart muscle. The muscle is mounted at the upper focus of the ellipsoid, which is in a glass tube projecting downward from the temperature-controlled muscle bath. The physiological salt solution is oxygenated in the chamber at the right of the bath, and circulated rapidly past the tissue by the pumping action of the gas stream. The muscle is held at the bottom by a miniature clamp containing stimulating electrodes and is tied at the top to a servo-controlled electromagnetic lever which senses (and controls) both force and length. A photographic bellows can be closed to make the apparatus completely light-tight. The muscle is positioned precisely at the focal point of the reflector by means of the vertical slide at the left. From Ref. 52, with permission of the publisher.

plier is vertical. Snowdowne and Borle have published a detailed description of an apparatus of the latter type in another volume of this series,[47] and also describe several methods that they have developed to keep cells in place as they are superfused.

The apparatus described by Allen and Kurihara[53] for use with heart muscle is optically very efficient, but has the disadvantage of not incorporating a shutter. A more recent version does incorporate a glycerol-filled shutter in the acrylic plastic light guide.[54] By having a thin gap between the two parts of the light guide to accommodate a shutter blade and by filling the gap with fluid having a refractive index equal to that of the plastic in the light guide, one can "fool" most of the light into passing through the shutter as if no gap were present.

The output of the photomultiplier can be processed in either digital (i.e., by photon counting) or analog form (i.e., by measurement of anode current). Photon counting is becoming increasingly widespread in photometry, but the equipment required is considerably more expensive than that for simple current recording. Photon counting certainly has "high-tech appeal" and probably has real advantages when very low light levels are to be compared over long periods of time or when signals will later be processed in digital form. On the other hand, it has distinct disadvantages as well. Two of these derive from the fact that multiple photons arriving within the particular dead-time characteristic of the system will be counted as one. This means that radioactive light sources cannot be used in a simple way to calibrate the system (each scintillation consists of many photons emitted nearly synchronously). It also means that corrections for photon coincidence must be applied when the system is used to measure relatively high light intensities. In analog systems one can reduce the voltage across the dynode chain at such times in order to avoid amplifier saturation or nonlinear performance of the photomultiplier and dynode chain. Other disadvantages of photon counting are the greater complexity and expense of the equipment required and the need to use photomultiplier housings especially shielded against radio-frequency interference.

Light signals from small cells injected with photoproteins are often very weak, which means that the signal of interest may be buried in background noise from other sources. If the signal is a repetitive one, signal averaging can sometimes be used to extract it. This approach has proved virtually essential in work with aequorin-injected heart muscle. There the light signals corresponding to individual Ca^{2+} transients are not always clearly discernible above the dark noise of the photomultiplier and the "resting glow" of the muscle. Even when they are, signal averaging is usually required to get a smooth record of the time course of the calcium transient. This would be true even if there were no dark current or resting glow at all, because of "shot noise" in the photomultiplier output. Shot

---

[53] D. G. Allen and S. Kurihara, *J. Physiol. (London)* **327**, 79 (1982).
[54] M. B. Cannell and D. G. Allen, *Pfluegers Arch.* **398**, 165 (1983).

noise is unlike the noise in most electrical measurements because it is inherent to the signal itself. It is an unavoidable consequence of the physical nature of light—of the fact that light consists of discrete photons and that weak light signals do not contain very many of them. The "noisy" photomultiplier record is a faithful representation of the real light signal, and its time course is greatly distorted by electronic filtering heavy enough to smooth the record. What is needed is a record of the time course of the statistical probability of the arrival of a photon. That can be provided by signal averaging if the signal is a repetitive one that may be assumed to be of fundamentally constant character for the duration of the averaging period. Signal averaging can be carried out only when the sweep of the averager can be synchronized with the signal, and for the most part (see Ref. 55 for an exception), this limits its use to electrically driven preparations. Depending on light intensity, anywhere from eight to several hundred successive signals from cardiac muscle may have to be averaged to give a satisfactorily smooth signal. Signal averaging can be carried out with appropriately programmed general-purpose computers or by special-purpose signal processors, of which there are now many on the market (e.g., from Nicolet Instrument Corp., 5225 Verona Road, Madison, WI 53711).

One of the advantages that photoproteins share with fluorescent indicators such as fura-2 is the readiness with which they lend themselves to the localization of Ca^{2+} within the system of interest. Because of the nature of the signal from the photoprotein (light emitted against a background of total darkness), microscopic image intensification can be applied to detect the spatial distribution as well as the time course of changes in Ca^{2+} within a preparation. The steepness of the Ca^{2+} concentration–effect curve (see below) is an advantage in this application, because it tends to increase the contrast between regions of high and low $[Ca^{2+}]$.

Commercial television systems intended for nighttime surveillance have been used successfully in some types of work with aequorin,[56] but we have found them insufficiently sensitive for most purposes. Methods of using more powerful image intensifiers for the localization of weak light sources through the microscope have been reviewed recently.[57,58] It should be emphasized, however, that this is a field of rapidly advancing

[55] A. Fabiato and A. O. Wist, *Am. J. Physiol.* **242**, H291 (1982).
[56] B. Rose and W. R. Loewenstein, *J. Membr. Biol.* **28**, 87 (1976).
[57] G. T. Reynolds and D. L. Taylor, *BioScience* **30**, 586 (1980).
[58] G. R. Bright and D. L. Taylor, *in* "Applications of Fluorescence in the Biomedical Sciences" (D. L. Taylor, A. S. Waggoner, F. Lanni, R. F. Murphy, and R. R. Birge, eds.), p. 257. Liss, New York, 1986.

technology, and an investigator who plans to set up an image intensification system would be wise to determine the best way of meeting his particular needs with currently available equipment, rather than to copy a particular setup described in the literature. Light signals from aequorin-injected cells must be fairly strong, as judged by photomultiplier records from the same preparation, before useful images can be obtained by simple image intensification. The main reason for this is that the light used by a microscope objective to form an image is collected either over a small solid angle (low-power objectives) or from a small volume of cytoplasm (high-power objectives), with the result that only a small fraction of the light emitted by the preparation contributes to the image. The use of reflectors, so helpful in the case of photomultiplier recording, is actually destructive in work with image intensification because it introduces a spurious diffuse elevation of the signal. Light arising from points above or below the plane of focus has a similar effect, so it is advantageous to record from thin portions of individually injected cells and to minimize the reflection of light into the microscope objective.

For the reasons just discussed, images recorded with high-gain image intensification from cells containing photoproteins tend to be formed from small numbers of photons, and are therefore inherently noisy. It seems likely that their quality will be greatly enhanced by the application of digital image processing techniques which permit the addition, subtraction, or averaging of images obtained at various times.[59] Such methods should make it possible to do for image intensification what digital signal averaging has done for photomultiplier recording.

Determination of Ca^{2+} Concentration–Effect Curves

Precise Ca^{2+} concentration–effect curves are important both to the development of an understanding of the mechanism of the reactions of photoproteins with Ca^{2+} and to the translation of light signals into absolute Ca^{2+} concentrations. The reaction should be carried out under conditions in which the binding of Ca^{2+} to the photoprotein never appreciably alters the Ca^{2+} concentration of the test medium. That is, there is always a substantial molar excess of Ca^{2+} over photoprotein, even at the lowest Ca^{2+} concentrations. Our usual method involves the rapid injection (by means of a Hamilton CR 700-20 spring-loaded syringe) of 10-μl aliquots of a dilute (<1 μM) solution of EDTA-free photoprotein into 1-ml volumes of the test solutions (for apparatus see Fig. 7). The peak light intensity is recorded, as is (in mid-to-high [Ca^{2+}]) the time course of decline of light

---

[59] J. E. Wampler, in "Bioluminescence and Chemiluminescence: Instruments and Applications" (K. Van Dyke, ed.), Vol. 2, p. 123. CRC Press, Boca Raton, Florida, 1985.

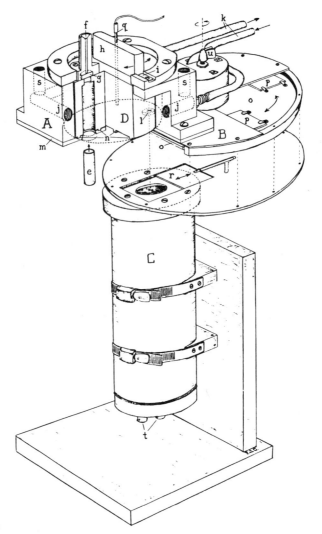

FIG. 7. Temperature-controlled reaction unit used for the determination of calcium concentration–effect curves. The apparatus is a more complex version of the assay apparatus illustrated in Ref. 1, differing primarily in that it has provision for temperature control of the sample, a filter wheel containing neutral density filters, and a more sensitive photomultiplier. Key: (A) temperature-controlled housing for cuvette holder, (B) housing for filter wheel, (C) photomultiplier housing, (D) cuvette rotor, (e) cuvette, (f) needle guide for sample injection, (g) rubber septum, (h) rotor handle, (i) detent for rotor, (j) channel for temperature-controlled water, (k) water tubing to temperature controller, (l) window in plastic bearing plate, (m) black plastic bearing plate, (n) concentric lightseals on bottom of rotor (mate with grooves in bearing plate), (o) filter wheel, (p) neutral density filters, (q) thermistor probe from temperature regulator, (r) shutter, (s) cuvette holders for preequilibrating cuvettes to temperature of housing, (t) photomultiplier cable connections, (u) handle for filter wheel. From Ref. 15, with permission of the publisher.

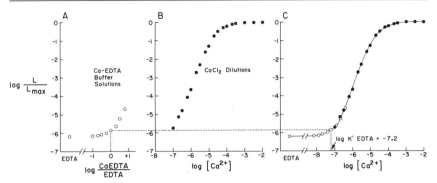

FIG. 8. Determination of a calcium concentration–effect curve for aequorin. Shown are double logarithmic plots; L/L_{max} indicates light intensity as a fraction of that achieved in saturating [Ca^{2+}]. (A) Experimental points determined with a series of Ca-EDTA (ethylenediaminetetraacetic acid) buffers (○). The point marked EDTA was determined in the EDTA solution used for the calcium buffers with no added Ca^{2+}. (B) Points determined with simple dilutions of $CaCl_2$ (●). In C the Ca^{2+}-buffer points have been shifted as a group to give the best fit to the points determined with $CaCl_2$ dilutions. This has the effect of establishing the Ca-EDTA binding constant appropriate to the particular conditions of the experiment, and the two sets of points together constitute a complete concentration–effect curve. All solutions in this experiment contained 150 mM KCl, 5 mM PIPES [piperazine-N,N'-bis(2-ethanesulfonic acid)], pH 7.0, that had been run through a column of chelating resin. No Mg^{2+} was present. The [EDTA] was 1 mM in the calcium-buffered solutions. All determinations were made at 21°.

intensity (reflecting the consumption of aequorin) and the time integral of the light signal (i.e., the total amount of light emitted). Full concentration–effect curves are generally determined with the combined use of simple dilutions of $CaCl_2$ for the higher concentrations and calcium buffers for the lower ones. This permits the determination of a curve covering a wider range of [Ca^{2+}] than would be possible with dilutions alone or one buffer system alone. The use of Ca^{2+} buffers not only makes it possible to extend the curve below the point at which it would be difficult or impossible to control Ca^{2+} contamination without the buffer, but also makes it possible to use concentrations of the photoprotein that otherwise might bind enough Ca^{2+} to have a significant impact on the free [Ca^{2+}] at the lower end of the curve. Furthermore, if the segments of the curve determined with $CaCl_2$ and with a buffer can be made to overlap, they provide a precise measure of the apparent binding constant of the chelator under the exact conditions of the experiment. The general method for making such measurements is illustrated in Fig. 8.

Although in principle the determination of Ca^{2+} concentration–effect curves is very simple, accuracy requires close attention to a number of practical details:

1. The photometer. Essential features are (a) high sensitivity combined with the ability to measure light intensities accurately over a range of at least seven orders of magnitude; (b) provision for injecting photoprotein into the reaction cuvette while light is being monitored without interruption; (c) electronics that provide signals proportional to both instantaneous light intensity and the total amount of light emitted; and (d) provision for temperature control of the reaction mixture. The comments made earlier (in the section on Assay of Photoproteins) about photometers apply here as well. However, it is important in this context to emphasize the absolute necessity of the first point in the list of features just above. Because a full Ca^{2+} concentration–effect curve typically spans more than a million-fold range of light intensities (see Figs. 8–11), it is absolutely essential that the photometer respond linearly to changes in light intensity over such ranges. The investigator must satisfy himself that this is so with his apparatus under the conditions of his experiments, and not rely on the claims of the manufacturer or designer of the apparatus. The photometer illustrated in Fig. 7 was designed specifically for use in determining Ca^{2+} concentration–effect curves for photoproteins, and performs very well. However, it should be noted that the selection and calibration of the neutral density filters is absolutely critical. Filters

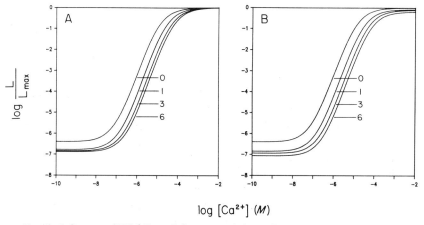

FIG. 9. Influence of $[Mg^{2+}]$ on Ca^{2+} concentration–effect curves for aequorin. Curves were determined as illustrated in Fig. 8, except that 1 mM EGTA [ethyleneglycol-bis-(β-aminoethyl ether)-N,N,N',N'-tetraacetic acid] was used as the calcium buffer. All curves were determined in 150 mM KCl, 5 mM PIPES buffer, pH 7.0, at 20°. In each case the aequorin was preequilibrated with 150 mM KCl, 5 mM PIPES, pH 7.0, containing the same concentration of $MgCl_2$ as the solutions with which it was subsequently to be mixed. The numbers beside the curves indicate the concentration of $MgCl_2$ (mM) in the reaction mixture. In A the results have been plotted with all light intensities expressed as fractions of the L_{max} determined in the corresponding $[Mg^{2+}]$. In B the same results have been replotted with all light intensities expressed as fractions of the L_{max} determined in the absence of Mg^{2+}.

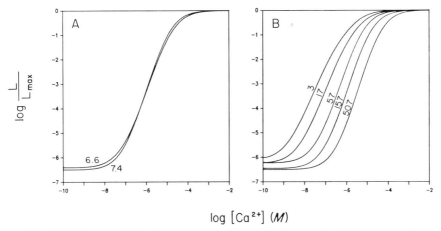

log [Ca^{2+}] (M)

FIG. 10. Influence of pH and monovalent salt concentration on the Ca^{2+} concentration–effect curve for aequorin. Curves were determined as in Fig. 8 except that both 1 mM EGTA and 1 mM CDTA (1,2-cyclohexylenedinitrilotetraacetic acid) were used as calcium buffers. Temperature, 21°. (A) Curves determined at pH 6.6 and 7.4; 150 mM KCl, 5 mM PIPES buffer. (B) The numbers beside the curves indicate the [K$^+$] in the test solution (mM). For the curve in 3 mM K$^+$, 1 mM PIPES was used as the pH buffer; in all other curves the concentration of PIPES was 5 mM (pH 7.0). The balance of the salt concentration was made up with KCl.

should be calibrated with the photoprotein as a light source, and should have the same apparent optical density over a wide range of light intensities. Reflective (metal film) filters should be avoided because of problems deriving from multiple reflections.

2. Clean solutions. Most reagents are contaminated with enough calcium to cause major problems in the determination of Ca^{2+} concentration–effect curves. Therefore, no matter how good the distilled water and in spite of all other precautions to avoid Ca^{2+} contamination, stock solutions of all the reagents to be used (except Ca^{2+}, Mg^{2+}, and chelating agents) must be run through chelating columns before use (see section on Control of Ca^{2+} Contamination, above) and stored in plastic containers. It should also be emphasized that Ag$^+$ contamination must be scrupulously avoided because traces of Ag$^+$ increase the Ca^{2+}-independent luminescence of aequorin and cause rapid loss of luminescent activity. In this connection it is pertinent to call the reader's attention to a potential hazard related to the use of pH electrodes. Even if a solution is destined for subsequent Chelex treatment, pH electrodes must not be put directly into it because the reference electrodes usually contain silver, and Chelex 100 has a low affinity for Ag$^+$. Instead, measurements of pH should be made in small samples of the solution that will be discarded afterward. Silver

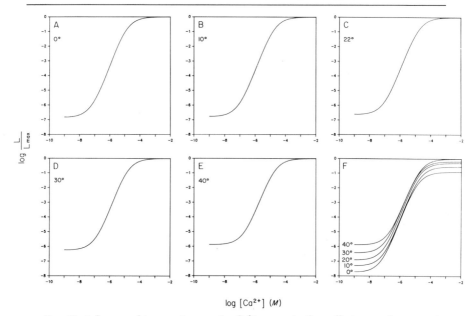

FIG. 11. Influence of temperature on the Ca^{2+} concentration–effect curve for aequorin. Curves were determined as in Fig. 8 except that 1 mM EGTA was used as the calcium buffer. In A–E the light intensities are expressed as fractions of L_{max} determined at the given temperature. In F, the same curves have been replotted with all light intensities expressed as fractions of L_{max} determined at 40°.

contamination can be checked for by adding CN^-, which binds Ag^+ very tightly and eliminates Ag^+-induced luminescence. (The cyanide solution must be Chelex-treated, however, or added in the presence of a chelator such as EGTA or EDTA.) When low calcium concentrations are used, the aequorin must be added to the test solution before the cuvette is put into the photometer. This avoids contamination that may be introduced into the tip of the needle when it is passed through the rubber septum in the injection port of the photometer. Fortunately, this contamination is not a problem when the [Ca^{2+}] is so high that the light intensity decays appreciably in the time required to insert the cuvette into the photometer.

3. *Calcium buffers.* With the use of Chelex-treated solutions, it is fairly easy to control Ca^{2+} concentrations down to about 10^{-6} M and possible to go as low as 10^{-7} M without the use of calcium buffers. Below this level, problems of Ca^{2+} contamination usually require that calcium buffer systems be used. Calcium buffers present numerous traps for the unwary, and their use should not be undertaken without a thorough understanding of the principles involved. A very readable account of these

principles can be found in Chapter 2 of Ref. 10. Recipes have been published for making a series of calcium buffers for the determination of aequorin calibration curves,[47] but they are mentioned here only to warn potential users against following them blindly. Some salient points with respect to the use of calcium buffers are as follows.

(a) Calcium buffers are similar in principle to pH buffers: a chelator plays a role analogous to that of a weak acid in a pH buffer system, and Ca^{2+} takes the place of the hydrogen ion. The apparent dissociation constant of a Ca^{2+}-chelator complex is defined as $K_D = [Ca^{2+}]$[free chelator]/ [Ca-chelator]. Therefore, $[Ca^{2+}] = K_D$ times the ratio of [Ca-chelator] to [free chelator] (frequently referred to as the buffer ratio). A convenient way of making calcium buffers of various $[Ca^{2+}]$ is to make up stock solutions that are as close to identical as possible in all respects except that one contains free chelator and the other contains exactly the same concentration of Ca-chelator. These can then be mixed in various ratios to get solutions with various buffer ratios.

(b) The control of pH in calcium buffer solutions is absolutely critical, because the apparent dissociation constants of most of the common chelators used to control $[Ca^{2+}]$ are strongly influenced by pH. (Within the physiological range, the log of the apparent dissociation constant for the Ca^{2+}-EGTA complex changes by a factor almost exactly twice the change in pH: thus, to keep the $[Ca^{2+}]$ of Ca-EGTA buffer solutions within 0.1 pCa unit of a particular level, one must keep the pH from changing more than 0.05 pH unit.) Protons are released when most chelators bind Ca^{2+}, and therefore it should not come as a surprise that when one prepares calcium buffers by mixing stock solutions that are equimolar in Ca-chelator and free chelator and identical in pH, the mixture will have a pH lower than that of the stock solutions. (This is especially true if there is a slight molar excess of Ca^{2+} over chelator in the Ca-chelator stock solution.) The stock solutions must be heavily pH-buffered to minimize the pH change on mixing, and for precise results one should check and adjust the pH of each buffer mixture, even though both stock solutions are known to have precisely the pH intended. It is also a good practice to check the pH of calcium buffer solutions again after they have been used in an experiment to make sure that the pH has not been altered beyond acceptable limits. Alterations may occur, of course, simply as a result of fluxes of hydrogen ions into or out of the system, in which case the $[Ca^{2+}]$ will change because of the influence of pH on the apparent dissociation constant of the chelator. However, pH changes may also result from changes in the amount of calcium bound to the chelator, and in this case there will be two reasons for unsatisfactory control of $[Ca^{2+}]$: a change in the buffer ratio and a change in the apparent dissociation constant of the chelator. Unfortunately, both changes will displace the [Ca^{2+}] in the same direction. That is, the addition of calcium to a calcium buffer solution will displace protons from the chelator, thus acidifying the solution and decreasing the affinity of the chelator for calcium; the extra calcium will also increase the free [Ca^{2+}] by increasing the ratio of Ca–chelator to free chelator. Finally, investigators using calcium buffers should be aware of recent reports that compact combination pH electrodes may give systematic errors if the salt concentration of the test solution is substantially different from that of the calibration buffer.[60]

(c) The apparent association constant for a chelator at a particular pH can readily be calculated, but until recently most of the calculations in the biological literature were seriously in error because of failure to recognize the need to use "mixed" association constants in conjunction with measurements of pH. The problem stems from the fact that glass electrodes give measurements of hydrogen ion activity, while the proton association constants usually tabulated for the chelators are determined from measurements of hydrogen ion concentration. To calculate the apparent association constant for a chelator at a particular pH, one must either convert pH values (hydrogen ion activities) to concentrations, or used "mixed" proton association constants, which are appropriate for use when hydrogen ions are measured as activities, but all other substances are measured as concentrations. Detailed discussions of this problem have been published elsewhere.[5,10] No one should attempt to use calcium buffers without a clear understanding of this matter.

(d) One should not have unreasonable expectations about the range of [Ca^{2+}] over which the buffer can be effective. To decide what that range is, one must first consider what it is that one wants the buffer to do. As Thomas[10] points out in his book, calcium buffers are used to achieve two distinctly different sorts of things, and how far one can push the buffer depends on what one expects to accomplish by using it.

(i) One of the things that calcium buffers are called upon to do is to minimize fluctuations in free Ca^{2+} concentrations brought about by uncontrolled changes in the total amount of calcium in the system (buffering function). As we were all taught in first-year chemistry, buffers do this most effectively when the pH is close to the pK_a of the buffer acid (or the [Ca^{2+}] is close to the apparent dissociation constant of the chelator). A common rule of thumb is not to ask a buffer to serve this function when it is more than a log unit away from its dissociation constant. (Of course, if the nature of the experiment permits it, one can often alter the effective

[60] J. A. Illingworth, *Biochem. J.* **195**, 259 (1981).

buffering range of Ca^2 buffers by altering the pH, because for many chelators the apparent dissociation constant is highly pH-dependent.)

(ii) Calcium buffers are also used to establish Ca^{2+} ion concentrations at levels below the range that can be set reliably by simple dilutions of calcium salts (calibrating function). (This application is analogous to the use of standardized pH buffers to calibrate pH electrodes.) Of course this function is not entirely separate from the first one . . . a calcium buffer system is useful in establishing known low Ca^{2+} concentrations because it buffers the increases in $[Ca^{2+}]$ that would otherwise be caused by accidental entry of Ca^{2+} into the solution. But there is another way of looking at its role in establishing a known low Ca^{2+} concentration: the chelator greatly increases the total amount of calcium that must be added to produce any given level of $[Ca^{2+}]$. Since at low buffer ratios the free Ca^{2+} concentration is essentially proportional to the total amount of calcium in the system, one can use calcium buffers to establish free Ca^{2+} concentrations at buffer ratios down to the point at which contaminating calcium becomes a significant fraction of the total calcium. If one defines a limit on the error that one is willing to tolerate in the $[Ca^{2+}]$ of a calibration solution, and can put an upper limit on the amount of contaminating calcium that is likely to be present, one can readily put a rational lower limit on the buffer ratio that one is willing to use in making up calibration solutions. One can use aequorin to estimate the Ca^{2+} present in stock solutions of many of the reagents used to construct calcium buffers, and if the solutions are Chelex-treated, it is not difficult to keep Ca^{2+} contamination well below $10^{-6}\,M$. EGTA that has been recrystallized by the addition of acid and protected from subsequent contamination contains virtually no calcium. Magnesium salts cannot be freed of calcium contamination by Chelex treatment, but reagent-grade magnesium salts regularly contain less than 0.01% calcium; therefore, if one adds 1 mM Mg^{2+} to a calcium buffer solution, one could expect to add no more than $10^{-7}\,M\,Ca^{2+}$ to the mixture by this means. Thus, with reasonable care, one should be able to keep the total contaminating calcium in a calcium buffer solution below $10^{-6}\,M$. If one were willing to tolerate a 10% error in the final $[Ca^{2+}]$ (0.1 pCa unit), this means that one could use a calcium buffer mixture in which the added calcium was $10^{-5}\,M$ (since at low buffer ratios the free Ca^{2+} concentration is directly proportional to the total amount of calcium in the system). Therefore, if the total chelator concentration is 1 mM, a buffer ratio of 1/100 could be used. Of course, this calculation depends on the assumption that all reagents but magnesium salts have been decalcified by one means or another, and that other sources of calcium contamination are negligible. If reagents have not been treated in this way, if one is not confident that other sources of calcium contamination have been eliminated, or if a 10% error in the final Ca^{2+} concentration cannot be tolerated, then the lower limit on the buffer ratio should be made accordingly higher. On the other hand, one could make the acceptable ratio even smaller by using a higher concentration of chelator.

For buffer ratios greater than one, the concern is not with calcium contamination (because contaminating calcium is always negligible in comparison with the large amount of calcium added deliberately), but with uncertainties in establishing precisely the relative amounts of chelator and of calcium that have been added to the system. Errors that would not have a serious impact at low buffer ratios loom large at high ratios. For example, a 5% overestimate of the total chelator concentration will cause only a 6% error in $[Ca^{2+}]$ when the buffer ratio is 0.1, but it will lead to an error of more than 120% when the buffer ratio is 10. As with low ratios, one can estimate how far the system can be safely pushed if one can put limits on the uncertainties involved in establishing the total concentrations of calcium and of chelator in the calcium buffer solutions. For example, let us assume that the concentrations of total calcium and chelator added can be established to within 1%. (As will be seen in the next paragraph it is difficult to do much better than that.) In the worst case, the chelator concentration will be 1% too low, and the calcium concentration 1% too high. This combination of errors would lead to a true buffer ratio of 12.8 when the nominal ratio was 10, and thus the $[Ca^{2+}]$ would be 28% higher than intended. At a nominal buffer ratio of 3, the true buffer ratio would be 3.26, and the error in $[Ca^{2+}]$ nearly 10%. Thus when calcium buffers are used for calibration purposes, it would seem prudent never to use buffer ratios above about 3 without checking the $[Ca^{2+}]$ of the solutions by an independent method.

(e) In making up calcium buffer solutions it is critical to establish the calcium and chelator concentrations with precision. As should be apparent from the previous paragraph, this is more critical the higher the buffer ratio. The most serious obstacle derives from the fact that most commercial preparations of EGTA are of unknown purity. Although most distributors of EGTA acknowledge that their products are less than 100% pure, those investigators who have taken the trouble to check them by titration methods such as those described by Moisescu and Pusch,[61] by Smith and Miller,[62] or by Bers,[63] have often found purities less than those claimed, and frequently in the vicinity of 95%.[62,64] Moore[65] recently tested the

[61] D. G. Moisescu and H. Pusch, *Pfluegers Arch.* **355**, R122 (1975).
[62] G. L. Smith and D. J. Miller, *Biochim. Biophys. Acta* **839**, 287 (1985).
[63] D. M. Bers, *Am. J. Physiol.* **242**, C404 (1982).
[64] D. J. Miller and G. L. Smith, *Am. J. Physiol.* **246**, C160 (1984).
[65] E. D. W. Moore, Ph. D. thesis, University of Minnesota, Rochester, 1986.

EGTA sold by a number of domestic suppliers (Aldrich, Baker, Eastman, and Sigma), and found that the Eastman product (batch A5A) was the purest (at 98.7%). He also found that the purity of all the preparations tested could be greatly improved (to better than 99%) by a fairly simple process of solvent extraction and recrystallization. Clearly, a reliably pure source of EGTA would be a great asset in this field. In the absence of that, it behooves the careful investigator to estimate the purity of his chelators by titration methods of the sort referred to above, or to avoid using calcium buffers to control [Ca^{2+}] at levels much above the apparent dissociation constant of the Ca^{2+}–chelator complex unless a Ca^{2+}-selective electrode is used to check freshly made calcium–buffer solutions against standard solutions of known [Ca^{2+}]. Ca^{2+} electrodes probably should be used much more widely for this purpose than they are. One wouldn't make up a pH buffer without checking it with a glass electrode. Why should a calcium buffer be any different?

Interpretation of Light Measurements

Calcium-regulated photoproteins have a number of properties that complicate the interpretation of light signals obtained from systems containing them. The following characteristics must be taken into consideration whenever plans are made to use these substances as Ca^{2+} indicators.

1. The photoprotein is consumed in the luminescent reaction. Corrections must be made for this if the [Ca^{2+}] in the compartment under study rises high enough or for long enough to discharge an appreciable fraction of the indicator present.

2. The speed with which light intensity follows sudden changes in [Ca^{2+}] is limited. When aequorin is mixed very rapidly with a solution of saturating [Ca^{2+}], light intensity rises with a halftime of about 6 msec at 20°.[66] Obelin responds more than twice as fast.[67] Over a wide range, the rate of rise is influenced very little by the concentration of Ca^{2+} with which the photoprotein is mixed,[66,67] but according to Moore[65] the rate of rise of aequorin luminescence does slow appreciably at Ca^{2+} concentrations as low as 1 μM. Exactly how the kinetics of the reaction are influenced by the concentrations of Ca^{2+} and aequorin when both are present in very low concentrations remains to be determined. Rapid-mixing studies are exceedingly difficult to carry out under such conditions because of the very low light intensities that must be measured.

[66] J. W. Hastings, G. Mitchell, P. H. Mattingly, J. R. Blinks, and M. van Leeuwen, *Nature (London)* **222,** 1047 (1969).
[67] D. G. Stephenson and P. J. Sutherland, *Biochim. Biophys. Acta* **678,** 65 (1981).

3. The rate of the luminescent reaction is independent of [Ca^{2+}] both at very high and at very low calcium concentrations (see Figs. 8–11). One would, of course, expect the reaction to saturate at high [Ca^{2+}], but we were somewhat surprised when we first observed that the calcium concentration–effect curve for aequorin also flattens out at very low [Ca^{2+}]. (The same is true for obelin,[67] halistaurin,[7] and phialidin.[7]) The fact that it does so imposes a lower limit on the range of [Ca^{2+}] detectable with aequorin that is independent of the sensitivity of the light-measuring apparatus or the amount of photoprotein present.

4. The relation between [Ca^{2+}] and light intensity is nonlinear. The calcium concentration–effect curve is sigmoid on a log–log plot, and has a maximum slope of about 2.5 under physiological conditions (Figs. 8–11). The steepness of this curve means that changes in light intensity tend to give an exaggerated impression of the changes in [Ca^{2+}] responsible for them. An additional consequence is that local regions of high [Ca^{2+}] tend to dominate patterns of light emission. In a system where the aequorin concentration is uniform, an amount of [Ca^{2+}] added at one point will give a much brighter signal when its concentration is locally high than after it has diffused to reach a uniform concentration. This behavior complicates the quantitative interpretation of photomultiplier records, but may be helpful in work with image intensification, since it tends to increase contrast.

5. The reactions of photoproteins with Ca^{2+} are influenced profoundly by a number of variables of physiological importance (Figs. 9–11). Temperature, ionic strength, and [Mg^{2+}] have marked effects on the relation of luminescence to [Ca^{2+}]; changes of pH within the physiological range have very little effect (Fig. 10). One must also be alert to the possibility that the properties of photoproteins might be altered by drugs or other substances that may be added during the course of an experiment, or even by interaction with normal constituents of the cell. It is known, for example, that the substitution of D_2O for H_2O greatly alters the properties of aequorin,[68] that many alcohols increase the quantum yield of its luminescent reaction,[69,70] that certain catecholamines increase its Ca^{2+}-independent luminescence,[71] and that urethane increases its sensitivity to Ca^{2+}.[5,72] Although it has been reported that other anesthetics (both local and general) also increase the Ca^{2+} sensitivity of aequorin,[72] most of these obser-

---

[68] D. G. Allen, J. R. Blinks, and R. E. Godt, *J. Physiol. (London)* **354**, 225 (1984).
[69] O. Shimomura, F. H. Johnson, and Y. Saiga, *J. Cell. Comp. Physiol.* **59**, 223 (1962).
[70] I. R. Neering and M. W. Fryer, *Biochim. Biophys. Acta* **882**, 39 (1986).
[71] M. Endoh and J. R. Blinks, *Circ. Res.* **62**, 247 (1988).
[72] P. F. Baker and A. H. V. Schapira, *Nature (London)* **284**, 168 (1980).

vations have been disputed.[5,73] While there is good evidence that aequorin binds to unknown structures in mechanically skinned cardiac muscle cells,[74] there is as yet no reason to believe that its properties are significantly changed as a result.

Calibration of Light Signals Recorded from Living Cells

Two general approaches have been used to translate measurements of light intensity from photoprotein-containing cells into absolute Ca^{2+} concentrations: for convenience I shall refer to them as the null method and the lysis method.

The null method depends on the injection or diffusion of calcium buffers into the space occupied by the photoprotein, with the goal of finding a buffer mixture (of known $[Ca^{2+}]$) that just does not alter the light emission from the cell. Obviously, the null method can be used in its most direct form only if the $[Ca^{2+}]$ to be estimated is constant; however, in principle, it should be possible to use it in conjunction with an appropriate $[Ca^{2+}]$ concentration–effect curve for the photoprotein to estimate Ca^{2+} concentrations from light levels considerably above or below the level at which the null determination was made. The most serious drawback of the null method is that there are not many situations in which it is technically feasible to use it. So far the method has been used only with the giant squid axon; calcium buffers have been introduced either by injection[75] or by perfusing them through an axially mounted dialysis capillary.[76]

The lysis method is called that because when it is used, the cell or cells containing the photoprotein are lysed in order to obtain a normalization factor required in the calibration.[77] If light signals recorded from living cells are to be related to Ca^{2+} concentration–effect curves determined *in vitro*, all of the light measurements must be expressed in the same units. Raw photomultiplier records are calibrated in terms of anode current or photon counts per second, and these units cannot be compared from one experimental system to another without information about the amount of photoprotein in the tissue, the optical geometry of the recording system, the quantum efficiency of the photocathode, and the gain of the photomul-

---

[73] S. Kurihara, M. Konishi, T. Miyagishima, and T. Sakai, *Pfluegers Arch.* **402**, 345 (1984).
[74] A. Fabiato, *J. Gen. Physiol.* **78**, 457 (1981).
[75] P. F. Baker, A. L. Hodgkin, and E. B. Ridgway, *J. Physiol. (London)* **218**, 709 (1971).
[76] R. DiPolo, F. Requena, F. J. Brinley, Jr., L. J. Mullins, A. Scarpa, and T. Tiffert, *J. Gen. Physiol.* **67**, 433 (1976).
[77] D. G. Allen and J. R. Blinks, in "Detection and Measurement of Free Ca^{2+} in Cells" (C. C. Ashley and A. K. Campbell, eds.), p. 159. Elsevier/North-Holland, Amsterdam, 1979.

tiplier. All of these factors can be ignored, and light signals from the tissue compared meaningfully with those recorded from photoproteins *in vitro* if all signals are expressed in terms of what we have referred to as "fractional luminescence."[77] To convert a light signal (L) into units of fractional luminescence, one divides it by the peak light signal (L_{max}) that would be obtained *under the optical conditions of the experiment* if all of the photoprotein present were instantly exposed to a saturating concentration of Ca^{2+}. This is simple enough in the case of Ca^{2+} concentration–effect curves determined *in vitro*: L_{max} is established automatically at the top of the curve, and by expressing all points on the curve as fractions of L_{max}, one normalizes for the concentration of photoprotein used in a particular experiment. Converting light signals recorded from living cells into units of fractional luminescence is a bit more complicated. L_{max} cannot be determined directly because it is not possible instantly to expose all of the photoprotein in the cell to a saturating concentration of Ca^{2+}. The best one can do is to discharge the photoprotein as rapidly as possible by lysing the cells containing it while measuring the total amount of light emitted under the same optical conditions used for the measurement of the light from the living cells. If the total amount of light emitted by a quantity of photoprotein is independent of the speed with which it is discharged (and we have found that to be true *in vitro*, at least over the range in which the question is readily testable), one can calculate the L_{max} corresponding to the measurements *in vivo* by multiplying the time integral of the light signal recorded during the lysis by the peak-to-integral ratio of light signals recorded *in vitro* when samples of the same batch of aequorin are rapidly mixed with a saturating concentration of Ca^{2+}. The peak-to-integral ratio is reduced by preequilibrating the photoprotein with Mg^{2+}, and must be determined with a sample of the photoprotein that has been preequilibrated with a concentration of Mg^{2+} appropriate to the intracellular environment. All measurements, including that of the Ca^{2+} concentration–effect curve, must be made at the same temperature. The concentration–effect curve must be determined at an ionic strength and $[Mg^{2+}]$ appropriate to the intracellular environment, and with photoprotein that has been preequilibrated with the concentration of Mg^{2+} used for the curve. The lysis must be carried out at the same temperature as the rest of the experiment because the quantum yield of the light-yielding reaction is highly temperature-dependent.[69] However, it is not essential that the ionic strength or $[Mg^{2+}]$ of the lysing solution be controlled, because these factors do not influence the quantum yield of photoprotein luminescence.

An example of the calibration method just described is illustrated in Figs. 12 and 13. In this particular instance the cell was an isolated frog

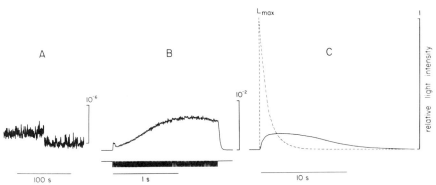

FIG. 12. Method used to estimate fractional luminescence (L/L_{max}) of frog muscle cells injected with aequorin. Tracing A shows photomultiplier output recorded from an intact aequorin-injected frog skeletal muscle fiber at rest (upper level) and the background signal recorded from the same fiber after the aequorin had been discharged by lysing the cell membranes with Triton X-100 (lower level). Tracing B shows the light output recorded from the fiber during a tetanic contraction (50 Hz). Tracing C shows the light signal recorded when 0.5% Triton X-100 was flushed into the bath to discharge all of the aequorin in the cell. All were recorded at 21° under the same optical conditions. The dashed line has been drawn to subtend the same area as the recorded curve, and shows the time course of the flash that would have been obtained if all of the aequorin in the cell had been exposed instantly to a saturating [Ca^{2+}] (see text). The peak of this curve is our best estimate of L_{max} for the amount of aequorin in the cell. The three measurements of light intensity must be made under identical optical conditions and at the same temperature and with the same batch of aequorin as the calibration curve (Fig. 13). Note the different scales of light intensity and time in tracings A, B, and C. From J. R. Blinks, W. G. Wier, and K. W. Snowdowne, in "Energy Transport, Protein Synthesis, and Hormonal Control of Heart Metabolism," NIH Publ. 80-2017, p. 13. National Institutes of Health, Bethesda, Maryland, 1980.

skeletal muscle fiber, and the lysis was carried out simply by exposing the cell to a physiological salt solution containing 0.5% of the detergent Triton X-100. This detergent has been found *in vitro* not to influence the quantum yield of the aequorin reaction, and as can be seen in Fig. 12, it completely discharges the aequorin contained in isolated frog skeletal muscle fibers in just a few seconds. Because the discharge is rapid, the intensity of the light signal recorded during lysis is very high, and steps must be taken to reduce the sensitivity of the light recording system before lysis is carried out. (It is necessary to do this to avoid exceeding the linear range of the photomultiplier–dynode chain combination.) Our approach to doing this has been to reduce the voltage across the dynode chain by an amount (e.g., from 1200 to 800 V) previously determined to reduce the sensitivity of the photomultiplier by a carefully measured factor (of the order of 40). In some situations it might be possible to use a carefully calibrated optical

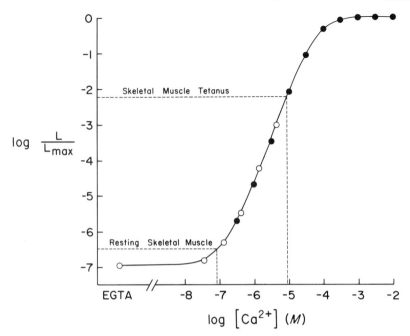

FIG. 13. Results of Fig. 12 applied to a Ca^{2+} concentration–effect curve determined *in vitro* under conditions thought to be appropriate to the intracellular environment (150 mM KCl, 2 mM Mg^{2+}, pH 7.0, 21°). As in Fig. 8, the filled circles indicate points determined by simple dilution of CaCl$_2$; the open circles indicate points determined with calcium buffers (1 mM EGTA). Horizontal dashed lines indicate estimates of fractional luminescence determined from the aequorin-injected muscle cell of Fig. 12. From Ref. 4, with permission of the publisher.

filter to reduce the light intensity recorded during lysis; our experimental setup would make this difficult.

The method of estimating L_{max} by lysing cells with Triton X-100 was first applied to single skeletal muscle fibers and atrial trabeculae from frogs,[33,77] and works exceedingly well with these tissues. However, it does not work well with some other kinds of tissue because the lysis is too slow. We have found the problem to be particularly serious with multicellular preparations of mammalian heart muscle in which multiple cells have been microinjected or chemically loaded with aequorin. These preparations often go on emitting small bursts of light for an hour or more after exposure to Triton X-100, and in such cases it is difficult or impossible to estimate the total light output with precision. Allen and Orchard[78] re-

[78] D. G. Allen and C. H. Orchard, *J. Physiol.* (*London*) **339**, 107 (1983).

ported that in cat, rat, and ferret papillary muscles the lysis could be accelerated by the use of a hypotonic solution (specifically 5% Triton X-100 in 10 mM CaCl$_2$). They did not state how quickly the lysis could be achieved, but Allen[79] has informed me that it was always complete within less than 5 min. For reasons that are not clear, it has been our experience that even with the solution recommended by Allen and Orchard, the lysis is usually too slow to give results that we consider reliable in most kinds of mammalian heart muscle (with the notable exception of dog Purkyně strands). Clearly, there is a pressing need for a more rapid and reliable method of lysing cells containing photoproteins without altering the quantum yield of the luminescent reaction.

Acknowledgments

Supported by USPHS Grant HL 12186. Use of the facilities of the Friday Harbor Laboratories, University of Washington, is gratefully acknowledged. The various pieces of apparatus illustrated in this chapter were constructed in the departmental workshop by Mr. Merlin Neher, Mr. Gary Harrer, and the author.

[79] D. G. Allen, personal communication (1987).

[13] Electron Probe X-ray Microanalysis of Ca^{2+}, Mg^{2+}, and Other Ions in Rapidly Frozen Cells

By A. V. Somlyo, H. Shuman, and A. P. Somlyo

Introduction

The unique contribution of electron probe X-ray microanalysis (EPMA) to biology is its capability to determine, *in situ* and quantitatively, the distribution of elements within organelles and in the cell cytoplasm.[1,2] EPMA is also a truly microanalytic method for biochemists, as it can be used to quantitate the elemental composition of microdroplets[3] and other very small samples, equivalent to or smaller than a single phage head or ferritin molecule. The subcellular compartmentalization and transport of elements within tissues in various functional states can be

[1] A. V. Somlyo, H. Gonzalez-Serratos, H. Shuman, G. McClellan, and A. P. Somlyo, *J. Cell Biol.* **90**, 577 (1981).
[2] A. P. Somlyo, *Cell Calcium* **6**, 197 (1985).
[3] C. P. Lechene and R. R. Warner, *in* "Microbeam Analysis in Biology" (C. P. Lechene and R. R. Warner, eds.), p. 279. Academic Press, New York, 1979.

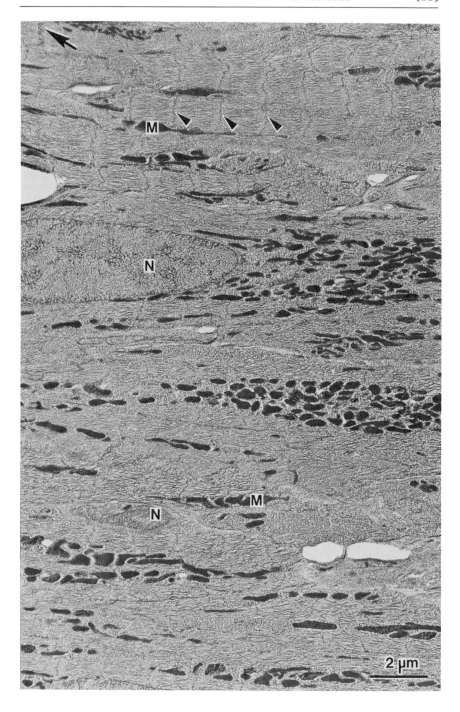

captured with this technique, on a millisecond time scale, by using ultrarapid freezing. Meaningful quantitation requires that the tissue preparatory techniques, including dissection, freezing, cryosectioning, and transfer to the microscope, as well as the analysis itself, preserve the elemental composition as it exists *in vivo*. This precludes the use of chemical fixatives and stains.

EPMA is based on the detection of signals due to atomic core shell excitations that are, for practical purposes, insensitive to chemical environment; therefore, it provides a measure of *total* elemental concentration. A single energy-dispersive X-ray spectrum contains information about all elements present from atomic number 11 (sodium) and higher, as well as about the total mass in which these elements are contained. This chapter will describe the tissue preparatory methods and the techniques for extracting quantitative information from X-ray spectra.

Specimen Preparation

Rapid Freezing

The distribution of diffusible elements in cells is preserved by ultrarapid freezing followed by cryoultramicrotomy. The freezing rate of biological samples is slowed by the insulating properties of ice, which retards the diffusion of heat from the sample. Therefore, even under optimal conditions, only the outer 5–20 μm will be free of ice crystal damage in most biological tissues that are highly hydrated. Cubed tissues are not suitable for EPMA, as the best frozen area on the surface will include damaged cells. Specimens should be small, for example, bundles of 15 frog semitendinosus fibers, or 200-μm diameter strips of smooth muscle attached to low-mass stainless steel mesh holders or single wires. In addition, 10-μl droplets of red blood cells, platelets, or mitochondrial pellets placed on tiny wooden splinters, which have a low heat capacity and can be directly clamped in the microtome chuck, have been frozen successfully. A cryosection from a 10-day-old rat papillary muscle attached to the septal wall and mounted on a small aluminum stub that is suitable for analysis is illustrated in Fig. 1. We have found that freezing by propelling tissues into subcooled Freon 22 ($-164°$) is most convenient for

FIG. 1. Typical cryosections showing several myocardial cells from a papillary muscle of a 10-day-old rat. The approximately 100-nm-thick section was cut at $-130°$ and freeze-dried at 10^{-6} Torr. The contrast arises from electron scattering due to differences in mass in the different regions. Nuclei (N), mitochondria (M), sarcomeres and Z-lines (small arrows), and intercalated disk (large arrow) are indicated.

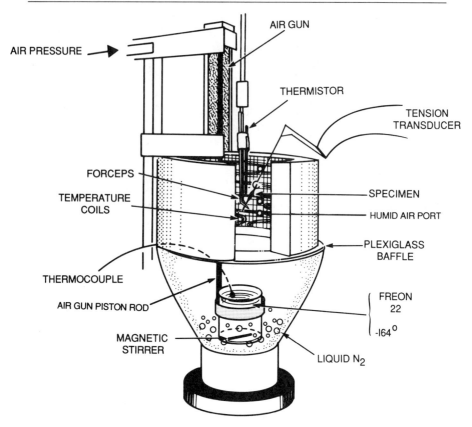

Fig. 2. Schematic view of the environmental chamber and freezing apparatus. The environmental chamber is isolated from the coolant by a plexiglass baffle that is removed the instant prior to triggering the air gun, which raises the beaker of supercooled Freon 22 at a speed of 80 cm/sec into the environmental chamber, freezing the preparation at the desired time during the physiological response. For mammalian tissues, the chamber is maintained at 37° with high humidity to prevent drying and cooling of evaporation.

physiological experiments designed to capture a functional state on a <1 sec time scale. The specimen is brought into contact with the rapidly stirred supercooled Freon 22 by triggering an air gun to which is attached either the specimen or the beaker of coolant, depending on whether the specimen is driven into the coolant or the coolant raised to the specimen as shown in Fig. 2. The shaft of the air gun travels at 80 cm/sec, and drives the specimen several centimeters into the coolant, resulting in good contact between the two surfaces and a reduction of nucleate boiling on the surface, which would otherwise slow the freezing rate. Another approach that has led to satisfactory freezing of the liver *in vivo* employs a spring-loaded clamp with two opposing cups of solidified Freon to capture a lobe of liver exposed through a small abdominal incision in the anesthetized animal.[4]

The handling of the specimen up to the moment of freezing is of critical importance in order to avoid cell swelling, sodium loading, depolarization, anoxia, and accumulation of mitochondrial calcium granules. Therefore, careful dissection is required. It is also desirable to use an environmental chamber (Fig. 2) positioned above the coolant and separated by a removable plexiglass baffle. The interior of the chamber is maintained at the desired temperature by stainless steel tubing wound within the plexiglass walls of the environmental chamber, and connected to a temperature-controlled recirculator. The chamber is humidified by air forced through air stones immersed in distilled water at the base of the walls of the chamber (not shown) and passed over the warmed coils into the interior of the chamber through multiple small openings on its interior wall. A small thermocouple positioned next to the specimen records the specimen temperature. The specimen is held by fine stainless forceps. The interior diameter of the chamber (15 cm) is sufficiently large so that other apparatus, such as a small organ bath, force transducer, or stimulating electrodes, can be mounted within the chamber. The solutions in an organ bath can be both oxygenated and exchanged within the chamber as desired. The organ bath, if used, and the styrofoam baffle are removed prior to activating the air gun that rapidly raises the coolant into the environmental chamber.

The "slammer" device or chilled copper pliers are also suitable for cryoultramicrotomy and EPMA.[5] The slammer, which brings the specimen into contact with a helium-cooled polished copper surface, has been perfected[5] to capture the twitch of a single skeletal muscle fiber. Using specially designed holders, this method yields specimens suitable for cryosection and freeze fracture from a single specimen sample. Other methods such as spray freezing of organelles or whole cells, plunging methods, and the use of other coolants such as liquid propane, are reviewed in Refs. 6 and 7.

[4] A. P. Somlyo, M. Bond, and A. V. Somlyo, *Nature (London)* **314**, 622 (1985).
[5] R. Nasser, N. R. Wallace, I. Taylor, and J. R. Sommer, in "Scanning Electron Microscopy—1986" (O. Johari, ed.), p. 309. SEM Inc., Chicago, 1986.
[6] H. Plattner and L. Bachman, *Int. Rev. Cytol.* **79**, 237 (1982).
[7] J. C. Gilkey and L. A. Staehelin, *J. Electron. Microsc. Tech.* **3**, 177 (1986).

Cryoultramicrotomy

Following freezing, the tissues mounted on the stainless steel mesh holders,[8] or droplets on the wooden splinters, can be clamped directly into a microtome chuck. The rather bulky liver samples frozen with the Freon clamp are cut into small pieces, under liquid nitrogen, with a hammer and a sharpened chisel. The small chunks (1–2 mm diameter) can then be clamped directly in the microtome chuck with the well-frozen surface exposed for sectioning and the chuck tightened to hold the tissue rigidly without crushing it. Other samples, such as free-standing thin strips of smooth muscle, can be cut into multiple samples and mounted on a specimen stub using toluene as a low-temperature cement.[9] This technique allows the specimen to be oriented for transverse or longitudinal sectioning. The remaining pieces of frozen tissue can be stored, for future sectioning, in Freon frozen solid in liquid nitrogen. For sectioning most materials the ambient temperature of the cryochamber is $-130°$, with the specimen temperature set at $-110°$. Thin (100–200 nm) sections are cut with a glass or diamond knife without facing or trimming the blockface, as this would remove the best frozen material. Well-frozen material yields glassy sections with interference colors, while poorly frozen material results in whitish crumbling sections that are discarded. The sections are picked up dry from the knife edge with precooled bamboo splinters, and placed on copper grids covered with thin (≈ 7.5 nm) carbon foils. The grids are held in a special grid holder positioned in front of the knife. The carbon foils are made on freshly cleaved mica using spectroscopically pure carbon rods. Each new lot of grids with carbon foils is checked for contamination by collecting several thousand seconds of EPMA spectra. With proper care, carbon foils free of calcium can be prepared. Sulfur and silicon contamination can generally be avoided by using S- and Si-free pump oils in the microscope, freeze-drying apparatus, and the vacuum evaporator for carbon coating and for the preparation of carbon foils. The sections are then sandwiched between two grids with carbon foils and pressed together with a prechilled polished brass rod. Alternatively, grids are held in a specially designed multiple grid holder that, following the deposition of sections, is covered with a lid containing small spring-loaded Teflon dowels that apply light pressure to the sections during drying. The grid holders containing the specimens are transferred from the cryochamber to the freeze-drying apparatus in a large liquid-nitrogen-chilled covered brass cup, and dried overnight at 10^{-6} Torr. The lid of the specimen holder can be removed with magnets attached to a lever within

---

[8] A. V. Somlyo, H. Shuman, and A. P. Somlyo, *J. Cell Biol.* **74**, 828 (1977).
[9] R. D. Karp, J. C. Silcox, and A. V. Somlyo, *J. Microsc.* (*Oxford*) **125**, 157 (1982).

the apparatus, without breaking the vacuum, and the sections are then carbon coated ($\simeq 1.5$ nm). If the copper grid specimen sandwich technique is used, then the vacuum must be broken and the grids opened before carbon coating. If done rapidly, this causes no detectable elemental translocation. The carbon-coated specimens are stored in a desiccator. We shall not deal here with special cases requiring the use of frozen hydrated specimens, such as analysis of the aqeuous lumen of secretory epithelia.

Electron Probe X-Ray Microanalysis

Principles, X-Ray Detectors, and Their Calibration

Electron probe X-ray microanalysis is based on the fact that ionization of a core shell atomic electron by an incident fast (high-energy) electron can give rise to the emission of an X-ray photon. The energy of the photon is characteristic of the element ionized, or more precisely, of the specific electronic transition leading to deexcitation of the ion. For example, following ionizaton of a K-shell orbit electron of a Ca atom, the filling of the ionized K shell by an L shell electron can create a 3.69-keV X-ray photon; a similar transition in a phosphorus atom can produce 2-keV X-rays. Therefore, bombardment of a thin specimen with the focused electron beam of an electron microscope will result in an X-ray spectrum containing a series of peaks that are representative of the elements present in the microvolume irradiated. The number of X-rays emitted, $I_x = I_e \omega_x Q_x N_x T_x$, where x denotes a given element, ω_x and Q_x are the respective fluorescent yields and ionization cross-sections, N_x is the number of atoms of x in the microvolume, I_e is the number of incident electrons, and T is the transmission function of the X-ray detector, taking into account geometric and other factors that determine whether an emitted X-ray is detected. The ionization cross-section is the probability of a given atomic shell being ionized. The fluorescence yield is the probability of an ionization being deexcited by the emission of an X-ray photon rather than, for example, an Auger electron.

The second component of an X-ray spectrum consists of the continuum X-rays that form the background underlying the characteristic peaks (Fig. 3). The background X-rays are generated, in a broad energy band up to the energy of the incident electron, through "bremsstrahlung" scattering of the incident electrons by atomic nuclei. In an ultrathin specimen, the number of background X-rays, B, generated is directly proportional to the *total* mass, M, of the microvolume irradiated. Therefore, the characteristic X-ray/background ratio (I_x/B) is proportional to elemental concentration (N_x/M), and the use of this ratio is the basis of the most

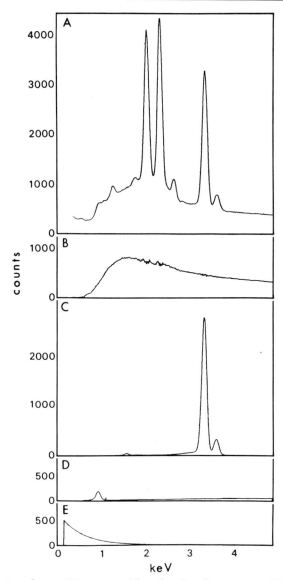

FIG. 3. Spectrum from a 100-nm cryosection of rat liver frozen *in vivo*. (A) raw spectrum; (B) thin target Bremsstrahlung after subtraction of elemental peaks, copper grid, and the $1/E$ low-energy tail. For most biological specimens, the statistically most suitable continuum region is the 1.34–1.64 keV region; (C) characteristic X-ray peaks of potassium, subtracted from A and composed of the potassium K_α and K_β as well as the Si escape peak at 1.6 keV. Escape peaks are due to potassium X-rays which have lost energy due to their exciting Si in the Si[Li] detector; (D) copper grid spectrum subtracted from A; (E) [$1/E$] tail subtracted from A and largely due to detector noise.

convenient and reliable method for quantitative electron probe X-ray microanalysis (EPMA) of ultrathin biological specimens.[10–14]

The X-rays emitted by a specimen can be detected by energy-dispersive or wavelength-diffractive detectors. For most biological studies and, in particular, for applications at high spatial resolution, energy-dispersive (EDS) detectors are now commonly employed. A pulse of electronically measured charged particles is created in the EDS Si(Li) crystal by each incident X-ray photon. The size of the pulse is proportional to the X-ray energy. In order to reduce the detector and amplifier noise, EDS X-ray detectors are continuously cooled with liquid nitrogen. The major advantage of these EDS detectors is their much better geometric efficiency and the parallel detection of a large number of characteristic X-rays (up to 40/keV). However, even EDS detectors detect, at most, 1% of the X-rays emitted by the specimen. The EDS detectors generally used in conjunction with transmission electron microscopes are equipped with thin ($\sim$7.5 μm) Be windows that absorb some of the lower energy X-rays (see Fig. 3) before they can reach the detector. This absorption reduces the sensitivity for elements between Na and P, and precludes the detection of elements with atomic numbers below Na (e.g., C, N, O, F, B). These low-atomic-number elements can be detected by wave-length-diffractive or "windowless" energy-dispersive X-ray detectors or by electron energy loss spectroscopy.[15,16]

The characteristic and continuum X-rays emitted from a microvolume of the specimen excited by a focused electron beam (Fig. 3) are computer processed, and the characteristic/continuum ratios are converted into elemental concentrations. This is sufficient if only X-rays originating from the specimen are detected. In practice, however, X-rays are also generated in the electron microscope column and specimen support grid, and noise in the detector electronics can add "counts" in the low-energy region. It is essential to minimize such extraneous peaks and continuum X-rays that are, respectively, not representative of the composition of the specimen (for example, copper from copper grids) and introduce errors in

---

[10] T. A. Hall, in "Physical Techniques in Biological Research" (G. Oster, ed.), Vol. 1A. Academic Press, New York, 1971.
[11] H. Shuman, A. V. Somlyo, and A. P. Somlyo, *Ultramicroscopy* **1**, 316 (1976).
[12] H. Shuman, A. V. Somlyo, and A. P. Somlyo, in "Scanning Electron Microscopy—1977" (O. Johari, ed.), Vol. 1, pp. 663–672. IIT Research Institute, 1977.
[13] T. Kitazawa, H. Shuman, and A. P. Somlyo, *Ultramicroscopy* **11**, 251 (1983).
[14] T. A. Hall and B. Gupta, *J. Microsc. (Oxford)* **136**, 193 (1984).
[15] A. P. Somlyo, "Recent Advances in Electron and Light Imaging in Biology and Medicine." New York Academy of Sciences, New York, 1986.
[16] H. Shuman and A. P. Somlyo, in "Analytical Electron Microscopy—1981" (R. H. Geiss, ed.), p. 202. San Francisco Press, San Francisco, California, 1981.

quantitation due to a reduction of the characteristic/continuum ratio. The detector noise that varies as E^{-1} (where E = energy) can be scaled to evaluate and subtract its contribution to the continuum region used for quantitation (Fig. 3). The continuum contribution by the copper grid can be scaled to the Cu L peak (Fig. 3), and also subtracted by the computer program. The instrumental elimination of extraneous X-rays has been met with varying success by different manufacturers through optimization of lens pole pieces, shielding of the microscope column, inclusion of a "spray aperture" above the specimen, and careful collimation of the X-ray detector. In addition to "parasitic" X-rays, scattered electrons can also reach and, eventually, damage the detector. This is a limiting factor in improving the geometric efficiency of detection by moving the detector closer than about 20 mm from the specimen, as electron scattering into the detector is difficult or impossible to eliminate.

The X-ray photons detected by the detector are pulse processed to determine the number of X-rays at a given energy, and stored in successive channels of a multichannel analyzer. Spectra can be stored at 10, 20, or 40 eV/channel, depending on the energy range to be explored and the memory capacity of the multichannel analyzer. The major portion of the biologically relevant spectrum (between Na and Ca) can be conveniently stored at 10 eV/channel in 512 of the (usually) 2048 channels of modern multichannel analyzers. The energy resolution of X-ray detectors varies with count rate and X-ray energy. The expected energy resolution (full width at half-maximum, FWHM) of a 30 mm^2 EDS detector (2000 cps) is approximately 110 eV at 1 keV energy and approximately 160 eV at 6 keV. The inherently narrow width (4 eV or less) of an X-ray line is very much broadened by the electronic noise of the X-ray detector. It is, however, still necessary and possible to calibrate the detectors to within 1 eV, when collecting both the reference and the experimental spectra. The locations of the peak centroid and the energy resolution of the detector are calibrated in a low- and in a high-energy region of the spectrum. Obviously, such calibration can not be obtained by visual inspection of spectra stored at 10 eV/channel, but is readily performed by a computer fitting program that determines the centroid positions and FWHM of two Gaussian peaks: one at a low, the other at a high energy. It is convenient to use a thin Cu film as a standard, and calibrate the detector each day with the Cu L line at 931 and the K_α line at 8044 eV.

Spectral Processing: Elemental Standards, Least Squares Fit, and Count Statistics

Once the X-ray spectrum has been collected and stored in the multichannel analyzer, it is necessary to determine the number of characteristic X-ray counts in each peak of the spectrum, as well as the number of the X-ray continuum counts to be used for thin film quantitation. The most widely used method for obtaining the number of counts in a given peak is to perform a multiple least-squares (ML) fit of the experimental spectrum to a library of primary reference files containing, as primary standards, "pure element" spectra of each of the elements expected to be present within the specimen.[11,13,17] A reference file used for the analysis of average biological specimens is "built" by collecting "noise-free" X-ray peaks of Na, Mg, Si, P, S, Cl, K, and Ca. (Si is included because it is an extremely common environmental contaminant, and may also appear due to excitation of the "dead layer" of the X-ray detector.) Primary reference files of other elements (e.g., Fe, Mn, Cu) may also be included, when necessary for analysis. To conserve computer memory and computation time, it is advisable to include in the ML fit primary references for only those elements that give rise to peaks within the energy region included in the ML fit and are expected to be present in the specimen. For example, biological spectra are commonly "fitted" between 500 and 5120 eV. Therefore, for specimens mounted on Cu grids, the Cu L line (931 eV) has to be included in the fit, but not the Cu K_α line (8044 eV). It must be emphasized that the exclusion of a given element from the reference library does not interfere with its recognition, if present. In addition to the appearance of an X-ray peak in the collected spectrum, failure to include in the ML fit an element that is present will result in a large χ^2, as the ML fitting routine (see below) is designed to minimize the χ^2 statistic and displays the χ^2 value. A large χ^2 signals an error due to a peak not included among the references and/or miscalibration.

Primary standards are collected for each element from specimens giving rise only to the peak generated from that one element: for example, with an EDS detector having the customary Be window, only the potassium $K_{\alpha,\beta}$ X-ray peaks will be present in a spectrum collected from $KHCO_3$, as H, C, and O do not give rise to detectable characteristic X-rays. Following subtraction of the associated X-ray continuum, the remaining K peaks are stored (as the "pure element" K standard), in the reference file, and represent the position and shape (width) of the potassium $K_{\alpha'\beta}$ peaks (Fig. 3). The number of characteristic X-rays in the peaks of the experimental spectra are obtained by ML fitting the experimental spectra to a reference file containing the primary standards.

A significant potential source of error in determining the number of characteristic counts is due to overlap of adjacent peaks, such as the potassium K_β and calcium K_α at, respectively, 3590 and 3690 eV as shown

---

[17] D. J. McMillan, G. D. Baughman, and F. H. Schamber, in "Microbeam Analysis—1985" (J. T. Armstrong, ed.). San Francisco Press, San Francisco, California, 1985.

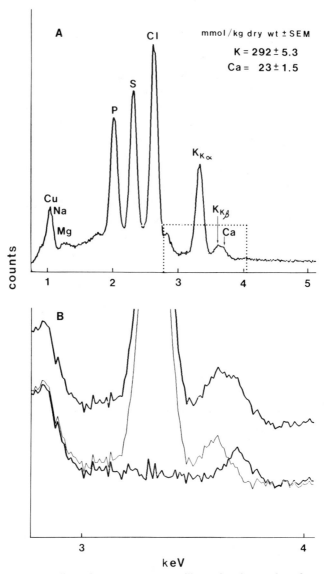

FIG. 4. (A) Energy dispersive X-ray spectrum illustrating the overlap of potassium K_β and calcium K_α peaks. The spectrum was recorded by defocusing the probe to cover a region containing high Ca and low K (extracellular space) as well as a region containing high K and low Ca (frog retinal rod outer segment). (B) The boxed area of A is expanded on the energy axis and at higher gain. The upper curve shows the original spectrum, the middle curve is the spectrum after multiple least-squares fit and subtraction of the Ca K_α, and the lowest trace shows the Ca K_α peak after subtraction of the K K_α and K_β peaks.

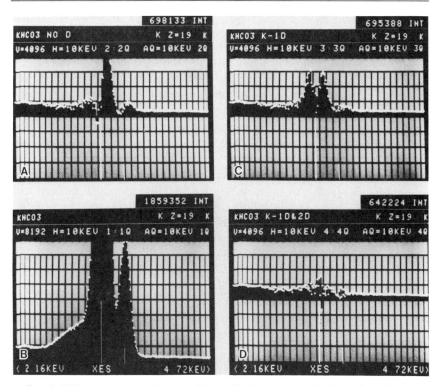

FIG. 5. X-Ray spectra showing the effects of including the derivatives of the K peak in the multiple least-squares fit of a spectrum collected in the presence of (intentional) detector miscalibration. (A) Spectrum after the characteristic X-rays, measured with the reference file that did *not* include the derivatives, have been stripped; (B) raw spectrum before stripping peaks (K $K_{\alpha,\beta}$ peaks overflowed); (C) same as B but with the inclusion of the first derivative of the $K_{\alpha,\beta}$ in the multiple least-squares fit; (D) same as C but also including the second derivative of the K $K_{\alpha,\beta}$ peaks in the fit. From Kitazawa *et al., Ultramicroscopy* **11**, 251 (1983).

in Fig. 4. Miscalibration of the EDS detector increases the errors in quantitation due to overlapping peaks, and for accurate quantitation with the ML fit, the calibration of the detector should be identical during the collection of the experimental spectra and of the primary reference files. Figure 5A demonstrates incomplete subtraction of the K $K_{\alpha,\beta}$ peaks which give rise to spurious Ca counts when the detector has been purposefully miscalibrated. Shifts in either the peak centroid or in the energy resolution can lead to significant errors in the measurement of low concentrations of Ca associated with much higher concentrations of K, generally encountered in biological specimens. The errors can be minimized by careful calibration of the X-ray detectors and by including in the ML reference files the first and second derivatives of the K peaks as shown in Figs. 4 and 5. Inclusion of the first derivative corrects any shift in the centroid position (Fig. 5C), while fitting to the second derivative corrects for changes in energy resolution (broadening or narrowing of the peaks) (Fig. 5D). It is possible to reliably measure, in this manner, low concentrations of Ca in biological specimens, such as the approximately 300 μmol/kg dry wt Ca in the outer segment of vertebrate retinal rods,[18] the cytoplasmic concentrations of Ca in smooth and in skeletal muscle or the Ca in mitochondria or rough endoplasmic reticulum of hepatocytes (Fig. 6). Other techniques suitable for overcoming the problem of overlapping peaks include the use of wave diffractive spectrometers that have significantly better energy resolution or electron energy loss spectroscopy that can conveniently resolve to 1 eV or better.

The X-ray continuum counts used to measure the mass of the microvolume for quantitative EPMA are best obtained from a statistically suitable region of the spectrum, between 1340 and 1640 eV, that contains no characteristic peaks in biological specimens. (Notable exceptions are specimens containing aluminum, bromine, or rubidium. Should these elements be present, this continuum region can still be used for quantitation, after subtraction of the characteristic peaks, but the statistical errors will be larger.) Continuum counts in the higher energy region of the spectrum are also used occasionally, but contain larger contributions of extraneous continuum originating from the electron microscope and supporting grid. For information on the optimal continuum region see Ref. 11. Since the number of continuum X-rays generated varies with atomic number Z ($\sim Z^2$), it is desirable to correct for local variations in average Z by subtracting the continuum contributed by inorganic constituents. If quantitation includes the appropriate correction (see Eq. 5 in Ref. 11), the concentrations are obtained as mmol/kg dry organic mass.

Counts, in general, are subject to Poisson statistics and, to a minimum "Poisson error" with a standard deviation, SD = $\sqrt{N}$, where N is the number of counts in the peak. In practice, the background counts also contribute to the statistical errors of concentration measurements. Therefore, while nine counts in a background free peak would theoretically be sufficient to detect the presence of an element with 99% probability (3 × SD = 3 × $\sqrt{9}$), a larger number of counts has to be collected to minimize the combined error of signal/background. An additional consequence of the Poisson statistics of the counting process is that the statistical error of the measurement can be reduced by increasing the probe current or the

---

[18] A. P. Somlyo and B. Walz, *J. Physiol.* (*London*) **358**, 183 (1985).

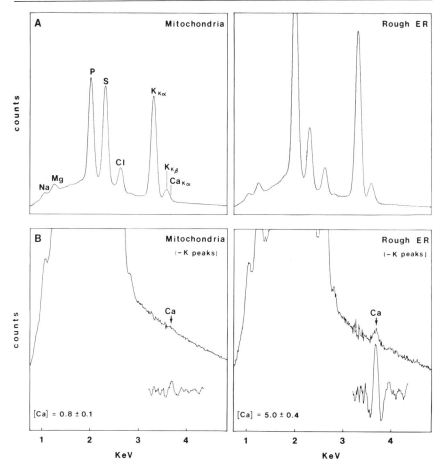

FIG. 6. X-Ray spectra of mitochondria and rough endoplasmic reticulum (ER). The Cu background was subtracted from all spectra by the computer. Peaks representing other elements are indicated. In the lower panels (8× higher gain) the potassium K_α and K_β peaks have also been removed to allow a comparison of the relative sizes of the calcium peaks originating from, respectively, mitochondria and ER. The insets show the Ca peaks in the filtered spectra. From Somlyo et al.[4]

collection time. However, because of the Poisson statistics, the measurement errors are linearly reduced with a quadratic increase of collection time or probe current: a 10-fold improvement of the detection sensitivity of a 100-sec count, for a given probe current and specimen, requires a 10,000-sec count.

Occasionally X-ray results are misinterpreted due to a lack of familiarity with least-squares analysis of noisy data. For example, when peak-to-background ratios are small or zero (i.e., low elemental concentrations), statistical fluctuations of the background will give rise to random variations in peak counts. Negative values of concentration will invariably occur, as expected on the basis of a normal statistical distribution centered about zero, and must be included in the values to be averaged. If this procedure is not followed, a finite concentration of an element will always be "found," even in specimens in which it is not present.

Quantitation of ultrathin cryosections is based on the linear relationship between characteristic/continuum ratio and elemental concentrations. Hence, the concentration of element $C_x = (I_x/B)W_x$, where I_x, as before, is the X-ray count in the peak generated by the ionization of element X, B is the X-ray continuum count, and the W_x is a proportionality constant that relates the X-ray ratios to concentrations. Therefore, once the number of characteristic and continuum X-rays in a given spectrum have been determined (Fig. 3), it is necessary to develop secondary standards to establish the relationship between the characteristic/continuum ratios and the respective concentrations.

Thin-film *secondary standards* that are homogeneous to at least 100 nm are generally difficult to prepare. However, it is not necessary to obtain absolute standards for each element of interest (see below). Some success has been met with polymers into which salts of Na, K, or Ca are introduced with Crown ethers.[11,19] Another, suitable procedure is to use, as a homogeneous secondary standard, a protein film containing a known, stoichiometric concentration of bound S within amino acid residues.[11,13] To avoid loss of S from the standard due to radiation damage, spectra from such specimens are collected at low temperature ($\leq -100°$) on cryogenic stages and with low beam currents. Given these precautions, the S K X-ray/continuum ratio obtained in a thin film of bovine albumin containing 600 mmol S/kg yields the W_x value for S. The other W_x values can then be obtained from spectra of "binary standards" having known stoichiometries. For example, having determined W_s with albumin, the W_K can be obtained from K_2SO_4, then the W_{Cl} (from KCl) and so forth. These W_x values have to be obtained for each detector and microscope configuration, because differences in the thickness of the detector's Be window and other factors can influence the proportion of low-energy X-rays (e.g., Na and Mg) reaching the active region of the detector, and the sensitivity of EDS detectors equipped with Be window is approximately sixfold better for K and Ca than for Na ($W_{Na} \simeq 6 \times W_{Ca}$). Fortunately, given the

[19] A. R. Spurr, in "Microprobe Analysis as Applied to Cells and Tissues" (T. Hall, P. Echlin, and R. Kaufman, eds.). Academic Press, New York, 1974.

biological prevalences of, respectively, Na and Ca, it is rarely necessary to measure Na with the same precision as Ca. In general, the largest source of "error" in measuring such common elements as Na and K is not statistical or instrumental, but the wide range of biological variability. Most high-resolution EPMA measurements are performed on dry cryosections, and even small (1–2%) *in vivo* differences in hydration can introduce very large specimen-to-specimen variations in the dry weight concentration of elements like K that are normally in solution.[20] The experimental design should include analysis of a sufficient number of cells and samples to adequately sample biological variations.

Spatial Resolution and Sensitivity

High spatial resolution with EPMA requires the use of ultrathin sections (i.e., ≤200 nm thick cryosection), as the diameter of an incident electron probe in thick specimens is increased by large-angle, elastic scattering in the specimen. Spatial resolution is primarily determined by the minimum probe diameter that can be produced with sufficient beam current for the generation of a statistically significant number of X-ray counts within a reasonable time period. The amount of current that can be focused into a small probe depends on the brightness of the electron source, and the small probes required for high-resolution analyses are achieved by using LaB_6 or field emission guns. A LaB_6 source that is considerably less expensive than a field emission gun is suitable for analysis with approximately 50-nm probes, perhaps smaller, depending on the lens geometry; for 10-nm resolution or better, a field emission gun is generally required. However, just as resolution in conventional microscopy must be considered in conjunction with contrast, spatial resolution in EPMA also depends on the "compositional contrast" between two points to be resolved. Thus, large differences in local concentrations of an element are considerably easier to resolve than small differences. The best spatial resolution formally documented to date, obtained with a field emission gun in X-ray maps of molybdenum-stained catalase crystals, is 8.7 nm. The minimal detectable concentration of Ca that can be measured in the presence of high concentrations of K, using the derivative fitting routine and collecting about 100 spectra for 400 sec each from a 100-nm section with a probe current of 1.0 nA in a 50-nm diameter spot, is approximately 0.3 mmol Ca/kg dry wt. The sensitivity could be further improved by better detector efficiency or longer counting time.

[20] A. P. Somlyo, A. V. Somlyo, and H. Shuman, *J. Cell Biol.* **81**, 316 (1979).

Independent Validations of EPMA

The computer-fitting routines and standards can be validated by preparing homogeneous films of standards such as proteins containing covalently linked elements (see Fig. 1 in Ref. 13) or films of albumin solutions with differing concentrations of, for example, Ca-EGTA. Ca quantitation of the films is carried out with EPMA using the same solutions measured with atomic absorption spectrophotometry. The two method were shown to be equally reliable.[13] It is also important that validation of specimen preparatory techniques be established, indicating that elemental concentrations and distributions are preserved. Examples in Table I compare concentrations of elements in a variety of tissues using EPMA with

TABLE I
COMPARISON OF TOTAL ELEMENTAL CONCENTRATIONS DETERMINED BY EPMA AND BY OTHER METHODS[a]

Tissue	Element	Concentration (mmol/kg dry wt ± SEM)		(Method)[b]
		EPMA	Other	
Erthrocytes				
Human	Ca	0.2 ± 0.3	0.03	(AA)
	Na	38 ± 2	33 ± 0.4	(AA)
	K	314 ± 9	268 ± 1.5	(AA)
	Cl	141 ± 5	132	(Chem)
Rat	Na	15 ± 1	14 ± 0.5	(AA)
	K	341 ± 4	378 ± 24	(AA)
Frog skeletal muscle				
Whole fiber	K	431 ± 8	411	(NAA)
I band	K	510 ± 7	516	(NAA, ILM)
I band (increase during tetanus)	Ca	3.5 ± 0.1	3.9	(CM)
I band in skinned fibers	Ca	9.0 ± 1.0	9	(Cal)
	Mg	16 ± 2	15.8	(Cal)
	P	31 ± 1	51.5	(Cal)
Liver				
Whole cell	K	422 ± 6	362 ± 10	(AA)
	Ca	3.4 ± 0.5	3.2 ± 0.5	(AA)

[a] References for values shown are given in A. P. Somlyo et al., in "Microbeam Analysis—1986" (A. D. Romig, Jr., and W. F. Chambers, eds.), pp. 199–204. San Francisco Press, San Francisco, California, 1986.

[b] AA, Atomic absorption spectroscopy; Chem, chemical analysis; NAA, neutron activation analysis; ILM, interference light microscopy; CM, computer modeling, based on free Ca^{2+} measurements and Ca^{2+}-binding affinities of troponin and parvalbumin; and Cal, calculated value, based on known stoichiometry of thin filaments.

results of other methods and show excellent agreement. Preservation of distribution can be maintained in small compartments within cells, such as the terminal cisternae of skeletal muscle (Table II) or the 30- to 40-nm atrial granules illustrated in the X-ray mapping section (Fig. 7).

Radiation Damage and Contamination

Radiation damage is the inevitable consequence of inelastic interactions of electrons with matter, and the production of X-rays used for EPMA is due to such inelastic interactions. Fortunately, the earliest signs of radiation damage, loss of high-resolution structure, crystallinity, and the removal of hydrogen, have little or no effect at the resolution and sensitivity of EPMA. On the other hand, with exposure of biological specimens to the electron doses required for EPMA at reasonably high spatial resolution, mass loss is inevitable. Fortunately, in dry cryosections, such mass loss is relatively constant (~15–25% of the total mass of the specimen) and reaches this maximum once a "terminal dose" has been delivered. This mass loss would decrease the continuum component of the spectrum due largely to loss of C, N, H, and O_2, and would lead to an overestimate of elemental concentrations measured with EPMA. Fortuitously, this is largely compensated for by the additional mass of the C support film and coat. Specific elements are removed by radiation only under special circumstances: S is lost during electron irradiation at room temperature, and Cl and other halogens are lost when irradiated by electrons while in crystal form. There is insufficient information about the radiation chemistry of electron-induced Cl loss to determine why this does not occur when Cl is within a protein matrix, but this phenomenon

TABLE II
ELEMENTAL COMPOSITION OF THE TERMINAL CISTERNAE OF FROG SKELETAL MUSCLE[a]

Tissue	n	Concentration: mean ± SD (mmol/kg dry wt)						
		Na	Mg	P	S	Cl	K	Ca
Control	229	56 ± 39	59 ± 22	415 ± 82	214 ± 40	43 ± 19 (17)	554 ± 138	117 ± 48
Tetanus	222	58 ± 37	72 ± 23*	413 ± 82	225 ± 56	42 ± 20 (16)	604 ± 103[b]	48 ± 20[b]

Effect of tetanus (mEq): -138Ca $+ 50$K $+ 26$Mg $= -62$ mEq[b]

[a] Modified from Somlyo *et al.*[1] The concentrations expressed as mmol/liter H_2O are shown in parentheses and are based on a value of 80% H_2O in the I band and 72% H_2O in the TC.
[b] $p < 0.001$, statistical comparisons between control and tetanus.

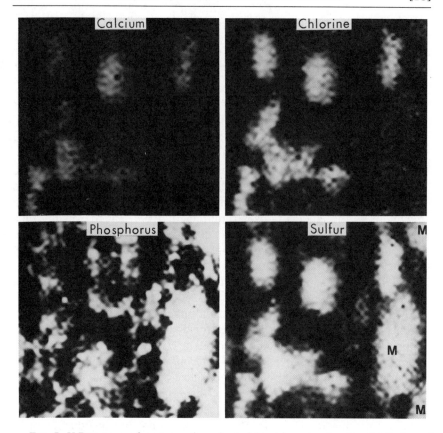

FIG. 7. X-Ray maps of a cryosection of rat right atrium showing the distribution of elements in a region of the nuclear pole containing several atrial granules known to contain atrial natriuretic factor (three discrete granules at the top and lower left of the image) and mitochondria (labeled M in the lower right-hand map). Maps of four elements, calcium, chloride, phosphorus, and sulfur, are shown individually. Note the presence of calcium in the granules but not in the mitochondria, the higher concentrations of sulfur and chlorine in the granules than in the surrounding cytoplasm, and the high phosphorus in the mitochondria compared with the atrial granules. The intensities of the four images were adjusted to represent the number of atoms of each type present in the section.

has permitted reliable EPMA of Cl in cryosections. It is customary to perform EPMA at $-100°$ or below,[11] to eliminate radiation-induced loss of S or Br (if present) and to minimize contamination. A very good ($<5 \times 10^{-8}$ Torr total pressure) microscope vacuum is important and specimens should be at a temperature that will not cause the deposition of ice, from residual water vapor in the column, on the specimen. Ice on the specimen can lead to an extremely severe form of radiation damage, etching. This process, thought to be due to active radicals produced by electrons in ice, can lead to total mass loss: the electron beam can drill a hole in a section covered with ice.[11,21] Deposition of ice can be prevented by interfacing a small residual gas analyzer with the electron microscope column to measure the H_2O pressure, and so maintain the specimen at a temperature at which the vapor pressure of ice is above that of the residual water vapor pressure in the column. The extreme radiation sensitivity of hydrated specimens also implies that frozen hydrated material is generally not suitable for EPMA at high spatial resolution, as the etching caused by the beam currents required for high-resolution analysis would be prohibitive. Inadvertent radiation damage can also occur when frozen hydrated specimens are transferred to be dried in the electron microscope, if the drying process is not properly controlled. EPMA of frozen hydrated specimens is feasible at low resolution with defocussed (low current) probes, particularly in the case of semithick hydrated specimens; this approach has provided considerable useful information about cellular and transcellular pathways of epithelial transport.[22]

Contamination due to the deposition of organic material from oils, vacuum greases, etc. in the microscope column is another potential source of error in quantitation, because the contaminant mass contributes to the X-ray continuum count. This particular problem has been serious with older microscopes, although even in those, contamination could be reduced by maintaining the specimen stage at cryogenic temperatures. The improved vacuum of modern electron microscopes is such that contamination of unstained cryosections, in a well-maintained microscope, should not add sufficient mass to be detectable by EPMA.

Biological Applications

Quantitative EPMA has been successfully applied to a variety of biological systems.[2,11,15,22-24] To illustrate the class of problems that are amenable to this technology, studies of the composition of the sarcoplasmic reticulum in skeletal muscle, at rest and during activation,[1,25] will be briefly summarized. Bundles of approximately 20 frog semitendinous fibers were dissected and frozen in the resting state, during a 1.2-sec teta-

[21] H. Shuman and A. P. Somlyo, *Proc. Natl. Acad. Sci. U.S.A.* **73**, 1193 (1975).
[22] T. A. Hall and B. Gupta, *J. Microsc. (Oxford)* **136**, 20449 (1984).
[23] R. B. Moreton, *Biol. Rev.* **56**, 409 (1981).
[24] T. E. Hutchinson and A. P. Somlyo, "Microprobe Analysis of Biological Systems." Academic Press, New York, 1981.
[25] A. V. Somlyo, G. McClellan, H. Gonzalez-Serratos, and A. P. Somlyo, *J. Biol. Chem.* **260**, 6801 (1985).

nus or at various time points during relaxation from tetanus. Tissue preparation, cryosectioning, and analysis were carried out as described above. Using the second condenser lens, the electron beam was focused to approximately 50 nm diameter for spot mode analysis of individual terminal cisternae (TC) or paired analysis of the adjacent I-band cytoplasm. Approximately 60–70% of the total fiber calcium was localized in the TC in resting frog muscle. During a 1.2-sec tetanus, 60% of the calcium content of the TC was released (Table II), enough to raise total cytoplasmic calcium concentration by approximately 1 mmol/liter of fiber water. This value is considerably higher than the micromolar ionized Ca^{2+} measured with calcium-sensitive dyes, and provided the first direct evidence that intracellular calcium buffers, in this case parvalbumin, play a functional role in muscle during a tetanus. The high spatial resolution of EPMA and the fast time resolution of rapid freezing established that the calcium released from the TC during tetanus is localized in the cytoplasm and not in the longitudinal reticulum. Good time resolution is attainable with EPMA, although very laborious, because tissues have to be frozen for each time point sampled. For example, in exploring further the role of parvalbumin as a calcium buffer in skeletal muscle, muscles were frozen at five different posttetanic intervals,[25] and the TC were analyzed to determine the movement of counterions and the posttetanic rate of return of calcium to the TC. At 400 msec following a 1.2-sec tetanus at room temperature, the force had returned to baseline, and 0.3 mmol Ca^{2+}/liter of cytoplasmic water had been pumped by the SR, indicating that the *in situ* pumping rate of the SR Ca^{2+}-ATPase is sufficiently high to account for the removal of Ca^{2+} from the Ca^{2+}-specific sites of troponin (0.18 mmol of Ca^{2+}-specific sites/liter of cytoplasmic H_2O) and for the rate of relaxation from a tetanus at room temperature. The half-time of the return of the total 1 mmol of Ca^{2+}/liter of cytoplasmic H_2O released during a tetanus was 1.1 sec: comparable to the slow K_{off} rate of Ca^{2+} from parvalbumin and consistent with the hypothesis that the return of this Ca^{2+} to the TC is rate limited by the Ca^{2+} off-rate from parvalbumin. Thus, EPMA is extremely useful in testing whether various processes measured *in vitro*, such as the calcium pumping rates of isolated SR fractions or the affinities of calcium-binding proteins, are also representative of physiological properties *in vivo*. More important, perhaps, is the fact that it is possible to measure with EPMA the concentration of diffusible ions that are translocated during cell fractionation, and the potential of detecting ion movements, such as Mg^{2+} fluxes in the SR, that were not predicted from studies of isolated cell fractions. Furthermore, the distribution of monovalent ions across organelle membranes also provides information about the electrical potential across organelle membranes. A major exam-

TABLE III
MITOCHONDRIAL CALCIUM CONTENT in Situ[a]

Animal	Tissue	Condition	Concentration (mmol/kg dry mitochondrion) ±SEM[b]
Rabbit	Portal vein	Relaxed	1.6 ± 0.2
	Portal vein	Contracted 30'	2.3 ± 0.4
	Portal vein	No Ca	0.1 ± 0.2[c]
	Portal vein	Na-loaded	2.0 ± 1.0
	Main pulmonary artery	Relaxed	2.6 ± 0.31
	Main pulmonary artery	Contracted	2.0 ± 0.32
	Heart (papillary)		0.5 ± 0.2
Guinea pig	Portal vein	Relaxed	0.7 ± 0.3
	Portal vein	Contracted	1.1 ± 0.3
Rat	Liver	In animal	0.8 ± 0.1
	Brain cortex	In animal	1.5 ± 0.26
Rana pipiens	Retinal rod	Dark adapted	0.0 ± 0.2
	Retinal rod	Illuminated	0.4 ± 0.2

[a] Values shown are from studies referenced in A. P. Somlyo and A. V. Somlyo, J. Cardiovasc. Pharmacol. **8,** 542 (1986).
[b] mmol/kg dry wt ≃ nmol/mg mitochondrial protein.
[c] Significantly different ($p < 0.01$) from relaxed rabbit portal vein.

ple of the power of EPMA in decisively resolving a question bearing on *in vivo* metabolic regulation was the demonstration of the low (<3 nmol/mg mitochondrial protein) Ca content of mitochondria in normal cells (Table III).

X-Ray Mapping

Ideally, the distribution of elements in cells could be determined directly by imaging the characteristic fluorescent X-rays emitted by each element, just as molecular fluorophores are imaged in light microscopes. However, the unavailability of large aperture X-ray lenses suitable for focusing X-ray photons precludes this approach altogether, with recent advances in precision microlithography, Fresnel zone plates have been made for focusing soft X-rays ($E < 400$ eV) to better than 1 μm resolution.[26] This is still insufficient for direct imaging of elemental distributions within organelles and for imaging with higher-energy (less thickness dependent) K shell fluorescent X-rays. The only current alternative to direct imaging is to collect, as in EPMA, the X-rays generated by a finely focused electron probe. In EPMA the ultimate attainable resolution is determined by the dimensions of the probe, rather than the X-ray detector. Instead of obtaining images globally, elemental maps are generated sequentially, by rastering the probe and collecting X-ray spectra from each point (pixel) in the specimen. In the simplest method of X-ray mapping, often used with metallurgical specimens, digital images of the elemental distribution are formed, by counting for each pixel of the specimen, the number of X-rays detected within a characteristic peak selected in a range of channels (a "window") of an energy-dispersive X-ray spectrum.[26] If several such windows are selected, multiple element maps and mass thickness maps can be created. This method works well for high-concentration specimens, and 8.7-nm resolution maps of molybdenum in negatively stained catalase crystals have been obtained.[27] However, the concentrations of physiologically relevant elements in normal cryosections of biological specimens and, therefore, the characteristic peak-to-background ratios, P/B, in energy-dispersive X-ray spectra, are generally small. For a Ca concentration of approximately 100 mmol/kg, the $P/B = 1$. In addition, low X-ray count rates are generated in thin specimens, with the low currents obtainable in high-resolution electron probes. For the spectrum shown in Fig. 4, the total X-ray count rate was 1000 counts/sec, with only 20 counts/sec in the Ca peak region. Yet high spatial resolution requires the use of small probes and thin specimens. Binary images, with small P/B and low peak counts, are difficult to interpret and quantitate. An alternative method for X-ray mapping has been developed to overcome these problems. The probe is still scanned across the specimen, but the beam is allowed to dwell on a pixel for a long enough time to collect and X-ray spectrum with sufficient statistics for accurate concentration measurements. The required dwell time can be estimated from the concentration sensitivity needed, the detector efficiency, and the current available in the probe. To detect a Ca concentration of 30 mmol/kg in a 60-nm-thick (carbon film equivalent) specimen, the required dwell time per pixel is estimated to be 5 sec with a 1.0% geometric collection efficency detector and a 5-nA beam current into a 10-nm-diameter probe. In order to map a 1 μm^2 area of specimen at 10 nm resolution, a 13-hr collection time is required.

If the X-ray spectra are quantitated while the beam is scanning, real time elemental maps can be generated in the same way dark- or bright-field scanning transmission electron microscope (STEM) images are produced. The ML fit used for quantitative EPMA, although slow (0.5 sec for

[26] A. P. Somlyo, *J. Ultrastruct. Res.* **83**, 135 (1985).
[27] A. P. Somlyo and H. Shuman, *Ultramicroscopy* **8**, 219 (1982).

a 12-element fit on a Vax 750 or 1.0 sec on a PDP 11/34) compared to the windowing method, accurately subtracts the background from beneath the peaks and, as noted earlier, also adjusts for small changes in detector calibration, as is required for quantitating Ca in the presence of high concentrations of K.

The basic instrumental requirement for the slow scan method is a computer-controlled scan generator, closely coupled to a conventional multichannel energy-dispersive X-ray detector, and a simple digital image display system. There are, of course, as many ways to implement this idea as there are microscopists willing to assemble the system. The microprocessor used is a matter of personal choice, cost, availability of the required interface hardware, and computer bus compatibility with preexisting X-ray multichannel analyzers. Suitable systems can be made using IBM PC, Dec LSI 11, or MicroVax, Apple II, and many other microprocessors. The interfaces needed are (1) microprocessor access to the memory of the multichannel analyzer, (2) a programmable clock to select the pixel dwell time, (3) a pair of digital-to-analog converters to position the electron beam with the microscopes scan coils, under computer control, (4) one or two analog-to-digital converters (or counters) to digitize STEM bright- or dark-field signals for further computer processing, and (5) a solid-state frame buffer to store and display slow-scan STEM images on a TV monitor. The hardware basis for our system is a raster scan LSI 11-based interface card that performs synchronous data collection and scan generation.[28] The board operates in three modes. In the image collection mode, the timing for the scan is provided by an internal programmable clock, which simultaneously steps the beam position in the microscope, reads the count rates from bright- and dark-field voltage-to-frequency converters, reads the counts in four windows of a multichannel analyzer, and supplies interrupt signals to the microprocessor (a MicroVax I) for end-of-pixel and end-of-line data transfer. In the spectral collection mode the automatic scanning is disabled, and the beam position is directly controlled by the MicroVax. In the focus mode no interrupts or data transfers occur and the images are displayed on a long-persistence CRT monitor as in analog scanning microscopes. Digital images are stored and diplayed with a Recognition Concepts frame buffer system on the MicroVax Q-bus. The MicroVax is coupled via a DecNet link to the multichannel analyzer processor (PDP 11/34) so that X-ray spectral analysis is also under its control.

Slow-scan X-ray mapping has been used to determine the distributions of Mg, Ca, and P in *E. coli* B cells.[29] Previous reports indicated that Ca is

---

[28] D. Kowarski, *J. Electron Microsc. Tech.* **1**, 175 (1984).
[29] C.-F. Chang, H. Shuman, and A. P. Somlyo, *J. Bacteriol.* **167**, 935 (1986).

bound to isolated *E. coli* outer membrane and it has also been suggested that Mg is localized to the nucleoid. The X-ray maps showed Ca uniformly distributed around the periphery of the cell, while Mg and P are distributed equally and uniformly throughout the cell, with no preferential localization of Mg. This experiment also revealed that, although this method can be quite powerful, the long-term mechanical instability of electron microscope stages limits the obtainable image quality by introducing large drift distortions in the maps. It would be difficult (and very expensive) to implement a "drift-free stage"; however, the drift can be compensated for with a cross-correlation feedback loop using STEM dark-field imaging.

The feedback needed is easily obtained with the computer-controlled raster generator, and follows the ideas already in use for following live cells in computer-controlled light microscopes. In our implementation a reference dark-field fiduciary image from an area of the specimen is collected before the long-term mapping is begun. At appropriate time intervals during mapping, additional fiduciary images are collected and cross-correlated with the reference image to compute the intervening drift, and a new set of rastering coordinates computed to continue the mapping.

The data collection proceeds as follows. After the areas of the specimen to be mapped are selected and centered in conventional TEM, the microscope is switched to STEM focus mode, where the various optical adjustments are made and specimen position refined. Two partial frames out of a maximum 512×512 pixels are then selected, one for a fiduciary region and the second for the map region. The partial frames are hardware preselected to be 32, 64, 128, or 256 pixel squares, although arbitrary-sized rectangles can be chosen with a computer-coupled joystick. The microprocessor switches to image collection mode, the reference dark-field fiduciary image is collected and transferred to a frame store. Although the absolute coordinates of the fiduciary and map regions can vary during data collection, their relative positions are fixed. For each pixel in the map region the computer system performs the following operations concurrently: (1) switches to spectral mode, positioning the beam on the next point of the specimen to be analyzed, waiting for a predetermined number of seconds while the Kevex 7000 spectrometer collects a spectrum, (2) the computer calculates the cross-correlation (with a fast Fourier transform) between the reference and the most recent fiduciary images, determines the relative drift during the last analysis point, stores the spectrum from the last pixel on a data file, does a ML fit to the last spectrum acquired, and stores the peak count and the computed concentrations in another file. In the next sequence of operations, the computer reads the new spectrum from the X-ray spectrometer, collects a new fiduciary image, clears the spectrometer, and repositions the beam using the last computed offset. Maps that are 128 × 128 pixels large, acquired for as long as 22 hr, have been collected with this software.

The X-ray maps illustrated in Fig. 7 were collected with this computer-controlled cross-correlation feedback loop. Shown are a cluster of atrial-specific granules, known to contain prohormones of atrial natriuretic peptides, and mitochondria at the nuclear pole in a cryosection from rat atrium. A complete X-ray spectrum was obtained from each 15-nm pixel, thus equivalent to a compositional resolution of 30 nm. The brightness of the image is proportional to the number of atoms of a given element. The granules, but not the mitochondria, contain a high concentration of Ca (71 mmol/kg dry wt) that may reflect Ca binding to grouped acidic residues of the atrial natriuretic peptide prohormone sequence. Based on the volume occupied by atrial granules, their Ca content represents approximately 17% of total cell Ca. The S content is compatible with the six S-containing residues of the atrial natriuretic peptide precursors. Based on the X-ray continuum measurements of the relative hydrations, the Cl concentration of granules is 6.1-fold higher than in the adjacent cytoplasm, which, if free, would lead to a +47 mV equilibrium potential. Incubation in NaBr and low Cl resulted in Br accumulation by the granules, demonstrating that the granules are anion permeable and can not only maintain, but also generate, an anion gradient, possibly through an inward-directed proton-pumping membrane ATPase similar to that in other secretory granules and in the Golgi system.

The computer-controlled scanning X-ray analysis system can also eventually be used to automate point mode data collection, so that once the biologist has selected areas of interest on the basis of specimen morphology, any further need for user intervention during analysis is eliminated. In a typical session, a large-area STEM image would be collected and displayed on the frame buffer and a correlation region selected for automatic repositioning. The user would then outline the analysis areas on the image with a joystick, with the appropriate coordinates and analysis times stored in computer memory. The computer then begins X-ray data collection from the specified areas automatically, thereby reducing the most labor intensive and tedious aspects of electron probe X-ray microanlaysis.

Acknowledgment

Supported by HL15835 to the Pennsylvania Muscle Institute.

[14] Measurement of Cytosolic Free Ca^{2+} with Quin2

By ROGER TSIEN and TULLIO POZZAN

Introduction: General Advantages and Disadvantages

The fluorescent Ca^{2+} indicator dye, quin2 (Fig. 1), is the basis of one of the most popular methods for measuring cytosolic free Ca^{2+} [$(Ca^{2+})_i$], especially in suspensions of mammalian cells. Briefer reviews of the use of quin2 have appeared previously.[1-3] Its widespread use probably stems from the following advantages that it has over other methods for measuring cytosolic free Ca^{2+}. (1) Quin2 can be loaded into the cytosol of millions of cells simultaneously, without any micromanipulations or any disruption of the plasma membrane.[4] Cells are simply incubated with quin2/AM, the tetra(acetoxymethyl) ester of quin2 (Fig. 1). This uncharged, hydrophobic derivative of quin2 diffuses freely across membranes but is gradually hydrolyzed by cytoplasmic esterases, eventually regenerating quin2 itself, which remains trapped in the cytosol as a relatively impermeant tetranion. All other methods, particularly those using ion-selective microelectrodes, aequorin, and arsenazo III, require breaching of the plasma membrane to enable the sensor molecules to contact the cytoplasm. Because of the enormous electrochemical gradient of Ca^{2+} across the plasma membrane, any threat to membrane integrity is bad for the cells and for the measurement. The nondisruptive method of loading quin2 via permeant hydrolyzable esters permits its use on cells too small or delicate to tolerate puncture or lysis and resealing. Also, the ability to measure an average $[Ca^{2+}]_i$ from large numbers of cells can be very helpful in establishing correlations of $[Ca^{2+}]_i$ with cell responses that must also be measured from whole populations, for example secretion, proliferation, or metabolism of cylic nucleotides, lipids, or proteins. (2) Quin2 is usually monitored with a conventional cuvette spectrofluorimeter. Both the dye and the instrumentation are widely available by purchase or loan from colleagues, whereas other methods tend to require much specialized or custom equipment. (3) Quin2 readings are usually easily calibrated at typical resting levels of $[Ca^{2+}]_i$ ($\sim 10^{-7}$ M) and below (to 10^{-8} M), whereas all other techniques do their best at detecting activated levels ($\sim 10^{-6}$ M)

---

[1] R. Y. Tsien, *Annu. Rev. Biophys. Bioeng.* **12**, 94 (1983).
[2] R. Y. Tsien, T. Pozzan, and T. J. Rink, *Trends Biochem. Sci.* **9**, 263 (1984).
[3] T. J. Rink and T. Pozzan, *Cell Calcium* **6**, 133 (1985).
[4] R. Y. Tsien, T. Pozzan, and T. J. Rink, *J. Cell. Biol.* **94**, 325 (1982).

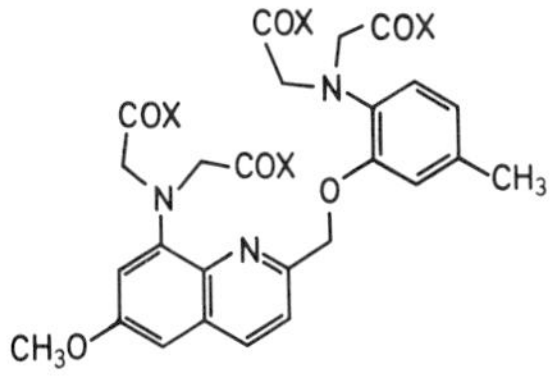

FIG. 1. Structure of Ca^{2+} indicator quin2 and its acetoxymethyl ester quin2/AM. Reproduced from Ref. 4, *The Journal of Cell Biology* (1982) 94:325–334, by copyright permission of the Rockefeller University Press.

and are nearly at their detection limits at resting levels. (4) Quin2 avoids many of the drawbacks particular to other $[Ca^{2+}]_i$ measurement techniques, for example, the irreversible destruction of aequorin by Ca^{2+}, the poor Mg^{2+} rejection and messy Ca^{2+} stoichiometry of arsenazo III and antipyrylazo III, the slow response time and need for voltage referencing of ion-selective electrodes, and the null point method's need to destroy large numbers of cells in low extracellular $[Ca^{2+}]$ for each data point.[1,5]

Of course, quin2 suffers from significant disadvantages, most of which were described in the first full articles[4,6] on the technique and which have been repeatedly rediscovered by subsequent authors. (1) The fluorescence of quin2 is not exceptionally bright, so that intracellular loading of several hundred micromolar is usually required for the quin2 signal to dominate cellular autofluorescence. This much quin2 can partially buffer $[Ca^{2+}]_i$ transients, though it generally does not alter $[Ca^{2+}]_i$ levels set by long-term homeostatic processes. This buffering can actually be extremely useful when examining whether $[Ca^{2+}]_i$ is really essential for a given cell response such as exocytosis or phosphoinositide breakdown.[6-9] $[Ca^{2+}]_i$ rises can often be suppressed to controllable extents by judicious use of varying quin2 loading, extracellular Ca^{2+} deprivation, and ionophore pretreatments. Comparisons of damped $[Ca^{2+}]_i$ signals and cell responses from the very same populations are a powerful tool to show whether $[Ca^{2+}]_i$ rises are really important, often revealing that they are

---

[5] R. Y. Tsien and T. J. Rink, *in* "Current Methods in Cellular Neurophysiology" (J. L. Barker and J. McKelvy, eds), pp. 249–312. Wiley, New York, 1983.
[6] T. Pozzan, P. Arslan, R. Y. Tsien, and T. J. Rink, *J. Cell Biol.* **94**, 335 (1982).
[7] F. DiVirgilio, D. P. Lew, and T. Pozzan, *Nature (London)* **310**, 691 (1984).
[8] T. J. Rink, S. W. Smith, and R. Y. Tsien, *FEBS Lett.* **148**, 21 (1982).
[9] L. M. Vicentini, A. Ambrosini, F. DiVirgilio, T. Pozzan, and J. Meldolesi, *J. Cell Biol.* **100**, 1330 (1985).

not as necessary as previously hypothesized. This seems an important and underused application of quin2, especially because other measurement techniques are much less amenable to variation in the extent of $[Ca^{2+}]_i$ buffering. But when rapid $[Ca^{2+}]_i$ transients are to be measured with minimum perturbation, an indicator working at lower loadings than quin2 would be highly desirable. (2) Obviously, quin2 is inappropriate for tissues that are highly opaque due to pigment or scattering or that are particularly sensitive to near-UV-light. (3) Because quin2 signals Ca^{2+} primarily by its fluorescence at a single set of excitation (339 nm) and emission wavelengths (490–500 nm), its signal becomes more difficult to calibrate in terms of absolute $[Ca^{2+}]_i$ values as one switches from suspensions to monolayers to single cells. A dye that shifted its preferred wavelengths rather than just changing its amplitude of fluorescence would be much easier to calibrate, since a mere change in dye content or instrumental sensitivity would not masquerade as a $[Ca^{2+}]_i$ change.[4,10] (4) Quin2 can be bleached by high illumination levels either in a cuvette or on a microscope.[11] (5) Quin2 does have some sensitivity to Mg^{2+}. Mg^{2+} by itself causes little change in the fluorescence with standard wavelengths of 339 nm excitation, but does act as a competitive inhibitor of Ca^{2+}-binding and associated fluorescence enchancement.[4,12] For a given error in estimating or guessing $[Mg^{2+}]_i$, quin2 at resting $[Ca^{2+}]_i$ levels would be much less affected than arsenazo III or an organophosphate-based ion-selective electrode, similarly or slightly less perturbed than aequorin would be, but more affected than an electrode using neutral carriers.[1,5] (6) A few tumor cell types have enough heavy metals in them to perturb the dye.[13] (7) Not all cells can be loaded by the gentle means of hydrolyzing membrane-permanent esters. Plant, bacterial, and a scattering of invertebrate cells load poorly or not at al,[14,15] whereas most vertebrate cells seem to load, with the possible exception of some but not all types of muscle. (8) In cells that do load by ester hydrolysis, the chemical by-products, formaldehyde and protons,[16] could have harmful side effects. (9) Quin2 is maximally sensitive to Ca^{2+} levels near its Ca^{2+} dissociation constant, $\sim 10^{-7} M$;

[10] G. Grynkiewicz, M. Poenie, and R. Y. Tsien, *J. Biol. Chem.* **260**, 3440 (1985).
[11] B. A. Kruskal, C. H. Keith, and F. R. Maxfield, *J. Cell Biol.* **99**, 1167 (1984).
[12] R. Y. Tsien, *Biochemistry* **19**, 2396 (1980).
[13] P. Arslan, F. DiVirgilio, M. Beltrame, R. Y. Tsien, and T. Pozzan, *J. Biol. Chem.* **260**, 2719 (1985).
[14] R. J. Cork, *Plant Cell Environ.* **9**, 157 (1986).
[15] J. I. Korenbrot, D. L. Ochs, J. A. Williams, D. L. Miller, and J. E. Brown, *in* "Optical Methods in Cell Physiology" (P. DeWeer and B. M. Salzberg, eds), pp. 347–363. Wiley, New York, 1986.
[16] R. Y. Tsien, *Nature (London)* **290**, 527 (1981).

much higher levels ($>10^{-6}$ M) nearly completely saturate the dye and cannot be distinguished from each other.[4] Moreover, if the dye resides in several compartments with different $[Ca^{2+}]_i$, a measurement that ignores the compartmentation will underestimate the deviations from the mean.[6]

Many of the above problems can be alleviated by more recently introduced relatives of quin2. These newer fluorescent indicators,[10] especially fura-2 and indo-1, are much brighter in fluorescence than quin2. Because so much less dye needs to be introduced, both the Ca^{2+} buffering and the potential for toxicity from hydrolysis products are greatly reduced. The new dyes shift wavelengths not just intensity upon binding Ca^{2+}, so calibration of signals from monolayers and single cells is greatly eased. Fura-2 is much more resistant to photobleaching[17] than quin2 and has better selectivity for Ca^{2+} over Mg^{2+} and heavy metals. Because fura-2 and indo-1 are sensitive enough to give large calibratable signals from single cells or regions of cells,[18-20] they allow direct measurement of compartmentalization and heterogeneity. For those reasons, we expect quin2 eventually to be replaced in most applications by fura-2, indo-1, or even newer relatives under development. This article, written in 1985–1986, concentrates mainly on quin2, the longer established indicator, whereas the newer dyes are discussed in more recent reviews.[21-25] Many of the protocols and cautions described below are similar for all the dyes. Moreover, to take full advantage of the advanced properties of the new dyes does require more elaborate equipment, so many newcomers to the field of $[Ca^{2+}]_i$ measurement will still want to get started using quin2 in a conventional cuvette fluorometer.

Dye Loading

The protocol for loading quin2 into intact cells was originally described in lymphocytes[4,26] and has been successfully applied with minor modifications to a variety of cell types, mostly mammalian. Basically,

---

[17] D. A. Williams, K. E. Fogarty, R. Y. Tsien, and F. S. Fay, *J. Gen. Physiol.* **86**, 37a (1985).
[18] R. Y. Tsien, T. J. Rink, and M. Poenie, *Cell Calcium* **6**, 145 (1985).
[19] D. A. Williams, K. E. Fogarty, R. Y. Tsien, and F. S. Fay, *Nature (London)* **318**, 558 (1985).
[20] M. Poenie, J. Alderton, R. Steinhardt, and R. Y. Tsien. *Science* **233**, 886 (1986).
[21] R. Y. Tsien and M. Poenie, *Trends Biochem. Sci.* **11**, 450 (1986).
[22] P. H. Cobbold and T. J. Rink, *Biochem. J.* **248**, 313 (1987).
[23] R. Y. Tsien, *Trends Neurosci.* **11**, 419–424.
[24] R. Y. Tsien, *Methods Cell Biol.* **30**, 127 (1989).
[25] R. Y. Tsien, *Annu. Rev. Neurosci.* **12**, in press.
[26] R. Y. Tsien, T. Pozzan, and T. J. Rink, *Nature (London)* **295**, 68 (1982).

quin2/AM from a stock solution in dimethyl sulfoxide (DMSO) is added to cell suspensions or monolayers. After some time the cells are washed, resuspended in fresh medium, and recording of fluorescence is initiated. Major variables in this protocol are (1) source of quin2/AM, (2) quin2/AM concentration and cell density, (3) incubation time and temperature, and (4) addition of serum or albumin.

Source

Quin2 and quin2/AM are now available from several commercial sources: Lancaster Synthesis, Amersham, Calbiochem-Behring, Sigma, Aldrich, and Dojindo Laboratories. The first five firms have branches both in the United States and Europe, while Dojindo is at 2861 Kengunmachi, Kumamotoshi (862), Japan. [It is our understanding that Amersham nonradioactive material and Sigma are merely repackagings of the Dojindo material.] We have not seen any consistent difference in quality between products from different companies, though there have been several isolated incidents of bad batches. Quin2 free acid is a freeflowing powder; its color is white or yellow, depending on purity and degree of protonation. Quin2/AM is a yellow gum or resin that should be fairly hard at refrigerator temperature but usually becomes sticky at room temperature. Routinely we keep both compounds desiccated over silica gel at $-20°$ for long-term storage, since the protonated free acid can gradually decarboxylate and the AM ester can hydrolyze and/or oxidize. (The only derivative known to be crystallizable and indefinitely stable at room temperature is the tetraethyl ester, which is still the best source for critical studies on quin2 tetranion.)

Tests for the quality of quin2, in order of increasing laboriousness and stringency, are determination of extinction coefficient ($\varepsilon = 5000\ M^{-1}\ cm^{-1}$ at 354 nm in Ca^{2+}- and Mg^{2+}-free medium above pH 7.0 with millimolar EDTA or EGTA to chelate contaminating Ca^{2+}), determination of fluorescence excitation spectra before and after addition of excess Ca^{2+}, and determination of K_D for Ca^{2+}. Measurement of the extinction coefficient tests for gross inert contamination. Fluorescence excitation spectra at micromolar dye concentrations in high Ca^{2+} should show a peak near 339 nm excitation, whose amplitude above blank background is five- to sevenfold greater than the amplitude above background of the same dye concentration at the same wavelength in zero Ca^{2+} with EDTA or EGTA. Examples of such excitation spectra at saturating and zero Ca^{2+} are shown in Fig. 2 as the two extremes of the series of spectra. If adequate signal amplitude or stability cannot be attained, either the dye or fluorometer could be deficient; if an adequate enhancement due to Ca^{2+} cannot be

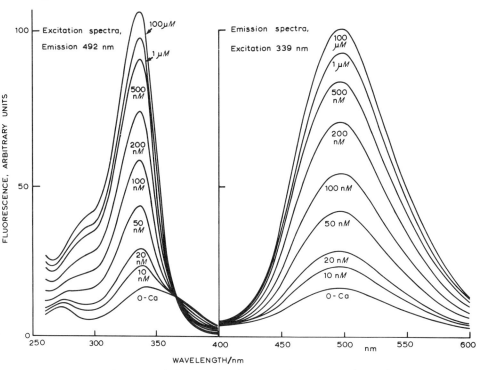

FIG. 2. Excitation and emission spectra of 20 μM quin2, with varying [Ca^{2+}] as shown against an ionic background of 120–135 mM K$^+$, 20 mM Na$^+$, 1 mM free Mg^{2+}, and pH 7.05 at 37°. Reproduced from Ref. 4, *The Journal of Cell Biology* (1982) 94:325–334, by copyright permission of the Rockefeller University Press. Further experimental details may be found in that article.

found with a small excitation bandpass (<5 nm), then bad dye should be suspected. Measurement of the K_d for Ca^{2+} using intermediate buffered Ca^{2+} values (as in Fig. 2 and the section Calibration Procedure: Quin2 Affinity for Ca^{2+}, below) tests that all the molecules have the same high-affinity binding site; if some molecules lack one or more carboxylates, they would have lower affinities, yet would show the same spectra at zero and very high Ca^{2+}.

Unlike quin2, the ester quin2/AM is much more soluble in organic solvents than in water and has no Ca^{2+}-binding properties. When one has access to a cell system known to hydrolyze quin2/AM, the latter is usually tested merely by its ability to load quin2 into the cells. If other tests are required, thin-layer chromatography on silica gel is convenient. A solution of the dye in chloroform or dichloromethane is spotted on a

Merck 5554 silica gel TLC plate, 10 cm long, developed with hexane–ethyl acetate (1 : 1, v/v), and visualized under 254 or 365 nm illumination. Quin2/AM runs with R_f = 0.27 in this system. Unfortunately DMSO perturbs this R_f. Minor fluorescent contaminants are often present and usually tolerable. Yet another test of quin2/AM is to hydrolyze the dye chemically. Thus a small aliquot of a 1–10 mM, quin2/AM solution in DMSO is mixed with an equal volume of 0.1–1 M aqueous NaOH or KOH. After waiting 5–10 min, the mixture should be a stock solution of quin2 tetranion which can be tested as described above. Of course, base hydrolysis does not test the completeness or correctness of esterification.

Quin2/AM stock solutions of 1–10 mM concentration are usually made in DMSO because of its ready miscibility with water, biological inertness in most systems, nonvolatility, and ease of microliter pipetting. The DMSO should be anhydrous, and stock solutions of quin2/AM should be kept frozen and desiccated. Because the melting point of DMSO (18°) is sharply depressed by water content, moisture pickup is easily seen as rapid melting of the solution upon removal from the refrigerator.

Quin2/AM Concentration and Cell Density

Typically, the 1–10 mM stock solution of quin2/AM in DMSO is diluted 100- to 1000-fold into the cell suspension while stirring or shaking. Depending on cell type and other variables discussed below, typically 5–40% of the total quin2/AM added will become trapped as quin2 inside the cells. Since the cells usually occupy 0.1–1% of the volume of the suspension, intracellular quin2 concentrations reach 0.1 to 5 mM, substantially higher than the 1–100 μM initial concentration of quin2/AM in the suspension.

Quin2/AM before hydrolysis is actually quite hydrophobic and poorly soluble in simple salines. Light-scattering measurements (T. Pozzan, unpublished observations) suggest that above a few micromolar, the ester forms colloidal suspensions or precipitates. Therefore loading efficiency, or the percentage of quin2/AM usefully trapped, is improved by squirting the dye solution directly into a dense and stirred suspension of cells.[4] High concentrations of dye ester without sufficient cells or protective proteins simply precipitate dye wastefully, some of which may then be endocytosed and generate undesired intracellular background fluorescence.

Though loading efficiency is improved by high initial cytocrits, nearer 1% than 0.1%, the cells soon acidify the medium. Also, the unavoidable death of a fraction of the cells may release enzymes that can harm other cells and hydrolyze extracellular quin2/AM. Therefore, dense suspensions should be diluted up to 10-fold a few minutes after the ester is added, since that interval seems sufficient to complete the initial association of the ester with cells, though not the full hydrolysis.

Incubation Time and Temperature

Quin2/AM is transformed to quin2 tetranion by four separate hydrolysis steps, one for each acetoxymethyl group. It is essential that this process be completed before measurements are begun. Any dye molecule with incompletely hydrolyzed ester groups contributes fluorescence yet is crippled in its Ca^{2+}-binding capability, leading to an underestimate of $[Ca^{2+}]_i$. If hydrolysis continues during the observation of fluorescence, the gradual increase in Ca^{2+} affinity will cause a rising curve that will simulate a progressive rise in $[Ca^{2+}]_i$. To aid completion of hydrolysis, it is important that the free concentration of the starting tetraester be reduced practically to zero well before observation begins. Otherwise the cells never get a chance to clear the metabolic pipeline of partially hydrolyzed intermediates. In dense suspensions of cells with high esterase activity, as in our initial studies with blood-forming cells, this exhaustion of quin2/AM is automatic as the cells take up and hydrolyze the ester. In some cases precipitation can help reduce the quin2/AM concentration. But if cell density or esterase activity is low, the experimenter should help by washing the cells with ester-free media after the initial contact period of a few minutes to 30 min. The cells should then be incubated in dye-free media until complete hydrolysis of the trapped intermediates which still bear one, two, or three ester groups. A quick and crude indication[4] of hydrolysis is a shift in the emission peak from 430 to 490–500 nm, matching that of the dye tetranion with or without Ca^{2+}. However, this widely used test is not fully rigorous, because an acetoxymethyl group lingering on the benzene ring portion of quin2 would not affect the fluorescence properties of the quinoline portion. The most rigorous test for full hydrolysis is to release the dye by lysing the cells, then titrate the dye with known buffered Ca^{2+} levels and quantitatively verify its responsiveness.[4] Obviously, such titration does not have to be done routinely, but should be checked during the establishment of a loading protocol for the particular tissue being studied.

We and most others have allowed about 1 hr incubation for full hydrolysis of the four ester groups. Some cells apparently hydrolyze the ester faster (for example, hepatocytes load well in just 2.5 min[27]), some more

[27] A. Binet, B. Berthon, and M. Claret, *Biochem. J.* **228**, 565 (1985).

slowly (i.e., some cultured cell lines), some not at all. In most cases the loading temperature has been 37°. There are a few reports, mainly in monolayers, where the loading was performed at room temperature. One of us (TP) has observed in brain synaptosomes, neutrophils, and the insulinoma cell line RINm5F, that loading at temperatures below 37° seems to result in trapping of quin2/AM in cell compartments different from the cytoplasm. In these cases, the emission spectrum retains a significant shoulder at 430 nm, indicating noncomplete hydrolysis. This shoulder surprisingly cannot be abolished either by prolonging the incubation time or by rewarming the cell to 37°. Another sign of such compartmentation is that digitonin treatment, which normally releases most cellular quin2, releases a smaller fraction (70–80%) of the dye; the residue requires Triton X-100 for release.

Albumin and Serum during Loading

Inclusion of either bovine serum albumin (0.5–1%) or fetal calf serum (1–5%) in the loading medium generally tends to increase the loading efficiency, probably because the proteins reduce the precipitation of quin2/AM and act as buffers for it. Moreover they increase the viability of most cell types. Very high concentrations of protein, as in platelets in their own plasma, may somewhat reduce loading efficiency by excessive competition for the dye, but the minimization of cell manipulation more than outweighs the waste of dye (T. J. Rink, personal communication). BSA and FCS can have some fluorescence of their own, and definitely reduce the potency of calcium ionophores,[28] so they are omitted or minimized in the final medium during actual measurements.

Recording the Fluorescence

Once dye loading is complete, the cells are typically washed once or twice to remove leaked or excess dye and hydrolysis by-products. Resuspension for fluorescence measurements should be in a medium containing as little as possible of such absorbing and fluorescing substances as phenol red, fetal calf serum, tryptophan, riboflavin, and pyridoxine. Cells that are not to be used immediately are resuspended in whatever medium at whatever temperature will best preserve their viability and dye content; often, room temperature is suitable. If the cells have been kept for a considerable time, a final wash just before use is helpful again to remove leaked dye and fluorescent additives from the medium.

[28] L. O. Simonsen, *J. Physiol. (London)* **318**, 34P (1981).

A wide variety of commercially available spectrofluorometers have been used successfully with quin2 in cell suspensions in cuvettes, so instrumental requirements are not stringent. However, it is important to scan the excitation and emission spectra to verify the quality and rate of dye loading. Therefore, filter fluorometers working at fixed wavelengths are not adequate unless the user is very experienced with quin2 and the tissue particularly tractable. Probably the most important characteristics of a good spectrofluorometer for quin2 studies are stability of output signal over the time scale of an experiment, and ability of the sample compartment to maintain whatever thermostatting, gassing, and/or stirring are needed for the particular cell type. Instrumental stability is easily tested by placing an unstirred solution of quin2 free acid in buffer of known high or low [Ca^{2+}] and looking for an unchanging signal amplitude. Assuming temperature equilibrium has been reached, any progressive decline in signal suggests a failing instrument, or more likely, dye bleaching due to excessive illumination intensity. Bleaching is confirmed if the signal stops declining or even partially recovers either upon stirring the solution or temporarily blocking the illumination. Bleaching can be minimized by reducing the monochromator slit settings or inserting filters, screens, or apertures in the excitation beam. Most commercial instruments have enough sensitivity if in good repair to reach a stable and sufficiently quiet signal with excitation intensity low enough to avoid significant bleaching over the 0.5–1 hr time scale. Very high sensitivity is not required, since intrinsic autofluorescence from the cells is usually readily detectable at moderate gain and one generally has to load enough dye (usually several hundred micromolar) to overcome autofluorescence by a factor of two or more.

Thermostatting and gassing of the cuvette are obvious requirements for many cell preparations. Continual stirring is often unnecessary with small and well-dispersed cells, which settle sufficiently slowly so that occasional manual stirring with a Pasteur pipette is adequate. However, larger cell or clumps of cells need continual stirring. Magnetic drive is better than an overhead paddle because addition of solutions is unobstructed.

Traditional 1 cm square quartz cuvettes are expensive, fragile, and very wasteful of solution volume and cells. We find that cheap round shell vials or test tubes made of borosilicate glass are usually adequate[3] except for the most precise physicochemical measurements. The glass walls are thin enough to absorb a negligible amount (a few percent) of the incident UV beam. The cylindrical shape reduces wasted volume in the corners, improves stirring, yet seems not to cause optical problems. Depending on precise dimensions, sample volumes one-fourth to one-half that of 1 cm^2

cuvettes are sufficient, reducing usage of tissue and dyes correspondingly. The vials are rugged and cheap enough to be centrifugable, a help when assessing leakage. The main disadvantage of the vials is the need to build a metal adapter to center the vial in a conventional square sample holder.

Quin2 has also been used in cell monolayers grown on glass coverslips.[29] The coverslip is clamped by spring tension in a holder that fits in a square cuvette and maintains the coverslip at a near-diagonal orientation to the fluorimeter beams. Coverslips obviously are advantageous for anchorage-dependent cells; also they permit complete *in situ* replacement of the solution, whereas one can add to but not subtract from a cuvette of suspended cells. However, stability of the fluorescence amplitude is more difficult with coverslips, since minute motions of the coverslips, bleaching in the fixed zone of illumination, and detachment of cells can cause severe artifacts. Some success has been reported[30] in growing anchorage-dependent cells on microcarrier beads and suspending them in cuvettes, but this variant has not achieved popularity, perhaps because the considerable light scattering from the beads tends to contaminate or obscure the fluorescence signal, and the vigorous stirring needed to keep the beads in suspension tends to scrape the cells off.

Calibration Procedure

The fluorescence signal recorded from loaded cells reflects not only some sort of mean cytosolic free Ca^{2+} concentrations, $[Ca^{2+}]_i$, but also many other factors. These necessarily include the effective affinity of quin2 for Ca^{2+}, the concentration of quin2 within the cells, the number of cells in the effective sample volume, and the sensitivity of the fluorometer. Furthermore, if any significant quantity of dye is present in noncytoplasmic compartments such as extracellular space or intracellular organelles, such dye will contaminate the overall fluorescence signal. The purpose of calibration is to give a quantitative estimate of $[Ca^{2+}]_i$ that is as independent of the other variables as possible. We first discuss the determination of the effective affinity of quin2 for Ca^{2+}, then the several methods available for handling uncertainties in dye content, location, and instrumental sensitivity. Operationally, calibration is a much simpler procedure than one might think from the length of the following discussion, which attempts to summarize the major alternative procedures and their implicit assumptions.

[29] W. H. Moolenaar, L. G. J. Tertoolen, and S. W. deLaat, *J. Biol. Chem.* **259**, 8066 (1984).
[30] J. D. R. Morris, J. C. Metcalfe, G. A. Smith, T. R. Hesketh, and M. V. Taylor, *FEBS Lett.* **169**, 189 (1984).

Quin2 Affinity for Ca^{2+}

Ideally, one would determine the dissociation constant (K_D) of the quin2-Ca^{2+} complex in actual cytoplasm. However, this would require a highly accurate method for setting or reading $[Ca^{2+}]_i$ independently of quin2 itself. Such methods are rarely available in the cell systems to which quin2 is most suited. Instead one has to determine the K_D *in vitro* in buffers believed to mimic the main ionic constituents of cytoplasm, as described below. Are these *in vitro* K_D values applicable to cytoplasm? Probably yes, because quin2 estimates of $[Ca^{2+}]_i$ are rather close to those obtained by quite independent techniques. For example, estimates of resting $[Ca^{2+}]$ in toad stomach smooth muscle[19] cells are 129 ± 4 nM, 137 ± 13 nM, and 148 ± 20 nM by quin 2, fura-2, and ion-sensitive microelectrodes, respectively. In hepatocytes, quin2 values of 200 nM[31] and 160 nM[27,32] compare well with 211 ± 9 nM and 190 ± 10 nM using aequorin[33] and null points,[34] respectively.

The *in vitro* affinity of quin 2 for Ca^{2+} is determined by setting free Ca^{2+} to a series of accurately known values, measuring the dye fluorescence at each value, and fitting the data to the theoretical equation for one-to-one binding.

$$F = F_{min} + (F_{max} - F_{min}) \left(\frac{[Ca^{2+}]}{[Ca^{2+}] + K_D} \right)$$

Examples of the fluorescence data and the fit are shown in Figs. 2 and 3. This titration was done against a constant background of 120 mM K^+, 20 mM Na^+, 1 mM free Mg^{2+}, pH 7.05, at 37°, an ionic milieu believed to correspond in major cation composition to lymphocyte cytoplasm.[4,35] Nearly all other investigators using quin2 have continued to use the K_D, 115 nM, obtained from Fig. 3. However, the effective K_D is definitely sensitive to ionic strength, temperature, free Mg^{2+}, and pH values if they fall below about 6.5–6.8. In addition, any deterioration of the dye sample or incompleteness of ester hydrolysis will generate chemical heterogeneity that can jeopardize the quantitative calibration. Therefore it is highly desirable, though admittedly tedious, to generate the equivalent of Figs. 2 and 3 on lysates of loaded cells from each new tissue system being studied, with the correct temperatures and ionic background for that system.

---

[31] R. Charest, P. F. Blackmore, B. Berthon, and J. H. Exton, *J. Biol. Chem.* **258**, 8769 (1983).

[32] A. P. Thomas, J. Alexander, and J. R. Williamson, *J. Biol. Chem.* **259**, 5574 (1984).

[33] N. M. Woods, K. S. R. Cuthbertson, and P. H. Cobbold, *Nature (London)* **319**, 600 (1986).

[34] E. Murphy, K. Coll, T. L. Rich, and J. R. Williamson, *J. Biol. Chem.* **255**, 6600 (1980).

[35] T. J. Rink, R. Y. Tsien, and T. Pozzan, *J. Cell Biol.* **95**, 189 (1982).

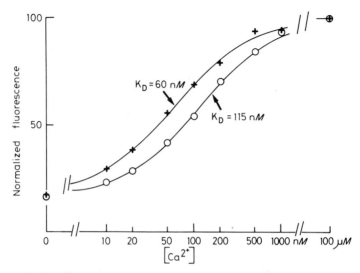

Fig. 3. Quin2 calibration curves in 120–135 mM K$^+$, 20 mM Na$^+$, pH 7.05, at 37°, and two different [Mg^{2+}] values. Normalized quin2 fluorescence (excitation 339 nm; emission 492 nm) plotted against free [Ca^{2+}] at 0 Mg (crosses) and 1 mM free Mg^{2+} (circles). The data points are fitted with curves corresponding to one-to-one stoichiometry and the indicated apparent dissociation constants. Reproduced from Ref. 4, *The Journal of Cell Biology* (1982) 94:325–334, by copyright permission of the Rockefeller University Press.

Ionic Strength and Monovalent Cation Concentration. So far, quin2 has been mostly used in mammalian cells, though a few reports of its application to other eukaryotic cells have also been published. In mammalian cells it is likely that the ionic strength and total monovalent cation concentration are normally rather constant. However, a change in ionic strength from that typical of a mammalian cell to that typical of marine invertebrate axoplasm increases the Ca^{2+} dissociation constant of related tetracarboxylate chelators by three- to fourfold.[10] Though we have not measured the quin2 K_D at high ionic strength, we would expect the same increase to apply.

Temperature. Increasing the temperature moderately increases the dissociation constant for Ca^{2+}, but a more significant effect is to decrease the fluorescence intensity of quin2 at fixed [Ca^{2+}]. A change from 20° to 37° decreases the fluorescence intensity of quin2 by a factor of approximately 2. This effect explains the common experience that cold solutions that are put in a warm thermostatted cuvette often decrease in fluorescence for the first few minutes before reaching a steady baseline level.

pH. Quin2 and other tetracarboxylate Ca^{2+} indicators are much less sensitive to pH changes around 7 than their parent EGTA is. This is because the highest pK_a of quin2 is ≤6.5 compared to 9.58 for EGTA. The effective K_D of quin2 for Ca^{2+} is increased by acidification of the medium below pH 6.8, while above this value pH changes are practically ineffective. In lymphocytes we measured[35] a resting pH_i of 7.05 for pH_o = 7.4, hence the choice of pH for Fig. 2. Since in most mammalian cells pH_i has been shown to range between 6.9 and 7.4, pH_i is probably not a major source of error in the calibration of $[Ca^{2+}]_i$. There are, however, some manipulations or drugs which can acidify pH_i considerably, which would complicate their effects on quin2 signals. Examples are mitochondrial uncouplers, anoxia, and permeant weak acids.

$[Mg^{2+}]_i$. There are many fewer direct measurements of free $[Mg^{2+}]_i$ than of pH_i. Our measurement of about 1 mM in lymphocytes[35] was the basis for the choice of $[Mg^{2+}]$ in Fig. 2. The affinity of quin2 for Mg^{2+} is in the 1–2 mM range, so an increase of Mg^{2+} can significantly increase the effective K_D of quin 2 for Ca^{2+}. Thus a change of Mg^{2+} from 0 to 1 mM results in a increase of the apparent K_D of quin 2 for Ca^{2+} from 60 to 115 mM, respectively.[4] Fortunately, the standard excitation wavelength of 339 nm not only maximizes the effect of Ca^{2+} on the fluorescence but also minimizes the direct effect of Mg^{2+} on the fluorescence, which is why 339 mn was chosen. At other wavelengths Mg^{2+} directly affects the spectrum as well as competing for Ca^{2+} (Fig. 4). There is little evidence up to now for changes of $[Mg^{2+}]_i$ upon cell stimulation, but the possibility of variability of the $[Mg^{2+}]_i$ between cell types has to be remembered.

Quin2 Localization in Cells

One of the most attractive features of the methodology is that most if not all of the trapped quin2 seems to reside in the cytoplasm or nucleus as opposed to other organelles. This generalization seems valid for the cells we have investigated, i.e., lymphocytes, platelets, neutrophils, Ehrlich and Yoshida carcinomas, and the cell lines PC12, RINm5F, and HL60. However, few workers have investigated in detail the localization of quin2 within other cell types, which may act differently.

Intracellular localization has been determined in various ways: (1) Direct morphological observation by fluorescence microscopy[11] reveals mainly diffuse rather than punctuate fluorescence, suggesting a cytoplasmic localization. (2) Detergents like digitonin, which preferentially permeabilize membranes with high-cholesterol content (like the plasma membrane), release practically all trapped quin2 in parallel to cytoplasmic

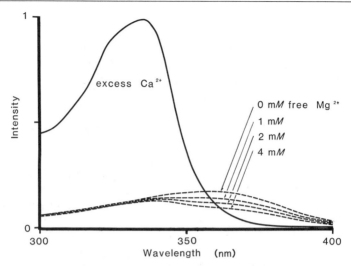

FIG. 4. Effect of Mg^{2+} on the excitation spectrum of quin2. Though Mg^{2+} has little effect at 335–340 nm, it significantly depresses the excitation amplitude at 350–370 nm, so that the ratio of the excitations at these two bands can be perturbed by physiological [Mg^{2+}]. The solution contained 10 μM quin2 in 132 mM KCl, 1 mM K_2H_2EGTA, 10 mM K-MOPS, pH 7.18, at 20°. $MgCl_2$ was added to reach 1, 2, and 4 mM free [Mg^{2+}], then 4 mM $CaCl_2$ was added to saturate the dye with Ca^{2+}. Excitation bandpass was 1.8 nm; emission was collected at 500 nm with 9.25 nm bandpass. Excitation spectra were corrected by a rhodamine B quantum counter.

markers, with minor release of enzymes from mitochondria, lysosomes, and endoplasmic reticulum.[4,13,36,37] (3) High voltage discharge, which makes small holes in the plasma membrane, releases most intracellular quin2.[4] (4) Incubation of quin2/AM with partially purified cellular fractions reveals that the relevant esterase activity is localized in the soluble cytosolic fraction.[4] (5) Measurement of extracellular quin2 after strong stimulation of secretion indicates that, in platelets[8] and neutrophils,[36] quin2 is not appreciably contained in secretory granules. However, it appears that mast cell granules can pick up substantial amounts of fluorescence.[38]

In lymphocytes we also found[4] that the nuclear membrane does not constitute a permeability barrier for quin2, so that the dye is neither concentrated nor excluded from the nucleus. Since most evidence had suggested that ionic activities are the same in the nucleus as in the cyto-

[36] T. Pozzan, D. P. Lew, C. Wollheim, and R. Y. Tsien, *Science* **221**, 1413 (1983).
[37] J. Meldolesi, W. B. Huttner, R. Y. Tsien, and T. Pozzan, *Proc. Natl. Acad. Sci. U.S.A.* **81**, 620 (1984).
[38] W. Almers and E. Neher, *FEBS Lett.* **192**, 13 (1985).

plasm, we felt that the nucleus could be considered as part of the cytoplasm as far as quin2 and Ca^{2+} were concerned. However, recent evidence suggests that, in toad stomach smooth muscle cells, the nucleus preferentially accumulates dye[39] and may regulate Ca^{2+} to a level different from the cytoplasm.[19] Obviously this question needs reexamination in other cell types.

Calibration by Lysis

This procedure has been adapted from the calibration commonly used[40] for the Ca^{2+}-sensitive photoprotein, aequorin. The cells are lysed at the end of the experiment, usually with a detergent, to release all the indicator into the medium where Ca^{2+} is known to be high (for example, 1 mM). The fluorescence F_{max} thus obtained represents the fluorescence intensity of the dye at saturating Ca^{2+}. Ca^{2+} in the medium is then chelated with excess EGTA and base to establish F_{min}, the fluorescence of the dye at zero Ca^{2+}. Any fluorescence value F intermediate between F_{min} and F_{max} corresponds to a free [Ca^{2+}] given by the following calibration equation:

$$[Ca^{2+}]_i = K_D \left(\frac{F - F_{min}}{F_{max} - F} \right) \tag{1}$$

Assumptions in this equation are that all the dye molecules share the same dissociation constant K_D for binding Ca^{2+} in a 1 : 1 complex, that the fluorescence is a linear function of the concentrations of Ca^{2+} bound and unbound dye, that any cellular autofluorescence has either been deducted or is unaffected by lysis, and that the instrument's sensitivity to dye fluorescence is not affected by lysis. This last assumption should be true in a dilute well-stirred cell suspension in a cuvette, since the average number of dye molecules in the macroscopic light path of the spectrofluorometer is the same before or after lysis. However, this assumption is insecure when a cell monolayer is held at the center of the cuvette, since lysis allows dye to spread into nonilluminated corners of the cuvette. The assumption is even worse when single cells are viewed by microscopy.

As discussed below in the section Correction for Extracellular Dye, Autofluorescence, and Heavy Metals, it is often advantageous to lower external Ca^{2+} to zero before lysis, then determine F_{min} before F_{max}.

Since F_{max} and F_{min} at 339 nm excitation are not affected by Mg^{2+}, pH above 6.8, K^+, or Na^+, it is not necessary that these variables are the

---

[39] D. A. Williams and F. S. Fay, *Am. J. Physiol.* **250**, C779 (1986).
[40] J. R. Blinks, W. G. Wier, P. Hess, and F. G. Prendergast, *Prog. Biophys. Mol. Biol.* **40**, 1 (1982).

same in the lysate as inside the cells. Thus to calculate $[Ca^{2+}]_i$, only F_{max}, F_{min}, and the K_D inside the cells need to be known. One detail that is sometimes neglected is that, to obtain a true F_{min} in Ca-EGTA mixtures, it is necessary to raise the pH of the medium above 8. This is required to lower the effective K_D of EGTA for Ca^{2+} to nanomolar levels.

Calibration with Ionophore and Mn^{2+}

A method for driving the quin2 to two extreme states without actually lysing the cells is to use ionomycin, a Ca^{2+} ionophore. This calibration[41] is particularly advantageous in cells attached to coverslips and in single cells viewed by fluorescence microscopy.[42] Another useful application of this calibration is when very low quin2 loadings are used, so that the autofluorescence artifacts due to cell lysis become large compared to the quin2 signal.

The actual procedure is the following: At the end of the experiment, in the presence of extracellular Ca^{2+}, a large dose of a Ca^{2+} ionophore is added to increase $[Ca^{2+}]_i$ to several micromolar. Normally ionomycin is used instead of A23187, because the latter is fluorescent at the quin2 wavelengths. Ionomycin is now commercially available from Calbiochem (Behring Diagnostics). The fluorescence level obtained with ionomycin is assumed to be F_{max}. Mn^{2+} at ~ 1 mM is then added to quench intracellular and extracellular quin2 fluorescence. The intracellular signal can be quenched because the ionophore transports Mn^{2+} into the cells. When a new steady level is reached, usually requiring a few minutes, the residual signal should correspond to cell autofluorescence. F_{min} is then calculated according to the equation $F_{min} = 0.16(F_{max} - \text{autofluorescence})$. Note that F_{min} does not equal the autofluorescence after Mn^{2+}. The factor 0.16 (see Refs. 41 and 42) derives from the observation that, in pure dye solutions *in vitro*, F_{min} is typically 16% of F_{max}. In different samples, F_{min}/F_{max} has varied from 0.12 to 0.20. Then Eq. (1) is applied. In order to verify that the dose of Ca^{2+} ionophore is sufficient to increase $[Ca^{2+}]_i$ to F_{max}, this calibration has to be checked with the lysis method. In some cells, for example, in the insulinoma cell line RINm5F, ionophore treatment does not approach F_{max} as measured by lysis unless the ionomycin concentration is high enough to cause lysis by itself. We have occasionally observed a paradoxical opposite, that the signal with the ionophore is higher than F_{max} obtained by lysis. There are two possible explanations

---

[41] T. R. Hesketh, G. A. Smith, J. P. Moore, M. V. Taylor, and J. C. Metcalfe, *J. Biol. Chem.* **258**, 4876 (1983).

[42] J. Rogers, T. R. Hesketh, G. A. Smith, M. A. Beaven, J. C. Metcalfe, P. Johnson, and P. B. Garland, *FEBS Lett.* **161**, 21 (1983).

for this result: quenching of extracellular quin2 by traces of heavy metals present in the medium (see the section Correction for Extracellular Dye, Autofluorescence, and Heavy Metals, below) or incomplete hydrolysis. In this latter case it is probable that the intensity of fluorescence of unhydrolyzed or partially hydrolyzed quin2 is higher inside the cells compared to the medium.

Calibration by the Ratio of Two Excitation Wavelengths

Excitation spectra of quin2 at varying Ca^{2+} levels (Fig. 2) show maximal sensitivity to Ca^{2+} at 335–340 nm and a crossover point somewhere between 355–360 nm. At the crossover point, the dye fluorescence is independent of Ca^{2+}. The ratio of intensity at 335–340 nm excitation to that at or above the crossover wavelength reflects the Ca^{2+} level but is independent of dye concentration and path length. In principle, such ratios can be calibrated to read Ca^{2+} without requiring any lysis of the cells or use of Ca^{2+} ionophore. Ratio calibration is particularly desirable when observing single cells by microscopy. In practice, however, quin2 is poorly suited to ratio calibration, despite attempts[11] to use it in that way. The wavelength and relative amplitude of the crossover point, but not the Ca^{2+} peak are quite sensitive to free Mg^{2+} (Fig. 4) and environmental viscosity, so that these parameters must be known and matched in the calibration medium. Quin2 fades rapidly at the high illumination intensities of fluorescence microscopy, but the photoproduct(s) retains some fluorescence whose excitation ratio is different from quin2 (R. Y. Tsien, unpublished observations). Autofluorescences at the two wavelengths must be accurately deducted before forming the ratio. Such subtraction of autofluorescence is particularly difficult at the longer excitation wavelength, at which quin2 is dim but cellular fluorophores such as NADH are peaking. Boosting the excitation intensity at the longer wavelength, for example, by using a mercury instead of xenon arc, can make the short- and long-wavelength quin 2 signals more nearly equal but cannot overcome the contamination by autofluorescence. Ratio calibration only becomes practical with dyes of much greater brightness and reduced Mg^{2+} sensitivity, such as fura-2 and indo-1 (see Refs. 10 and 18–20).

Correction for Extracellular Dye, Autofluorescence, and Heavy Metals

Whichever calibration procedure one decides to use, often corrections for (1) autofluorescence, (2) extracellular dye, and (3) intracellular heavy metals need to be performed before $[Ca^{2+}]_i$ is calculated.

Autofluorescence. Correction for autofluorescence or scattered stray light due to the cells themselves is critical if one uses the ratio mode. With the two other calibration procedures, any constant background level of autofluorescence is cancelled out in $F - F_{min}$ or $F_{max} - F$. Only changes of autofluorescence need to be deducted. Such changes are simply assessed by subjecting an aliquot of unloaded cells, at the same cytocrit as used for the loaded cells, to the entire sequence of manipulations in the cuvette. Usually the largest change in autofluorescence is observed upon cell lysis. Its effect on the calibration is greatest with low loadings of intracellular quin2.

There are a number of compounds which fluoresce at the wavelength of quin2, such as the Ca^{2+} ionophor A23187, anticalmodulin drugs such as trifluperazine, etc. Their fluorescence has to be subtracted before calibrating the signal. Other compounds are not fluorescent but quench the signal of quin2. Examples include the mitochondrial uncouplers FCCP, CCCP, and 2,4-dinitrophenol, and the microtubule inhibitor colchicine. These UV-absorbing drugs probably just attenuate the excitation beam before it reaches the zone from which the emission is collected. Excessive turbidity due to too dense a cell suspension can have a similar effect. Such inner-filter effects can be assessed by control experiments in which drugs or unloaded cells are added to quin2 in buffer. Inner filtering can be reduced by either reducing the optical density, offsetting the cuvette, decreasing its path length, or collecting emission from its front face rather than at right angles to the excitation.

Extracellular Dye. Ideally, the quin2 fluorescence signal measured should originate only from dye trapped in the cell cytoplasm. In practice, it is common to find some extracellular dye even after repeated washings. Its contribution to the overall fluorescence signal needs thus to be subtracted before calculating $[Ca^{2+}]_i$. Extracellular quin2 can derive from (a) carryover from the loading incubation, (b) cell breakage during centrifugation and resuspension of the pellet, or (c) leakage during the experiment. In this latter case, quin2 can either leak from intact cells or be released as a consequence of cell death.

Carryover (a) is a trivial problem and can be easily solved by more thorough washing. Several maneuvers can be tried to reduce extracellular quin2 to effects of (b) and (c), but it has proved impossible in our hands to eliminate it completely in cell suspensions. Extracellular quin2 due to (b) varies considerably between cell types. It is quite large (20–40%) in cell types which require trypsinization during preparation, such as in hepatocytes and pancreatic acini and in the two ascites carcinoma lines Ehrlich and Yoshida. On the other hand, extracellular quin2 is of minor relevance in lymphocytes and practically negligible in platelets and neutrophils. Compared to extracellular quin2 found immediately after spinning and resuspension, quin2 leakage during the experiment has usually proved to be less important. We calculated a leakage rate of 10–15% of total content in 1 hr at 37° in mouse thymocytes and up to 30% under the same conditions for Ehrlich ascites carcinomas. The leakage rate is very temperature sensitive, being negligible at room temperature. Thus, after the loading we usually keep the stock suspension of cells at room temperature or even in a water bath at 10–20°.

The most obvious way to determine the extent of extracellular quin2 is to centrifuge an aliquot of the cell suspension and measure the amount of quin2 in the supernatant. This procedure is simple but tedious to repeat on every sample. Alternatively, the amount of extracellular quin2 can be determined *in situ* either by quenching with Mn^{2+} or by chelation of extracellular Ca^{2+} with EGTA. Because the Mn^{2+}–quin2 complex has negligible fluorescence, and quin2 binds Mn^{2+} about 500 times more avidly than it binds Ca^{2+},[41] Mn^{2+} concentrations $> 10^{-5}$ M nearly completely quench the fluorescence signal of quin2 even in the presence of millimolar concentrations of Ca^{2+}. Because Mn^{2+} has a very low permeability through the plasma membrane of resting cells in the absence of ionophore, 50–100 μM external Mn^{2+} will quench external quin2 immediately, while intracellular dye is only slowly affected. To continue the experiment, Mn^{2+} can then be chelated with a slight excess of diethylenetriaminepentaacetic acid (DTPA), a chelator whose $Mn^{2+}:Ca^{2+}$ preference, 63,000,[43] greatly exceeds that of quin2 or EGTA. This should restore fluorescence to the initial value. Sometimes the fluorescence restoration due to DTPA is larger than the quenching due to Mn^{2+}, because traces of heavy metals were already present in the medium. The opposite discrepancy, only partial restoration of fluorescence by DTPA, can occur when the membrane permeability to Mn^{2+} has been increased so that Mn^{2+} could relatively rapidly quench the intracellular signal as well. However, DTPA itself, with five negative and two positive charges, is very unlikely to enter any cell still capable of trapping quin2. Therefore the fluorescence restoration due to DTPA is still a good index of extracellular quin2. The quenching of intracellular and extracellular signal by Mn^{2+}, and their different sensitivities to DTPA can be turned into a powerful experimental tool when it is necessary to distinguish whether a rise in quin2 fluorescence is due to a rise in $[Ca^{2+}]_i$ or due to quin2 release into the medium,[41] and when studying influx through "Ca^{2+} channels" that also let Mn^{2+} or Ni^{2+} through.[44] In the first case, including Mn^{2+} in the medium before adding the stimulus will give a decrease of fluorescence if

---

[43] A. E. Martell and R. M. Smith, "Critical Stability Constants," Vol. 1. Plenum, New York, 1974.
[44] T. Hallam and T. J. Rink, *FEBS Lett.* **186**, 175 (1985).

quin2 leakage occurs, which will be completely reversed on DTPA addition. In the second case, the entrance of Mn^{2+} into the cells will cause a quenching of the intracellular quin2 fluorescence that cannot be reversed by DTPA.

Another way of determining extracellular quin2 is to chelate all extracellular Ca^{2+} with EGTA and a base such as Tris to raise the pH of the medium to >8. Extracellular quin2, or quin2 in leaky cells, responds immediately to a drop in extracellular Ca^{2+}, whereas intracellular quin2 should respond only gradually. Thus the immediate drop of fluorescence upon addition of EGTA + Tris is taken as a measure of extracellular quin2. Because of the change in pH required by this method, it is usually performed at the end of the experiment, just before lysis. Obviously, extracellular dye must not have been quenched by traces of heavy metals, so it is advisable to include $10^{-5}-10^{-4}$ M DTPA in the startup medium.

Figure 5 gives an example of how $[Ca^{2+}]_i$ can be calibrated in the

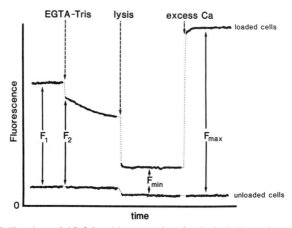

FIG. 5. Calibration of $[Ca^{2+}]_i$ with correction for leaked dye using the EGTA-Tris method. These traces are fictitious but representative of typical real records. The top trace indicates that fluorescence of quin2-loaded cells subjected to sudden chelation of extracellular Ca^{2+}, lysis, then readdition of excess Ca^{2+}. The bottom trace represents the autofluorescence of a matching aliquot of unloaded cells undergoing the same manipulatons, though in practice it would not be necessary to make the timings between additions exactly parallel to the run with loaded cells. In this example, the autofluorescence is shown as decreasing upon lysis; this actually varies with cell type. F_1, F_2, F_{min}, and F_{max} each are fluorescence amplitudes measured above autofluorescence levels. F_1 can be measured at any time desired as long as no increase in leakage happens between F_1 and F_2; F_2 should be measured as soon as possible after mixing, with back extrapolation if necessary; F_{min} and F_{max} are measured whenever their amplitudes have stabilized. Then $[Ca^{2+}]_i$ corresponding to F_1 is $K_D(F_2 - F_{min})/F_{max} - F_1)$ where K_D is the quin2 effective dissociation constant for Ca^{2+}, typically 115 nM under conditions representative of mammalian cytoplasm.

presence of extracellular dye, using the EGTA–Tris protocol to measure leakage. The top trace starts with intact quin2-loaded cells in medium with >10 μM Ca^{2+}. The lower trace represents the signal from a matching sample of unloaded cells, which will be treated with the same sequence of manipulations so that the autofluorescence contribution can be deducted at each stage. Let x be the fraction of leaked dye, and $(1 - x)$ the fraction still in the cells. The composite fluorescence F_1 due to the dye (autofluorescence having been subtracted) is $xF_{max} + (1 - x)([Ca^{2+}]_i F_{max} + K_D F_{min})/([Ca^{2+}]_i + K_D)$, where F_{max} and F_{min} as usual denote the fluorescence that the dye would have if all of it were exposed to >5 μM or <2 nM Ca^{2+}, respectively. Then external $[Ca^{2+}]$ is abruptly lowered to <2 nM by addition of EGTA and Tris base, then the dye fluorescence is immediately remeasured as F_2. The section How to Calculate and Prepare Ca-EGTA Buffers (see below) discusses how to compute the necessary pH rise and EGTA addition; it may be valuable to rehearse with blank medium to determine in advance how much base is needed for the pH change. If one can assume that x and $[Ca^{2+}]_i$ have not yet changed, then $F_2 = xF_{min} + (1 - x)([Ca^{2+}]_i F_{max} + K_D F_{min})/([Ca^{2+}]_i + K_D)$. The fluorescence will probably then drop gradually as x increases or $[Ca^{2+}]_i$ decreases by pumping. Upon complete lysis of the cells by any means, all the dye is exposed to very low Ca^{2+}, yielding a direct measure of F_{min}. F_{max} then results from readdition of enough Ca^{2+} to the lysate to give a reliable excess (say 1 mM) of total Ca^{2+} over EGTA. Solution of the above expressions for F_1 and F_2 gives the simple equations:

$$x = (F_1 - F_2)/(F_{max} - F_{min})$$
$$[Ca^{2+}]_i = K_D(F_2 - F_{min})/(F_{max} - F_1)$$

The latter equation is remarkably like Eq. (1) except that F_2 and F_1 replace F in numerator and denominator, respectively.

Intracellular Quenching by Heavy Metals. Some cell types, notably dedifferentiated tumor cell lines, already contain enough exchangeable heavy-metal ions to quench intracellular quin2.[13,40] Heavy metals such as Mn^{2+}, Zn^{2+}, Fe^{2+}, and Cu^{2+} not only directly suppress quin2 fluorescence but also competitively reduce Ca^{2+} binding.[36] Such quenching is not constant but tends to be relieved when $[Ca^{2+}]$ is raised very high to determine F_{max} in either the lysis or ionophore calibration procedures. It is also removed when F_{min} is attained by alkaline EGTA addition to lysates. Since intracellular heavy metals depress F more than they affect F_{max} or F_{min}, they lead to underestimation of $[Ca^{2+}]$. The error is greatest at low quin2 loadings, because the exchangeable heavy-metal pool is of limited size.

Testing for dye quenching by endogenous heavy metals is now very simple with a membrane-permeant, heavy-metal-specific chelator N,N,N',N'-tetrakis(2-pyridylmethyl)ethylenediamine (TPEN).[13] TPEN has much higher affinity for heavy metals but lower affinity for Ca^{2+} than quin2 has. Unlike DTPA, TPEN is almost uncharged at physiological pH and readily crosses membranes as the free base. TPEN can effectively strip heavy metals from quin2 both in homogeneous solution and when applied externally to liposomes containing trapped quin2 and heavy metals. A similar effect has also been demonstrated in intact cells both on endogenous as well as deliberately introduced heavy metals.[13] The amount of intracellular quenching by heavy metals varies significantly between cell types. Among the cells tested by us, the highest quenching was observed in two ascites carcinomas (Ehrlich and Yoshida) and the lowest in human neutrophils. We thus suggest that TPEN should always be tested in cell preparations when absolute values of $[Ca^{2+}]_i$ are of importance. TPEN is now commercially available from Molecular Probes and Calbiochem-Behring.

Direct evidence that tetracarboxylate Ca^{2+} indicators bind Zn^{2+} inside tumor cells has been obtained[45] with 5FBAPTA, an analog of quin2 which can be monitored by fluorine nuclear magnetic resonance. The NMR chemical shifts of 5FBAPTA, Ca-5FBAPTA, and several heavy metal–5FBAPTA complexes are different. Ehrlich ascites tumor cells showed a peak for Zn–5FBAPTA whose intensity was about 30% of the Ca–5FBAPTA peak.

Measurement of Dye Content

Measurement of the intracellular dye loading is particularly important when considering Ca^{2+} buffering or other potential perturbing effects of the dye or the loading process. Some workers present such data only in terms of initial ester concentration in the suspension, but the final cellular content of de-esterfied dye is obviously more directly relevant. The latter needs actually to be measured because it depends not only on the nominal ester concentration but on the purity of the ester, the exact loading protocol, and any subsequent dye leakage.

Intracellular dye content is most simply determined by measuring $(F_{max} - F_{min})$ from the cells and comparing it to the $(F_{max} - F_{min})$ obtained from one or more samples of quin2 free acid of known concentration, analogously exposed to saturating versus zero Ca^{2+}. The comparison establishes the concentration of quin2 in the cell lysate or suspension. One

---

[45] J. C. Metcalfe, T. R. Hesketh, and G. A. Smith, *Cell Calcium* **6**, 183 (1985).

then needs to know the cytocrit, or ratio of cell volume to total suspension volume. Typically, this is estimated by hemacytometer or Coulter counting of the cells before lysis, together with some estimate for the average volume per cell. The prelysis concentration of dye in the cells is then just the lysate or suspension concentration divided by the cytocrit. Thus if lysate ($F_{max} - F_{min}$) corresponds to 1 μM dye, and the cells had occupied 0.1% of the suspension volume, the dye content was 1 mM. In this example and all of our work, such concentrations are with respect to total cell volume, though they could be expressed in moles per liter of cytosolic volume or intracellular water if one had the appropriate values per cell.

Some authors have preferred to estimate loadings using radioactively labeled dye and scintillation counting.[41] This approach could be valuable in tissues that are too pigmented or autofluorescent for the standard method. Otherwise, tracer labeling would seem to offer little advantage, since the radioactive material is much more costly than ordinary dye and requires additional data on radiochemical purity, specific activity, and counting efficiency.

Side Effects of Quin2 Loading

Quin2 loading often affects cell function, especially at higher levels of dye content. Therefore it is always desirable to check the relevant physiological functions of the tissue, especially acute responses that can be measured in the same time scale as the quin2 fluorescence signals. Most known effects of quin2 loading are probably due to one of three obvious mechanisms, in order of increasing prevalence: (1) heavy metal chelation by quin2; (2) toxicity or acidification induced by the formaldehyde and acetic acid by-products of acetoxymethyl (AM) ester hydrolisis; (3) Ca^{2+} buffering by quin2.

1. Just as intracellular quin2 fluorescence can be perturbed by heavy metals, the dye could be chelating heavy metal ions that are potentially important in cell function.[13] The obvious test for such a mechanism would be to see if the membrane-permeant heavy metal chelator TPEN has similar effects on the cell. Because TPEN has higher affinities than quin2 has for heavy metals, and TPEN can probably reach all compartments of the cell, TPEN efficacy should exceed that of quin2. Failure of TPEN to affect cell function probably rules out heavy metal chelation as a mechanism for quin2, whereas a finding of TPEN activity is more ambiguous. Because of this ambiguity and the only recent commercial availability of TPEN, heavy metal chelation has not yet been proven as the mechanism for any particular side effect of quin2 loadings.

2. Each molecule of quin2/AM eventually generates four molecules of

formaldehyde, four of acetate, and eight protons upon full hydrolysis.[16] The acetate ions are probably innocuous and carry away four of the protons as they diffuse out of the cell. The remaining protons ought still to be easily managed by the normal pH-regulatory mechanism of the cell, though instances have been reported[46] in which a small depression of internal pH persists. Such acidification may really be an indirect consequence of the formaldehyde, which is undoubtedly the most worrisome side product. Fortunately, the formaldehyde is generated gradually and may partly be able to diffuse out of the cell. The most clearly analyzed consequence of the formaldehyde produced by AM ester hydrolysis is inhibition of glycolysis in human red cells due to formaldehyde depletion of NAD and formation of NADH.[47] A recommended remedy[48] for this effect is to include several millimolar pyruvate in the medium. Pyruvate can not only reoxidize NADH but also serve as an alternative substrate for oxidative phosphorylation in cells with mitochondria. A further defense against other reactions of formaldehyde is ascorbate at several hundred micromolar. Ascorbate is known to block for example, the reactions of formaldehyde with lysine residues.[49] Pyruvate and ascorbate together seem sufficient to protect even retina, a tissue notoriously sensitive to formaldehyde (G. Ratto, W. G. Owen, and R. Y. Tsien, unpublished results; see also Ref. 15). Further agents that might be beneficial include thiols such as mercaptoethanol, dithiothreitol, or permeant glutathione esters, though such agents alone were ineffective in erythrocytes.[47]

An additional test for toxicity due to AM ester by-products is to use an AM ester of a carboxylic acid with negligible Ca^{2+} affinity. A particularly realistic model is the acetoxymethyl ester of o-anisidine-N,N'-diacetic acid ("anis1/AM"[5] or "APDA AM ester"[41]). Because anis1/AM represents one isolated half of the quin2/AM molecule, it should be given at twice the concentration of quin2/AM. Indeed, anis1/AM can reproduce the lowering of cellular ATP sometimes caused by quin2/aM. Unfortunately, anis1/AM is not yet commercially available. Less realistic but more accessible agents for comparison are methylene diacetate and extracellular formaldehyde itself.

An ideal solution to the formaldehyde problem would be to replace the acetoxymethyl ester group by another α-acyloxyalkyl group. For example, α-acetoxyethyl esters would release acetaldehyde, whose lesser toxicity than formaldehyde is reflected in our ability to imbibe ethanol not

---

[46] D. C. Spray, J. Nerbonne, A. Campos de Carvalho, A. L. Harris, and M. V. L. Bennett, *J. Cell Biol.* **99**, 174 (1984).

[47] T. Tiffert, J. Garcia-Sancho, and V. L. Lew, *Biochim. Biophys. Acta* **773**, 143 (1984).

[48] J. Garcia-Sancho, *J. Physiol. (London)* **357**, 60P (1984).

[49] L. Trezl, I. Rusznak, E. Tyihak, T. Szarvas, and B. Szende, *Biochem. J.* **214**, 289 (1983).

methanol. α-Acetoxyethyl esters of simple carboxylic acids and even of anis1 do hydrolyze satisfactorily, but when applied to all four carboxylates of quin2, the extra carbon atoms reduce solubility to an unacceptable degree and prevent significant loading.

3. Ca^{2+} buffering, the most common and most inherent side effect of quin2 loading, has already been discussed in the Introduction. Obscure side effects can result from Ca^{2+} buffering, for example, inhibition of Na_i/Ca_o exchange in squid axons.[50] This effect is equally well produced by EGTA and appears to be a consequence of slowing a rise in $[Ca^{2+}]_i$. The normal mode of operation of the exchanger in resting cells, Na_o/Ca_i exchange, is not affected by quin2 or EGTA.

The Ca^{2+} buffering caused by quin2 can be highly useful if properly controlled and varied. Valuable information can be deduced concerning the origin and the role of the $[Ca^{2+}]_i$ rises induced by agonists. For example, if a $[Ca^{2+}]_i$ rise originates only from an intracellular pool of limited capacity, increasing intracellular quin2 should change the amplitude of the rise but hardly affect its kinetics.[4,6] Furthermore, by measuring the amplitude of the $[Ca^{2+}]_i$ rises at different intracellular quin2 concentrations, it should also be possible to calculate the actual amount of Ca^{2+} released and the endogenous Ca^{2+} buffering capacity of the cell.[5] However, if the $[Ca^{2+}]_i$ rise is due to the establishment of a new steady-state level of Ca^{2+} influx and efflux at the plasma membrane, then quin2 should increase the time required to attain the steady state, but not affect the level eventually achieved. Most agonists often act both on the influx and on the release of Ca^{2+} from stores, and the effect of quin2 loading is a mixture of the two. Another effect sometimes observed with high concentrations of quin2 is that the release from stores and the influx from the extracellular medium become well separated kinetically, since the release from stores is completed in less than 10 sec while the influx continues for several minutes. Varying the amount of quin2 in the cell is not only useful to study the mechanism of the $[Ca^{2+}]_i$ rises induced by agonists, but also to study the relationships between Ca^{2+} and cell activation. In fact, if other side effects can be excluded, then the effect of Ca^{2+} buffering on cell responses shows whether Ca^{2+} was important in that response. A further effective manipulation to study the causal relationship between $[Ca^{2+}]_i$ and cellular responses is to load the cell with quin2 in the absence of Ca^{2+} in the extracellular medium. Under these conditions, the cells have no Ca^{2+} with which to titrate the dye to the normal resting level, so quin2 loading will result in a dramatic reduction of the resting $[Ca^{2+}]_i$.[4,6-8] Surprisingly, this use of quin2, i.e, as a Ca^{2+} buffer as well as a Ca^{2+} indica-

---

[50] T. J. A. Allen and P. F. Baker, *Nature (London)* **315**, 755 (1985).

tor, has found little attention among investigators other than the present authors and their collaborators. Although newer Ca^{2+} indicators such as fura-2 and indo-1 may supersede quin2 as a Ca^{2+} indicator, quin2 may remain important as an intracellular Ca^{2+} buffer because it is easier than the new indicators to load into cells at high concentrations. In conclusion, quin2 loading can inherently alter the kinetics and/or the amplitude of $[Ca^{2+}]_i$ changes, but this perturbation can be turned into an invaluable tool if used appropriately.

How to Calculate and Prepare Ca-EGTA Buffers

Practically all methods for measuring or controlling intracellular free Ca^{2+} require reference media with $[Ca^{2+}]$ well buffered at low levels. Usually, such buffers are based on the Ca^{2+} chelator EGTA, ethylene glycol bis(2-aminoethyl)ether N,N,N',N'-tetraacetic acid. Other buffers are known[12] with the same Ca:Mg selectivity as EGTA but lesser pH sensitivity, faster kinetics, UV indicator properties, and a range of Ca^{2+} affinities, but they are not used for routine reference or extracellular solutions because they are very much more expensive than EGTA and are not as indefinitely stable at room temperature. Despite the central importance of accurate EGTA buffers in work on cellular Ca^{2+}, beginning workers often have difficulty finding a coherent description of how to make them. Many are under the misapprehension that elaborate computer programs[51,52] are necessary to prepare buffer recipes. In fact, such programs are rarely necessary. Far more important is a basic understanding of the chelation chemistry of EGTA and the possible sources of systematic error in solution preparation.

EGTA is widely available from most chemical suppliers and is an amorphous white powder practically insoluble in water. We have used the "puriss." grade of Fluka, with which we have had no difficulties. (For detailed comparison of EGTAs from different manufacturers, see Refs. 53 and 54.) None of the commercial products are >99.5% pure, but traditionally no one purifies EGTA further. Convenient stock solutions of EGTA are 0.1–1.0 M and are simply prepared by adding sufficient base [NaOH, KOH, $Me_4N^+OH^-$, tris(hydroxymethyl)aminomethane, or whatever base provides the desired cation] to dissolve all the EGTA and reach a pH of 7–8, then diluting to the appropriate final volume. With a strong base such as KOH, approximately two equivalents are required, resulting in a solu-

[51] D. D. Perrin and I. G. Sayce, *Talanta* **14**, 833 (1967).
[52] A. Fabiato and F. Fabiato, *J. Physiol. (Paris)* **75**, 463 (1979).
[53] D. M. Bers, *Am. J. Physiol.* **242**, C404 (1982).
[54] D. J. Miller and G. L. Smith, *Am. J. Physiol.* **246**, C160 (1984).

tion of K_2H_2EGTA since EGTA starts out with four titratable protons. The dissolution should proceed smoothly as base is stirred in; if granules appear that dissolve only very slowly even after the supernatant pH is >7, impurities are suspected.

Obviously the above Ca-free stocks of EGTA are convenient for diluting into the other constituents of physiological media. To prepare a given Ca-EGTA mixture with defined free $[Ca^{2+}]$, the simplest approach is merely to add the correct amount of $CaCl_2$, whose amount is calculated as described below. This procedure has practical drawbacks. Mixing $CaCl_2$ with K_2H_2EGTA gives $K_2CaEGTA + 2H^+ + 2Cl^-$. The large acid production and pH sensitivity of EGTA require that solutions be retitrated to the correct pH after mixing. An even more serious problem concerns the accuracy of the $[Ca^{2+}]$ setting when the amount of Ca begins to approach that of the EGTA. Small errors in the precise titer of either the Ca or EGTA stock can cause enormous swings in final free $[Ca^{2+}]$. So, if one wishes to prepare a range of Ca-EGTA buffers, it is usually worthwhile adopting a different approach, the preparation of a concentrated stock solution of $K_2CaEGTA$ in which the Ca and EGTA contents are verified to be within 0.5% of each other. This approach is much more accurate than the first for preparing final solutions where Ca is a large fraction of EGTA. It relies on the availability of sensitive methods for assessing whether total Ca is in excess or deficit compared to EGTA in a concentrated solution. One obvious method is to use a precalibrated Ca^{2+}-selective electrode; concentrated EGTA is titrated with $CaCl_2$ until an inflection point is seen in the plot of $\log[Ca^{2+}]$ versus $CaCl_2$ added. At this inflection point, $[Ca^{2+}]$ should be close to the geometric mean of the total EGTA concentration and the effective K_D of EGTA for Ca^{2+} at the existing pH.

A less obvious but often more convenient method for titrating Ca against EGTA was first described by Moisescu and Pusch.[55] Their "pH-metric" method requires only a pH electrode, not a Ca^{2+}-selective electrode. It relies on the fact that if EGTA at pH 6–9 is in excess over Ca, an addition of Ca will then acidify the solution. Once Ca is in excess over EGTA, adding further Ca negligibly affects the pH. The following is a typical recipe for 10 ml of 1.0 M stock $K_2CaEGTA$, in which excess KCl is minimized: 3.84 g EGTA free acid (1% over the theoretical weight of 10 mmol, to allow for the typical purity of 99%), about 0.95 g $CaCO_3$ (analytical grade, 9.5 mmol), and 19 mmol solid KOH (about 1.20 to 1.25 g of 85% pellets) and 6 ml H_2O are mixed in a beaker, stirred, heated to 90–100° until CO_2 evolution ceases, then returned to room temperature. Concen-

---

[55] D. G. Moisescu and H. Pusch, *Pfluegers Arch.* **355**, R122 (1975).

trated aqueous KOH (the commercially available 45% solution is convenient) is cautiously added in 10-μl portions until the pH reaches 7 to 8. Almost all the solids should have dissolved. Then the solution pH is noted before and after addition of $CaCl_2$ in 0.01- to 0.05-mmol aliquots, for example, 10–50 μl of a 1 M solution. Assuming the pH has dropped, KOH is then added to restore the pH; about 2 mmol per mmol Ca should be required. Precise adjustment of pH is not essential so long as it remains between 6.5 and 8. The process of Ca addition and pH restoration is repeated. It is worth keeping a log of how many Ca additions have been made and what pH change each produced. Eventually, the pH decrement per unit of Ca added decreases fairly abruptly. We stop adding Ca when ΔpH/ΔCa falls below one-half of its original value. The mixture is transferred to a 10-ml volumetric flask. Several small portions of water are used to rinse the pH electrode and beaker and are added to the volumetric flask to complete the transfer and reach the 10-ml mark. After mixing the flask contents, residual haziness is removed by membrane filtration.

For most reasonable buffer compositions, a few simple steps on a hand calculator are sufficient to calculate how much Ca^{2+}, Mg^{2+}, and EGTA should be mixed to achieve a desired free [Ca^{2+}] and [Mg^{2+}]. A simple understandable procedure is preferable to a black-box computer program whose operations cannot be checked. Our first step is to obtain the effective dissociation constants of EGTA for Ca^{2+} and Mg^{2+} at the desired pH and temperature. This discussion refers to dissociation constants instead of association constants merely because units of real concentration are more easily remembered and interpreted than values in liters/mole, which are prone to errors in carrying exponents. The values used in our laboratory are listed in Table I at 0.05 − 0.1 pH unit intervals from 6.50 to 8.20 at 20° and 37° in 0.1 M ionic strength, and at 18° in 0.25 M ionic strength.

Once the effective dissociation constants K_D(Ca) and K_D(Mg) at the relevant pH are known, Ca^{2+}-EGTA buffers are governed by the following equations, where [$EGTA_f$] is defined as [all forms of EGTA not bound to Ca^{2+} or Mg^{2+}] = [$EGTA^{4-}$] + [$H \cdot EGTA^{3-}$] + [H_2EGTA^{2-}]. Likewise [MgEGTA] is defined as [all MgEGTA complexes] = [$MgEGTA^{2-}$] + [$MgHEGTA^-$].

$$[CaEGTA] = [Ca^{2+}][EGTA_f]/K_D(Ca) \qquad (2)$$
$$[MgEGTA] = [Mg^{2+}][EGTA_f]/K_D(Mg) \qquad (3)$$
$$\text{Total EGTA} = [EGTA_f] + [CaEGTA^{2-}] + [MgEGTA] \qquad (4)$$
$$\text{Total Ca} = [Ca^{2+}] + [CaEGTA^{2-}] \approx [CaEGTA^{2-}] \qquad (5)$$
$$\text{Total Mg} = [Mg^{2+}] + [MgEGTA] \qquad (6)$$

The key to easy calculation is to decide first on [$EGTA_f$], then work out what total EGTA, Ca, and Mg are required. This contrasts with the traditional procedure of fixing total EGTA then calculating total Ca and Mg, a masochistic exercise in simultaneous nonlinear equations that is largely responsible for the fear and trembling with which these equilibria are often viewed.

Suppose we desire a buffer with 100 nM free Ca^{2+} and 1 mM free Mg^{2+} at 37°, pH 7.05. From Table I, $K_D(Ca)$ is 213 nM and $K_D(Mg)$ is 9.65 mM. Suppose we choose [EGTA$_f$] = 2 mM. Then [CaEGTA] = 100 nM × 2 mM/213 nM = 0.939 mM. Also, [MgEGTA] = 1 mM × 2 mM/9.65 mM = 0.207 mM. Therefore the total EGTA required is 2 mM + 0.939 mM + 0.207 mM = 3.146 mM. The total Ca required is 0.939 mM + 100 nM = 0.9391 mM. The total Mg required is 0.207 mM + 1 mM = 1.207 mM.

Suppose we now wish to raise free Ca^{2+} in the preceding solution to 1 μM while maintaining pH and free Mg^{2+} constant. Obviously we need to increase [CaEGTA] by 900 nM × 2 mM/213 nM = 8.45 mM. This is best added from a concentrated stock solution of $K_2CaEGTA$ whose Ca and EGTA contents are accurately balanced and whose pH is already near 7–8. If instead one tried to add 8.45 mM $CaCl_2$ and 8.45 mM K_2H_2EGTA, a small percentage error in either concentration would lead to a much larger percentage error in [EGTA$_f$] and the actual [Ca^{2+}] achieved. Also, about 16.9 mM H^+ would be generated, requiring additional careful tritration to restore pH to 7.05.

Note that by adding Ca in conjunction with EGTA, we preserve [EGTA$_f$] and avoid disturbing the Mg^{2+} equilibria. If one added $CaCl_2$ without EGTA, [EGTA$_f$] would fall and [free Mg^{2+}] would rise, though not above 1.207 mM.

It is no more difficult to calculate how much EGTA to add to reduce Ca^{2+} from extracellular to intracellular levels. Suppose the starting medium has 1.00 mM Ca^{2+} and 0.50 mM Mg^{2+}. What do we add to reach 200 nM free Ca^{2+} and 0.8 mM free Mg^{2+} at pH 7.00, 20°? Once we add the EGTA, essentially all the 1 mM Ca^{2+} (actually, 1 mM–200 nM = 0.9998 mM) will become CaEGTA. From Table I, $K_D(Ca)$ and $K_D(Mg)$ are 376 nM and 33.9 mM, respectively. From Eq. (2), [EGTA$_f$] needs to be $K_d(Ca)$[CaEGTA]/[free Ca^{2+}] = 376 nM × 1 mM/200 nM = 1.88 mM. Then [MgEGTA] will be 0.8 mM × 1.88 mM/33.9 mM = 0.044 mM. Therefore total EGTA to be added is 1.0 mM + 1.88 mM + 0.044 mM = 2.924 mM, while total Mg is 0.844 mM or 0.344 mM extra. Of course, pH must be checked and adjusted *after* the above additions have been made.

This type of analysis is readily extended to buffers containing other ligands as well as EGTA, provided that the input concentration values refer to free cation, free ligand, or a specific complex, not to total cation or ligand. For example, suppose we want 500 nM free Ca^{2+}, 2 mM free

TABLE I
DISSOCIATION CONSTANTS OF EGTA FOR Ca^{2+} AND Mg^{2+a}

pH	$K_D(Ca)$ (nM)			$K_D(Mg)$ (mM)		
	20°, 0.1 M	37°, 0.1 M	18°, 0.25 M	20°, 0.1 M	37°, 0.1 M	18°, 0.25 M
6.5	3728	2646	5273	119.3	41.7	339
6.6	2354	1672	3331	93.5	32.4	267
6.70	1487	1057	2104	73.1	25.1	209
6.75	1182	841	1673	64.5	22.0	185.3
6.80	940	669	1330	56.9	19.25	163.9
6.85	747	532	1057	50.1	16.84	144.9
6.90	594	423	841	44.0	14.70	127.9
6.95	472	337	668	38.7	12.80	112.9
7.00	376	268	532	33.9	11.13	99.4
7.05	299	213	423	29.7	9.65	87.5
7.10	238	170.0	336	26.0	8.35	76.9
7.15	189.1	135.4	268	22.7	7.21	67.5
7.20	150.5	170.9	213	19.73	6.21	59.2
7.25	119.8	86.0	169.6	17.15	5.34	51.7
7.30	95.4	68.6	135.0	14.87	4.57	45.2
7.35	76.0	54.7	107.5	12.86	3.91	39.4
7.40	60.5	43.7	85.7	11.10	3.33	34.3
7.45	48.2	34.9	68.3	9.52	2.83	29.7
7.50	38.5	27.9	54.4	8.20	2.40	25.8
7.6	24.5	17.88	34.7	6.00	1.709	19.2
7.7	15.61	11.49	22.1	4.34	1.207	14.2
7.8	9.99	7.42	14.1	3.11	0.846	10.35
7.9	6.41	4.82	9.08	2.20	0.590	7.49
8.0	4.13	3.15	5.85	1.547	0.409	5.37
8.1	2.68	2.08	3.79	1.080	0.284	3.82
8.2	1.75	1.39	2.47	0.751	0.197	2.70

[a] $K_D(Ca) = [1 + \text{antilog}(9.58 - pH) + \text{antilog}(18.54 - 2pH)] (10^{-10.97} M)$ at 20°, ionic strength $(I) = 0.1\ M$. $K_D(Ca) = [1 + \text{antilog}(9.34 - pH) + \text{antilog}(18.06 - 2pH)] (10^{-10.64} M)$ at 37°, $I = 0.1\ M$. $K_D(Ca) = [1 + \text{antilog}(9.537 - pH) + \text{antilog}(18.498 - 2pH)] (10^{-10.777} M)$ at 18°, $I = 0.25\ M$ (0.2 M KCl + 0.05 M NaCl) (see Ref. 10). $K_D(Mg) = [1 + \text{antilog}(9.58 - pH) + \text{antilog}(18.54 - 2pH)] (10^{-5.21} M)/[1 + \text{antilog}(7.73 - pH)]$ at 20°, $I = 0.1\ M$. $K_D(Mg) = [1 + \text{antilog}(9.34 - pH) + \text{antilog}(18.06 - 2pH)] (10^{-5.41} M)/[1 + \text{antilog}(7.49 - pH)]$ at 37°, $I = 0.1\ M$. $K_D(Mg) = [1 + \text{antilog}(9.537 - pH) + \text{antilog}(18.498 - 2pH)] (10^{-4.567} M)/[1 + \text{antilog}(7.885 - pH)]$ at 18°, $I = 0.25\ M$ (0.2 M KCl + 0.05 M NaCl) (see Ref. 10). The parameters for 0.1 M ionic strength are derived from the fundamental binding constants listed in the critical compilation of Martell and Smith.[43] The effective dissociation constant of EGTA for Ca^{2+}, $K_D(Ca)$, is defined as [free Ca^{2+}][all forms of EGTA not bound to Ca^{2+} or Mg^{2+}]/[CaEGTA complex]. At pH >4, there are three relevant forms of metal-free EGTA: the tetraanion $EGTA^{4-}$, the singly protonated species $(H \cdot EGTA)^{3-}$, and the doubly protonated species

Mg^{2+}, 4 mM MgATP, 2 mM [EGTA$_f$] at pH 7.10, 20°. Table I gives K_D(Ca) = 238 nM, K_D(Mg) = 26.0 mM, and the analogous effective values for ATP binding to Ca^{2+} and Mg^{2+} are 226 and 116 μM, respectively, from the absolute values in the Martell and Smith compendium.[43] [CaEGTA] = 500 nM × 2 mM/238 nM = 4.20 mM; [MgEGTA] = 2 mM × 2 mM/26.0 mM = 0.154 mM; [ATP$_f$] = [MgATP] × 116 μM/[Mg^{2+}] = 4 mM × 116 μM/2 mM = 232 μM; [CaATP] = 500 nM × 232 μM/226 μM = 513 nM. Total EGTA = 2 mM + 4.20 mM + 0.154 mM = 6.354 mM; total ATP = 0.232 mM + 4 mM + 513 nM = 4.2325 mM; total Ca = 4.20 mM; total Mg = 2 mM + 0.154 mM + 0.232 mM = 2.386 mM.

Complex programs are needed only if total cation or ligand is prespecified. The main occasion when this is necessary is the evaluation of the effect of a given error in the assumed binding constant or in the composition of a buffer. Thus if total Ca, total Mg, and total EGTA are given as Ca$_T$, Mg$_T$, and L$_T$, free Ca^{2+} and Mg^{2+} are obtained from the coupled quadratic equations:

(H$_2$EGTA)$^{2-}$. These protonations are described by conventional Henderson–Hasselbach equations, so that [H · EGTA^{3-}] = [EGTA^{4-}]$10^{(pK_1 - pH)}$ and [H$_2$EGTA^{2-}] = [H · EGTA^{3-}]$10^{(pK_2 - pH)}$ = [EGTA^{4-}]$10^{(pK_2 + pK_1 - 2pH)}$, where pK_1 and pK_2 are the pK_a values for the first and second protons, respectively. Therefore, K_D(Ca) = [Ca^{2+}][EGTA^{4-} + HEGTA^{3-} + H$_2$EGTA^{2-}]/[CaEGTA^{2-}] = [1 + $10^{(pK_1 - pH)}$ + $10^{(pK_2 + pK_1 - 2pH)}$] [EGTA^{4-}][Ca^{2+}]/[CaEGTA^{2-}] = [1 + $10^{(pK_1 - pH)}$ + $10^{(pK_2 + pK_1 - 2pH)}$]/K_{Ca}. In the above, K_{Ca} ≡ [CaEGTA^{2-}]/[Ca^{2+}][EGTA^{4-}] is the "absolute affinity" of EGTA for Ca^{2+}, achieved at very high pH where the protonations occur to a negligible extent. K_{Ca} is $10^{10.97}$ M^{-1} at 20°, 0.1 M ionic strength; pK_1 is 9.58 and pK_2 is 8.96. Note that the values for pK_1 and pK_2 are each 0.11 unit higher than the values actually listed in the tables for EGTA. The need for the 0.11 correction is explained on pp. XI–XII of Martell and Smith.[43] In brief, the original value of 9.47 for pK_1 meant that [H · EGTA^{3-}] = [EGTA^{4-}][H$^+$]$10^{9.47}$ M^{-1}. Now, pH as operationally used by biologists is not exactly $-$log[H$^+$] as stated in elementary texts. Rather, pH = $-$log[H$^+$ activity] = $-$log(0.78[H$^+$]) at 0.1 M ionic strength by National Bureau of Standards convention. Therefore [H$^+$] = (10^{-pH})/0.78 = $10^{0.11 - pH}$, so [H · EGTA^{2-}] = [EGTA^{4-}]($10^{9.58 - pH}$). The same argument applies for pK_2 = 8.96 corrected from 8.85.

The effective dissociation constant for Mg^{2+} is slightly more complicated because the complex (MgH · EGTA)$^-$ is significant as well as (MgEGTA)$^{2-}$. From Martell and Smith[43] again, [MgHEGTA$^-$] = [Mg · EGTA^{2-}]$10^{(7.73 - pH)}$ at 20°, and K_{Mg} ≡ [MgEGTA^{2-}]/[Mg^{2+}][EGTA^{4-}] = $10^{5.21}$ M^{-1}. Therefore K_d(Mg) = [1 + $10^{(pK_1 - pH)}$ + $10^{(pK_2 + pK_1 - 2pH)}$]/K_{Mg}[1 + $10^{(7.73 - pH)}$]. Values at 37° were calculated by adding $(\Delta H)(2.303R)^{-1}[(293°K)^{-1} - (310°K)^{-1}]$ = (0.041 mol/kcal)(ΔH) to the log stability constants at 20°. Data for the behavior of EGTA at other ionic strengths (reviewed in Ref. 10; see also Ref. 57) are scanty, and theoretical extrapolations using Debye–Hückel theory are of dubious value for a complicated zwitterion like EGTA.

$$0 = [Ca^{2+}]^2 + (L_T - Ca_T + K_D(Ca) + K_D(Ca)[Mg^{2+}]/K_D(Mg))[Ca^{2+}]$$
$$- (K_D(Ca))(Ca_T)(1 + [Mg^{2+}]/K_D(Mg))$$
$$0 = [Mg^{2+}]^2 + (L_T + Mg_T + K_D(Mg) + K_D(Mg)[Ca^{2+}]/K_D(Ca))[Mg^{2+}]$$
$$- (K_D(Mg))(Mg_T)(1 + [Ca^{2+}]/K_D(Ca))$$

We are indebted to V. L. Lew for showing that these equations are efficiently solved by iterative refinement. Make a guess at [Mg^{2+}], solve the first quadratic equation for [Ca^{2+}], substitute that in the second equation, solve for [Mg^{2+}], resubstitute that in the first equation again, etc. Repeat this cycle until successive rounds give negligible change in the values of Ca^{2+} and Mg^{2+}. Convergence is usually rapid.

Because EGTA is so pH-dependent, pH should be checked in the final solutions, not assumed to stay as initially set by pH buffers in the medium. Also, the pH electrode combination should be stable[56] and should have been calibrated with buffers thermostatted at the same temperature as the test solution, since the temperature-compensating knob on most pH meters only corrects the slope, not the potential offset of an electrode. Commercial pH buffers tend to be of indefinite age and extent of CO_2 pickup, sometimes disagree with one another, and are not always specified at other than room temperature, so we prefer to prepare fresh pH standards from the well-known Natural Bureau of Standards recipes.

[56] J. A. Illingworth, *Biochem. J.* **195**, 259 (1981).
[57] R. DiPolo, H. Rojas, J. Vergara, R. Lopez, and C. Caputo, *Biochim. Biophys. Acta* **728**, 311 (1983).

[15] Analyzing Transport Kinetics with Desk-Top Hybrid Computers

By HAROLD G. HEMPLING

Introduction

Modern technology has converted the powerful and rapid calculating resources of the modern computer into a desk-top unit, easily accessible to the bench scientist. The purpose of this chapter is to describe several programs which may be used on a routine basis to acquire kinetic data, to manipulate the data, and to optimize parameters. These programs are based on well-established kinetic equations from the literature and serve to integrate these equations into paradigms for analysis. Access to this type of resource may encourage workers in the field of membrane transport to extend the limits of their experimentation beyond measurements of initial fluxes or equilibrium distributions and to get at the more difficult problems of transport among multiple compartments. Those who are interested in volume regulation and the consequences of solute displacement on the movements of water may be interested in those programs which calculate permeability coefficients for water and solute from osmotic data and volume changes.

Optimizing Solutions for Three-Compartment Kinetics

Series Model

This first section takes the equations which describe the exchange of isotope between two cell compartments and the environment in a series model and in a parallel model and programs them for the analog computer. The two options are diagrammed in Figs. 1 and 2.

The kinetics of iostope distribution among the components of a three-compartment system in a series is described by

$$\dot{Q} = k_{12}P - k_{21}Q - \dot{R} \tag{1}$$
$$\dot{R} = k_{23}Q - k_{32}R \tag{2}$$

where P, Q, and R are the plasma compartment (or external medium, as the case may be), the first-cell compartment, and the second-cell compartment, respectively, all with dimensions of radioactivity in counts per minute (cpm). The rate constants k_{12}, k_{21}, k_{23}, and k_{32} describe the exchange from P to Q, Q to P, Q to R, and R to Q, respectively, with dimensions of reciprocal time (t^{-1}).

If P, the radioactivity in the external medium, remains constant because the compartment is made so much larger than Q or R, then at infinity:

$$\frac{P}{p} = \frac{(Q + R)}{(q + r)} \tag{3}$$

$$P = (Q + R)\left(\frac{p}{q + r}\right) \tag{4}$$

where q and r represent the compartments of exchangeable substance in the cell, and p is the medium compartment, in micromoles/compartment size.

If we scale $(Q + R)$ to be equivalent to 10 V, the maximal output of the analog computer, let us scale P as

$$\left[\frac{P}{\frac{p}{q+r}}\right]$$

so that the scaled variables of P and $(Q + R)$ will be equal in volts at ∞. The scaled equation is then

$$[\dot{Q}] = k_{12}\left(\frac{p}{q+r}\right)\left[\frac{P}{\left(\frac{p}{q+r}\right)}\right] - k_{21}[Q] - [\dot{R}] \quad (5)$$

$$[\dot{R}] = k_{23}[Q] - k_{32}[R] \quad (6)$$

If we use as the scaled variable not radioactivity, but specific activity, normalized to the total ion content of the cell, $(q + r)$, we can divide both sides of both equations by $(q + r)$.

$$\left[\frac{\dot{Q}}{q+r}\right] = k_{12}\left(\frac{p}{q+r}\right)\left[\frac{P}{\left(\frac{p}{q+r}\right)(q+r)}\right] - k_{21}\left[\frac{Q}{q+r}\right] - \left[\frac{\dot{R}}{q+r}\right] \quad (7)$$

$$\left[\frac{\dot{R}}{q+r}\right] = k_{23}\left[\frac{Q}{q+r}\right] - k_{32}\left[\frac{R}{q+r}\right] \quad (8)$$

Scaling is completed by the insertion of β into the denominator of all terms in order to convert machine time to real time according to the equation:

$$M = \beta t \quad (9)$$

where M is machine time and t is real time.

The circuit diagram for this analog program is shown in Fig. 1. Table I defines the potentiometers and Table II lists the amplifier outputs. Note that amplifier 6 gives

$$\frac{Q}{q+r} + \frac{R}{q+r} = \frac{Q+R}{q+r}$$

or the specific activity of the cell as a function of time. The output from this amplifier was used to fit the computer solution to the experimental data.

The selection of a scale factor for P such that this compartment would equal $(Q + R)$ at ∞ has made easier the task of fitting the machine solution to experimental data to arrive at values for the rate constants and fluxes.

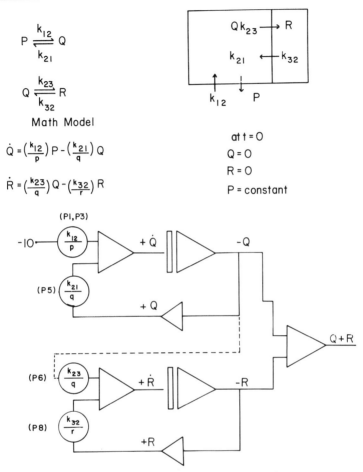

FIG. 1. Analog computer program of a series model of a three-compartment closed system.

Consider that at ∞, $[\dot{Q}]$ and $[\dot{R}] = \phi$

$$k_{12}\left(\frac{p}{q+r}\right)\left[\frac{P}{\frac{p}{q+r}}\right] = k_{21}[Q] \qquad (10)$$

and

$$\frac{k_{12}}{k_{21}}\left(\frac{p}{q+r}\right)\left[\frac{P}{\frac{p}{q+r}}\right] = [Q] \qquad (11)$$

TABLE I
POTENTIOMETER SETTINGS FOR ANALOG
PROGRAMS WHICH MODEL FLUXES ACROSS A
THREE-COMPARTMENT SYSTEM

Potentiometer	Series	Parallel
1	$\dfrac{k_{12}p}{\beta(q+r)}$	$\dfrac{k_{12}p}{\beta(q+r)}$
2	$\dfrac{k_{21}}{\beta}$	$\dfrac{k_{21}}{\beta}$
3	$\dfrac{k_{23}}{\beta}$	$\dfrac{k_{13}p}{\beta(q+r)}$
4	$\dfrac{k_{32}}{\beta}$	$\dfrac{k_{31}}{\beta}$

TABLE II
AMPLIFIER OUTPUTS FOR ANALOG PROGRAMS
WHICH MODEL FLUXES ACROSS A
THREE-COMPARTMENT SYSTEM

Amplifier	Output variable	
	Series	Parallel
1	$-\dfrac{Q}{(q+r)}$	$-\dfrac{Q}{(q+r)}$
2	$\dfrac{R}{(q+r)}$	$-\dfrac{R}{(q+r)}$
3	$-\dfrac{R}{(q+r)}$	$\dfrac{\dot{Q}}{(q+r)}$
4	$\dfrac{\dot{Q}}{(q+r)}$	$\dfrac{\dot{R}}{(q+r)}$
5	$-\dfrac{R}{(q+r)}$	$\dfrac{k_{21}}{\beta}\dfrac{Q}{(q+r)}$
6	$\dfrac{Q+R}{(q+r)}$	$\dfrac{k_{31}}{\beta}\dfrac{R}{(q+r)}$
7	—	$\dfrac{Q+R}{(q+r)}$

Also, from Eq. (8),

$$k_{23}[Q] = k_{32}[R] \tag{12}$$

and

$$\left(\frac{k_{23}}{k_{32}}\right)[Q] = [R] \tag{13}$$

Now, let us define $(Q + R) = T$, the total radioactivity, properly scaled, in the cells at ∞.
Then

$$[Q] + \left(\frac{k_{23}}{k_{32}}\right)[Q] = T \tag{14}$$

$$[Q]\left(1 + \frac{k_{23}}{k_{32}}\right) = T \tag{15}$$

and, from Eq. (11),

$$\left(\frac{k_{12}}{k_{21}}\right)\frac{p}{q+r}\left[\frac{P}{\frac{p}{q+r}}\right]\left[1 + \left(\frac{k_{23}}{k_{32}}\right)\right] = T \tag{16}$$

Since we arranged by scaling to have $(Q) + (R) = [P/(p/q + r)]$ then

$$\frac{T}{\left(\frac{P}{\frac{p}{q+r}}\right)} = 1$$

and

$$1 + \frac{k_{23}}{k_{32}} = \frac{k_{21}}{k_{12}\left(\frac{p}{q+r}\right)} \tag{17}$$

This relation, by defining the constraints on the rate constants, simplifies the selection of the proper rate constants for fitting the machine solution to experimental data.

Finally, from amplifiers 2 and 4 we obtain, at ∞, the fraction of total cell sodium which is in each compartment, since:

From amplifier 4 we have $Q/(q + r)$. At ∞, $Q/q = (Q + R)/(q + r)$. Then $Q/(q + r) = [(Q + R)/(q + r)][q/(q + r)]$. Since $(Q + R)/(q + r)$ equal specific activity of the total cell, when normalized to the value of 1, amplifier 4 is equivalent to $q/(q + r)$ or the fraction of the total cell sodium in compartment q.

Parallel Model

For a system of three compartments arranged in parallel, the appropriate equations are

$$\dot{Q} = k_{12}P - k_{21}Q \qquad (18)$$
$$\dot{R} = k_{13}P - k_{31}R \qquad (19)$$

When scaled in a fashion similar to that described above, the equations are

$$\left[\frac{\dot{Q}}{q+r}\right] = k_{12}\left(\frac{p}{q+r}\right)\left[\frac{P}{\left(\frac{p}{q+r}\right)(q+r)}\right] - k_{21}\left[\frac{Q}{q+r}\right] \qquad (20)$$

$$\left[\frac{\dot{R}}{q+r}\right] = k_{13}\left(\frac{p}{q+r}\right)\left[\frac{P}{\left(\frac{p}{q+r}\right)(q+r)}\right] - k_{31}\left[\frac{R}{q+r}\right] \qquad (21)$$

The circuit diagram for this analog program is shown in Fig. 2. Likewise, Tables I and II contain listings of potentiometers and of amplifier outputs.

Reasoning in a manner similar to the series analysis, we can put constraints on the rate constants, since

$$\frac{k_{12}\left[\frac{p}{q+r}\right]}{k_{21}} + \frac{k_{13}\left[\frac{p}{q+r}\right]}{k_{31}} = 1 \qquad (22)$$

Optimization of Kinetic Parameters

Figure 3 describes how the analog voltages are sampled by an A/D interface (Mountain Computer, Inc., Santa Cruz, CA) and stored for use in a BASIC program written for the Apple II+ microprocessor (The Apple Corporation, Cupertino, CA).

The BASIC program, HYBRIP.BA (parallel model) (Fig. 4 and HYBRIS.BA (series) sample the analog solutions at those time intervals equal to the experimental data points, compare them to the experimental points, calculate deviations, absolute variances, fractional variances, and then by D/A interface (Fig. 5) change the values for rate coefficients, and repeat the analog cycle with new set of solutions. The following strategy is used to optimize the analog solution and to achieve the minimum fractional variance.

1. The fractional variance is obtained by taking the difference between the value from the model and the value for the experimental point and normalizing to the experimental point. This ratio is squared and the fractional variance is defined as the sum of squared values for all data points.

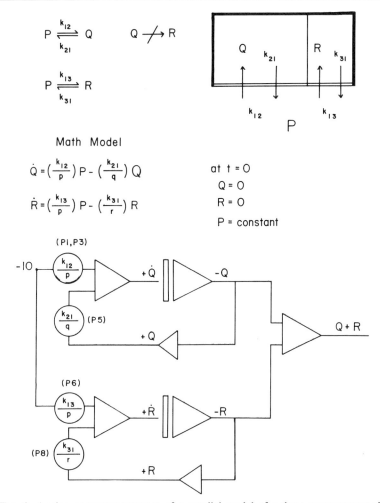

FIG. 2. Analog computer program of a parallel model of a three-compartment closed system.

2. Constraints are put on k_{12}, the rapid influx coefficient so that the value chosen is determined by the earliest experimental value. This procedure is identical to the procedure most investigators use when they wish to calculate the initial influx, but do not have the data or the inclination to do a complete kinetic analysis.

3. Then a first estimate for k_{21}, the rapid efflux coefficient, is chosen.

4. Once k_{21} is set, the ratio k_{23}/k_{32} for a series analysis or k_{13}/k_{31} for a

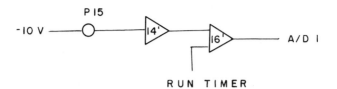

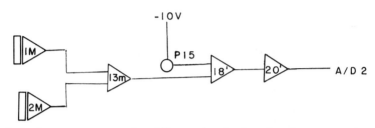

Fig. 3. Analog computer network to deliver analog signals to the microcomputer by way of A/D interfaces.

```
10    REM    HYBRIP.BA EDITION
11    REM    HYBRID PROGRAM TO MODEL
12    REM    TWO CELL COMPTS IN PARALLEL
15    OQ1 =  INT (( PEEK (A2) / 255) * 100)
35    D$ =   CHR$ (4): REM   CTRL D
55    DIM Y1(15),Q(15),C$(72):V = 0:N3 = 0
57    DIM A(15),B(15):N4 = 0
58    X1 = 1
60    GOSUB 300
70    GOSUB 390
80    FOR I = 2 TO 4
90    ADDR = 49280 + (4 * 16) + I
92    ADDR = 49280 + (4 * 16) + I
93    REM    IF I>2 THEN 900 OPTION
95    POKE ADDR,K(I) * 255
97    POKE ADDR,K(I) * 255
100   NEXT I
110   A1 = 49280 + (4 * 16) + 1
115   A1 = 49280 + (4 * 16) + 1
120   A2 = 49280 + (4 * 16) + 2
125   A2 = 49280 + (4 * 16) + 2
130   GOSUB 600
180   IF Y < A(1) THEN 190
185   GOSUB 600: GOTO 180
190   FOR I = 1 TO N
200   IF Y > A(I) THEN 210
205   GOSUB 600: GOTO 200
210   Y1(I) = Y:Q(I) = Q1
230   NEXT I: GOSUB 410
```

Fig. 4. Listing in BASIC of the program HYBRIP.BA designed to optimize rate constants for the parallel model of a three-compartment closed system.

```
231 N4 = N4 + 1: IF N4 < 3 THEN 255
235 N4 = 0: IF T4 > 0 THEN 250
240 K(2) = K(2) - ABS (T4) * K(2)
245  GOTO 255
250 K(2) = K(2) + 0.05 * K(2)
255  IF (S4 - T4) > 0 THEN 270
260 K(4) = K(4) + ABS (S4 - T4) * K(4)
265  GOTO 275
270 K(4) = K(4) - ABS (S4 - T4) * K(4)
275 K(3) = K(4) * (1 - (K(1) / K(2)))
280 V = S2 / F
285  GOTO 70
300  PRINT D$;"OPEN DATA"
305  PRINT D$;"READ DATA": INPUT A$
310  PRINT A$
325  INPUT PO,QO,C,N,N1
361  PRINT D$;"CLOSE DATA"
362  PRINT "MACHINE TIME TO REAL TIME "
364  INPUT B3
365  PRINT D$;"OPEN ";A$
366  PRINT D$;"READ ";A$
367  FOR I = 1 TO N
370  INPUT A(I),B(I)
375 A(I) = B3 * A(I):B(I) = B(I) * 1000
377  PRINT A(I),B(I)
380  NEXT I
382  PRINT D$;"CLOSE";A$
385  PRINT "K1 ": INPUT K(1)
386  PRINT "K2 ": INPUT K(2)
387  PRINT "K3 ": INPUT K(3)
388  PRINT "K4 ": INPUT K(4)
390 S2 = 0:S4 = 0:F = 0:S6 = 0
400  RETURN
410  FOR I = 1 TO N
412  IF Q(I) = 0 THEN 470
415 B = B(I) / 1000
420 Q = Q(I) / 1000
425 S1 = (Q - B) * (Q - B)
430 S3 = (Q - B) / Q
435 S5 = S3 * S3:S6 = S6 + S5
440 S4 = S4 + S3:F = F + 1
450 S2 = S2 + S1
455  IF I = N1 THEN 465
460  GOTO 470
465 T4 = S4
470  NEXT I
475 S7 = SQR (S6 - (S4 * S4 / F) / (F - 1))
480 S8 = S7 / (F - 1)
481  IF X1 < (S6 / F) THEN 487
482  IF X1 > (S6 / F) THEN 483
483 X1 = S6 / F:X2 = 2: GOTO 489
487 X2 = 1
488  GOSUB 700
489  FOR I = 1 TO 3: PRINT : NEXT I
```

FIG. 4. (*continued*)

```
490  PRINT "FRACTIONAL VARIANCE = ";S6 / F
495  IF X2 = 1 THEN 592
500  PRINT "TOTAL FRACTIONAL ERROR = ";S4
505  PRINT "NO. OF POINTS = ";F;"AVG ERROR = ";S4 / F
507  PRINT "S.E.M = ";S8
510  A1 = K(1) / K(2)
515  A2 = 1 - K(1) / K(2)
520  R1 = K(1) * (QO / PO) * B3
525  R2 = K(2) * B3
530  R3 = K(3) * (QO / PO) * B3
535  R4 = K(4) * B3
540  F1 = (R1 * PO) / C
545  B1 = (R2 * A1 * QO) / C
550  F2 = (R3 * PO) / C
555  B2 = (R4 * A2 * QO) / C
556   PRINT "K12= ";
557   PRINT  INT (R1 * 10000) / 10000;
558   PRINT "K21= ";
559   PRINT  INT (R2 * 10000) / 10000;
560   PRINT "K13= ";
561   PRINT  INT (R3 * 100000) / 100000
562   PRINT "K31= ";
563   PRINT  INT (R4 * 100000) / 100000
565   PRINT "K1= ";K(1);
566   PRINT "K2= ";K(2);
567   PRINT "K3= ";K(3);
568   PRINT "K4= ";K(4);
570   PRINT "    COMPARTMENT 1   "
575   PRINT "   SIZE = ";A1
576   PRINT "   FLUX IN = ";F1,"OUT = ";B1
580   PRINT "    COMPARTMENT 2   "
585   PRINT "   SIZE = ";A2
586   PRINT "   FLUX IN = ";F2,"OUT = ";B2
592  N3 = N3 + 1: IF N3 > 25 THEN 2000
593   PRINT X2,N3
595   RETURN
600  Y =   PEEK (A1) / 255 * 100 * B3
610  Y =   PEEK (A1) / 255 * 100 * B3
620  Q1 =   PEEK (A2) / 255 * 1000
630  Q1 =   PEEK (A2) / 255 * 1000
650   RETURN
700   PRINT D$;"PR#1"
770   RETURN
900   IF K(I) > 0.0999 THEN 1500
910   POKE ADDR,K(I) * 255 * 10
920   GOTO 100
1000   PRINT : PRINT D;"PR#0": PRINT
1100   RETURN
1500   PRINT "RAISE GAIN ON K";I
1510   POKE ADDR,K(I) * 255
1520   POKE ADDR,K(I) * 255
1530   GOTO 100
2000   END
```

FIG. 4. (*continued*)

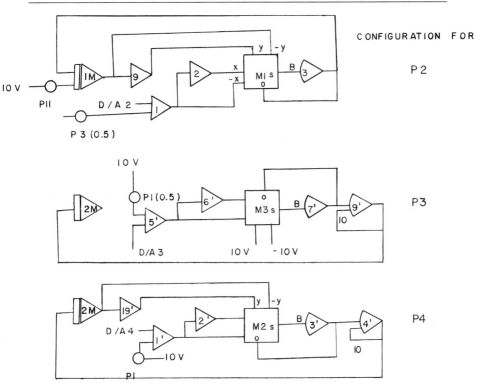

FIG. 5. Analog network to generate settings for rate constants P2, P3, and P4 as defined in Table I. Optimized values for rate constants from the microcomputer are transferred by D/A interfaces to multipliers which generate potentiometer values and apply them to the appropriate integrators.

parallel analysis is set either by Eq. (17) or Eq. (22). Usually low absolute values for k_{23} and k_{32} (series) or for k_{13} and k_{31} (parallel) are chosen.

6. After each run, the total fractional error is used to set a new absolute value for k_{23} (and therefore k_{32} since the ratio is constant for the series solution) or k_{13} (and therefore k_{31}) for the parallel solution.

7. The absolute values for the individual components of the ratio are optimized for three to five iterations by this process.

8. The k_{21} is adjusted on the fractional error between early experimental points and the model solution. This procedure works best because k_{21} is most influential on the early part of the uptake of isotope.

9. With k_{21} reset and k_{12} still held constant, a new ratio for

$$\frac{k_{23}}{k_{32}} \quad \text{(series) or} \quad \frac{k_{13}}{k_{31}} \quad \text{(parallel)}$$

is calculated and another series of iterations is carried out on their absolute values.

10. With this strategy, the experimental curve is optimized to the minimum fractional variance, using about 25 iterations.

Rapid Calculation of Permeability Coefficient to Water, Using Desk-Top Hybrid Computers

In this section, a computer program is described which interfaces a desk-top analog computer with a desk-top microprocessor to collect data from a particle-size analyzer and to store it on floppy disks. The data are readily available to calculate the permeability of single-cell suspensions to water by methods which are described in the following paragraphs.

In 1977[1] we described how we combined the attributes of a particle-size analyzer with an analog computer (EAI Miniac, Long Branch, NJ) to measure changes in cell volume at intervals of 100 msec or less. We have now added the digital computing power of a small microprocessor (APPLE II, Cupertino, CA) for data storage and processing.

The patching program for the analog computer is shown in Fig. 6. The interface between the Miniac and the APPLE II uses an A/D board from Mountain Computer Inc., Santa Cruz, CA). The program written in BASIC has been named MCV VERSUS TIME (Fig. 7). The analog data from the particle-size analyzer are sampled at an interval which may be set in response to the query: DURATION. The value entered is used to define the range of an iterative loop which puts in a delay between successive PEEKS at two Miniac Track/Stores.

Track/Store 34 stores the sum of integrated spikes from cells which passed through the aperture of the Coulter Counter Model B. Track/Store 24 stores the voltage from an integrator which integrates a constant voltage whenever an electronic switch is activated. This switch is activated whenever Comparator 33 goes HIGH. Comparator 33 goes HIGH whenever the peak spike voltage exceeds a preselected value (P33). Therefore the voltage from TRACK/STORE 24 is proportional to the number of cells which passed through the orifice during the sampling period. The voltage from TRACK/STORE 34 is proportional to the summed volumes of these cells during the same period. The ratio, Voltage from TRACK/STORE 34 to Voltage from TRACK/STORE 24, is proportional to the mean corpuscular volume. The division is performed by the microprocessor after PEEKing T/S 34 and T/S 24 at A/D 1 and A/D 2, respectively. Amplifiers 14′, 16′ and 18′ are used to condition outputs so that $0 = -5$ V

---

[1] H. G. Hempling, *Acta Cytol.* **21**, 96 (1977).

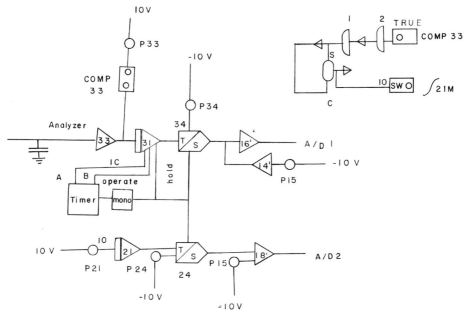

FIG. 6. Analog network to count signals from a particle-size analyzer and to transmit them to the microcomputer via A/D 2 or to integrate the signals and to transmit the summed values via A/D 1.

```
10   REM     DURATION=(MSEC-58.239)/1.69347
20   D$ =    CHR$ (4): REM    CTRL D
30   I = 1
40   DIM MCV(1000)
50   A1 = 49280 + (4 * 16) + 1
60   A2 = 49280 + (4 * 16) + 2
70   A3 = 49280 + (4 * 16) + 3
80   A4 = 49280 + (4 * 16) + 4
110  INPUT "DURATION  ";R
120  INPUT "NO. OF POINTS   ";K
130  PRINT "WAITING TO START"
140  GOSUB 510: IF V > 50 THEN 160
150  GOTO 140
160  PRINT "STARTED"
170  FOR I = 1 TO K
180  FOR J = 1 TO R: NEXT J
190  GOSUB 510
200  MCV(I) = V / C
210  NEXT I
250  GOTO 380
260  PRINT "SAVE DATA PRESS Y"
265  ONERR GOTO 260
```

FIG. 7. BASIC program, MCV VERSUS TIME designed to accept A/D signals from a particle-size analyzer and to convert them to equivalent units of volume for storage on floppy disks.

```
270   INPUT Q$: IF Q$ = "Y" THEN 285
280   GOTO 110
285   FOR I = 1 TO 20: PRINT MCV(I): NEXT
290   INPUT "NAME OF FILE   ";A$
300   INPUT "START ";L: INPUT "END ";M
310   PRINT D$;"OPEN";A$
320   PRINT D$;"WRITE";A$
330   FOR I = L TO M
340   PRINT MCV(I)
350   NEXT I
360   PRINT D$;"CLOSE";A$
370   GOTO 110
380 HI = MCV(5):LO = MCV(5)
390   FOR I = 1 TO K
400   IF MCV(I) < LO THEN LO = MCV(I)
410   IF MCV(I) > HI THEN HI = MCV(I)
420   NEXT I
425   POKE A3,0: POKE A3,0: POKE A4,255: POKE A4,255
430   PRINT HI,LO
432   INPUT "ARE HI-LO SATISFACTORY?Y OR N   ";A$
434   IF A$ = "N" THEN 8000
436   PRINT "TO PLOT PRESS RETURN": GET X$
438   POKE A1,0: POKE A1,0
440   FOR I = 1 TO K
450   FOR J = 1 TO R: NEXT J
460   POKE A3,(I / K) * 255
470   POKE A4,((MCV(I) - LO) / (HI - LO)) * 255
480   POKE A4,((MCV(I) - LO) / (HI - LO)) * 255
490   NEXT I
495   POKE A1,255: POKE A1,255
500   GOTO 260
510 V =   PEEK (A1) / 255 * 10000
520 V =   PEEK (A1) / 255 * 10000
530 C =   PEEK (A2) / 255 * 10000
540 C =   PEEK (A2) / 255 * 10000
550   RETURN
560   END
8000  INPUT "HI ";HI: INPUT "LO   ";LO
8100  GOTO 436
```

FIG. 7. (*continued*)

and maximum output = +5 V. After completion of the run, the data may be stored on a floppy disk.

Once the data are stored, they may be recalled for subsequent processing. A program, CALCULATING LP (Fig. 8) uses the data to calculate an osmotic permeability coefficient for water, using the equation of Lucké et al.[2]

We used two approaches to test our methodology. The first method was to generate a curve of cell volume versus time from the analog model, with a preset value for the permeability coefficient to water, L_p, using the analog program described in 1967.[3] Treating the output from the model as

[2] B. Lucké, H. K. Hartline, and M. McCutcheon, *J. Gen. Physiol.* **14**, 405 (1931).
[3] H. G. Hempling, *J. Gen. Physiol.* **70**, 237 (1967).

```
10    DIM X(1000),Y(1000),YS(1000): GOSUB 110
20    X1 = 0:Y1 = 0
30    INPUT "ANY AVERAGING? ";Q$
40    IF Q$ = "Y" THEN 1110
50    INPUT "CALCULATE LP? ";Q$
60    PRINT D$;"PR#1"
70    IF Q$ = "Y" THEN 280
80    PRINT D$;"PR #0"
90    GOSUB 910
100   END
110   D$ = CHR$ (4): REM  CONTROL D
120   INPUT "INPUT FILE   ";Q$
130   INPUT "NO. OF POINTS   ";N
140   INPUT "Y FACTOR   ";F
150   INPUT "X FACTOR ";F1
160   PRINT D$;"OPEN";Q$
170   PRINT D$;"READ";Q$
180   FOR I = 1 TO N
190   INPUT Y(I): NEXT I
200   PRINT D$;"CLOSE";Q$
210   FOR I = 1 TO N:Y(I) = Y(I) * F
220   YS(I) = Y(I)
230   X(I) = I * F1: NEXT I
240   PRINT D$;"PR #1": PRINT : PRINT Q$
250   PRINT D$;"PR #0"
260   ONERR  GOTO 30
270   RETURN
280   PRINT "ISOSMOLARITY   ": INPUT P0
290   ONERR  GOTO 30
300   P0 = P0 * 0.082 * (273 + TP)
310   PRINT "TEMP CENTIG ": INPUT TP
320   PRINT "ISOSMOTIC CELL VOLUME ": INPUT V0
330   PRINT "START VOLUME ": INPUT VS
340   PRINT "START APPLE VOLUME ": INPUT AS
350   AX = AS
360   PRINT "B VALUE ": INPUT B
370   PRINT "EQUILIB CELL VOL ": INPUT VE
380   PRINT "EQUIL APPLE VOL ": INPUT AE
390   PRINT D$;"PR #0"
400   REM :CF=CELL FACTOR
410   CF = (VS - VE) / (AS - AE)
420   DENOM = (36 * 3.1416) ^ (1 / 3) * P0 * (V0 - B)
430   S0 = (VE - B) / DENOM
440   S1 = 1 - B / VE
450   INPUT "ANY OMISSIONS Y OR N";O$
460   IF O$ = "Y" THEN 1180
470   INPUT "STARTING POINT   ";L
480   INPUT "ENDING POINT   ";M
490   FOR J = L TO M
500   IF Y(J) = 0 THEN 630
510   IF Y(J) > AS THEN Y(J) = AX
520   IF Y(J) < AE THEN Y(J) = AX
530   AX = Y(J)
540   Y(J) = (Y(J) - AE) * CF + VE
550   QUAD = VE ^ (2 / 3) + (VE * Y(J)) ^ (1 / 3) + Y(J) ^ (2 / 3)
560   T1 = 2 * Y(J) ^ (1 / 3) + VE ^ (1 / 3)
```

FIG. 8. BASIC program, CALCULATING LP, designed to sample stored data of cell volumes and elapsed times and to calculate the hydraulic coefficient, L_p, from the equation of Lucké et al.[2] (see text).

```
570 T2 = 3 ^ (1 / 2) * VE ^ (1 / 3)
580 S3 = 3 ^ (1 / 2) *   ATN (T1 / T2)
590 DF = (VE ^ (1 / 3) - Y(J) ^ (1 / 3)) ^ 2
600 S2 = (1 / 2) *   LOG (QUAD / DF)
610 KT = S0 * (S1 * VE ^ (1 / 3) * (S2 + S3) - 3 * Y(J) ^ (1 / 3))
620  PRINT J,Y(J):Y(J) = KT
630  NEXT J
640  PRINT "START    ";L
650  PRINT "END    ";M
660 N = 0:S = 0:S1 = 0:S2 = 0:S3 = 0:S4 = 0
670  INPUT "STARTING POINT ";L
680  INPUT "END POINT   ";M
690  FOR I = L TO M
700  IF Y(I) = 0 THEN 760
710  IF Y(I) = Y(I - 1) THEN 760
720  PRINT X(I),Y(I)
730 S = S + X(I):S1 = S1 + Y(I):S3 = S3 + X(I) ^ 2
740 S2 = S2 + Y(I) ^ 2:S4 = S4 + X(I) * Y(I)
750 N = N + 1
760  NEXT I
770  PRINT D$;"PR#1"
780  PRINT : PRINT
785  PRINT "INTERVAL ";L;" - ";M;"   N = ";N
790 C = S4 - ((S * S1) / N)
800 N1 = S3 - ((S * S) / N)
810 B1 = C / N1
820 N2 = S2 - ((S1 * S1) / N):B2 = C / N2
830 YD =   SQR ((S2 / (N - 1)) - ((S1 * S1) / (N * (N - 1)))
840 YSEM = YD /   SQR (N)
850   PRINT "PEARSON R   ";C / (  SQR (N1) *    SQR (N2))
860   PRINT "Y =   ";B1;" X +   ";
870   PRINT (S1 / N) - ((S / N) * B1)
880   PRINT "AVG = ";S1 / N;"   SEM = ";YSEM
890   PRINT D$;"PR#0"
900   GOTO 30
910   INPUT "OUTPUT FILE? ";Q$
920   INPUT "X AXIS FACTOR?   ";J
930   INPUT "Y AXIS FACTOR?   ";K
940   INPUT "RANGE:LOWER   ";L
950   INPUT "RANGE:UPPER   ";M
960   INPUT "POINTS:GRAPH 1   ";O
970   INPUT "POINTS:GRAPH 2   ";P
980   PRINT D$;"OPEN";Q$
990   PRINT D$;"DELETE";Q$
1000   PRINT D$;"OPEN";Q$
1010   PRINT D$;"WRITE";Q$
1020   PRINT O
1030   PRINT P
1040   FOR I = L TO M
1050   IF Y(I) = Y(I - 1) THEN 1080
1060   PRINT X(I) * J
1070   PRINT Y(I) * K
1080   NEXT I
1090   PRINT D$;"CLOSE";Q$
1100   RETURN
1110   INPUT "STARTING POINT   ";L
1120   INPUT "ENDING POINT   ";M
1130   FOR I = L TO M
```

FIG. 8. *(continued)*

```
1140    Y(I) = YS(I)
1150    NEXT
1160    N = 0:S = 0:S1 = 0:S2 = 0:S3 = 0:S4 = 0
1170    GOTO 690
1180    INPUT "START ";L1
1190    INPUT "END   ";M1
1200    FOR I = L1 TO M1
1210    Y(I) = 0: NEXT
1220    GOTO 470
```

FIG. 8. (*continued*)

if it were experimental data, we sampled the data with the program MCV VERSUS TIME. We then recalled the data from the disk and used the equation of Lucké et al.[2] to calculate a value of L_p, the permeability coefficient. In their article, it requires the calculation of an algebraic equation, $K_t = f$ (cell volume) where K is identical to L_p and t is equal to time. A plot of K_t versus t was linear, with a slope equal to K (or L_p). Using the BASIC Program CALCULATING LP in a typical run, the correlation coefficient was 0.9857 and the value of K calculated from the slope was 0.511. The value used in the analog model was 0.494. Therefore the error was 3%. This result satisfied us that an analog solution agreed with a digital solution within the limits of our analog model and was satisfactory for our needs, since biological variation was likely to be greater than 3%.

With the second approach, we measured the permeability coefficient to water of a line of rat megakaryocytopoietic cells in culture as described, using the analog computer to calculate L_p.[3] We also calculated

TABLE III
COMPARISON OF PERMEABILITY COEFFICIENT OF MEGAKARYOCYTOPOIETIC CELLS IN CULTURE TO WATER[a]

Experiment	Digital[b]	R value[c]	Analog[b]
1	0.373	0.87	0.402
2	0.540	0.93	0.528
3	0.387	0.94	0.395
4	0.528	0.94	0.491
5	0.554	0.68	0.587
6	0.400	0.87	0.421

[a] Calculated by a digital solution to the equation of Lucké et al.[2] with an analog solution.[3]
[b] Permeability coefficient to water (μM min^{-1} atm^{-1}).
[c] R is correlation coefficient of least squares fit to a straight line generated in the solution.

the permeability coefficient from the equation of Lucké et al.[2] from the data stored on the disks, using the program CALCULATING LP. The results are summarized in Table III. Again, we found agreement between the two methods quite satisfactory.

The advantage of the digital method lay in its ease and speed of solution. In addition, we could replace the criterion of best fit by eye, used in arriving at an analog solution, with a statistic, the correlation coefficient. This statistic characterized the best fit to a straight line which is the outcome of the algebraic solution of Lucké et al.[2] The slope of such a straight line was equal to the permeability coefficient to water.

The present method eliminates a time-consuming curve-fitting procedure which had been used in the analog method.[3] In its place is a rapid algebraic solution, using the stored data to calculate permeability coefficients to water. This type of calculation makes it easier to compare permeability coefficients statistically since each calculated value has its best estimate of fit to the data. In addition, each permeability run is stored on a disk for future reference.

Modeling Solute Movement and Volume Changes with the Desk-Top Microcomputer

Using principles from irreversible thermodynamics, Kedem and Katchalsky[4] derived a pair of equations which predicted how cells would change volume when placed in an environment containing combinations of permeable and impermeable solutes.

$$\dot{V}_{H_2O} = 4.8(V)^{2/3} \left[\frac{N_1}{V_{H_2O}} - C_i + \sigma \left(\frac{N_s}{V_{H_2O}} C_s \right) \right] \quad (23)$$

$$\dot{n}_s = 4.8(V)^{2/3} (\omega RT) \left[C_s - \frac{N_s}{V_{H_2O}} \right] + \frac{(1-\sigma)}{2} \left[\frac{N_s}{V_{H_2O}} - C_s \right] \dot{V}_{H_2O} \quad (24)$$

These equations have been applied successfully to calculate the permeability coefficient of the membrane for the permeable solute (ωRT) and the interaction of permeable solute with the solvent or with the membrane as defined by the Staverman reflection coefficient, σ. Table IV contains a parameter list.

Initially an analog computer solution was prepared. Since then, a digital solution has been written using a simple Euler iteration procedure and adapted to the microcomputer (APPLE II+ or APPLE IIe). It has been used with equal success to calculate the phenomenological coefficients

[4] O. Kedem and A. Katchalsky, *Biochim. Biophys. Acta* **27**, 229 (1958).

TABLE IV
PARAMETER LIST FOR THE EQUATIONS OF KEDEM AND KATCHALSKY

Equation variable	Meaning	Program variable	Remarks
$\dot{V}_{H_2O}$	Net water movement	DV	Liters/time
$4.8V^{2/3}$	Surface area	A	
N_i	Osmoles impermeable solute	N1	Set at 0.3 osmol but may be changed
V	Cell volume	V	$V = V_1 + B$
V_{H_2O}	Cell water	V1	Set at 1 liter at time = 0
C_i	Osmolarity of permeable solute in medium	C1	Chosen by user
b	Volume of osmotically inactive material	B	Set at 0.2 but may be changed
σ	Reflection coefficient (values anywhere between 0 and 1 may be set	SG	Set at 1.0 but useful to change
N_s	Osmole permeable solute	S	Chosen by user
C_s	Osmolarity of permeable solute in medium	CS	Chosen by user
L_p	Permeability coefficient for water	DS	Chosen by user
$\dot{n}_s$	Net solute movement	DS	Osmoles/time
ωRT	Permeability coefficient for solute	O	Chosen by user

from data or to predict change sin cell volume and solute transport when selected coefficients are entered. Two versions are available: PERMEA R2V, which is listed in Fig. 9, and PERMEA/R2V AND PLOT, which is listed in Fig. 10.

In the first instance, the display of volume versus time appears on the screen and is handy for teaching, demonstrations, and quick modeling. The second version is adapted for use with an $x-y$ plotter. It is designed to convert the calculated values of volume or solute through D/A interfaces (Mountain Computer, Inc., Santa Cruz, CA) to voltage signals as input to the $x-y$ plotter. This program is particularly useful to estimate ωRT and σ by fitting solutions to experimental data in an iterative mode.

The user may set the following values: (1) duration of the experiment; (2) volume of the cell in cubic microns; (3) volume of osmotically inactive material, the so-called "b" value from a Boyle–van't Hoff plot in cubic microns; (4) the intracellular osmolarity of impermeable solute; (5) the intracellular osmolarity of permeable solute; (6) the extracellular osmolarity of impermeable solute; (7) the extracellular osmolarity of permeable solute; (8) the permeability of the membrane to water, L_p, in μ/min atm;

```
100  REM                           DEPARTMENT OF PHYSIOLOGY
110  REM     MEDICAL UNIV. OF S.C.
120  :
130  REM     CHARLESTON, S.C. 29425
140  :
150  REM     COPYRIGNT RESERVED
160  :
170  :
180  REM     =========================
190  REM             PERMEA/R2V
200  REM     =========================
210  :
220  :
230  :
240  V = 0:V1 = V:B = V:A = V:W1 = V:W2 = V
245  V0 = V:X = V:H1 = V:XF = V:XIN = V:Y = V
247  YIN = V:YF = V:DV = V:DS = V:LP = V:N1 = V
248  C1 = V:S = V:CS = V:O = V:H = V
250  D$ = CHR$ (4)
260    TEXT : HOME
270    DIM A$(25)
280    PRINT D$;"BLOAD HI-RES CHARACTER GEN $6000"
290    PRINT D$;"BLOAD CHARACTER TABLE $6800"
300    GOSUB 330
310    GOSUB 1180
320    GOTO 900
330    HCOLOR= 2: HGR
340    HPLOT 260,150 TO 10,150
350    HPLOT 10,150 TO 10,10
360    HTAB 1: VTAB 1:A$ = "VOLUME": GOSUB 710
370    HTAB 33: VTAB 19:A$ = "TIME": GOSUB 710
380    VTAB 24: RETURN
390  S = S1
400  W1 = 4.76:W2 = .67:W3 = 2
410  H1 = 0:V = V1 + B
420  V0 = V
430    HCOLOR= 1
440    FOR I = 1 TO XHIGH * 100
450  V = V1 + B:A =. W1 * V ^ W2
460    IF V / V0 > YHIGH THEN   HTAB X / 7: VTAB Y / 7.5: GOTO 700
470  X =    INT (H1 * XF) - XIN:Y = YIN -    INT (V / V0 * YF): HPLOT X,Y
475    PRINT H1,V
480  DV = A * LP * ((N1 / V1 - C1) + SG * (S / V1 - CS))
490  DS = A * O * (CS - S / V1) + (1 - SG) / 2 * (CS + S / V1) * DV
500  V1 = V1 + DV * H
510  S = S + DS * H
520  H1 = H1 + H
530    NEXT I
535    VTAB 24: PRINT "MINUTES","CUBIC MICROMETERS"
538    VTAB 24: PRINT "PRESS RETURN TO CONTINUE"
539    GET X$
540  A$ =  STR$ ( INT ((V / V0) * 100))
550    VTAB 24
560    PRINT "H20 PERMEA = ";L1;
570    PRINT  TAB( 20);"SOL PERMEA = ";O
580    PRINT "IMP SOL"; TAB( 12);"PERM SOL(OUT)";
590    PRINT  TAB( 28);"PERM SOL(IN)"
```

FIG. 9. BASIC program, PERMEA/R2V, designed to model the equations of Kedem and Katchalsky[4] (see text) for graphic display.

```
600   PRINT " ";C1 / 10 ^ - 15;" OSM"; TAB( 13);
610   PRINT CS / 10 ^ - 15;" OSM"; TAB( 29);S1 / 10 ^ - 15;" OSM"
620   HTAB X / 7: VTAB Y / 7.5
630   GOSUB 710
640   HTAB (X / 7) + 5
650   IF C1 = 0 GOTO 700
660   V = (B + N1 / C1)
663   IF V / V0 > YHIGH THEN  VTAB Y / 7.5: GOTO 700
665   GOSUB 1280: VTAB Y / 7.5
670   A$ = STR$ ( INT ((B + N1 / C1) / V0 * 100))
680   GOSUB 710
690   GOTO 770
700   A$ = "LYSIS": GOSUB 710: GOTO 770
710   P1 = PEEK (54):P2 = PEEK (55)
720   PR# Q: POKE 54,0: POKE 55,96
730   PRINT A$
740   POKE 54,P1: POKE 55,P2
750   POKE - 16301,0
760   RETURN
770   VTAB 24
780   GET X$
790   PRINT : PRINT "FOR ANOTHER RUN,    PRESS RETURN"
800   PRINT : PRINT "FOR OTHER OPTIONS,    PRESS Y   ";
810   INPUT Y$: IF Y$ = "Y" THEN 880
820   PRINT : PRINT : PRINT : VTAB 24
830   PRINT "ERASE DATA (Y OR N)   ";
840   INPUT Q$: IF Q$ = "Y" THEN 300
850   PRINT : PRINT : PRINT
860   VTAB 24: PRINT "EXIT (Y OR N) "
865   INPUT Q$
867   IF Q$ = "Y" THEN  PRINT D$;"RUN HELLO2"
870   PRINT : PRINT : PRINT : VTAB 24: GOTO 320
880   TEXT : HOME
890   PRINT D$;"RUN PERMEABILITY OPTION"
900   V = 0:S = 0
905   VTAB 24: INPUT "VH2O ";V1
910   VTAB 24: INPUT "CELL IMP. OSMOLARITY ";N1
915   N1 = N1 * V1 * 10 ^ - 15
920   H = 0.01
925   VTAB 24: INPUT "B VALUE ";B
930   PRINT "IMPERMEABLE SOLUTE,ENVIRONMENT"
950   INPUT C1:C1 = C1 * 10 ^ - 15
960   PRINT : PRINT : VTAB 24
970   PRINT "PERMEABILITY TO WATER"
990   INPUT L1:LP = L1 * 0.082 * 298 * 10 ^ 15
1000  PRINT : PRINT : VTAB 24
1002  PRINT "REFLECTION COEFFICIENT"
1004  PRINT "CHOOSE A VALUE BETWEEN 0 AND 1   ";
1010  INPUT SG
1015  PRINT : PRINT : VTAB 24
1020  PRINT "PERMEABLE SOLUTE IN ENVIRONMENT"
1040  INPUT CS:CS = CS * 10 ^ - 15
1050  PRINT : PRINT : VTAB 24
1060  PRINT "INTERNAL CONC OF PERM SOLUTE"
1080  INPUT S1:S1 = S1 * 10 ^ - 15
1090  PRINT : PRINT : VTAB 24
1100  PRINT "PERMEABILITY TO SOLUTE"
1110  INPUT O
```

FIG. 9. (*continued*)

```
1120    PRINT : PRINT : PRINT : VTAB 24
1170    GOTO 390
1180    VTAB 24: INPUT "TIME IN MINUTES ";XHIGH
1190    XLOW = 0
1200    XD = XHIGH - XLOW
1210    XF = 200 / XD
1220    XIN =   INT (XLOW * XF) - 10
1230    YHIGH = 1.67:YLOW = 0.3
1240    YD = YHIGH - YLOW
1250    YF = 130 / YD
1260    YIN = 20 +  INT (YHIGH * YF)
1270    RETURN
1280    X =   INT (H1 * XF) - XIN
1290    Y = YIN -  INT (V / V0 * YF)
1300    RETURN
1310    S = S1
1320    W1 = 4.76:W2 = .67:W3 = 2
1330    H1 = 0:V1 = 1:V = V1 + B
1340    V0 = V
1350    HCOLOR= 1
1360    FOR I = 1 TO 300
1370    V = V1 + B:A = W1 * V ^ W2
1380    IF V / V0 > YHIGH THEN   HTAB X / 7: VTAB Y / 7.5: GOTO 700
1390    X =   INT (H1 * XF) - XIN:Y = YIN -  INT (V / V0 * YF): HPLOT X,Y
1400    DV = A * LP * ((N1 / V1) + SG * (S / V1 - CS))
1410    DS = A * O * (CS - S / V1)
1420    V1 = V1 + DV * H
1430    S = S + DS * H
1440    H1 = H1 + H
1450    NEXT
1460    GOTO 540
1470    S = S1
1480    W1 = 4.76:W2 = .67:W3 = 2
1490    H1 = 0:V1 = 1:V = V1 + B
1500    V0 = V
1510    HCOLOR= 1
1520    FOR I = 1 TO 300
1530    V = V1 + B:A = W1 * V ^ W2
1540    IF V / V0 > YHIGH THEN   HTAB X / 7: VTAB Y / 7.5: GOTO 700
1550    X =   INT (H1 * XF) - XIN:Y = YIN -  INT (V / V0 * YF): HPLOT X,Y
1560    DV = A * LP * ((N1 / V1 - C1) + SG * (S / V1))
1570    DS = A * O * ( - S / V1)
1580    V1 = V1 + DV * H
1590    S = S + DS * H
1600    H1 = H1 + H
1610    NEXT
1620    GOTO 540
1630    S = S1
1640    W1 = 4.76:W2 = .67:W3 = 2
1650    H1 = 0:V1 = 1:V = V1 + B
1660    V0 = V
1670    HCOLOR= 1
1680    FOR I = 1 TO 300
1690    V = V1 + B:A = W1 * V ^ W2
1700    IF V / V0 > YHIGH THEN   HTAB X / 7: VTAB Y / 7.5: GOTO 700
1710    X =   INT (H1 * XF) - XIN:Y = YIN -  INT (V / V0 * YF): HPLOT X,Y
1720    DV = A * LP * (N1 / V1 - C1)
1730    V1 = V1 + DV * H
```

FIG. 9. (*continued*)

```
1740    H1 = H1 + H
1750    NEXT
1760    GOTO 540
1770 S = S1
1780 W1 = 4.76:W2 = .67:W3 = 2
1790 H1 = 0:V1 = 1:V = V1 + B
1800 V0 = V
1810    HCOLOR= 1
1820    FOR I = 1 TO 300
1830 V = V1 + B:A = W1 * V ^ W2
1840    IF V / V0 > YHIGH THEN   HTAB X / 7: VTAB Y / 7.5: GOTO 700
1850 X =    INT (H1 * XF) -  XIN:Y = YIN -    INT (V / V0 * YF):  HPLOT X,Y
1860 DV = A * LP * ((N1 / V1 - C1) + SG * (S / V1 - CS))
1870 V1 = V1 + DV * H
1880 H1 = H1 + H
1890    NEXT
1900    GOTO 540
```

FIG. 9. (continued)

```
100  REM
110  REM      MEDICAL UNIV. OF S.C.
120  :
130  REM      CHARLESTON, S.C. 29425
140  :
150  REM      COPYRIGNT RESERVED
160  :
170  :
180  REM      ========================
190  REM           PERMEA/R2 V AND PLOT
200  REM      ========================
210  :
220  :
230  :
240  V = 0:V1 = V:B = V:A = V:W1 = V:W2 = V
244  V0 = V:X = V:H1 = V:XF = V:XIN = V
246  Y = V:YIN = V:YF = V:DV = V:DS = V
247  LP = V:N1 = V:C1 = V:S = V:CS = V
248  O = V:H = V
250  D$ =    CHR$ (4)
260     DIM A$(25),V(1000),S2(1000)
265     TEXT : HOME : FOR I = 1 TO 5: PRINT : NEXT I
270     PRINT "WHEN REQUESTED ENTER  "
275     PRINT
280     PRINT "VOLUMES IN CUBIC MICROMETERS"
281     PRINT
282     PRINT : PRINT "LP IN MICROMETERS/MIN ATM"
283     PRINT
284     PRINT "CONCENTRATIONS IN OSMOLES/LITER"
285     PRINT
286     INPUT "ENTER TIME IN MINUTES ";ZZ
295  ZZ = ZZ * 1000
320     GOSUB 900
390  S = S1
400  W1 = 4.76:W2 = .67:W3 = 2
410  H1 = 0:V = V1 + B
```

FIG. 10. BASIC program, PERMEA/R2V AND PLOT, designed to model the equations of Kedem and Katchalsky[4] (see text) and to plot results on an x–y plotter.

```
440    FOR I = 1 TO ZZ
450    V(I) = V1 + B:A = W1 * V(I) ^ W2
460    S2(I) = S
480    DV = A * LP * ((N1 / V1 - C1) + SG * (S / V1 - CS))
490    DS = A * O * (CS - S / V1) + (1 - SG) / 2 * (CS + S / V1) * DV
500    V1 = V1 + DV * H
510    S = S + DS * H
520    H1 = H1 + H
530    NEXT I
535    PRINT D$;"PR #1"
540    PRINT : PRINT
560    PRINT "H20 PERMEA = ";LP;
570    PRINT  TAB( 20);"SOL PERMEA = ";O
580    PRINT "IMP SOL"; TAB( 12);"PERM SOL(OUT)";
590    PRINT  TAB( 28);"PERM SOL(IN)"
600    PRINT " ";C1;" OSM"; TAB( 13);
610    PRINT CS;" OSM"; TAB( 29);S1;" OSM"
620    PRINT : PRINT
630    PRINT D$;"PR #0"
700    GOSUB 4000
710    GOSUB 2000
900    V = 0:S = 0
910    H = 0.001: INPUT "VH20 ";V1
915    INPUT "B ";B: INPUT "N ";N1
920    N1 = N1 * V1 * 10 ^  - 15
925    VTAB 24
930    PRINT "IMPERMEABLE SOLUTE,ENVIRONMENT"
950    INPUT C1:C1 = C1 * 10 ^  - 15
960    PRINT : PRINT : VTAB 24
970    PRINT "PERMEABILITY TO WATER"
990    INPUT LP:LP = LP * 0.082 * 298 * 10 ^ 15
1000   PRINT : PRINT : VTAB 24
1002   PRINT "REFLECTION COEFFICIENT"
1004   PRINT "CHOOSE A VALUE BETWEEN 0 AND 1  ";
1010   INPUT SG
1015   PRINT : PRINT : VTAB 24
1020   PRINT "PERMEABLE SOLUTE IN ENVIRONMENT"
1040   INPUT CS:CS = CS * 10 ^  - 15
1050   PRINT : PRINT : VTAB 24
1060   PRINT "INTERNAL CONC OF PERM SOLUTE"
1080   INPUT S1:S1 = S1 * V1 * 10 ^  - 15
1090   PRINT : PRINT : VTAB 24
1100   PRINT "PERMEABILITY TO SOLUTE"
1110   INPUT O
1120   PRINT : PRINT
1125   PRINT "CALCULATIONS STARTED. PLEASE WAIT"
1130   RETURN
2000   A3 = 49280 + (4 * 16) + 3
2100   A4 = 49280 + (4 * 16) + 4
2200   A1 = 49280 + (4 * 16) + 1
2400   INPUT "DURATION  ";R
2500   PRINT HI,LO
2550   INPUT "ARE HI-LO SATISFACTORY? Y OR N ";A$
2560   IF A$ = "N" THEN 8000
2600   INPUT "START PLOT AT ";L
2700   INPUT "END PLOT AT ";M
2750   INPUT "TOTAL ELAPSED TIME UNITS ";M1
2760   IF TV = 2 THEN 6000
```

FIG. 10. (*continued*)

```
2762    POKE A1,255: POKE A1,255
2765    POKE A3,0: POKE A3,0
2770    POKE A4,((V(1) - LO) / (HI - LO)) * 255
2780    POKE A4,((V(1) - LO) / (HI - LO)) * 255
2790    PRINT "TO PLOT PRESS RETURN": GET X$
2795    POKE A1,0: POKE A1,0
2800    FOR I = L TO M
2900    FOR J = 1 TO R: NEXT J
3000    POKE A3,(I / M1) * 255
3100    POKE A3,(I / M1) * 255
3200    POKE A4,((V(I) - LO) / (HI - LO)) * 255
3300    POKE A4,((V(I) - LO) / (HI - LO)) * 255
3400    NEXT I
3450    POKE A1,255: POKE A1,255
3500    INPUT "PLOT SATISFACTORY? Y OR N ";X$
3600    IF X$ = "Y" THEN 10000
3700    GOTO 700
4000    HOME : PRINT "PLOT VOLUME ENTER 1"
4050    PRINT "PLOT SOLUTE ENTER 2  "
4070    INPUT TV: IF TV = 2 THEN 5000
4100    HI = V(1):LO = V(1)
4200    FOR I = 2 TO ZZ
4400    IF V(I) < LO THEN LO = V(I)
4500    IF V(I) > HI THEN HI = V(I)
4600    NEXT I
4650    IF HI = LO THEN 7000
4700    RETURN
5000    HI = S2(1):LO = S2(1)
5100    FOR I = 2 TO ZZ
5200    IF S2(I) < LO THEN LO = S2(I)
5300    IF S2(I) > HI THEN HI = S2(I)
5400    NEXT I
5450    IF HI = LO THEN 7000
5500    RETURN
5550    POKE A1,255: POKE A1,255
6000    POKE A4,((S2(1) - LO) / (HI - LO)) * 255
6100    POKE A4,((S2(1) - LO) / (HI - LO)) * 255
6150    POKE A3,0: POKE A3,0
6160    PRINT "TO PLOT PRESS RETURN": GET X$
6170    POKE A1,0: POKE A1,0
6200    FOR I = L TO M
6300    FOR J = 1 TO R: NEXT J
6400    POKE A3,(I / M1) * 255
6500    POKE A3,(I / M1) * 255
6600    POKE A4,((S2(I) - LO) / (HI - LO)) * 255
6700    POKE A4,((S2(I) - LO) / (HI - LO)) * 255
6800    NEXT I
6850    POKE A1,255: POKE A1,255
6900    GOTO 3500
7000    PRINT "NO CHANGE IN VARIABLE"
7100    GOTO 3500
8000    INPUT "HI   ";HI: INPUT "LO   ";LO
8100    GOTO 2600
10000   D$ = CHR$ (4): REM  CTRL D
10010    PRINT "SAVE DATA PRESS Y"
10020    INPUT Q$: IF Q$ = "Y" THEN 10040
10030    GOTO 270
10040    INPUT "NAME OF FILE  ";A$
```

FIG. 10. (*continued*)

```
10060    INPUT "START ";L: INPUT "END ";M
10070    PRINT D$;"OPEN";A$
10080    PRINT D$;"WRITE";A$
10090    FOR I = L TO M
10095    IF TV = 2 THEN 10500
10100    PRINT V(I)
10110    NEXT I
10120    PRINT D$;"CLOSE";A$
10400    GOTO 700
10500    PRINT S2(I)
10600    GOTO 10110
```

FIG. 10. (*continued*)

(9) the permeability of the membrane of one of the solutes, ωRT, in μm/min; (10) the reflection coefficient, σ between 0 and +1.

In summary, this chapter has provided several programs and their listings. The descriptions are general enough so that those who are familiar with the BASIC language and with analog computer networks can adapt them to their own microcomputers and hybrid analog/digital instruments. Floppy disks carrying these programs are available from the author at cost.

[16] Synthesis and Properties of Caged Nucleotides

By JEFFERY W. WALKER, GORDON P. REID, and DAVID R. TRENTHAM

Introduction

Investigation of organized biological systems such as cells (intact or permeabilized), active transport proteins, muscle fibers, protein crystals, or biological membranes frequently requires rapid introduction of effector molecules with known concentration and localization. Time-resolved measurements with such systems are hampered by delays due to diffusion of effector into the preparation and complications arising from degradation of the effector or desensitization of the biological response. One way to circumvent these problems is by introducing biological substrates rapidly and uniformly by pulse photolysis of precursor compounds that are biologically inactive.[1] Photosensitive precursors have been termed caged compounds.[2] We are concerned here with caged compounds that on pho-

[1] A. M. Gurney and H. A. Lester, *Physiol. Rev.* **67**, 583 (1987).
[2] J. H. Kaplan, B. Forbush III, and J. F. Hoffman, *Biochemistry* **17**, 1929 (1978).

tolysis yield nucleotides or nucleotide analogs as shown for caged ATP in Eq. (1).

$$\underset{NO_2}{\underset{|}{\bigcirc}}\!\!-\!\!\underset{CH_3}{\underset{|}{CH}}\!\!-\!\!O\!-\!\underset{\underset{O^-}{|}}{\overset{\overset{O}{\|}}{P}}\!\!-\!\!O\!-\!\underset{\underset{O^-}{|}}{\overset{\overset{O}{\|}}{P}}\!\!-\!\!O\!-\!\underset{\underset{O^-}{|}}{\overset{\overset{O}{\|}}{P}}\!\!-\!\!O\!-\!\text{[ribose-adenine]} \xrightarrow{h\nu} \underset{NO}{\underset{|}{\bigcirc}}\!\!-\!\!\underset{CH_3}{\underset{\|}{C}}\!\!=\!\!O + \text{ATP} \quad (1)$$

In this approach, a biological preparation is equilibrated with a caged nucleotide and it is then illuminated by a near-UV light pulse to liberate the nucleotide that initiates the biological response. Such a strategy has been used to make time-resolved measurements of the actions of ATP and ATPγS on muscle fibers,[3-7] the Na^+/K^+ ion pump ATPase,[8] and sarcoplasmic reticulum vesicles[9] and of GTP analogs on sensory neurons.[10] These nucleotides have in common the fact that they contain weakly acidic phosphate groups and the photolabile precursors used are 1-(2-nitrophenyl)ethyl phosphate esters as in Eq. (1). The esters have advantageous photochemical properties,[2,11] and, when applied in an appropriate concentration range, generally have no detectable biological activity before photolysis (but see Ref. 8).

We describe here the synthesis and characterization of 1-(2-nitrophenyl)ethyl phosphate esters of nucleotides using as representative examples caged ATP, ATPβ,γNH, and GTPγS.[12]

[3] Y. E. Goldman, M. G. Hibberd, J. A. McCray, and D. R. Trentham, *Nature (London)* **300**, 701 (1982).
[4] M. G. Hibberd and D. R. Trentham, *Annu. Rev. Biophys. Biophys. Chem.* **15**, 119 (1986).
[5] G. Rapp, K. J. V. Poole, Y. Maeda, K. Guth, J. Hendrix, and R. S. Goody, *Biophys. J.* **50**, 993 (1986).
[6] A. V. Somlyo, Y. E. Goldman, T. Fujimori, M. Bond, D. R. Trentham, and A. P. Somlyo, *J. Gen. Physiol.* **91**, 165 (1988).
[7] J. A. Dantzig, J. W. Walker, D. R. Trentham, and Y. E. Goldman, *Proc. Natl. Acad. Sci. U.S.A.* **85**, 6716 (1988).
[8] B. Forbush III, *Proc. Natl. Acad. Sci. U.S.A.* **81**, 5310 (1984).
[9] D. H. Pierce, A. Scarpa, M. R. Topp, and J. K. Blasie, *Biochemistry* **22**, 5254 (1983).
[10] A. C. Dolphin, J. F. Wootton, R. H. Scott, and D. R. Trentham, *Pfluegers Arch.* **411**, 628 (1988).
[11] J. A. McCray, L. Herbette, T. Kihara, and D. R. Trentham, *Proc. Natl. Acad. Sci. U.S.A.* **77**, 7237 (1980).
[12] Abbreviations: caged ATP, 1-(2-nitrophenyl)ethyl P^3-ester of ATP; S-caged ATPγS and O-caged ATPγS, 1-(2-nitrophenyl)ethyl P^3-esters of ATPγS in which the ester linkage is through S and O atoms, respectively; caged ATPβ,γNH, 1-(2-nitrophenyl)ethyl P^3-ester of ATPβ,γNH; caged ADP, 1-(2-nitrophenyl)ethyl P^2-ester of ADP; 1-(2-nitrophenyl)ethyl esters of guanosine nucleotides are similarly abbreviated.

$$\text{2-nitroacetophenone} \xrightarrow{H_2N-NH_2/H^+} \text{hydrazone} \xrightarrow{MnO_2} \text{diazoethane}$$

(2)

$$\text{diazoethane} + Y-\underset{O^-}{\underset{|}{\overset{O}{\overset{\|}{P}}}}-OH \xrightarrow{CHCl_3/H_2O} \text{caged nucleotide} + N_2$$

(3)

The synthesis is in two stages as illustrated by Eq. (2) and (3). In the first stage, a relatively stable diazoethane is prepared, and in the second the diazoethane reacts selectively in a mixed water/organic phase with YPO_3H^-, a nucleotide containing a weakly acidic phosphate group. The synthesis has general applicability and may be used for a wide variety of nucleotides and nucleotide analogs,[13] as well as other phosphate-containing compounds.[14] Advantages of this approach are the selectivity of the reaction, the high yield, and the ease of isolation of the caged nucleotide. Nucleotides that do not contain a weakly acidic phosphate group such as cyclic AMP have to be prepared in a nonaqueous solvent. Dimethyl sulfoxide is generally used.[15,16]

Characterization of caged nucleotides has three main components. First, there is structure determination for which NMR spectroscopy is the primary approach. Second, there are the photochemical properties of the caged nucleotides for which measurements of quantum yield and kinetics of photolysis are required. Finally, there are the general parameters of concentration, extend of photolysis, and purity of caged nucleotides. These analyses depend principally on UV spectroscopy and HPLC.

A by-product of the photolysis [eq. (1)] is 2-nitrosoacetophenone, a compound that is reactive toward thiols and so may interfere with biological activity.[2,13,17,18] Practical considerations concerning 2-nitrosoace-

[13] J. W. Walker, G. P. Reid, J. A. McCray, and D. R. Trentham, *J. Am. Chem. Soc.* **110**, 7170 (1988).
[14] J. W. Walker, A. V. Somlyo, Y. E. Goldman, A. P. Somlyo, and D. R. Trentham, *Nature (London)* **327**, 249 (1987).
[15] J. Engels and E.J. Schlaeger, *J. Med. Chem.* **20**, 907 (1977).
[16] J. M. Nerbonne, S. Richard, J. Nargeot, and H. A. Lester, *Nature (London)* **310**, 74 (1984).
[17] J. W. Walker, J. A. McCray, and G. P. Hess, *Biochemistry* **25**, 1799 (1986).
[18] Y. E. Goldman, M. G. Hibberd, and D. R. Trentham, *J. Physiol. (London)* **354**, 577 (1984).

tophenone and the effect of laser and flash lamp illumination on the measurement of the biological responses are briefly discussed.

Syntheses

The Hydrazone of 2-Nitroacetophenone: Step 1 of Eq. (2)

To 8.26 g 2-nitroacetophenone (0.05 mol) in 100 ml of 95% ethanol are added 5.62 g hydrazine hydrate (0.112 mol) and 3.2 ml glacial acetic acid (0.05 mol) and the mixture is heated under reflux for 3 hr. The ethanol is evaporated under vacuum and the residue partitioned between chloroform and water to remove hydrazine acetate. After drying the chloroform phase with anhydrous $MgSO_4$, chloroform is evaporated off and the hydrazone is distilled (118° at 0.22 mm Hg) from the mother liquor and stored at −20°. The yield from the ketone approaches 100%. Characterization: UV spectrum in ethanol, ε 9400 M^{-1} cm^{-1} at $\lambda_{248\,nm}$ (see Fig. 1). The NMR spectrum in $CDCl_3$ shows a mixture of the two hydrazones [syn (Z) and anti (E) isomers] in a ratio (that depends on the solvent) of 6:1

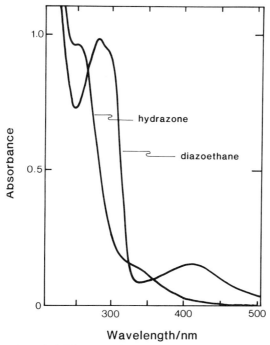

FIG. 1. Near-UV and visible absorption spectra of 100 μM hydrazone of 2-nitroacetophenone and 112 μM 1-(2-nitrophenyl)diazoethane in ethanol with a 1-cm path length cell.

(25°); δ 7.30–8.20 (4H, aromatic), 5.54 and 4.91 [6:1; broad singlet, 2H, amine (exchangeable with D_2O)], 2.03 and 2.18 (6:1; singlet, 3H, methyl).

1-(2-Nitrophenyl)diazoethane[19]: Step 2 of Eq. (2)

Reproducible oxidations of the hydrazone to the diazoethane are obtained in high yield (95% by ^1H NMR analysis of a reaction run in $CDCl_3$) with an activated grade of MnO_2 from BDH Merck[20] (product number 805958; used without further purification). It is normally appropriate to make just sufficient diazoethane for immediate needs as the compound decomposes spontaneously in chloroform with a $t_{1/2}$ of 2.3 days at 21°. One hundred and seventy seven milligrams (1 mmol) hydrazone and 500 mg MnO_2 are added to 5 ml chloroform or diethyl ether and stirred very vigorously for 5 min with the solution protected from daylight. MnO_2 and $Mn(OH)_2$ are removed by filtration with a Büchner flask using gentle suction. Progress of the reaction, which should be complete after the 5 min stirring, may be followed from the absorption spectrum of the product in ethanol (see Fig. 1; the peak at 405 nm is specific to the diazoethane). If the reaction is incomplete, the MnO_2 oxidation can be repeated. Spectra: UV in ethanol, ε 1300 M^{-1} cm^{-1} at $\lambda_{405\,nm}$ and ε 8700 M^{-1} cm^{-1} at $\lambda_{272\,nm}$; IR, 2050 cm^{-1} (>C=N$^+$=N$^-$); NMR in $CDCl_3$, δ 6.95–7.60 (4H, aromatic) and 2.10 (singlet, 3H, methyl).

Caged Nucleotides: General Considerations

An aqueous solution of the nucleotide is first acidified with Dowex 50, H$^+$ form, or with dilute HCl if the amount to be esterified is less than 100 μmol, and then stirred with diazoethane in chloroform or diethyl ether at room temperature.[21] The progress of the reaction is followed by HPLC of the aqueous phase.

Caged nucleotides are purified by a one- (DEAE-cellulose) or two-step (C_{18} reversed-phase preparative HPLC plus DEAE-cellulose) column

---

[19] In general, caution is needed in handling diazoethanes. We have experienced no explosions in 3 years of frequent use. However, we have not experimented with more than 4 mmol diazoethane and have only rarely evaporated off the organic solvent, and then with caution.

[20] Other grades of MnO_2 activated for oxidation (e.g., Aldrich 21, 746-6) may be used, but yield of diazoethane in our hands were variable even after base treatment and washing to neutrality.

[21] On the suggestion of R. S. Goody, we have used diethyl ether rather than chloroform in some instances. (However, note the caution in Ref. 19.) Esterification of the nucleotides proceeds more rapidly in an ether/water mixed solvent. This is an advantage when the nucleotide is relatively unstable at pH 4, as is the case with GTPγS and ATPγS. The rate of hydrolysis of ATPγS in water at pH 4.0 and 21° is 0.22 hr^{-1}.

procedure depending on the purity required.[22] The reversed-phase column retards compounds containing the 1-(2-nitrophenyl)ethyl group, so that caged nucleotides are well resolved from uncaged nucleotides. The column also resolves caged nucleotides from one another on the basis of their difference in charge. The DEAE-cellulose column separates nucleotides on the basis of charge and also permits desalting of the caged nucleotides.[23]

Caged ATP. The cations of 1 mmol ATP are exchanged with protons by treatment with 5 g Dowex 50, H^+ form, and the solution (10 ml) adjusted to pH 4.0 with NaOH. Then 1.5 mmol 1-(2-nitrophenyl)diazoethane dissolved in 10 ml chloroform is added and the mixture stirred vigorously at room temperature in a flask protected from daylight. Progress of the reaction may be followed by analysis of the aqueous phase by HPLC (Fig. 2, Table I). Maximum yields are obtained within 24 hr. Yields may be increased toward 100% by readjusting the pH to 4.0 with dilute HCl at intervals and, if necessary, by replacement of the chloroform solution with one containing freshly prepared diazoethane. At the end of the reaction the aqueous phase is washed twice with chloroform.

Batches of 250 μmol of the caged ATP solution are purified at room temperature on a preparative reversed-phase C_{18}-μBondpak HPLC column (2.5 cm × 30 cm, packed with 55–105 μm beads, Waters Associates) with 10 mM KH_2PO_4/K_2HPO_4 at pH 5.5 and methanol (85:15, v/v) as eluting solvent. As with the analytical reversed-phase column (Fig. 2), uncaged nucleotide elutes first, followed by caged ATP, and then any contaminating caged ADP. Caged ATP fractions are analyzed for purity by HPLC (Table I). Further batches of caged ATP may be similarly purified. The combined extracts of caged ATP are adjusted to pH 7.5 before being loaded onto a DEAE-cellulose column (2.2 cm × 50 cm) and eluted with a linear gradient of 10–500 mM triethylammonium bicarbonate (TEAB) at pH 7.5 (2 liters total volume) and 5°. TEAB is prepared by bubbling CO_2 into an ice-cooled stirred solution of triethylamine in water until the pH reaches 7.5. TEAB is removed from the caged ATP fractions by rotary evaporation under reduced pressure with four methanol washes, keeping the water bath at <40°. Caged ATP may be stored as a frozen aqueous solution of its triethylammonium salt at −20°.

---

[22] Typically, we have found that, using the one-step procedure, caged nucleotides can be obtained with <2% nucleotide contamination. With the two-step procedure, levels of contaminating ADP and ATP in caged ATP preparations are <0.02%.

[23] The preparative reversed-phase column procedure introduces phosphate ions that are readily removed by the DEAE column in the case of caged ATP purification. However, this is not necessarily the case when the caged molecule carries less charge, for example, in the case of caged ADP purification.

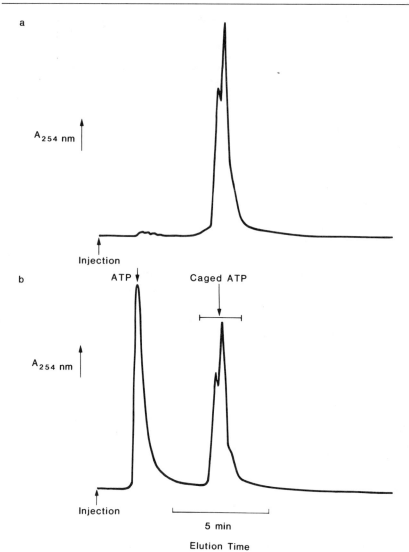

FIG. 2. Analytical HPLC traces of the aqueous phase of a reaction between ATP and 1-(2-nitrophenyl)diazoethane. Aliquots were injected onto a C_{18} reversed-phase HPLC column equilibrated in 10 mM KH_2PO_4, pH 5.4, methanol (9:1, v/v) and eluted at 1.5 ml/min. (a) Aqueous phase after 15 hr of reaction is composed of >95% caged ATP. The complex shape of the caged ATP peak arises because caged ATP exists as two diastereoisomers. (b) Same solution after exposure to 300–350 nm light. 2-Nitrosoacetophenone binds tightly to the reversed-phase HPLC column under the conditions used.

TABLE I
HPLC RETENTION TIMES[a]

Nucleotide	Retention time (min)	
	PartiSphere SAX column (Whatman)	LiChrosorb RP-8 column (BDH Merck)
ADP	0.7[a]	vv[b]
Caged ADP	2.5[a]	6.2[b]
ATP	1.8,[a] 3.7[c]	vv[b]
Caged ATP[d]	5.1,[a] 0.2[c]	1.3 and 1.9[b]
ATPγS	24.5[e]	0.2[b]
S-caged ATPγS	10.5[e]	2.6[b]
O-caged ATPγS[d]	11.0 and 13.5[e]	2.6[b]
ATPβ,γNH	4.8[f]	0.9[g]
Caged ATPβ,γNH	6.3[f]	5.5[g]
GDP	0.8[a]	
Caged GDP	1.5[a]	
GTP	1.9[a]	
Caged GTP	3.7[a]	
GTPγS	5.7[a]	
S-caged GTPγS	4.3[a]	
O-caged GTPγS	4.1[a]	
GTPβ,γNH	1.4[a]	
Caged GTPβ,γNH	6.5[a]	

[a] pH adjustments with HCl or KOH. Flow rates were 1.5 ml/min except *f* which were 2.0 ml/min. vv Void volume. Solvent: 0.7 M $(NH_4)_2HPO_4$, pH 4.0 unless noted.
[b] 10 mM KH_2PO_4, pH 5.5, + 15% methanol by volume.
[c] 0.25 M $(NH_4)_2HPO_4$, pH 5.7, + 13% methanol by volume.
[d] Two retention times for O-caged ATPγS in solvent *e* and caged ATP in solvent *b* are given because of resolution of diastereoisomers. Elution peaks of caged nucleotides are frequently broad because the mixtures of diastereoisomers are partially resolved.
[e] 0.4 M $(NH_4)_2HPO_4$, pH 4.0, + 8% methanol by volume.
[f] 0.25 M $(NH_4)_2HPO_4$, pH 5.4, + 8% methanol by volume.
[g] 20 mM KH_2PO_4, pH 6.5, + 12% methanol by volume.

Caged GTPγS. Esterification of GTPγS is complicated because the γ-phosphate containing the weakly acidic group possesses two different atoms and the P^3 atom is at a prochiral center. Consequently, treatment of GTPγS with 1-(2-nitrophenyl)diazoethane results in S- and O-caged GTPγS in which the bridging atom between the 1-(2-nitrophenyl)ethyl moiety and the P^3 atom is S and O, respectively. Like caged ATP, S-caged GTPγS exists as two diastereoisomers, but the asymmetric center at the P^3 atom in O-caged GTPγS means that this compound has two pairs of diastereoisomers.

To 30 μmol of GTPγS (Boehringer-Mannheim, containing some contaminant GDP) that is dissolved in 5 ml water and adjusted to pH 4.3 with aqueous HCl is added 0.75 mmol diazoethane in 6 ml diethyl ether.[21] The mixture is vigorously stirred in a flask protected from daylight. HPLC analysis (Table I) of the reaction mixture after 20 min shows 20% unreacted nucleotide. The pH is restored to 4.3 with HCl and, after a further 20 min, the reaction is essentially quantitative. The ether phase is removed. The aqueous phase is concentrated to about 3 ml by rotary evaporation under vacuum, taking care to avoid excess frothing due to dissolved ether. The caged GTPγS solution may be clarified by centrifugation if necessary and then loaded onto the preparative reversed-phase C_{18}-μ Bondapak column and eluted at room temperature with 10 mM KH_2PO_4/K_2HPO_4 at pH 5.5 and methanol (100:20, v/v). After a small initial peak absorbing at 254 nm, the main peak containing caged GTPγS is eluted, followed by caged GDP. The pooled fractions of caged GTPγS are adjusted to pH 7 with KOH and subjected to chromatography on DEAE-cellulose (1.5 cm × 35 cm column) with a linear gradient of 10–800 mM TEAB (1.2 liters total volume) at 5°. The fractions between a and d (Fig. 3, 360–520 mM TEAB) contain pure caged GTPγS in almost 100% yield. Between a and b and between c and d (Fig. 3) the nucleotide is >95% S-caged GTPγS and >95% O-caged GTPγS, respectively, as determined by ^{31}P NMR spectroscopy (see below, Table II). The caged GTPγS

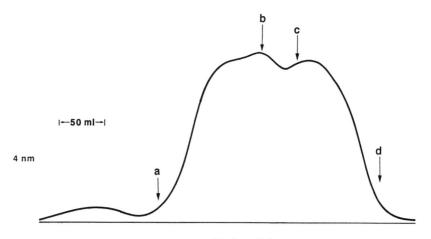

FIG. 3. Elution profile of caged GTPγS from a DEAE-cellulose column (see text). S-Caged GTPγS is eluted between a and b and O-caged GTPγS between c and d. The complex shape of the peak is associated with the multiple diastereoisomers present. On photolysis in the presence of 1 mM dithioerythritol, all fractions between a and d yield GTPγS as the only nucleotide. If dithioerythritol is omitted, then GTPγS is slowly modified during the photolysis.

TABLE II
NMR PROPERTIES OF CAGED NUCLEOTIDES[a]

Nucleotide[b]	^{31}P NMR[c]: Chemical shift (ppm)		
	α	β	γ
ATP	7.9 (d)	19.5 (t)	2.8 (d)
Caged ATP	8.3 (d)	20.3 (t)	9.4 (d)[d]
ATPγS	7.9 (d)	19.8 (dd)	−36.7 (d)
S-caged ATPγS	8.2 (d)	20.5 (dd)	−9.6 (sd)[d]
O-caged ATPγS	8.3 (d)	21.0 (dd)	−45.2 (sd)[d]
ATPβ,γNH	7.3 (d)	3.5 (dd)	−2.9 (d)
Caged ATPβ,γNH	7.8 (sd)	8.2 (dd)	−1.6 (sd)

[a] Reproduced by permission of the American Chemical Society.
[b] In our experience, ^{31}P NMR spectra of caged nucleoside triphosphates and their analogs are independent of the nucleoside.
[c] Samples at 21° are in 20% D$_2$O, 1 mM EDTA, pH 9.5 (pH 11 for ATPβ,γNH and caged ATPβ,γNH). Except for caged ATPβ,γNH, spectra were recorded on a Bruker 200 MHz instrument at 81 MHz using a broadband highfield probe and broadband proton decoupler. Chemical shifts are relative to 85% H$_3$PO$_4$ using the sign convention of Jaffe and Cohn.[28] Symbols in parentheses indicate the observed peak patterns: d, doublet; t, triplet; dd, doublet of doublets (due to spin coupling to two nuclei with different coupling constants); sd, split doublet (due to presence of diastereoisomers with different chemical shifts). For caged ATPβ,γNH the spectrum was recorded on a Bruker 400 MHz instrument. This enabled the chemical shifts of the α and β phosphorus atoms to be resolved. It also permitted observation of the split doublet in the phosphorus signal.
[d] Esterification gives rise to diastereoisomers which are expected to have slightly different chemical shifts. For caged ATP the chemical shift differences are too small for separate peaks to be observed. For S-caged ATPγS the chemical shifts are different by 14 Hz, giving rise to a split doublet. O-Caged ATPγS represents four different compounds since a chiral center is formed at the γ-phosphate as well as at the benzyl carbon. Four doublets are possible but only two are observed with a chemical shift difference of 8 Hz.

fractions may be concentrated and stored as aqueous solutions at −20° as for caged ATP.

Characterization

The purity of caged nucleotides and characterization of the nucleotide formed on photolysis are established by HPLC monitoring absorption at 254 nm. A combination of anion-exchange and reversed-phase chromatography is desirable. Table I lists retention times of several caged nucleotides and nucleotides. Retention times vary considerably with the age of columns and can be decreased for caged nucleotides by increasing the percentage of methanol in the eluting solvent (cf. retention times of ATP and caged ATP in solvents a and b, Table I).

Structure

The primary technique of establishing the structure of caged nucleotides is ^{31}P NMR spectroscopy. ^{31}P NMR is used to determine which phosphate group is esterified and, in the case of ATPγS and GTPγS, through which atom (i.e., O or S) the 1-(2-nitrophenyl)ethyl group is attached.

The main feature of ^{31}P-NMR spectra of caged nucleotides (Table II) is that the chemical shifts of the terminal (or γ) phosphate groups of ATP and ATPγS are the most sensitive to esterification but that this is not so for ATPβ,γNH. This difference in sensitivity is similar to that associated with protonation of ATPβ,γNH; unexpectedly, protonation of the phosphate group causes a larger chemical shift of the β compared to the γ phosphorus atom.[24,25] All nucleotides so far examined are esterified specifically on weakly acidic residues of the phosphate moiety.

The concentration of caged nucleotide is most conveniently measured from its UV spectrum.[2] In addition to the chromophore from the purine or pyrimidine base there is absorption due to the 1-(2-nitrophenyl)ethyl group. This has a rather broad band extending to 400 nm with its peak absorption, ε 4240 M^{-1} cm^{-1}, at $\lambda_{265\,nm}$.[2] The spectrum of caged ATP from $\lambda_{230\,nm}$ to $\lambda_{400\,nm}$ within experimental error equals the sum of those of ATP plus 1-(2-nitrophenyl)ethyl phosphate. The same result is obtained for the spectrum of caged GDP. On this basis, for caged adenine nucleotides, ε = 19,600 M^{-1} cm^{-1} at $\lambda_{260\,nm}$ and, for caged guanine nucleotides, ε = 16,900 M^{-1} cm^{-1} at $\lambda_{255\,nm}$. The maximum errors in these extinction coefficients are estimated to be ±5%. These estimates can be checked by photolysis of a solution of caged nucleotide in an aqueous solvent and extraction of the 2-nitrosoacetophenone by-product into chloroform. The absorption and the extent of photolysis in the aqueous solution are measured, from which the extinction coefficient of the caged nucleotide is then compared with that of the parent nucleotide.

Photochemical Properties

Two photochemical properties of most immediate interest when working with caged nucleotides are the extent of photolysis on illumination and

[24] S. Tran-Dinh and M. Roux, *Eur. J. Biochem.* **76**, 245 (1977).
[25] M. A. Reynolds, J. A. Gerlt, P. C. Demou, N. J. Oppenheimer, and G. L. Kenyon, *J. Am. Chem. Soc.* **105**, 6475 (1983).

the rate of release of nucleotide on pulse photolysis. The extent of photolysis depends on the quantum yield, Q_p. Following steady-state illumination, Q_p equals the ratio of the number of molecules undergoing photolysis to the number absorbing a photon. Q_p can be measured for caged nucleotides by comparison with known Q_p values of other 1-(2-nitrophenyl)ethyl phosphates such as caged P_i ($Q_p = 0.54$)[2] or caged ATP ($Q_p = 0.63$).[13] The method is illustrated for caged ATPβ,γNH in Fig. 4.

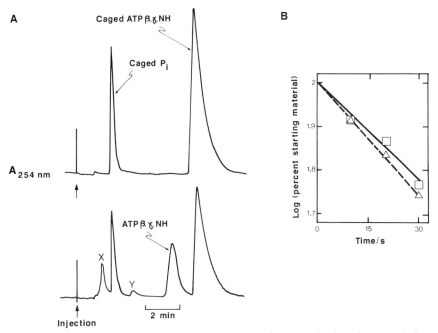

FIG. 4. Steady-state photolysis of caged ATPβ,γNH for determination of Q_p. A solution containing a mixture of 1 mM caged ATPβ,γNH and caged P_i in 100 mM TES, pH 7, is exposed to 300–350 nm irradiation and samples are analyzed by analytical HPLC on a SAX anion-exchange column (see Table I). In A is shown the elution profile initially and after 20 sec illumination. X and Y are probably 2-nitrosoacetophenone and a further by-product derived from it. In B is shown a semilogarithmic plot. The negative of the slopes is proportional to Q_p. --△--, caged ATPβ,γNH photolysis: —□—, caged P_i photolysis. The decrease of the peak size of caged P_i and caged ATPβ,γNH gives their extents of photolysis. Alternatively, in the case of caged ATPβ,γNH, the ratio of the ATPβ,γNH to caged ATPβ,γNH peaks may be used to measure the extent of photolysis, noting that allowance must be made for the ratio (= 0.81) of the extinction coefficients at 254 nm of the two compounds. 1-(2-Nitrophenyl)ethyl phosphate (caged P_i) may be synthesized by the methods described here, except that the aqueous phase for the reaction with the diazoethane is adjusted to pH 9 so that HPO_4^{2-} rather than $H_2PO_4^-$ is present. This maximizes the yield of the singly esterified compound. Alternative syntheses of caged P_i are described in Refs. 2 and 26.

In pulse photolysis, the extent of photolysis depends not only on Q_p but also on the number of times an individual molecule may be excited during the pulse, that in turn depends on the lifetime of the excited state.[26,27] However, it is found in practice that the same amounts of caged compounds are photolyzed relative to caged P_i or caged ATP in pulse as in steady-state photolysis.[13]

The rate of release of nucleotides from their caged precursors on pulse photolysis equals the rate of decay of a presumed *aci*-nitro intermediate. A spectrophotometer designed to measure rate constants of the exponential decay process has been described.[13] The values for caged ATP, S-caged ATPγS, O-caged ATPγS, and ATPβ,γNH are 86, 35, 105, and 250 sec^{-1}, respectively, at 21° in 140 mM KCl, 3 mM MgCl$_2$, and 100 mM N-tris(hydroxymethyl)methyl-2-aminoethanesulfonic acid (TES) adjusted to pH 7.1 with KOH. The observed rates are proportional to H$^+$ concentration between pH 6 and 9.

The rate of decay of the *aci*-nitro intermediate is normally measured in the presence of millimolar dithiothreitol to prevent precipitation of the photolysis by-product, 2-nitrosoacetophenone, which is relatively insoluble in water. One must take into account that the 2-nitrosoketone–dithiothreitol adduct absorbs light in the near-UV, giving rise to an additional spectral signal. At 406 nm the extinction coefficient (= 9100 M^{-1} cm^{-1}) of the *aci*-nitro intermediates of the nucleotides described here[13,17] is an order of magnitude greater than that of the 2-nitrosoketone–dithiothreitol adduct.

Technical Aspects in Application

We conclude with a few technical points concerning the application of caged nucleotides. As mentioned above, 2-nitrosoacetophenone is insoluble in water. It also may react with thiols in proteins. Problems due to these properties can be averted by having a hydrophilic thiol such as dithiothreitol or glutathione in solution at a concentration (generally 1 to 10 mM) that depends on the system under investigation.[2,18]

Laser light sources are preferable when millimolar concentrations of photolysis products are needed due to their high intensity and because it is easy to direct, shape, and focus laser beams. However, release of millimolar ATP from caged ATP can be achieved with a flash lamp.[5] When photolysis is only required to the extent of 100 μM, then flash

[26] M. A. Ferenczi, E. Homsher, and D. R. Trentham, *J. Physiol.* (*London*) **352**, 575 (1984).
[27] B. I. Greene, R. M. Hochstrasser, R. B. Weisman, and W. A. Eaton, *Proc. Natl. Acad. Sci. U.S.A.* **75**, 5255 (1978).

lamps in which 300–350 nm light is selected by use of bandpass filters are generally adequate.

Careful consideration needs to be given to obtaining uniform photolysis throughout the sample under investigation.[18] To achieve this, there should be at least 50% transmission of the photolyzing light through the sample. It is therefore an advantage that the extinction coefficient of caged nucleotides is relatively low at the wavelengths at which they are photolyzed (at 347 nm, the wavelength of a frequency-doubled ruby laser, $\varepsilon = 660\ M^{-1}\ cm^{-1}$).

Biological systems and measuring devices (electrical, mechanical, and optical) are generally not damaged by intense light pulses, though transient artifacts of the order of a millisecond are common. Heating of the samples exposed to 300–350 nm light pulses may be calculated[26] and is less than 0.5° with typical experimental setups. However, with focused laser beams onto small samples, temperature rises can be as high as 4°.[26]

In studies of biological systems involving caged nucleotides, it is necessary to know the extent of photolysis and frequently to follow the biochemical fate of the released compound (e.g., enzyme-catalyzed hydrolysis[26]). This can be accomplished by HPLC analysis of nucleotides in test samples or tissue extracts following the light pulse. The synthetic method described here can be used to prepare radiolabeled caged nucleotides which may aid analysis.

Acknowledgments

Supported by NIH grant HL15835 to the Pennsylvania Muscle Institute, the Muscular Dystrophy Association of America, EEC Stimulation Action No. 85200162UK04PUJUI, and by the Medical Research Council, United Kingdom.

[28] E. K. Jaffe and M. Cohn, *Biochemistry* **17**, 652 (1978).

[17] Measurement of Ion Fluxes in Membrane Vesicles Using Rapid-Reaction Methods

By HERBERT S. CHASE, JR., MARK D. GELERNT, and MARC C. DEBELL

Introduction

Our ability to examine the molecular details of membrane transport phenomena was greatly advanced with the introduction of the techniques for isolation of plasma membrane vesicles. Questions regarding specific mechanisms and modulators of ion transport could be examined directly by manipulating the intravesicular and extravesicular environment. In a typical experiment, the transport of ions across a membrane vesicle is examined over time and the rate coefficient[1] serves as a macroscopic measure of the membrane permeability. Using this method, investigators have examined directly many important properties of ion transporters as well as identified and reconstituted the proteins which underlie the transport function.

In theory, any transport process in membrane vesicles can be studied, provided that the flux is measured over a time period fast enough to generate an accurate rate constant. The isotopic rate coefficient for ion transport via a typical carrier is sufficiently low so that experiments can be performed easily without special equipment. That is, the initial rate flux occurs over seconds to minutes. The broad success and wide appeal of the membrane vesicle technique is due, in part, to the fact that experiments can be performed by hand with a simple filtration apparatus. However, for ion channels with turnover rates orders of magnitude faster than carriers, experiments cannot be performed by hand. The initial rate flux across a vesicle containing a channel may well be within a time frame of milliseconds rather than seconds—considerably faster than a traditional carrier. Thus, measurement of ion permeability in vesicles which contain channels cannot be determined easily by simple "hand-held" methods and usually requires rapid-reaction methods.[2]

There was great impetus to develop a method to study ion channels in membrane vesicles because reconstitution, photolabeling, and identification of the channel proteins would be possible if reliable transport assays were available. Investigators solved the problem of obtaining fast (under 1 sec), initial rate fluxes by adapting rapid-reaction methods, which were originally designed to examine biochemical reactions in order to study ion fluxes. Fluxes are performed over 5 to 1000 msec using a machine which rapidly mixes solutions and initiates an ion flux, measured either isotopi-

---

[1] The rate coefficient k may be determined from the relationships: $\ln[1 - (S_t/S_i)] = -kt$, for influx and $\ln(S_t/S_0) = -kt$ for efflux, where S is vesicle tracer content at time t, 0, and infinity i. Measuring k may be tedious because of the need to obtain multiple time points. One may measure a flux over two time points, provided the points chosen are very early in the reaction.

[2] The flux through a channel in a vesicle can be slowed. If the vesicle is made larger by freezing and thawing, the rate coefficient will decline because k is indirectly proportional to the vesicular radius (see footnote 3). Also, establishing a membrane potential in a direction which will augment the flow of the ion permeating the channel will lengthen the time period within which an initial rate flux can be measured. [See H. Garty, B. Rudy, and S. Karlish, *J. Biol. Chem.* **258**, 13094 (1983) and Kleyman *et al.*, *J. Biol. Chem.* **261**, 2839 (1986).]

cally or with an ion-sensitive dye. In the following report we will describe two uses of a rapid-reaction apparatus to study ion transport. First, we will describe flow–quench, in which an ion flux is terminated with a high-affinity, specific inhibitor of the transport process under study. The time of the reaction is fixed and the ion flux is calculated after the vesicles are harvested. Second, we will describe the use of stopped-flow, a method which circumvents the need for an inhibitor. In stopped-flow, reactants are rapidly mixed and the reaction is observed in real time.

Before mounting the considerable effort to build a rapid-reaction apparatus, one must first determine whether these methods are required. Unfortunately, there is no reliable way to estimate the rate constant of a vesicular flux. Because the rate coefficient is inversely related to the radius of the vesicle,[3] the marked size heterogeneity of an average vesicle preparation makes it difficult to estimate a single value for k. Even if an average radius can be determined, it will not be clear if channels are present in all of the vesicles, or only the small (or large) ones. Furthermore, the membrane permeability of a vesicle is likely to be different from that *in vivo*. The permeability depends on the number of channels per vesicle, the relative purity of the preparation, as well as the turnover number and open time of the channel in the vesicle, parameters that are either unmeasurable or unknown. Eventually, one must take a practical approach and measure the flux within a time frame of seconds to minutes to determine if the initial rate is faster than the earliest points which can be obtained by hand.

Measurement of Isotope Fluxes Using Flow–Quench

Definition of Flow–Quench

The term flow–quench is used to describe how the isotopic flux is performed. Isotope-loaded vesicles are diluted into isotope-free buffer, initiating the flux, and flow to a mixing chamber where they meet an inhibitor to terminate the flux. After inhibition, the vesicles are harvested and the flux calculated by examining the vesicular tracer content as a function of time.

Clearly, the *sine qua non* for the success of the experiments is the availability of a high-affinity inhibitor, and finding a suitable inhibitor for quenching the reaction is the first priority in setting up flow–quench experiments. Even if an inhibitor known to work *in vivo* is available, it may

---

[3] $k = P$ (area/volume) $= 3 P/r$, where P is the membrane permeability and r is the vesicle radius.

not terminate the reaction. Drugs which act at the outside face of a channel may fail to quench a flux across an inside-out vesicle. On the other hand, an inhibitor might permeate an inside-out vesicle fast enough to block transport through the channel from the inside of the vesicle. That is, the rate coefficient for inhibitor entry may be an order of magnitude greater than that of the isotopic efflux. Due to the uncertainties regarding vesicle preparation, orientation, and inhibitor action, it is not possible to predict *a priori* whether an inhibitor will succeed in quenching the reaction. One must perform experiments using the potential quenching solution as the diluent (see below) to see if the flux is inhibited.

Typical Experiment

In tight, sodium-transporting epithelia such as the toad urinary bladder, the entry of sodium through the amiloride-sensitive[4] sodium channel in the luminal membrane is the rate-limiting step. Because the luminal sodium permeability is affected by hormones and is under the control of intracellular ions such as sodium, calcium, and H^+, investigators have concentrated their efforts on defining the properties of the channel. An ideal way to examine directly these issues is to measure membrane permeability in luminal membrane vesicles. In our preliminary studies, which we performed by hand, we found that the sodium flux into luminal membrane vesicles was complete in less than 2 sec, a result which demonstrated the need to use rapid-reaction methods.

We built a flow–quench apparatus, depicted in Fig. 1, based on a prototype described in considerable detail by Thayer and Hinkle.[5] In this particular model there are three syringes which deliver solutions to two mixing chambers. Isotope-loaded (^{22}Na) vesicles are placed in syringe 3 and isotope-free buffer is placed in syringe 2. When the syringes are emptied by the force of the air-driven piston, the two solutions are mixed in mixing chamber 1. This results in dilution of the vesicles and reduction in the extravesicular specific activity, causing isotope to leave the vesicle. The diluted vesicles meet the quench solution, amiloride, at mixing chamber 2, delivered in syringe 1. Addition of amiloride terminates the reaction by blocking the flow of sodium ions out of the vesicle. After the vesicles have been mixed with the inhibitor amiloride, extravesicular isotope is removed either by collecting the vesicles on filters and washing away isotope or by running the vesicles down an ion-exchange column,[6] which

[4] P. J. Bentley, *J. Physiol. (London)* **195**, 317 (1968).
[5] W. S. Thayer and P. C. Hinkle, this series, Vol. 56, p. 492.
[6] O. D. Gasko, A. F. Knowles, H. G. Shetzer, E.-M. Suolinna, and E. Racker, *Anal. Biochem.* **72**, 57 (1976).

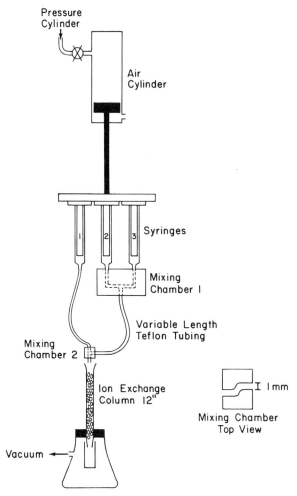

FIG. 1. Flow–quench apparatus. The apparatus was made of plexiglass and the holes drilled to suit our specific needs (see Ref. 9). Because we were interested in using as small a volume of vesicles as possible, mixing chamber 1 was drilled so that a 200-μl gas-tight syringe could empty into the chamber. The dead space from the tip of the syringe insert to the start of the tubing was only 10–20 μl, meaning that we lost only 10% of our vesicles during the flux. The descent time of the syringes could be accurately measured on an oscilloscope by installing a potentiometer to measure the movement of the piston as it depressed the syringes. Reproduced with permission of the Rockefeller University Press.

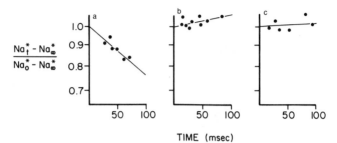

FIG. 2. Typical flux in membrane vesicles using flow–quench. We measured the fractional loss of ^{22}Na in (a) luminal, (b) basolateral, (c) mitochondrial membrane vesicles prepared from the toad bladder (see Ref. 9 for details). The slope of the line drawn through the points is the efflux rate coefficient. Reproduced with permission of the Rockefeller University Press.

removes the extravesicular isotope from the vesicles. The vesicles are then harvested either on a filter or in a vial (see Fig. 1) and counted.

To obtain a rate coefficient, one performs a number of fluxes, varying the time of each flux. Because the time of the flux is the time it takes for the diluted vesicles to go from mixing chamber 1 to 2, the reaction time is easily varied by changing either the length of tubing that connects the two mixing chambers, the volume of the syringes, or the flow rate.[7] To calculate the flux time, one divides the volume of the tubing connecting mixing chambers 1 and 2 by the flow rate through the tubing. The flow rate is calculated by dividing the volume of the syringes by the emptying time for the syringes, which can be measured on an oscilloscope (see legend, Fig. 1). Once a number of time points have been obtained, the rate constant is calculated from the plot of the natural log of vesicle isotope content versus time.[8]

A typical experiment, using membrane vesicles from the toad urinary bladder, is depicted in Fig. 2. We measured the fractional loss of sodium in luminal, basolateral, and mitochondrial membrane vesicles.[9] There is a fast sodium efflux within 100 msec from luminal membrane vesicles, which are expected to contain sodium channels, and not from either the basolateral or mitochondrial membranes. The $t_{1/2}$ of the flux in the luminal

---

[7] The flow rate through the syringes is determined by the total resistance of the system which is composed of the resistance of the syringe plungers and that of the tubing. Thus, both the type of the syringes and the caliber of the tubing will affect the flow rate, each of which can be adjusted separately.

[8] The apparent rate coefficient is affected by the degree to which the isotope-loaded vesicles are diluted. k obtained by diluting vesicles 1 : 20 will be higher than that from diluting 1 : 1. The true efflux rate coefficient can be calculated using corrected fluxes: $((S_t - S_i)/(S_0 - S_i)) = e^{-kt}$.

[9] H. Chase, Jr., and Q. Al-Awqati, *J. Gen. Physiol.* **81**, 643 (1983).

membrane vesicles was ~175 msec, close to that predicted from the macroscopic measurements of the luminal membrane permeability of the intact bladder.[10] Amiloride, when present on both on the inside and outside of the vesicles, inhibited the sodium flux in the membrane vesicles in a concentration similar to that found *in vivo*. This result suggested that the fast sodium flux was likely to be through the amiloride-sensitive sodium channel present in the luminal membrane of the intact tissue (see below). We next studied the effect of calcium on luminal sodium permeability and showed that calcium directly inhibited sodium transport in a range well within the measured values for intracellular calcium,[11] suggesting that changes in cell calcium could have an effect on luminal sodium permeability. This direct examination of the channel, as well as the precise buffering of calcium, could not have been performed *in vivo*, and demonstrates the utility of the vesicle studies.

Identification of the Flux

How did we know that the rapid flux was through the sodium channel rather than through some "leak" pathway? To a certain extent the experiments are rigged so that only the flux through the sodium channel is measured. This conclusion follows from the fact that we used a potent inhibitor of the channel to quench the reaction. If isotope escapes quickly from a vesicle by a route other than the sodium channel, then at the time of harvest these vesicles will not contain any isotope, thus having no effect on the calculated rate constant. In other words, amiloride will terminate the isotopic flux and trap tracer only in those vesicles in which the only route of exit is through the sodium channel. For a vesicle with a slow leak via a sodium carrier, the rate of isotopic flux will be dominated by the faster process, the flux through the sodium channel. Thus, the presence of either slow or fast leaks will have no effect on the rate coefficient when measured by using the channel inhibitor amiloride in the quenching solution. Thus, the observed flux is literally defined by the inhibitor chosen to terminate the flux, provided the inhibitor is specific for the particular channel.

A more convincing demonstration that the observed flux is through the channel is afforded by kinetic studies using the inhibitor. If the flux *in vitro* is inhibited by a drug with the same affinity as *in vivo*, then it is likely that the flux in both the vesicle and intact epithelium is through the same pathway. If the environment *in vitro* is sufficiently different from *in vivo*, however, then the measured affinities (K_i) may be different.

[10] H. S. Chase, Jr., and Q. Al-Awqati, *J. Gen. Physiol.* **77**, 693 (1981).
[11] S. M. E. Wong and H. S. Chase, Jr., *Am. J. Physiol.* **250**, C841 (1986).

Another approach to demonstrate that the vesicular flux is through the channel under study is to measure ion selectivity of the flux and demonstrate that it is the same as it is for the channel *in vivo*, if known. For example, we found that the luminal membrane vesicles were not permeable to either ^{86}Rb or ^{42}K, consistent with the studies *in vivo* demonstrating an ion selectivity of Na $\gg$ K, Rb.[12] In summary, one must demonstrate that the isotopic flux measured in the vesicles is indeed through the channel of interest.

Stopped-Flow Methodology

Definition of Stopped-Flow

In stopped-flow experiments, the flux across a vesicle is recorded in real time. The movement of ions is monitored either directly, with ion-selective electrodes or indirectly by measuring the light intensity of dyes whose fluorescent properties or absorptive intensity are a function of ion composition, pH, or membrane potential. The utility of the method is that it eliminates the need for an inhibitor of the transport process under study.

Typical Experiment

There has been a great deal of attention focused on the potassium channel present in the basolateral membrane in epithelia. The rate of potassium efflux from the cell, across this membrane, is coupled to potassium influx via the sodium–potassium pump.[13,14] To date, there has been little success in examining directly the properties of this channel(s), in part because of the lack of specific, high-affinity inhibitors. Complicating the issue is the fact that there is probably more than one type of potassium channel in the basolateral membrane, each with differing ion selectivities.[15]

We sought to examine the properties of the potassium channel in basolateral membrane vesicles, but could not use flow–quench due to the lack of a suitable inhibitor. We could have looked for an inhibitor by screening many drugs and measuring their effects on ^{86}Rb, which might permeate the potassium channel at a slower rate, thus enabling us to measure the flux by hand. Unfortunately, this flux would not identify those potassium channels whose selectivity excluded rubidium.

[12] L. G. Palmer, *J. Membr. Biol.* **67,** 91 (1982).
[13] S. G. Schultz, *Am. J. Physiol.* **241,** F579 (1981).
[14] H. Chase, Jr., *Am. J. Physiol* **247,** F869 (1984).
[15] W. J. Germann *et al., J. Gen. Physiol.* **88,** 237, 253 (1986).

To circumvent these problems we used stopped-flow methods[16,17] using thallium as a substitute for potassium. Thallium, able to permeate all potassium channels in which it has been measured,[18] quenches fluorescent dyes such as ANTS (8-aminonaphthalene-1,3,6-trisulfonic acid) and PTS (1,3,6,8-pyrenetetrasulfonic acid). The thallium flux can thus be observed after mixing dye-loaded vesicles with thallium and measuring the decay in fluorescence intensity over time as the thallium moves into the vesicle.

We prepared basolateral membrane vesicles from the rat kidney,[19] and measured a thallium flux across the vesicles. We preloaded vesicles with the fluorescent dye PTS, ran them down an anion-exchange column to remove extravesicular PTS, and placed the dye-loaded vesicles in syringe 3. The thallium containing diluent was placed into syringe 2. We connected a small tube from mixing chamber 1 to a small capillary tube (50 μl volume) fixed to the stage of our fluorescence microscope.[20] We focused a 40× lens on the capillary tube, excited the sample at 360 nm using a mercury light source, and recorded fluorescence at 405 nm using a monochrometer and photometer attached to the microscope. To start the reaction, the piston was pushed down manually at a rate sufficiently fast for the diluted vesicles to be in view of the microscope lens in less than 10 msec.

The experiment depicted in Fig. 3 demonstrates the movement of thallium into the vesicles in real time. PTS-loaded vesicles were diluted into a potassium glutamate buffer containing thallium (tracing A in Fig. 3) and the gradual decline in fluorescence represents the movement of thallium into the vesicles.[21] That the decay of the signal is due to thallium quenching intravesicular PTS rather than extravesicular dye is demonstrated by tracing B in Fig. 3. Vesicles were pretreated with Triton X-100 before the mixing with the thallium in order to release dye trapped in the

---

[16] H.-P. H. Moore and M. A. Raftery, *Proc. Natl. Acad. Sci. U.S.A.* **77,** 4509 (1980).
[17] A. M. Garcia and C. Miller, *J. Gen. Physiol.* **83,** 819 (1984).
[18] B. Hille, in "Ion Channels of Excitable Membranes." Sinauer, Sunderland, Massachusetts, 1984.
[19] J. E. Scoble, S. Mills, and K. A. Hruska, *J. Clin. Invest.* **75,** 1096 (1985).
[20] One drawback to stopped-flow is that the volume of vesicles and diluent must be sufficient to be read in a conventional spectrofluorimeter, usually 1 ml. A great deal of membrane protein, usually at a premium, may be required to fill up a cuvette. By using a microscope to make fluorescence measurements, one needs only a trivial volume (10 μl) to measure fluorescence decay.
[21] The $t_{1/2}$ of the flux was ~4 sec, slower than one would expect for a single, large 200-psec potassium channel in a small (0.3 μm) vesicle. As mentioned earlier in the text, based on the uncertainties regarding the purity of the vesicle preparation, the channel permeability *in vitro*, as well as a host of other unknown parameters, it is not possible to state with any confidence that the flux is as fast as, faster, or slower than predicted.

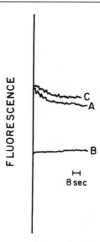

FIG. 3. Typical flux in membrane vesicles using the stopped-flow method. Basolateral membrane vesicles were suspended in potassium glutamate, 100 mM; calcium glutamate, 1 mM; MOPS, 10 mM; N-methyl-D-glucamine, 10 mM; pH 7.5, and loaded with 10 μM PTS by freeze–thaw in dry ice–acetone. Immediately before performing the flux, the extravesicular PTS was removed by passing the vesicles down an ion-exchange column (Dowex 1-X4-200, chloride form) which was prewashed with potassium glutamate buffer. Vesicles were drawn into a 100-μl gas-tight, glass syringe and placed in position 3 (see Fig. 1). Thallium–glutamate buffer, made by replacing potassium with thallium, was drawn into a 200-μl syringe and placed in position 2. In tracing A, vesicles were diluted into thallium; in B, vesicles were first pretreated with 0.1% Triton X-100; in C vesicles were pretreated with 50 μM phencyclidine and diluted into a thallium buffer also containing the inhibitor.

intravesicular space. Upon meeting the thallium, the signal was immediately quenched (<20 msec) and to a much greater extent than the untreated vesicles.[22] The difference between tracings A and B in Fig. 3 thus represent the intravesicular dye. It is of interest that greater than half of the vesicles seem not to contain a pathway for thallium entry, as indicated by the fact that the steady-state fluorescence in tracing A remains higher than that in tracing B in Fig. 3. The thallium flux observed in real time can thus be used to measure the potassium permeability of the vesicle membrane.

Identifying the Flux

How does one identify the flux as going through a potassium channel? Inhibitor experiments clearly would suffice, but there are few specific high-affinity inhibitors of any of the different potassium channels.[23] For example, as demonstrated in tracing C in Fig. 3, the thallium flux was

[22] Triton X-100 does not quench PTS.
[23] R. Latorre and C. Miller, *J. Membr. Biol.* **71**, 11 (1983).

inhibited by phencyclidine,[24] a known inhibitor of potassium channels. Although the flux was also inhibited by the potassium channel blockers barium and quinidine, they are not specific for a particular type of potassium channel. In fact, quinidine can inhibit a variety of channels, in the appropriate doses.[25] Thus, although good inhibitors are not required to perform the stopped-flow experiments, as they are in flow–quench, they are ultimately needed to define the flux.

Proof that a flux is through a particular channel may be obtained in selectivity experiments. For example, the thallium flux may be measured in the presence of counterions such as potassium, rubidium, or lithium. Thallium flow is slowed in the presence of impermeant counterions because of charge buildup, a result of the asymmetric thallium distribution across the membrane vesicle (see Fig. 3). Thus, demonstration that the thallium flux is fastest in the presence of potassium rather than another cation is indirect evidence that the thallium flux is through a potassium channel. However, even defining the selectivity of the flux will not identify the particular potassium channel, especially where these data are not available from *in vivo* studies. Defining types of potassium channels in the basolateral membrane in epithelia will require a concerted effort using a variety of techniques such as patch-clamp and reconstitution, either in vesicles or bilayers.

Other Uses of Rapid-Reaction Apparatus

The rapid-reaction apparatus is flexible and can be built to suit one's needs. For example, Karlin and associates[26] have used the apparatus to photolabel the acetylcholine receptor and compare labeling of receptor in its resting or desensitized states to that in its active state. The success of their experiment depended on their ability to time precisely channel activation and photolabeling. This is accomplished by adjusting either the flow rate or the length of tubing connecting the mixer, at which the channels are activated, and the light source, where the vesicles are photolyzed.

Reliability of the Apparatus

The heart of the rapid-reaction apparatus is the rapid mixing capability. One must test the reliability and speed of mixing before performing

[24] M. P. Blaustein and R. K. Ickowicz, *Proc. Natl. Acad. Sci. U.S.A.* **80,** 3835 (1983).
[25] A. Hermann and A. L. F. Gorman, *J. Gen. Physiol.* **83,** 919 (1984).
[26] R. N. Cox, R. Kaldany, P. W. Brandt, B. Ferreu, R. A. Hudson, and A. Karlin, *Anal. Biochem.* **136,** 476 (1984).

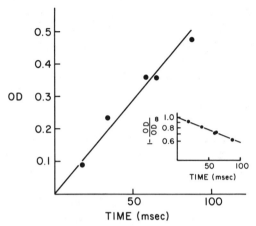

FIG. 4. Measurement of hydrolysis of DNPA using flow–quench.[9] Reproduced with permission of the Rockefeller University Press.

any experiments with membrane vesicles. To accomplish this, we examined the alkaline hydrolysis of dinitrophenyl acetate (DNPA). Upon hydrolysis the colorless DNPA is converted to the yellow DNP. In syringe 2 (see Fig. 1) was placed DNPA dissolved in ethyl acetate and in syringe 3 was NaOH. The reaction was quenched by HCl in syringe 1. After collecting the effluent of the reaction, the pH of the solution was brought to 4.5 and absorption measured in a spectrophotometer at 405 nm. The optical density (OD) of the solution is plotted as a function of time in Fig. 4. The reaction is clearly linear up to 100 msec and the fact that the line drawn through the points intersects at the origin demonstrates that mixing is uniform and nearly instantaneous. Had there been a delay in mixing, one would have expected the line to intersect on the x-axis, whose value represented the time delay mixing.

Conclusions

Rapid reaction methods, originally developed to study enzyme reactions, have been adapted for membrane transport studies. The method is particularly suited to the study of ion channels in which the time course of the ion flux is usually less than 1 sec. The rapid-reaction apparatus method can be designed and adjusted to fit a variety of needs.

Acknowledgments

This chapter is dedicated to the memory of James V. Bone, an exceptional young man, whose intense hard work and dedication were responsible for the success of the flow–

quench experiments. We sincerely thank Heinz Rosskothen, who used his considerable talents to build the flow–quench apparatus and Dr. Qais Al-Awqati, whose counsel and support made this work possible. (H. Chase is an Established Investigator of the American Heart Association.)

[18] Tracer Studies with Isolated Membrane Vesicles

By ULRICH HOPFER

Introduction

Vesicles of isolated membranes are well suited for the characterization of functional properties of transport systems. Both native as well as reconstituted membranes can be used to study properties of specific transport systems, such as substrate specificity, species of cosubstrates and countersubstrates, inhibitors, activators, pH dependence, and temperature dependence. A prerequisite is that the isolated membranes reseal to form vesicles, thereby forming an effective barrier between an intravesicular aqueous space and the outside medium. The major advantages of vesicular systems are high mechanical stability and large surface areas for solute flux. The major disadvantage is that measurements are made on an ensemble of parallel compartments that may not behave identically. Vesicles may differ with respect to size and shape or to presence and/or concentration of transporters. The amount of retrievable information about specific transporters will therefore depend on the complexity and heterogeneity of the vesicle population and the extent to which the experimental protocols for measurements and analyses of transport data take into consideration any heterogeneity.

This chapter will cover the basis for transport experiments with radiolabeled isotopes in homogeneous and heterogeneous vesicle populations. The use of radioactive solutes has the advantage that the uptake of any solute of interest is measured directly. Radiolabeled compounds and nucleotides are widely available and the actual measurements of uptake (or release) of radiolabeled solutes are straightforward. They consist of incubation of aliquots of the membrane fraction with labeled substrate, quenching of any solute movement across the membrane, separation of substrate in the medium from that associated with the membranes, and finally counting of the radioactivity. The major difficulties are in the area of design and interpretation of transport experiments because inferences about properties of transporters in the membrane are made indirectly from the dependence of solute transport on experimental conditions. Rate and extent of transport, however, are determined not solely by transporters, but also by other variables, such as solute concentrations and gradients of substrate and cosubstrates, on both sides of the membrane.

Design and Interpretation of Transport Experiments

General Considerations

Osmotic Forces. To observe transport of solutes across the membrane, the availability of an intravesicular space is crucial. This is accomplished by trapping relatively impermeant solutes into the intravesicular space during formation of vesicles, i.e., during homogenization and fractionation of cells. Sucrose or mannitol at concentrations of 0.1–0.3 M are commonly employed. Isolated plasma membrane vesicle preparations generally behave like perfect osmometers over a considerable range when exposed to hypertonic solutions of impermeant solutes, but not when exposed to hypotonic solutions. Under the first condition membranes shrink, while under the latter condition membrane vesicles tend to burst and the membrane looses its ability to act as a barrier separating two compartments. This osmotic behavior has consequences for the design of transport experiments: Conditions that would lead to swelling and bursting of vesicles have to be avoided. Such conditions are replacement of the solution with which the membrane vesicles have been equilibrated by (1) a hypotonic solution, or (2) an isotonic solution in which *permeant* solutes replace relatively impermeant ones. In the latter case, vesicle swelling and bursting will follow a time course that is tied to the movement of the permeant solute into the vesicles, water moving along secondarily.

Vesicle Size. Major underlying assumptions in transport experiments with vesicles are that the intravesicular aqueous phase is a well-mixed compartment with bulk phase properties and that boundary layer effects do not make a major contribution to solute uptake. These assumptions can be true only for vesicles of a certain size range. The reasons are: (1) mixing of the intravesicular space occurs by diffusion and is therefore slower the greater the volume, setting an upper limit on size for which the assumption is fulfilled; (2) the magnitude of boundary layer effects is directly related to the surface area and therefore the fraction of the intravesicular volume that is influenced by the boundary increases with decreasing vesicle size, setting a lower limit. The time constant of equilibration (τ) for any solute, assuming a spherical compartment that is well stirred by diffusion, can be calculated from vesicle size as follows:

$$\tau = r^2/D \tag{1}$$

where r is the vesicle radius and D is the diffusion constant of the solute in the intravesicular medium. Fortunately, τ is less than a millisecond for typical vesicles with radii of 0.1–0.3 μm and small solutes, such as glucose, with diffusion constants of 10^{-5} cm^2/sec in aqueous media. The extent of the boundary layer can be approximated for electrical effects from the Debye length, estimated, for example, on the basis of the Gouy–Chapman theory.[1] Significant Donnan effects that may have their origin in surface charges have been observed in isolated intestinal and respiratory plasma membranes.[2,3]

Uptake versus Transport. The design of transport experiments strongly depends on the extent to which uptake (or release) of labeled solutes reflects transport across the membrane, the type of transport system under investigation, and the level of information sought. Interesting information can be obtained either from the movement of solutes against their concentration gradients or from the kinetics of solute transport, i.e., from the dependence of transport rates on experimental variables such as substrate concentration, pH, or temperature. Since the primary data measure time-dependent increases (or decreases) of radiolabeled solute uptake by membrane suspensions, the first question that needs to be addressed is whether and to what extent any uptake is related to transport across the membrane. The observed uptake constitutes transport if it is due to movement from the medium into the intravesicular space and/or to binding sites on the inside of the membranes (or in the reverse order). The observed uptake does not constitute transport if it is due to partitioning of the solute into the lipid phase of the membrane (applies to hydrophobic solutes) and/or binding to the outside surface of the membranes. The underlying factors for solute uptake have to be sorted out for each solute/membrane pair. Criteria for solute uptake into an intravesicular space are: (1) the equilibrium uptake of the solute of interest is directly proportional to the intravesicular space, which can be manipulated by increasing the medium osmolarity with impermeant solutes (valid only for hyperosmotic media) (Fig. 1); (2) the equilibrium uptake is proportional to solute concentration (no saturation); (3) the normalized uptake (solute uptake divided by its concentration in the medium) is identical for all solutes that are taken up into the same intravesicular space. However, this latter criterion does not necessarily hold for electrolytes; their equilibrium uptake is also influenced by Donnan effects, i.e., electrical effects due to fixed charges on the inside of the vesicles. These effects can be quanti-

[1] S. McLaughlin, *Curr. Top. Membr. Transp.* **9**, 71 (1977).
[2] C. M. Liedtke and U. Hopfer, *Am. J. Physiol.* **242**, G263 (1982).
[3] J. E. Langridge-Smith and W. P. Dubinsky, *Am. J. Physiol.* **249**, C417 (1985).

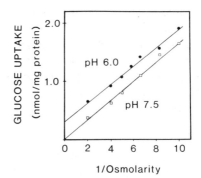

FIG. 1. Effect of medium osmolarity on D-glucose uptake in isolated intestinal brush border membranes. The data demonstrate that D-glucose uptake at pH 7.5 (squares) can be accounted for completely by transport into an osmotically active space. In contrast, for D-glucose at pH 6.0 (circles), about 15% of the total uptake appears to be insensitive to vesicle volume (intercept of y-axis relative to uptake under isosmotic conditions [1/osmolarity = 10]). The membranes were prepared in 100 mM cellobiose, 1 mM HEPES adjusted to pH 7.5, and 0.1 mM MgSO$_4$. Uptake was measured from a medium containing 25 mM NaSCN, 1 mM Tris–HEPES at pH 7.5 or pH 6.0, 0.1 mM MgSO$_4$, 1 mM D-glucose, and cellobiose. The concentration of cellobiose was adjusted to give the indicated osmolarity (plotted as inverse osmolarity) disregarding other solutes. Incubation: 10 min at 25°. Redrawn with permission from U. Hopfer and K. Sigrist-Nelson, *Nature (London)* **252**, 422 (1974).

tated in terms of the Donnan potential from the dependence of the equilibrium uptake of different electrolytes on nature and valency of their charge and on the ionic strength of the intravesicular solution.[4]

The most difficult situation to sort out is that of binding to sites on the inside, which must be distinguished from binding to the outside. In the former case, the observed uptake represents transport across the membrane and thus interesting information on transport can possibly be obtained from uptake measurements. The question of sidedness of binding can be resolved with kinetic experiments if transport across the membrane is rate-limiting and the transport kinetics dominate the overall uptake kinetics. For example, saturation kinetics for isotope exchange at equilibrium provide one criterion.[5]

Once it has been established that uptake of a particular solute represents transport, it is possible to probe the characteristics of this transport and, in the case of mediated transport, of the transporter(s). The characterization can be accomplished by using kinetic and/or thermodynamic approaches for experimental design and interpretation. The thermody-

[4] E. Heinz, "Electrical Potentials in Biological Membrane Transport." Springer-Verlag, Berlin, Federal Republic of Germany, 1981.
[5] W. A. Muir, U. Hopfer, and M. King, *J. Biol. Chem.* **259**, 4896 (1984).

namic approach is particularly useful for investigations of co- and countertransport systems.

Approach Based on Thermodynamics

Overshooting Uptake. Thermodynamics provides a framework for the design and interpretation of experiments concerned with coupled transport, either co- or countertransport. It has the advantage that it is independent of specific molecular models for the transporter which brings about the coupled transport. The treatment can be either in terms of thermodynamics of equilibria (see below) or of irreversible processes.[6]

In general, for a coupled transport reaction between two compartments, one can state at constant temperature and pressure and in the absence of other reactions

$$dG = \sum \Delta\mu_i \, dn_i \qquad (2)$$

where dG is the variation in Gibbs free energy (G) for the entire system, $\Delta\mu$ the difference in electrochemical potential of solute i in the two compartments, and n the number of moles of solute i which have been transported.[7]

This relationship allows one to investigate coupled transport by setting up experimental conditions in which one solute is transported uphill against its electrochemical gradient (driven solute) while others supply the necessary energy by moving downhill (driving solutes). This situation appears as active transport for the driven solute and is therefore termed secondary active.

The exact form of Eq. (2) depends on the number of different solutes involved, the stoichiometry at which they are transported, and whether the solute movement constitutes co- or countertransport. It can be derived easily from the corresponding transport reaction. For example, for the tightly coupled transport of two solutes A and B, which are cotransported from compartment 1 to compartment 2 at a ratio of $n_a : m_b$, the transport reaction is as follows:

$$n_a A_1 + m_b B_1 \rightarrow n_a A_2 + m_b B_2 \qquad (3)$$

The corresponding Gibbs standard free energy change associated with the net movement of n_a moles via the transporter is

$$\Delta G = n_a \, \Delta\mu_a + m_b \, \Delta\mu_b \qquad (4)$$

---

[6] H. Murer, H. Lücke, and R. Kinne, in "Secretory Diarrhea" (M. Field, J. S. Fordtran, and S. G. Schultz, eds.), p. 31. Am. Physiol. Soc., Bethesda, Maryland, 1980.

[7] D. Walz and S. R. Caplan, *Biochim. Biophys. Acta* **859,** 151 (1986).

This equation allows predictions of the direction of the coupled transport via the transporter on the basis of the magnitudes of $\Delta\mu_a$ and $\Delta\mu_b$. The coupled transport spontaneously proceeds from left to right if $\Delta G < 0$, and reaches equilibrium when $\Delta G = 0$. Thus movement of solute A from compartment 1 to 2 against an electrochemical gradient ($\Delta\mu_a > 0$) can occur if $\Delta\mu_b < 0$ and $|n_a \Delta\mu_a| < |m_b \Delta\mu_b|$ (for more details, see, e.g., Ref. 8).

Equation (4) can be transformed into a more useful format by replacing $\Delta\mu$ in terms of concentrations and electrical potential. For dilute aqueous solutions (activity coefficients = 1) and at constant temperature and pressure:

$$\Delta\mu_i = RT \ln([i]_2/[i]_1) + zF \Delta\Psi \tag{5}$$

where R is the gas constant, T the absolute temperature, F the Faraday constant, z the electrical charge on solute i, and $\Delta\Psi$ the transmembrane potential. A relationship between the concentrations of the driven solute A and the driving solute B can be derived for cotransport from left to right ($\Delta G \leq 0$) from Eqs. (4) and (5) as follows:

$$\ln([A]_2/[A]_1) \leq (m_b/n_a) \ln([B]_1/[B]_2) + [(m_b/n_a)z_b - z_a]F \Delta\Psi \tag{6}$$

The equality sign applies to equilibrium ($\Delta G = 0$) with no flow through the transporter, including a constrained equilibrium with $\Delta\mu_a > 0$. Equation (6) predicts that solute A can be transported against an electrochemical gradient by an appropriate electrochemical gradient of solute B. The existence of sodium–glucose cotransport in the small intestinal brush border membranes was initially shown on the basis of such considerations.[9] For example, Fig. 2 shows glucose uptake by vesicles after an electrochemical Na gradient had been established at time zero. The Na^+ gradient supports glucose uptake into the vesicles against a concentration gradient; however, with time, as the Na^+ gradient dissipates the excess glucose leaves the vesicles, giving the appearance of a so-called overshoot. It is important to stress that observations of overshooting solute uptake are evidence for secondary active transport only if other possible explanations can be excluded. These are (1) initial swelling of vesicles and subsequent shrinkage due to osmotic forces when impermeant solutes in the medium are replaced by more *permeant* ones at the start of an incubation, or (2) breakdown of the membrane barrier with time, or (3) in the case of electrolytes, an electrical potential that draws ions into the in-

---

[8] E. Heinz, "Mechanics and Energetics of Biological Transport." Springer-Verlag, Berlin, Federal Republic of Germany, 1978.

[9] H. Murer and U. Hopfer, *Proc. Natl. Acad. Sci. U.S.A.* **71**, 484 (1974).

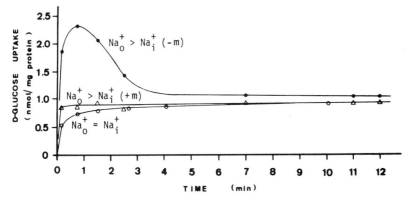

FIG. 2. Overshooting uptake of D-glucose by intestinal brush border membranes. D-Glucose uptake was measured either in the presence of a NaSCN gradient without further addition (filled circles; $Na_o^+ > Na_i^+$; $-m$), or in the presence of a NaSCN gradient plus 20 μg/ml monactin (open triangles; $Na_o^+ > Na_i^+$; $+m$), or in the presence of NaSCN, but absence of a gradient (open circles; $Na_o^+ = Na_i^+$). The membranes were prepared in 0.1 M mannitol, 0.1 mM MgSO$_4$, and 1 mM Tris-HEPES (pH 7.5). For the first two experiments ($Na_o^+ > Na_i^+$), D-glucose uptake was measured after simultaneous addition of labeled D-glucose and NaSCN plus other constituents of the membrane isolation buffer so that the final concentrations of D-glucose and NaSCN were 1 mM and 100 mM, respectively, and the concentrations of the other constituents did not change. Note that the incubation medium is hypertonic relative to the membrane isolation medium. In the experiment ($Na_o^+ = Na_i^+$), the membranes were preincubated with 100 mM NaSCN for 10 min before addition of D-glucose. The overshooting part represents concentrative uptake since osmotic considerations eliminate initial vesicle swelling. The energy for the overshooting uptake comes from the electrochemical Na$^+$ gradient established at time zero. It is eliminated either by the inclusion of the ionophore monactin, equilibrating the electrochemical Na$^+$ gradient, or dissipation of the Na$^+$ gradient by preincubation. Redrawn with permission from U. Hopfer, *Proc. Natl. Acad. Sci. U.S.A.* **72**, 2027 (1975).

travesicular space. If these three possibilities can be excluded on the basis of experimental design and controls, observations of overshooting uptake prove the existence of coupled transports. Determination of the types of solutes and electrical potentials that produce overshoots can be used to obtain information on co- or countersubstrates in coupled transport.

Constrained Equilibrium (Static Head). Equation (6) also provides the basis for a quantitative treatment of transport data to evaluate the stoichiometry ($m_b : n_a$) of coupled transport.[10] One of the conditions of no flow through the transporter is that of constrained equilibrium in which $\Delta G = 0$ and the force inherent in $n_a \Delta \mu_a$ ($\Delta_{\mu a} > 0$) is balanced by that in $m_b \Delta \mu_b$. The experimental design calls for a fixed concentration gradient of solute

[10] J. Turner, *Ann. N.Y. Acad. Sci.* **456**, 10 (1985).

A, which is established at the beginning of the incubation period, and then determination of the electrochemical gradient of solute B at which there is no flux of solute A through the coupling transporter (Fig. 3). The experiments can be carried out with either just a concentration gradient of solute B (voltage clamped to zero) or an electrical potential (no concentration gradient of solute B). If the concentrations of the solutes in the medium and the intravesicular space and the electrical potential are known, the stoichiometry ($m_b : n_a$) can be calculated from Eq. (6).

Kinetic Approaches

General Considerations. Kinetic approaches to transport in vesicles can provide substantial information about the dynamic properties of transporters. It is important to keep in mind that kinetic data are inter-

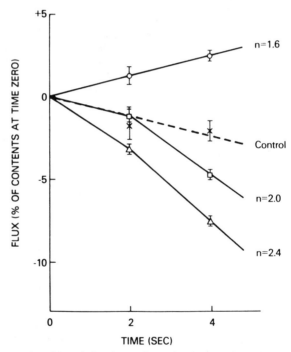

FIG. 3. The results of a static head experiment for the Na^+-dependent glucose transport in renal outer medullary brush border membrane vesicles. Vesicles were loaded with 20 mM NaCl, 50 mM choline chloride, and 0.25 mM labeled D-glucose and then diluted 1:5 into isosmotic media containing various concentrations of Na^+. The values of n indicated on the figure are the Na^+ : glucose stoichiometries that would be predicted if that run were to lie on the control (static head condition). Reproduced with permission from J. Turner and A. Moran, *J. Membr. Biol.* **70,** 37 (1982).

preted on the basis of a model and that therefore the adequacy of the model constitutes a critical aspect of obtaining new information. For vesicle systems, not only molecular models for the transporter of interest have to be considered, but also the complexity of the vesicle population.

Solute flow across a membrane is influenced by many factors. Some of the obvious ones are cis or trans concentrations of the substrate, cosubstrate, or countersubstrate, pH, membrane potential, ionic strength, number of transporters per vesicle, intravesicular volume. A general mathematical description accounting for the dependencies of solute flux on all these factors and valid for all possible experimental conditions would be quite complex. In practice, analysis of flux data can be carried out only if the dependence of uptake on time is simple. It is particularly important to avoid all complex time dependencies that can arise from changes of transmembrane gradients of co- or countertransported substrates, of the membrane potential, or of the environment which could affect the transporter of interest, such as intravesicular pH and ionic strength. Two types of experimental conditions lend themselves particularly well to kinetic experiments and analyses because of a simple time dependence: (1) initial steady-state rate measurements (linear dependence of uptake on time), (2) conditions with exponential dependence of uptake on time, such as isotope exchange at equilibrium. The types of information that are obtained with these two approaches are different and usually complementary.

Initial Rates. The determination of initial steady-state rates can be carried out under a wide variety of conditions. This approach therefore can provide considerable information about the dynamic behavior of transporters. In principle, it is valid for all transport processes that are determined by time-independent rate constants and for all types of vesicle populations, i.e., even for heterogeneous ones. In this method, flux measurements are carried out over short periods while the transport rates are constant and transmembrane gradients are not significantly disturbed by any fluxes. Parallel incubations are set up which differ only with respect to an independent experimental variable such as cis and trans concentrations of substrate or of cosubstrates. This protocol yields the dependence of the initial rate (the dependent variable) on an independent variable, which can then be interpreted in terms of molecular models (e.g., simple diffusion and/or properties of transporters). Often a Michaelis–Menten type of relationship between rate and an independent variable of interest is observed.

The initial rate approach requires that all transmembrane gradients and the electrical potential are "clamped" during the time course of the flux measurements, i.e., that they are not substantially influenced by any fluxes. More importantly, the transmembrane gradients and the potential, except for the independent variable itself, have to be identical in the parallel incubations that measure the dependence of the rate on the independent variable. Clamping of gradients at specific values, including zero (identical solute concentrations on both sides of the membrane), can be accomplished by appropriate buffering systems and/or solute concentrations that are high relative to the transmembrane fluxes. The membrane potential can be clamped to the Nernst potential of a particular ion by the presence of this ion in the intra- and extravesicular spaces provided the membrane also has a high conductance for this ion (usually established by an ionophore, such as valinomycin).

Although widely used, initial rate kinetics can have considerable problems, particularly when used with heterogeneous membrane vesicles. Major kinds of problems are (1) linearity of uptake of the solute of interest with time under all experimental conditions, (2) achievement of a steady state for transport in all vesicles, (3) effects of the experimental variable on transmembrane solute gradients and/or the potential so that *several* factors that determine the transport of interest change simultaneously, (4) dominance of the initial rate kinetics by a subfraction of the membrane population. Some of these problems can be experimentally controlled. For example, linearity of uptake can be measured for the different experimental conditions of interest and the sampling period restricted to the linear range. However, considerable diligence is required since with discrete sampling, which is the method of choice for radiolabeled solutes, changes in rate do not immediately become obvious in the uptake domain, but only after having persisted for some time (uptake is the time integral of rate). The other problems are more difficult to correct for, since useful criteria are lacking for vesicle populations of unknown complexity on how the existence of a steady state and the magnitude of transmembrane gradients can be determined in the vesicle fraction that is responsible for solute uptake. Particularly in the case of solutes that are concentrated by secondary active transport, the measured initial rate can originate from only a small fraction of the vesicles, so that bulk measurements on the entire population may not be informative about the state of a particular subpopulation. For instance, contamination of basolateral plasma membranes with brush border membranes containing a secondary active transport system for a specific solute will essentially determine initial rate kinetics of the entire fraction if the basolateral membrane transports the solute only very poorly. The contamination may show up only under some experimental conditions if the signal from the contaminating fraction is particularly high due to active transport.

There are many examples in the literature in which initial rate kinetics proved crucial for the discovery of new transport mechanisms or isolation of transporters.[10–12] On the other hand, there are also examples in which the characterization of transporters in terms of kinetic parameters (e.g., Michaelis constant and maximal velocity) appears uninformative because transmembrane gradients were not controlled and therefore cannot be reproduced.[13] An indication for this type of problem is that many investigators consider different values for the Michaelis constant of a particular transporter as being in "close agreement" if they fall within an order of magnitude, even though the difference may be outside the statistical error.

Equilibrium Exchange. The rationale behind measurements of tracer exchange at equilibrium is quite different from that behind initial rate kinetics. The method consists of preincubation of vesicles until chemical equilibrium is reached, and then tagging of the solute of interest in the medium with a different label (e.g., replacement of an unlabeled solute by one labeled with a radioisotope). The time course of tracer exchange between a small and a large compartment is known *a priori* to be an exponential, again provided the rate constants involved are time independent. Thus, the tracer uptake (or release) of a single vesicle is characterized by a time constant and an amplitude. All parallel transport pathways in a vesicle give rise to a single lumped time constant. The amplitude is proportional to the intravesicular space occupied by the solute of interest. Measurements of an entire time curve of tracer exchange for an entire vesicle population can therefore be used to determine (1) the fractional intravesicular volume associated with subpopulations of vesicles characterized by their transport rates, and (2) initial rates. The equilibrium exchange approach has then not only the advantage of well-defined experimental conditions for tracer transport and initial rate estimates from entire time curves of exchange, but, more importantly, has the ability to deal explicitly with the problems of heterogeneity (see below for a quantitative treatment). Using initial rate estimates, equilibrium exchange allows similar kinetic experiments as described above for the conventional initial rate method, i.e., evaluation of the dependence of the transport rate of a solute on independent experimental variables (see Fig. 4). However, the conditions are restricted to equilibrium states with equal cis and trans concentrations of all solutes.

Heterogeneity of vesicles in functional terms can easily be demonstrated for membrane vesicles with *time-independent* rate constants by

[11] J. Turner and A. Moran, *Am. J. Physiol.* **242,** F406 (1982).
[12] B. I. Kanner and R. Radian, *Ann. N.Y. Acad. Sci.* **456,** 153 (1985).
[13] U. Hopfer, *Am. J. Physiol.* **234,** F89 (1978).

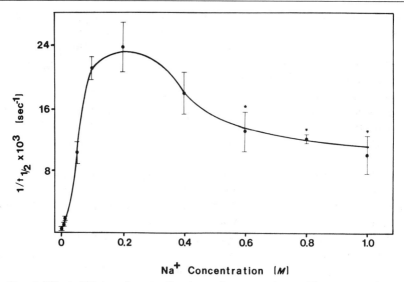

Fig. 4. Effect of Na$^+$ on the rate of D-glucose isotope exchange. The rates are observed $(t_{1/2})^{-1}$ values for labeled isotope uptake. The membranes were preincubated in buffers containing 1 mM unlabeled D-glucose, 0.1 M mannitol, 1 mM Tris/HEPES, pH 7.4, 0.1 mM MgSO$_4$, and Na$^+$ and K$^+$ as chloride salts at a combined concentration of 1 M. The isotope exchange was initiated by adding a small amount of identical buffer with labeled D-glucose. Asterisks indicate significantly ($p < 0.05$) different values from the maximal rate at 0.2 M Na$^+$ (bars = ±1 SD). Reproduced with permission from U. Hopfer and R. Groseclose, *J. Biol. Chem.* **255**, 4453 (1980).

measuring isotope flux under equilibrium exchange conditions. The finding that more than one apparent rate constant is required to describe the macroscopic uptake as a function of time suffices to conclude that some degree of heterogeneity exists in the vesicle population[14] (see below). Heterogeneity in vesicle preparations can have two different origins: (1) contamination of the preparation with types of membranes which are different from the bulk and are not of interest; (2) differences in vesicle size or shape and/or in the surface concentration of the transporter(s) of interest. In the latter case, the vesicles, although not identical, are of the same type and respond similarly to factors that affect transporter activity, such as substrate concentration, pH, or temperature. Membrane preparations that exhibit the latter type of heterogeneity have therefore been labeled as "similar." Kinetic experiments that test the effects of indepen-

---

[14] U. Hopfer, *Fed. Proc., Fed. Am. Soc. Exp. Biol.* **40**, 2480 (1981).

dent variables on transport rates in similar vesicles can be carried out not only using initial rate estimates, but also other types that are proportional to the initial rates, such as reciprocal half-times of exchange, facilitating experimental design.[14]

Gated Ion Transport. The above discussion of kinetics assumed the existence of time-*independent* rate constants that determine membrane transport. Another order of magnitude of complexity exists in membranes with gated channels for which the elementary transport process is described by time-*dependent* constants. Such gated ion transport exists in excitable membranes (e.g., acetylcholine receptor-mediated Na^+ channels). Bernhardt and Neumann[15–17] have analyzed experimental conditions that may be suitable for extracting information from vesicles with gated channels. They were particularly interested in information about kinetic constants of receptor reactions that control channel opening and closing, and showed that such an analysis can be accomplished for vesicle populations that fall into the category of similar membranes (i.e., all membranes contain one or more of the excitable channels). In essence, the information that is needed for analysis consists of measurements of tracer flux and of the internal volumes of size subclasses. The quantitative interpretation of the events in vesicles in terms of kinetic constants of receptor-mediated gating depends on information about rates as well as amplitude of neurotransmitter-controlled tracer fluxes.

Transport in Nonhomogeneous Vesicle Populations

Quantitation of Heterogeneity of Transport. The experimental conditions that allow quantitative inferences about heterogeneity of transport in membrane vesicles become obvious when one considers the basic equations relating macroscopic flux measurements to events in individual vesicles. For discrete sampling of solute uptake (or release) by (from) vesicles, the time-dependent macroscopic uptake is the sum of the events in the individual vesicles:

$$\text{Uptake }(t) = \sum \int_0^t v_i \, dt + \text{uptake }(0) \quad (7)$$

where v_i is the transport velocity in an individual vesicle and the summation is carried out over all the vesicles in the measured sample. The membrane properties (P_m) can only be inferred from the velocities, v_i,

---

[15] J. Bernhardt and E. Neumann, *Biophys. Chem.* **14**, 303 (1981).
[16] J. Bernhardt and E. Neumann, *Biophys. Chem.* **15**, 317 (1982).
[17] J. Bernhardt and E. Neumann, *Biophys. Chem.* **15**, 327 (1982).

after accounting for effects of surface area, substrate concentration, and appropriate driving force:

$$v_i = P_m \text{ (surface area)} \times f \text{ ([substrate], driving force)} \quad (8)$$

where $f(\ )$ means some "function of."

In general, v_i can take on positive or negative values. For example, for solute–nonelectrolyte cotransport, the experimental conditions are often designed to give overshooting uptake (see above). If the time of maximum uptake differs in different vesicles, a time period must exist when net uptake occurs in some vesicles and net release in others. A consequence of Eq. (7) is therefore that the v_i values cannot be evaluated without *a priori* knowledge of their time dependence. In addition, if the velocities (v) are taken as normalized (specific) values (e.g., per unit of surface area) the weights or contributions of individual vesicles (with a particular time dependence of v) to the macroscopic uptake need to be known or determinable.

An experimental condition for which the general time course of solute flux is known *a priori* is isotope exchange between a small and large compartment in equilibrium with each other. For single-walled vesicles with a time-independent solute permeability (i.e., absence of gated channels), the isotope flux is determined by a single exponential characterized by an apparent rate constant of exchange, whether solute flux is carrier-mediated or not. All parallel pathways for solute flux across a given membrane give rise to a combined rate constant. Under these experimental conditions it is possible to make inferences about the distribution (frequency) of rate constants in the population.

To demonstrate mathematically the relationship between the frequency of apparent rate constants of exchange in the population and the macroscopic uptake data, it will be useful to express the experimental data in terms of a complementary exchange function $\text{cxf}(t)$:

$$\text{cxf}(t) = 1 - \text{fractional exchange} \quad \text{[dimensionless]} \quad (9)$$

$\text{cxf}(t)$ is a time-dependent function, covering both the situations of isotope uptake and of release. For an individual vesicle, the time dependence of tracer exchange between an infinitely large and a small compartment is given by a single exponential:

$$\text{cxf}(t) = \exp(-kt) \quad (10)$$

where k is a function of membrane-specific properties, such as surface, volume, and permeability, as well as of experimental condition-dependent variables, such as substrate concentration, temperature, and pH. k is assumed to be time independent. For a vesicle population, the macroscopic complementary exchange function is then given by

$$\text{CXF}(t) = \sum g(k_i) \exp(-k_i t) \tag{11}$$

where the summation is carried out over all the vesicles in the ensemble. $g(k_i)$ measures the contribution that each set of vesicles with a particular k_i makes to the macroscopic flux and is proportional to the intravesicular volume associated with a particular rate constant. Equation (11) implies that $\sum g(k_i) = 1$.

For samples with large numbers of vesicles (and isolated membrane preparations typically contain more than 10^{10} vesicles per milligram protein), the time dependence of CXF can be approximated as the integral:

$$\text{CXF}(t) = \int_0^\infty g(k) \exp(-kt)\, dk \tag{12}$$

with $\int_0^\infty g(k)\, dk = 1$. Equation (12) illustrates that the exchange function $\text{CXF}(t)$ is the Laplace transform of the frequency distribution $g(k)$ of the apparent rate constants. The Laplace transform is a well-understood operation, akin to the Fourier transform. The interesting information about the heterogeneity of the vesicle population is contained in $g(k)$, which can be obtained from the experimentally measured $\text{CXF}(t)$ by an inverse Laplace transformation. A practical inversion method is provided by the computer program CONTIN (Fig. 5). With sufficiently precise transport data, the frequency distribution $g(k)$ may reveal distinct subpopulations.

Mean Transport Properties of a Population. Information on the frequency distribution of the apparent rate constant of a population allows calculation of a defined, mean apparent rate constant and initial rate which can both be used to characterize and compare vesicle populations, e.g., during purification of transporters. In addition, for spherical vesicles it is possible to calculate a mean permeability (i.e., essentially a rate constant normalized to surface area), provided the size distribution is known and is independent of the permeability distribution.

The mean rate constant for a population is defined in general [i.e., even if $\sum g(k_j)$ is not normalized to 1]:

$$\langle k \rangle = \sum [g(k_j) k_j] \Big/ \sum g(k_j) \quad (\text{units: time}^{-1}) \tag{13}$$

where k_j corresponds to the rate constants of the grid points j for which the numerical inverse Laplace transformation have been carried out. Thus, $\langle k \rangle$ can be calculated in a straightforward manner from $g(k_j)$. Interestingly, the mean rate constant gives a reliable estimate of the initial steady-state rate of the isotope exchange as is seen when Eq. (12) is differentiated with respect to time.

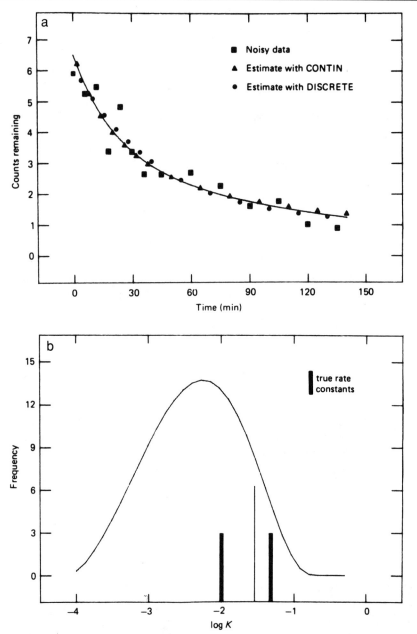

FIG. 5. Simulation of isotope exchange (a) and analysis in terms of continuous distribution or a discrete number of exponentials (b). Comparison of possible solutions with CONTIN[19-21] and DISCRETE (S. W. Provencher, Technical report. Max-Planck-Institute

The initial rate is of interest when comparing different membrane populations as it should be dependent only on the surface involved and thus can be used to define a specific activity. The initial velocity of isotope exchange $V_{t=0}$ is given by

$$V_{t=0} = \langle k \rangle \text{ [substrate](volume)} \quad \text{(units: mass flux/time)} \quad (14)$$

[substrate] represents the substrate concentration at which the isotope exchange is carried out and volume refers to the intravesicular volume involved (i.e., the small compartment where the concentrations of the tagged solute are changing). If the intravesicular volume is expressed per unit protein (typically, 1–5 μl/mg protein) then a specific initial velocity is obtained. Because the initial velocity is dependent only on surface area, Eq. (14) is applicable to vesicles of any shape and can be used to compare the activity of different populations to evaluate, for example, enrichment of transporters.

While the determination of the specific transport activity on a protein basis is straightforward [Eq. (14)], conversion to transport activity on a surface area basis is more difficult. Such information is desirable in many cases, particularly if activity in planar membrane systems is to be compared with that in vesicular ones. The conversion is possible in the case of spherical vesicle populations if the size distribution is determined by independent measurements, such as photon correlation spectroscopy (see below), and the size distribution is independent of the permeability distribution. This situation is probably encountered frequently, although the alternative is conceivable. For example, when cells are disrupted and fractionated, the size of the membrane fragments and of the resulting resealed vesicles could depend on the carrier density so that vesicle size

---

for Biophysical Chemistry, Goettingen, Federal Republic of Germany, 1976). (a) Time-dependent efflux of isotope. The simulation of isotope efflux was carried out by using the built-in feature of CONTIN assuming two subpopulations each with a rate constant indicated as solid bars in b. The two subpopulations contribute equally to the signal, i.e., each subpopulation contributes 50% of the intravesicular volume. The two rate constants from b were used to calculate "data without noise" (continuous line in a), and with added noise of about 20% of maximal counts remaining (squares in a). For clarity only every third point is plotted. (b) The noisy data were then used to estimate a distribution of rate constants with CONTIN (continuous thin line in b) or discrete exponentials with DISCRETE (thin bar in b). The estimated rate constants in turn were used to calculated "fitted data" in a. The high level of noise, which was added for demonstration purposes, makes retrieval of the true rate constants difficult with either a continuous distribution or a discrete number of exponentials. The scales of the ordinate in b are arbitrary and are different for solutions with CONTIN and DISCRETE. The mean estimated rate constants were 22.9 and 35.1 msec^{-1} by CONTIN and DISCRETE, respectively, as compared with the true mean one of 30 msec^{-1}. Reproduced with permission from Hopfer.[18]

and permeability are no longer independent. Therefore, the assumption of independence of the two parameters needs to be stressed and may need to be evaluated experimentally for particular membranes.

With the above precautions in mind, it can be shown that Eq. (15) holds[18]:

$$\langle P \rangle = \langle k \rangle/(3 \langle 1/R \rangle) \quad \text{(units: length/time)} \quad (15)$$
$$\langle P \rangle = \langle k \rangle_{\text{vol}} \langle R \rangle_{\text{sur}}/3 \quad (16)$$

$\langle P \rangle$ is the surface area fraction-averaged permeability and $\langle 1/R \rangle$ is the weight fraction-averaged inverse radius. The determination of the latter parameter requires knowledge of the size distribution of the vesicle population. Equation (16) is equivalent to Eq. (15). The subscripts vol and sur indicate volume (weight) and surface area averaged values, respectively.

Vesicle Size Distribution from Photon Correlation Spectroscopy

Photon correlation spectroscopy (PCS) (also termed quasielastic light scattering) is well suited to determine size distribution and mean sizes of spherical vesicles in heterogeneous membrane populations for sizes up to about 0.3 μm in radius, based on the author's experience. Newer, more stable instruments are capable of extending the range. The advantages of this method are (1) size can be determined in the medium and at the temperature of interest without fixation or dehydration, (2) need for little material (microgram quantities of membranes), and (3) the measurements can be accomplished within a few minutes. It should be stressed that the method provides data that are readily interpretable in terms of size only in the case of spherical vesicles.

The method requires experimental determination of three parameters: (1) autocorrelation curve of the scattered light from the membrane vesicles in the medium of interest, (2) refractive index of the medium, and (3) viscosity of the medium. Because of the strong scattering by membrane vesicles, sufficient scattering signals from the membranes can be accumulated in a few minutes. In addition, only very low concentrations of the membranes are required.

Analysis of the autocorrelation data in terms of size distribution of the membrane vesicles is most easily achieved with the CONTIN program by Provencher.[19-21] CONTIN is a general-purpose inversion program written

[18] U. Hopfer, *Biochem. Soc. Symp.* **50**, 151 (1985).
[19] S. W. Provencher, *Comp. Phys. Commun.* **27**, 213 (1982).
[20] S. W. Provencher, *Comp. Phys. Commun.* **27**, 229 (1982).
[21] S. W. Provencher, "CONTIN User's Manual," Version 2, Eur. Mol. Biol. Lab. Tech. Rep. EMBL-DA05. EMBL, Heidelberg, Federal Republic of Germany, 1982.

in FORTRAN, with specific switch settings for analysis of PCS data in terms of vesicle (hollow sphere) size distributions. The autocorrelation data, medium viscosity, refective index, as well as other machine-specific parameters (e.g., wavelength of the laser light, angle) are entered into the program, which calculates possible solutions, a probability of the solutions, fitted curves with standard deviations, and mean sizes. (For an in-depth analysis of size distributions of membrane vesicles, see Ruf et al.[22])

The question of shape of membrane vesicles can be experimentally approached by measuring the apparent size distribution from data obtained at several angles between incident and scattered light. The apparent size distribution will be independent of angle if the vesicles have a spherical shape.

The program can be obtained:
Title: CONTIN
Catalogue number: AAOB (double precision version)
AAOC (single precision version)
CPC Program Library, Queen's University of Belfast, Northern Ireland

Useful information for running the program is contained not only in the two articles describing its characteristics,[19,20] but also in the accompanying "User's Manual."[21]

Acknowledgment

The author's research is supported by grants from the National Institutes of Health.

[22] H. Ruf, Y. Georgolis, and E. Grell, this volume [21].

[19] Electron Paramagnetic Resonance Methods for Measuring pH Gradients, Transmembrane Potentials, and Membrane Dynamics

By DAVID S. CAFISO

Chemical and electrical gradients of ions across membranes are of fundamental importance to many bioenergetic, transport, and signaling processes. As a result, a wide range of molecular probe techniques has been developed to study these gradients. This chapter describes the use of several electron paramagnetic resonance (EPR) techniques to estimate membrane electrical potentials and pH gradients in model, reconstituted, or biological membrane systems. These techniques offer some unique features when compared to other probe methodologies. Because the synthetic methods used to manipulate nitroxides are well developed, a wide range of probe structures is possible. The mechanisms by which these probes function are well understood, a feature that permits their use in a quantitative fashion, without the need for an empirical calibration. Frequently, a simple line-shape analysis will provide additional data on the probe environment and membrane dynamics, information that can greatly aid in the interpretation of probe behavior. Finally, because the sensitivity of detection of the nitroxyl radical is high, measurements can be made on relatively small samples. This sensitivity permits membrane currents in vesicles to be accurately measured to less than 1 pA/cm^2.

The paramagnetic probes described here partition between aqueous and membrane phases as a function of membrane potentials or ΔpH. Quantitation of this probe phase partitioning is easily accomplished using EPR spectroscopy. Two categories of probe will be discussed: paramagnetic hydrophobic ions and paramagnetic secondary amines. The use of paramagnetic amphiphiles to determine membrane surface potentials is discussed elsewhere in this series.[1]

Hydrophobic ions (such as tetraphenylphosphonium and tetraphenylboron) appear to bind to bilayers in two regions, termed "boundary" regions, located below the level of the membrane–solution interface.[2-5] These ions also experience a lower energy barrier to transport, compared to simple inorganic ions, and readily transit membranes. The free-energy profile for a hydrophobic ion in a hydrocarbon membrane is depicted in Fig. 1. In practice, the binding of the probes discussed here is adjusted so that significant membrane-bound and aqueous populations of probe are present, As a result, these ions equilibrate between the four regions labeled o, m_o, m_i, and i in Fig. 1, and their distribution is a function of the relative electrostatic potentials in each region. Three potentials are referred to here: $\Delta\psi$, the transmembrane potential difference between regions i and o; ψ_o and ψ_i, the differences in potential between each boundary region (m_o or m_i) and its corresponding bulk aqueous phase (o or i). ψ_o and ψ_i are termed the external and internal boundary potentials, respec-

---

[1] D. Cafiso, A. McLaughlin, S. McLaughlin, and T. Winiski, this series, Vol. 171 [16].
[2] B. Ketterer, B. Neumcke, and P. Lauger, *J. Membr. Biol.* **5**, 225 (1971).
[3] D. A. Haydon and S. B. Hladky, *Q. Rev. Biophys.* **5**, 187 (1972).
[4] S. B. Hladky, *Curr. Top. Membr. Transp.* **12**, 53 (1979).
[5] P. Lauger, R. Benz, G. Stark, E. Bamberg, P. C. Jordan, A. Fahr, and W. Brock, *Q. Rev. Biophys.* **14**, 513 (1981).

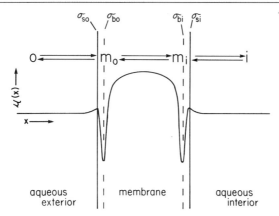

FIG. 1. Potential energy profile, $\mu(x)$, for a hydrophobic ion across a thin hydrocarbon membrane (see Ketterer et al.[2]). There are free-energy minima for these ions at "boundary" regions lying below the membrane–solution interface. The hydrophobic ion probes I–IV (see Fig. 3) are designed to phase partition between the membrane boundary regions, m_o and m_i, and the interior and exterior vesicle volumes, i and o, respectively. These ions transit membranes rapidly compared to inorganic ions, thus establishing a transmembrane equilibrium. The phase partitioning of these probes is a function of ψ_i and ψ_o, the boundary potentials, and $\Delta\psi$, the transmembrane potential (see text).

tively. Boundary potentials can be altered by changes in the charge densities, labeled σ_{bo} and σ_{bi} in Fig. 1, or changes in the surface charge densities, σ_{so} and σ_{si}.[6]

It is readily apparent that the partitioning of hydrophobic ion probes will depend on the potentials ψ_o or ψ_i. For example, if these potentials become more positive, the aqueous concentration of a positive hydrophobic ion will increase. In vesicles, the partitioning of these probes is also dependent on $\Delta\psi$, the transmembrane potential difference. This dependence arises because of the large difference between the internal and external aqueous volumes of vesicles. For the secondary amine labels, the dependence of probe phase partitioning on ΔpH also results because of this volume asymmetry. In this case, the chemical gradient of protons, rather than $\Delta\psi$, determines the distribution of probe across the membrane.

A detailed description of the behavior and synthesis of hydrophobic

---

[6] Changes in the charge density within other membrane regions can also alter the potentials ψ_o and ψ_i. For example, the movement of charge across the membrane interface to a fixed point within the membrane will also alter the boundary potential, provided the charge is localized deeper than the membrane boundary region [see D. S. Cafiso and W. L. Hubbell, Biophys. J. **30**, 243 (1980)].

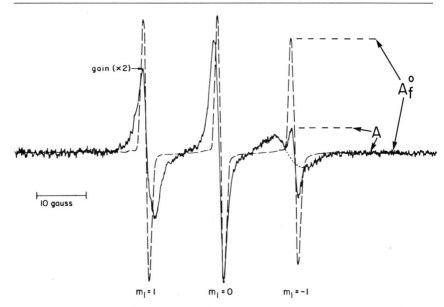

Fig. 2. EPR spectra of the paramagnetic phosphonium ion I(3) in buffer (---) and in the presence of sonicated egg PC vesicles (——). In the presence of vesicles, the spectrum is a simple sum of an aqueous and membrane-bound spectrum. A_f^o is the total signal amplitude of the $m_I = -1$ resonance in buffer and A is the amplitude in the presence of membranes. From A_f^o and A the phase partitioning of the probe, λ, can be determined.

ion and secondary amine labels is given elsewhere.[7–9] Here, experimental procedures for obtaining and interpreting data from these labels will be discussed.

Quantitation of Label Phase Partitioning

The determination of potentials and pH gradients using paramagnetic labels is based on measurements of phase partitioning. Here, the partitioning of the label is given by λ, which is defined as the ratio of membrane associated to aqueous probe populations. Shown in Fig. 2 are spectra of a spin-labeled phosphonium ion in aqueous solution and in a lipid vesicle suspension. In the vesicle suspension, the EPR spectrum of the hydrophobic ion has contributions arising from both aqueous and membrane-

[7] D. S. Cafiso and W. L. Hubbell, *Biochemistry* **17**, 187 (1978).
[8] D. S. Cafiso and W. L. Hubbell, *Biochemistry* **17**, 3871 (1978).
[9] D. S. Cafiso and W. L. Hubbell, *Annu. Rev. Biophys. Bioeng.* **10**, 217 (1981).

bound probe. This "composite" spectrum is also seen for the secondary alkylamines.

Several approaches can be taken to quantitate λ.[10] One general approach involves obtaining the bound and aqueous line shapes for the label and applying computer methods to determine the contributions of each population to the composite spectrum. Another simpler technique that is frequently utilized involves calibrating the amplitude of the high-field resonance line ($m_l = -1$ in Fig. 2) in terms of the aqueous spin population. Because the bound and free resonance amplitudes are proportional to the number of bound and free spins (N_b and N_f, respectively), the phase partitioning, λ, in the presence of membranes can be written as[9]:

$$\lambda = N_b/N_f = \frac{A_f^\circ - A}{A - (\beta/\alpha)A_f^\circ} \quad (1)$$

Here, A_f° is the amplitude of the $m_l = -1$ resonance for the total number of aqueous spins in the absence of membranes, and A is the amplitude of this resonance in the presence of membranes (see Fig. 2). The ratio β/α is used to correct for the contribution that bound spin makes to the amplitude A. It is the ratio of amplitudes for the total number of spins bound versus free measured at the field setting of the aqueous $m_l = -1$ resonance. Although the amplitudes can be measured peak to peak, a baseline-to-peak amplitude measurement is used here since it readily permits time-dependent measurements of λ (see below). Since the $m_l = -1$ resonance for bound spin is usually very broad, the quantity β/α is seldom large. Often, β/α can be ignored, especially in cases where the aqueous population of spin exceeds that of the bound spin population.

Experimental Considerations

Several limitations must be kept in mind when calculating λ from absolute spectral amplitudes. The measurement requires reproducible tuning of the spectrometer and consistency in the total spin population. Clearly, processes such as spin reduction will present a problem for this simpler method. If spin reduction occurs at a rate that is rapid compared to the measurement of λ, rapid scan methods or a computer analysis of line shapes should be employed. The measurement of spectral amplitudes also requires that changes in aqueous or bound line shapes do not occur;

---

[10] Descriptions of procedures to quantitate λ, the phase partitioning of the probe, can be found in J. D. Castle and W. L. Hubbell [*Biochemistry* **15**, 4818 (1976)] and in Ref. 9 above.

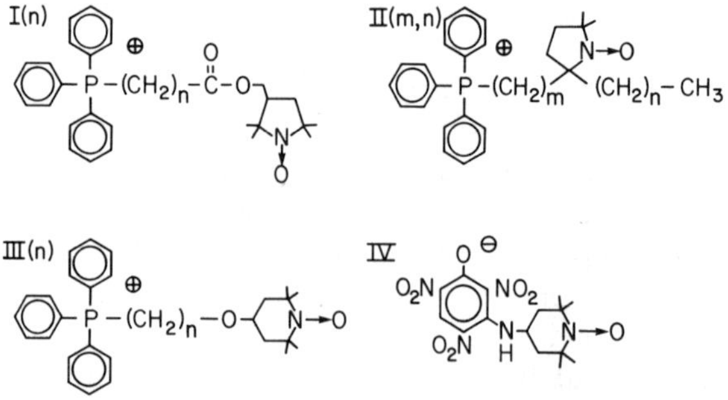

FIG. 3. Examples of spin-labeled hydrophobic ions. The phosphoniums **I** and **III** have been characterized in vesicle systems and their synthesis has been previously described.[7,8] Probe **I** suffers from the presence of an ester linkage which leads to a slow hydrolysis of the label in aqueous solution (stock solutions must be made up weekly). Label **III** is much more stable; however, the bound line shape is not as broad as **I**, making determinations of λ slightly more difficult. The synthesis of the phosphonium **II** is described by Keana and LaFleur.[11] The synthesis of the negative hydrophobic trinitrophenol label **IV** has been accomplished by Flewelling and Hubbell.[12]

this is usually not a major concern since aqueous line shapes rarely change and the bound $m_I = -1$ resonances are typically very broad.

Time-Dependent Measurements of λ

Time-dependent measurements of the probe phase partitioning, λ, can be made simply be setting the magnetic field on the high-field ($m_I = -1$) resonance and monitoring the amplitude A with time. While a field-frequency lock is desirable for these measurements, standard EPR magnets usually provide adequate field stability. As discussed below, data on membrane currents can be obtained in this manner. The response time of most commercial spectrometers permits events to be measured on the order of a few milliseconds.

Quantitation of Membrane Potentials $\Delta\psi$, ψ_o, ψ_i

Paramagnetic hydrophobic ions have been used to estimate membrane potentials in vesicles and several of these labels are shown in Fig. 3.[11,12] These labels include a series of positive phosphonium ions (**I–III**) and a negative trinitrophenol label (**IV**). It should be noted that $\partial\lambda/\partial\Delta\psi$ is greatest for the positively charged labels when $\Delta\psi < 0$ and for the negative probe (**IV**) when $\Delta\psi > 0$. From the equilibria shown in Fig. 1, the phase

[11] J. W. F. Keana and L. E. LaFleur, *Chem. Phys. Lipids* **23**, 253 (1979).
[12] R. Flewelling and W. L. Hubbell, unpublished observations.

partitioning of these labels as a function of the potentials $\Delta\psi$, ψ_o, and ψ_i can be determined. If K_i' and K_o' are the binding constants of the probe to the inner and outer boundary regions from the aqueous phase in the absence of membrane potentials, then $\lambda(\psi_o, \psi_i, \Delta\psi)$ can be written as

$$\lambda = \frac{V_{mi}}{V_i} \left(\frac{K_i' e^{-zF\psi_i/RT} + V_{mo}/V_{mi} K_o' e^{-zF\psi_o/RT} e^{zF\Delta\psi/RT}}{1 + V_o/V_i e^{zF\Delta\psi/RT}} \right) \quad (2)$$

Here, V_{mo} and V_{mi} are the effective thermodynamic volumes of the external and internal boundary regions, and V_o and V_i are the volumes of the external and internal aqueous phases, respectively; z, F, R, and T are the valence of the probe, the Faraday constant, gas constant, and temperature, respectively. For cases where the boundary potentials do not change, we can write the effective binding constants as $K_i = K_i' \exp(-zF\psi_i/RT)$ and $K_o = K_o' \exp(-zF\psi_o/RT)$. If symmetrical binding constants are assumed, so that $K_o = K_i \equiv K$, $\lambda(\Delta\psi)$ can be written as[7]

$$\lambda = \frac{KV_{mi}(1 + V_{mo}/V_{mi} e^{zF\Delta\psi/RT})}{V_i(1 + V_o/V_i e^{zF\Delta\psi/RT})} \quad (3)$$

In sonicated egg phosphatidylcholine (PC) vesicles, Eq. (3) has been tested using K^+-valinomycin-induced potentials.[7] Excellent agreement is obtained between the experimental values of λ and λ determined from Eq. (3) using the values of $\Delta\psi$ expected from a K^+ electrochemical equilibrium. Experimentally, the measurement of $\Delta\psi$ from λ using Eq. (3) requires that three quantities reflecting intrinsic properties of the vesicles and probe be obtained: V_o/V_i, V_{mo}/V_{mi}, and KV_{mi}/V_i. The ratio of internal to external aqueous volumes, V_o/V_i, can be readily determined from a trapped volume measurement as described previously.[7] If $\bar{V}_i$ and $\bar{V}_\ell$ are the trapped volume and hydrated volume per gram of lipid where m_ℓ is the mass of lipid and V_t is the total vesicle suspension volume,

$$V_o/V_i = \frac{V_t}{m_\ell \bar{V}_i} + \left(\frac{\bar{V}_\ell}{\bar{V}_i} + 1 \right) \quad (4)$$

If the vesicle suspension is well characterized, V_o/V_i could also be estimated from the vesicle concentration and vesicle size. The ratio of the external to internal binding volumes for hydrophobic ions, V_{mo}/V_{mi}, can be estimated simply from a measure of the vesicle size. Here, a simple estimate of the ratio of internal to external surface areas provides an adequate estimate of this volume ratio. The values of KV_{mi}/V_i can be determined experimentally by measuring λ for vesicle suspensions where no transmembrane potential is present. If $\Delta\psi = 0$, KV_{mi}/V_i is obtained from Eq. (3) using $\lambda(\Delta\psi = 0)$. In practice, KV_{mi}/V_i can be determined by making one measurement on the vesicle sample of interest (at $\Delta\psi = 0$). A more accurate procedure is to construct a plot of λ^{-1} versus V_t/m_ℓ at $\Delta\psi = 0$; the slope of this line is then $1/\bar{V}_i\beta$ where $\beta = KV_{mi}/V_i(1 + V_{mo}/V_{mi})$. In most vesicle systems depolarization of the membrane (so that $\Delta\psi = 0$) can be accomplished using ionophores or by eliminating ion gradients.

Measurement of Internal and External Probe Binding Constants

The procedure described above to quantitate $\Delta\psi$ works well in most simple model membrane systems despite the assumption that $K_i = K_o$. In sonicated lipid vesicles, careful measurements reveal small differences in these binding constants, for example, in egg PC vesicles, K_i for probe **I**(3) is about 17% larger than K_o.[13] At values of $\Delta\psi$ near -60 mV this binding difference will lead to an error of only a few millivolts. In other cases, however, much more dramatic errors are encountered. In egg PC vesicles containing 10% cholesterol with a K^+ gradient of 10/1 (inside/outside), a potential of -45 mV is measured with probe **III**(4) by assuming $K_i = K_o$ [Eq. (3)]. This value is, of course 14 mV, less than that expected based on a simple K^+ electrochemical equilibrium. In these vesicles an independent measurement of K_i and K_o yields a ratio for K_i/K_o of 0.55 (presumably due to the asymmetric packing of sterol in these small vesicles). Using the experimentally determined values for K_i and K_o, a value of -58 mV is obtained.[14]

A determination of the asymmetry in the phosphonium binding is readily accomplished by measuring the time-dependent partitioning of the probe following mixing with vesicles. Shown in Fig. 4 are recordings of the high-field $m_I = -1$ resonance amplitude when probe **III**(3) and egg PC vesicles are mixed at $t = 0$. This time-dependent decay results from the transmembrane migration of label and can be used to determine the values of K_i and K_o.[13] The label rapidly equilibrates with the external surface (faster than this measurement can resolve) and at $t = 0$ the phase partitioning is given by

$$\lambda(t = 0) = K_o V_{mo}/V_o \tag{5}$$

If $\Delta\psi = 0$, the equilibrium partitioning $\lambda(t = \infty)$ is given by Eq. (2) with $\Delta\psi = 0$. The ratio of K_i/K_o is readily derived as

$$K_i/K_o = (V_{mo}/V_{mi})[(\lambda(t = \infty)/\lambda(t = 0)(1 + V_i/V_o) - 1] \tag{6}$$

From Eqs. (5) and (6) a determination of K_i and K_o can be made.

In Fig. 4, time-dependent amplitudes for both egg PC and egg PC + 10 mol% ergosterol are shown and the ratios of K_i/K_o are approximately 1.25

[13] D. S. Cafiso and W. L. Hubbell, *Biophys. J.* **39**, 263 (1982).
[14] S. Dudley and D. S. Cafiso, unpublished observations.

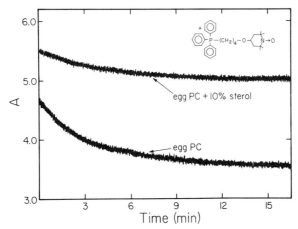

FIG. 4. Tracings of the high-field line, $m_I = -1$, for label **III**(4) following mixing with sonicated egg PC vesicles; curves for egg PC at 53 mg/ml and egg PC + 10% ergosterol at 49 mg/ml are shown. The relaxation amplitudes can be used to calculate K_i and K_o, the internal and external binding constants for the label (see text). In sterol-containing vesicles K_i is much less than K_o; hence, the signal change upon mixing is much less. The high-field resonance amplitude, A, is calibrated so that 20 μM free label has an amplitude of 15 units.[14]

and 0.45, respectively. The decreased amplitude change following mixing in ergosterol containing vesicles is a reflection of the decreased value of K_i.

Kinetics of Transmembrane Equilibration

While the equilibration of phosphoniums with membrane surfaces is fast, their transmembrane migration (as indicated above) can be slow. In native membrane systems these rates are much faster; in ROS disk membranes, for example, $t^{1/2}$ is less than 1 sec. If this kinetic limitation becomes a problem, several approaches can be taken. The addition of small amounts of tetraphenylborate ($\phi_4 B^-$) can be used to dramatically enhance the transmembrane equilibration rate of phosphonium labels.[11] Shown below in Fig. 5 are data indicating the rates of transmembrane movement of **I**(3) in the presence of $\phi_4 B^-$. The highest levels of $\phi_4 B^-$ used correspond to a $\phi_4 B^-$/PC ratio of less than 1/10,000, levels that do not appear to effect other ionic properties of the vesicles. Alternatively, the trinitrophenol label **IV** could be used since its transmembrane migration rate in egg PC vesicles has a $t^{1/2}$ of less than 10 msec.[15]

[15] The transmembrane migration of label **IV** is sufficiently rapid so that mixing measurements capable of resolving events on the order of ~10 msec cannot resolve this transmembrane event.

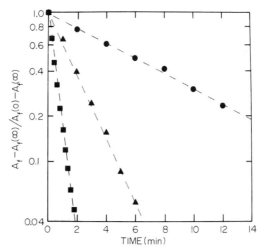

FIG. 5. A plot showing the fractional change in the $m_I = -1$ resonance amplitude following mixing with egg PC vesicles. The decay in the signal intensity, which is due to the transmembrane migration of the label, is shown in the absence of $\phi_4 B^-$ (●) and in the presence of 1 μM (▲) and 5 μM (■) $\phi_4 B^-$. Egg PC vesicles are a concentration of approximately 40 mg/ml. Reproduced from the *Biophysical Journal*[11] by copyright permission of the Biophysical Society.

Effects of Vesicle Size Heterogeneity

As shown above, the potential dependent phase partitioning of probes I–IV, $\lambda(\psi_o, \psi_i, \Delta\psi)$, is determined by the volume ratios V_o/V_i and V_{mo}/V_{mi}, which vary with vesicle size. Equations (2) and (3) assume a uniform vesicle size. While excellent agreement of these expressions is found for sonicated systems, these systems have a narrow distribution of sizes. We consider here the quantitation of $\lambda(\psi_o, \psi_i, \Delta\psi)$ in vesicle suspensions having a greater range of sizes. Experimentally, these probes work well for other more heterogeneously sized lipid systems, such as reverse phase, reconstituted, or biological membrane systems. The reason for this can be seen as follows. Consider a suspension of n vesicles where a vesicle, j, has volumes v_{ij}, v_{mij}, and v_{moj}, corresponding to the trapped internal volume and the inner and outer boundary region volumes, respectively. In this case, V_o/V_i and V_{mo}/V_{mi} in Eqs. (2) and (3) become

$$V_{mo}/V_{mi} = \sum_{j=1}^{n} v_{moj} / \sum_{j=1}^{n} v_{mij} \quad \text{and} \quad V_o/V_i = V_o / \sum_{j=1}^{n} v_{ij} \qquad (7)$$

The quantity V_o/V_i is still described by a measure of the effective internal/external volume ratio for the vesicle suspension. Note, however, that the ratio V_{mo}/V_{mi} would not be given by v_{mo}/v_{mi} for a vesicle of average radius. Because the effective volumes appear to occupy only a thin shell at the membrane interface, they can be approximated by surface areas. In this case V_{mo}/V_{mi} can be written as

$$V_{mo}/V_{mi} \cong \sum_{r=r_1}^{r_2} f(r) r^2 / \sum_{r=r_1}^{r_2} f(r)(r-d)^2 \qquad (8)$$

where $f(r)$ is the frequency of finding a vesicle of outer radius r, d is the bilayer thickness, and r_1 and r_2 are a range of vesicle outer radii. The distribution of vesicle sizes $f(r)$ could be determined by electron microscopy, and in principal would be needed for an accurate determination of V_{mo}/V_{mi}. In practice, simply taking V_{mo}/V_{mi} for the average radius is likely to be adequate; more heterogeneously sized vesicles such as those found in biological systems are usually large and V_{mo}/V_{mi} will have a value near 1, regardless of vesicle size. In egg PC vesicles prepared by reverse phase evaporation techniques, Eq. (3) works well when V_{mo}/V_{mi} for an average vesicle size is taken.[16]

Measuring Boundary Potential Changes

The partitioning of labels **I–IV** will be affected by changes in transmembrane or boundary potential. These can be readily distinguished by the use of ionophores that reduce or eliminate $\Delta\psi$ but do not affect ψ_i and ψ_o. As defined here, ψ_i and ψ_o are made up of a surface potential component (ψ_{si} and ψ_{so}, respectively) and an interfacial potential drop form the interface to the boundary region. A change in ψ_i or ψ_o due to the surface potentials can be examined either by a study of ψ_i and ψ_o versus ionic strength or by independently measuring ψ_{si} and ψ_{so} directly with probes of surface potential.[17]

Estimation of Ionic Currents in Phospholipid Vesicles

In addition to the determination of equilibrium potentials, the paramagnetic hydrophobic ion probes described above can be used to obtain measurements of ionic currents in vesicle systems.[13,17,18] These measurements are made by recording λ as a function of time where experimental

---

[16] W. R. Perkins and D. S. Cafiso, unpublished observations.
[17] D. S. Cafiso and W. L. Hubbell, in "Molecular Basis of Drug Action" (T. Singer and R. Ondarza, eds.), p. 253. Elsevier/North-Holland, Amsterdam, 1981.
[18] D. S. Cafiso and W. L. Hubbell, *Biophys. J.* **44**, 49 (1983).

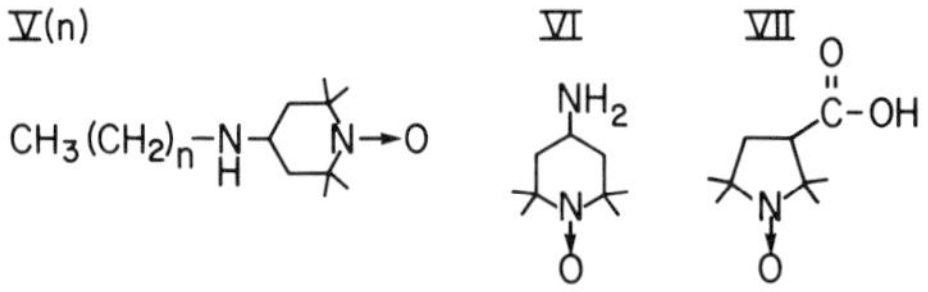

FIG. 6. Paramagnetic probes for the determination of ΔpH. Probe **V** functions by changing its phase partitioning as a function of ΔpH; its synthesis and behavior has been previously described.[8] Probes **VI** and **VII** do not bind strongly to membranes and their distribution across the membrane is revealed using a spin-broadening agent, such as $K_3Fe(CN)_6$.[19,20] Probes **VI** and **VII** are commercially available.

conditions permit a single ion conductance to predominate. From $\lambda(t)$, $\Delta\psi(t)$ is obtained. The ionic current, i, can be determined by $i = (\partial\Delta\psi/\partial t)c$, where c is the membrane capacitance. One must, of course, ensure that the transmembrane migration rate of the probe is not rate limiting (particularly when probes **I–III** are used) and that the levels of probe are small relative to the total ionic charge that is transferred. These data can also be used to construct a current–voltage curve for the ionic species in question. In egg PC vesicles, H^+/OH^- currents and a current–voltage curve were obtained in this fashion using the phosphonium **I(3)**.[18]

Quantitation of Transmembrane pH Gradients, ΔpH

In Fig. 6 several nitroxide probes are shown that have been used to quantitate transmembrane pH gradients. Probes **V–VII** are weak acids or bases and distribute across membranes as a function of ΔpH. Two general approaches to detect the distribution of these molecules can be employed. For probes **VI** and **VII**, which do not bind strongly to membranes, a spin-broadening agent such as $K_3Fe(CN)_6$ can be added to the external vesicle volume; the EPR resonance from spin located externally is broadened and only the internal population of spin is detected.[19,20] For these measurements, one must ensure that the broadening agent is not permeable during the time period of the experiment. Possible effects of this agent on the system being studied should be considered since 30 mM $Fe(CN)_6$ is typically used in these measurements. Probe **V** will bind to membranes and its transmembrane distribution can be detected, as was the case for probes **I–IV**, by monitoring its phase partitioning. It has been shown that[8]

$$\lambda = \frac{V_{mo}}{V_i}\left(\frac{K_i + K_o V_{mo}/V_{mi} \cdot 10^{\Delta pH}}{1 + V_o/V_i \cdot 10^{\Delta pH}}\right) \quad (9)$$

[19] A. T. Quintanilha and L. Packer, *Arch. Biochem. Biophys.* **190**, 206 (1978).
[20] A. T. Quintanilha and R. J. Melhorn, *FEBS Lett.* **91**, 104 (1978).

Here K_i and K_o are the binding constants of the charged form of the probe to the internal and external vesicle surfaces, respectively. The volume ratios are calculated as described above for labels **I–IV**. Equation (9) accurately predicts values for λ that are measured in egg PC vesicles with known values of ΔpH. The charged form of this probe was also shown to be impermeable in these vesicles and is not expected to act as a proton carrier.[8] It should be noted that because label **V** is present in its charged form, it will also sense surface potentials. Because of this dependence, measurements using probes **VI** or **VII** may provide a simpler measure of ΔpH in complex membrane systems. For each of these molecules there is a pH range over which they can be used, since the probes must be present predominantly in a charged form. The alkylamine **V**(5), for example, can be used up to pH 8.5.

Additional Experimental Considerations

Sensitivity Considerations

The probes described above can be used under a wide range of experimental conditions. One can vary both the concentration of the vesicle suspension (which alters V_o/V_i) and the probe-binding constant. Generally, two factors must be taken into account when designing these experiments. First, the phase partitioning of the probe, λ, must be adjusted so that significant bound and aqueous label populations are present. At the concentrations of label that are typically employed, 20 μM, roughly equal membrane versus aqueous distributions of probe are desirable. If λ is very large, the amplitude A may be too small to detect (see Fig. 2). If λ is small, A will undergo little change with potential. Second, since the labels work by equilibrating with existing potentials, it is important that they be used at levels that do not alter these potentials. For example, probes **I–IV** can themselves alter boundary potentials. At a probe/lipid ratio of 1/100 the phosphoniums reach electrostatic saturation in the boundary region.[21] Measurements in systems where large ion gradients maintain transmembrane potentials or pH gradients can be made with higher levels of probe. Ideally, the vesicle concentration to be used is determined by deciding upon an appropriate probe/lipid ratio. A probe-binding constant that will produce reasonable values of λ can then be chosen.

In biological membranes, the sensitivity over a large range of potentials will generally be lower than that for model membrane systems. This is a result of the larger sizes expected for these vesicles; biological sys-

[21] R. Flewelling and W. L. Hubbell, *Biophys. J.* **49**, 531 (1986).

tems, utilized at optimal vesicle concentrations, will have smaller values of V_o/V_i. It can be seen from Eq. (3) that λ becomes independent of $\Delta\psi$ as V_o/V_i decreases and approaches V_{mo}/V_{mi}.

For a more rigorous analysis, the sensitivity, S, of probes I–V above can be defined as the rate of change of A (the amplitude of the $M_I = -1$ resonance) with either $\Delta\psi$ or ΔpH. For probes I–IV, S is described by[9]

$$S = \partial A/\partial\Delta\psi = (\partial A/\partial\lambda)(\partial\lambda/\partial\Delta\psi) \tag{10}$$

This function can be determined by substituting Eqs. (1) and (3) into Eq. (10); for probe V, $\Delta\psi$ is replaced by ΔpH and Eq. (9) is used in place of Eq. (3). S can then be optimized for each experimental variable in Eq. (10).

Probe and Membrane Dynamics

Spin-labeled derivatives of fatty acids, phospholipids, and sterols have, for a number of years, provided an important methodology to study membrane dynamics. Order parameters, which describe the average angular fluctuations of high-frequency alkyl chain motions, lipid lateral diffusion, and rotational diffusion, can be examined using spin-labeled probes. These techniques have been widely utilized and have been previously discussed in great detail.[22,23] An important aspect of the voltage-sensitive probes described here is that information on probe dynamics can be obtained. This is an important feature since clues to any anomolous behavior may be found in the line shape of the bound probe population. As an example, the behavior of the phosphonium probes has been examined in membranes containing the acetylcholine receptor. In this system, these paramagnetic labels interact with the receptor and apparently bind to channel-blocking sites. This binding is revealed in the EPR spectra of these labels and is seen as a broadening of the bound line shape.[24] Thus, the spectrum of the bound probe can provide diagnostic information regarding the source of changes in λ. In other biological systems we have studied, such as the disk membrane from vertebrate photoreceptors, probes I–III are well behaved and do not appear to interact specifically with membrane proteins.[9]

In general, the binding constants of potential or ΔpH-sensitive probes are expected to be altered by nonelectrostatic events, such as changes in

---

[22] L. Berliner, "Spin-Labeling Theory and Applications," Vols. 1 and 2. Academic Press, New York, 1979.
[23] G. I. Likhtenshtein, "Spin Labeling Methods in Molecular Biology." Wiley, New York, 1974.
[24] D. S. Cafiso, *Biophys. J.* **45**, 6 (1984).

membrane dynamics or lipid order. Several approaches are possible if the electrostatic nature of changes in λ is in question. First, the phase partitioning of probes such as TEMPO or alkylamide nitroxides could be examined since λ for these labels should be sensitive to these nonelectrostatic events. Membrane-associated fatty acid or lipid spin-labels can also reveal changes in the physical state of the bilayer. Finally, if oppositely charged probes are utilized, demonstration of opposite changes in λ provides strong evidence for the electrical nature of changes in λ.

Membranes are not necessarily homogeneous in the plane of the membrane and these labels may not distribute uniformly. Charges on membrane proteins, for example, could affect the lateral distribution of probe. Since probe localized near a protein will yield a different spectrum than probe in bulk lipid, a nonuniform probe distribution would likely be revealed in its bound EPR spectrum. While this behavior could clearly complicate potential measurements, it would also present an opportunity to obtain additional information about the lateral distribution of charge in the membrane.

Summary

In this article, we have described the use of paramagnetic hydrophobic ions and secondary alkylamines to estimate pH gradients, and transmembrane and interfacial boundary potentials in membrane vesicle systems. Several features of these probes make them powerful tools for the investigation of membrane potentials. The mechanisms by which they function are well understood, thereby facilitating their use in a quantitative fashion. They have a relatively high sensitivity and are not affected by turbidity. While they are ideally suited for use in model membrane systems, they have also been used successfully in biological and reconstituted membrane systems. Finally, the synthetic flexibility of the nitroxide permits a much wider range of probes to be synthesized than has been described here. Future work will likely bring additional developments in the application and information obtainable from this methodology.

Acknowledgments

I would like to thank Drs. W. L. Hubbell and S. C. Hartsel for their comments and reading of this manuscript. Some of the work reported by the author was supported by NSF grant BNS 8302840.

[20] Transport Studies with Renal Proximal Tubular and Small Intestinal Brush Border and Basolateral Membrane Vesicles: Vesicle Heterogeneity, Coexistence of Transport Systems

By HEINI MURER, PIOTR GMAJ, BRUNO STIEGER, and BRUNO HAGENBUCH

Introduction

The membrane mechanisms involved in small intestinal and renal proximal tubular transcellular transport of various solutes can be explored in great detail in isolated brush border and basolateral membrane vesicles.[1-7] Besides the description of the properties of a transport system, studies with isolated membrane vesicles are also useful for the determination of asymmetric distribution of transport pathways between the brush border and the basolateral membrane. However, the heterogeneity of vesicle preparations presents problems which must be considered in the interpretation of experiments on the localization of specific transport pathways.

It must be assumed that preparations of membrane vesicles are heterogeneous, i.e., membranes with different transport pathways even for the same substrate might be present.[8] This heterogeneity can be related to (1) cross-contamination of basolateral membranes with brush border membranes and vice versa and (2) a heterogeneous epithelial cell population used for vesicle isolation. For example, in the small intestine tissue heterogeneity results from longitudinal differences in transport functions (proximal versus distal small intestine) and/or to the villus tip/villus crypt axis. In membrane preparations isolated from renal cortex, vesicle heterogeneity is due to a complex anatomical structure of the kidney tissue. If only proximal tubular nephron segments are considered, the outermost

---

[1] H. Murer and R. Kinne, *J. Membr. Biol.* **55**, 81 (1980).
[2] H. Murer and G. Burckhardt, *Rev. Physiol. Biochem. Pharmacol.* **96**, 1 (1983).
[3] H. Murer, J. Biber, P. Gmaj, and B. Stieger, *Mol. Physiol.* **6**, 55 (1984).
[4] H. Murer and B. Hildmann, in "Handbook of Experimental Pharmacology" (T. Z. Csaky, ed.), p. 157. Springer-Verlag, Berlin, Federal Republic of Germany, 1984.
[5] H. Murer and P. Gmaj, *Kidney Int.* **30**, 171 (1986).
[6] U. Hopfer, *Am. J. Physiol.* **233**, E445 (1977).
[7] U. Hopfer, *Am. J. Physiol.* **234**, F89 (1978).
[8] U. Hopfer, *Fed. Proc., Fed. Am. Soc. Exp. Biol.* **40**, 2480 (1981).

cortex contains mainly proximal convoluted tubules with early and late segments of superficial nephrons. The juxtamedullary cortex contains the proximal tubules of deep nephrons (early and late) and the pars recta contains proximal tubules of the superficial nephrons. The outermost medulla consists mainly of pars recta of both nephron populations. Thus, isolation procedures starting with epithelial cells from the entire small intestine and from the entire kidney cortex yield heterogeneous vesicle preparations. The degree of heterogeneity can be influenced by a prefractionation of the starting tissue for vesicle isolation. By using a crude anatomical separation of kidney cortex before the isolation of brush border membrane vesicles, a tubular heterogeneity of transport systems for D-glucose,[4-12] phosphate,[13,14] lactate,[15] L-amino acids, and Na^+-H^+ exchange[16] has been demonstrated in several laboratories.

In the present chapter, a few suggestions will be offered regarding the problems related to tissue heterogeneity and/or cross-contaminations. The complexity of the problems constrains us to select a few representative examples. We would also like to present some experimental approaches which allow documentation of the different transport pathways coexisting in one membrane vesicle.

It must be emphasized that the examples of the experimental procedures outlined in this chapter are only illustrations of the type of experiments that could be performed rather than precise experimental protocols. Thus, modifications or adaptations will probably be required in further application of these procedures.

Cross-Contamination of Basolateral and Brush Border Membranes

The separation of brush border and basolateral membranes is followed by the determination of enzyme activities known to be associated with one of the two membranes. Na^+,K^+-ATPase and hormone-stimulated adenylate cyclase are used as marker enzymes for the basolateral membrane; alkaline phosphatase, γ-glutamyltransferase, animopeptidase M,

[9] R. J. Turner and A. Moran, *J. Membr. Biol.* **67**, 73 (1982).
[10] R. J. Turner and A. Moran, *Am. J. Physiol.* **242**, F406 (1982).
[11] R. J. Turner and A. Moran, *J. Membr. Biol.* **70**, 37 (1982).
[12] U. Krag-Hansen, H. Roigaard-Petersen, C. Jacobsen, and M. J. Sheikh, *Biochem. J.* **220**, 15 (1984).
[13] M. G. Brunette, M. Chan, U. Maag, and R. Beliveau, *Pfluegers Arch.* **400**, 356 (1984).
[14] S. T. Turner and T. P. Dousa, *Kidney Int.* **27**, 879 (1985).
[15] K. E. Jørgensen and M. I. Sheikh, *Biochem. J.* **223**, 803 (1984).
[16] M. J. Sheikh, U. Krag-Hansen, and H. Roigaard-Petersen, *Proc. Eur. Colloq. Renal Physiol.*, 5th Abstr. 215 (1985).

and different disaccharidases are used as marker enzymes for the brush border membrane.[1,3–5]

Although marker enzymes are highly enriched in standard preparations of brush border membranes and basolateral membranes, respectively, the activities of marker enzymes characteristic of other membrane structures are always measurable in the "final" membrane fractions. Therefore, in order to localize a particular transport system in either of the two membranes, the influence of cross-contaminating membranes must be very carefully considered. Various procedures have been employed as functional tests for cross-contamination. For example, the cross-contamination of basolateral membrane preparations with brush borders is usually evaluated by measuring some transport processes that are characteristic for the brush border membrane, e.g., Na^+-dependent D-glucose uptake.[17] Such controls, however, have only a relative value, since the brush border membranes are not homogeneous (see above) and the properties of brush borders copurified with the basolateral membranes might not necessarily be the same as the properties of the bulk brush border fraction. Futhermore, unequal sensitivity of the test system due to different transport rates in the contaminating membrane and in the membranes of interest can lead to an incorrect conclusion, i.e., an apparent "functional" absence of cross-contamination.

The selection of a method for membrane isolation is dictated by the aim of the study. For most of the studies on the properties of transport systems (kinetics, energetics, stoichiometry, biochemical modification/identification) a fast isolation procedure providing high yields of sealed vesicles is of primary importance. For localization studies, on the other hand, a method which effectively separates all the membrane structures is mandatory. Thus, one has to distinguish between analytical and preparative isolation procedures. The analytical procedures usually separate membranes on the basis of at least two different criteria, e.g., density and surface properties. The preparative procedures separate on the basis of one criterion only. The localization of a transport system in either the brush border or the basolateral membrane can be analyzed best by a preparative procedure followed by a second analytical separation. A careful comparison of the activity to be located with sensitive markers of well-established localization is needed to determine precisely the location of the mechanism under study.

In the following we will describe two examples from our laboratory related to cross-contamination of brush border with basolateral membranes and of basolateral membranes with brush border membranes. To

---

[17] B. Sacktor, I. L. Rosenbloom, C. T. Liang, and L. Cheng, *J. Membr. Biol.* **60**, 63 (1981).

simplify, only the mutual cross-contamination and not contamination by endomembranes, e.g., by endoplasmic reticulum, mitochondria, and Golgi will be considered.

Contamination of Brush Border Membranes with Basolateral Membranes

The earlier methods to isolate brush border membranes from rat small intestinal mucosal scrapings and from renal cortex were based on an isolation of large brush border fragments—the brush border caps—followed by a separation of the membrane from cytoskeletal structures.[18–24] At present, the most frequently used methods for brush border membrane isolation are based on a differential precipitation of subcellular organelles with either calcium or magnesium.[25–33] Although the brush border marker enzymes are highly enriched, the brush border membrane fractions are always contaminated with basolateral membranes, as indicated by measurable activities of Na^+,K^+-ATPase.

The experimental example to be discussed below was prompted by a report describing carrier-mediated transport of glucose in the presence and absence of sodium in brush border vesicles isolated from rat small intestine.[34] In our study, brush border membrane vesicles from the small intestine of male Wistar rats are obtained by a divalent cation precipitation method according to Hauser et al.[32] The final vesicle fraction is

[18] S. J. Berger and B. Sacktor, J. Cell Biol. **47**, 637 (1970).
[19] R. Kinne and E. Kinne-Saffran, Pfluegers Arch. **308**, 1 (1969).
[20] P. Miller and R. K. Crane, Biochim. Biophys. Acta **52**, 293 (1961).
[21] R. F. Wilfong and D. M. Neville, Jr., J. Biol. Chem. **245**, 6106 (1970).
[22] A. Eichholz and R. K. Crane, J. Cell Biol. **26**, 687 (1965).
[23] G. G. Forstner, S. M. Sabesin, and K. J. Isselbacher, Biochem. J. **106**, 381 (1968).
[24] U. Hopfer, K. Nelson, I. Perrotto, and K. J. Isselbacher, J. Biol. Chem. **248**, 25 (1973).
[25] A. G. Booth and A. J. Kenny, Biochem. J. **142**, 575 (1974).
[26] S. A. Hilden and B. Sacktor, Kidney Int. **14**, 279 (1978).
[27] P. Malathi, H. Peiser, P. Fairclough, P. Mallett, and R. K. Crane, Biochim. Biophys. Acta **554**, 259 (1979).
[28] C. Evers, W. Haase, H. Murer, and R. Kinne, Membr. Biochem. **1**, 203 (1978).
[29] J. Biber, B. Stieger, W. Haase, and H. Murer, Biochim. Biophys. Acta **647**, 169 (1981).
[30] J. Schmitz, H. Preiser, D. Maestracci, B. K. Ghosh, J. J. Cerda, and R. K. Crane, Biochim. Biophys. Acta **323**, 98 (1973).
[31] M. Kessler, O. Acuto, C. Storelli, H. Murer, M. Müller, and G. Semenza, Biochim. Biophys. Acta **506**, 136 (1978).
[32] H. Hauser, K. Howell, R. M. C. Dawson, and D. E. Boyer, Biochim. Biophys. Acta **602**, 567 (1980).
[33] B. Hildmann, C. Storelli, W. Hasse, M. Barac-Nieto, and H. Murer, Biochem. J. **186**, 169 (1980).
[34] K. J. Ling, W. B. Im, and R. G. Faust, Int. J. Biochem. **13**, 693 (1981).

loaded with 400 mM mannitol, 20 mM HEPES–Tris, pH 7.4. For further fractionation, the vesicle fraction is applied to a linear sucrose gradient [20–50% (w/w)] in a TST 41.13 swinging bucket rotor from Kontron at 41,000 rev/min [214,600 g_{av}] for 2 hr. After centrifugation, two protein peaks and a shoulder on the low-density peak are observed (data not shown). According to this preliminary result, a discontinuous sucrose density gradient is chosen. The vesicle suspension—obtained by the divalent cation precipitation method—is layered on the following sucrose gradient (w/w): 10 ml 50%, 12 ml 35%, 12 ml 30%. All the sucrose solutions are buffered with 10 mM Tris-HCl, pH 7.4, and the density is checked with a Bausch and Lomb refractometer. The centrifugation is performed for 5 hr in an AH-627 swinging bucket rotor with 36-ml buckets from Sorvall at 27,000 rev/min [96,700 g_{av}]. The bands at the interfaces are removed with a syringe and diluted with the buffer needed for the experiments to follow. The vesicles are collected by a 30-min centrifugation at 20,000 rev/min [31,180 g_{av}] in a SS-34 rotor. The vesicles are washed and recentrifuged for 30 min at 20,000 rev/min. The pellets are then resuspended in a small volume of buffer and homogenized with a syringe with a thin needle. Designation of the fractions was (I) overlay/30%, (II) 30%/35%, (III) 35%/50%.

The activities of marker enzymes for the brush border and basolateral membranes are measured in the different fractions and compared with that in the starting homogenate (enrichment factors, Table I). After su-

TABLE I
SPECIFIC ACTIVITIES AND ENRICHMENT FACTORS OF MARKER ENZYMES IN BRUSH BORDER MEMBRANE VESICLES[a]

Enzyme and property	Starting vesicles	Sucrose gradient		
		Fraction I	Fraction II	Fraction III
Aminopeptidase M ($n = 10$)				
Specific activity (mU/mg protein)	860 ± 265	85 ± 15	282 ± 161	1276 ± 285
Enrichment	10.4	0.9	3.2	14.5
Maltase ($n = 11$)				
Specific activity (mU/mg protein)	642 ± 212	50 ± 11	121 ± 90	949 ± 352
Enrichment	9.4	0.6	1.8	14.5
Na$^+$,K$^+$-ATPase ($n = 11$)				
Specific activity (mU/mg protein)	207 ± 50	793 ± 281	893 ± 270	55 ± 28
Enrichment	3.2	12.2	14.6	0.8

[a] Vesicles were isolated from rat jejunum by magnesium precipitation and analyzed further by sucrose gradient centrifugation. Modified from Stieger and Murer.[35] The data are means ± SD from the number of experiments given in parentheses.

TABLE II
TRANSPORT PROPERTIES OF SUBFRACTIONS OF VESICLES OBTAINED BY SUCROSE
DENSITY GRADIENT CENTRIFUGATION[a]

		Equilibrium uptake (%) based on D-glucose uptake in:			
Conditions	Medium	Whole vesicle fraction	Fraction I	Fraction II	Fraction III
Control	Na^+	31.2	20.3	50.3	58.3
	K^+	24.2	23.6	52.0	15.8
Phlorizin (100 μM)	Na^+	25.2	21.8	42.0	30.0
	K^+	16.1	22.8	47.5	11.4
Phloretin (100 μM)	Na^+	20.5	5.2	5.6	53.5
	K^+	5.3	4.9	5.6	6.8

[a] Brush border membrane vesicles were resuspended after density gradient centrifugation and equilibrated in 300 mM mannitol, 20 mM HEPES–Tris, pH 7.4, 0.1 mM D-glucose, 100 mM NaCl or 100 mM KCl. Uptake experiments were performed in the same buffers containing radioactively labeled D-glucose. The concentration of phlorizin and phloretin was 0.1 mM. The data are expressed as a percentage of labeled D-glucose uptake after 8 sec with respect to equilibrium uptake after 90 min. The values are the means of five determinations with an experimental scatter not larger than 5% (before normalization). From Stieger and Murer.[35]

crose density gradient centrifugation, the vesicles show an increased enrichment of brush border membrane markers in fraction III and a significant enrichment of basolateral membrane markers in fractions I and II. Thus, it is possible to fractionate further the apparently purified brush border and basolateral membranes.[35]

To characterize further the different membrane fractions, a membrane-specific transport system is analyzed. We have chosen D-glucose transport for this purpose, since it shows characteristic properties at the brush border side of the enterocytes: it is sodium-dependent, electrogenic, and sensitive to phlorizin, a competitive inhibitor of this transport system.[24] At the basolateral side, transport of D-glucose is sodium-independent and inhibited by phloretin.[36,37] Table II shows the properties of D-glucose transport of the vesicles obtained by the fractionation on a sucrose gradient; the membranes are equilibrated with the salts and

[35] B. Stieger and H. Murer, *Eur. J. Biochem.* **135,** 95 (1983).
[36] H. Murer and R. Kinne, in "Biochemistry of Membrane Transport" (G. Semenza and E. Carafoli, eds.), p. 292. Springer-Verlag, Berlin, Federal Republic of Germany, 1977.
[37] H. Murer, U. Hopfer, E. Kinne-Saffran, and R. Kinne, *Biochim. Biophys. Acta* **345,** 170 (1974).

D-glucose, and the uptake measurements are started by the addition of radioactively labeled D-glucose (tracer exchange). Membranes obtained in fraction III show better sodium stimulation of the D-glucose tracer exchange than the control. Fractions I and II show no difference between the sodium and the potassium medium. In conjunction with the enzyme enrichment factors, this observation suggests that brush border membranes are concentrated in fraction III. Similar conclusions are obtained from the inhibition experiments. It can be seen that phlorizin is a better inhibitor of the exchange in fraction III than in the control, whereas there is no inhibition in fraction I. The inhibition in fraction II can be explained by the presence of small amounts of brush border membrane vesicles (compare enrichment factors between fractions I and II in Table I). In fractions I and II, the D-glucose tracer exchange is almost completely abolished in the presence of phloretin, whereas in fraction III there is only a minor inhibition. Thus, D-glucose transport in membranes of fractions I and II shows properties of basolateral D-glucose transport, i.e., sodium-independent, phloretin-sensitive D-glucose transport.

Contamination of Basolateral Membranes with Brush Order Membranes

Basolateral membranes are usually isolated from a crude plasma membrane fraction obtained by differential centrifugation from homogenates of renal cortical slices or small intestinal epithelial cells. Density gradient centrifugation in different media and free-flow electrophoresis are used for further membrane separation.[1,36-60] Although basolateral membrane

[38] A. K. Mircheff, *Am. J. Physiol.* **244**, G347 (1983).
[39] R. Kinne, H. Murer, E. Kinne-Saffran, M. Thees, and G. Sachs, *J. Membr. Biol.* **21**, 375 (1975).
[40] M. S. Medow, K. S. Roth, K. Ginkinger, and S. Segal, *Biochem. J.* **214**, 209 (1983).
[41] R. Hori, M. Takano, T. Okano, S. Kitazawa, and K. Inui, *Biochim. Biophys. Acta* **692**, 97 (1982).
[42] H. E. Ives, V. J. Yee, and D. G. Warnock, *J. Biol. Chem.* **258**, 13513 (1983).
[43] A. M. Kahn, S. Branham, and E. J. Weinman, *Am. J. Physiol.* **246**, F779 (1984).
[44] L. H. Lash and D. P. Jones, *Biochim. Biophys. Res. Commun.* **112**, 55 (1983).
[45] J. R. Del Castillo, R. Marin, T. Proverbio, and F. Proverbio, *Biochim. Biophys. Acta* **692**, 61 (1982).
[46] D. W. Windus, S. Klahr, and M. R. Hammerman, *Am. J. Physiol.* **247**, F403 (1984).
[47] P. Gmaj, H. Murer, and E. Carafoli, *FEBS Lett.* **144**, 226 (1982).
[48] M. P. Van Heeswijk, J. A. Geertsen, and C. H. Van Os, *J. Membr. Biol.* **79**, 19 (1984).
[49] V. Scalera, Y. K. Huang, B. Hildmann, and H. Murer, *Membr. Biochem.* **4**, 49 (1981).
[50] E. F. Boumendil-Podevin and R. A. Podevin, *Biochim. Biophys. Acta* **735**, 86 (1983).
[51] C. T. Liang and B. Sacktor, *Biochim. Biophys. Acta* **466**, 474 (1977).
[52] R. D. Mamelok, S. S. Tse, K. Newcomb, C. L. Bildstein, and D. Liu, *Biochim. Biophys. Acta* **692**, 115 (1982).

marker enzymes are highly enriched, marker enzymes for the brush border membranes are always measurable in basolateral membrane preparations, indicating a contamination with brush border membranes.

The experiments to be discussed below were prompted by a report showing both sodium-dependent and sodium-independent transport of inorganic phosphate in basolateral membranes isolated from dog kidney cortex.[61,62] Therefore, it was of interest to analyze whether sodium-dependent and sodium-independent phosphate transport could be separated by further membrane fractionation.

In our study, basolateral membrane vesicles are isolated from rat kidney cortex by Percoll gradient centrifugation according to the method described by Scalera et al.[49] The separation of the microsomal fraction into brush border and basolateral membranes is obtained in a Sorvall SS-34 rotor with a Percoll concentration of 10.8% (v/v) at 31,180 g_{av} for 30 min (integrator setting 8×10^9 rad^2/sec). The basolateral membranes are collected between 3.5 and 11 ml (from the top of the gradient). Percoll is removed by centrifugation in a Sorvall T-865 rotor for 60 min at 100,00 g_{av}. The membrane layer on top of the glassy Percoll pellet is resuspended in chamber buffer (see below) and used for further separation. For this purpose, a DESAGA FF48 free-flow electrophoresis apparatus is used at the following separation conditions: electrode buffer, 100 mM triethanolamine, 100 mM acetic acid adjusted to pH 7.4 with NaOH; chamber buffer, 100 mM mannitol, 10 mM triethanolamine, 10 mM acetic acid adjusted to pH 7.4 with NaOH; current, 150 mA at 1000 V; chamber buffer flow rate, 200 ml/hr; sample injection, 0.7 ml/hr, membrane vesicles suspended in chamber buffer at a concentration of 5 mg protein/ml. The membranes elute from the electrophoresis apparatus between fractions 4 and 19 as a symmetrical peak with a shoulder at the higher fractions (Fig. 1, OD 280 nm). The basolateral membranes are found in a

---

[53] M. J. Sheikh, U. Krag-Hansen, K. E. Jørgensen, and H. Roigaard-Petersen, Biochem. J. **208,** 377 (1982).
[54] H. Murer, E. Ammann, J. Biber, and U. Hopfer, Biochim. Biophys. Acta **433,** 509 (1976).
[55] A. K. Mircheff and E. M. Wright, J. Membr. Biol. **28,** 309 (1976).
[56] M. Fujita, H. Ohta, K. Kawai, and H. Matsui, Biochim. Biophys. Acta **247,** 336 (1972).
[57] A. K. Mircheff, G. Sachs, S. D. Hanna, C. S. Labiner, E. Rabon, A. P. Douglas, M. W. Walling, and E. M. Wright, J. Membr. Biol. **50,** 343 (1979).
[58] V. Scalera, C. Storelli, C. Storelli-Joss, W. Haase, and H. Murer, Biochem. J. **186,** 177 (1980).
[59] J. D. Quigley and G. S. Gotterer, Biochim. Biophys. Acta **173,** 456 (1969).
[60] B. A. Lewis, A. Alkin, R. H. Michell, and R. Coleman, Biochem. J. **152,** 71 (1975).
[61] S. J. Schwab, S. Klahr, and M. R. Hammerman, Am. J. Physiol. **246,** F663 (1984).
[62] S. J. Schwab, S. Klahr, and M. R. Hammerman, Am. J. Physiol. **247,** F543 (1984).

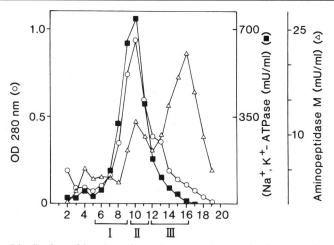

FIG. 1. Distribution of basolateral membranes and contaminating brush border membranes during separation by free-flow electrophoresis. Basolateral membranes were isolated by a Percoll centrifugation method and further separated by free-flow electrophoresis as outlined in the text.

symmetrical peak coinciding with the major portion of the membrane material (OD 280 nm) as indicated by Na^+,K^+-ATPase activity, whereas the brush border membranes, indicated by the animopeptidase M activity, are largely separated from the basolateral membrane and coincide with the shoulder observed in the OD 280 measurements (Fig. 1). The fractions are collected in three different pools (I/II/III in Fig. 1), marker enzyme activities are measured, and the membranes are frozen in liquid nitrogen for further measurements. Na^+,K^+-ATPase is enriched in fraction I, aminopeptidase M in fraction III (Table III). Sodium-dependent D-glucose is concentrated in fraction III; ATP-dependent Ca^{2+} transport is absent in fraction III (Hagenbuch and Murer, manuscript in preparation, data not shown).

In the experiments presented in Table III, we analyze phosphate transport at 1 min of incubation. The membranes obtained in the pools I–III are resuspended in 300 mM mannitol, 20 mM HEPES–Tris, pH 7.4. Uptake is initiated by mixing 20 μl of vesicle suspension with 100 μl of incubation medium containing 100 mM mannitol, 100 mM NaCl or KCl, 20 mM HEPES–Tris, pH 7.4, and 0.05 mM $KH_2{}^{32}PO_4$. Sodium stimulates the rate of phosphate transport in the basolateral membranes obtained from the Percoll gradient. However, in the electrophoretic fractions with the highest activity of Na^+,K^+-ATPase (fraction I) sodium stimulation is absent, whereas fractions with highest brush border membrane marker

TABLE III
SEPARATION OF MEMBRANES CONTAINING Na^+-P_i COTRANSPORT SYSTEMS FROM THE BULK OF BASOLATERAL MEMBRANES[a]

Step	Basolateral membranes	Electrophoretic fractions		
		I	II	III
Na^+-P_i cotransport (% stimulation by Na^+ of initial uptake)	106 ± 23	26 ± 23	87 ± 3	162 ± 17
Na^+,K^+-ATPase (mU/mg protein)	1247 ± 380	1793 ± 28	1233 ± 28	761 ± 86
Aminopeptidase M (mU/mg protein)	125 ± 11	39 ± 3	61 ± 6	251 ± 51

[a] Basolateral plasma membranes were isolated from rat kidney cortex by Percoll gradient centrifugation, and separated further into three fractions by free-flow electrophoresis. Values are means ± SE, $n = 4$. From Hagenbuch and Murer (unpublished observations).

activity show highest sodium stimulation of phosphate transport. This observation documents that, at least in rat cortical membrane preparations, sodium-dependent phosphate transport is located exclusively in the brush border membrane and that Na^+-dependent phosphate transport observed in "standard" basolateral membrane preparations might be attributed to cross-contamination by the brush border membranes.

The procedures described above are two selected examples for the type of experiments which must be performed if the distribution of transport systems between the apical and the basolateral cell surface is to be documented. Several misinterpretations which have occurred in the past indicate that the measurements of transport activity in membrane vesicles obtained by preparative procedures are not sufficient to prove the localization of a transport system. Analytical separation based on at least two different criteria is indispensable if unequivocal results are to be obtained.

Heterogeneity of Vesicles Related to Tissue Heterogeneity

In this section, we discuss vesicle heterogeneity which results from cell/tissue heterogeneity of small intestinal and renal proximal tubular epithelium. We describe an experimental example related to brush border membranes isolated from kidney cortex.

Small Intestine

The transport functions are different in the proximal and distal parts of small intestine. For example, sodium-dependent phosphate transport is most active in the proximal parts whereas sodium-dependent sulfate transport is most active in distal parts.[63,64] Another complication is illustrated by the behavior of sodium-dependent transport of D-glucose. Following the nomenclature introduced by the authors of a recent study,[65] there exists serial and parallel heterogeneity in rat intestinal D-glucose transport. A kinetic analysis of sodium-dependent D-glucose transport in brush border vesicles isolated from proximal small intestine of 7- to 8-week-old rats demonstrated a curvilinear Wolf–Augustin–Hofstee plot consistent with multiple (at least two) transport mechanisms. In contrast, only one transport component was seen in the distal segment. It is easy in an experiment to take into account that the longitudinal (serial) heterogeneity membrane fractions can be isolated from the shortest possible segments and then compared to those isolated from other segments. It is more difficult to analyze a parallel heterogeneity in detail. If two transport components are revealed by kinetic analysis, then the question arises whether the two transport systems are present in the same membrane, or rather in membranes originating from different cells, e.g., those from the tip of the villus and those from the crypt (crypt–tip axis). The problem of heterogeneity of brush border membranes isolated from a well-defined intestinal segment can only be resolved by a separation of the epithelial cells according to their location on the villus.[66,67] For this purpose, cells can be liberated by vibration from everted and slightly extended small intestinal segments mounted on a vibrating tube. The cells released first into the bathing solution are those from the top of the villus, the cells liberated after prolonged vibration are those from lower regions of the villus. Thus, to analyze the "axial" heterogeneity, the vesicles must be isolated from cells obtained from various regions of the villus; preliminary experiments in our laboratory with rabbit jejunum suggest that brush border membranes can be isolated by the standard divalent cation procedure from cells collected by the vibration method.

---

[63] G. Danisi, H. Murer, and R. W. Straub, *Am. J. Physiol.* **246**, G180 (1984).
[64] H. Lücke, G. Stange, and H. Murer, *Gastroenterology* **80**, 22 (1981).
[65] H. J. Freeman and G. A. Quamme, *Am. J. Physiol.* **251**, G208 (1986).
[66] M. M. Weiser, *J. Biol. Chem.* **248**, 2536 (1973).
[67] D. Gratecos, M. Knibihler, V. Benoit, and M. Semeriva, *Biochim. Biophys. Acta* **512**, 508 (1975).

Renal Proximal Tubules

As mentioned in the introduction, renal proximal tubules are functionally heterogeneous, i.e., the transport of a given substrate can be different in early as compared to late proximal tubule or in proximal tubules of superficial and deep nephrons.

The differences in transport function between superficial and deep nephrons need to be taken into account (see below). However, it is difficult if not impossible to account for the longitudinal heterogeneity, i.e., for different transport properties of vesicles originating from early versus late proximal tubules. For this purpose, a preparative procedure for the isolation of epithelial cells or tubular fragments from early and late segments of either superficial or deep proximal tubules would be required; such a high-yield procedure, which would provide sufficient starting material for vesicle isolation, is not available at present. In the following, we will outline an attempt at resolving the heterogeneity of brush border membrane D-glucose transport.[9-12] Other examples related to heterogeneity of phosphate transport,[13,14] lactate transport,[15] L-amino acid transport,[12,68,69] and Na^+-H^+ exchange[16] have been published.

The experimental example to be discussed has been developed by Turner and Moran.[10] The kidneys of New Zealand White rabbits (1.5–3 kg) are used. "Outer cortical" tissue is obtained by taking slices of 0.5 mm thickness with a Stadie–Riggs microtome; 1–2 g of tissue could be obtained from one rabbit. "Outer medullary" tissue (0.5–1.0 g/rabbit) is obtained by first sectioning the kidney transversely into 2-mm slices. Under a dissecting microscope the cortex is trimmed completely away from these slices and a strip of outer "medulla" 1–2 mm thick is then removed. Brush border membrane vesicles are then prepared from outer cortical or outer medullary tissue using one of the modifications of the divalent cation precipitation method.[25-29] In our laboratory the modification published by Biber et al.[29] is preferred, which is very similar to that described by Hauser et al.[32] for intestinal brush border membranes.

The enzymatic characterization of the brush border membranes of outer cortical and outer medullary vesicle preparations is given in Table IV. The data indicate that brush border membranes of identical purity are obtained from the two different starting materials.

The kinetics of D-glucose transport into brush border membrane vesicles isolated from outer renal cortex are different from those of inner renal

---

[68] U. Kragh-Hansen and M. I. Sheikh, *J. Physiol. (London)* **354**, 55 (1984).
[69] H. Roigaard-Petersen and M. I. Sheikh, *Biochem. J.* **220**, 25 (1984).

TABLE IV
ENZYMATIC CHARACTERIZATION OF OUTER CORTICAL AND OUTER MEDULLARY BRUSH BORDER VESICLE PREPARATIONS[a]

Enzyme	Outer cortex			Outer medulla		
	Activity in homogenate (μmol/mg protein/hr)	Enrichment (vesicles/ homogenate)	Recovery (%) (vesicles/ homogenate)	Activity in homogenate (μmol/mg protein/hr)	Enrichment (vesicles/ homogenate)	Recovery (%) (vesicles/ homogenate)
Maltase	1.1 ± 0.2	11.1 ± 0.3	50 ± 5	0.62 ± 0.31	13.4 ± 4.6	66 ± 21
Alkaline phosphatase	5.7 ± 0.7	8.1 ± 0.8	36 ± 5	6.8 ± 1.3	6.3 ± 1.2	32 ± 6
Na^+,K^+-ATPase	2.6 ± 0.3	0.22 ± 0.06	0.9 ± 0.2	1.8 ± 0.5	1.5 ± 0.7	8.0 ± 4.4

[a] Values are means ± SD for four or five independent determinations. Activities are given as micromoles substrate consumed per milligram protein per hour. Data from Turner and Moran.[10]

cortex (Fig. 2). In the outer cortical preparation, the sodium-dependent component of D-glucose transport behaves as a relatively low-affinity system with an apparent K_m of about 6 mM as measured under zero trans conditions at 40 mM NaCl and at 17°. In contrast under the same conditions, the sodium-dependent component of D-glucose flux in the outer medullary preparation is characterized by a high-affinity system with an apparent K_m of about 0.35 mM. The D-glucose transport in vesicles from the outer cortex has a high capacity as compared to that from the inner cortex. Furthermore, qualitative differences in D-glucose transport can also be observed: in outer cortical brush border membranes stoichiometry between D-glucose and sodium flux is 1 : 1 and the system is highly sensi-

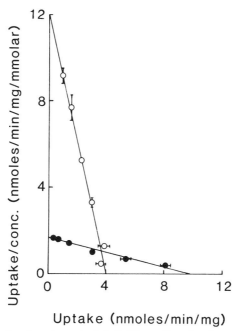

FIG. 2. Kinetics of sodium-dependent, initial linear (4 sec) D-glucose flux into (●) outer cortical and (○) outer medullary brush border membrane vesicles isolated from rabbit kidney. D-Glucose transport was measured under zero trans sodium and glucose conditions at 17°. The membranes were suspended in a buffer containing 10 mM Tris–HEPES, pH 7.4, 100 mM KSCN, 300 mM D-mannitol, and 12.5 μg valinomycin/mg protein. The incubation medium had a similar composition but contained in addition 40 mM NaCl or 40 mM choline chloride and labeled D-glucose at concentrations between 0.2 and 20 mM for outer cortical preparations and 0.1 and 8.0 mM for outer medullary preparations. Least square fits yield: apparent K_m = 5.7 ± 1.3 mM, apparent V_{max} 9.7 ± 1.7 nmol min^{-1} mg protein^{-1} for outer cortex; and apparent K_m = 0.34 ± 0.05 mM, apparent V_{max} = 4.1 ± 0.3 nmol min^{-1} mg protein^{-1} for outer medulla. Data from Turner and Moran.[10]

tive to phlorizin, whereas in brush border membrane vesicles from outer medulla stoichiometry between D-glucose and sodium flux is 1:2 and the system is less sensitive to phlorizin.[10,11]

It is obvious that the brush border membranes isolated from the whole kidney cortex represent a mixture of membranes originating from different nephron populations and/or different segments of proximal tubules. Their transport properties are accordingly heterogeneous, which may result in curvilinear kinetic plots.[10] Therefore, the presence of two components in a kinetic analysis of a transport system provides no evidence that the two systems coexist in the same membrane. In some cases a careful anatomical separation of the tissue prior to membrane isolation may help to resolve the heterogeneity of transport systems. When, however, different transport systems are present in cells located in close proximity within the renal tubule, such separation may fail, and different methods must be applied in order to prove the coexistence of transport systems in the same membrane.

Experimental Evidence for Coexistence of Transport Systems

In a membrane vesicle preparation several transport systems for the same substrate can usually be identified. This, however, does not mean that they coexist in one membrane, i.e., they could be present in different vesicles (heterogeneity, see previous section). A first experimental suggestion for a coexistence in the same membrane is obtained when, in analytical separation procedures (see above), the different transport systems comigrate during all fractionation steps. However, the possibility exists that two different types of membranes are not separable by the methods used. Thus, it would certainly be helpful to have experimental tests which determine the physical coexistence of transport pathways. Again we can only outline the experimental strategies which have been applied so far, illustrate one example in some detail, and mention other examples in the literature.

The strategies involved in such experiments are outlined in Fig. 3. In all the experiments to be described, the same membrane exists for distinct transport systems which share one common substrate. In example 1 of Fig. 3, solute X_1 is transported together with sodium via a cotransport mechanism, and an increased intravesicular concentration of X_1 permits the transport of X_2 via an exchange mechanism; ideally, a concentrative (overshooting) uptake of X_2 should be observed. An example of this experimental situation is that of the anion-exchange properties in dog renal microvillus membrane vesicles.[70] In the presence of an inwardly directed

[70] S. E. Guggino, G. J. Martin, and P. S. Aronson, *Am. J. Physiol.* **244**, F612 (1983).

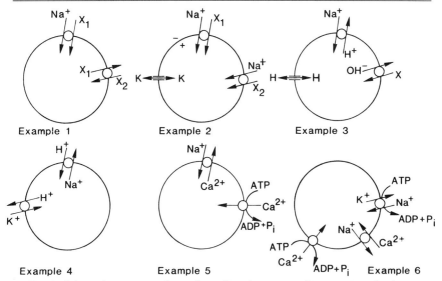

FIG. 3. Schematic representations of coupling phenomena of transport mechanisms coexisting in the same vesicular membrane.

Na^+ gradient, lactate stimulates the uphill movement of urate, indicating that a Na^+–lactate cotransport mechanism coexists in these vesicles with a lactate–urate exchange. In example 2, two sodium-dependent transport pathways coexist. The sodium-dependent transport of X_1 influences sodium-dependent transport of X_2—usually an inhibition is observed—due to competition for a common driving force. This mutual inhibition is especially pronounced for potential-dependent (electrogenic) transport systems. The inhibition is reduced or prevented by an analysis of the transport pathways in sodium-equilibrated conditions and in the presence of a sodium ionophore or for potential-dependent transport systems by short-circuiting the membrane potential, e.g., by valinomycin and equilibrated potassium concentrations. Examples from the literature to be mentioned are the mutual inhibition of D-glucose and neutral amino acid cotransport in intestinal brush border membranes[71] or the inhibition of phosphate transport by D-glucose or neutral amino acids in renal membranes.[72,73] In example 3, an inwardly directed sodium gradient leads via the operation of Na^+/H^+ exchange to an alkalinization of the intravesicular space. Intravesicular alkalinization is then the driving force for the influx of solute X via $OH(HCO_3)/X$ exchange. An example from the litera-

[71] H. Murer, K. Sigrist-Nelson, and U. Hopfer, *J. Biol. Chem.* **250,** 7392 (1975).
[72] P. Q. Barrett and P. S. Aronson, *Am. J. Physiol.* **242,** F126 (1982).
[73] J. Thierry, P. Poujeol, and P. Ripoche, *Biochim. Biophys. Acta* **647,** 203 (1981).

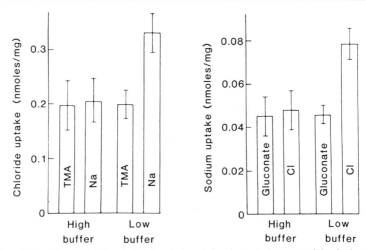

FIG. 4. Coupling of Na^+ and Cl^- flux in brush border membrane vesicles isolated from rabbit ileum. The uptake of 1 mM Cl^- (15 sec) was determined with 150 mM TMA–gluconate inside, 100 mM TMA–gluconate and 50 mM Na–gluconate outside, or 150 mM TMA–gluconate on both sides, in presence of high buffer (30 mM Tris, 14 mM HEPES, 90 mM MES, 0.9 mM choline–HCO_3, pH 6.0, and 50 mM mannitol inside and outside) or low buffer (0.3 mM Tris, 0.14 mM HEPES, 0.9 mM MES, 0.9 mM choline–HCO_3, pH 6.0, and 60 mM mannitol inside and outside). All solutions were gassed with 95% Na_2–5% CO_2. The uptake of 1 mM Na^+ (15 sec) by membrane vesicles in low or high buffered media (as described above) was determined in the presence (50 mM TMA–CL, 100 TMA–gluconate outside) and absence (150 mM TMA–gluconate outside) of chloride. All solutions were gassed with 95% N_2–5% CO_2. From Knickelbein et al.[74]

ture for this situation is the documentation of a coexistance of Na^+/H^+ exchange and $OH(HCO_3)/Cl^-$ exchange as the elements of "coupled" electroneutral NaCl absorption in rabbit ileal brush border vesicles.[74] This experiment is illustrated in Fig. 4 and the experimental details are given in the legend to Fig. 4. It is evident that, in this situation, coupling between Na^+ and Cl^- flux depends on the Na^+ flux- or Cl^- flux-dependent development of a transmembrane ΔpH. Thus, "coupling" can be prevented by high buffer capacities (Fig. 4) or by an ionophore-dependent pH clamp, e.g., nigericin or valinomycin/FCCP in the presence of equilibrated potassium. Example 4 of Fig. 3 illustrates a similar situation. An outwardly directed sodium gradient leads via Na^+/H^+ exchange to intravesicular acidification which can be measured, e.g., by the quenching of acridine orange fluorescence.[75] The quenching of acridine orange, i.e., the release

[74] R. Knickelbein, P. S. Aronson, C. M. Schron, J. Seifter, and J. W. Dobbins, Am. J. Physiol. **249**, G236 (1985).
[75] G. Cassano, B. Stieger, and H. Murer, Pfluegers Arch. **400**, 309 (1984).

of intravesicular acidification can be reversed by the addition of a substrate for a second exchanger. Recently, in a combination of tracer experiments and fluorescence quench experiments we have been able to prove coexistence of separate Na^+/H^+ and K^+/H^+ exchange in rat ileal brush border membranes.[76] The last two examples are related to the coexistence of a primary active pump and an equilibrating system in basolateral membranes isolated either from small intestinal or proximal tubular epithelial cells. In example 5, the Ca^{2+}-ATPase promotes intravesicular Ca^{2+} uptake in inside-out oriented vesicles; ATP-dependent Ca^{2+} uptake can be "inhibited," or the accumulated Ca^{2+} can be released by the addition of Na^+, due to the existence of a Na^+/Ca^{2+} exchange system in the same membrane as the primary active Ca^{2+} pump. Examples from the literature for this situation are the studies on Ca^{2+} transport in basolateral membranes isolated either from rat renal cortex or rat small intestinal epithelial cells.[77,78] Example 6 illustrates coexistence of Na^+,K^+-ATPase, Ca^{2+}-ATPase, and Na^+/Ca^{2+} exchange. The primary active Na^+ pump promotes intravesicular, ouabain-sensitive Na^+ accumulation which, at least at low extravesicular sodium concentrations, promotes an additional driving force for Ca^+ influx via Na^+/Ca^+ exchange in addition to that mediated by the Ca^{2+}-ATPase. At high extravesicular Na^+ concentrations, Na^+ will reduce ATP-dependent Ca^{2+} uptake as the sodium is always present under inwardly directed gradient conditions, i.e., it promotes Ca^{2+} efflux via Na^+/Ca^{2+} exchange. An example for this rather complex situation is given in the literature by Van Heeswijk et al.[48] for the coexistence of Na^+,K^+-ATPase, Na^+/Ca^{2+} exchange, and Ca^{2+}-ATPase in basolateral membranes from rat kidney cortex.

Conclusions

We have outlined above some experimental strategies that have been applied to resolve the problems resulting from vesicle heterogeneity (cross-contamination/tissue heterogeneity). In view of the complexity of such experiments, detailed experimental protocols cannot be offered, but nevertheless some basic strategies can be developed. The experimental examples given are representative of the type of experiments to be performed. It should be noted that the problems can not be avoided, but they can certainly be experimentally reduced. This article has fulfilled its purpose if the problems discussed in some detail are taken into account in the interpretation of transport studies with isolated vesicles. It must be real-

---

[76] H. J. Binder and H. Murer, *J. Membr. Biol.* **91**, 77 (1986).
[77] P. Gmaj, H. Murer, and R. Kinne, *Biochem. J.* **178**, 549 (1979).
[78] B. Hildmann, A. Schmidt, and H. Murer, *J. Membr. Biol.* **65**, 55 (1982).

ized that vesicle preparations are heterogeneous and that a precise localization of transport pathways to either the apical or the basolateral membrane needs several control experiments. Similarly, the documentation of coexistence of transport systems in one membrane requires careful control experiments, and ideally an experimental evidence for the "functional" cooperation of the different transport pathways should be obtained.

Acknowledgment

The work presented from this laboratory was supported by the Schweiz. Nationalfonds.

[21] Dynamic Laser Light Scattering to Determine Size Distributions of Vesicles

By Horst Ruf, Yannis Georgalis, and Ernst Grell

Phospholipid vesicles, which are widely used as model membranes in reconstitution experiments and as vehicles for drug targeting, can be prepared by a variety of methods[1-18] that in general yield vesicles of different size and structure. Depending on whether a single lipid bilayer or a multi-

---

[1] A. D. Bangham, M. M. Standish, and J. C. Watkins, *J. Mol. Biol.* **13**, 238 (1965).
[2] P. Pagano and T. E. Thompson, *Biochim. Biophys. Acta* **144**, 666 (1967).
[3] C. Huang, *Biochemistry* **8**, 344 (1969).
[4] W. D. Seufert, *Biophysik* **7**, 60 (1970).
[5] Y. Kagawa and E. Racker, *J. Biol. Chem.* **246**, 5477 (1971).
[6] S. Batzri and E. D. Korn, *Biochim. Biophys. Acta* **298**, 1015 (1973).
[7] J. Brunner, P. Skrabal, and H. Hauser, *Biochim. Biophys. Acta* **455**, 322 (1976).
[8] E. Skriver, A. B. Maunsbach, and P. L. Jørgensen, *J. Cell Biol.* **86**, 746 (1980).
[9] D. Deamer and A. D. Bangham, *Biochim. Biophys. Acta* **443**, 629 (1976).
[10] J. M. H. Kremer, M. W. J. v. d. Esker, C. Pathmamanoharan, and P. H. Wiersma, *Biochemistry* **16**, 3932 (1977).
[11] F. Szoka and D. Papahadjopoulos, *Proc. Natl. Acad. Sci. U.S.A.* **75**, 4194 (1978).
[12] F. Olson, C. A. Hunt, F. C. Szoka, W. Vail, and D. Papahadjopoulos, *Biochim. Biophys. Acta* **557**, 9 (1979).
[13] Y. Barenholz, S. Amselem, and D. Lichtenberg, *FEBS Lett.* **99**, 210 (1979).
[14] O. Zumbuehl and H. G. Weder, *Biochim. Biophys. Acta* **640**, 252 (1981).
[15] L. T. Mimms, G. Zampighi, Y. Nozaki, C. Tanford, and J. A. Reynolds, *Biochemistry* **20**, 833 (1981).
[16] S. Kim and G. M. Martin, *Biochim. Biophys. Acta* **646**, 1 (1981).
[17] H. Hauser and N. Gains, *Proc. Natl. Acad. Sci. U.S.A.* **79**, 1683 (1982).
[18] N. E. Gabriel and M. F. Roberts, *Biochemistry* **23**, 4011 (1984).

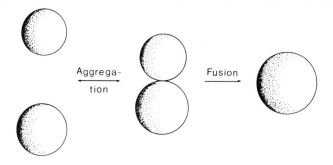

+ MICELLES

FIG. 1. Schematic illustrations of particles which can be present in a thermodynamically metastable vesicle suspension: vesicles of different size, aggregates of vesicles which may fuse to form larger vesicles, and small micelles, e.g., due to the presence of detergents or lysolipids in the lipid dispersion (aggregation products between vesicles and micelles may also exist but are not shown here). The processes which may occur between these particles, i.e., aggregation and fusion, will generally change the size distribution with time.

layer is forming the shell of a vesicle, these are called unilamellar vesicles or multilamellar vesicles (MUV). Unilamellar vesicles, in addition, are classified on the basis of their size; vesicles under 100 nm are considered small unilamellar vesicles (SUV), whereas vesicles with a greater diameter are large unilamellar vesicles (LUV). It is important to note that many physical and biological studies on vesicles require not only the knowledge of the average size but also of the size distribution for the final interpretation of results. From the size distribution it can be decided whether vesicles are homogenous in size. Its characteristics can be used for controlling the stability of a sample, and for studying processes like aggregation and fusion (Fig. 1). The average size is a measure which is often too insensitive for indicating such changes. Size, shape, and size distribution can be determined by various methods such as light microscopy, electron microscopy, analytical ultracentrifugation, NMR spectroscopy, gel chromatography, and static or dynamic light scattering techniques. The nonmicroscopic methods requiring rather homogeneous populations with a well-defined shape are most appropriately applied to small unilamellar vesicles. Electron microscopy is the method used for vesicles having an extended size range. The vesicles, however, have to be treated additionally for this method, which may influence their shape and size. Furthermore, to quantify a distribution by electron microscopy is a cumbersome task. Among the techniques appropriate for small unilamellar vesicles, the method of dynamic light scattering or intensity fluctuation spectroscopy (IFS), by which the hydrodynamic radius of the vesicles can be determined, has received much attention in recent years. The relatively short times (about 30 min) required to perform a measurement, the fact that the vesicles are measured directly in the suspension practically undisturbed by the measuring process, and that it is also possible to determine with some confidence the size distribution of spherical vesicles are features that make this method very attractive for the characterization and control of vesicle suspensions.

However, in contrast to the simplicity of carrying out an experiment, the theory is quite involved and comprises a number of different topics. An understanding requires an acquaintance with the theories of light scattering, the statistics of light fields, principles and practice of light detection, the methods of data evaluation, and some knowledge of fast electronic circuitry. The method is based on the property of macromolecules dissolved in a liquid to scatter light from an incident light beam in all directions of space. Since the macromolecules perform random walk movements due to the Brownian motion of the molecules, the total scattered light field at a given point in space resulting from superposition of all the wavelets scattered from the individual particles fluctuates with time. The analysis of the fluctuations yields the lateral diffusion coefficient of the particles and, by means of the Stokes–Einstein relation, the hydrodynamic radius for the case of spherical particles. The relationship between the measured quantities and the quantities to be determined, however, is complicated by the fact that the intensity instead of the field is measured with common light detectors and that light is measured by quantum processes whose statistics have to be considered.

The fundamentals of these topics have been treated in detail in a number of articles and reviews.[19-31] Unfortunately, the terminology used

---

[19] M. Kerker, "The Scattering of Light and Other Electromagnetic Radiation." Academic Press, New York, 1969.
[20] H. Z. Cummins and E. R. Pike (eds.), *NATO Adv. Study Inst. Ser., Ser. B* **3** (1974).
[21] H. Z. Cummins and E. R. Pike (eds.), *NATO Adv. Study Inst. Ser., Ser. B* **23** (1977).
[22] S.-H. Chen, B. Chu, and R. Nossal (eds.), *NATO Adv. Study Inst. Ser., Ser. B* **73** (1981).
[23] J. C. Earnshaw and M. W. Steer (eds.), *NATO Adv. Study Inst. Ser., Ser. A* **59** (1983).
[24] B. Chu, "Laser Light Scattering." Academic Press, New York, 1974.
[25] B. Berne and R. Pecora, "Dynamic Light Scattering." Wiley, New York, 1976.
[26] E. O. Schulz-Dubois (ed.), *Springer Ser. Opt. Sci.* **38** (1983).
[27] B. E. Dahneke (ed.), "Measurements of Suspended Particles by Quasi-Elastic Light Scattering." Wiley, New York, 1983.
[28] B. Chu, *Annu. Rev. Phys. Chem.* **21**, 145 (1970).
[29] R. Pecora, *Annu. Rev. Biophys. Bioeng.* **1**, 257 (1972).
[30] P. N. Pusey and J. M. Vaughan, in "Dielectric and Related Molecular Processes" (M. Davies, ed.), Vol. 2, p. 48. The Chemical Society, London, 1975.
[31] F. D. Carlson, *Annu. Rev. Biophys. Bioeng.* **4**, 243 (1975).

in discussing intensity fluctuation experiments is quite confused and no standard notation exists. The terms in use include dynamic light scattering, optical-mixing spectroscopy, light-beating spectroscopy, quasielastic or elastic light scattering, Rayleigh line width studies, Doppler-shifted light scattering, intensity fluctuation spectroscopy (IFS), and photon correlation spectroscopy (PCS). The last two terms will probably be used for standard notation: IFS to cover the general class of experiments where intensity fluctuations are analyzed, regardless of the signal-processing technique actually used, and PCS when intensity fluctuation experiments are processed by digital correlation techniques. A concise presentation of the basic principles of the method can be found in the literature.[24-26,28-31] However, an article comprising all the relevant aspects of the method required for its application to vesicles is not yet available. Here, an attempt is made to fill this void using a presentation with a minimum number of mathematical equations. By presenting a survey of the underlying principles with regard to vesicle suspensions, in particular to small unilamellar vesicles with sizes up to about 200 nm, and by giving special attention to the various data evaluation methods including the most recent and powerful developments, we hope to reduce the difficulties facing the new researcher in this field, who should also be aware of all the preconditions that have to be fulfilled for obtaining reliable results.

Scattering Theory

The scattering of light through the interaction with molecules can be explained by the classical theory of electrodynamics, as well as by the more sophisticated quantum field theory. A short review of the classical approach (see, e.g., Refs. 19, 24, and 25) is sufficient to demonstrate the physical basis of the method described in this chapter.

The scattering geometry is shown in Fig. 2. A monochromatic plane light wave of angular frequency ω_o, which for the purposes of this discussion is assumed to be polarized perpendicular to the plane of the diagram, impinges on the scattering medium, being a transparent liquid suspension of macromolecules such as small unilamellar phospholipid vesicles. The incident beam is characterized by the wave vector $\mathbf{k}_i$, of which the magnitude is related to the wavelength *in vacuo*, λ_o, by $k_i = 2\pi n/\lambda_o$ with n being the refractive index of the medium. The scattered light, detected at a distance R apart from the scattering medium by a detector, set at an angle θ with respect to the incident direction, is represented by the wave vector $\mathbf{k}_f$. The wavelength of the light is usually changed very little due to the interaction with the molecules of the medium, so that the magnitudes of the two wave vectors are essentially the same, $k_f \cong k_i$. These two vectors,

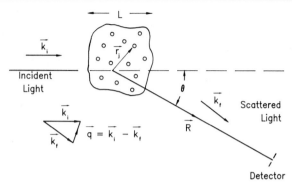

FIG. 2. Typical geometry of a light-scattering experiment.

enclosing the scattering angle θ, define the scattering plane. The difference vector

$$\mathbf{q} = \mathbf{k}_i - \mathbf{k}_f$$
$$q = |\mathbf{q}| = (4\pi/\lambda)\sin(\theta/2), \quad (1)$$

characterizes the deflection from the incident direction, and is called the scattering vector, where λ denotes the wavelength of the light in the scattering medium. The volume V, from which the scattered radiation is detected, is given from the intersection of the incident beam and the scattered beam intercepted by the detector. The characteristic dimension of the scattering volume, L, is, for typical light-scattering experiments, about 0.1 mm.

The incident light field exerts a force on the charges of the atoms in the scattering volume, and thereby induces electric dipoles. These dipoles oscillate with the frequency of the incident light, and consequently radiate light in all directions. The atoms in a volume element dV, small compared with λ^3, at the position $\mathbf{r}$ in the illuminated region, will see at any instant essentially the same incident electric field. Therefore, the instantaneous electric dipole moment induced in this subregion can be written as

$$\mathbf{P}(\mathbf{r}, t) = \alpha(\mathbf{r}, t)\, \mathbf{E}(\mathbf{r}, t)\, dV \quad (2)$$

where $\alpha(\mathbf{r}, t)$ is the polarizability tensor. The total scattered light field at a given position $\mathbf{R}$ in space then is obtained from the superposition of the fields radiated by each subregion of the scattering volume. At large distances from the scattering volume ($R = |\mathbf{R}| \gg L \gg \lambda$), the scattered field has the form of a spherical wave

$$\mathbf{E}(\mathbf{R}, t) = \frac{E_o \omega_o^2}{c^2 R} \exp[i(k_f R - \omega_o t)] F(t) \quad (3)$$

where E_o is the maximum field amplitude of the incident light, c the velocity of light, and $F(t)$ is a function modulating the amplitude in time. For the case of scalar polarizability, this reads

$$F(t) = \int_V \alpha(\mathbf{r}, t) \exp(i\mathbf{q}\mathbf{r}) \, dV \tag{4}$$

This integral over the scattering volume, which may be interpreted as an instantaneous interference factor, contains the time-dependent spatial distribution of the polarizability and the corresponding phase shifts due to the different optical paths taken by light scattered from the different parts of the illuminated region for reaching the detector. If the subregions are optically identical, that is, each has the same polarizability or dielectric constant, the scattered wavelets will be identical except for the phase. Ignoring surface effects, we then can always pair two wavelets scattered from different subregions in the scattering volume having phase shifts differing by $\lambda/2$ and which, therefore, interfere in a destructive manner. Thus, there will be no light scattered in other than the forward direction. In order to observe a scattering intensity in directions deviating from the incident one, the local polarizability or the local dielectric constant has to vary in space. This semimacroscopic view of light scattering being the result of local fluctuations of the dielectric constant of the medium was originally introduced by Einstein.[32]

In general, the tensor α will be a fluctuating quantity. In a pure liquid, the density will undergo spontaneous, thermally induced, fluctuations. If in a macromolecular solution the polarizability of the macromolecules differs from that of the solvent, fluctuations in α will in particular arise from fluctuations of the local concentration. Because of the high polarizability of a macromolecule compared with that of a small solvent molecule, and because of the stronger spatial fluctuations, the contribution of the solvent to the scattered light can usually be treated as weak background. The polarizability can be split into an averaged and a fluctuating part, $\alpha(\mathbf{r}, t) = \langle\alpha\rangle + \Delta\alpha(\mathbf{r}, t)$. Since the average polarizability will not contribute to the scattered intensity, we can confine ourself to the fluctuating part. For a dilute suspension, containing in the scattering volume N noninteracting particles that are small compared to the wavelength of the incident light, the fluctuating part of the polarizability can be written as

$$\Delta\alpha(\mathbf{r}, t) = \alpha_o \sum_{j=1}^{N} \delta[\mathbf{r} - \mathbf{r}_j(t)] \tag{5}$$

[32] A. Einstein, *Ann. Phys. (Leipzig)* **33**, 1275 (1910).

where α_o is the polarizability difference between the particle and the volume of solvent that it replaces and $\mathbf{r}_j(t)$ is the position if the jth particle at time t (small in practice means that all dimensions of the particles must be less than $\lambda/20$). The term $\delta[\mathbf{r} - \mathbf{r}_j(t)]$ is the Dirac delta function, which contributes one to the polarizability if $\mathbf{r}_j(t)$ is in neighborhood of $\mathbf{r}$ or zero otherwise. In this approximation each particle represents an elementary dipole radiator. Unilamellar lipid vesicles, in general, are too large to be approximated by dipoles of polarizability α_o. Wavelets scattered from various subregions inside a vesicle interfere destructively to some extent. Since the relative positions of these scattering elements are given from the vesicle's geometry, the intravesicular interferences can be summarized in a so-called scattering form factor.[33,34] For unilamellar spherical vesicles of outer wall radius R_o that satisfy the Rayleigh–Debye condition, $(4\pi/\lambda)R_o(n/n_o - 1) << 1$ (n_o is the index of refraction of the liquid into which the vesicles are suspended), this reads[19,35]

$$P(q, R_i, R_o) = \left\{ \left[\frac{3}{x^3(1 - y^3)} \right] [(\sin x - x \cos x) - (\sin xy - xy \cos xy)] \right\}^2 \quad (6)$$

with R_i being the inner wall radius, $x = qR_o$, and $y = R_i/R_o$. The form factor describes the reduction of the scattered intensity in dependence of the vesicle radius and the scattering angle θ. In practice, the vesicles to be investigated have to be smaller than about $\lambda/2$. Such vesicles can be considered as elementary dipole radiators possessing a reduced ability to scatter light. For vesicles of very small size, or for small scattering angles, the form factor, P, approaches one. For vesicles with dimensions greater than $\lambda/2$ the intensity diminishes very rapidly with increasing θ, and P may become an oscillating function with respect to both size and scattering angle. Furthermore, for larger vesicles the distortion of the electric field of the incident radiation due to the higher refractive index of the lipid phase becomes appreciable, and has to be considered too. This makes the complete treatment exceedingly complicated, and so far has been solved only for the simplest shapes such as full spheres, as done by Mie[36] almost 80 years ago, and for oriented cylinders.[37]

The polarizability of the vesicle, α_o, is also a function of its size. By

---

[33] Lord Rayleigh, *Proc. R. Soc. London, Ser. A* **84**, 25 (1910); see also *Proc. R. Soc. London, Ser. A* **90**, 219 (1914) and *Proc. R. Soc. London, Ser. A* **94**, 256 (1918).
[34] P. Debye, *Ann. Phys. (Leipzig)* **46**, 809 (1915).
[35] R. Pecora and S. R. Aragon, *Chem. Phys. Lipids* **13**, 1 (1974).
[36] G. Mie, *Ann. Phys. (Leipzig)* **25**, 377 (1908).
[37] H. C. van de Hulst, "Light Scattering by Small Particles." Wiley, New York, 1957.

introducing α_v as the polarizability per unit volume of lipid phase, the polarizability may be written as

$$\alpha_o(R_i, R_o) = (4\pi/3)(R_o^3 - R_i^3)\alpha_v \tag{7}$$

The field scattered by a dispersion of unilamellar lipid vesicles of uniform size is then given by

$$\mathbf{E}(\mathbf{R}, t) = \frac{\mathbf{E}_o \omega_o^2}{c^2 R} \exp[i(k_f R - \omega_o t)]\alpha_o(R_i, R_o)$$

$$\times [P(q, R_i, R_o)]^{1/2} \sum_{j=1}^{N} \exp[i\mathbf{q}\mathbf{r}_j(t)] \tag{8}$$

where $\mathbf{r}_j$ is now the position of the center of mass of the jth vesicle. Accordingly, the strength of the scattered field is determined by the strength of the incident light, the scattering angle, and the size and instantaneous spacial distribution of the vesicles.

Time-Dependent Fluctuations of the Scattered Field due to the Motion of the Scattering Particles

Macromolecules suspended in a solvent are in permanent motion, i.e., they undergo vibrational, rotational, and translational movements. The light scattered by these moving particles, therefore, is shifted in frequency and in phase (Doppler shifts). This means that the line width of the incident light becomes broadened upon scattering and the amplitude of the scattered field fluctuates with time. These two phenomena are directly related, i.e., a spectral feature with a finite frequency width is always associated with temporal fluctuations of the amplitude. The exact relationship is given by the Wiener–Khintchine theorem,[38,39] which states that the power spectrum (the dependence of power on frequency) of any stationary, ergodic, random signal is the Fourier transform of the first-order time correlation function of the corresponding amplitude. The characteristics of the frequency distribution of the scattered light, or the period of fluctuations, thus reflect the type of movement of the macromolecules. For random-walk diffusion of macromolecules under the influence of Brownian motion, the spectrum of the scattered light is described by a Lorenzian curve whose line width can be directly related to the translational diffusion coefficient.[40] For the analysis of the fluctuations in the

---

[38] N. Wiener, *Acta Math.* **55**, 117 (1930).
[39] O. Khintchine, *Math. Ann.* **109**, 604 (1934).
[40] R. Pecora, *J. Chem. Phys.* **40**, 1604 (1969).

time domain by means of correlation functions, there exists a corresponding relation between the translational diffusion coefficient and the correlation time typifying the fluctuations, which is given below. The first-order field autocorrelation function given by

$$G^{(1)}(\tau) = \langle \mathbf{E}^*(t)\mathbf{E}(t + \tau) \rangle \tag{9}$$

calculates the ensemble average of the products of the values of the scattered field, measured at times t and $t + \tau$, at position $\mathbf{R}$. The angle brackets indicate time or ensemble average for stationary ergodic systems. The autocorrelation functions of such systems depend only on the temporal separation of the events, τ, which is also called the delay time, and therefore the time t can be set arbitrarily to zero. For random fluctuation $G^{(1)}$ is a decaying function. If τ is small compared with the times typifying the fluctuations, the field will not change remarkably. The autocorrelation function then assumes values near to $\langle \mathbf{E}^*(0)\mathbf{E}(0) \rangle = \langle I \rangle$, i.e., to the average intensity of the scattered field. On the other hand, for delay times which are large compared with the period of the fluctuations the values of the scattered field will be completely independent or uncorrelated, and $G^{(1)}$ will be zero.

For small unilamellar vesicles, which can be modeled by rigid spherical shells, it is the translational motion that is of special interest. In a sufficiently dilute suspension, the vesicles encounter each other so rarely that their positions can be assumed to be statistically independent. By introducing Eq. (8) into Eq. (9) we obtain

$$G^{(1)}(q, \tau) = \frac{E_0^2 \omega_0^4}{c^4 R^2} \exp(i\omega_0 \tau) \alpha_0^2(R_i, R_o)$$

$$\times P(q, R_i, R_o) \left\langle \sum_{j=1}^{N} \exp\{i\mathbf{q}[\mathbf{r}_j(\tau) - \mathbf{r}_j(0)]\} \right\rangle \tag{10}$$

Thus, the field autocorrelation function is proportional to the autocorrelation function of the interference factor including all the displacements from the initial positions that the vesicles suffered during the time interval τ. Since these displacements are the result of random walk processes, the autocorrelation function can be obtained from the solution of a diffusion problem. For the derivation of this relation, we refer to the literature (e.g., Ref. 25); here only the results are presented. Usually the autocorrelation function $G^{(1)}$ is normalized by division through $G^{(1)}(0) = \langle I \rangle$. Then, quantities such as the magnitude of the incident field E_o, the distance of the detector from the scattering volume R, and the fourth power of the wavelength, $\lambda_o^4 = (2\pi c/\omega_o)^4$ are canceled out. For particles of uniform size, the form factor P, the polarizability α_o, and the number N, of independent scatterers in the illuminated region are canceled too. The normalized first-order field autocorrelation function, for the case of a monodisperse suspension, reads

$$g^{(1)}(q, \tau) = \exp(i\omega_0\tau) \exp(-q^2D\tau) \qquad (11)$$

As will be shown later, it is the magnitude of the field autocorrelation function

$$|g^{(1)}(q, \tau)| = \exp(-q^2D\tau) = \exp(-\Gamma\tau) \qquad (12)$$

which can be determined experimentally, and which under these circumstances decays as a single exponential. The diffusion coefficient D is obtained from the characteristic decay rate $\Gamma = q^2D$. The corresponding hydrodynamic radius then can be calculated by means of the Stokes–Einstein relation

$$R_o = kT/(6\pi\eta D) \qquad (13)$$

where η is the viscosity of the medium, T the absolute temperature, and k is Boltzmann's constant.

However, in general the vesicles in a suspension are not uniform in size. The actual size distribution may be influenced by the method of preparation, the type of lipid used, and the ionic composition of the solvent, etc. The first-order field autocorrelation function, in this case, consists of a series of exponentials representing different vesicle sizes, each of which decays with its characteristic decay time, and is weighted according to the relative number of vesicles of radius R_o. If vesicles in the size range between R_o and $R_o + dR_o$ are grouped together, the normalized first-order field autocorrelation can be approximated by the integral equation

$$|g^{(1)}(q, \tau)| = \frac{\int_0^\infty N(R_o)I(q, R_i, R_o) \exp[-q^2D(R_o)\tau] \, dR_o}{\int_0^\infty N(R_o)I(R_o, q) \, dR_o} \qquad (14)$$

with $N(R_o)$ being the number of vesicles in the size range between R_o and $R_o + dR_o$, and $I(q, R_i, R_o)$ the intensity scattered by one of these vesicles given by

$$I(q, R_i, R_o) = \frac{E_0^2 \omega_0^4}{c^4 R^2} \alpha_0^2(R_i, R_o) P(q, R_i, R_o) \qquad (15)$$

Measurement of the Fluctuations of the Scattered Light

The measurement of light with available detectors such as photomultipliers is based on quantum processes, i.e., light is detected in discrete portions of energy which are called photons. The detection of a photon, leading to the emission of an electron into the vacuum tube of the photomultiplier, appears (after amplification, discrimination, and standardization) as a charge pulse at the output of the detector. The principles of light detection have some important consequences. The output signal is related to the intensity of light, i.e., to the square of the field amplitude averaged over a period of oscillation, and not to the field itself. Accordingly, the detector is able to record the temporal fluctuations of the scattered intensity (Fig. 3a) but not the fast oscillations of the light field (the corresponding frequency for visible light of wavelength 500 nm being about 6×10^{14} Hz). In addition, the absorption (or annihilation) of a photon from the incident light field is a statistical event, and consequently the photodetection rate is not simply proportional to the incident intensity but is related to it by the statistics of the detection process.

The classical technique to measure light spectra is to filter spectrally the light prior to its detection using interferometric instruments such as diffraction gratings or the Pérot–Fabry etalon. The line broadening, $\Delta\omega_0/\omega_0 \lesssim 10^{-8}$, of the scattered light arising from the diffusional motion of macromolecules,[40] however, is too small to be resolved by classical methods. Since the frequency shifts manifest themselves in the fluctuations of the scattered intensity, they can be determined from postdetection analysis. Postdetection methods, which in this field are sometimes called optical-mixing or light-beating spectroscopy, are known from radio and radar technology. The mixing of electromagnetic waves reflected from moving objects with a reference wave is used to extract the velocities of the objects from difference or beat frequencies. In analogy, depending on whether the scattered light is mixed with a portion of a shifted (or unshifted) reference laser beam derived from the incident laser source, or is studied alone without reference beam, the optical methods are called heterodyne, homodyne, or self-beating spectroscopy. The latter is also referred to as the direct intensity fluctuation method.

The intensity fluctuations can be analyzed either in the frequency domain by determining the frequency distribution, as done in the pioneering work of Cummins et al.,[41] or in the time domain by calculating the second-order intensity autocorrelation function

$$G^{(2)}(q, \tau) = \langle I(q, \tau)I(q, 0)\rangle \tag{16}$$

With the development of fast, hard-wired, digital correlators, the latter has become the preferred technique for the study of macromolecular suspensions.

[41] H. Z. Cummins, N. Knable, and Y. Yeh, *Phys. Rev. Lett.* **12**, 150 (1964).

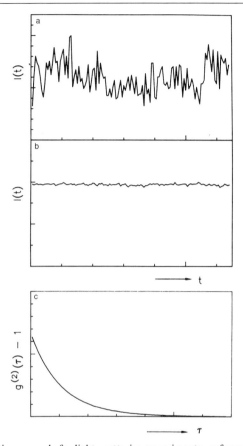

FIG. 3. Illustrative example for light-scattering experiments performed with Dow Chemical latex particles of 85 nm nominal diameter at 4×10^{-5} g/g aqueous solution ($\theta = 90°$). (a) Fluctuations of the scattered light intensity as recorded by a multichannel analyzer after a small number of sweeps (136 channels, sample time 0.1 msec, 20 sweeps). (b) Scattering intensity obtained from averaging about 10^4 sweeps. The values in each of the channels become very similar, which indicates that the mean intensity is measured when the number of samples becomes large. (c) Intensity autocorrelation function as measured by a correlator by taking about 10^8 samples (sample time 0.01 msec).

In many cases, this function can be simply related to the field correlation function and hence to the spectrum. In situations where the scattering volume contains a large number of independent scatterers performing random walk diffusion, the scattered field obeys Gaussian statistics.[42] For these conditions, the intensity autocorrelation function measured in a

---

[42] Lord Rayleigh, *Philos. Mag.* **10,** 73 (1880).

homodyne experiment is related to the field autocorrelation function by[43,44]

$$G^{(2)}(q, \tau) = |G^{(1)}(q, 0)|^2 + |G^{(1)}(q, \tau)|^2$$

or by (17)

$$g^{(2)}(q, \tau) = 1 + |g^{(1)}(q, \tau)|^2$$

which represents the normalized form of this relation being obtained from the division by $|G^{(1)}(q, 0)|^2 = \langle I(q) \rangle^2$.

At the low light-scattering levels typical for macromolecular suspensions, it becomes advantageous to use the technique of photon counting to improve the signal-to-noise ratio. Fast photomultipliers, capable of resolving charge pulses of a few nanoseconds width, allow one to record individual photons under these conditions. The signal output in the form of standardized pulses can be processed as such by digital correlators. As mentioned before, the detection of photons is a statistical process. The number of photons n, counted for a given sampling time interval Δt, have a simple Poisson probability distribution (see, e.g., Ref. 24), which for the case of constant intensity I, reads

$$P(n) = \frac{(\sigma I \, \Delta t)^2}{n!} \exp(-\sigma I \, \Delta t) \quad (18)$$

where σ is the quantum yield of the photocathode. The mean photocount per sample time $\langle n \rangle = \sum_{n=1}^{\infty} nP(n)$, which can be identified as $\sigma I \, \Delta t$, is thus proportional to the intensity. An extension of this relation to the situation where the intensity fluctuates is given by Mandel's formula[45]

$$P(n) = \int_0^\infty \frac{(\sigma I \, \Delta t)^2}{n!} \exp(-\sigma I \, \Delta t) P(I) \, dI \quad (19)$$

where $P(I)$ is the probability distribution of the incident intensity. The mean photocount, $\langle n \rangle = \sigma \langle I \rangle \Delta t$, now is proportional to the mean intensity of the radiation field (it is the mean photocount rate, which appears as current output of detectors of low time resolution, or under conditions of high illumination, yielding what is an analog signal for the intensity of the incident light). This can be demonstrated by sampling the incoming photocounts over a longer period in consecutive channels of a multichannel analyzer, all of which are set to the same sample time Δt (Fig. 3b). The tendency for each channel to register after 10^4 sweeps a similar number of

---

[43] A. J. F. Siegert, *Mass. Inst. Technol. Lab. Rep.* **465** (1943).
[44] R. J. Glauber, *Phys. Rev.* **131**, 2766 (1963).
[45] L. Mandel, *Proc. Phys. Soc.* **74**, 233 (1959).

counts is clearly indicated. Despite the indirect relationship between intensity and photon statistics it can be shown (see, e.g., Ref. 46) that the normalized photocount autocorrelation function is equal to the normalized intensity autocorrelation function except for the zeroth channel, or

$$\frac{C(q, \tau)}{\langle n(q)\rangle^2} \equiv \frac{\langle n(q, \tau)n(q, 0)\rangle}{\langle n(q)\rangle^2} = \frac{\langle I(q, \tau)I(q, 0)\rangle}{\langle I(q)\rangle^2} \quad (\tau \neq 0) \quad (20)$$

The extra term contributing to the zeroth channel, often called the shot noise term, results from the statistical nature of the photon detection. This is one of the reasons why this channel does not exist in most of the available correlators. As an illustrative example, the photocount autocorrelation function for a sample of calibrated polystyrene latex particles is shown in Fig. 3c.

In the Gaussian approximation, the photocount autocorrelation function can be related to the first-order field autocorrelation function by[47]

$$C(q, \tau) = \langle n(q)\rangle^2[1 + f(A)|g^{(1)}(q, \tau)|^2] \quad (21)$$

where $f(A)$ is a spatial coherence factor. In a real experiment, the scattered light is detected on a finite area, A, of the multipliers photocathode, and therefore is not described by a single **k** vector but by a range of **k** vectors. The superposition of these different waves reduces the correlation values. By choosing the area to cover about one coherence area, the reduction can be taken into account by a single factor. In practice, $f(A)$ is treated as an adjustable parameter, and is determined from fitting of Eq. (21) to the data. For a suspension of lipid vesicles of uniform size, the normalized photocount or intensity autocorrelation function is a single exponential decay function decaying from two to one with the double of the characteristic decay rate Γ of the corresponding first order field autocorrelation function, which decays from one to zero. Deviations of the experimentally determined normalized photocount autocorrelation function from the initial value of two indicate that the detection area of the detector comprises more than one coherence area, and from the final value of one that a certain level of background scattering occurs, which may be due to dust. A digital correlator, in general, samples the signal for set periods Δt, and calculates the correlation function for values of τ given by $\Delta t, 2 \Delta t, \ldots, m \Delta t$, where m is the number of channels, being typically of the order of 10^2. The baseline can be calculated either from a small group of extra channels with delay times large compared with $m \Delta t$, called

---

[46] E. O. Schulz-Dubois, *Springer Ser. Opt. Sci.* **38**, 1 (1983).
[47] E. Jakeman and E. R. Pike, *J. Phys. A* **1**, 128 (1968); see also *J. Phys. A* **2**, 115 and 411 (1969).

the delay channels, or from the total number of counts. For the specifications of the various correlator devices, which are now available, we refer to the literature[24,48-51] and the articles cited therein. The run time of an experiment and the sample time are normally choosen to cover many fluctuations of the light under study. By taking about 10^6–10^8 samples, in general, the contents of the channels represent, within a certain statistical error, to a good approximation the photocount autocorrelation function $C(q, \tau)$. For measuring a correlogram of light scattered by a sample to be investigated, however, special attention has to be given to the range of delay time, location of delay channels, and importance of baseline. A proper choice of these parameters is important for obtaining reliable results from the methods of data evaluation, which are presented in the next section. A detailed treatment of the experimental parameters can be found in several articles.[24,48-53]

Methods for the Evaluation of PCS Data of Polydisperse Samples

By introducing $\Gamma = Dq^2$, the field autocorrelation function of polydisperse systems (Eq. 14) may be recognized as the Laplace transform of the distribution function $S(\Gamma)$ with respect to Γ, which reads

$$g^{(1)}(q, \tau) = \int_0^\infty S(\Gamma) \exp(-\Gamma\tau) \, d\Gamma \quad (22)$$

where $S(\Gamma)$ is related to the distribution function representing the amount of light scattered by the particles of size R_o by

$$S(\Gamma) = \tilde{N}(R_o)I(R_o, q)(dR_o/d\Gamma) \quad (23)$$

where $\tilde{N}(R_o) \, dR_o$ denotes the number fraction of particles having radii between R_o and $R_o + dR_o$. The distribution function $S(\Gamma)$ now may be recovered from the inversion of the Laplace transform. The inversion process, however, is an ill-conditioned problem in the mathematical sense that small changes in the experimental data can produce large changes in the solution, which means that the solution in general is not stable for

---

[48] R. Foord, E. Jakeman, C. J. Oliver, E. R. Pike, R. J. Blagrove, E. Wood, and A. R. Peacocke, *Nature (London)* **227**, 242 (1970).
[49] C. J. Oliver, *NATO Adv. Study Inst. Ser., Ser. B* **3**, 151 (1974).
[50] V. Degiorgio, *NATO Adv. Study Inst. Ser., Ser. B* **23**, 142 (1977).
[51] C. J. Oliver, *NATO Adv. Study Inst. Ser., Ser. B* **73**, 87 and 121 (1981).
[52] B. Chu, *NATO Adv. Study Inst. Ser., Ser. A* **59**, 53 (1983).
[53] N. C. Ford, in "Measurements of Suspended Particles by Quasi-Elastic Light Scattering" (B. E. Dahneke, ed.), p. 32. Wiley, New York, 1983.

small variations of the data (relevant illustrative discussions to this point can be found elsewhere[54-59]). Experimental data are nearly always subjected to errors that, in quasi-elastic light-scattering experiments, may arise from a variety of sources. For example, unwanted modes of oscillations in the laser, so called off-axis modes, may occasionally produce spurious components to the correlation function. The detection system is also imperfect in that all components such as the power supplies delivering the necessary voltages, the photomultiplier, and the amplifier–discriminator are contributing noise to the signals. Furthermore, unwanted stray light from various parts of the optical system due to misalignments, and in particular light scattered from impurities of the sample such as dust can be sources for serious errors (these critical points have been considered in detail elsewhere[53]). Hence, the experimentally determined autocorrelation function $g^{(1)}$ measured for a given set of delay times τ_k has to be described rather by

$$|g_k^{(1)}| = |g^{(1)}(q, \tau_k)| + \varepsilon_k + B \qquad (24)$$

where ε_k represents the unknown noise contribution, and B allows counting for a constant background. Accordingly, the exact formulas for Laplace inversion or iterative algorithms converging to them cannot be directly applied to experimental data because there always exists a large set of possible solutions which can have arbitrarily large deviations from each other, all of which fit the data to within experimental error.

In order to extract useful information from data of polydisperse systems, a series of alternative methods has yet to be devised. We shall give a short overview to show how the inherent problems have been tackled on various levels of complexity. This, however, is far from being complete, and for details we refer to the corresponding literature. The common feature of these approaches consists in approximating Eq. (22) by functions containing a smaller number of independent parameters that has to be determined from the fitting procedure. This can be achieved either by approximating the distribution $S(\Gamma)$ or the kernel, $\exp(-\Gamma\tau)$, in the integral equation, or both.

The cumulants method, described by Koppel[60] in 1972, characterizes

---

[54] D. L. Phillips, *J. Assoc. Comput. Mach.* **9**, 84 (1962).
[55] D. R. Wiff, *J. Polym. Sci., Polym. Symp.* **43**, 219 (1963).
[56] S. W. Provencher, *J. Chem. Phys.* **64**, 2772 (1976).
[57] J. G. McWhirter and E. R. Pike, *J. Phys. A: Math., Nucl., Gen.* **11**, 1729 (1978).
[58] S. W. Provencher, *Comp. Phys. Comm.* **27**, 213 and 229 (1982).
[59] S. Bott, in "Measurements of Suspended Particles by Quasi-Elastic Light Scattering" (B. E. Dahneke, ed.), p. 129. Wiley, New York, 1983.
[60] D. E. Koppel, *J. Chem. Phys.* **57**, 4814 (1972).

the distribution by its moments. By expanding the exponential term of Eq. (22) into a MacLaurin series about its mean value $\exp(-\langle\Gamma\rangle\tau)$, where $\langle\Gamma\rangle$ is the mean decay rate, it can then be shown that[60,61]

$$\ln[f^{1/2}(A)|g^{(1)}(\tau)|] = \frac{1}{2}\ln[f(A)] - \langle\Gamma\rangle + \frac{1}{2!}\mu_2\tau^2$$
$$- \frac{1}{3!}\mu_3\tau^3 + \frac{1}{4!}(\mu_4 - 3\mu_2^2)\tau^4 + \ldots \quad (25)$$

The mean decay rate represents the first moment of the distribution

$$\langle\Gamma\rangle = \int_0^\infty S(\Gamma)\, d\Gamma \quad (26)$$

and the μ's are the moments of $S(\Gamma)$ about the mean decay rate

$$\mu_2 = \int_0^\infty (\Gamma - \langle\Gamma\rangle)^2 S(\Gamma)\, d\Gamma, \text{ etc.} \quad (27)$$

Data analysis thus reduces to the problem of fitting the logarithm of the experimentally determined field autocorrelation function to a polynomial of appropriate order. In practice, however, no more than the first two or three cumulants can be determined with any reasonable degree of confidence, which means that only the mean, the variance, and maybe the skewness of the distribution can be obtained. When calculating the field autocorrelation function from the measured intensity autocorrelation function the constant background is subtracted, so it has to be determined with fairly high accuracy. Otherwise, contributions arising from the background might seriously distort the logarithm of the experimentally determined autocorrelation function and thus might indicate details of the distribution that actually do not exist. This method, which does not require assumptions about the shape of the distribution, proved to be very useful for cases of narrow monomodal distributions, and it is also often used for preliminary analysis of more complex systems. Its advantages are that it can be realized easily even with small computers and that the evaluation can be carried out in a very short time.

All the other approaches use approximations for the distribution function $S(\Gamma)$. If the shape of the distribution of a polydisperse sample is known, the number of parameters that has to be determined can be reduced considerably. Accordingly, for systems for which such prior knowledge exists, parameterized distributions such as the Schulz,[62] the

---

[61] P. N. Pusey, D. E. Koppel, D. W. Schaefer, R. D. Camerini-Otero, and S. H. Koenig, *Biochemistry* **13**, 952 (1974).

[62] S. R. Aragon and R. Pecora, *J. Chem. Phys.* **64**, 2395 (1976).

log-normal,[63] or a Pearson distribution[64] can be used. Bimodal distributions have also been analyzed by fitting a double exponential to the experimental data.[64] The evaluation, in general, has to be performed with nonlinear least squares techniques, whose details can be found in the corresponding articles. In order to improve the resolution, the cumulant technique has been used together with several parameterized distribution functions, i.e., Gaussian[65] and various Pearson distributions.[66]

No special shape for the distribution has to be assumed if the actual distribution function $S(\Gamma)$ is approximated by an interpolation function identical with the distribution function for a given set of grid points. This, however, increases the number of parameters to be determined. The histogram method[67] uses a staircase type of function which is constant within the selected intervals of Γ. A different approach takes a piece-wise linear function called a first-order spline[68] or a cubic spline function,[69] which are constructed from a set of basic spline functions with knots at selected values of Γ, respectively. In both cases, the integrals can be readily evaluated for each interval, and the autocorrelation function can be approximated by a sum of discrete values. The only adjustable parameters then are the heights of the steps, or in the second case, the amplitudes at the grid points. Since the experimental autocorrelation function consists of a set of discrete data, this approach reduces to solve a system of linear equations. Actually, in order to include nonnegativity for the amplitudes and thus avoiding physically unrealistic solutions, these interpolation methods use systems of quadratic equations to be solved by nonlinear least squares techniques. The distributions obtained from these methods, however, depend strongly on the selection of the knots or grid points, both in spacing and in number. The most efficient combination has to be determined empirically, and how this can be achieved is described in some detail elsewhere.[67-69]

All the above methods do not consider explicitly the contribution of noise to the experimental data. Upon inversion of Eq. (22), noise is transformed into actually nonexisting components of a distribution which might appear with arbitrarily large magnitudes. This prevents recovery of such details of a distribution which contribute to the transform with magnitudes below the noise level, and thus only a limited amount of informa-

---

[63] D. S. Thompson, *J. Phys. Chem.* **75**, 789 (1971).
[64] B. Chu and A. DiNapoli, in "Measurements of Suspended Particles by Quasi-Elastic Light Scattering" (B. E. Dahneke, ed.), p. 81. Wiley, New York, 1983.
[65] C. B. Bargeron, *J. Chem. Phys.* **61**, 2134 (1974).
[66] R. L. McCally and C. B. Bargeron, *J. Chem. Phys.* **67**, 3151 (1977).
[67] E. Gulari, E. Gulari, Y. Tsunashima, and B. Chu, *J. Chem. Phys.* **70**, 3965 (1979).
[68] J. H. Goll and G. Stock, *Biophys. J.* **19**, 256 (1977).
[69] A. N. Lavery and J. C. Earnshaw, *J. Chem. Phys.* **80**, 5438 (1984).

tion can be extracted from the transform. For example, this prevents the determination of the higher moments of the cumulant method with some confidence.

There are presently methods available which allow the extraction of a maximum of information from the experiments in view of the noisy nature of the data. In order to elicit the maximum extractable information content, the Laplace transform has been investigated for its ability to transmit the various components of a given distribution. This might be illustrated by the approaches[56,57] whereby the distribution function $S(\Gamma)$ is expanded into a complete set of eigenfunctions of the Laplace transform operator $\exp(-\Gamma\tau)$. Eigenfunctions retain their identity under the action of the integral operator but are scaled in magnitude by their corresponding eigenvalues. The eigenfunctions of the Laplace transform operator are characterized by a frequency (in terms of the size coordinate), where eigenfunctions of low frequency represent coarse and those of high frequency represent fine details of the distribution. The larger the eigenvalue the more efficient is the transmission of the corresponding eigenfunction through the integral operator. By investigating the dependence of the magnitude of the eigenvalues on frequency it was found that, above a certain limiting frequency, the magnitude of the eigenvalues becomes so small that the corresponding transmitted eigenfunctions are lost in noise, and therefore cannot be recovered. It is a special property of the Laplace transform that it is acting as a low-pass filter, damping down those components representing the very fine details of a distribution. In the method developed by McWhirter and Pike[57] and by Ostrowsky et al.[70] those eigenfunctions which are transmitted below the noise level are truncated, and thus a "filtered" Laplace transform is used for evaluating the experimental data. Details for the determination of the cutoff frequency and the evaluation technique can be found elsewhere.[57,70]

The other nearly equivalent approach developed by Phillips,[54] Twomey,[71] Provencher,[58,72,73] and Wiff[55] involves regularizing the solution by imposing smoothness. This can be achieved, for example, by minimizing the total curvature[54,58,73] of the solution. Such a constraint penalizes against overoscillatory solutions. Smoothing, however, tends to increase the sum of squared residuals whose minimum is searched for in the least squares analysis. The optimal solution thus is found by minimizing the sum of these two counteracting minimum conditions. A regularized inver-

---

[70] N. Ostrowsky, D. Sornette, P. Parker, and E. R. Pike, *Opt. Acta* **28,** 1059 (1981).

[71] S. Twomey, *J. Assoc. Comput. Mach.* **10,** 97 (1963).

[72] S. W. Provencher, J. Hendrix, L. DeMaeyer, and N. Paulussen, *J. Chem. Phys.* **69,** 4273 (1978).

[73] S. W. Provencher, *Macromol. Chem.* **180,** 201 (1979).

sion of Eq. (14) can be performed using the method developed by Provencher.[58,73] The method, for which a computer program CONTIN[74] is available, allows inversion of the data represented by linear algebraic equations or by Fredholm integral equations of the first kind such as Eq. (14) or the Laplace transform [see Eq. (22)]. The program CONTIN, in addition, includes *a priori* knowledge such as nonnegativity for the values of the distribution function, $S(\Gamma) > 0$, by which the accuracy and the resolution of the solution obtained from the inversion are increased considerably. The optimal solution is found by searching for the smoothest (most parsimonious) solution that is consistent with the data. This is achieved by inverting the integral equation for different degrees of smoothing characterized by a factor, α, which gives the fraction in which the "regularizor" acts in the minimum condition, and then by choosing the optimal α using a criterion involving the Fisher test. The SIPP package,[75] dedicated to solve ill-conditioned problems by employing regularization techniques as well, bears certain similarities to Provencher's algorithm.

The last two methods, which face the ill-conditioned nature of the problem, implicitly assume that the true solution is fairly smooth and as such take into consideration the low resolution obtained from light-scattering experiments. Both methods are mathematically quite involved. However, their ability to extract a maximum of useful information, for example, about the particle size distribution in a sample, makes them extremely powerful in light-scattering data analysis. Nowadays, complete experimental setups for dynamic light-scattering experiments, where the described evaluation procedures are partly incorporated, including the last two powerful methods, are commercially available.

Determination of Size Distributions of Lipid Vesicles in Suspensions by Intensity Fluctuation Spectroscopy

Size distributions of vesicles originally were determined by electron microscopy.[4] However, this well-established technique, which gives detailed insight into the nature of a sample, is quite time consuming, in spite of the improved evaluation techniques[76] available. This method is less suited for investigations where many samples have to be measured or for rapid control measurements. In contrast, intensity fluctuation spectroscopy (IFS) permits a complete measurement within 30–60 minutes. Fur-

---

[74] S. W. Provencher, "CONTIN User's Manual" Version 2, Eur. Mol. Biol. Lab. Tech. Rep. EMBL-DA05. EMBL, Heidelberg, Federal Republic of Germany, 1982.
[75] G. R. Danovich and I. N. Serdyuk, *Springer Ser. Opt. Sci.* **38**, 315 (1983).
[76] M. Belanger and W. D. Seufert, *Biophysik* **9**, 39 (1972).

thermore, it does not require extensive preparations which might alter the characteristics of the sample. Therefore, IFS is being used more extensively for the study of vesicles where the average size and the size distribution in a dispersion are of importance. The reliability of this method has been demonstrated by comparing size distributions of lipid vesicles[77,78] or biological membrane vesicles[79] evaluated by electron microscopy with those obtained from IFS.

The type of distribution obtained from the two techniques as well as from the various evaluation procedures may differ and this should be taken into consideration when comparing results from different sources. The size of individual vesicles is determined from electron micrographs. An approximation to the size distribution is then obtained by counting the number of vesicles $N(R_j)$ having radii in the range between R_j and $R_j + dR_j$, where $j = 1, 2, 3, \ldots, n$ designates the n selected radii. Accordingly, the size distribution obtained from electron microscopy is the number distribution $N(R)$. The distribution function $S(\Gamma)$ involved in the autocorrelation function [Eqs. (14) and (22)] represents the mass square, form factor-weighted distribution of sizes since vesicles of radius R contribute to the field autocorrelation function according to the corresponding scattered light intensity. This is not only proportional to the number $N(R)$ but also to the square of the vesicle mass $M(R)^2$ and additionally is weighted by the corresponding form factor $P(q, R)$. This distribution function generally is included in the cumulant analysis, yielding the mean decay rate $\langle \Gamma \rangle$ from which the average diffusion coefficient $\langle D \rangle = \langle \Gamma \rangle / q^2$ and an average hydrodynamic radius $\langle R \rangle$ can be determined according to[80]

$$\langle R \rangle = \frac{kT}{6\pi\eta\langle D \rangle} = \frac{kTq^2}{6\pi\eta} \frac{\int_0^\infty N(R)M^2(R)P(q, R) \, dR}{\int_0^\infty N(R)M^2(R)P(q, R)D(R) \, dR} \quad (28)$$

where $D(R)$ is given in Eq. (13). For the case of $P(q, R) = 1$, which is valid either for sufficiently small particles ($R \leq \lambda/20$), or for particles of any size in the limit $\theta \to 0$, this average represents the z-average diffusion coefficient.[60,80] The second parameter that can be determined from the cumulant analysis for characterizing a distribution is the polydispersity $Q = \mu_2 / \langle \Gamma \rangle^2$, which is also a form factor-weighted z-average quantity.

Assuming a given shape for the distribution (Gaussian, etc.), the eval-

---

[77] R. A. Schwendener, M. Ansager, and H. G. Weder, *Biochem. Biophys. Res. Commun.* **100**, 1055 (1981).

[78] P. Schurtenberger and H. Hauser, *Biochim. Biophys. Acta* **778**, 470 (1984).

[79] G. Perevucnic, P. Schurtenberger, D. D. Lasic, and H. Hauser, *Biochim. Biophys. Acta* **821**, 169 (1985).

[80] J. C. Brown, P. N. Pusey, and R. Dietz, *J. Chem. Phys.* **62**, 1136 (1975).

uation methods yield number distributions. However, in methods which include inversion of the Laplace transform or use interpolation formula to approximate the distribution, different distribution functions may be obtained. Upon inversion of the Laplace transform the kernel in Eq. (14) may be weighted by the scattering form factor, the mass, or the squared mass of the vesicles. Therefore different distribution functions, $F_k(R)$, can be obtained: the number distribution, $N(R)$, the mass distribution, $N(R)M(R)$, the distribution of the squared mass, $N(R)M^2(R)$, or the distribution of the squared mass weighted by the scattering form factor, $N(R)M^2(R)P(R)$. Correspondingly, number-averaged (n), mass-averaged (m), z-averaged (z), or form factor-weighted z-averaged (pz) radii can be calculated from the relation

$$\langle R \rangle_k = \frac{\int_0^\infty F_k(R) R \, dR}{\int_0^\infty F_k(R) \, dR} \tag{29}$$

where $k = n, m, z, pz$ designates the various distributions and averages. The integrals in Eq. (28) are the zeroth and first moment of the distribution $S(\Gamma)$. Generally, the moments of the distribution function $F_k(R)$ can be written as

$$M_{k,1} = \int_0^\infty F_{k,1}(R) R^1 \, dR \tag{30}$$

Thus, by means of the second, first, and zeroth moments the mean squared deviation from the average radius can be calculated by

$$\sigma_k^2 = \langle R^2 \rangle_k - \langle R \rangle_k^2 = \frac{M_{k,2}}{M_{k,0}} - \left(\frac{M_{k,1}}{M_{k,0}}\right)^2 \tag{31}$$

which, for monomodal distributions, provides a useful measure for the width.

Phospholipid vesicles are, in general, thermodynamically metastable structures. Therefore, the preparation technique, the lipids used, the composition of the solution, and other factors can have a strong influence on the size of the vesicles. IFS is applied to characterize vesicles with respect to their size and to study the temporal stability of such size distributions. In addition, processes like coagulation and fusion, or the effect of size on the incorporation of proteins into vesicular membranes are studied.[81] In agreement with studies employing other methods, the results show that small rather uniform-sized vesicles can be obtained from differ-

---

[81] K. C. Aune, J. G. Gallagher, A. M. Gotto, and J. D. Morriset, *Biochemistry* **16**, 2151 (1977).

ent procedures such as sonication[68,82–88] with subsequent application of a separation technique like gel chromatography[3] or high-speed ultracentrifugation[82]; the ether injection technique[89]; or the various methods of detergent removal from solutions containing mixed lipid–detergent micelles.[77,90–93] For small unilamellar vesicles, the existence of a lower radius limit due to packing constraints has been postulated[94] to be $R = 9.9$ nm. This has been experimentally confirmed from IFS measurements.[86] Vesicles prepared from lipids having saturated hydrocarbon chains like α,β-dimyristoylphosphatidylcholine (DMPC) or α,β-dipalmitoylphosphatidylcholine (DPPC) undergo a phase transition from gel state to a liquid–crystalline state at a characteristic temperature, which is essentially dependent on the chain length of the lipid. Above the transition temperature, T_m, the average size of vesicles seems to be quite stable with time, but the vesicles tend to aggregate or to fuse below T_m.[59,95,96] The time scale on which aggregation and fusion of the vesicles occur seems to be dependent on factors such as the presence of impurities. From experiments on DMPC and DPPC vesicle preparations,[87–89] where special care has been taken to remove fusogenics such as Ca^{2+} ions no aggregation or fusion was observed, and quite narrow monomodal size distributions

[82] Y. Barenholz, D. Gibbes, G. B. Litman, J. Goll, T. E. Thompson, and F. D. Carlson, *Biochemistry* **16**, 2806 (1977).
[83] S. R. Aragon and R. Pecora, *J. Colloid Interface Sci.* **89**, 170 (1982).
[84] E. L. Chang, B. P. Graber, and J. P. Sheridan, *Biophys. J.* **39**, 197 (1982).
[85] B. A. Cornell, G. C. Fletcher, J. Middlehurst, and F. Separovic, *Biochim. Biophys. Acta* **690**, 15 (1982).
[86] J. Goll, F. D. Carlson, Y. Barenholz, B. J. Litman, and T. E. Thompson, *Biophys. J.* **38**, 7 (1982).
[87] B. A. Cornell, G. C. Fletcher, J. Middlehurst, and F. Separovic, *Biochim. Biophys. Acta* **642**, 375 (1981).
[88] E. L. Chang, J. P. Sheridan, and B. P. Gaber, in "Biomedical Applications of Laser Light Scattering" (D. B. Satelle, W. I. Lee, and B. R. Ware, eds.), p. 167. Elsevier, Amsterdam, 1982.
[89] A. Milon, J. Ricka, S.-T. Sun, T. Tanaka, Y. Nakatani, and G. Ourisson, *Biochim. Biophys. Acta* **777**, 331 (1984).
[90] P. Schurtenberger, N. Mazer, S. Waldvogel, and W. Kaenzig, *Biochim. Biophys. Acta* **75**, 111 (1984).
[91] M. H. W. Milsmann, R. A. Schwendener, and H.-G. Weder, *Biochim. Biophys. Acta* **512**, 147 (1978).
[92] E. Stelzer, H. Ruf, and E. Grell, *Springer Ser. Opt. Sci.* **38**, 329 (1983).
[93] E. Stelzer and H. Ruf, *Stud. Phys. Theor. Chem.* **24**, 37 (1983).
[94] C. Huang and J. T. Mason, *Proc. Natl. Acad. Sci. U.S.A.* **75**, 308 (1978).
[95] M. Wong and T. E. Thompson, *Biochemistry* **21**, 4133 (1982).
[96] D. Sornette and N. Ostrowsky, *NATO Adv. Study Inst. Ser., Ser. B* **73**, 351 (1981).

were reported for all temperatures in the transition range. In some cases these vesicles change their size reversibly with alterations in temperature within the transition range.

For most of the investigations cited above the IFS data were analyzed by the cumulants method, which provides only the mean size and the width of a distribution with some confidence. This restricts this technique essentially to monomodal distributions, although it has been applied in a special way to cases where bimodal distributions due to coagulation or fusion were assumed.[95] In some of these cases distributions of higher modality were also evaluated by fitting two or three exponentials to the data, implying the existence of two or three exact discrete radii. Vesicle dispersions, however, are generally polydisperse and often polymodal, especially when coagulation or fusion occurs. Such distributions would be more readily investigated by using the methods involving the inversion of the Laplace transform.[59,92,93,96–98] The power of these methods to analyze complex distributions occurring in dispersions of aggregating or fusing vesicles has been demonstrated by applying the eigenfunction approach where a truncated version of the Laplace transform is inverted,[96] or the method of searching for a regularized solution (CONTIN).[59] The size distributions obtained from these two studies showed an intrinsic bimodality with peaks at about 16 and 55 nm, or 11 and 40 nm, where the smaller component was initially dominating. The temporal behavior of the data reflecting the formation of larger particles was interpreted as fusion in the first case, and as reversible aggregation and fusion in the second. However, these processes which were presumably studied under slightly different conditions are not yet fully understood, and are under investigation.

Some illustrative results from our laboratory on small lipid vesicles employing IFS will be shown and discussed. The equipment for photon correlation measurements comprises a light-scattering spectrometer (ALV Laser Ges., Langen, FRG) equipped with a He–Ne Laser (20 mW, Spectra Physics, Darmstadt, FRG) and a 4-bit correlator (BI 2020, Brookhaven Instruments, Ronkonkoma, NY), which is connected to a minicomputer (HP 1000, Hewlett Packard, Frankfurt, FRG). The IFS data sampled with the correlator and transferred to the minicomputer by means of specific transfer programs[99] were analyzed with the program CONTIN, made available by Dr. S. Provencher (Max-Planck-Institut für

[97] A. Flamberg and R. Pecora, *J. Chem. Phys.* **88**, 3026 (1984).
[98] U. Hopfer, *Biochem. Soc. Symp.* **50**, 151 (1986).
[99] E. Stelzer, thesis, University Frankfurt, Federal Republic of Germany, 1982.

Biophysikalische Chemie, Göttingen, FRG). Pure egg lecithin is prepared by the method of Singleton et al.,[100] and its fatty acid composition is determined. Vesicles are prepared from a solution containing mixed lecithin–cholate micelles (2% lecithin; 1.3% sodium cholate) by fast removal of the detergent at 6° employing a lipid-saturated Sephadex G-50 column.[7] Prior to IFS measurements performed at 25°, the samples were filtered through 0.2-μm Nuclepore filters. Further details concerning instrumentation and lecithin and vesicle preparation are given elsewhere.[101] In order to allow us to compare our data with those from other sources, the radii and widths given in the text are form factor-weighted z-average quantities.

From the analysis of IFS data with CONTIN, the average radius of the main fraction of chromatographically separated vesicles (in buffer 20 mM PIPES–NaOH, pH 7.2, containing 65 mM NaCl) is about 15 nm and the width as characterized by the standard deviation over the mean, s_R, is ±0.25 nm (Fig. 4b) [s_R is the square root of Eq. (31) divided by the mean radius]. It should be noted that the average quantities given in the text are obtained by averaging the results of different samples measured in some cases with different sampling times, and thus are not necessarily identical with the parameters given for a single experiment documented in Figs. 4 and 5. This average radius agrees well with data obtained by electron microscopy and other methods.[7] It is surprising that this simple preparation yields vesicles of rather uniform size. Dilution of this sample by a factor of three does not essentially change the characteristics of the distribution. This indicates that in this range vesicle concentration has no effect on the measured photocount autocorrelation function and that vesicle aggregation or fusion can be largely excluded. Dialysis of the samples after the chromatographic separation for periods up to 2 days as well as concentration of dilute vesicle samples by ultrafiltration (XM50 filter; Amicon, Witten, FRG) also does not cause any significant change of the size distribution. However, if the lecithin content of the original vesicle-forming solution is reduced from the usual value of 2% to 1.3% (w/v) and the sodium cholate content is kept constant at 1.3%, a decrease in the mean radius to 12 nm (s_R = ±0.25) is generally observed (Fig. 4a). A similar trend is observed under different conditions, namely by reducing the ionic strength. The vesicles prepared in 20 mM PIPES–NaOH, pH 7.2 (Fig. 4c), have a mean radius of 12.5 nm (s_R = ±0.25). In this particular case a fractionation of a vesicle dispersion is also performed by Sephacryl S-400 chromatography. Analysis of the IFS measurements of the fractions

[100] W. S. Singleton, M. S. Gray, M. L. Brown, and J. L. White, *J. Am. Oil Soc.* **42**, 53 (1965).
[101] Y. Georgalis, H. Ruf, and E. Grell, manuscript in preparation.

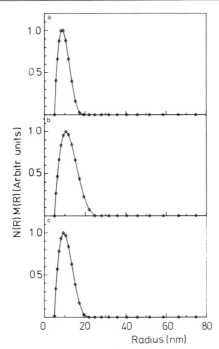

FIG. 4. Mass-weighted size distributions of small unilamellar lecithin vesicles prepared (a) in 20 mM PIPES–NaOH, pH 7.2, containing 65 mM NaCl (1.7 mM P$_i$) with $\langle R \rangle_w$ = 9.1 nm, s_w = ±0.26, and $\langle R \rangle_{pz}$ = 12.3 nm, s_{pz} = ±0.19 (vesicle-forming solution containing in this case 1.2% lecithin and 1.3% w/v sodium cholate); (b) in the same buffer (7.6 mM P$_i$) with $\langle R \rangle_w$ = 11.0 nm, s_w = ±0.32, and $\langle R \rangle_{pz}$ = 15.8 nm, s_{pz} = ±0.22; and (c) in 20 mM PIPES–NaOH, pH 7.2 (1.3 mM P$_i$), with $\langle R \rangle_w$ = 9.9 nm, s_w = ±0.29, and $\langle R \rangle_{pz}$ = 13.7 nm, s_{pz} = ±0.20 [s denotes the normalized standard deviation; see Eq. (31)]. Measurements carried out at 20°.

investigated indicates that no relevant differences exist with respect to the size distributions. Thus, no further improvement of vesicle size homogeneity is achieved by this type of chromatographic separation. Although the size distributions of samples prepared under apparently identical conditions are all fairly narrow and generally very similar. In a few cases, larger mean radii are evidently observed. As an example, the most deviating preparation in the medium given for Fig. 4b has led to a mean radius of 19 nm (s_R = ±0.3). Figure 5 illustrates the effect of ionic strength on the parameters of vesicle size distributions. In 20 mM PIPES–LiOH buffer, pH 7.2, a mean radius of 11 nm (s_R = ±0.15) is found (Fig. 5a). The distribution is as narrow as those of Dow latex particles, which are generally used for particle size calibration purposes. If the preparation is performed in the same medium but containing in addition 65 mM LiCl (Fig.

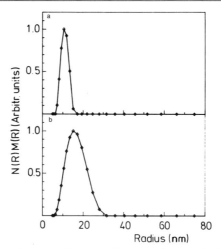

FIG. 5. Mass-weighted size distributions of small unilamellar lecithin vesicles prepared (a) in 20 mM PIPES–LiOH, pH 7.2 (3.2 mM P_i), with $\langle R \rangle_w$ = 10.7 nm, s_w = ±0.16, and $\langle R \rangle_{pz}$ = 12.0, s_{pz} = ±0.13; and (b) in the same buffer containing 65 mM LiCl (2.7 mM P_i) with $\langle R \rangle_w$ = 15.4 nm, s_w = ±0.29, and $\langle R \rangle_{pz}$ = 20.1 nm, s_{pz} = ±0.21. (Measurements carried out at 20°.)

5b), an increase of the mean radius to 19 nm (s_R = ±0.3) is detected. Furthermore, the stability of this sample has been investigated as a function of time. After 4 days, the mean radius is 20 nm, which indicates that no appreciable amount of vesicle fusion has occurred during this period of time.

Unilamellar vesicles prepared according to the method of detergent removal by dialysis[8] are larger in size. The results of similar studies using IFS together with the evaluation technique CONTIN are given elsewhere.[101]

From the study employing CONTIN for the analysis of IFS data of lipid vesicles, it is concluded that generally a single characterization of a particular preparation with respect to its size distribution is not sufficient. As a consequence of the remarkable sensitivity of the parameters to the size distribution of these vesicles based on the composition of the medium, it is important to characterize and to analyze each vesicle preparation.

Section III

Membrane Analysis and Characterization

[22] Sensitive Protein Assay in Presence of High Levels of Lipid

By RONALD S. KAPLAN and PETER L. PEDERSEN

Introduction

The accurate determination of low amounts of protein in samples containing a large excess of lipid has proven to be a rather difficult task. Several different experimental strategies have been utilized.[1-7] One approach has been to remove interfering lipid by extraction with organic solvents.[1] However, since certain proteins display a limited solubility in such solvents,[2] this strategy often fails. Another, widely used approach,[3-6] has involved the inclusion of SDS[8] in a modified Lowry[9] procedure in order to reduce lipid (and detergent) interference. However, since oxidized lipid continues to react to produce a substantial amount of color in the Lowry assay,[4,10] and since most lipid samples are at least partially oxidized, this procedure is not suitable for accurate measurements of protein in samples containing a large excess of lipid.[4] Finally, Mokrasch[2,7] has used 2,4,6-trinitrobenzenesulfonic acid to quantify 50–500 µg of protein in the presence of up to 10 mg of lipid. However, the accuracy of this procedure decreases when the quantity of sample protein is less than 50 µg.

In this chapter, we describe a method we have developed,[11] that is capable of accurately measuring low amounts of protein (i.e., 2–24 µg) in the presence of very high levels of lipid (i.e., 20–40 mg). Our procedure

---

[1] H. Tornqvist and P. Belfrage, *J. Lipid Res.* **17**, 542 (1976).
[2] L. C. Mokrasch, this series, Vol. 35, p. 334.
[3] M. A. K. Markwell, S. M. Haas, L. L. Bieber, and N. E. Tolbert, *Anal. Biochem.* **87**, 206 (1978).
[4] M. B. Lees and S. Paxman, *Anal. Biochem.* **47**, 184 (1972).
[5] G. L. Peterson, *Anal. Biochem.* **83**, 346 (1977).
[6] G. L. Peterson, this series, Vol. 91, p. 95.
[7] L. C. Mokrasch, *Anal. Biochem.* **36**, 273 (1970).
[8] The abbreviations used are: HEPES, 4-(2-hydroxyethyl)-1-piperazineethanesulfonic acid; MES, 4-morpholineethanesulfonic acid; MOPS, 4-morpholinepropanesulfonic acid; BSA, bovine serum albumin; TCA, trichloroacetic acid; SDS, sodium dodecyl sulfate; BCA, bicinchoninic acid.
[9] O. H. Lowry, N. J. Rosebrough, A. L. Farr, and R. J. Randall, *J. Biol. Chem.* **193**, 265 (1951).
[10] J. Eichberg and L. C. Mokrasch, *Anal. Biochem.* **30**, 386 (1969).
[11] R. S. Kaplan and P. L. Pedersen, *Anal. Biochem.* **150**, 97 (1985).

was developed from the Amido Black 10B methods of Schaffner and Weissmann[12] and Newman et al.[13] and incorporates several critical modifications that enable the assay to be performed with lipid-containing samples without interference.

Method for Protein Determination in the Presence of Lipid

The following method is routinely used in this laboratory for measuring protein in the presence of a large excess of lipid [a protein-to-lipid ratio of up to 10,000/1 (w/w) has been tested]. All operations are performed at room temperature. Proteolipid samples (containing 2–24 μg of protein) are diluted to 2.0 ml with deionized water. Then 0.2 ml of 10% (w/v) SDS, 0.3 ml of 1 M Tris/1% (w/v) SDS, pH 7.5, and 0.6 ml of 104% (w/v) TCA are added sequentially. After each addition, the reaction mixtures are vigorously vortexed. Then, following the TCA addition, the samples are incubated for at least 3 min. Each sample is subsequently transferred via a Pasteur pipette onto a Millipore filter (HAWP 024 00; 0.45 μm pore size, 24 mm diameter) that had been previously numbered near the top in pencil and fastened into a Millipore sampling manifold (12 place). Each sample is then filtered under vacuum and the filter is immediately washed with 2 ml of 6% (w/v) TCA. When all samples have been processed in this manner, the filters are removed and placed into a beaker which contains approximately 200 ml of a staining solution [0.1% (w/v) Amido Black 10B (Bio-Rad) dissolved in methanol/glacial acetic acid/deionized water, 45/10/45, v/v/v]. The filters are stained for 3 min (with gentle stirring). The stain is then decanted (and saved; the stain can be reused for several weeks) and the filters are rinsed once with approximately 200 ml of deionized water. The filters are then washed with three successive 200-ml portions of destaining solution (methanol/glacial acetic acid/water, 45/1/4, v/v/v) for 1 min per wash (with gentle stirring). They are subsequently washed with approximately 200 ml of deionized water for 2 min (again with gentle stirring) and are then placed on a paper towel and blotted with Kimwipes in order to remove excess water. The blue spot on each filter (or the equivalent sized area on the blank) is then excised, cut into smaller pieces, and placed into a test tube that contains 0.7 ml of 25 mM NaOH/0.05 mM EDTA/50% (v/v) ethanol. The dye is then eluted from the filters during a 20-min incubation with occasional vortexing. The absorbance of the eluate is read at 630 nm in a 1.0-ml

[12] W. Schaffner and C. Weissmann, *Anal. Biochem.* **56,** 502 (1973).
[13] M. J. Newman, D. L. Foster, T. H. Wilson, and H. R. Kaback, *J. Biol. Chem.* **256,** 11804 (1981).

cuvette against air. Reagent blanks, which contain all components except protein, are routinely run. The blank absorbance values (i.e., 0.05–0.08) are then subtracted from the sample values. These corrected absorbance values are stable (to within approximately 5%) for several hours. Unless indicated otherwise, the absorbance values presented below have been corrected for in this way.

When determining the effect of lipid on the assay, partially purified asolectin (soybean phosphatidylcholine, Type IV-S; Sigma Chemical Co.), which comprises some of the lipids that are frequently encountered in biological samples as well as in reconstitution studies (i.e., phosphatidylcholine, phosphatidylethanolamine, etc.), has been employed as the lipid source. When present, the lipid is added as a dispersion[11] (approximately 111 mg lipid/ml) in buffer A (120 mM HEPES, 50 mM KCl, 1 mM EDTA, pH 7.4) before the sample volume is brought to 2.0 ml. For comparison, samples without lipid contain the equivalent volume of buffer A.

General Properties of the Protein Assay and the Effect of Lipid

The Amido Black 10B protein assay, when performed as described above, yields a linear standard curve with 2–24 μg of BSA in both the absence (Fig. 1A) and the presence (Fig. 1B) of 20 mg of lipid. Furthermore, lipid does not significantly change the magnitude of the observed absorbance values. For example, in the absence of lipid, 20 μg of BSA results in a mean absorbance value (21 determinations) of 0.744 ± 0.054

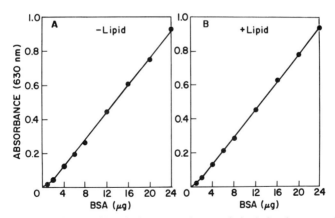

FIG. 1. Linearity of the Amido Black 10B protein assay in both the absence and presence of lipid. Samples contained either 180 μl of buffer A (A) or 20 mg of lipid that had been dispersed in 180 μl of buffer A (B). In addition, tubes contained varying amounts of BSA. Each datum point represents a mean of at least three determinations. From Kaplan and Pedersen,[11] with permission of the publisher.

(SD), whereas in the presence of 20 mg of lipid, the value obtained is 0.775 ± 0.048 (SD). The low standard deviation values point to the high degree of reproducibility that we observe with this assay. We have also tested this procedure employing mitochondria as the protein source (i.e., a mixture of hydrophobic and hydrophilic proteins) and obtain similar results (i.e., a linear standard curve and noninterference by 20 mg of lipid). Finally, Fig. 2 shows that, with either 4 or 20 μg of BSA, the assay is unaffected by as much as 40 mg of lipid (the highest amount tested). Thus, the assay is capable of accurately measuring protein even in the presence of a 10,000-fold excess (w/w) of lipid. These conditions are quite demanding indeed and we know of no other spectrophotometric protein assay that is capable of accurately estimating protein in the presence of such a large excess of lipid.

We have assessed the susceptibility of this procedure to interference by a number of the reagents that are typically encountered with isolated protein fractions as well as with reconstituted proteoliposomal systems. We find that in the presence of 20 μg of BSA and 20 mg of lipid, 200 mM of the following salts and buffers (in the original 2.0 ml sample) do not interfere with this procedure: NaCl, KCl, NaP$_i$, KP$_i$, HEPES, Tris, MES, and MOPS. Furthermore, the assay can tolerate up to 0.47% (v/v) Triton X-100, 0.47% (v/v) Genapol X-080 (American Hoechst Corp.,

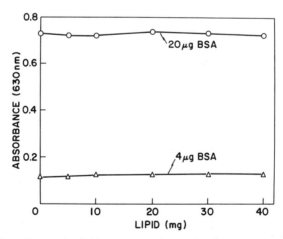

FIG. 2. Effect of increasing lipid on the absorbance due to BSA in the Amido Black 10B protein assay. Reagent blanks (i.e., samples without BSA) were prepared for each amount of added lipid. These absorbance values were subtracted from the values obtained with samples containing either 4 or 20 μg of BSA. All samples contained a total of 360 μl of buffer A. Each datum point represents a mean of at least four determinations. From Kaplan and Pedersen,[11] with permission of the publisher.

Somerville, NJ), and at least 1.55% (w/v) octylglucoside in the 2.0-ml sample. Thus, this procedure can be used to measure protein in the presence of high levels of both lipid and nonionic detergent.

Precautions

Several precautions should be mentioned. First, any reduction in the volumes of the various assay components should be made with extreme caution. The rather high volumes that we employ appear to be important for maintaining the lipid in solution (probably by decreasing the final lipid concentration). Second, the final SDS concentration in the assay is important. We obtain optimal results (i.e., maximal standard–blank absorbance values together with minimal blank values) in the presence of an SDS concentration (after TCA addition) of at least 0.5% (w/v) [up to 1.7% (w/v) has been tested]. For routine use, we recommend 0.7% (w/v) SDS, as the susceptibility of the assay to interference by other detergents has been examined at this SDS level. Finally, it is necessary to employ a fairly high TCA concentration [i.e., 16–20% (w/v)] in order to maximally precipitate the protein present [6–20% (w/v) has been tested].

Effect of Lipid on Several Alternative Protein Assay Methods

We have compared the degree of lipid interference with our protein assay relative to that observed with several popular alternative protein estimation methods.[1,5,6,12] With our Amido Black 10B procedure (Fig. 3A) neither the standard (i.e., 20 μg of BSA) nor the blank (no BSA) absorbance values are affected by 0–40 mg of lipid. In contrast, with the original Amido Black 10B method of Schaffner and Weissmann[12] (Fig. 3B), as little as 0.5 mg of lipid causes substantial interference in the standard absorbance values. Figure 3C depicts results obtained with the modified Lowry method of Peterson,[5,6] in which SDS is added in order to promote solubilization of a deoxycholate–TCA-precipitated sample, as well as to minimize lipid interference. We find that lipid causes an approximate parallel rise in both the standard and blank absorbance values. The procedure appears to be able to tolerate up to approximately 4 mg of lipid while still maintaining a reasonably low blank value. However, as the lipid level is increased further, the blank absorbance becomes rather high. For example, with 10 mg of lipid, the standard (20 μg of BSA) yields an absorbance increase of 0.27 over a blank value of 0.74. Figure 3D depicts results obtained with a procedure[1] that removes sample lipid by extraction of a TCA precipitate with diethyl ether/ethanol (3/1, v/v) followed by protein estimation via a scaled-down[1] Lowry procedure.[9] The data indicate that

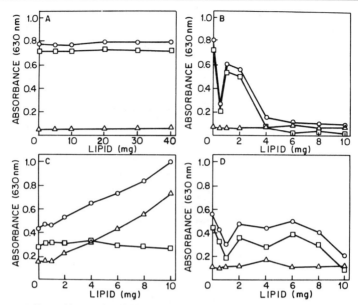

FIG. 3. Effect of increasing lipid on several alternative protein determination methods. The absorbance due to BSA was measured via our Amido Black 10B protein assay (A), the original Amido Black 10B protein assay of Schaffner and Weissmann[12] (B), a modified Lowry assay[5,6] that incorporates a deoxycholate–TCA precipitation step as well as the subsequent addition of SDS (C), and a procedure[1] involving the extraction of lipid from a TCA precipitate with diethyl ether/ethanol (3/1, v/v) followed by protein estimation according to a modified[1] Lowry assay[9] (D). Experimental conditions for A were as described for Fig. 2. Modifications in the procedures followed for B–D have been described previously.[11] ○, Standard incubations containing 20 μg of BSA; △, blank incubations that do not contain BSA; □, the difference between the standard and the blank absorbance values (i.e., the corrected values). Each datum point represents a mean of at least triplicate determinations. From Kaplan and Pedersen,[11] with permission of the publisher.

extraction of the added lipid causes a decrease and scatter in the standard absorbance values. Finally, we find that as little as 0.5 mg of lipid causes substantial interference in the new BCA protein assay[14] (data not shown).

Final Comments

It is noteworthy that protein quantification via the Amido dye technique often yields somewhat lower values than those obtained by other spectrophotometric procedures. For example, we find that with a sample of mitochondrial protein, our Amido dye procedure yields a protein esti-

[14] Pierce Chemical Co., Technical Bulletin 23225. Rockford, Illinois. 1984.

mate that is approximately 40% lower than that obtained with either modified Lowry[9,15] or biuret[11,16] procedures (BSA was used as the standard throughout). Similarly, Newman et al.[13] report that with E. coli membrane preparations, the Lowry procedure yielded values that were 1.5-fold higher than those obtained via the original Amido dye procedure of Schaffner and Weissmann[12] (additional SDS included). They also indicate that, with the purified lactose carrier, the accuracy of the Amido dye procedure was confirmed by amino acid analysis. However, it is important to note that the Amido dye, the Lowry, and the biuret procedures are all relative methods of protein estimation. They compare the staining capacity of sample protein(s) with that of a standard protein such as BSA.[17] Hence, the accuracy of these methods will vary depending on the similarity of the amino acid composition of the sample and standard proteins. When the goal is to determine accurately the absolute amount of protein in a sample, a given method should be calibrated either with known quantities of the sample protein or with a reliable alternative method such as quantitative amino acid analysis.[6,17]

Acknowledgment

This work was supported by National Science Foundation Grant PCM 8300772 to PLP.

[15] R. S. Kaplan and P. L. Pedersen, *J. Biol. Chem.* **260**, 10293 (1985).
[16] A. G. Gornall, C. J. Bardawill, and M. M. David, *J. Biol. Chem.* **177**, 751 (1949).
[17] W. H. Peters, A. M. M. Fleuren-Jakobs, K. M. P. Kamps, J. J. H. H. M. de Pont, and S. L. Bonting, *Anal. Biochem.* **124**, 349 (1982).

[23] Orienting Synthetic and Native Biological Membranes for Time-Averaged and Time-Resolved Structure Determinations

By L. G. Herbette and J. K. Blasie

Overview

Knowledge of the structure of biological membranes can provide a basis for understanding their functional role at the molecular level. The detailed enzymatic steps involved in such processes as active transport of ions across otherwise impermeable membrane bilayers, the passive gating actions of ion channels, energy transduction processes (oxidative phosphorylation), light energy ion gradient transduction, and receptor/channel regulatory processes all require knowledge of both the biochemical and biophysical properties of the membrane. The correlation of a specific biochemically characterized event with an associated structural event would provide the most basic understanding of how biological membranes function.

There are few techniques which can provide the detailed structural information that is required to make this correlation. Diffraction techniques (optical, X-ray, and neutron), spectroscopy (fluorescence, nuclear magnetic and electron spin resonance), and microscopy (optical and electron) are a few such techniques. These approaches can provide structural information with resolutions spanning the molecular (ångstrom) to morphological (micron) resolution range.

However, in order to capitalize on the potential of these structural techniques, it is highly favorable to introduce into the biological membrane sample *orientation* along a direction perpendicular to the membrane plane. Other than the retinal rod outer segment membrane,[1-4] which can occur naturally as a "stack" of membrane bilayers, most membranes must be artificially oriented along the direction normal to the membrane plane. This orientation provides for a macroscopic sample composed of membrane stacks (lamellae) forming a one-dimensional liquid crystal that is referred to as a multilayer and which is particularly suited for certain structural studies. For example, an oriented membrane multilayer is usually crucial to the design of a diffraction experiment, since it provides for enhanced interference of scattered photons, improving the ability to detect a diffraction/scattering pattern. Several standard and some relatively new techniques for orienting biomembranes are described below in Methods for Orienting Synthetic and Native Membranes, with particular attention paid to orienting biomembranes *without destroying their functionality*.

Likewise, it is unusual to find isolated biomembranes with a natural high degree (crystalline) in-plane *order* for the protein components. For a few isolated examples, notably, the purple membrane,[5] the acetylcholine receptor,[6] and the gap junction membrane,[7] the major protein components

---

[1] N. G. Webb, *Nature (London)* **235**, 44 (1972).
[2] W. J. Gras and C. R. Worthington, *Proc. Natl. Acad. Sci. U.S.A.* **63**, 233 (1969).
[3] M. Chabre, *Biochim. Biophys. Acta* **382**, 322 (1975).
[4] S. Schwartz, J. E. Cain, E. A. Dratz, and J. K. Blasie, *Biophys. J.* **15**, 1201 (1975).
[5] R. Henderson and P. N. T. Unwin, *Nature (London)* **257**, 28 (1975).
[6] M. J. Ross, M. W. Klymkowsky, D. A. Agard, and R. M. Stroud, *J. Mol. Biol.* **116**, 635 (1977).
[7] L. Makowski, D. L. D. Caspar, W. C. Phillips, and D. A. Goodenough, *J. Cell Biol.* **74**, 629 (1977).

within these membranes occur in an ordered array within the plane of the membrane, making them particularly amenable to certain structural techniques. However, for the majority of biomembranes, the protein within the plane of the membrane exists in a random or liquid crystalline arrangement such that artificial means must be used to establish the required degree of order. This crystalline-like order for a specific protein component within the plane of the membrane, in principle, may be artificially introduced, as speculated below in the section Future Directions—In-Plane Membrane Protein Order.

Obviously, the combination of orientation of an ordered two-dimensional array of proteins where there exists layer-to-layer registration of the two-dimensional crystalline arrays would describe a three-dimensional crystalline lattice structure for the membrane. Such a biological membrane sample, if it could be prepared, would provide the maximum amount of structure information from the above-mentioned physical techniques.

Methods for Orienting Synthetic and Native Membranes

Several techniques have been developed for orienting a biological membrane to make it suitable for structural studies. First, orientation may be introduced in the stacking direction by placing membranes on a substrate using one of several techniques (methods I–V); these stacks of biomembranes then comprise an oriented array of lamellae along this direction, referred to as a multilamellar or multilayer structure. Second, it may be feasible to artificially introduce order within the plane of the membrane (i.e., perpendicular to the stacking direction) by substrate binding and reconstitution techniques.

Method I: Centrifugation/Partial Dehydration

Dispersions of membranous vesicles in an appropriate buffer (sucrose-free preferred) can be prepared to a final volume of 2.5 ml and centrifuged in a special Lucite sedimentation cell (see Fig. 1) onto aluminum foil, Mylar, Saran, etc. strips for 1–3 hr at 85,000 g in a Beckman SW-28 swinging bucket rotor. For most native biological membranes, the temperature during centrifugation is usually controlled in the range of 5–15°, whereas higher temperatures, if required, can be used for pure lipid dispersions. As shown, the cell is designed with a lower inner diameter of 5 mm and sealed at the bottom by the strip onto which the sedimented membranes adhere. These cells can be designed to provide larger diameter samples (up to 1.5 cm), depending on the application. The strip is then removed from the sedimentation cell and glued to a more permanent

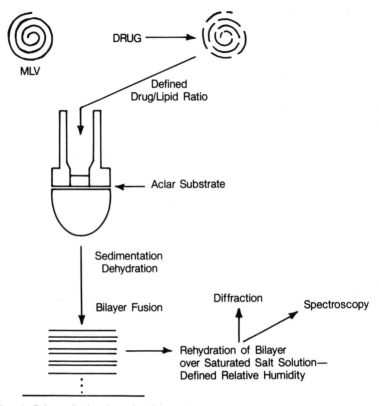

FIG. 1. Schematic drawing of sedimentation cell insert used to prepare oriented membrane multilayers for diffraction and spectroscopy. This cell is designed to fit an SW-28 swinging bucket. For studies employing the spin-dry method, the cap to the swinging bucket has a small hole so that drying and sedimentation occur together (see Fig. 3).

substrate depending on the application (for example, glass or quartz) and suspended in sealed vials over one of several saturated salt solutions, defining a relative humidity (RH) in the range of 10–100% usually at 5–15° for 10–30 hr.

After *slow* partial dehydration, specimens can then be placed in a temperature-regulated chamber preequilibrated and maintained throughout the X-ray or spectroscopy experiment. Specimens can be subjected to controlled humidity conditions within the chamber by being supported in a mount near a cup of saturated salt solution producing a specified relative humidity. Options include flowing helium through the chamber saturated at the same relative humidity as provided by the enclosed cup of saturated salt. Alternatively, the specimen chamber may be sealed completely with the cup of saturated salt. Experience indicates that sample lifetime and stability are markedly improved when the latter method (sealed sample holder) is employed. An example of a membrane oriented by these procedures and subsequently fixed and stained[8] for observation in the electron microscope is provided in Fig. 2. This procedure has been found to preserve the functionality of a variety of biomembranes and is the method of first choice.

Method II: Centrifugation/Complete Dehydration/Rehydration

The techniques described by Clark *et al.*[9] can be used as described or modified slightly for simplification. This technique takes advantage of the centrifuge vacuum to dehydrate a sample during centrifugation so as to produce highly ordered multilayer samples. As such, precise control can be exerted over membrane (or added substrate, ligand, small molecule, etc.) recovery since, as usually found, >95% of the added material is typically recovered in the pellet. In addition, other components that may be present (e.g., amphiphiles, ions, etc.) can be quantitatively recovered. In a more simplified application of the originally described procedure,[9] the Lucite sedimentation cells (see Fig. 1) are used which provide a flat substrate (aluminum foil, Mylar, etc.) surface onto which the membranes may be pelleted. A second set of rotor bucket caps ("spin-dry caps") has been tapped with a 1-mm hole which is overlayed with a Mylar window containing a 100-μm diameter hole. The size and condition of each hole is readily assessed and cleaned (with a microsyringe wire, <100 μm diameter) on a dissecting microscope.

Typically, 50 μl of the membrane dispersion in buffer is added to the Lucite sedimentation cells and sedimented onto the substrate at 85,000 g for 30 min in an SW-28 rotor using the normal SW-28 caps without a pinhole. The normal bucket caps are then replaced with the spin-dry caps and the pelleted vesicles are then dried at 3° under centrifuge vacuum at 65,000 g for up to 5 hr. On completion of the spin-dry process, the samples can be mounted as above and rehydrated over saturated salt solutions which define specific relative humidities. This procedure produces highly ordered pure lipid multilayers but should be used cautiously for native biomembranes because the dehydration step can introduce irreversible protein/lipid phase separation.

Method III: Isopotential Spin Dry

The isopotential spin-dry process also described by Clark *et al.*,[9] is similar to that described above with the advantage of maintenance of

[8] L. G. Herbette, A. Scarpa, J. K. Blasie, C. T. Wang, A. Saito, and S. Fleischer, *Biophys. J.* **36**, 47 (1981).
[9] N. A. Clark, K. J. Rothschild, D. A. Luippold, and B. A. Simon, *Biophys. J.* **31**, 65 (1980).

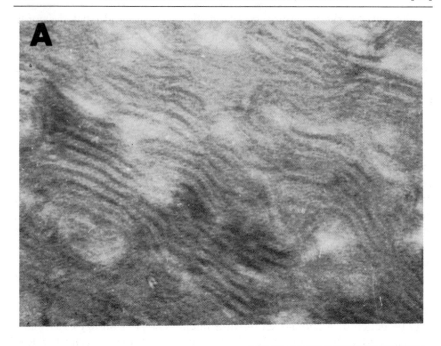

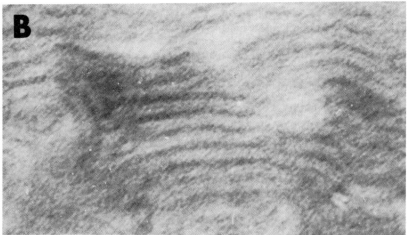

FIG. 2. Fixed and stained oriented multilayers of sarcoplasmic reticulum membranes used in diffraction and spectroscopy studies. Magnification: (A) ×166, 500; (B) ×279, 800.

initial solute (buffer, salt, etc.) concentrations during the drying process. This is a substantial advantage over the spin-dry technique where solute concentration must be adjusted (or solutes excluded) to allow for concentration increases during dehydration. The isopotential spin-dry method requires a modification of the design of the Lucite sedimentation cells (Figs. 3 and 4) which provides for two chambers, an inner sample sedimentation chamber and an outer remote evaporation chamber, which are connected by a semipermeable (or porous) membrane. Evaporation of solvent from the sample chamber is limited by a cap containing a pressure equilibration hole. The isopotential spin-dry process depends on centrifugation-induced hydrostatic pressure flow of solvent (and solute) from the inner sample chamber into the outer remote evaporation chamber. Evaporation from this chamber during centrifugation provides a hydrostatic pressure head in the sample chamber which maintains solvent flow in this direction. Provided that the hydrodynamic flow from the inner chamber is always greater than back diffusion, constant solute concentrations can readily be maintained during dehydration.

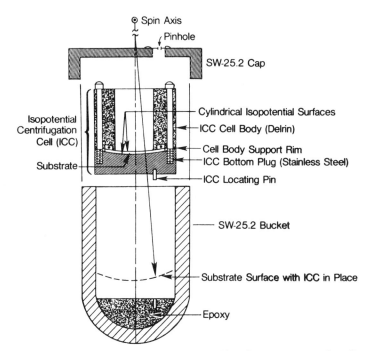

FIG. 3. Schematic drawing of the isopotential centrifugation apparatus made to fit an SW-25.2 swinging bucket. This figure and Fig. 4 were reproduced from the literature[9] with permission. A modified design can be inserted into an SW-28 swinging bucket (see Fig. 1).

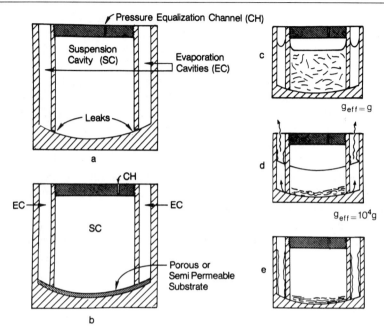

FIG. 4. Remote evaporation isopotential centrifugation cells for sedimenting and compacting a membrane array while maintaining a constant salt concentration. Evaporation takes place in cavities (EC) connected to the sedimentation cavity by (a) leaks or (b) a porous membrane; evaporation from the sedimentation cavity is suppressed by a cap with a small channel (CH) for pressure equalization; (c)–(e) schematic of the operating cycle for these isopotential centrifugation cells.

Method IV: Magnetic Orientation

Order may be introduced into biological samples by a reasonably strong magnetic field (10–20 kilogauss). This method could be applicable to dilute solutions and concentrated suspensions of macromolecular assemblies, fibers, and mono- and multilayers of biomembranes. Specifically, the orientation of biomembranes in a magnetic field must be due to their anisotropic magnetic susceptibility.[10,11]

Different approaches may be used but they have not been fully compared as to their usefulness. The multilayer (or single layer) biomembrane sample can be subjected to a strong magnetic field during the partial dehydration process and then removed from the field and utilized in the

---

[10] M. J. Glucksman, R. D. Hay, and L. Makowski, *Science* **231**, 1273 (1986).
[11] J. M. Pachence, R. Knott, I. S. Edelman, B. P. Schoenborn, and B. A. Wallace, *Ann. N.Y. Acad. Sci.*, in press.

appropriate diffraction or spectroscopy experiment. Alternatively, the mono- or multilayer sample could be kept in a magnetic field (10–20 kilogauss) during the course of the diffraction or spectroscopy experiment.

An interesting application of this approach may be the orientation of small lipid-soluble molecules (e.g., drugs) in mono- or multilayer lipid structures since the small molecule could possess a similar or higher degree of anisotropy than the lipid into which it has partitioned. Small molecules will insert into a lipid bilayer structure with a well-defined location along the bilayer axis (i.e., perpendicular to the bilayer plane) and usually with reasonable orientation. However, if the orientation is not well defined, a small organic molecule/lipid mono- or multilayer sample could be placed in a strong magnetic field, such that the small molecule may be aligned in the plane of the membrane bilayer, producing a rotationally averaged (about the bilayer normal) orientation of the small molecule (see Fig. 5). This pseudocrystalline/liquid crystalline composite may allow relatively high-resolution structural information to be obtained if the contrast between the small molecule and its lipid environment is sufficient (e.g., deuteration of the drug in a fully protonated lipid environment to be used in either a neutron diffraction or nuclear magnetic resonance experiment). This oriented monomolecular dispersion of the small molecule in a lipid bilayer could be compared to the crystal structure of the small organic molecule to test the relevance of small molecule crystal structure determinations of lipophilic substances.

Method V: Langmuir–Blodgett Film Technique

Numerous physical techniques have been employed to measure the forces of interaction between lipids. One of the oldest of these is the measurement of force–area characteristics of monomolecular films. A small amount of substance is allowed to spread on the surface of water in a Langmuir trough and its compressibility measured by a suitable apparatus. Films of long-chain saturated fatty acids at an air–water interface are condensed at relatively low pressures because the hydrocarbon chains can pack together as in a crystalline substance near its melting point. The hydrocarbon chains are then almost normal to the surface. Films of highly charged substances reflect the repulsive effects of the charged groups and the films are more expanded. Similar effects are brought about by introducing double bonds into the hydrocarbon chain. In this case, the kink in the chain does not allow such close packing.

This approach has been exploited by dipping a flat plate surface through the air–water interface to prepare monomolecular films of lipids

FIG. 5. Schematic of a rotationally averaged monomolecular dispersion of a small molecule in a lipid monolayer using a magnetic field in one direction.

for structural studies. A single monolayer, a bilayer, or a multibilayer of lipids can be oriented on a suitable substrate. The so-called Langmuir–Blodgett films are highly oriented and very suitable for a number of biophysical applications. An extensive literature has been established on the preparation, conditions, and composition of such oriented films. Future potential applications of this approach to biological membrane proteins are described in the next section.

Future Directions—In-Plane Membrane Protein Order

In contrast to the more proven methods for orienting biological membranes, the introduction of in-plane ordering of the protein components is more speculative. Apart from some naturally occurring membrane proteins which crystallize into two-dimensional arrays within the membrane, one possible artificial means of introducing the required order would involve structured substrate binding of the protein followed by reconstitution of the lipid bilayer. Specifically, if a substrate surface could be prepared with a covalently bound ligand which, in turn, would bind to a particular protein, the substrate surface could be decorated with bound protein, especially if it had a high affinity for the ligand. If the density of the ligand was high enough, a protein could be made to condense into a two-dimensional crystalline structure. The bilayer could effectively then be reconstituted using the Langmuir–Blodgett technique, which would result in a structure forming a monomolecular deposition of the crystalline protein in the bilayer matrix.

Alternatively, some membrane proteins, such as bacteriorhodopsin, are known to crystallize into two-dimensional arrays within membranes. If the protein molecule did not protrude excessively from the plane of the bilayer, a monolayer containing an appropriate fraction of the protein and lipid spread in a Langmuir trough could be used with the above protein-coated substrate to build up a three-dimensionally ordered membrane protein structure utilizing the Langmuir–Blodgett technique. Thus, it is possible that the monomolecular dispersion of proteins bound to ligands on a particular substrate, which would form the first two-dimensional crystalline layer, might then be used as a starting substrate to layer additional monolayers of the appropriate fraction of protein and lipid which should deposit in an ordered fashion on this modified substrate surface. The hope here is that the first layer, where a two-dimensional crystalline array of protein is formed, would promote the next layers deposited by the Langmuir–Blodgett technique to be registered into a three-dimensional membrane crystalline structure.

Conclusions

Ordering of protein or other constituents within membranes as well as orientation of membranes relative to one another usually is a requirement for structural studies. In general, methods for ordering and orienting membranes should: (1) preserve the functionality of the membrane, in particular, the enzymatic or biological processes it performs; (2) not alter the inherent structure of the membrane, particularly, the "conformation" of proteins; and (3) where possible, be verified by other techniques to ensure that the proper order and/or orientation has been achieved.

As our structural techniques continue to advance, the need for artificially introducing the requisite orientation and order in isolated and synthetically prepared biomembrane materials may be lessened.

Acknowledgments

The authors would like to thank Dr. B. P. Schoenborn, Dr. A. Saxena, and associated staff at the High Flux Beam Reactor, Brookhaven National Laboratory, Upton, NY, for their assistance in making it possible to carry out neutron diffraction studies. This work was supported by research grants HL-33026, HL-18708 from the National Institutes of Health, by the Whitaker Foundation and the Patterson Trust Foundation, and a gift from RJR Nabisco, Inc. Dr. Herbette would like to thank his colleagues, Dr. Chester and Dr. Rhodes, who were previously supported by NIH HL-07420, for their valuable interactions. Sincere thanks to Mrs. T. Wojtusik for her dedication in preparing this manuscript for submission. Dr. Herbette would like to acknowledge his affiliation as an Established Investigator of the American Heart Association. Development of membrane orientation procedures has taken place in part, in the Biomolecular Structure Analysis Center at the University of Connecticut Health Center. We would like to thank the staff of the Structure Center for their dedication in keeping the facilities in optimal running condition. The Biomolecular Structure Analysis Center acknowledges support from the State of Connecticut Department of Higher Education's High Technology Programs.

[24] Radiation Inactivation of Membrane Components and Molecular Mass Determination by Target Analysis

By E. S. KEMPNER and SIDNEY FLEISCHER

Introduction

Membranes are common to all cells and carry out a diversity of functions. The basic structural feature is the phospholipid bilayer into which a variety of different proteins are inserted. There is only limited information regarding the orientation and association of protein constituents in the membrane. In principle, protein components within the membrane can associate (1) with one another to form oligomers or complexes[1]; and/or (2) with compartmental components at either surface, i.e., with membrane-associated proteins or the microskeletal system. Such associations may change during the exercise of function. Target inactivation analysis is one of the most powerful approaches to characterize protein–protein associations of membrane constituents. The method is directly applicable to the determination of molecular size in the native membrane and does not depend on purification of the component to be studied. It requires only the measurement of a functional characteristic such as binding of a ligand or enzymatic activity after the sample has been irradiated. The target size can then be obtained from the slope of a plot of the log activity as a function of radiation dose.

Mechanism of Radiation Action and Principles of Target Analysis

The technique of molecular weight determination from radiation inactivation of a biological function depends on certain assumptions. Important among these are that (1) action of radiation directly on biologically active structures completely destroys that activity, and (2) radiation action elsewhere is without effect on that activity. Proper use of this technique involves finding those conditions where these assumptions are valid; they will be discussed in this section and throughout the remainder of this chapter.

Physics of Radiation Action

Radiation occurs in several forms: nonparticulate electromagnetic waves of different wavelengths (energies), and particles, both uncharged (neutrons, neutrinos) and charged (electrons, protons, and heavier ions). All of these deposit energy in any matter with which they interact. The mechanisms of these interactions depend on the nature and energy of the radiation involved. For the purpose of molecular weight determination, only ionizing radiation in the form of gamma rays and high-energy electrons is useful.

In practice, gamma rays are obtained almost exclusively from ^{60}Co or ^{137}Cs sources, and various accelerators are used to provide electrons with energies of 1 to 20 MeV. In general these radiations have similar mecha-

---

[1] We use the term "oligomer" to refer to a multimer of identical subunits, and "complex" to refer to an association of nonidentical polypeptides. The complex can also be oligomeric, as $(a, b, c)_n$.

nisms of action. There is an initial "primary ionization" of an orbital electron; the electron is displaced with a certain amount of kinetic energy, usually quite small, and a "positive hole" is left in the original atom. It is important to note that these primary ionizations occur randomly throughout the mass of irradiated matter, and also that each primary ionization is associated with the transfer of significant amounts of energy to the target material. On the average there is a deposition of approximately 65 eV (~1500 kcal/mol).[2,3] This is sufficient to break many covalent bonds. Ultimately, this energy is absorbed by a variety of different mechanisms. Among these are vibrational and rotational energy level changes, excitations, ionizations, free radical formation, and others. The only mechanisms of concern to this technique are those which result in irreversible molecular changes which alter the biochemical activity under study. Mechanisms of energy absorption which lead to covalent bond breakage are among the most significant. In proteins, those in a polypeptide backbone are especially important. In general, the total energy absorbed is so large that the polypeptide is split into several separate fragments which may themselves be chemically altered in other ways. The damage is observed throughout the structure of a covalently bound polymer and to a first approximation it is completely random throughout that structure. Damage can also spread to closely associated polypeptide chains in a multimer by way of energy transfer mechanisms (see section General Principles). Furthermore, the destruction of a polypeptide chain could be expected also to result in conformational changes in adjacent subunits of an oligomer in those situations where the subunit interaction stabilizes a unique conformational state.

As a result of the structural changes there is a total loss of biological activity, and this is the basis for target analysis to determine molecular weights.

Mathematics of Target Analysis

Since ionizations caused by gamma rays and high-energy electrons occur at random throughout the mass of irradiated material, the probability of such "hits" in a molecule is described by a Poisson distribution[4]

$$P(n) = \frac{e^{-x} x^n}{n!} \quad (1)$$

---

[2] A. M. Rauth and J. A. Simpson, *Radiat. Res.* **22**, 643–661 (1964).
[3] G. R. Kepner and R. I. Macey, *Biochim. Biophys. Acta* **163**, 188–203 (1968).
[4] R. B. Setlow and E. C. Pollard, "Molecular Biophysics." Addison-Wesley, Reading, Massachusetts, 1962.

where n is the number of primary ionizations per molecule when x is the average number of hits. Since the only biological activity which remains after irradiation is from molecules which have not been hit, i.e, for which $n = 0$,

$$P(0) = \frac{e^{-x}x^0}{0!} = e^{-x} \tag{2}$$

The average number of hits is given by

$$x = kD \tag{3}$$

where D is the radiation dose in rads (1 rad = 100 ergs energy absorbed per gram of matter) and k is a constant related to the mass of the structure and also containing conversion factors to change units from ergs per gram to hits per dalton.

After radiation exposure, each sample is assayed for surviving biochemical activity, which is expressed as a fraction of the activity in unirradiated samples. Since each hit in a biochemically active unit destroys that unit completely, the probability of no hits (obtained from the Poisson distribution) will also be the probability of surviving activity. Therefore

$$\frac{A}{A_0} = e^{-kD} \tag{4}$$

The graphical representation of data as $\ln(A/A_0)$ versus radiation dose (D) is called the "inactivation curve" (Fig. 1[5]). For simple systems this will be a straight line with a slope equal to k. Estimation of this slope permits calculation of the molecular weight from

$$MW = 6.4 \times 10^{11} k \tag{5}$$

which was determined empirically by Kepner and Macey.[3] A theoretical argument has been presented[6] which leads to a similar equation. When there is an average of one primary ionization (p.i.) per active unit (i.e., $x = 1$),

$$A = A_0 e^{-x} = A_0 e^{-1} \simeq 0.37 A_0$$

Thus, when the measured activity is reduced to 37% of the control, $x = 1 = kD_{37}$.

---

[5] M. E. Lowe and E. S. Kempner, *J. Biol. Chem.* **257**, 12478 (1982).
[6] E. S. Kempner and H. T. Haigler, in "Growth and Maturation Factors" (G. Guroff, ed.), Vol. 3, pp. 149–173. Wiley, New York, 1985.

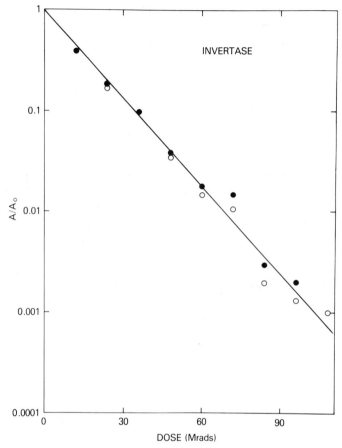

Fig. 1. Loss of invertase activity after exposure to high-energy electrons.[5] Native yeast enzyme (solid circles) and carbohydrate-depleted invertase (open circles) irradiated in buffer at −135°.

The dose in rads can be expressed in different units:

$$1 \text{ rad} = \frac{100 \text{ ergs}}{\text{g}} \times \frac{6.24 \times 10^{11} \text{ eV}}{\text{erg}} \times \frac{1 \text{ p.i.}}{65 \text{ eV}} \times \frac{1.66 \times 10^{-24} \text{ g}}{\text{Da}}$$

where p.i. refers to primary ionizations. Thus,

$$1 \text{ rad} = 1.59 \times 10^{-12} \text{ p.i./Da}$$

A dose of D rads will produce $1.59 \times 10^{-12} D$ primary ionizations per dalton. For a molecule whose mass is MW daltons there will be $1.59 \times 10^{-12} (D)(\text{MW})$ primary ionizations. Since the radiation dose which causes an average of one p.i. per molecule is D_{37}, $1 = 1.59 \times 10^{-12}(D_{37})$ (MW), or MW $= (1.59 \times 10^{-12} \times D_{37})^{-1} = 6.3 \times 10^{11}/D_{37} = 6.3 \times 10^{11}\, k$.

Ideally, the biological process should be examined over a range of radiation exposures sufficient to reduce the activity by several orders of magnitude, to 10^{-2} or 10^{-3} of the control value. If the inactivation curve then demonstrates a single exponential decay (Fig. 1), some confidence can be placed in the application of Eq. (4). The greater the range of surviving activity, the more accurate is the determination of k, and therefore also the target molecular weight.

Arithmetically, k is equal to the reciprocal of the radiation dose which reduces activity to 37% of the control. Some authors have been misled by this relationship and have incorrectly assumed that radiation exposure which destroy two-thirds of the initial activity is a sufficient range for an inactivation study.

When the experimental measurement is that of biochemical activity, the molecular weight "target size" is an estimate only of the sizes of those structures needed for the biological assay. Structures present in the sample but which are not involved in the biochemical process will not contribute to the estimated size. This leads to the idea that the target size is a measure of the "functional unit." Evaluation of the amount of polypeptide remaining provides additional insight and sometimes indicates limitations in the simplistic interpretation of the functional unit (see General Principles).

Membranes containing the LDL receptor were irradiated with high-energy electrons; subsequent measurements of LDL binding yielded the accurate molecular weight of the LDL receptor protein.[7,8] The presence of other proteins and lipids in the membrane had no effect on the measurement. Similarly, irradiation of the enzyme complex tryptophan synthase (an $\alpha_2\beta_2$ protein) revealed a target size of the functional unit for the enzymatic conversion indole + serine → tryptophan was that of β_2 alone.[9] Nonessential portions of the structure are invisible in the assay. From this derives one of the most useful attributes of this approach: the ability to use impure samples. As long as the assay is specific, the presence of other molecules of unrelated function is without effect. For example, enzyme molecular weights have been determined from irradiation of intact yeast

---

[7] T. L. Innerarity, E. S. Kempner, D. Y. Hui, and R. W. Mahley, *Proc. Natl. Acad. Sci. U.S.A.* **78**, 4378 (1981).

[8] T. Yamamoto, C. G. Davis, M. S. Brown, W. J. Schneider, M. L. Casey, J. L. Goldstein, and D. W. Russell, *Cell* **39**, 27 (1984).

[9] E. S. Kempner, J. H. Miller, W. Schlegel, and J. Z. Hearon, *J. Biol. Chem.* **255**, 6826 (1980).

or bacterial cells, and receptor sizes from irradiation of frozen cells from brain and liver. A sequela is that different functions will be destroyed individually and independently; therefore, separate measurements of these other functions can be performed on the same irradiated samples, yielding the target sizes for each.

In the above discussion, it is assumed that the radiation damage is directly on the functional molecular structures. This appears to be true when frozen samples are irradiated at low temperature. It is clearly not true for irradiation of samples in liquid solutions. The bulk of samples is water and the vast majority of primary ionizations occur in this solvent. At room temperature, the radiation products from water [H_3O^+, H_2O_2, etc., but primarily OH^-] can diffuse from their site of production until they chemically react with proteins and inactivate them. At very low temperatures the bulk of the ionizations are still in the aqueous (ice) phase, but the products are not free to diffuse. Generally, they react with molecules in their immediate vicinity and are thereby neutralized.

Other Types of Data and Their Interpretation

The analysis described above assumes the simplest situation where all the active molecules have the same size and for which a simple exponential decay of activity with radiation exposure is observed. More complex cases have been reported which are detected because of a nonexponential loss of activity.

Improper assay conditions may result in such nonexponential losses. Therefore, it is vital that the concentration of every constituent be tested in both control and irradiated samples to ensure that the only rate-limiting substituent is the irradiated enzyme (receptor, etc.) under study. This artifact causes almost all the "nonexponential" observations and proper adjustment of the assay converts these to the simple case described above.

Sometimes a simple exponential loss of activity is observed only in irradiated samples, i.e., the nonirradiated control value does not fit the remaining data; usually the control shows considerably greater activity than predicted from the rest of the inactivation curve. This is caused most often by an inappropriate control which was treated differently from the irradiated samples. In other cases, such complexities can sometimes be eliminated by addition of thiol reagents or EDTA to the original enzyme preparation.

Complex inactivation curves which decrease from the control value at relatively low doses of radiation but then decrease much more slowly after greater radiation doses (as shown in Fig. 2) have been seen with

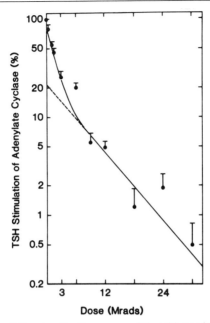

FIG. 2. Complex radiation inactivation curve of thyroid membranes. The assay is the stimulation of adenylate cyclase by thyroid stimulating hormone.[10] Data indicate a continuous loss of activity with increasing radiation exposure, but loss is not a simple exponential function of dose. Final exponential slope permits estimation of the smallest active unit. Initial curvature may indicate the presence of one or more independent larger functional structures.

enzymes and receptors.[10] They can be described mathematically as a sum of independent functions, at least one of which is assumed to be an exponential. Two situations can be described which would lead to these complex results. One is artifactual in the sense that it is due to assay conditions; the presence of a nonenzymatic reaction creating the measured product, or a nonspecific binding to small structures in the receptor preparation, would each contribute activity which would be diminished by radiation at a considerably different rate than the specific function of interest. More specific assay conditions which avoided this "background" reaction could obviate this difficulty. A second model involves the presence of two or more different-sized functional units, i.e., two different proteins, each of which independently catalyzes the same enzymatic reaction, or two receptors of differing molecular weights but binding the same ligand. The radiation inactivation of such a system would be

[10] T. B. Nielsen, Y. Totsuka, E. S. Kempner, and J. B. Field, *Biochemistry* **23**, 6009 (1984).

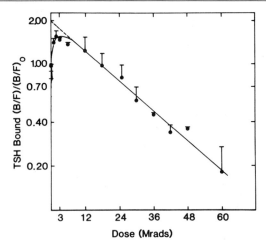

FIG. 3. Complex radiation inactivation curve for the irradiation of thyroid membranes. The binding of TSH to the receptor[10] was assayed. There is an initial increase in measurable activity after low radiation exposure, followed by a simple exponential decrease in activity. Final slope yields size of the active structure.

properly described as a sum of two or more exponentials:

$$A = B_0 e^{-bD} + C_0 e^{-cD} + \ldots \qquad (6)$$

If the radiation inactivation experiment is conducted over a sufficiently wide range of exposure, it is sometimes possible to resolve two exponential components; both their individual target sizes and their relative contributions to the initial activity can be calculated. It is unlikely that three components could be resolved in most experimental conditions.

Lastly, the complexity shown in Fig. 3 is indicated by inactivation curves that show an initial rise in activity after low radiation doses, followed by an exponential decrease in measureable function.[10] This has been interpreted as destruction of a larger-molecular-weight inhibitor[11] or activation of a silent unit.[12] Several theoretical analyses predict such curves in nonequilibrium and other conditions.[13-15]

---

[11] J. T. Harmon, C. R. Kahn, E. S. Kempner, and W. Schlegel, *J. Biol. Chem.* **255**, 3412 (1980).

[12] R. L. Kincaid, E. Kempner, V. C. Manganiello, J. C. Osborne, Jr., and M. Vaughan, *J. Biol. Chem.* **256**, 11351 (1981).

[13] A. S. Verkman, K. Skorecki, and D. A. Ausiello, *Proc. Natl. Acad. Sci. U.S.A.* **81**, 150 (1984).

[14] S. Swillens, *Biochem. J.* **222**, 273 (1984).

[15] M. Potier and S. Giroux, *Biochem. J.* **226**, 797 (1985).

The target size, obtained from radiation inactivation experiments, represents the mass of all the components necessary for the assayed function. Further interpretation of these numbers requires information obtained by other techniques; the most powerful adjunct comes from kinetic analysis of enzymes (double reciprocal plots of rates[16]) and receptor binding (Scatchard analysis of ligand binding[7]). The assumptions of complete loss of activity in radiation-damaged structures and lack of effect on other molecules lead to the predictions that a single enzyme will show a lowered V_{max} and unchanged K_m after radiation exposure, while a receptor will show a lowered B_m with unchanged K_D. These predictions have been confirmed experimentally.[7,16–18]

Multicomponent systems revealed by complex inactivation curves should be reflected by complex kinetic analyses in control samples and simpler analyses in irradiated samples after destruction of larger components. The presence of two independent, catalytically active proteins should be revealed by nonlinear Scatchard plots in the case of ligand binding, or nonlinear double reciprocal plots (v^{-1} versus $[S]^{-1}$) in the case of enzymes. The nonlinearity would be detectable if the two K_D or K_m values are sufficiently different and there are sufficient numbers of molecules of each type. When the high- and low-activity enzymes (receptors) are of different molecular weights, they will be affected differentially by radiation; the larger structures will be lost more rapidly than the smaller ones. Thus, the ratio of large to small units will decrease as radiation exposure increases, and at sufficiently high radiation doses the only surviving activity will be that due to the smaller target. Kinetic measurements in these highly irradiated samples will display only a single K_m or K_D, the one associated with the smaller unit.

Methodology

Sample Preparation

In principle, target analysis can be applied to any biological system provided the actions of radiation are confined to the molecule in which the primary ionization occurs. Indirect actions of radiation, such as inactivation of an enzyme by reaction with a radiation product produced elsewhere in the sample, make the biological activity more sensitive to radia-

---

[16] M. D. Suarez, A. Revzin, R. Narlock, E. S. Kempner, D. A. Thompson, and S. Ferguson-Miller, *J. Biol. Chem.* **259**, 13791 (1984).
[17] D. J. Fluke, *Radiat. Res.* **51**, 56 (1972).
[18] K. J. Angelides, T. J. Nutter, L. W. Elmer, and E. S. Kempner, *J. Biol. Chem.* **260**, 3431 (1985).

tion by three orders of magnitude. The main limitation comes from the presence of liquid water in the sample during irradiation, but this can be overcome by freezing the samples and in some cases by lyophilization.

Samples should contain sufficient activity so that reliable measurements of surviving function can be made after reduction to 1 or 5% of the initial value. For intact frozen cells this can require an inconveniently large mass of sample. A membrane preparation is often used in order to achieve a higher specific activity. However, it should be realized that this and other preparative procedures might alter the nature of the functional unit. Proteolysis during sample preparation is a major concern, since it could lead to the presence of several different target sizes. There is also concern when detergents are used to solubilize samples, even as a postirradiation procedure.[19] An attempt has been made to correct observed target sizes for the presence of detergents,[20] but it is not yet clear whether the same effect is involved in detergent treatment of all biological materials. Additional data and knowledge of mechanisms are needed.

Degradative and denaturing processes in sample preparation can sometimes be detected by comparison of the activity in fresh material with that in the extracted, frozen, and thawed sample. If the K_D and B_m for ligand binding are the same in fresh membranes and in the unirradiated frozen control, some confidence may be placed in the preparative procedures. As in all freezing and thawing procedures, the speed of these processes may be an important factor that is readily tested.

The choice of buffer or other solutions for sample preparation can be important in survival of activity through the freezing and thawing procedure. The measurement of transport requires special care since freezing and thawing could give rise to increased leakiness of the vesicle. Also, indirect damage from radiation products can lead to leakiness of the membrane.[21,22] The damage is separate from that on the pump molecules and appears to be on the membrane lipids. Cryoprotective agents[21–24] and reagents which neutralize the secondary radiation products have been used to minimize membrane leakiness so that transport can be measured (see 2b in the section Analysis of Samples after Irradiation). In studies of erythrocyte membrane acetylcholinesterase, lyophilization from phos-

---

[19] C. J. Steer, E. S. Kempner, and G. Ashwell, *J. Biol. Chem.* **256**, 5851 (1981).
[20] G. Beauregard and M. Potier, *Anal. Biochem.* **140**, 403 (1983).
[21] L. Hymel, A. Maurer, C. Berenski, C. Y. Jung, and S. Fleischer, *J. Biol. Chem.* **259**, 4890 (1984).
[22] B. K. Chamberlain, C. J. Berenski, C. Y. Jung, and S. Fleischer, *J. Biol. Chem.* **258**, 11997 (1983).
[23] J. Cuppoletti, C. Y. Jung, and F. A. Green, *J. Biol. Chem.* **256**, 1305 (1981).
[24] S. M. Jarvis, D. A. Fincham, J. C. Ellory, A. R. P. Paterson, and J. D. Young, *Biochim. Biophys. Acta* **772**, 227 (1984).

phate buffer was found to yield a target size twice as large as that obtained when Tris buffer was used,[25] and phosphate buffer has been found to be a problem in unpublished studies of frozen samples (Kempner).

There are some special parameters in sample preparation that are unique to radiation inactivation studies; the most important of these is sample thickness. As a beam of gamma rays or electrons transverses matter, the individual rays/particles lose energy and some stop completely. Thus, the dose delivered to the proximal layer of sample is greater than that deposited deep inside. The thickness of the sample must be kept thin enough so that there is no major change in dose delivered to any portion of the material. For ^{60}Co gamma rays, a maximum sample thickness of 1.5 cm was recommended,[26] but this value must be reduced to adjust for absorption by the sample container and any other intervening absorbers. For electrons, 0.15 cm for 1 MeV electrons and 2 cm for 10 MeV electrons would be appropriate. The container wall thickness must also be considered.

A sample parameter important to radiation studies is protein concentration. It has been observed that irradiation of samples in glass containers may lead to a loss of total protein. Charges induced on the irradiated glass are capable of binding otherwise-undamaged protein molecules so that the protein concentration in the withdrawn fluid is appreciably reduced.[27] It is not known whether this phenomenon extends to plastic and metal containers. The problem can be minimized by the inclusion of high concentrations of protein (5 mg/ml or more), either by increased quantities of sample or by the addition of some inert protein which does not interfere in the biochemical assay. Transferrin has been used for this purpose.

Addition of other proteins has been used for the purpose of providing an internal standard such as an enzyme of known radiation target size. Endogenous enzymes (or other active molecules) native to the preparation are even better markers. These offer the possibility of an independent measure of radiation dose within each sample and also a detector for any abnormal conditions to which it may have been exposed.

Most of the early studies of radiation analysis utilized lyophilized samples containing enzymes, viruses, or receptors. Subsequently, the use of frozen samples became more common, but discrepancies between results of membranes that were irradiated either frozen or lyophilized appeared in the literature. The membrane-bound Na$^+$,K$^+$-ATPase was

---

[25] D. Parkinson and B. A. Callingham, *Radiat. Res.* **90,** 252 (1982).
[26] J. T. Harmon, T. B. Nielsen, and E. S. Kempner, this series, Vol. 117, p. 65.
[27] G. C. Ness, M. J. McCreery, C. E. Sample, M. Smith, and L. C. Pendleton, *J. Biol. Chem.* **260,** 16395 (1985).

reported to yield very different inactivation curves under the two conditions.[28,29] A comparison of receptor target analyses showed that lyophilized samples usually yielded target sizes considerably larger than those from frozen samples.[18,30-34] In all cases where the protein monomer size was independently known, good agreement was obtained with the target size observed in frozen samples. Lai et al.[33] have suggested that receptor aggregation may occur on lyophilization and lead to aberrantly large target sizes. Therefore, in any radiation experiment lyophilized membrane preparations should be used only with caution.

Physical Conditions for Irradiation

Several parameters dictate the physical conditions for radiation exposure. The most important is the nature of the radiation source. Different arrangements of samples will be needed when gamma rays or electrons are provided as a collimated beam, as from an accelerator, or if they are uniformly distributed in space, as in some ^{60}Co sources. Temperature control of samples and maintenance of a specific atmosphere surrounding the samples impose further restrictions on the experimental arrangement. The sample containers should be of uniform thickness and of known radiation absorption characteristics. The radiation absorption of all materials which intervene between the radiation source and the experimental material must be accounted for.

Radiation exposure of air leads to formation of considerable amounts of ozone which reacts strongly with organic materials and could therefore be classed as a secondary effect of radiation. It is prudent to irradiate samples under conditions where there is only limited exposure to potential ozonolysis. This is partially obtained by controlling the gas phase in the sample containers. Vacuum,[35] nitrogen,[35,36] or restricted volume of air[26] have been used.

---

[28] P. Ottolenghi and J. C. Ellory, *J. Biol. Chem.* **258**, 14895 (1983).
[29] S. J. D. Karlish and E. S. Kempner, *Biochim. Biophys. Acta* **776**, 288 (1984).
[30] L.-R. Chang and E. A. Barnard, *J. Neurochem.* **39**, 1507 (1982).
[31] S. M. Paul, E. S. Kempner, and P. Skolnick, *Eur. J. Pharmacol.* **76**, 465 (1981).
[32] R. D. Schwartz, J. W. Thomas, E. S. Kempner, P. Skolnick, and S. M. Paul, *J. Neurochem.* **45**, 108 (1985).
[33] F. A. Lai, E. L. Newman, E. Peers, and E. A. Barnard, *Eur. J. Pharmacol.* **103**, 349 (1984).
[34] S. Ott, T. Costa, B. Hietel, W. Schlegel, and M. Wüster, *Arch. Pharmacol.* **324**, 160 (1983).
[35] S. M. Jarvis, J. D. Young, and J. C. Ellory, *Biochem. J.* **190**, 373 (1980).
[36] C. Y. Jung, *in* "Molecular and Chemical Characterization of Membrane Receptors" (C. Venter, ed.), pp. 193-208. Liss, New York, 1984.

Dose Delivery and Measurement

Since all target analyses are dependent on measurements of biological function and of radiation dose, both are crucially important. The measurement of radiation dose is somewhat unfamiliar to most biochemists and needs to be considered carefully. Physical and chemical radiation detectors each have specificity for the type of radiation and its energy. Among the more common dosimeters are calorimetric devices, ionization chambers, X-ray film, thermoluminescent dosimeters (TLDs), poly(methyl methacrylate) (Perspex), and radiochromic dye films. Usually these are compared with the ferrous sulfate chemical (Fricke) dosimeter which is used as a standard. It has been suggested that the inactivation of experimental biochemical processes can be studied in comparison to the inactivation of known viruses or enzymes[37-39] and thereby avoid direct measurement of radiation dose.[39] This approach is arduous, but valid provided that the radiation target sizes of the standards are well known and correct.

The total dose delivered per sample is usually varied by the length of time of exposure in a uniform radiation field. An alternative technique is to utilize the variation in dose rate at different positions in an inhomogeneous radiation field while exposing all samples for the same length of time.[40,41] This procedure is less facile and popular since it requires more extensive dosimetry than the time-variation method.

Further details about sample preparation and physical conditions can be found in articles by Beauregard *et al.*,[42] Jung,[36] and Harmon *et al.*[26]

Temperature Control and Corrections

Target size determinations vary with the temperature during irradiation.[43] The same temperature coefficient has been observed for 25 enzymes,[44] including membrane components and viruses. The quantitative temperature variation is accurately known and appropriate correction (S_t) can be made; Eq. (5) is modified

$$\text{MW} = 6.4 \times 10^{11} k S_t$$

---

[37] J. W. Preiss, M. Belkin, and W. G. Hardy, *J. Natl. Cancer Inst.* **27**, 1115 (1961).
[38] D. J. Fluke, *Radiat. Res.* **28**, 336 (1966).
[39] M. M. S. Lo, E. A. Barnard, and J. O. Dolly, *Biochemistry* **21**, 2210 (1982).
[40] Lübbecke, D. R. Ferry, H. Glossman, E.-L. Sattler, and G. Doell, *Naunyn-Schmeideberg's Arch. Pharmacol.* **323**, 96 (1983).
[41] S. Uchida, K. Matsumoto, K. Takeyasu, H. Higuchi, and H. Yoshida, *Life Sci.* **31**, 201 (1982).
[42] G. Beauregard, S. Giroux, and M. Potier, *Anal. Biochem.* **132**, 362 (1983).
[43] D. J. Fluke, *Radiat. Res.* **28**, 677 (1966).
[44] E. S. Kempner and H. T. Haigler, *J. Biol. Chem.* **257**, 13297 (1982).

When the temperature data are presented as an Arrhenius plot, a curved line is always seen.[38] This indicates that even in an individual protein the temperature effect is not due to a unique process with a single energy of activation. One underlying mechanism is the creation of free radicals.[45]

When corrected for this temperature dependence, target sizes give molecular weight values which accurately correspond to known polypeptide sizes. Thus, it is important to understand and account for this parameter. Two alternative procedures are widely used. Stringent measurement and control of temperature during irradiation[26,36,43] and subsequent correction of data[44] is the most direct approach and has been described in detail.[26] Some workers have used temperature-dependent dosimeters, either the above-mentioned use of enzymes as dosimeters[39] or the use of conventional dosimeters which are temperature-sensitive. In the latter case it is necessary to establish that the temperature variation of the dosimeter is the same as that for the biological materials under study.[36]

Use of Internal Standards

The use of internal standards is highly recommended as a control that the proper radiation dose has been used. It is important to ascertain that the enzyme which is added to the sample remains the proper size throughout the manipulations of target irradiation. Also, it is desirable to match the size of the internal standard to be in the same range as the unknown. Glucose-6-phosphate dehydrogenase from *Leuconostoc mesenteroides* has been found to be a reliable internal standard for radiation inactivation studies of membranes in the frozen state.[46]

Analysis of Samples after Irradiation

1. It has been our experience that frozen samples after irradiation can be stored for days at liquid nitrogen temperature. However, once the sample has been thawed, the most labile activities should be assayed right away with the remainder of the sample promptly refrozen after removal of the aliquot. Storage of the sample on ice for even a few hours can result in loss of function. Sample instability is variable and probably reflects the action of products of ionizing radiation such as free radicals. The addition of sulfhydryl reagents or metal chelators can help to stabilize the sample, e.g., 1 mM dithiothreitol or EDTA. As for the assay of nonirradiated

---

[45] E. S. Kempner, R. Wood, and R. Salovey, *J. Polym. Sci., Polym. Phys. Ed.* **24**, 2337 (1986).
[46] J. O. McIntyre and P. Churchill, *Anal. Biochem.* **147**, 468 (1985).

samples, the addition of any reagent must be checked with regard to its effect on functional characteristics. For example, dithiothreitol can dissociate disulfide bridges and has been found to alter nitrendipine binding,[47] and EDTA at pH 8 and higher can increase the permeability of the membrane system.[48] The choice of assay should take into consideration the possible lability of the irradiated sample.

2. Any functional characteristic can be used to obtain the target size of a membrane component (although see the section General Principles regarding general principles and interpretation). See Table II.

a. Enzymatic activity: This is the most generally applied assay. Due to the problem of sample instability after irradiation, it is appropriate to select a rapid assay and to confirm that the measurement is linear as a function of time. A rapid loss of rate which would otherwise be linear is a hallmark of damage from free radicals.

b. Transport: The assay of transport has thus far been limited to only several studies. The problem with this assay is that irradiated membranes can become leaky and therefore the transport measurement is no longer a characteristic of the transporter, per se. The problem has been overcome by solubilizing the sample after irradiation and reconstituting a nonleaky membrane.[49]

c. Binding assay: The binding assay has been especially useful for measuring receptors and channels. A variety of ligands, both agonists and antagonists, are available for assay of some channels.

d. Polypeptide integrity: The amount of polypeptide remaining can be measured readily by SDS–PAGE. The quantitation is then carried out by densitometry on gels stained with either Coomassie Brilliant Blue or silver staining. It is important to load different amounts of sample onto the gel and to establish the concentration range that gives a linear response. The application of this method by necessity has been limited to native membranes where a constituent is present in substantial amount, or to the purified and reconstituted system. More recently, immunoprecipitation with specific antibody has been used prior to electrophoresis for quantitation of a specific polypeptide remaining after irradiation.[50] In principle, antibodies also can be used to quantitate a minor component of the membrane directly on the gel. Such procedures now permit target analysis of the structure of even minor components in a complex membrane.

---

[47] W. R. Brandt, R. W. Kawamoto, and A. H. Caswell, *Biochem. Biophys. Res. Commun.* **127,** 205 (1985).

[48] P. F. Duggan and A. Martonosi, *J. Gen. Physiol.* **56,** 147 (1970).

[49] T. Goldkorn, G. Rimon, E. S. Kempner, and H. R. Kaback, *Proc. Natl. Acad. Sci. U.S.A.* **81,** 1021 (1984).

[50] P. A. Edwards, E. S. Kempner, S.-F. Lan, and S. K. Erickson, *J. Biol. Chem.* **260,** 10278 (1985).

General Principles

The membrane component may be part of a multicomponent system which carries out a complex function. The assay of a simple characteristic may involve only one polypeptide, whereas measurement of the holofunction may involve several constituents of the complex. Therefore, the specific assay(s) employed could lead to different target sizes, reflecting the number of constituents involved in partial versus holofunction. The target size of a multimer consisting of same (oligomer) or different subunits (complex) is a measure of the combined mass of the polypeptides required to carry out the measured function.

Analysis of Target Inactivation Data

Two types of measurements can be made: (1) biological activity and (2) integrity of the polypeptide. These form the basis of functional versus structural analysis of size (see Table I). In this context, it should be recognized that the functional size (active unit) is not necessarily equivalent to the structural size, and vice versa.

(1) Functional analysis: When all subunits are essential to the function of a multimer, then destruction of any polypeptide would lead to loss of function and the target size would be that of the multimer. However, a monomeric size can be indicated for function even though the structural unit may be multimeric.

(2) Structural analysis from polypeptide fragmentation: A single hit of a polypeptide could lead via energy transfer to the fragmentation of other subunits of the complex.[51] This type of result would mean that the structural unit is a multimer, but it would not necessarily be relevant to the size of the functional unit. In actuality, measurement of a target size by functional analysis could be invalid when energy transfers occurs, since destruction of other polypeptides in the structural unit obtains.

Monomeric versus Oligomeric Target Size

Four permutations are possible based on measurement of polypeptide and function remaining (see Table I).

---

[51] The precise basis for energy transfer between polypeptides is not yet understood. Clearly, there must be sufficient intimate association between polypeptides for energy transfer to occur. This might find explanation in terms of surface complementarity of the polypeptides and/or the nature of the noncovalent association. It is known in the case of ricin that the transfer of energy between polypeptides is by way of dithiol linkage.[52] However, transfer of energy between polypeptides can occur in the apparent absence of such association.[21,22,53–56]

[52] H. T. Haigler, D. J. Woodbury, and E. S. Kempner, *Proc. Natl. Acad. Sci. U.S.A.* **82**, 5357 (1985).

[53] J. O. McIntyre, P. Churchill, A. Maurer, C. J. Berenski, C. Y. Jung, and S. Fleischer, *J. Biol. Chem.* **258**, 953 (1983).

TABLE I
PERMUTATION OF MONOMERIC VERSUS OLIGOMERIC SIZE BY TARGET ANALYSIS[a]

Assay	Monomeric or oligomeric target size			
	Class: 1	2	3	4
Function	Monomer	Monomer	Oligomer	Oligomer
Polypeptide remaining (PAGE)	Monomer	Oligomer	Monomer	Oligomer

[a] Based on measurement of function and polypeptide remaining after irradiation.

Class 1: A monomeric target size is indicated by functional assay as well as by the polypeptide remaining. The functional unit is the monomer although the structural unit could be an oligomer. An example is lactate dehydrogenase,[57] which structurally is a tetramer.

Class 2: A monomeric target by activity and oligomeric size structurally. There is no instance of a monomeric target size based on function and oligomer by polypeptide remaining. This would not be likely, since fragmentation of the polypeptides generally would be expected to result in loss of function.

Class 3: An oligomeric size is indicated by enzyme assay, whereas a monomer is obtained by assay of polypeptide remaining. This case, where energy transfer does not occur or lead to fragmentation, gives a unique interpretation. The functional unit is the oligomer and the subunits are assembled as an oligomer. This type of result could be explained by (a) instability of the other subunits of the complex when a single subunit has been hit, or (b) some energy transfer from the target polypeptide which is sufficient to inactivate but not to fragment the other subunits. Glutamate dehydrogenase[58] and hydroxymethylglutaryl (HMG)-CoA reductase[50] are examples of this class.

Class 4: An oligomeric target size is obtained by measurement of both function and polypeptide remaining. Examples include a glucose transporter,[59] two types of calcium pump protein,[21,22] the proton pump of *Neurospora* plasma membrane,[54] the K^+-H^+ pump of the gastric mucosa,[55] and D-β-hydroxybutyrate dehydrogenase.[53] Such a result is obtained when energy from a single hit on one polypeptide has been transferred to other polypeptides of the multimer. Members of this class are structurally

---

[54] J. Bowman, C. J. Berenski, and C. Y. Jung, *J. Biol. Chem.* **260**, 8726 (1985).
[55] G. Saccomani, G. Sachs, J. Cuppoletti, and C. Y. Jung, *J. Biol. Chem.* **256**, 7727 (1981).
[56] J. O. McIntyre, E. S. Kempner, and S. Fleischer, unpublished observations.
[57] D. J. Fluke and P. W. Hochachka, *Radiat. Res.* **26**, 395 (1965).
[58] E. S. Kempner and J. H. Miller, *Science* **222**, 586 (1983).
[59] M. Takahashi, P. Malathi, H. Preiser, and C. Y. Jung, *J. Biol. Chem.* **260**, 10551 (1985).

oligomeric, albeit the oligomer is not necessarily the functional unit. Separate kinetic evidence then would be required to assess cooperative behavior of the subunits of the oligomer.

A similar analysis has been developed independently by Potier and Beauregard.[60]

Complexes (Association of Nonidentical Polypeptides)

We next consider the case of a complex consisting of nonidentical polypeptides. Similar interpretation would pertain to a complex as for an oligomer. If only a single subunit polypeptide is essential for function, its destruction would result in a target size equivalent to the subunit molecular weight. If all (or a portion of) the subunits were essential, the molecular weight of the total (or portion of the) complex would be obtained. The target size of the complex also could be an oligomer, depending on which is the functional unit. Cytochrome oxidase,[16] adenylate cyclase,[61] and the Na^+-K^+ pump[29] are examples of studies with complexes. For cytochrome oxidase from bovine heart, which is isolated as a complex consisting of 10 or more different polypeptides, the target size obtained by measuring electron transport was equivalent to only a portion of the size of the complex. This result means that not all of the peptides in the complex are essential for the function that was assayed. One example of analysis (by SDS–PAGE) of polypeptide remaining in a complex is the Na^+-K^+ pump. By this technique, the target size was found to be equal to the α-subunit.

Domains

In a few cases, target sizes have been reported which were smaller than the monomeric polypeptide. This means that after radiation exposure, there remain fragments of the original polypeptide which can carry out these activities. It is not yet known whether this represents a radiation-resistant domain or an ensemble of different-sized fragments which retain activity. The measured functions include the binding of antibodies[62,63] or ligands[64-66] and also enzymatic reactions.[67-70] The antibody reac-

[60] M. Potier and G. Beauregard, *Methods Biochem. Anal.* **32**, 313 (1987).
[61] W. Schlegel, E. S. Kempner, and M. Rodbell, *J. Biol. Chem.* **254**, 5168 (1979).
[62] C. Fewtrell, E. Kempner, G. Poy, and H. Metzger, *Biochemistry* **20**, 6589 (1981).
[63] T. Olivecrona, G. Bengtsson-Olivecrona, J. C. Osborne, Jr., and E. S. Kempner, *J. Biol. Chem.* **260**, 6888 (1985).
[64] D. E. Richards, J. C. Ellory, and I. M. Glynn, *Biochim. Biophys. Acta* **648**, 284 (1981).
[65] A. S. Verkman, K. L. Skorecki, C. Y. Jung, and D. A. Ausiello, *Am. J. Physiol.* **251**, C541 (1986).
[66] S. G. McGrew, R. J. Boucek, Jr., J. O. McIntyre, C. Y. Jung, and S. Fleischer, *Biochemistry* **26**, 3183 (1987).
[67] L. P. Solomonson and M. J. McCreery, *J. Biol. Chem.* **261**, 806 (1986).

tions apparently detect epitopes, and an analogous explanation would fit the ligand binding. Among the enzymes, the measured parameter apparently represents a partial function. Target sizes for monomer destruction, and where known, the target size for holofunction correlate with one or more complete polypeptide subunits.

Tabulation of Data in Literature

Table II summarizes some of the key literature on target analysis of membrane proteins. Growth factors and hormone receptors have been recently reviewed[6,71] and will not be repeated here. We have been somewhat selective in the data presented. Some values at variance with those in Table II have been excluded when it appeared that they were technically flawed. Table II includes values obtained on samples prepared by lyophilization as well as freezing. Sometimes the same target values were obtained. Most often the data obtained on the lyophilized samples exhibited large scattering and yielded much larger target sizes. It is our view that the values obtained using frozen samples are more reliable (see Sample Preparation in the Methodology section, above).

While most analyses concerned biochemical activity (enzymatic activity and ligand binding), fragmentation of the polypeptide has also been reported for a number of different systems. The loss of staining intensity of the monomer band after different radiation exposures was used to determine the target size based on polypeptide remaining. Of eight experiments listed, three showed target sizes equal to that of the subunit polypeptide and five showed higher multiples. These latter results show that a primary ionization in one subunit can lead to energy transfer and destruction of other polypeptides (see General Principles, above and Ref. 51). Such results have been obtained for a number of membrane proteins including the calcium pump protein from skeletal muscle and cardiac SR, and the H^+-K^+ pump of the gastric mucosa plasma membrane as well as D-β-hydroxybutyrate dehydrogenase, a lipid-requiring enzyme. In a number of instances, the purified and reconstituted preparations were also studied and gave the same target size as that in the native membrane.[21,53,59]

---

[68] M. J. Barber, L. P. Solomonson, and M. J. McCreery, *Arch. Biochem. Biophys.* **256**, 260 (1987).

[69] P. Anataki, N. Vigneault, G. Beauregard, M. Potier, and K. D. Roberts, *Biol. Reprod.* **37**, 249 (1987).

[70] T. Kume and I. Ishigaki, *Biochim. Biophys. Acta* **914**, 101 (1987).

[71] W. Schlegel and E. S. Kempner, in "Investigation of Membrane-Located Receptors" (E. Reid, G. M. W. Cook, and D. J. Morre, eds.), pp. 47–57. Plenum, New York, 1984.

TABLE II
RADIATION STUDIES OF MEMBRANE PROTEINS[a]

Protein	Source	Monomer size(s) (in thousands)	Assay	Target size (kDa) Frozen	Lyophilized	Ref.
I. Transporters (primary pumps)						
A. ATP-driven pumps						
1. H^+-ATPase	*Neurospora* PM	104	H^+-ATPase	230	—	54
2. H^+,K^+-ATPase	Pig gastric mucosa PM	100	H^+,K^+-ATPase	270	320	55
	Pig gastric mucosa PM		H^+,K^+-ATPase	—	444	73
	Pig gastric mucosa PM		Gel band	250	—	55
3. Calcium pump protein (Ca^{2+}-ATPase)	Rabbit SkM SR	115	Ca^{2+}-ATPase	214	—	21
	Rabbit SkM SR		Ca loading	251	—	21
	Rabbit SkM SR		Gel band	213	—	21
	Rabbit SkM SR reconstituted		Ca^{2+}-ATPase	224	—	21
	Rabbit SkM SR reconstituted		Gel band	230	—	21
4. Calcium pump protein	Rabbit SkM SR		Ca^{2+}-ATPase	240	—	54
	Dog cardiac SR	115	Ca^{2+}-ATPase	213	—	22
	Dog cardiac SR		Ca loading	223	—	22
	Dog cardiac SR		CPP gel band	229	—	22
5. Calcium pump protein	Human RBC PM	138	Ca^{2+}-Mg^{2+}-ATPase	—	290	74
	Human RBC PM		Ca^{2+},Mg^{2+}-ATPase	—	251	75
	Human RBC PM		Ca^{2+},Mg^{2+}-ATPase	290	—	76
	Human RBC PM	145	Ca^{2+}-ATPase (calmodulin dependent)	317	—	76
6. Mg^{2+}-ATPase[b]	Human RBC PM	—	Mg^{2+}-ATPase	224	—	76

Enzyme	Source	(value)	Form	(value)	(value)	(value)
7. Na+,K+-ATPase	Rabbit kidney PM	α = 115; β = 55	Na+,K+-ATPase	—	332	55
	Purified from pig kidney microsomes		Na+,K+-ATPase	250	—	77
	Purified from pig kidney microsomes		Gel band	(α)184	—	77
	Guinea pig kidney microsomes		Na+,K+-ATPase	—	262[c]	78
	Guinea pig kidney microsomes		Na+,K+-ATPase	—	264[c]	28
	Guinea pig kidney microsomes		Na+,K+-ATPase	184[c]	—	29
	Guinea pig kidney microsomes		Gel band	(α)128	—	29
	Guinea pig kidney microsomes		ATPase	—	190	3
	Human RBC ghosts		ATPase	200	300	3
B. Reversible ATP synthetase-proton pump (F_1F_0)						
1. H+-ATPase (F_1)	E. coli PM	382	ATPase	320	—	49
	Beef heart mitochondria		ATPase	260	270	79
II. Electron transport machinery and related redox enzymes						
1. Cytochrome oxidase	Purified from beef heart and rat liver	5 to 57 (total 200)	EA	114	—	16
2. NADH oxidase	Micrococcus PM		EA	47	—	80
NADH dehydrogenase	Micrococcus PM	50	EA	49	—	80
	Micrococcus PM		EA	70	—	81
3. D-Lactate dehydrogenase	E. coli PM	65	EA	58	—	49

(continued)

TABLE II (continued)

Protein	Source	Monomer size(s) (in thousands)	Assay	Target size (kDa) Frozen	Target size (kDa) Lyophilized	Ref.
4. D-β-Hydroxybutyrate dehydrogenase	Beef heart mitochondrial Ves	31	EA	106	—	53
	Beef heart mitochondrial Ves, purified and reconstituted		EA	119	—	53
	Beef heart mitochondrial Ves, purified and reconstituted		Gel band	104	—	53
5. Malate dehydrogenase	*Micrococcus* PM	—	EA	73	—	81
III. Other transporters						
1. Glucose carrier	Human RBC ghosts	46	Cytochalasin B	220	—	82
	Human RBC		D-Glucose transport	185	—	23
2. Glucose carrier–Na$^+$ symport	Rabbit kidney BBM	—	D-Glucose transport	1000	—	59
	Rabbit kidney BBM		D-Glucose transport, purified/reconstituted	328/358	—	59
	Rabbit kidney BBM	165	Gel band	298	—	59
	Calf kidney M Ves		D-Glucose exchange	345	—	83
	Calf kidney M Ves		Phlorizin B	230	—	83
	Rabbit kidney BBM		Phlorizin B (2 sites)	>4000 and 110	—	84
3. Lactose carrier	*E. coli* PM Ves reconstituted liposomes	47	Lactose transport (not energized/energized)	50/100	—	49

4. L-Amino acid transporter	Human RBC		Leucine uptake	100	—	85
5. L-Alanine transporter	Rabbit kidney BBM	—	Alanine uptake	1200	—	59
6. Nucleoside transporter	Human RBC	55?	Uridine influx and exchange	122	—	86
	Human RBC ghosts		Nitrobenzyl-thioinosine B	—	122	35
IV. Ion channels						
1. Calcium channel	Rabbit SM TT	213 (α = 130, β = 50, γ = 33)	Nitrendipine B	—	210	87
	Guinea pig SM M		Nitrendipine B	278	—	88
	Guinea pig brain M		Nimodipine B	185	—	89
	Guinea pig SM TT		Nimodipine B	179	—	90
	Guinea pig SM TT		Diltiazem B	130	—	90
	Guinea pig SM TT		Desmethoxy-verapamil B	107	—	90
	Guinea pig SM TT		Verapamil B	110	—	91
	Rat brain synaptosomes		Saxitoxin B	224	240	18
2. Sodium channel (PM)	Electric eel, pig brain	α = 190; β_1 = 39; β_2 = 37	Tetrodotoxin B	—	229	92
	Rat brain synaptosomes		Tetrodotoxin B	—	260	93
	Rat brain synaptosomes		CSS toxin II B	—	266	93
	Rat brain synaptosomes		CSS toxin II B	45	190; 360	18
	Rat brain synaptosomes		TSS toxin γ B	—	266	93
	Rat brain synaptosomes		Lqq toxin V B	263	—	18
	Rat brain synaptosomes		Batrachotoxin B, (2 sites)	287/51	472	18

(*continued*)

TABLE II (continued)

Protein	Source	Monomer size(s) (in thousands)	Assay	Target size (kDa) Frozen	Target size (kDa) Lyophilized	Ref.
3. Potassium channel, (Ca-dependent)	Rat brain synaptosomes	—	Apamin B	—	250	94
4. Acetylcholine receptor (PM)	Electric eel and muscle	255 [α_2 = (40)$_2$, β = 50, γ = 60, δ = 65]	α-Bungarotoxin B	—	300	39
V. Membrane-bound proteins						
1. Acetylcholinesterase[d]	Electric eel, purified	70	EA	240	255	95
	Electric eel, purified		EA?	248	—	96
	Electric eel, purified		EA	260	—	55
	Electric eel, purified		Gel band	220	—	55
	Electric eel M		EA	77	75; 72	92, 95
	Rat brain synaptosomes		EA	70	—	18
	Bovine RBC	72	EA	—	73	97
	Llama RBC ghosts		EA	71	75	95
	Human RBC M		EA	—	74	98
	Human RBC ghosts		EA	71	—	76
	Human RBC ghosts (2 buffers)	80	EA	—	67(Tris)/130 (PO$_4$)	25
	Human RBC ghosts		EA	—	70	99
2. HMG-CoA reductase[e]	Rat liver cells and microsomes	97	EA	200	—	50
	Rat liver cells		Gel band	100	—	50
	Rat liver microsomes (normal/dietary stim.)		EA	200/100	—	27

#	Enzyme	Source		Method/Substrate			Ref
3.	Alkaline phosphatase	Rabbit kidney BBM Ves		EA	67	—	84
		GP kidney microsomes		EA	70	—	3
		Calf intestine, purified		EA	72	—	100
		Human liver plasma M	29	EA	129	—	101
		Human fibroblast		EA	85	—	7
4.	5′-Nucleotidase (PM)	Rabbit kidney BBM Ves	85	EA	58	—	84
		Rat brain synaptosomes		EA	90	—	18
5.	Adenylate cyclase (catalytic unit)	Rat liver PM	—	EA	—	160	102
6.	γ-Glutamyltransferase	Rat liver PM		EA	150	—	61
		Turkey RBC PM		EA	92	—	103
		Bovine thyroid PM		EA (2 sites)	150/85	—	10
		Rat liver microsomes	40 + 20	Transferase/hydrolase	—	99/92	104
		Rat hepatoma microsomes		Transferase/hydrolase	—	102/107	104
7.	N-Acetylhexosaminidase	Rat kidney BBM		EA	—	93(112^f)	20
		Calf kidney BBM		EA	74	—	83
		Human spleen M	—	EA	—	101(99^f)	105
		Var tissue culture cells		EA	—	110	106
8.	Sphingomyelinase	Human spleen M	70	EA	—	86(90^f)	105
		Var tissue culture cells		EA	—	105	106
9.	β-Glucosidase	Human spleen M	66	EA	—	98(53^f)	105
		Human spleen M		EA	—	78(72^f)	107
		Var tissue culture cells		EA	—	120	106
10.	UDP-glucuronyltransferase	Rat liver microsomes	—	EA (different substrates)	—	41 to 175	108
11.	Monoamine oxidase	Rat liver mitochondria	55 and 60	EA (5HT/benzylamine)	—	143/203	109
		Rat liver mitochondria, outer M		EA (phenylethylamine/serotonin)	—	300/300	110

(*continued*)

TABLE II (continued)

Protein	Source	Monomer size(s) (in thousands)	Assay	Target size (kDa) Frozen	Target size (kDa) Lyophilized	Ref.
12. Steroid sulfatase	Human placenta microsomes	72	EA	—	533(79^f)	111
13. Neuraminidase	Rat liver lysosomal M	—	EA	—	56	112
	Human placenta M	—	EA	—	63	113
	Human leucocyte M	—	EA	—	240/203	114
14. Phosphodiesterase	Bovine retinal rod, outer segment	88+85+11	EA	—	176	98
15. Rhodopsin	Bovine retinal rod, outer segment	39	ΔOD_{500}	—	20	115
	Bovine retinal rod, outer segment		Opsin survival (dark/light)	—	47/44	115
16. Lipoprotein lipase	Rat adipose homogenate	—	EA	127	—	116

^{a} B, binding; BBM, brush border plasma membrane; CPP, calcium pump protein; EA, enzyme activity; M, membrane; PAGE, polyacrylamide gel electrophoresis; PM, plasma membrane; RBC, erythrocytes; SkM, skeletal muscle; SM, smooth muscle; SR, sarcoplasmic reticulum; TT, transverse tubules; var, various; Ves, vesicles. Monomer size from SDS gels is exclusive of carbohydrate and lipids.
^{b} This activity has not been associated as yet with a pump molecule.
^{c} A recent study finds that the Na^+-K^+ pump, as the $\alpha + \beta$ monomer solubilized in detergent, has good Na^+,K^+-ATPase activity [P. L. Jørgenson and J. P. Andersen, *Biochemistry* **25**, 2889 (1986)].
^{d} Membrane-bound acetylcholinesterase displays a monomer target size, but purified preparations show a target size four times larger.
^{e} Dietary manipulation has been found to modify the kinetic behavior of hydroxymethylglutaryl-CoA reductase so that the Hill coefficient is shifted from two in the normal to one [J. Roitelman and I. Schechter, *J. Biol. Chem.* **261**, 5061 (1986)].
^{f} In Triton.

A number of observations based on target analysis are worth noting: (1) Membrane proteins (perhaps most) have been documented to exist as oligomers in the native membrane (compare target size with monomer size). (2) Carbohydrate was found not to contribute to the target size of glycoproteins.[72] (3) Phospholipid was found not to contribute to the target size (see especially Ref. 53). (4) Two studies have been carried out to correlate oligomeric size with functional state. The calcium pump protein of skeletal muscle sarcoplasmic reticulum was found to be a dimer in the membrane in three different states of the calcium pump cycle.[21] In another study, the lactose carrier from *E. coli* membrane vesicles (item III-3 in Table II[73–116]) gave a target size of a dimer when the vesicles were

[72] E. S. Kempner, J. H. Miller, and M. J. McCreery, *Anal. Biochem.* **156,** 140 (1986).
[73] J. J. Schrijen, W. A. H. M. van Groningen-Luyben, H. Nauta, J. J. H. H. M. de Pont, and S. L. Bonting, *Biochim. Biophys. Acta* **731,** 329 (1983).
[74] A. M. Minocherhomjee, G. Beauregard, M. Potier, and B. D. Roufogalis, *Biochem. Biophys. Res. Commun.* **116,** 895 (1983).
[75] J. D. Cavieres, *Biochim. Biophys. Acta* **771,** 241 (1984).
[76] L. Hymel, M. Nielsen, and K. Gietzen, *Biochim. Biophys. Acta* **815,** 461 (1985).
[77] J. Jensen and P. Ottolenghi, in "The Sodium Pump" (I. M. Glynn and J. C. Ellory, eds.), pp. 77–81. Company of Biologists, Cambridge, England, 1985.
[78] P. Ottolenghi, J. C. Ellory, and R. A. Klein, *Curr. Top. Membr. Transp.* **19,** 139 (1983).
[79] Y. Kagawa, *J. Biochem. (Tokyo)* **65,** 925 (1969).
[80] M. E. Zinov'eva, A. S. Kaprel'yants, and D. N. Ostrovskii, *Biokhimiya* **48,** 1319 (1983).
[81] D. N. Ostrovskii, I. M. Tsfasman, and N. S. Gelman, *Biokhimiya* **34,** 993 (1969).
[82] C. Y. Jung, T. L. Hsu, J. S. Hah, C. Cha, and M. N. Haas, *J. Biol. Chem.* **255,** 361 (1980).
[83] J.-T. Lin, K. Szwarc, R. Kinne, and C. Y. Jung, *Biochim. Biophys. Acta* **777,** 201 (1984).
[84] R. J. Turner and E. S. Kempner, *J. Biol. Chem.* **257,** 10794 (1982).
[85] D. A. Fincham, J. D. Young, and J. C. Ellory, *Biochem. Soc. Trans.* **13,** 229 (1985).
[86] S. M. Jarvis, D. A. Fincham, J. C. Ellory, A. R. P. Paterson, and J. D. Young, *Biochim. Biophys. Acta* **772,** 227 (1984).
[87] R. I. Norman, M. Borsotto, M. Fosset, M. Lazdunski, and J. C. Ellory, *Biochem. Biophys. Res. Commun.* **111,** 878 (1983).
[88] J. C. Venter, C. M. Fraser, J. S. Schaber, C. Y. Jung, G. Bolger, and D. J. Triggle, *J. Biol. Chem.* **258,** 9344 (1983).
[89] D. R. Ferry, A. Goll, and H. Glossman, *Arch. Pharmacol.* **323,** 292 (1983).
[90] A. Goll, D. R. Ferry, J. Streissnig, M. Schober, and H. Glossmann, *FEBS Lett.* **176,** 371 (1984).
[91] A. Goll, D. R. Ferry, and H. Glossmann, *Eur. J. Biochem.* **141,** 177 (1984).
[92] S. R. Levinson and J. C. Ellory, *Nature (London), New Biol.* **245,** 122 (1973).
[93] J. Barhanin, A. Schmid, A. Lombet, K. P. Wheeler, M. Lazdunski, and J. C. Ellory, *J. Biol. Chem.* **258,** 700 (1983).
[94] H. Schmid-Antomarchi, M. Hugues, R. Norman, C. Ellory, M. Borsotto, and M. Lazdunski, *Eur. J. Biochem.* **142,** 1 (1984).
[95] S. R. Levinson and J. C. Ellory, *Biochem. J.* **137,** 123 (1974).
[96] C. M. Fraser and J. C. Venter, *Biochem. Biophys. Res. Commun.* **109,** 21 (1982).
[97] G. Beauregard, M. Potier, and B. D. Roufogalis, *Biochem. Biophys. Res. Commun.* **96,** 1290 (1980).
[98] S. M. Hughes and M. D. Brand, *Biochemistry* **22,** 1704 (1983).

energized by electron transport (protonmotive force was imposed).[49] When the electrochemical gradient was collapsed, the target size obtained was the monomer.

Concluding Remarks

This is the first overview for the use of target inactivation analysis to determine the size of membrane proteins. The application of this method has already become an important part of the literature of membrane proteins. Two special features of the method are that it permits (1) analysis of a function of a component without its isolation, and (2) the study of the multimeric state of constituents in the native membrane. For optimal interpretation, it is desirable to correlate both biological activity and polypeptide remaining with radiation dose. Quantitation of polypeptide remaining by SDS–PAGE is now possible even for a minor constituent in a

---

[99] B. R. Martin, J. M. Stein, E. L. Kennedy, C. A. Dorberska, and J. C. Metcalfe, *Biochem. J.* **184**, 253 (1979).

[100] R. J. Pollet, E. S. Kempner, M. L. Standaert, and B. A. Haase, *J. Biol. Chem.* **257**, 894 (1982).

[101] A. L. Schwartz, C. J. Steer, and E. S. Kempner, *J. Biol. Chem.* **259**, 12025 (1984).

[102] M. D. Houslay, J. C. Ellory, G. A. Smith, T. R. Hesketh, J. M. Stein, G. B. Warren, and J. C. Metcalfe, *Biochim. Biophys. Acta* **467**, 208 (1977).

[103] T. B. Nielsen, P. M. Lad, M. S. Preston, E. Kempner, W. Schlegel, and M. Rodbell, *Proc. Natl. Acad. Sci. U.S.A.* **78**, 722 (1981).

[104] J. L. Ding, G. D. Smith, A. Searle, and T. J. Peters, *Biochim. Biophys. Acta* **707**, 164 (1982).

[105] A. Maret, M. Potier, R. Salvayre, and L. Douste-Blazy, *Biochim. Biophys. Acta* **799**, 91 (1984).

[106] G. Dawson and J. C. Ellory, *Biochem. J.* **226**, 283 (1985).

[107] A. Maret, M. Potier, R. Salvayre, and L. Douste-Blazy, *FEBS Lett.* **160**, 93 (1983).

[108] W. H. M. Peters, P. L. M. Jansen, and H. Nauta, *J. Biol. Chem.* **259**, 11701 (1984).

[109] B. A. Callingham and D. Parkinson, in "Monoamine Oxidase: Structure, Function, and Altered Functions" (T. Singer, R. von Korff, and D. Murphy, eds.), pp. 81–86. Academic Press, New York, 1979.

[110] K. F. Tipton and T. J. Mantle, *Mod. Pharmacol.–Toxicol.* **10**, 559 (1977).

[111] H. Noël, G. Beauregard, M. Potier, G. Bleau, A. Chapdelaine, and K. D. Roberts, *Biochim. Biophys. Acta* **758**, 88 (1983).

[112] G. Beauregard and M. Potier, *Anal. Biochem.* **122**, 379 (1982).

[113] D. McNamara, G. Beauregard, H. V. Nguyen, D. L. S. Yan, M. Belisle, and M. Potier, *Biochem. J.* **205**, 345 (1982).

[114] V. Nguyen Hong, G. Beauregard, M. Potier, M. Belisle, L. Mameli, R. Gatti, and P. Durand, *Biochim. Biophys. Acta* **616**, 259 (1980).

[115] S. M. Hughes, G. Harper, and M. D. Brand, *Biochem. Biophys. Res. Commun.* **122**, 56 (1984).

[116] A. S. Garfinkel, E. S. Kempner, O. Ben-Zeev, J. Nikazy, S. J. James, and M. C. Schotz, *J. Lipid Res.* **24**, 775 (1983).

membrane with the use of specific antibody. The more that is known about the system, the more powerful is the interpretation. A conceptual basis for the interpretation of the data is presented. We can look to the continued application of target analysis to membrane proteins in the future, especially with emphasis on correlation of changes in multimeric size with functional state.

[25] Prediction of Bilayer Spanning Domains of Hydrophobic and Amphipathic Membrane Proteins: Application to the Cytochrome b and Colicin Families

By J. W. Shiver, A. A. Peterson, W. R. Widger, P. N. Furbacher, and W. A. Cramer

Introduction

It has been noted that knowledge of the structure of membrane proteins is limited because X-ray crystallography and other biophysical tools are not so easily applied.[1] At that time bacteriorhodopsin was the only membrane protein whose structure was known to moderate (~7 Å) resolution.[2] Since then, there has been a major step forward, the solution of the X-ray structure of the four-polypeptide reaction center complex from the photosynthetic bacterium *Rhodopseudomonas viridis* to 3 Å resolution,[3] and of the reaction center complex from *Rhodopseudomonas sphaeroides*.[4,5] Although high-resolution crystals exist for some other membrane proteins (e.g., porin[6]), and the methods developed for the solution of the reaction center structures are being disseminated, as yet only the latter structures have been solved to high resolution. The techniques of membrane biochemistry and molecular biology have progressed rapidly in recent years so that there exists a plethora of membrane proteins characterized *in vitro* whose primary sequences are known, and for

[1] D. Eisenberg, *Annu. Rev. Biochem.* **53**, 595 (1984).
[2] R. Henderson and P. N. T. Unwin, *Nature (London)* **257**, 28 (1975).
[3] J. Deisenhofer, O. Epp, K. Miki, R. Huber, and H. Michel, *Nature (London)* **318**, 618 (1985).
[4] J. Allen, G. Feher, T. O. Yeates, D. C. Rees, J. Deisenhofer, H. Michel, and R. Huber, *Proc. Natl. Acad. Sci. U.S.A.* **83**, 8589 (1986).
[5] C. H. Chang, D. Tiede, U. Smith, J. Norris, and M. Schiffer, *FEBS Lett.* **205**, 82 (1986).
[6] R. M. Garavito, T. Jenkins, J. N. Jansonius, R. Karlsson, and J. P. Rosenbusch, *J. Mol. Biol.* **164**, 313 (1983).

which *in vitro* reconstitution has yielded some structure information. In these cases, it is often desirable in the absence of X-ray analysis to obtain better working hypotheses for the folding pattern of the protein in the membrane.

Studies of the structure of these proteins in the membrane bilayer have been pursued (1) by topographical approaches using hydrophilic, hydrophobic, and antigenic probes (see, e.g., Ref. 7 for application to bacteriorhodopsin), and (2) by complementary theoretical considerations. The latter are the subject of this article and are based on (a) sequence comparisons, (b) the linear distribution of hydrophobic amino acids in a polypeptide chain and identification of possible transmembrane hydrophobic segments, (c) consideration of turn regions of the sequences, and (d) the prediction of transmembrane amphipathic α-helices or β-strands through the periodic distribution of polar and nonpolar residues.

Folding Pattern of Hydrophobic Membrane Proteins

Hydropathy ("Feeling about Water") Plots

The distribution of hydrophobic residues in an intrinsic membrane polypeptide can be an indicator of its folding pattern in the membrane.[8] Data bases for the hydrophobicity indices of the amino acids presented here were chosen to represent a range of viewpoints and/or results for the hydropathy plots. These data are based on the surface accessibility of amino acids in soluble proteins,[9] combinations of parameters including surface accessibility and free energy of transfer from water to a hydrophobic phase,[1,8] free energy of transfer from water to ethanol and side-chain interactions within proteins,[10] a weighted five-parameter algorithm refined by the fit to bacteriorhodopsin,[11] and bilayer partition free energy of amino acids in the α-helical conformation.[12] Considering the different methods by which the data bases were derived, each can result in somewhat different predictions.

The distribution of mean hydrophobicity in a polypeptide, a hydropathy plot, is generated by averaging the hydrophobicity index over an

---

[7] Y. A. Ovchinnikov, N. G. Abdulaev, R. G. Vasilov, I. Y. Uturina, A. B. Kuryatov, and A. V. Kiselev, *FEBS Lett.* **179,** 343 (1985).

[8] J. Kyte and R. F. Doolittle, *J. Mol. Biol.* **157,** 105 (1982).

[9] C. Chothia, *J. Mol. Biol.* **105,** 1 (1976).

[10] W. R. Krigbaum and A. Komoriya, *Biochim. Biophys. Acta* **576,** 204 (1979).

[11] P. Argos, J. K. M. Rao and P. A. Hargrove, *Eur. J. Biochem.* **128,** 565 (1982).

[12] D. M. Engelman, T. A. Steitz, and A. Goldman, *Annu. Rev. Biophys. Chem.* **15,** 321 (1986).

interval of amino acids, and then incrementing the calculation of this average, residue by residue, along the polypeptide. For the proteins discussed in this chapter, an averaging interval of at least seven residues is needed to avoid excessive noise. The maximum value of this interval is kept small enough to prevent distortion of the peaks and to allow detection of the transition from the hydrophobic membrane-spanning domains to the polar aqueous-soluble regions. It may be larger when searching for transmembrane peptides in the α-helical, as opposed to the β-sheet conformation (see below, Effect of Averaging Window on Cytochrome f Hydropathy Plot). The length of membrane-spanning domains would typically be 20–25 residues for an α-helix or 9–12 residues for a β-sheet, since an α-helix has a pitch of 1.5 Å/turn, a β-sheet occupies about 3.3 Å/residue, and the span of the bilayer is ~30–40 Å. The average length of the transbilayer α-helical domains that are known to exist in the bacteriorhodopsin and photosynthetic reaction center structures is approximately 25 residues. The overall shape of the hydropathy functions obtained for a protein using various data bases is similar, although transitions between hydrophobic and hydrophilic regions of the polypeptide, as well as amplitudes of the peaks of the function, may differ. Hence, for different protein classes (i.e., globular, membrane, and membrane-channel), different data bases may be of more utility.

It is useful to test the predictive value of the hydropathy plot on a family of proteins with similar functions and extensive sequence homology. The b cytochromes of energy-transducing membranes constitute such a family. The hydropathy plot using the Kyte and Doolittle data base (Table I) for bacteriorhodopsin[13] is shown in Fig. 1A, for chloroplast-encoded cytochrome b_6[14] in Fig. 1B, and for cytochrome b from complex III of maize mitochondria[15] in Fig. 1C. It has been shown that the hydropathy plots for cytochrome b_6 are similar to that shown in Fig. 1B when alternative amino acid hydrophobicity data bases are used,[16] such as those derived by Argos et al.,[11] Chothia,[9] Eisenberg,[1] and Kyte and Doolittle.[8] The plot is shifted upward toward increased hydrophobicity by 1–2 units, relative to the plot of Fig. 1B obtained using the Kyte and Doolittle scale,[8] when the Chothia[9] data base is used and downward toward increased hydrophobicity by about one unit with the Argos et al. scale.[11]

---

[13] R. Dunn, T. McCoy, M. Simsen, A. Majumdar, S. H. Chany, U. L. RajBhandary, and H. D. Khorana, *Proc. Natl. Acad. Sci. U.S.A.* **78**, 6744 (1981).
[14] W. Heinemeyer, J. Alt, and R. G. Herrmann, *Curr. Genet.* **8**, 543 (1984).
[15] A. J. Dawson, V. P. Jones, and C. J. Leaver, *EMBO J.* **3**, 2107 (1984).
[16] W. R. Widger, W. A. Cramer, and A. Trebst, "Molecular Biology of the Photosynthetic Apparatus" (K. Steinback, S. Bonitz, C. Arntzen, and L. Bogorad, eds.), pp. 89–97. Cold Spring Harbor Laboratory, Cold Spring Harbor, New York.

TABLE I
Cross-Correlation Coefficients between the Hydropathy Function of the Chloroplast Cytochrome b_6 and Cytochrome b Polypeptides from Mitochondria or Chromatophores, or Unrelated Proteins, Using Five Different Amino Acid Hydrophobicity Scales

	Cross-correlation coefficients				
			Data base		
Cytochrome b_6[14] versus[a]	Argos et al.[11]	Chothia[9]	Eisenberg[1]	Engelman[12]	Kyte and Doolittle[8]
A. nidulans mitochondria b[17]	0.63	0.71	0.69	0.68	0.78
Bovine mitochondria b[18]	0.58	0.72	0.70	0.73	0.72
Human mitochondria b[19]	0.61	0.74	0.72	0.72	0.76
Maise b[15]	0.67	0.80	0.74	0.66	0.80
Mouse mitochondria b[20]	0.58	0.73	0.71	0.72	0.76
Photosynthetic bacteria b[23]	0.52	0.65	0.67	0.67	0.75
Trypanosome b[22]	0.65	0.46	0.59	0.52	0.54
Yeast mitochondria b[21]	0.80	0.72	0.70	0.70	0.76
Cytochrome b_6	1.00	1.00	1.00	1.00	1.00
		Maximum coefficient[b]			
Bovine cytochrome oxidase II[29]	−0.085	0.26	0.17	0.24	0.24
Yeast cytochrome oxidase II[28]	0.34	0.31	0.22	0.20	0.30
Spinach cytochrome f[30]	0.37	0.24	0.20	0.22	0.13
Bacteriorhodopsin[14]	0.32	0.50	0.45	0.56	0.52

[a] The cytochrome sequences were aligned by the first two His residues characteristically separated by 14 residues (e.g., His-82 and His-96 in the chloroplast cytochrome b_6); no changes were made to the original sequences; the correlation was made over first 211 residues, or 210 for mouse, bovine, human cytochrome b, and 190 for the trypanosome cytochrome. The most recent sequence compilation includes 22 cytochrome b sequences including 14 from mitochondria.[27]

[b] In the absence of a rationale for alignment of the cytochrome b_6 with sequences from membrane proteins, other than cytochrome b, these correlation values are the maximum obtained by allowing the hydropathy function of cytochrome b_6 to orient with the second hydropathy function in all possible ways independent of phase, so that alignment with bacteriorhodopsin starts after its first α-helix.

Fig. 1A–C shows the existence for both bacteriorhodopsin and cytochrome b from chloroplasts and maize mitochondria of a number of hydrophobic domains separated by hydrophilic regions. To determine whether these plots are quantitatively similar and whether the degree of similarity of hydropathy plots is greater between two proteins from the same family relative to dissimilar proteins, the hydropathy functions can be mathematically compared. The amino acid sequences are known for a family consisting of seven analogous mitochondrial-encoded b cyto-

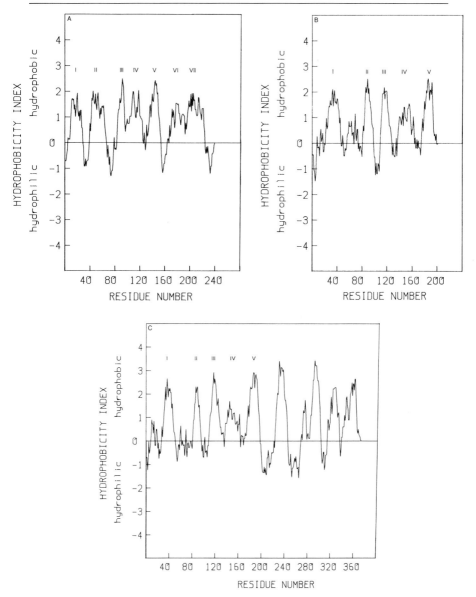

FIG. 1. Hydropathy plots for bacteriorhodopsin[13] (A), the chloroplast cytochrome b_6[14] (B), and the cytochrome b of mitochondrial complex III from maize[15] (C), using the amino acid hydrophobicity data base of Ref. 8, and an averaging interval of 11 residues.

chromes from maize,[15] the fungus *Aspergillus nidulans*,[17] as well as from bovine,[18] human,[19] mouse,[20] yeast,[21] and trypanosome[22] sources, and from chloroplasts[14] and photosynthetic bacteria.[23,24] The four conserved histidine residues needed for heme coordination are found at positions 82, 96, 183, and 197 (198 for chloroplasts) in the amino acid sequences of the mitochondrial cytochromes (numbering system for yeast; if the numbering is done only for the chloroplast polypeptide, the four conserved His residues would be residues numbered 85, 99, 186, and 201). The first five hydrophobic regions of the mitochondrial protein define the extent of the heme-binding domain.[25] There is a high degree of primary sequence identity in the region of the four histidine ligands and over the entire heme-binding domain, suggesting that the hydropathy functions for the respective polypeptides should also be alike.

Correlation of Hydropathy Functions

Homology in a family of proteins such as that of the b cytochromes can be quantitated by calculation of a cross-correlation coefficient[26] between (1) pairs of functions derived from two members of the family and (2) between functions derived from a b cytochrome and unrelated intrinsic membrane proteins of approximately the same size. The cross-correlation coefficient C is defined as[26]:

$$C = \frac{\sum_{j=1}^{n}(X_j - \bar{X})(Y_j - \bar{Y})}{\left[\sum_{j=1}^{n}(X_j - \bar{X})^2 \sum_{j=1}^{n}(Y_j - \bar{Y})^2\right]^{1/2}}$$

[17] R. B. Waring, R. W. Davies, S. Lee, E. Grisi, M. McPhail Berks, and C. Scazzocchio, *Cell* **27**, 4 (1981).
[18] S. C. Anderson, M. H. L. de Bruijn, A. R. Coulson, I. C. Eperon, F. Sanger, and I. G. Young, *J. Mol. Biol.* **156**, 683 (1982).
[19] S. Anderson, A. T. Bankeir, B. C. Barrell, M. H. L. de Bruijn, A. R. Coulson, J. Dronin, I. C. Eperon, D. P. Nierlich, B. A. Roe, F. Sanger, P. H. Schrier, A. J. H. Smith, R. Staden, and I. G. Young, *Nature (London)* **290**, 457 (1981).
[20] M. J. Bibb, R. A. Van Etten, C. T. Wright, M. W. Walbert, and D. A. Clayton, *Cell* **26**, 167 (1981).
[21] F. G. Nobrega and A. Tzagoloff, *J. Biol. Chem.* **255**, 9828 (1980).
[22] B. J. B. Johnson, G. C. Hill, and J. E. Donelson. *Mol. Biochem. Parasitol.* **13**, 135 (1984).
[23] N. Gabellini and W. Sebald, *Eur. J. Biochem.* **154**, 569 (1986).
[24] E. Davidson and F. Daldal, *J. Mol. Biol.* **195**, 25 (1987).
[25] W. R. Widger, W. A. Cramer, R. G. Herrmann, and A. Trebst. *Proc. Natl. Acad. Sci. U.S.A.* **81**, 674 (1984).
[26] D. D. Jones, *J. Theor. Biol.* **50**, 167 (1975).

for comparison of the functions over n residues of polypeptides X and Y, without any lag or delay between the functions. $\bar{X}$ and $\bar{Y}$ are the average values of the hydropathy function over the length, n, of the polypeptide, corrected for the averaging interval, m. X_j and Y_j are the average values of the hydrophobicity index over the sampling interval for the two functions, X and Y, so that

$$\bar{X} = \frac{1}{n} \sum_{j=1}^{n} X_j; \qquad \bar{Y} = \frac{1}{n} \sum_{j=1}^{n} Y_j$$

$$X_j = \frac{1}{m} \sum_{i=1}^{m} h_i; \qquad Y_j = \frac{1}{m} \sum_{i=1}^{m} h_i$$

in which h_i is the hydrophobicity of residue i. The above cross-correlation function is normalized to a value of 1.0 for autocorrelation (the hydropathy function compared to itself). Values of the correlation coefficient close to 1.0 and 0.0 are obtained for functions that are, respectively, similar in amplitude and phase, or not related to each other in phase and/or amplitude. The correlation coefficients between the chloroplast cytochrome b_6 and other members of this family along with a set of unrelated proteins are tabulated in Table I. The correlation coefficient measures the similarity between hydropathy functions and thus is dependent on the interval size and the data base. The values in Table I use an averaging interval of 11 and are higher than those previously reported,[16] where the hydrophobicity of the individual residues in the cytochrome b sequences was correlated.[28]

The correlation coefficients of the hydropathy functions between cytochrome b_6 and the heme-binding domain of the other b cytochromes are between 0.5 and 0.8 for five data bases (Table I), whereas the maximum values obtained with unrelated proteins, independent of phasing constraints, range from -0.08 to $+0.6$, the latter values obtained only if bacteriorhodopsin was aligned so that its first helix was omitted. The b cytochromes could be subdivided into groups of even greater internal similarity: *A. nidulans* and yeast; mouse, bovine, and human; and maize and photosynthetic bacterium (not shown). These coefficients were calculated from an alignment or phasing of the cytochrome hydropathy functions at the first pair of histidines. The calculation of the cross-correlation between cytochrome b_6 and an unrelated protein has the problem of sequence alignment. The correlation coefficients shown in Table I for the unrelated proteins with cytochrome b_6 are the maximum values calculated

---

[27] G. Hauska, W. Nitschke, and R. G. Herrmann, *J. Bioenerg. Biomembr.* **20**, 211 (1988).
[28] R. M. Sweet and D. Eisenberg, *J. Mol. Biol.* **171**, 479 (1983).

for all possible alignments of the cytochrome b sequence with that of the unrelated proteins: cytochrome oxidase subunit II from yeast[29] and bovine[30] sources, the chloroplast cytochrome f,[31] and bacteriorhodopsin.[13] These maximum coefficients are all much smaller than those obtained between the cytochrome pairs, with the exception of bacteriorhodopsin, aligned without its first helical domain.

This comparison of the correlation coefficients between hydropathy functions of two members of the b cytochrome family and those between a b cytochrome and an unrelated protein shows that the hydropathy function for the cytochrome b family is a well-conserved structural parameter. It can then be used as a reasonable basis for prediction of the folding pattern of the heme-binding domain of the cytochrome b polypeptide in the membrane. The ~20-residue-long hydrophobic regions I–V in Fig. 1A–C were hypothesized to span the bilayer in an α-helical conformation[25,32] (Fig. 2A). The choice of the bilayer–polar interface is dictated in the case of cytochrome b_6 by the presence of charged amino acids (e.g., R, arginine) at the termini of the nonpolar sequences (Fig. 3). The fourth helix of cytochrome b_6 in the model of Fig. 2A has a lower hydrophobicity index than the other four, as seen in Fig. 1B and C. The origin of the lower average hydrophobicity is apparent in Fig. 2A, where it can be seen by inspection that helix IV contains two acidic residues, three prolines, and is amphipathic (NB: hydrophobic residues on the left and more polar residues on the right, in helix IV of Fig. 2A). Together with data on the distribution of inhibitor-resistant mutants (properties of inhibitors summarized in Ref. 37) of cytochrome b in yeast[33] and mouse[34] mitochondria, as well as *Rhodopseudomonas capsulata*,[35] this suggested that helix IV may not span the bilayer, but is surface-bound or extrinsic.[33–36] Schematic diagrams comparing four- and five-helix models for cytochrome b_6, including the charged residues that might be used as sites for protease (e.g., trypsin, TR) probes of topography, are shown in Fig. 2B and C.

Using the models of Fig. 2, the four conserved histidine residues are distributed as two pairs on spanning peptides II and V (or IV) with one His from each pair on the p-side (positive proton electrochemical poten-

---

[29] G. Coruzzi and A. Tzagoloff, *J. Biol. Chem.* **254**, 9324 (1979).
[30] G. J. Steffens and G. Buse, *Hoppe-Seyler's Z. Physiol. Chem.* **360**, 613 (1979).
[31] J. Alt and R. G. Herrmann, *Curr. Genet.* **8**, 551 (1984).
[32] W. A. Cramer, W. R. Widger, R. G. Herrmann, and A. Trebst, *Trends Biochem. Sci.* **10**, 125 (1985).
[33] J. P. di Rago, and A.-M. Colson, *J. Biol. Chem.* **263**, 12564 (1988).
[34] N. Howell and K. Gilbert, in "Cytochrome Systems: Molecular Biology and Bioenergetics" (S. Papa, B. Chance, and L. Ernsten, eds.), pp. 79–86. Plenum Press, New York, 1987.
[35] F. Daldal, personal communication.
[36] A. R. Crofts, personal communication.

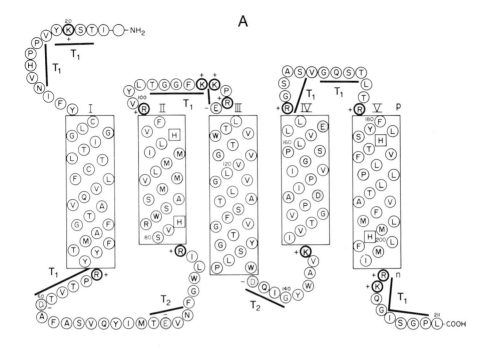
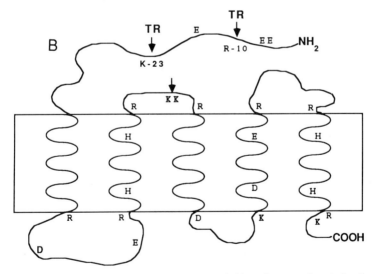

FIG. 2. (A) Proposed membrane-folding pattern of chloroplast cytochrome b_6; five helix model regions with high propensity for turns (T_1 and T_2) are marked. The number and percentages of residues participating in T_1 segments within the proposed connecting and helical regions are 44 (44.4%) and 9 (8.0%), respectively. Sequences identified as T_1s lying within the membrane but not shown are 130–133, and 148–152. (B and C) Schematic models for five and four transmembrane helices, showing distribution of charges, and the four conserved His residues on helices II and V (IV). Also shown are potential trypsin (TR) sites on the extrinsic segments that could be used as probes of the topography.

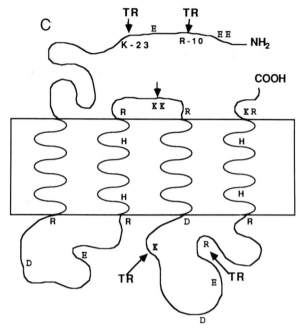

FIG. 2. (continued)

tial) and the other on the n-side of the membrane. This hypothesis for cytochrome polypeptide folding led to a model[25,32,37,38] for coordination of the two hemes whose orientation, approximately perpendicular to the membrane plane (Fig. 3), is in agreement with experiment.[39,40] This model provides a structural framework for transmembrane electron transport in the cytochrome b_6–f and b–c_1 complexes. With the residue numbering for the yeast cytochrome, the two pairs of heme-coordinating His residues are (His-82–His-198, His-96–His-183) in the five-helix model with antiparallel helices II and V, and (His-82–His-183, His-96–His-198) for a four-helix model with parallel helices II and IV.

Effect of Averaging Window on Cytochrome f Hydropathy Plot

From the existence of a 20-residue nonpolar sequence near the COOH-terminus of the sequence of the cytochrome f protein,[31,41] and the

[37] W. A. Cramer, W. R. Widger, M. T. Black, and M. E. Girvin, in "The Light Reactions" (J. Barber, ed.) Vol. 8, pp. 447–493. Elsevier, Amsterdam, 1987.
[38] M. Saraste, *FEBS Lett.* **166**, 367 (1984).
[39] M. Erecinska, D. F. Wilson, and J. F. Blasie, *Biochim. Biophys. Acta* **501**, 63 (1978).
[40] J. Bergström, *FEBS Lett.* **183**, 87 (1985).
[41] D. L. Willey, A. D. Auffret, and J. C. Gray, *Cell* **36**, 556 (1984).

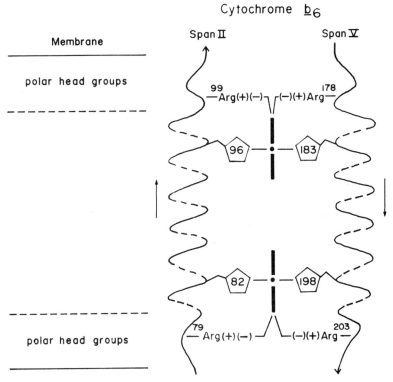

FIG. 3. Transmembrane arrangement of the two hemes of cytochrome b_6 cross-linking membrane-spanning peptides II and V, inferred from the folding pattern shown in Fig. 2A, the resulting position of the four conserved His residues, and the requirement of His–His ligation for both hemes.[32] Numbering system is according to the yeast sequence. The second (right) helix would be parallel to the first (left) in a four-helix model, and the His pairs on the two sides of the membrane would be (His-82–His-198) and (His-96–His-183) using the yeast numbering system. Reprinted with permission of Elsevier.

presence of a prominent broad hydrophobic peak in this region of the hydropathy plot (Fig. 4A and B), it has been inferred that cytochrome f contains only the one α-helical membrane-spanning domain near the COOH-terminus, with the heme and most of the polypeptide on the lumenal side of the membrane.[31,41] It can be seen, however, that additional narrow hydrophobic peaks of about 10 residues in width appear in the hydropathy plot as the averaging window is decreased to 9–11 residues (Fig. 4C and D). A further decrease in the averaging interval does not result in improved resolution because of additional "noise." These plots suggest the possibility of short membrane surface-binding domains between residues 50 and 125. The use of averaging intervals of 13 residues or

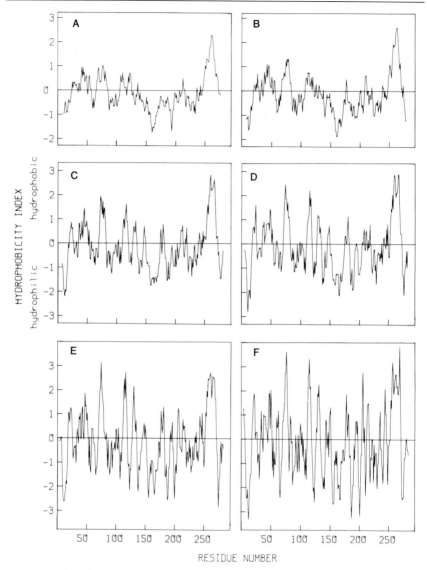

FIG. 4. Hydropathy plots of the mature cytochrome f^{31} polypeptide as a function of the length of the averaging interval. Interval length: (A) 19 residues, (B) 15 residues, (C) 11 residues, (D) 9 residues, (E) 7 residues, (F) 5 residues.

more biases the hydropathy plots against shorter hydrophobic segments of possible significance.

Turn Regions

Turns are presumed to exist in transmembrane proteins at the membrane–aqueous phase interface, so that determination of turn regions may assist in identifying membrane-spanning regions. Turns are made of polar amino acids and may be predicted by identifying hydrophilic regions of proteins in which the likelihood of turns is high. In addition, methods for assigning secondary structure may also be used to independently predict turns. Table II summarizes one method[42] in which sequences of 4–9 residues are classified into three groups according to their turn propensity, with the greatest probability for turn $T_1 > T_2 > T_3$. For soluble proteins, T_2 and T_3 turns are neglected unless the distance between T_1 turns is greater than 19 residues. For transmembrane proteins this number should be greater, because longer segments without turns may be required to transverse the membrane.

The accuracy of using turn regions to define the edges of membrane-spanning segments is illustrated by comparing membrane-spanning segments of the L and M polypeptides of the *R. viridis* reaction center determined experimentally[3] with those predicted from hydropathy plots and turn regions (Table III). These prediction methods correctly identify all five membrane-spanning helices of both the L and M subunits. Application of hydropathy plots alone appear to be less accurate in the determination of the ends of the membrane-spanning helices.[45]

The presence of turn regions in the polar phases linking the membrane-spanning peptides of cytochrome b_6 is indicated in Fig. 2A. These turn regions predicted by the algorithm of Cohen *et al.*[42] were mostly confirmed by the algorithm of Garnier *et al.*[43] While two segments within the proposed transmembrane helices are also identified as possible turns, Fig. 2A shows that the occurrence of turns is much greater in the extramembranous regions. The presence of a turn within a transmembrane region is possible. Transmembrane segments need not be straight, rigid helices, but may have bends or kinks. Proline and glycine, often considered helix breakers or turn promoters, are commonly found in transmem-

---

[42] F. E. Cohen, R. M. Abarbanel, I. D. Kuntz, and R. J. Fletterick, *Biochemistry* **22**, 4894 (1983).
[43] J. Garnier, D. J. Osguthorpe, and B. Robson, *J. Mol. Biol.* **120**, 97 (1978).
[44] M. Levitt, *Biochemistry* **17**, 4277 (1978).
[45] H. Michel, K. A. Weyer, H. Gruenberg, I. Dunger, D. Oesterhelt, and F. Lottspeich, *EMBO J.* **5**, 1149 (1986).

TABLE II
SEQUENCE CRITERIA FOR PROTEIN TURN REGIONS WITH DIFFERENT TURN PROPENSITY[a]

Label	Description	Residues in turns	Turn algorithms
T_1	High density of hydrophilic residues	t_1 = D, E, G, H, K, N, P, Q, R, S, or T $t_2 = t_1 + Y$ $t_3 = t_2 + A$	$(3t_1, 1t_2)$[b] $(4t_1, 1*)$[c] $(5t_1, 2*)$
T_2	High density of hydrophilics, but less than T_1	As above	$(3t_1, 1*)$ $(4t_1, 2t_3, 1*)$
T_3	High density of hydrophilics, but less than T_2	As above	$(3t_1, 4t_3)$ $(4t_1, 4*)$ $(5t_1, 4*)$ $(4t_1, 2t_3, 3*)$

[a] From Cohen et al.[42]
[b] Expressions of the form (3x,2y) mean that for a sequence of five residues, three are of type x, and two are of type y.
[c] *, Any amino acid.

TABLE III
PREDICTED TRANSMEMBRANE SPANNING REGIONS OF THE *Rhodopseudomonas viridis* PHOTOSYNTHETIC REACTION CENTER[a]

Subunit	Helix	Experimental[3]	Hydropathy[b]	Prediction method		
				Cohen et al.[42]	Garnier et al.[43]	Levitt[44]
L	A	32–55	29–40	24–53	30–53	30–53
	B	84–112	88–98	85–102	85–109	85–113
	C	115–140	116–136	115–139	119–139	119–139
	D	170–199	177–196	177–198	174–198	173–198
	E	225–251	223–247	219–251	222–250	224–250
M	A	52–78	46–75	46–75	53–79	53–79
	B	110–139	111–128	113–138	112–131	112–140
	C	142–167	140–165	144–166	137–158	143–162
	D	197–225	199–226	194–225	210–225	210–225
	E	259–285	267–289	268–297	268–289	256–285

[a] Transmembrane segments were determined by crystallographic analysis[3] and predicted from the sequences[46] through use of the hydropathy plots and application of the secondary structure algorithm of Refs. 42–44 to identify turn regions.
[b] End points were determined from hydropathy plots, using the data base of Ref. 8 as the transition between hydrophilic and hydrophobic regions.

brane segments of some proteins.[45,46] It should be noted that many secondary structure methods assign probabilities at each residue for turn, α-helix, and β-sheet,[43,44] so that the probability of each residue being in a nonturn secondary structure should be taken into account when assigning turns.

Transmembrane Channel Proteins: Colicins

The channel-forming colicins (E1, A, Ib, Ia, K, B) produced by the ColE1 plasmid of *E. coli* exert a lethal effect on sensitive *E. coli* by depolarizing and deenergizing the cell.[47] Depolarization is a consequence of colicin insertion into the cytoplasmic membrane and formation of an ~8 Å diameter channel[48] whose nonspecific ionic conductance is sufficient to depolarize the inner membrane. When measured as a function of colicin concentration, the lethal activity as well as activity in artificial membranes is one-hit or first order, suggesting that it is mediated by a single colicin molecule.[49,50] However, predictions of the potential colicin channel structure have raised the question of whether the lethal unit of the channel-forming colicin may be a dimer or oligomer instead of a monomer.[51,52]

The essential channel-forming portion of the colicin molecule is contained in the COOH-terminal third of the polypeptide.[53] Active channel-forming COOH-terminal peptides can be made by treatment of colicin with protease[53-56] or cyanogen bromide.[57] These COOH-terminal peptides have an α-helical content of 50–60% in aqueous detergent solutions,[52,58] large enough to indicate a possible role for this conformation in the structure. *In vitro* measurements of colicin ionophoretic activity with model

---

[46] C. J. Brandl and C. M. Deber, *Proc. Natl. Acad. Sci. U.S.A.* **83**, 917 (1986).
[47] W. A. Cramer, J. R. Dankert, and Y. Uratani, *Biochim. Biophys. Acta* **737**, 173 (1983).
[48] L. Raymond, S. L. Slatin, and A. Finkelstein, *J. Membr. Biol.* **84**, 173 (1985).
[49] A. A. Peterson and W. A. Cramer, *J. Membr. Biol.* **99**, 197 (1987).
[50] S. L. Slatin, *Biophys. J.* **53**, 155a (1988).
[51] H. R. Guy, *Biophys. J.* **41**, 363a (1983).
[52] F. Pattus, F. Heitz, C. Martinez, S. W. Provencher, and C. Lazdunski, *Eur. J. Biochem.* **152**, 681 (1985).
[53] J. R. Dankert, Y. Uratani, C. Grabau, W. A. Cramer, and M. Hermodson, *J. Biol. Chem.* **257**, 3857 (1982).
[54] V. L. Davidson, K. R. Brunden, F. S. Cohen, and W. A. Cramer, *J. Membr. Biol.* **79**, 105 (1984).
[55] Y. Ohno-Iwashita and K. Imahori, *J. Biol. Chem.* **257**, 6446 (1982).
[56] C. Martinez, C. Lazdunski, and F. Pattus, *EMBO J.* **2**, 1501 (1983).
[57] B. M. Cleveland, S. Slatin, A. Finkelstein, and C. Levinthal, *Proc. Natl. Acad. Sci. U.S.A.* **80**, 3706 (1983).
[58] K. R. Brunden, Y. Uratani, and W. A. Cramer, *J. Biol. Chem.* **259**, 7602 (1984).

membranes such as liposomes and planar lipid bilayers have demonstrated that acidic pH conditions are required for protein insertion,[54] implying that the transition from a soluble to membrane-associated protein requires protonation of acidic amino acids.[59]

Comparison of the amino acid sequences of the four colicins for which primary structures are available (E1, Ia, Ib, and A)[60–64] shows that the ~200 residues of the COOH-terminus containing the channel-forming domain have primary sequence identity between pairs of the colicin sequences of 24–44% (Fig. 5). The COOH-terminal sequences of colicins E1, Ia, and Ib align from the COOH-terminus forward without manipulation, whereas the colicin A sequence (70 residues longer than E1) is realigned by 12 residues (e.g., residue 450 of A corresponds to residue 392 of E1, Fig. 6). Table IV shows that the average cross-correlation values of the hydropathy functions of the channel-forming peptides are also relatively high. This trend begins after residue 345 for colicin E1, suggesting that the channel domain may start near this region. It is concluded that the overall distribution of hydrophobic residues is similar in the channel-forming domain of these colicins (sequence homology of individual residue hydrophobicity is 50–70%).

Hydropathy plots of the colicins and visual inspection indicate that there is only one long highly hydrophobic region, between residues 474 and 508 in the colicin E1 polypeptide (Fig. 6) and at a similar position in the other colicins (Fig. 5), which is likely to be a hydrophobic anchor region for the channel structure. The hydropathy plot of the COOH-terminal channel domain of colicin E1 shows seven relative hydrophobic maxima which may potentially be transmembrane segments (Fig. 6). The maximum hydrophobicity of peaks I–V (Fig. 6) for colicin E1 is smaller than that of hydrophobic membrane proteins, as expected because channel proteins have a higher content of charged and polar amino acids.

The most hydrophobic region (residues 474–508 in colicin E1) near the COOH-terminus has a minimum in the hydropathy function near its center that divides peaks VI and VII in Fig. 6, suggesting the presence of two transmembrane regions separated by a short "hairpin" turn or bend.[42,65]

---

[59] V. L. Davidson, K. R. Brunden, and W. A. Cramer, *Proc. Natl. Acad. Sci. U.S.A.* **82,** 1386 (1985).

[60] Y. Yamada, Y. Ebina, T. Miyata, T. Nakazawa, and A. Nakazawa, *Proc. Natl. Acad. Sci. U.S.A.* **79,** 2827 (1982).

[61] P. T. Chan, H. Ohmori, J. Tomizawa, and J. Leibowitz, *J. Biol. Chem.* **260,** 8925 (1985).

[62] J. Mankovich, C. H. Hsu, and J. Konisky, *J. Bacteriol.* **168,** 228 (1986).

[63] J. M. Varley and G. J. Boulnois, *Nucleic Acids Res.* **12,** 6727 (1984).

[64] J. Morlon, R. Lloubes, S. Varenne, M. Chartier, and C. Lazdunski, *J. Mol. Biol.* **110,** 271 (1983).

[65] G. D. Rose, *Nature (London)* **272,** 586 (1978).

```
            CNBr
              ↓            390                      415
    El KYSKMAQELADKSKGKKIGNVNEALAAFEKYKDVLNKKFSKADRDAIFNA
    Ia KAEQLAREMAGQAKGKKIRNVEEALKTYEKYRADINKKINAKDRAAIAAA
    Ib QASELAKELASVSQGKQIKSVDDALNAFDKFRNNLNKKYNIQDRMAISKA
    A  AIAKDIADNIKNFQGKTIRSFDDAMASLNKITANPAMKINKADRDALVNA


                         440                        465
    El LASVKYDDWAKHLDQFAKYLKITGHVSFGYDVVSDILKIKDTGDWKPLFL
    Ia LESVKLSDISSNLNRFSRGLGYAGKFTSLADWITEFGKAVRTENWRPLFV
    Ib LEAINQVHMAENFKLFSKAFGFTGKVIERYDVAVELQKAVKTDNWRPFFV
    A  WKHVDAQDMANKLGNLSKAFKVADVVMKVEKVREKSIEGYETGNWGPLML


                 ┌       490                        505
    El TLEKKAADAGVSYVVALLFSLLAGT          TLGIWGIAIVTGILC
    Ia KTETIIAGNAATALVALVFSILTGS          ALGIIGYGLLMAVTG
    Ib KLESLAAGRAASAVTAWAFSVMLGT          PVGILGFAIIMAAVS
    A  EVESWVLSGIASSVALGIFSATLGAYALSLGVPAIAVGIAGI LLAAVVG
                                    └


         ┐            522
    El SYIDKNKLNTINEVLGI
    Ia ALIDESLVEKANKFWGI
    Ib ALVNDKFIEQVNKLIGI
    A  ALIDDKFADALNNEIIRPAH
         ┘
```

FIG. 5. Alignment of the sequences of channel-forming colicins E1, Ia, Ib, and A.[52,60–64] The cyanogen bromide (CNBr) cleavage site at residue 370 of E1 is identified and the long hydrophobic sequences of the colicins are bracketed. The sequences are enumerated according to E1.

The additional residues within colicin A that do not align with the other three colicins occur at this predicted turn. The occurrence of a turn in this region is also indicated by the conservation of glycines located at positions 489, 493, and 496 (colicin E1) in the sequences of all four colicins (Fig. 5). The hydrophobic segments VI and VII in the postulated hairpin loop would be no longer than 15–17 residues in colicin E1 if they were in an α-helical conformation. Therefore, the hairpin loop would not completely span the membrane as α-helices for colicins E1, Ia, and Ib, but could do so if these segments were β-sheet. The additional amino acids in this region on colicin A would, however, allow two spanning α-helical peptides. Further analysis of the primary sequences of the four colicins indicates turn regions between all of the hydrophobic maxima of plots

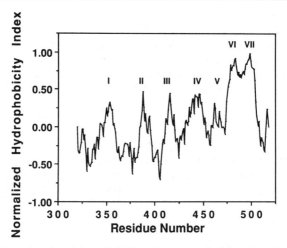

FIG. 6. Hydropathy plot of the colicin E1 carboxy-terminal domain calculated using the normalized consensus hydrophobicity data base of Eisenberg.[1] The proposed transmembrane segments are designated as I, II, III, IVa,b, V, VI, and VII.

such as that of Fig. 6 and also in the middle of peak IV (forming IVa and IVb). The breadth of peak IV relative to the other hydrophobic maxima indicates that two transmembrane peptides, IVa and IVb, could be present in the β-sheet conformation. A role for the region around segment

TABLE IV
AVERAGE CROSS-CORRELATION COEFFICIENTS
(C) OF HYDROPATHY PLOTS OF COLICIN
CHANNEL-FORMING DOMAINS

Comparison	C^a
E1/Ia	0.88
E1/Ib	0.87
Ia/Ib	0.87
E1/A	0.66
Ia/A	0.72
Ib/A	0.72

[a] C values for aligned hydropathy plots beginning at CNBr site of E1 (Fig. 6) using the data base of Ref. 10 and an averaging window of 11 residues.

IV in channel function was implied by the decrease of single channel conductance caused by chemical modification of histidines 427 and 440.[66]

Hydrophobic Moment

Most of the polypeptide chains forming a membrane channel are expected to be amphipathic, with the side chains of hydrophilic amino acids facing the channel lumen and those of the hydrophobic amino acids oriented toward the lipid. A search can be made for amphipathic segments using the periodic nature of the segregation between polar and nonpolar residues. The periodic difference in residue hydrophobicity creates a hydrophobic moment,[67-69] μ_H, defined by

$$\mu_H = \sum_{n=1}^{N} h_n S_n$$

for a polypeptide with N amino acids, where h_n is the hydrophobicity of the nth amino acid and S_n is a unit vector extending from its α carbon to the center of the side chain. For an α-helix or β-sheet, the distribution of the hydrophobic moment through the sequence can be written as a Fourier sum,[67,69] for which the magnitude of the moment may be calculated using the relationship:

$$\mu_H = \left\{ \left[\sum_{n=1}^{N} h_n \sin(\delta n) \right]^2 + \left[\sum_{n=1}^{N} h_n \cos(\delta n) \right]^2 \right\}^{1/2}$$

where δn is the angle, in radians, at which successive side chains emerge with a periodicity, δ, of 100° for α-helix, or 160°–180° for twisted and straight β structures. A hydrophobic moment angle (μ_H versus δ) profile in which μ_H is plotted versus the periodicity, using values of h_n obtained from Eisenberg's normalized consensus data base[1] (Fig. 7), indicates the preferred secondary structure of a segment if the putative amphipathic peptide q length N folds to maximize μ_H.[67]

Candidate regions for amphipathic transmembrane peptides in the colicin COOH-terminal domain were identified by calculating the moment of the segments around the local maxima I–V in the hydropathy plot (Fig. 6) that are conserved in the family of colicin sequences. A search was made

---

[66] L. J. Bishop. F. S. Cohen, V. L. Davidson, and W. A. Cramer, *J. Membr. Biol.* **92**, 237 (1986).
[67] D. Eisenberg, R. M. Weiss, and T. C. Terwilliger, *Proc. Natl. Acad. Sci. U.S.A.* **81**, 140 (1984).
[68] D. Eisenberg, E. Schwarz, M. Komaromy, and R. Woll, *J. Mol. Biol.* **179**, 125 (1984).
[69] J. Finer-Moore and R. M. Stroud, *Proc. Natl. Acad. Sci. U.S.A.* **81**, 155 (1984).

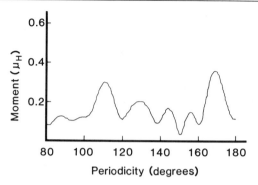

Fig. 7. Hydrophobic moment[67] as a function of periodicity for the 35 amino acids extending over residues 434–468 in the colicin E1 channel domain. Pronounced peaks are seen for a periodicity, $\delta = 110°$ and $160°–180°$, corresponding to a possible α-helical and β-sheet conformation, respectively.

for amphipathic segments overlapping regions I–V, with segment lengths of 19 residues used for α-helix and 9–12 residues for β-sheet. Hydrophobic moments for these segments were calculated for periodicities of 100°, 160°, and 180°, and the endpoints of segments exhibiting an amphipathic character adjusted by ±2 residues to maximize their moments. The threshold values for putative amphipathic transmembrane domains were assumed to be ≥0.35 for α-helical segments and ≥0.25 for β-sheet.[67,68] μ_H versus δ plots such as that shown in Fig. 7, together with plots and printouts of μ_H as a function of residue number for a given value of the periodicity, δ (e.g., Fig. 8, for $\delta = 180°$) were used to identify segments whose moments exceeded the threshold values. Peaks I–IVa in Fig. 8 show amphipathicity maxima for $\delta = 180°$ that overlap with the corresponding hydrophobicity maxima (Fig. 6). Data similar to those of Fig. 8, but for $\delta = 160°$ (twisted β-sheet) and $\delta = 100°$ (α-helix) were also used. Amphipathicity maxima for β-sheet and α-helix that exceed the threshold values and overlap significantly with the hydrophobicity maxima are italicized in Table V (segments denoted by prime are α-helical). Three amphipathic α-helices (I', II', IV') and four to five β-sheet segments (I, III, IVa, IVb, and V) could be proposed (β-sheet segment II seems less likely because its position does not correspond to a hydrophobicity maximum of colicin Ia). As a hybrid α–β channel seems unlikely because of the problem of saturating the hydrogen bonds of the β-sheet segments, the colicin channel for E1, Ia, and Ib would contain either four α-helices (three amphipathic and one hydrophobic) or six or seven β-sheet segments (four or five amphipathic, two hydrophobic). Since at least five to six α-helical transmembrane segments are needed to define an 8 Å channel, the present

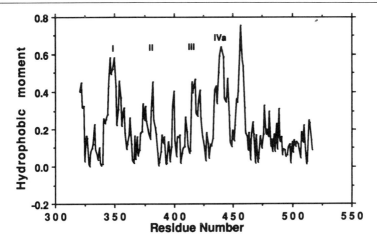

FIG. 8. Hydrophobic moment for periodicity, $\delta = 180°$, as a function of residue position in the colicin E1 channel polypeptide. Interval for calculation of moment, μ_H, was 11 residues. Maxima labeled I, II, III, IVa overlap with local hydrophobicity maxima of Fig. 6.

TABLE V
Hydrophobic Moments of Proposed Transmembrane α-Helical or β-Sheet Segments of the Colicin Channel

Colicin, segment number	Membrane spanning segment[a]	Hydrophobic moment[b] periodicity		
		100° (α)	160° (β)	180° (β)
E1				
I	344–355	—	0.13	*0.61*
I'	341–359	*0.41*	—	—
II	384–392[c]	—	0.22	*0.31*
II'	385–403	*0.40*	—	—
III	410–421	—	0.15	*0.44*
IVa	434–442	—	0.48	*0.65*
IVb	446–454	—	*0.44*	0.10
IV'	422–440	*0.47*	—	—
V	460–468	—	*0.43*	0.04
VI (hydrophobic)	475–488	0.25	0.06	0.17
VII (hydrophobic)	494–508	0.13	0.13	0.06
Ia				
I	448–459	—	0.29	*0.38*
I'	447–465	*0.54*	—	—
II	489–498[c]	*0.54*	0.24	*0.38*
II'	488–506	*0.49*	—	—

(*continued*)

TABLE V (continued)

Colicin, segment number	Membrane spanning segment[a]	Hydrophobic moment[b] periodicity		
		100° (α)	160° (β)	180° (β)
III	513–521	—	0.26	*0.57*
IVa	538–546	—	*0.31*	*0.33*
IVb	550–558	—	0.28	0.04
IV'	528–546	*0.53*	—	—
V	564–572	—	*0.53*	0.04
VI (hydrophobic)	579–592	0.25	0.11	0.12
VII (hydrophobic)	598–612	0.04	0.12	0.00
Ib				
I	448–459	—	0.28	*0.53*
I'	447–465	*0.65*	—	—
II	488–497[c]	—	0.27	0.18
II'	488–506	*0.56*	—	—
III	513–521	—	0.27	*0.42*
IVa	538–546	—	*0.36*	*0.33*
IVb	549–558	—	*0.44*	*0.33*
IV'	528–546	*0.42*	—	—
V	564–572	—	*0.64*	0.10
VI (hydrophobic)	579–592	0.11	0.16	0.15
VII (hydrophobic)	598–612	0.11	0.14	0.10
A				
I	417–427	—	0.19	*0.43*
I'	413–431	*0.55*	—	—
II	441–452[c]	—	0.08	*0.43*
II'	441–459	*0.47*	—	—
III	468–479	—	0.03	*0.30*
IVa	493–501	—	*0.51*	0.01
IVb	504–512	—	*0.68*	*0.34*
IV'	503–521	*0.35*	—	—
V	518–526	—	*0.41*	*0.40*
VI (hydrophobic)	533–546	0.14	0.24	0.01
VII (hydrophobic)	562–575	0.05	0.06	0.04

[a] Lengths of colicin E1, Ia, Ib, and A polypeptides are 522, 626, 626, and 592 residues, respectively. Approximate positions of membrane-spanning segments were identified from peaks I–VII in hydropathy plots in Fig. 6, the turn algorithms, and homology of sequence alignment; values in italics exceed threshold moment values (0.25 for β and 0.35 for α).

[b] Moments were calculated for each segment length using the normalized data base of Ref. 1. Values indicating a significant amphipathic character[67,68] are italicized. Unprimed and primed segments have lengths of 9–12 and 19 residues appropriate for transmembrane β-sheet and α-helix, respectively.

[c] This peptide region in colicin Ia does not have a hydrophobicity maximum.

analysis indicates that an α-helical channel for colicin E1, Ia, and Ib could not be monomeric. The data showing that colicin E1 channel formation is first order in channel peptide concentration[49,50] would then argue that the channel structure must have a β-sheet conformation, presumably forming some kind of β-barrel. Segments I–III would not be utilized and this kind of β-barrel model would also be impossible if full activity of the channel did not require any residue before that at position 428.[70,71]

The present analysis might allow five α-helices for colicin A (three amphipathic plus two hydrophobic, because of the larger extent of segments VI and VII). However, in an article[52] that preceded this one and which is similar to the present discussion in its treatment of potential amphipathic α-helical peptides, it was argued that the hydrophobicity maxima of most of the amphipathic helices are too small to be representative of membrane proteins.[68] Therefore, a dimer or trimer channel model with a total of six α-helical transmembrane segments was proposed.[52] These hypotheses await additional data on the structure of the colicin peptide in the membrane to test their validity. Inspection of Table V also suggests that there may be some differences in the structure of the channels of colicin E1, Ia, Ib in one group and colicins B (not discussed here)[72] and A in another.

A preliminary and abbreviated version of the work described in this chapter has already been published.[73]

Acknowledgment

The research described in this article was supported by grants from the NIH (GM-18457 and GM-38323) and NSF (DMB-84-03308). We thank Dr. V. L. Davidson for contributions to an initial stage of this project, Dr. J. Konisky for communication of the colicin Ia sequence prior to publication, and Sheryl Kelly and Janet Hollister for their dedicated efforts in the preparation of this manuscript. Computer programs, written in BASIC for an IBM PC, are available upon request.

---

[70] L. Raymond, S. L. Slatin, A. Finkelstein, Q.-R. Liu, and C. Levinthal, *J. Membr. Biol.* **92**, 255 (1986).

[71] Q.-R. Liu, V. Crozel, F. Levinthal, S. L. Slatin, A. Finkelstein, and C. Levinthal, *Proteins* **1**, 218 (1986).

[72] E. Schramm, J. Mende, V. Braun, and R. M. Kamp, *J. Bacteriol.* **169**, 3350 (1987).

[73] J. W. Shiver, W. R. Widger, A. A. Peterson, and W. A. Cramer, *in* "Membrane Proteins" (S. C. Goheen, ed.), p. 235. BioRad Laboratories, Richmond, CA, 1987.

[26] Order and Viscosity of Membranes: Analysis by Time-Resolved Fluorescence Depolarization

By MAARTEN P. HEYN

Intrinsic membrane proteins are embedded in a lipid matrix. Knowledge of the properties of this hydrocarbon chain environment is of critical importance in understanding membrane function. The lipid bilayer is a highly anisotropic system and rotational diffusion of acyl chains or rodlike probe molecules is expected to be anisotropic and restricted. Two parameters characterizing the rotational diffusion are of particular interest, and are accessible with time-resolved fluorescence depolarization methods: (1) the *rate* of rotational diffusion, which provides information on the size and shape of the rotating group as well as on the viscosity (dynamics), and (2) the *range* or extent of the restricted motion, which is a measure of the order of the anisotropic environment (structure).

Time-resolved fluorescence depolarization is a convenient method to determine these parameters. Its main advantages are (1) high sensitivity: nanomolar probe concentrations with milliliter samples are sufficient; (2) straightforward extraction of the information from the decay of the fluorescence anisotropy: the information on the rate of motion is obtained from the initial slope of the decay, and the information on the range of the motion is obtained from the nonzero asymptotic value which the fluorescence anisotropy reaches for long times, and (3) information is obtained on the nanosecond time scale: during the fluorescence lifetime almost no lateral diffusion occurs, and the observed signal is thus the ensemble average over labels in various fixed environments.

Some of the disadvantages are (1) the required fluorescent chromophore might perturb the environment that it is supposed to probe, (2) the depth in the bilayer and the average orientation of the probe molecules are not always well defined, (3) the required size of the chromophore precludes a high-resolution statement about local order and dynamics, (4) heterogeneity in probe environments is difficult to analyze.

The physical principles of the method are briefly as follows. An isotropic suspension of membranes in a cuvette is excited at time $t = 0$ by a nanosecond flash of linearly polarized light (this direction will also be referred to as vertical). Only those chromophores in the membranes with transition dipole moments approximately parallel to this polarization direction will be able to absorb. In this way, a highly ordered anisotropic distribution of excited chromophores is selected at $t = 0$ out of an originally isotropic distribution of unexcited chromophores (photoselection). The excited chromophores fluoresce and the fluorescence from this oriented population is highly polarized at $t = 0$, the intensity $I_\parallel$ observed with analyzer parallel to the excitation polarization direction being much larger than $I_\perp$ (analyzer perpendicular to excitation polarization direction). The degree of polarization is appropriately expressed by the fluorescence anisotropy $r(t)$:

$$r(t) = \frac{I_\parallel(t) - I_\perp(t)}{I_\parallel(t) + 2I_\perp(t)} \qquad (1)$$

The anisotropic angular distribution of excited chromophores created at $t = 0$ is not the equilibrium distribution. It will relax in time by rotational diffusion to the less-ordered equilibrium distribution which, in the case of membranes, is not expected to be isotropic. As a consequence, $r(t)$, which had a high initial value $r(0)$, will decay with a rate determined by the viscosity and the size of the chromophore to a nonzero asymptotic value r_∞ reached for long times, which is a measure of the degree of orientational constraint in the equilibrium state.

Measurement of Time-Resolved Fluorescence Anisotropy

Since this is a well-established method,[1] only those aspects will be discussed that are of particular relevance for work with membranes.

Membrane lipids do not fluoresce, so it is necessary to introduce a suitable fluorescent label into the membrane that reports the local order and viscosity. A wide choice of probe molecules is currently available for this purpose. Whereas for 1,6-diphenyl-1,3,5-hexatriene (DPH) the depth and orientation of the label are not well-defined, this is not the case for its various charged "surface-anchored" analogs such as DPH-propionic acid, trimethylammonium-DPH, and phospholipids incorporating the DPH moiety in the 2-position.[2-4] The second class of widely used probes are the parinaric acids, naturally occurring fatty acids, and their phospholipid analogs.[5] Many of these compounds are commercially available from Molecular Probes (Junction City, OR 97448). It is to be expected that the phospholipid analogs of DPH and parinaric acid interfere little with the

[1] R. B. Cundall and R. E. Dale, "Time-Resolved Fluorescence Spectroscopy in Biochemistry and Biology." Plenum, New York, 1983.
[2] R. A. Parente and B. R. Lentz, *Biochemistry* **24**, 6178 (1985).
[3] M. Cranney, R. B. Cundall, G. R. Jones, J. T. Richards, and E. W. Thomas, *Biochim. Biophys. Acta* **735**, 418 (1983).
[4] F. G. Prendergast, R. Haugland, and P. J. Callahan, *Biochemistry* **20**, 7333 (1981).
[5] P. K. Wolber and B. S. Hudson, *Biochemistry* **20**, 2800 (1981).

acyl chain-packing arrangement of the bilayer and provide at the same time fluorophores at a well-defined depth in the bilayer.

In order to perturb the bilayer as little as possible a low label-to-lipid ratio is desirable. Due to its high quantum yield, molar DPH-to-lipid ratios as low as 1 to 500 are sufficient for time-resolved measurements. At these low ratios the shape of the lipid-phase transition curve is unaffected.[6] Methods for incorporating the label into the membrane depend on the label. For DPH this problem has been recently discussed.[7]

Suspensions of membranes scatter light. The main effect in fluorescence anisotropy measurements is that the scattering leads to depolarization of the emitted light. Dilute samples are advisable for this purpose. Corrections for these effects have been discussed.[7]

Time-resolved fluorescence depolarization measurements are usually carried out with a single-photon counting apparatus using as a pulsed light source a nanosecond-flash lamp,[8] a laser, or a synchrotron at high repetition rates. To obtain a reliable value for the viscosity, it is important to measure the initial decay of $r(t)$ accurately. It is thus necessary to have good temporal resolution. With a properly tuned flash lamp of pulse width of 1.3–1.5 nsec (full width at half-height), a resolution of 0.2 nsec can be achieved which is sufficient to determine accurately the initial slope of the decay. The long time stability of the lamp pulse shape is of even greater importance because anisotropy measurements with membranes require several hours. Since there are always small changes in lamp profile during measurements, a compromise has to be found between good counting statistics in the two decay curves $I_\parallel(t)$ and $I_\perp(t)$ and lamp stability. The standard error of the anisotropy in a particular channel is determined by the number of counts in that channel. Good accuracy can be achieved if the count level at the end of the decay curves is larger than 100 per channel, or ~20000 counts in the maximum, or 5000–10000 counts in the difference between $I_\parallel$ and $I_\perp$. This can usually be achieved within 2 hr. The excitation profile should be measured before and after the anisotropy decay and the sum of the two pulses used to deconvolute the data. As a test of lamp stability the anisotropy data collected in 2 hr should be analyzed separately with the lamp pulse before and at the end of the data collection period. The parameters describing the anisotropy decay should be identical within the error bars. A high pulse repetition rate is desirable to collect data rapidly. Of course, the lamp pulse frequency should be stable. To average out the effects of lamp frequency fluctuations and

---

[6] M. P. Heyn, A. Blume, M. Rehorek, and N. A. Dencher, *Biochemistry* **20**, 7109 (1981).
[7] B. J. Litman and Y. Barenholz, this series, Vol. 81, p. 678.
[8] M. Rehorek, N. A. Dencher, and M. P. Heyn, *Biochemistry* **24**, 5980 (1985).

slight pulse shape instabilities, the analyzer may be rotated every 5×10^6 counts in the start channel (~100 sec at a frequency of 50 kHz).

The sensitivity of the detection system is usually different for $I_\parallel$ and $I_\perp$. To correct for this, the G-factor[1] should be determined several times before and after the measurement for each sample under the exact conditions of the experiment, except that the polarizer is set to the horizontal position. The G-factor can be conveniently calculated from the ratio of the total number of counts after the time to amplitude converter for vertically and horizontally polarized emission with the polarizer set to horizontal polarization.

Analysis and Interpretation of Membrane Fluorescence Anisotropy Data

Since the lamp pulse is always of nonzero width, the experimental data have to be deconvoluted in order to get the response to a δ-function excitation. This can be done in a number of ways.[1] A convenient way is to interface the multichannel analyzer to a DEC LSI 11/23 minicomputer and to analyze the data using a model for anisotropy decay described by Wahl[9] using a Marquardt nonlinear least-squares algorithm.[8] In the first part of the computer program, the convolution integral (*) of the excitation function, $L(t)$, with a sum of one to three exponentials is fitted to the experimental fluorescence decay curve:

$$S_{calc}(t) = L(t)*f(t) \tag{2}$$

with

$$f(t) = \sum_i \alpha_i e^{-t/\tau_i} \quad \text{and} \quad S_{exp}(t) = I_\parallel(t) + 2GI_\perp(t) \tag{3}$$

$S_{calc}(t)$, the parameters α_i and τ_i, and the time shift between $L(t)$ and $S_{exp}(t)$ are stored for later use. In the second part of the program, the convolution of $L(t)$ with the product of $f(t)$ and $r(t)$, divided by $S_{calc}(t)$, is fitted to $r_{exp}(t)$, with the parameters α_i and τ_i and the time shift kept constant:

$$r_{calc}(t) = L(t)*[f(t)r(t)]/S_{calc}(t) \tag{4}$$
$$r_{exp}(t) = [I_\parallel(t) - GI_\perp(t)]/S_{exp}(t) \tag{5}$$

For the analysis of membrane anisotropy data, it is necessary to use a sum of two exponentials plus a constant term.[8] The following model for $r(t)$ is then appropriate:

$$r(t) = r_\infty + (r_0 - r_\infty)[Re^{-t/\phi_1} + (1 - R)e^{-t/\phi_2}] \tag{6}$$

[9] P. Wahl, *Biophys. Chem.* **10**, 91 (1979).

with $R = r_1/(r_0 - r_\infty)$ and r_0 fixed. This expression contains four parameters. At $t = 0$, $r(0) = r_0$, which is a number that can be measured for any given chromophore and that depends on the angle between the absorption and emission transition dipole moment.[1] Its theoretical maximum value is 0.4. r_0 can be accurately determined by measuring $r(t)$ under conditions where the rotational diffusion of the chromophore is completely frozen on the nanosecond time scale. In this way we obtained for DPH in glycerol at $-5.6°$, a value of 0.394. Others have found $r_0 = 0.395$ for DPH.[10] In order to observe a large $r(t)$ signal, pick a chromophore with a large r_0 value. For parinaric acid r_0 is also very close to the theoretical maximum.[5] Rather than make the fit with r_0 fixed as in Eq. (6), it is of course also possible to make a five-parameter fit including r_0. The proper determination of r_0 is difficult, especially if ϕ_1 is short. The large errors in the amplitude of the fast component (0.2–0.5 nsec) determine the errors in r_0. Since the fitted value of r_0 for DPH is always within the error bars equal to 0.4, we recommend using a four-parameter fit with r_0 fixed. In order to determine the parameters of the fast component properly, it is advisable to start the fit of $r(t)$ at the channel number where the rising edge of the lamp pulse is at 50% of its maximal value. The following example illustrates the necessity of using a two-exponential fit for $r(t)$.[8] For DPH in DMPC vesicles at 35°, a single exponential fit with $r_0 = 0.4$ gave an unacceptably high χ_ν^2 of 3.55. Including r_0 in the fit led to a better fit ($\chi_\nu^2 = 1.08$) but r_0 was underestimated (0.302) and the residuals showed systematic deviations. A double-exponential analysis with $r_0 = 0.4$ yielded $\chi_\nu^2 = 0.928$ with $\phi_1 = 0.23 \pm 0.03$ nsec, $\phi_2 = 2.14 \pm 0.10$ nsec, $R = 0.42 \pm 0.03$, and $r_\infty = 0.035 \pm 0.001$. The two exponentials are of comparable amplitude. The standard deviations of the parameters can be calculated from the diagonal elements of the error matrix. The relative errors of the parameters of $r(t)$ are typically[8] ϕ_1: 15%, ϕ_2: 7%, R: 6%, and r_∞: 3%. A proper analysis of the initial decay is of special importance in connection with the determination of the viscosity (see below).

Having measured and fitted the anisotropy according to Eq. (6), it is of course possible to calculate from it the expected steady-state anisotropy $\langle r(t) \rangle$ by integration[8]:

$$\langle r(t) \rangle = \frac{\int_0^\infty r(t)f(t)\,dt}{\int_0^\infty f(t)\,dt}$$

$$= r_\infty + \frac{r_0 - r_\infty}{\sum_i \alpha_i \tau_i} \left[R\phi_1 \sum_i \frac{\alpha_i \tau_i}{\phi_1 + \tau_i} + (1 - R)\phi_2 \sum_i \frac{\alpha_i \tau_i}{\phi_2 + \tau_i} \right] \quad (7)$$

---

[10] S. Kawato, K. Kinosita, and A. Ikegami, *Biochemistry* **16**, 2319 (1977).

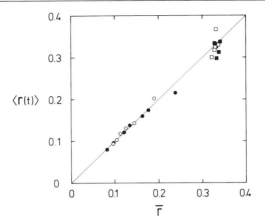

FIG. 1. Comparison of the time average of the anisotropy decay $\langle r(t) \rangle$, calculated according to Eq. (7), with the corresponding steady-state anisotropy $\bar{r}$ for DPH in bacteriorhodopsin/DMPC vesicles of various lipid-to-protein ratios. Ideally, all the points should be on the diagonal. (■, □) Data taken at 5°, (●, ○) data taken at 35°. Open symbols: the retinal chromophore removed by bleaching.

It is a good test for the performance of the instruments and the program to compare the $\langle r(t) \rangle$ values so obtained with steady-state values $\bar{r}$ measured on the same samples with a steady-state fluorometer. Ideally, the two values should be equal. In Fig. 1 such a comparison is made for a set of DMPC/bacteriorhodopsin vesicles of various $\bar{r}$ values.

The way to extract the information on order and dynamics from the parameters describing the decay of $r(t)$ is straightforward. One approach assumes that the chromophores carry out a wobbling diffusion within a cone of opening angle θ_c.[11] The cone angle θ_c is a measure of the orientational constraint of the equilibrium distribution. θ_c can be calculated from the experimental ratio r_∞/r_0 by solving[11]:

$$\frac{r_\infty}{r_0} = \left[\frac{\cos \theta_c (1 + \cos \theta_c)}{2} \right]^2 \quad (8)$$

The rotational diffusion constant D for wobbling in the cone can be calculated from the ϕ_i's but depends in addition on the cone angle.[11,12] When the decay of $r(t)$ is single exponential,

$$D_{\text{int}} = \langle \sigma \rangle / \phi_1 \quad (9)$$

[11] K. Kinosita, Jr., S. Kawato, and A. Ikegami, *Biophys. J.* **20**, 289 (1977).
[12] G. Lipari and A. Szabo, *Biophys. J.* **30**, 489 (1980).

where $\langle\sigma\rangle$ is a known function of θ_c.[11,12] When the decay of $r(t)$ is double exponential [Eq. (6)], the appropriate mean correlation time which measures the area under the decay curve is[8,13]

$$\bar{\phi}_{int} = \int_0^\infty \frac{r(t) - r_\infty}{r_0 - r_\infty} dt = R\phi_1 + (1 - R)\phi_2 \quad (10)$$

and the rotational diffusion constant D_{int} is calculated from

$$D_{int} = \langle\sigma\rangle/\bar{\phi}_{int} \quad (11)$$

The second approach for analyzing the data is model-free.[11,14,15] It turns out[14,15] that the quantity $(r_\infty/r_0)^{1/2}$ equals the second moment of the normalized orientational distribution function $f(\theta)$:

$$\left(\frac{r_\infty}{r_0}\right)^{1/2} = \int_{-1}^{+1} \left(\frac{3\cos^2\theta - 1}{2}\right) f(\theta) \, d\cos\theta = S \quad (12)$$

This is usually called the order parameter S. From r_∞ we can thus calculate this moment of the unknown distribution function $f(\theta)$ which is an appropriate measure of the order. To learn more about $f(\theta)$ we need to measure the higher moments, which is, in principle, possible using oriented samples. The rotational diffusion constant in this approach is calculated from the initial slope of the normalized anisotropy decay.[11,13] The appropriate mean correlation time $\bar{\phi}_{der}$ is

$$\bar{\phi}_{der} = -\left[\frac{\partial}{\partial t}\left(\frac{r(t) - r_\infty}{r_0 - r_\infty}\right)\right]_{t=0}^{-1} = \left(\frac{R}{\phi_1} + \frac{1-R}{\phi_2}\right)^{-1} \quad (13)$$

and the rotational diffusion constant D_{der} is computed from $\bar{\phi}_{der}$ using

$$D_{der} = (1 - r_\infty/r_0)/6\bar{\phi}_{der} \quad (14)$$

In both approaches, the viscosity η of the membrane bilayer is calculated from

$$\eta = \frac{kT}{6DV_e f} \quad (15)$$

using for D either D_{int} or D_{der}. V_e and f denote the effective volume and shape factor for the probe, approximated by a prolate ellipsoid. Based on 17 determinations, the average literature value of $V_e f$ for DPH is $(1.63 \pm 0.34) \times 10^{-22}$ cm^3.[8]

[13] K. Kinosita, Jr., S. Kawato, and A. Ikegami, *Adv. Biophys.* **17**, 147 (1984).
[14] M. P. Heyn, *FEBS Lett.* **108**, 359 (1979).
[15] F. Jähnig, *Proc. Natl. Acad. Sci. U.S.A.* **79**, 6361 (1979).

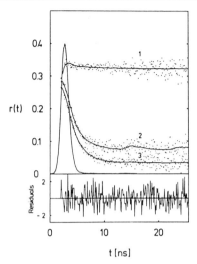

FIG. 2. Experimental (·····) and calculated (——) anisotropy decay curves for pure DMPC vesicles at 35° (curve 3), and for bacteriorhodopsin/DMPC vesicles of molar protein-to-lipid ratio 1 : 115 at 35° (curve 2) and at 0.5° (curve 1). The weighted residuals for curve 3 are shown at the bottom. The solid curves represent the anisotropy decay curves $r_{calc}(t)$ *after* convolution with the lamp pulse, calculated according to Eq. (4). The bump in curve 2 around 15 nsec is due to the corresponding bump in the lamp pulse. The bump shows only up in curve 2, since strong energy transfer from DPH to retinal leads to a rapid decay of the fluorescence intensity in this case.

Examples, Comparison with Steady-State Experiments, and Outlook

With all the membrane systems studied so far, it has been found that $r_\infty \neq 0$, indicating hindered reorientational motion. Increasing the protein-to-lipid or cholesterol-to-lipid ratio in reconstituted systems has invariably led to an increase in r_∞ (order) without much effect on D (viscosity). An example is shown in Fig. 2. The anisotropy decay of DPH at 35° in pure DMPC vesicles (curve 3) is compared with that in reconstituted DMPC/bacteriorhodopsin vesicles at a molar protein-to-lipid ratio of 1 : 115 (curve 2). Bacteriorhodopsin does not aggregate under these conditions.[16] The initial slope of the two curves is about the same, showing that the viscosity is not much affected by the incorporation of the protein. The end value r_∞, however, is about twice as large in curve 2, indicating a protein-induced increase in order. In Table I the values of r_∞ and η_{der} are collected for a set of DMPC/bacteriorhodopsin vesicles of various lipid-

[16] M. P. Heyn, R. J. Cherry, and N. A. Dencher, *Biochemistry* **20**, 840 (1981).

TABLE I
Viscosity of Bacteriorhodopsin/DMPC Vesicles[a]

Molar bacteriorhodopsin/DMPC ratio	r_∞	$\bar{r}$	$\eta_{MV}(P)$	$\eta_{der}(P)$
0	0.034	0.081	0.471	0.122 ± 0.019
1/393	0.045	0.095	0.618	0.143 ± 0.017
1/185	0.049	0.103	0.721	0.156 ± 0.015
1/174	0.059	0.115	0.881	0.133 ± 0.014
1/115	0.065	0.128	1.077	0.207 ± 0.018
1/97	0.070	0.144	1.353	0.132 ± 0.017

[a] η_{MV} calculated according to Eq. (18) in Ref. 17. Temperature at 35°. Viscosity in poise (P).

to-protein ratios at 35° (above the phase transition for DMPC). Whereas η_{der} is roughly independent of this ratio, r_∞ is not. Also included is a column with the steady-state values $\bar{r}$. These allow the so-called "microviscosity" to be calculated according to a standard recipe.[17] These values are collected in the column labeled η_{MV}. The chromophore of bacteriorhodopsin was removed in these experiments in order to eliminate the energy transfer from DPH to bacteriorhodopsin which shortens the DPH lifetime, leading to artificially high $\bar{r}$ values.[18] A large discrepancy between η_{MV} and η_{der} already exists for the pure lipid vesicles, which increases when more bacteriorhodopsin is incorporated. For a molar bacteriorhodopsin-to-lipid ratio of 1:97, which is a typical value for membranes, η_{MV} is a factor 10 larger than η_{der} in the liquid crystalline phase. In the microviscosity recipe, r_∞ is assumed to be zero (unrestricted isotropic rotation) and the large increase in $\bar{r}$ with increasing protein-to-lipid ratio thus has to be interpreted as a major increase in viscosity leading to the gross overestimate. The time-resolved measurements show on the other hand that the viscosity remains roughly constant, but that the proteins lead to an increase in order. The data of Table I show clearly that the claim that the error in η_{MV} above the phase transition of the lipids is rather small[7] is incorrect at realistic protein-to-lipid ratios.

The range of suitable fluorescent membrane probes is increasing steadily. In particular, the availability of phospholipid analogs of DPH and parinaric acid is a significant step forward. In the future, we expect that improved commercially available instrumentation will make the measure-

---

[17] M. Shinitzky and Y. Barenholz, *Biochim. Biophys. Acta* **515**, 367 (1978).
[18] M. Rehorek, N. A. Dencher, and M. P. Heyn, *Biophys. J.* **43**, 39 (1983).

ment of time-resolved fluorescence anisotropy of membranes a routine matter. Many agents and factors affecting the order and viscosity of membranes will be investigated. Usage in the field of lipid–protein interactions, as described above, will continue. The combination of fluorescence anisotropy with fluorescence energy transfer may well be very useful for studying lateral inhomogeneities in order and viscosity.[8] If the proteins carry a suitable acceptor for the probe's fluorescence, then those donor probes in the immediate neighborhood of the protein will be strongly quenched and will contribute little to the fluorescence anisotropy, which will be dominated by contributions from the bulk lipids. By removing the acceptors the previously masked contributions of the labels in the boundary region will be observed as well. In this way differences in dynamics and order in the bulk and boundary region may be investigated. By suitable choice of the donor–acceptor pair (Förster radius) different areas around the proteins may be sampled in this way.

[27] Rotational and Translational Diffusion in Membranes Measured by Fluorescence and Phosphorescence Methods

By THOMAS M. JOVIN and WINCHIL L. C. VAZ

Introduction

The dynamic nature of biological membranes is well established.[1] The lipid bilayer provides a compartmentalized matrix within and on which embedded proteins and other macromolecules are in a constant state of motion and redistribution, processes that are essential to cellular mechanisms such as metabolism, endocytosis and secretion, differentiation, locomotion, and signal transduction. The lipid bilayer in natural and reconstituted membranes imposes constraints[2,3] which restrict the larger segmental and global motions of constituent molecules to the microsecond to millisecond time domain. We consider in this chapter methods for analyzing such rotational and translation displacements based on the time-resolved emission of light in the form of fluorescence and phospho-

---

[1] M. D. Houslay and K. K. Stanley, "Dynamics of Biological Membranes." Wiley, New York, 1982; see also A. S. Perelson, C. DeLisi, and F. W. Wiegel (eds.), "Cell Surface Dynamics." Dekker, New York, 1984.
[2] M. Shinitzky (ed.), "Physiology of Membrane Fluidity," Vols. 1 and 2. CRC Press, Boca Raton, Florida, 1984.
[3] R. J. Cherry, *Biochim. Biophys. Acta* **559**, 289 (1979).

rescence. These techniques offer the sensitivity and selectivity required for studies of relatively sparse cell-surface receptors or other membrane components in preparations examined in the microscope and in suspension. Thus, we do not consider less-sensitive methods based on magnetic resonance (NMR, ESR) or other spectroscopic phenomena (transient absorption,[4-6] fluorescence correlation, IR, Raman, X-ray).

The physical–analytical techniques for the quantitation of macromolecular diffusion in the microsecond to millisecond range have been reviewed previously with respect to biological membranes[4-14] and, in the case of lateral diffusion, to artificial lipid bilayer membranes.[15-17] The vast literature associated with the motions of small fluorescence probes in membranes, a topic not treated in this chapter, can be accessed via recent articles and monographs.[14,18-21]

Rotational Diffusion

The rotational motions of a macromolecule are very sensitive to its size, shape, and rigidity, and to the nature of the immediate environment, which can exert a viscous retardation or impose anisotropic barriers to diffusion.[22] In the case of integral membrane proteins which span the lipid

[4] R. J. Cherry, this series, Vol. 54, p. 47; see also this series, Vol. 88, p. 248.
[5] S. Kawato and K. Kinosita, Jr., *Biophys. J.* **36**, 277 (1981).
[6] K. Kinosita, Jr., S. Kawato, and A. Ikegami, *Adv. Biophys.* **17**, 147 (1984).
[7] W. Hoffmann and C. J. Restall, in "Biomembrane Structure and Function" (D. Chapman, ed.), p. 257. Verlag Chemie, Weinheim, Federal Republic of Germany, 1984; see also E. K. Murray, C. J. Restall, and D. Chapman, p. 119, in Ref. 14.
[8] M. Edidin, *Annu. Rev. Biophys. Bioeng.* **3**, 179 (1974).
[9] W. W. Webb, *Ann. N.Y. Acad. Sci.* **366**, 300 (1981).
[10] K. Jacobson and J. Wojcieszyn, *Comments Mol. Cell. Biophys.* **1**, 189 (1981).
[11] R. Peters, *Cell Biol. Int. Rep.* **5**, 733 (1981).
[12] J. Schlessinger and E. L. Elson, *Methods Exp. Phys.* **20**, 197 (1982).
[13] D. Axelrod, *J. Membr. Biol.* **75**, 1 (1983).
[14] P. M. Bayley and R. E. Dale (eds.), "Spectroscopy and the Dynamics of Molecular Biological Systems." Academic Press, London, 1985.
[15] W. L. C. Vaz, Z. I. Derzko, and K. A. Jacobson, *Cell Surf. Rev.* **8**, 83 (1982).
[16] W. L. C. Vaz, F. Goodsaid-Zalduondo, and K. Jacobson, *FEBS Lett.* **174**, 199 (1984).
[17] R. M. Clegg and W. L. C. Vaz, in "Progress in Protein–Lipid Interactions" (A. Watts and J. J. H. H. M. de Pont, eds.), p. 173. Elsevier, Amsterdam, 1985.
[18] R. B. Cundall and R. E. Dale (eds.), "Time-Resolved Fluorescence Spectroscopy in Biochemistry and Biology," NATO Adv. Study Inst. Ser. Plenum, New York, 1983.
[19] A. G. Szabo and L. Masotti (eds.), "Excited State Probes in Biochemistry and Biology," NATO Adv. Study Inst. Ser. Plenum, New York, in press.
[20] J. R. Lakowicz, in "Spectroscopy in Biochemistry" (J. E. Bell, ed.), Vol. 1, p. 195. CRC Press, Boca Raton, Florida, 1981.
[21] A. Szabo, *J. Chem. Phys.* **81**, 150 (1984).
[22] G. Weber, *Adv. Protein Chem.* **8**, 415 (1953).

bilayer, intermolecular interactions (liganding, association, aggregation) within the plane of the membrane, on its surface, or extending into the surrounding medium are of functional importance, for example, in the transduction of energy or information (signaling). The paramount experimental challenge in studies of such systems is how to achieve the required: (1) selectivity (exploit specific intrinsic or extrinsic probes), (2) time resolution (use photochemical processes induced by pulsed or modulated light sources in the desired time domain), and (3) sensitivity (detect emitted light either with photomultipliers or intensified two-dimensional sensors). As indicated, these requirements can be met by adaptation of the classical[22] spectroscopic approach based on photoselective absorption and time-dependent polarized emission. The following alternative strategies currently in use are based on excitation of a suitable covalent or noncovalent probe to the first excited singlet state which is then photoselectively "photobleached" by either an irreversible or a reversible process (i.e., by generation of a long-lived triplet or ground-state photochromic intermediate). One has then the following options[23,24]: (1) observe the transient excited triplet state directly via its polarized emission (phosphorescence, delayed fluorescence), (2) monitor the depleted ground-state population of the probe by steady-state polarized emission (fluorescence anisotropy), and/or (3) monitor ground-state transient intermediates by steady-state polarized fluorescence. The photophysical relationships are shown schematically in Fig. 1 and the various techniques based on them are summarized in Table I.

The method of transient absorption (dichroism) based on the triplet state of halogenated fluorescein probes (e.g., eosin, tetrabromofluorescein) as well as natural chromophores has been exploited extensively. Cherry has given accounts in prior volumes of this series.[4] The sensitivity of the absorption measurement, however, is inherently limited such that probe concentrations must lie in the micromolar range. Let us examine this requirement in more detail. Consider a eukaryotic cell expressing 10^5 molecules of a surface antigen or receptor. A suspension of such cells at a density of 10^7 per milliliter (a practical upper limit due to the depolarizing effects of turbidity) corresponds to a concentration of the protein in question of only about 2 nM. While one may devise probes (hormones, toxins, antibodies) with greater than one conjugated chromophore, the attainable enhancement of sensitivity is necessarily modest. A quantitative assessment can be made by expressing the absorption of excitation energy (pho-

---

[23] E. Blatt and T. M. Jovin, in "Photophysical and Photochemical Tools in Polymer Science" (M. Winnik, ed.), NATO Adv. Study Inst. Ser. C, p. 351. de Reidel, Dordrecht, The Netherlands, 1986.

[24] J. Matkó and T. M. Jovin, manuscript in preparation.

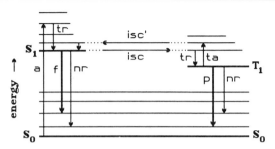

FIG. 1. Photophysical cycle leading to luminescence. Light is absorbed from the ground state S_0 (a), drops to the first excited singlet state S_1 (tr), decays by prompt fluorescence (f) and nonradiative pathways (nr), or crosses (isc) to the first excited triplet state (T_1). From there deactivation can occur by emission (phosphorescence p) or by nonradiative means (nr). Alternatively, thermal reactivation (ta) and intersystem crossing (isc') back to the singlet state lead to delayed fluorescence (f). Not shown is the possible intervention of fluorescent photochromic ground state species exploited in the TRF technique.

tons) by an individual chromophore as a probability given by its absorption cross-section $\sigma_\lambda = 10^3 \ln(10) \varepsilon_\lambda/N_A$ (cm^2 molecule^{-1}) where ε_λ is the molar absorption coefficient (M^{-1} cm^{-1}), and N_A is Avogadro's number. For a 2 nM solution or suspension, an illuminated volume of ~20 μl (a typical value in our experiments) corresponding to 2×10^{10} molecules and an ε_λ of 8.3×10^4 M^{-1} cm^{-1} (eosin[25]), only 10^{-5} of the excitation will be absorbed, a value well below the normal limits of detectability by light transmission. Consider, however, the hypothetical experiment in terms of luminescence detection and using an excitation pulse of 2 mJ (5×10^{12} photons at 520 nm). The number of absorbed photons is $\sim 4 \times 10^7$. This absorbed energy is converted into the various forms of emission, each with its characteristic quantum yield. In the case of delayed fluorescence or phosphorescence, both the quantum yield and the net detection efficiency may be on the order of 0.01[26,27]; thus, in our example, ~4000 photons/decay are available for generating signals at the detector. With suitable averaging (e.g., 10^3–10^4 repetitive excitations), the time-resolved polarized emission can be defined. [For such experiments, erythrosin (tetraiodofluorescein)[27] is a better choice than eosin; see below.] Note that this example corresponds to the regime of low absorption and relatively low irradiance such that saturation of the chromophore (and an accompanying depolarization[28]) are avoided.

[25] R. J. Cherry, A. Cogoli, M. Oppliger, G. Schneider, and G. Semenza, *Biochemistry* **15**, 3653 (1976).
[26] T. M. Jovin, M. Bartholdi, W. L. C. Vaz, and R. H. Austin, *Ann. N.Y. Acad. Sci.* **366**, 176 (1981).
[27] P. B. Garland and C. H. Moore, *Biochem. J.* **183**, 561 (1979).
[28] A. Szabo, R. M. Clegg, and T. M. Jovin, manuscript in preparation.

TABLE I
METHODS FOR MEASURING SLOW ROTATIONAL DIFFUSION USING FLUORESCENCE AND PHOSPHORESCENCE

Method	Advantages[a,b]	Disadvantages[a,b]
DLA (delayed luminescence anisotropy)[c] Phosphorescence	Dark reference state, sensitivity	Low quantum yield, low anisotropy, gating required
Delayed fluorescence	High anisotropy, increases with temperature	Interference by prompt emission (gating required)
FRA (fluorescence recovery anisotropy)[c,d]	Useful in the microscope as well as with cuvettes	
Reversible bleaching (triplet state)	High sensitivity, high anisotropy, long lifetimes	Bleaching, differential measurement, critical optics
Irreversible bleaching	Aerobic	Single-shot, limited to longer time range
TFA (transient fluorescence anisotropy)[b]	Aerobic, intrinsic probes, dark reference state	Lifetime environment sensitive, bleaching, unproven
FCS (fluorescence correlation spectroscopy)[d]	Independent of excited state lifetime	Technically demanding, unproven
FPR (fluorescence photobleach recovery)[e]	Independent of excited state lifetime	Requires 2π collection, unproven

[a] See also P. B. Garland and P. Johnson, in "Spectroscopy and the Dynamics of Molecular Biological Systems" (P. M. Bayley and R. E. Dale, eds.), p. 95. Academic Press, London, 1985.
[b] E. Blatt and T. M. Jovin, in "Photophysical and Photochemical Tools in Polymer Science" (M. Winnik, ed.), NATO Adv. Study Inst. Ser. C, p. 351. de Reidel, Dordrecht, The Netherlands, 1986; see also J. Matkó and T. M. Jovin, manuscript in preparation.
[c] Other designations: FD, fluorescence depletion; FRS, fluorescence recovery spectroscopy; see text for references.
[d] R. Rigler, in "Spectroscopy and the Dynamics of Molecular Biological Systems" (P. M. Bayley and R. E. Dale, eds.), p. 35. Academic Press, London, 1985.
[e] W. A. Wegener and R. Rigler, Biophys. J. **46**, 787 (1984).

DLA (Delayed Luminescence Anisotropy): Phosphorescence and Delayed Fluorescence

Probes. Three of the methods cited in Table I have been used most extensively since their introduction in 1979, i.e., the polarized emission originating directly (phosphorescence[29,30]) or indirectly (delayed fluorescence[31]) from the long-lived first excited triplet state (T_1; Fig. 1). The extrinsic probes of choice have been members of the fluorescein family substituted with the heavy atoms Br and I, which greatly potentiate intersystem crossing to the triplet state. Thus, the tetrahalogenated eosin and erythrosin are available as noncovalent ligands or with reactive groups (isothiocyanates, iodoacetamides, maleimides, thiosemicarbazides, and hydrazides) designed for conjugation to the amino and sulfhydryl side groups of peptides and proteins and to alcohol and carboxylic groups, e.g., of polysaccharides (Table II; a good primary source for these reagents is Molecular Probes, Inc., Eugene, OR). In the absence of oxygen, both eosin and erythrosin show delayed fluorescence at 500–600 nm and phosphorescence at 620–760 nm.[26,27] Erythrosin has triplet and phosphorescence yields significantly greater than those of eosin. Furthermore, its fluorescence quantum yield (0.02) is only about one-tenth that of eosin,[26,27] thereby alleviating the severe problem posed by the need to suppress (by gating the photomultiplier) the intense prompt emission elicited by pulsed excitation (see below). The triplet lifetimes, however, are generally shorter, i.e., a maximum of 0.5 and 1.8 msec, respectively, for erythrosin and eosin immobilized in plastic[27,32] or glycerol.[30] Diiodofluorescein is useful in measurements of polarized delayed fluorescence[26,33] but not of phosphorescence.[26] In general, the photophysical properties (spectra, lifetimes, quantum yields) of the halogenated fluoresceins change on binding and/or conjugation.[25–27,29] The generally more complex decay behavior can present problems in the analysis of the time-dependent polarization (see below). An additional complication is a pronounced tendency for nonspecific interactions of the free chromophores (e.g., erythrosin) with cell surfaces[34] and lipid vesicles[35] such that the specific signals may be masked by an intense phosphorescence with a

---

[29] C. Moore, D. Boxer, and P. B. Garland, *FEBS Lett.* **108**, 161 (1979).
[30] R. H. Austin, S. L. Chan, and T. M. Jovin, *Proc. Natl. Acad. Sci. U.S.A.* **76**, 5650 (1979).
[31] R. Greinert, H. Staerk, A. Stier, and A. Weller, *J. Biochem. Biophys. Methods* **1**, 77 (1979).
[32] A. F. Corin, E. Blatt, and T. M. Jovin, *Biochemistry* **26**, 2207 (1987).
[33] R. Greinert, S. A. E. Finch, and A. Stier, *Xenobiotica* **12**, 717 (1982).
[34] S. Damjanovich, L. Trón, J. Szöllösi, R. Zidovetzki, W. L. C. Vaz, F. Regateiro, D. J. Arndt-Jovin, and T. M. Jovin, *Proc. Natl. Acad. Sci. U.S.A.* **80**, 5985 (1983).
[35] W. Reed, D. Lasic, H. Hauser, and J. H. Fendler, *Macromolecules* **18**, 2005 (1985).

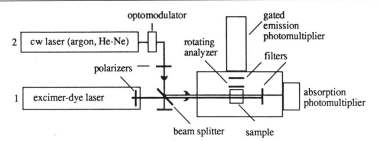

FIG. 2. Instrumentation for polarized luminescence and absorption measurements. See text for details. The two beams emanating from lasers 1 and 2 (used for FRA and TRF) are shown displaced for clarity; in practice, they are maintained in close coaxiality. Not shown are the microprocessor control and data acquisition elements.

time-dependent anisotropy. Thus, it is essential to rigorously exclude free dye from conjugated ligands or proteins by using a combination of dialysis, gel permeation and affinity chromatography, exposure to absorbent beads[36] (Bio-Beads SM-2, Bio-Rad, Richmond, CA), butanol extraction,[37] and other appropriate procedures. Numerous other chromophores should be suitable for triplet emission studies of biological membranes; a program for the systematic exploitation of intrinsic probes and the development of new extrinsic probes is clearly required; the photophysical literature is rich in possibilities.[38]

Instrumentation. The apparatus currently used in our laboratory for the measurement of time-resolved polarized delayed luminescence is shown schematically in Fig. 2 and described in more detail elsewhere.[23,26,30,32,39,40] It includes provisions for some of the other techniques summarized in Table I. The intense (actinic) source of pulse light energy for generation of the triplet state consists of an excimer laser (Model EMG50E, LambdaPhysik, Göttingen, FRG) operated with XeCl at 308 nm and driving a companion dye laser (Model FL2000). With the laser dye coumarin 307, 3–10 mJ in a 5- to 10-nsec vertically polarized pulse can be generated at 515 nm; the pulse-to-pulse variation can be significant, i.e., up to 20%. The diameter of the unfocused beam impinging on the sample

[36] E. Spack, Jr., Ph.D. thesis. The Johns Hopkins University, Baltimore, Maryland, 1985.
[37] G. D. Niswender, D. A. Roess, H. R. Sawyer, W. J. Silvia, and B. G. Barisas, *Endocrinology* **116**, 164 (1985).
[38] H. E. Lessing and A. von Jena, in "Laser Handbook" (M. L. Stitch, ed.), p. 753. North-Holland, Amsterdam, 1976.
[39] E. D. Matayoshi, A. F. Corin, R. Zidovetzki, W. H. Sawyer, and T. M. Jovin, in "Mobility and Recognition in Cell Biology" (H. Sund and C. Veeger, eds.), p. 119. de Gruyter, Berlin, Federal Republic of Germany, 1983.
[40] A. F. Corin, E. D. Matayoshi, and T. M. Jovin, p. 53, in Ref. 8.

FIG. 3. Cuvette used for rotational diffusion measurements involving triplet states. A nylon plunger with a circular Teflon disk at its tip fits loosely in the Teflon stopper and is used to gently agitate the solution (by hand or machine) under continuous argon flow so as to effect deoxygenation of the sample. The plunger is drawn up during measurement.

compartment at a distance of 1 m is ~2 mm. The sample is contained in a fused quartz fluorescence cuvette (Hellma, Müllheim, FRG), typically with inner dimensions of 5 × 5 mm (Fig. 3). The solution of suspension (350 μl) is deoxygenated by repetitive, gentle actuation of a Teflon plunger in the presence of a continuous flow of hydrated oxygen-free argon (99.998% purity, passed through an Oxisorb O_2 scrubber, Messer-Griesheim, Kassel, FRG, at an initial rate of 25 ml/min). This process requires about 5 min, as judged by the progressive increase in the amplitude and decay times of the delayed luminescence, and can be effected by hand or with a three-position, automatic, thermostatted, motorized driver. Other physical and biochemical procedures for creating anaerobic conditions have been described.[4,41–44] The interaction of certain ligands with cell-surface receptors can lead to a relative insensitivity to environmental O_2 such that extensive deoxygenation is not required.[45] During measurements, the cuvette is thermostatted in a metal holder. Collection and collimating lenses direct the emission (at 90° to the excitation path)

[41] K. Kinosita, Jr., S. Ishiwata, H. Yoshimura, H. Asai, and A. Ikegami, *Biochemistry* **23**, 5963 (1984).
[42] P. Johnson and P. B. Garland, *FEBS Lett.* **132**, 252 (1981).
[43] P. Johnson and P. B. Garland, *FEBS Lett.* **153**, 391 (1983).
[44] T. M. Eads, D. D. Thomas, and R. H. Austin, *J. Mol. Biol.* **179**, 53 (1984).
[45] R. Zidovetzki, Y. Yarden, J. Schlessinger, and T. M. Jovin, *Proc. Natl. Acad. Sci. U.S.A.* **78**, 6981 (1981).

through a sheet polarizer (analyzer), appropriate filters, and into a photomultiplier with an S20 quartz cathode (9817QAG, EMI, Ruislip, England). Mirrors can be inserted about the sample chamber in order to enhance the excitation and emission collection efficiencies. The analyzer is rotated in 90° steps by a stepping motor under microprocessor control, thereby generating records corresponding to positions of the emission analyzer parallel ($s_\parallel$) and perpendicular ($s_\perp$) to the excitation. The number of repetitions at each setting (typically 32–50) and the number of ($s_\parallel$, $s_\perp$) cycles (typically 16–40) collected for signal averaging are also determined by software. Thus, a usual data set consists of 1000–2000 summed decay curves (for each of the two polarized components) collected at a repetition rate of about 10 Hz. The averaging not only improves the signal-to-noise ratio in accordance with photon statistics but also compensates for the inherent laser pulse energy instability such that the final $s_\parallel$ and $s_\perp$ decay curves are accurately matched.

Electronic gating of the photomultiplier (by driving the focusing electrode negative to the photocathode) is essential in order to suppress the intense burst of prompt fluorescence that reflects the immediate radiative deactivation of the singlet state (Figs. 1 and 4); this process has the typical lifetimes of fluorescence (nanoseconds) and is essentially coincident with the laser pulse. With turbid samples, scattered light and induced filter luminescence also contribute to this perturbation signal, which may be orders of magnitude greater in intensity than the ensuing delayed fluorescence and/or phosphorescence. At the gain settings required for the latter, the prompt signal in the absence of gating can destroy the photocathode or at the very least generate long-lasting transients which obscure the true course of the delayed emission. The gating requirement remains the most vexing technical problem in this form of time-resolved transient spectroscopy; more refined technical solutions (other circuits, other detectors) would permit the extension of the technique into the nanosecond domain, i.e., well below the current practical limit of a few microseconds. It is worth noting that our instrument employs a fast (100 MHz) transient recorder for quantitation of the analog photomultiplier output; other designs incorporate photon-counting techniques.[27,31]

For the reasons cited, the selection of emission filters is rather critical. They should have sharp cutoff characteristics and exhibit minimal induced emission. We have found the KV series of fluorescence filters (Schott, Mainz, FRG) very useful (particularly the KV550) followed in the optical train (in the direction of the photomultiplier) by RG filters at wavelengths > 600 nm. Liquid filters[46] in 5-cm square quartz cuvettes

---

[46] S. L. Murov, "Handbook of Photochemistry." Dekker, New York, 1973.

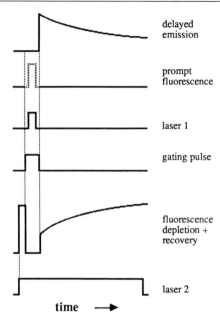

FIG. 4. Temporal sequence of excitation and emission. Laser 1 populates the triplet state which decays by delayed emission. The prompt fluorescence during the laser flash is suppressed at the detector photocathode by electronic gating (<4 μsec). With the two-laser techniques, a second continuous laser (2) is used to evoke a steady-state fluorescence signal from the ground state. This too is subject to gating. See also Ref. 32.

(Hellma) and dielectric interference filters can also be used. It is often necessary to attenuate the excitation energy in order to avoid saturation. Broad-band neutral density filters (e.g., the ND series of Schott) selected for this purpose must be resistant to the high power levels of the actinic laser.

A second laser (2 in Fig. 2) supplies an additional source for lower energy excitation of programmable amplitude and polarization state using electrooptic and/or acoustooptic modulators. The latter can be driven at megahertz rates and with any desired degree of amplitude modulation. Laser 2 is used (1) as an alternative pumping light source with which one can minimize the prompt fluorescence artifact by extending the excitation from a few nanoseconds to the microsecond–millisecond range. The triplet lifetimes can be obtained from both the "on" and the "off" phases of excitation; the determination of rotational depolarization, of course, is limited to times longer than those selected for excitation; (2) to achieve rapid repetition frequencies, i.e., greater than the <1 kHz rates achievable with the excimer laser; and (3) as a monitoring beam for assessing the depletion of the ground state or generation of a photophysical intermediate (e.g., the triplet state; Table I) by steady-state absorbance or fluorescence (see also section on fluorescence recovery anisotropy). One can cite alternative choices for laser 1 (Fig. 2), particularly the N_2 laser used in many installations as well as the Nd:YAG pulsed laser.[47] The latter is used directly by harmonic generation of visible and UV frequencies. In all cases, combinations of pump and tuneable dye lasers provide the greatest flexibility. In the case of delayed luminescence, net pulse energies of 10–100 μJ suffice for "macroscopic" samples (10–200 μl and probe concentrations of >10 nM); in depletion experiments based on pulsed excitation (Table I), millijoules and fairly tight focusing are required in order to achieve a significant depletion of the ground state.[40]

The relative timing of the various excitation and emission processes is shown in Fig. 4 (see also below).

Sample Preparation. The following is a typical protocol for cellular studies involving external ligands with conjugated probes[45,48]: Attached monolayer tissue culture cells are removed from the substrate by incubation for 10 min with 2 mM EDTA in a Tris–saline buffer followed by gentle agitation. Ca^{2+} and Mg^{2+} are replaced and the cells washed in a balanced salt solution twice before suspension at a concentration of 2 × 10^7 per milliliter in neutral saline buffer or growth medium containing 1 mg/ml bovine serum albumin. Labeling with monoclonal antibodies is carried out for 30–60 min, with peptide hormones for 5–10 min. Aggregates are removed from the ligands by prior centrifugation at >10^5 g, e.g., with a Beckman (Palo Alto, CA) Airfuge. The cells are separated from excess unbound ligand by centrifugation through fetal calf serum and washed again by one or two cycles of centrifugation in phosphate-buffered saline. Cells are resuspended at a density of 10^7 per milliliter for measurement. Viability after labeling is checked with the permeant fluorogenic substrate fluorescein diacetate (FDA) and typically exceeds 95%. After being purged with argon and subjected to the rotational diffusion measurements, cells generally retain >90% viability according to the FDA assay. Cells grown in suspension culture are treated in a similar fashion.

Dense cell suspensions scatter light and exhibit different degrees of autoluminescence. It is essential, therefore, to obtain parallel records taken with the following alternative control preparations: (1) unlabeled

---

[47] M. Hogan, J. Wang, R. H. Austin, C. L. Monitto, and S. Hershkowitz, *Proc. Natl. Acad. Sci. U.S.A.* **79**, 3518 (1982).
[48] R. Zidovetzki, M. Bartholdi, D. Arndt-Jovin, and T. M. Jovin, *Biochemistry* **25**, 4397 (1986).

but otherwise normally processed cells, (2) cells exposed to the label in the presence of an excess of competitive unlabeled ligand,[48] and (3) labeled cells exposed to a triplet quencher such as cyclooctatetraene.[26] These "blank" records are substracted from those obtained with intact labeled cells prior to generating the intensity and anisotropy decay curves.

In the case of synthetic lipid bilayers and other reconstituted systems, the protein in question is labeled with the desired extrinsic probe either *in situ* in its native membrane of origin, followed by extraction and processing, or after purification. There exist numerous strategies for reconstitution of membrane proteins[49]; experimental protocols are discussed below in the section on translational diffusion. Under fortunate circumstances, integral membrane proteins can be labeled covalently with a sufficiently high degree of specificity; prototypical examples are the two major surface glycoproteins of human erythrocytes, the band 3 anion transporter[4,5,26,50] and glycophorin.[51]

Specific ligands with conjugated probes (antibodies, hormones, toxins) are routinely examined alone in order to establish baseline values for the triplet excited state yields and lifetimes. The emission of these comparatively small ligands should be unpolarized in the microsecond time domain.[45] The same molecules can be examined in highly viscous solutions of glycerol or sucrose in order to confirm the integrity of the probe conjugation and to obtain additional information about the rotational relaxation modes of the macromolecular carrier.[30,48]

Data Processing and Analysis. Consider an idealized cylindrical integral membrane protein embedded in a lipid bilayer and able to undergo the canonical diffusional motions of (1) uniaxial rotation about a normal to the membrane surface, (2) translation in the plane of the membrane, and (3) wobbling, perhaps even flip-flop excursions. These possibilities are shown schematically in Fig. 5A. Polarized light is directed on a randomly oriented ensemble of such membranes (Fig. 5B). The probability of absorption has the usual cosine square dependence on the angle between the excitation vector and absorption transition moment (μ_a). The sequence of photophysical reactions ensues (Fig. 1), culminating in delayed emission with an anisotropy $r(t)$ defined as a function of time in terms of the two components polarized parallel ($I_\parallel$) and perpendicular ($I_\perp$) to the excitation vector.

$$r(t) \equiv [I_\parallel(t) - I(t)]/[I_\parallel(t) + 2I_\perp(t)] \tag{1}$$

[49] M. Montal, A. Darszon, and H. Schindler, *Q. Rev. Biophys.* **14**, 1 (1981).
[50] E. D. Matayoshi, W. H. Sawyer, and T. M. Jovin, manuscript in preparation.
[51] R. Cherry *et al. Proc. Natl. Acad. Sci. U.S.A.* **77**, 5899 (1980).

in which the denominator represents the total polarization-independent emission. In order to construct $r(t)$ from the experimental polarized signals $s_\parallel(t)$ and $s_\perp(t)$, we introduce two factors correcting for gain inequality (g) and for the depolarizing effect of the (relatively) large emission collection aperture (h). We also use difference $d(t)$ and sum $S(t)$ functions corresponding formally to the numerator and denominator of Eq. (1).

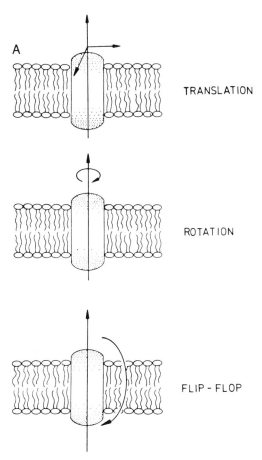

FIG. 5. Model for motions of an integral membrane protein. (A) The three canonical motions for a rigid protein spanning the membrane bilayer. Not depicted are wobbling excursions around the normal to the plane of the membrane. (B) Vectorial representation of the molecular axis (μ) and the two transition moments for absorption (μ_a) and emission (μ_e). The normal to the plane of the membrane is **d**, about which the protein undergoes axial rotation with possible wobbling motions (shown as an ellipse). In a randomly distributed sample (e.g., in suspension) the exciting light can be in any direction.

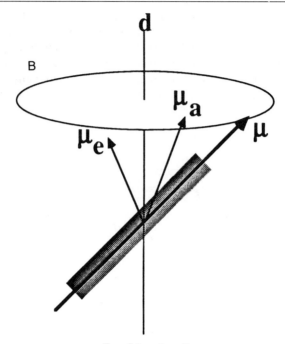

Fig. 5 (continued)

$$d(t) \equiv s_\|(t) - gs_\perp(t) \quad S(t) \equiv s_\|(t) + ghs_\perp(t) \quad (2a)$$
$$r(t) = d(t)/S(t) \quad (2b)$$

The factor g is determined experimentally with samples of free eosin or erythrosin in aqueous media (i.e., fully depolarized in the microsecond time domain). The factor h is given by an expression corresponding to the indicated geometry,[52] $h = 1 + \cos\varphi(1 + \cos\varphi)/2$, where φ is the effective cone angle for collection of the emission; thus, $h = 2$ for the limit of very narrow apertures and $h = 1$ for $\varphi = 90°$. For our apparatus (Fig. 2), $g = 1.031$ and $h = 1.79$; the latter value was determined by measuring samples of known anisotropy, and alternatively, by combining two measurements, one made with a pinhole interposed in the emission path as a limiting aperture and the other with the full collection aperture (which from the above value of h corresponds to an $\varphi = 30°$ or an $f/\#$ of 1.9). In the analysis of experimental records, the functions $d(t)$, $S(t)$, and $r(t)$ are calculated from $s_\|(t)$ and $s_\perp(t)$ after subtracting the pretrigger levels from

[52] T. M. Jovin, in "Flow Cytometry and Sorting" (M. Melamed, P. Mullaney, and M. Mendelsohn, eds.), p. 137. Wiley, New York, 1979.

the latter in order to establish the appropriate reference points corresponding to the condition of zero (or steady-state) signal.

The total emission $S(t)$ is a function of the excited state lifetime(s) τ_i.

$$S(t) = \sum \beta_i \exp(-t/\tau_i) \tag{3}$$

In the case of a sphere rotating in an isotropic medium, the emission anisotropy $r(t)$ decays monoexponentially.

$$r(t) = r_o \exp(-t/\phi) = r_o \exp(-6\mathbf{D}_r t) \tag{4}$$

in which ϕ, the rotational correlation time, is equal to $V\eta/kT$ (V is the molecular volume, η the viscosity of the medium, k Boltzmann's constant, and T the absolute temperature). From Eq. (4), it is seen that ϕ is also inversely related to the rotational diffusion constant $\mathbf{D}_r$, i.e., $\phi = 1/6\mathbf{D}_r$. r_o is the fundamental anisotropy in the absence of rotation (i.e., immediately after excitation) and is related in a simple way to the angle δ between the effective absorption and emission transition moments.

$$r_o = (2/5)P_2(\mu_a \cdot \mu_e) \equiv (2/5)P_2 \cos \delta \tag{5}$$

where $P_2(x) = (3x^2 - 1)/2$ is the second Legendre polynomial. It follows that r_o reflects intrinsic properties of the chromophore and not of the macromolecular carrier to which it is conjugated; furthermore, r_o is different for fluorescence and phosphorescence since different transition moments are involved (Fig. 5B). For eosin and erythrosin immobilized in polymethylmethacrylate, r_o has been reported to be 0.17–0.30 (phosphorescence)[27,32] and 0.26–0.33 (fluorescence),[32,42] corresponding to values of δ [Eq. (5)] of 24°–38° and 20°–29°, respectively. In experiments with membranes, the observed initial anisotropy r_{in} is in general significantly smaller in magnitude than r_o, a phenomenon we and others have generally attributed to local probe and segmental protein motions which lead to depolarization in the nanosecond domain, i.e., prior to the measurement of delayed luminescence.

For all rigid bodies other than spheres, the $r(t)$ function is more complex than that given by Eq. (4).[53] Three components are required in the case of a freely rotating ellipsoid of revolution, and this number reduces to 2 if the environment (membrane, Fig. 5) restricts the motion to a uniaxial rotation (the first model assumed for intrinsic membrane proteins[5]). In the latter case, the orientational constraints imposed by the anisotropic membrane structure (Fig. 5) lead to an observed finite limiting anisotropy r_∞ at long experimental times, expressions for which can be

---

[53] R. Rigler and M. Ehrenberg, *Q. Rev. Biophys.* **6**, 139 (1973).

derived in a model-independent manner, i.e., for any form of restricted motion.[21,54]

$$r_\infty = (2/5)P_2(\mu_a \cdot \mu)P_2(\mu_e \cdot \mu)[\langle P_2(\mu \cdot \mathbf{d})\rangle]^2 \quad (6)$$

in which the correlation function in angular brackets denotes the final equilibrium distribution and is equivalent to the order parameter used in other spectroscopic techniques.[5,6,54,55]

We now formulate a generalized multicomponent decay function $r(t)$ in a form which permits the correlation between experimentally determined parameters and theoretical models (see also Refs. 5, 6, 21).

Isotropic medium

$$r(t) = (r_{in}/r_o)\sum_{i=1}^{j} \alpha_i \exp(-t/\phi_i) \quad (7a)$$

Anisotropic medium

$$r(t) = (r_{in} - r'_\infty)\sum_{i=1}^{i\leq j} \alpha'_i \exp(-t/\phi_i) + r'_\infty \quad (7b)$$

in which ϕ_i are the rotational correlation times, $(r_{in}/r_o)\alpha_i$ the corresponding amplitudes, and α'_i the experimental fractional amplitudes [$\sum \alpha'_i = 1$; also, $\alpha'_i = \alpha_i/(r_o - r_\infty)$]. Note that the experimental value of the limiting anisotropy r'_∞ corresponds to r_∞ [Eq. (6)] depressed in magnitude by the common depolarization coefficient (r_{in}/r_o); i.e., $r'_\infty = r_\infty(r_{in}/r_o)$.

For the rigid ellipsoid of revolution, $j = 3$; the other factors in Eq. (7a) are tabulated below.

i	α_i	$\alpha < 0$ for	ϕ_i^{-1}
1	$(3/10)\sin^2\theta_a \sin^2\theta_e \cos 2\nu$	$\nu > 45°$	$4\mathbf{D}_\parallel + 2\mathbf{D}_\perp$
2	$(3/10)\sin 2\theta_a \sin 2\theta_e \cos \nu$	$\nu > 90°$	$\mathbf{D}_\parallel + 5\mathbf{D}_\perp$
3	$(2/5)P_2(\cos\theta_a)P_2(\cos\theta_e)$	θ_a or $\theta_e > 54.7°$	$6\mathbf{D}_\perp$
	$\sum \alpha_i \equiv r_o$	$\delta > 54.7°$	

(7c)

The angle ν denotes the separation between the projections of the vectors μ_a and μ_e on the plane normal to μ, and is given by the expression $\cos \nu = (\cos \delta - \cos \theta_a \cos \theta_e)/\sin \theta_a \sin \theta_e$. The diffusion constants $\mathbf{D}_\parallel$ and $\mathbf{D}_\perp$ correspond to axial and transverse motions, respectively, and are analytical functions of the molecular dimensions[56] as well as of the properties of

[54] G. Lipari and A. Szabo, *Biophys. J.* **30**, 489 (1980).
[55] F. Jähnig, *Proc. Natl. Acad. Sci. U.S.A.* **76**, 6361 (1979).
[56] E. W. Small and I. Isenberg, *Biopolymers* **16**, 1907 (1977).

the medium (membrane). Equation (7) is necessarily more elaborate than that applicable to transient dichroism based on the singlet state alone[5] because the orientations of two transition moments are involved. In addition, we note [third column, Eq. (7c)] that any or all of the amplitudes α_i in the anisotropy decay can be negative, depending upon the geometry of the system. That is, the anisotropy of delayed emission can decay to *higher* values ($r'_\infty > r_{in}$) and can even be negative given a large enough angle δ (>54.7°); both phenomena have been observed.[26,34,35,45,57-61]

Inasmuch as biological macromolecules are rarely rigid, additional complications arise in the appropriate formulation or the $r(t)$ function. The effects of librational motion,[5,6,21,62,63] segmental flexibility,[64-66] and combinations of torsional and bending motions[66-68] have been treated. In general, however, the usual practice has been to apply Eq. (7) to membrane proteins assuming a uniaxial mode of rotation about the normal to the membrane surface (**d** and μ parallel, Fig. 5B). In this case, the reorientation about axes in the plane of the membrane is presumed not to occur such that the effective $\mathbf{D}_\perp$ is 0. Thus, in Eq. (7b), there are two rotational correlation times ($i = 1, 2$) and r_∞ is equivalent to the coefficient α_3, as can also be seen from Eq. (6) by noting that the order parameter is unity for this case. The experimental rotational correlation times (most often two distinct components are not resolved experimentally) can be compared to the $\mathbf{D}_\parallel$ estimated by the equation elaborated by Saffman and Delbrück for a cylinder embedded in a homogeneous viscous medium.[69,70]

$$\mathbf{D}_\parallel = kT/(4\pi a^2 h \eta) \tag{8}$$

where a is the radius of the cylinder, h the length constrained within the membrane, and η the effective membrane (micro)viscosity. It is instruc-

---

[57] R. Zidovetzki, Y. Yarden, J. Schlessinger, and T. M. Jovin, *EMBO J.* **5**, 247 (1986).
[58] M. Bartholdi, F. J. Barrantes, and T. M. Jovin, *Eur. J. Biochem.* **120**, 389 (1981).
[59] M. M. Lo, P. B. Garland, J. Lamprecht, and E. A. Barnard, *FEBS Lett.* **111**, 407 (1983).
[60] A. Speirs, C. H. Moore, D. H. Boxer, and P. B. Garland, *Biochem. J.* **213**, 67 (1983).
[61] S. C. Harvey and H. C. Cheung, *Biochemistry* **16**, 5181 (1977).
[62] W. van der Meer, H. Pottel, W. Herreman, M. Ameloot, H. Hendrickx, and H. Schröder, *Biophys. J.* **46**, 515 (1984).
[63] C. J. Restall, R. E. Dale, E. K. Murray, C. W. Gilbert, and D. Chapman, *Biochemistry* **23**, 7665 (1984).
[64] W. A. Wegener, *J. Chem. Phys.* **76**, 6425 (1982).
[65] W. A. Wegener, *Biopolymers* **21**, 1049 (1982).
[66] V. Bloomfield, p. 1, in Ref. 14.
[67] M. Hogan, J. LeGrange, and R. Austin, *Nature (London)* **304**, 752 (1983).
[68] H. Yoshimura, T. Nishio, K. Mihashi, K. Kinosita, Jr., and I. Ikegami, *J. Mol. Biol.* **179**, 453 (1984).
[69] G. Saffman and M. Delbrück, *Proc. Natl. Acad. Sci. U.S.A.* **72**, 3111 (1975).
[70] P. G. Saffman, *J. Fluid Mech.* **73**, 593 (1976).

tive to compare Eqs. (8) and (4). In using this relationship, it is usually assumed that the external domains of the membrane protein and/or bound ligand which reside in the aqueous medium do not contribute an appreciable viscous retardation compared to that due to the segments of the molecule within the membrane, for which η is approximately 10^2 that of water.[2,3] More sophisticated models are required,[71] as further discussed below.

The above discussion has considered a single homogeneous molecular and chromophoric component. Photophysical complexity and macromolecular heterogeneity characterize the more general situation, however, such that the observed decay may comprise numerous superimposed components.

$$S(t) = \sum S_i(t) \qquad d(t) = \sum d_i(t) = \sum S_i(t)r_i(t) \qquad r(t) = d(t)/S(t) \qquad (9)$$

The $d(t)$ function is now the sum of exponential functions with time constants equal to the harmonic means of excited state lifetimes and rotational correlation times; it acquires additional complexity if the individual $S_i(t)$ and $r_i(t)$ functions are multicomponent [Eqs. (3 and 7)]. The resultant $r(t)$ may exhibit a spurious time dependence even in the absence of rotation[39] (Fig. 6). The rigorous solution to this problem consists of analyzing the $d(t)$ and $S(t)$ functions separately and then attempting a unique decomposition into τ_i and ϕ_i. This approach fails in the case of noisy records and one has to perform independent controls in order to establish the validity of the $r(t)$ decay curve, particularly in the presence of multiple excited states lifetimes (τ_i). One experimental strategy is to systematically vary the latter by introducing graduated fixed levels of oxygen or another triplet quencher into the system. An invariant $r(t)$ implies that the $S_i(t)$ coefficients in $d(t)$ [Eq. (8)] can be successfully factored out such that the anisotropy is not a function of the τ_i.

Fluorescence Recovery Anisotropy (FRA)

The technique introduced by Garland and colleagues[42,72,73] with the designation *fluorescence depletion* combines the greatly enhanced sensitivity of continuous fluorescence detection with the convenient "spectroscopic clock" of the triplet excited state (Table I, Figs. 1 and 4). Photoselective reversible bleaching into the triplet state via a laser light pulse generated with optomodulators depletes the ground state to a sufficient degree such that its subsequent interrogation by a low-level monitoring

[71] B. D. Hughes, B. A. Pailthorpe, and L. R. White, *J. Fluid. Mech.* **110**, 349 (1981).
[72] P. Johnson and P. B. Garland, *Biochem. J.* **203**, 313 (1982).
[73] P. B. Garland and P. Johnson, p. 95, in Ref. 14.

beam yields a polarized prompt fluorescence recovery curve similar in nature to that described above. A systematic study of different chromophores has been carried out in order to identify those which combine the properties of a good fluorescence quantum yield and a finite triplet yield.[72] Another microscope-based system has been described,[74] which shares the capability of performing translation diffusion measurements by fluorescence photobleaching recovery (FPR). The use of distinct actinic and monitoring beams introduces significant complications with respect to the analysis of the rotational relaxation[28,74,75] and to the experimental application of the method to solutions and suspensions.[32] The combination of two lasers[32] as opposed to a single[42,72–74] modulated light source serves to extend the time resolution into the region (a few microseconds) required for cellular studies (Table II).

Despite the problems, FRA offers the proven potential[42,72,73] for performing rotational diffusion measurements on single cells and other small oriented membrane structures in the optical microscope, a challenge beyond the capabilities of the triplet emission methods outlined in the previous section.

Another version of the FRA technique is based on single-shot, aerobic, irreversible photobleaching,[76] but has not been applied to processes occurring below the millisecond range.

Other Techniques

Table I summarizes other methods of actual or potential utility in rotational diffusion studies of membrane systems. In one case, designated TRA (transient fluorescence anisotropy),[23,24] a transient photochromic intermediate instead of the triplet state is exploited. The spectroscopic clock is no longer determined by the photophysics of an excited electronic state but rather by chemical reequilibration in the ground state. Photoselection with an actinic beam and interrogation with a second source provide in principle the same sensitivity of FRA but with some additional virtues: the spectral region for measurement is shifted from that of the initial state, the monitored emission is referenced to a dark state, there is no *a priori* requirement for anaerobic conditions, and the optical design is uncritical. The applicability of TRF to membrane systems, however, remains to be established. The same applies for the last two entries in Table I.

All of the techniques described above require a light pulse to photose-

---

[74] T. M. Yoshida and B. G. Barisas, *Biophys. J.* **50,** 41 (1986).
[75] W. A. Wegener, *Biophys. J.* **46,** 795 (1984).
[76] L. M. Smith, R. M. Weiss, and H. M. McConnell, *Biophys. J.* **36,** 773 (1981).

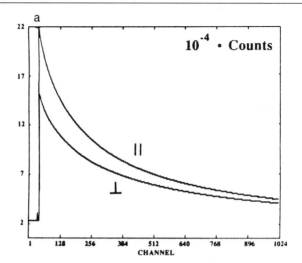

FIG. 6. Fluorescence decay of uranyl glass GG17. Excitation is at 515 nm and emission is >550 nm. (a) The two polarized emission components are shown; 1000 cycles were accumulated. (b) The derived ln $S(t)$ signal is "wrapped around" 30 times in order to emphasize the multicomponent decay (τ = 11, 61, and 265 μsec with corresponding fractional amplitudes of 0.17, 0.33, and 0.50). (c) The anisotropy decay was fit with two components (ϕ = 19 and 232 μsec and relative amplitudes 0.13 and 0.87).

lect a subpopulation of chromophores. There are distinct advantages to time-resolved measurements of this kind, but one pays the price of the severe prompt fluorescence transient and corresponding need for gating. The problem can be greatly alleviated by alternative methods based on modulated excitation and phase-sensitive detection, which have been used primarily in (prompt) fluorescence spectroscopy[77] but also have been proposed for the longer time regimes.[73] Finally, we cite the possibility (and need) to extend the spectral range to cover the intrinsic chromophores of proteins, for which steady-state phosphorescence anisotropy is of great utility.[78]

Representative Data

The application of the delayed luminescence technique according to the preceding description is illustrated in Fig. 6, which depicts the properties of a fluorescent uranyl glass GG17 supplied (formerly) by Schott (Mainz, FRG). The complex decays of intensity and anisotropy provide a

[77] E. Graton, D. M. Jameson, and R. D. Hall, *Annu. Rev. Biophys. Bioeng.* **13**, 205 (1984).
[78] G. B. Strambini and W. C. Galley, *Nature (London)* **260**, 554 (1976).

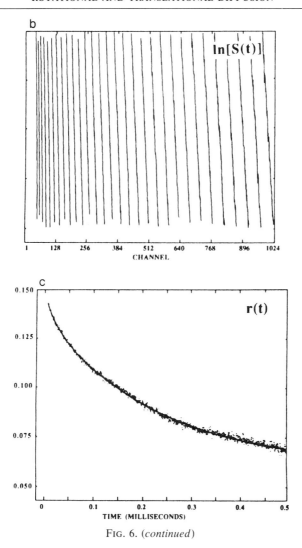

Fig. 6. (*continued*)

standard data set with which the performance of the instrument (Fig. 2) can be checked periodically. Note that the time-dependent anisotropy reflects excited state processes, which probably include energy transfer but obviously neither atomic nor molecular rotational diffusion [Eq. (7)].

Table II is a summary of some of the studies in which the phosphorescence and fluorescence techniques of Table I have been applied. They include (1) natural and synthetic membrane systems; (2) intact living cells;

TABLE II
MEMBRANE AND CELLULAR SYSTEMS STUDIED BY DELAYED LUMINESCENCE ANISOTROPY AND FLUORESCENCE RECOVERY ANISOTROPY

Target	Membrane/cells	Reagent/probe[a]	Method[b]	ϕ (μsec)[c]	Ref.
Cytochrome P-450	Reconstituted into vesicles	diIoFl-IA,Eo-M	D	13–12[g]	d
Rhodopsin	Bovine rod outer segments	Er-IA	P	24–39	e
	Reconstituted into vesicles	Er-ITC, diIoF-IA	P, D	29–43	f
Glycophorin	Reconstituted into vesicles	Er-ITC	P	1–15	g
Band 3 anion transporter	Human RBC ghosts	Eo, Er-ITC, Eo-M, diIoF-ITC	P, D, F	25–>10³	g–j
	Human RBC	Eo-M	P	28–330	i, k
	Human RBC ghosts	Eo-glyceraldehyde-PO₄-DH	P	$r = 0, >0$	k
IgE-Fc receptor	Rat basophil leukemia cells	Er-IgE	P	23–65	l
Acetylcholine-R (AChR)	Postsynaptic membranes	Er-bungarotoxin	P	13–26; 200–500; ∞	m
	Reconstituted into vesicles	Eo-AChR (monomer, dimer)	P	45–80	n
EGF-receptor (EGF-R)	Epidermoid carcinoma cells	Er-EGF (epidermal growth factor)	P	20–350	o
	Cellular membranes	Er-EGF	P	12–120	p
H-2K[k] antigen	Murine lymphoma cells	Er-anti-H-2K[k] antibodies (MAb)	P	17–210	q
H-2L[d] antigen	Transformed murine L cells	Er-anti-H-2L[d] antibody (MAb)	P	6–25	r
Con A receptors	Murine erythroleukemia cells	Eo-concanavalin A (Con A)	P, F	∞	g, s
lac permease	Vesicles	Eo-M	P	19–39	t
Ca²⁺-ATPase	Sarcoplasmic reticulum vesicles	Er-ITC, Er-IA	P	10–100	u
Na⁺,K⁺-ATPase	Membranes	Eo-M	P	∞	i
Lipid probe	Multilamellar vesicles	Eo-fatty acids (n = 11, 15)	P	16–68	v
	Lipid bilayers	diI	F	0.6–800 sec	w
Synthetic vesicles	(Un)polymerized vesicles	Er	P	2–1100	x

[a] Eo, Eosin; Er, erythrosin; diIoF, diiodofluorescein; ITC, 5-isothiocyanate; IA, 5-iodoacetamide; M, 5-maleimide; diI, [bis-2-(N-octadecyl-3,3-dimethyl-1-benzo[b]pyrrole]trimethincyanine iodide. If the reagent alone is indicated, then it has been used to label the designated target.

[b] P, Phosphorescence anisotropy; D, delayed fluorescence anisotropy; F, fluorescence recovery anisotropy.
[c] In many cases, discrete multicomponent relaxations were observed; in others, the range reflects changes in temperature, etc. Immobile probe is indicated by ∞.
[d] R. Greinert, H. Staerk, A. Stier, and A. Weller, *J. Biochem. Biophys. Methods* **1**, 77 (1979); see also R. Greinert, S. A. E. Finch, and A. Stier, *Xenobiotica* **12**, 717 (1982).
[e] M. Coke, C. J. Restall, C. M. Kemp, and D. Chapman, *Biochemistry* **25**, 513 (1986).
[f] E. M. Matayoshi and G. Schoellmann, unpublished data.
[g] R. H. Austin, S. L. Chan, and T. M. Jovin, *Proc. Natl. Acad. Sci. U.S.A.* **76**, 5650 (1979).
[h] T. M. Jovin, M. Bartholdi, W. L. C. Vaz, and R. H. Austin, *Ann. N.Y. Acad. Sci.* **366**, 176 (1981).
[i] E. D. Matayoshi, A. F. Corin, R. Zidovetzki, W. H. Sawyer, and T. M. Jovin, in "Mobility and Recognition in Cell Biology" (H. Sund and C. Veeger, eds.), p. 119. de Gruyter, Berlin, Federal Republic of Germany, 1983.
[j] C. Moore, D. Boxer, and P. B. Garland, *FEBS Lett.* **108**, 161 (1979); see also P. Johnson and P. B. Garland, *FEBS Lett.* **132**, 252 (1981).
[k] E. D. Matayoshi, W. H. Sawyer, and T. M. Jovin, manuscript in preparation.
[l] R. Zidovetzki, M. Bartholdi, D. Arndt-Jovin, and T. M. Jovin, *Biochemistry* **25**, 4397 (1986).
[m] M. Bartholdi, F. J. Barrantes, and T. M. Jovin, *Eur. J. Biochem.* **120**, 389 (1981); see also M. M. Lo, P. B. Garland, J. Lamprecht, and E. A. Barnard, *FEBS Lett.* **111**, 407 (1983).
[n] T. M. Jovin and M. Criado, unpublished data.
[o] R. Zidovetzki, Y. Yarden, J. Schlessinger, and T. M. Jovin, *Proc. Natl. Acad. Sci. U.S.A.* **78**, 6981 (1981).
[p] R. Zidovetzki, Y. Yarden, J. Schlessinger, and T. M. Jovin, *EMBO J.* **5**, 247 (1986).
[q] S. Damjanovich, L. Trón, J. Szöllösi, R. Zidovetzki, W. L. C. Vaz, R. Regateiro, D. J. Arndt-Jovin, and T. M. Jovin, *Proc. Natl. Acad. Sci. U.S.A.* **80**, 5985 (1983).
[r] M. Edidin, R. Zidovetzki, and T. M. Jovin, manuscript in preparation.
[s] A. F. Corin, E. Blatt, and T. M. Jovin, *Biochemistry* **26**, 2207 (1987).
[t] K. Dormier, A. F. Corin, J. Keith, and F. Jaehnig, *EMBO J.* **4**, 3633 (1985).
[u] C. Moore, D. Boxer, and P. B. Garland, *FEBS Lett.* **108**, 161 (1979); see also A. Speirs, C. H. Moore, D. H. Boxer, and P. B. Garland, *Biochem. J.* **213**, 67 (1983); C. J. Restall, R. E. Dale, E. K. Murray, C. W. Gilbert, and D. Chapman, *Biochemistry* **23**, 7665 (1984).
[v] E. Blatt and A. F. Corin, *Biophys. Biochim. Acta* **857**, 85 (1986).
[w] L. M. Smith, R. M. Weiss, and H. M. McConnell, *Biophys. J.* **36**, 73 (1981).
[x] W. Reed, D. Lasic, H. Hauser, and J. H. Fendler, *Macromolecules* **18**, 2005 (1985).

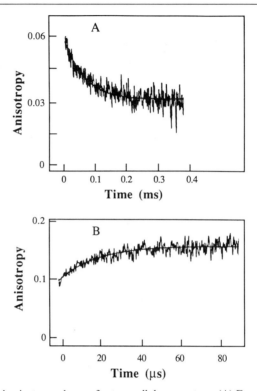

FIG. 7. Typical anisotropy decays for two cellular receptors. (A) Fc receptor for IgE on 2H3 rat basophil leukemia cells. Decay of bound erythrosin-labeled IgE at 5.0°. Data reproduced from Zidovetzki et al.[48] (B) Epidermal growth factor receptor (EGFR) on membrane fragments derived from A-431 epidermoid carcinoma cells. Decay of bound erythrosin-labeled EGF at 4°. The negative amplitude presumably reflects the particular orientation of the hormone (and conjugated erythrosin) to the rotation axis [see Eq. (7c) and associated text]. Data reproduced from Zidovetzki et al.[57]

(3) specific ligands and reconstituted proteins; and (4) receptors, antigens, and enzymes. It is particularly gratifying that living cells can be subjected to these anaerobic techniques without drastic loss of viability. Illustrative examples are the cell-surface receptors for immunoglobulin IgE, which triggers exocytosis of vasoactive amines in mast cells and basophils, and for EGF (epidermal growth factor) (Fig. 7). These integral membrane proteins exert their signaling function through mechanisms involving initial intermolecular cross-linking (microclustering) mediated by the binding of the specific external agents. The data of Fig. 7 provide evidence for an initial dispersed distribution of the receptors on the cell surface. Cross-linking of the tethered IgE antibodies[48] or, in the case of the EGF recep-

tor,[45,57] temperature-dependent activation of presumed conformational changes in the hormone-receptor complex, induces oligomerization and immobilization, which are detected by an increase in the rotational correlation times (slower diffusion) of the specific ligands (IgE, EGF) labeled with the triplet probe erythrosin. It is interesting that the corresponding translational diffusion of the EGF receptor (Table IV) is more *rapid* at the higher temperatures, indicating the relative insensitivity of that process to molecular size. The challenge to the biophysicist is to make the accurate quantitative estimations of the equilibrium and kinetic constants, specifying coupled reactions and conformational transitions on the cell surface.[1,79,80]

Translational Diffusion

The lateral transport of membrane components in the plane of the lipid bilayer is not only of interest from biological and biochemical viewpoints but is of intrinsic physicochemical interest[17,81] due to the possibility that many membrane-linked multimolecular biochemical reactions are diffusion-limited.[82,83] Since the earliest attempts to assess the movement of membrane components,[84-86] several methods have been developed for the study of translational diffusion based on optical spectroscopy (Table III), magnetic resonance spectroscopy,[85-92] electron microscopy,[93] electrophoresis of cell surface components,[94] and cell fusion.[84]

The various methods study the translational diffusion process over different characteristic distances in the plane of the membrane. Displacements less than two orders of magnitude larger than the radius of a lipid molecule can be studied by most techniques involving bimolecular reac-

[79] T. M. Jovin, *Mol. Immunol.* **21,** 1147 (1984).
[80] T. L. Hill, "Cooperativity Theory in Biochemistry." Springer-Verlag, New York, 1985.
[81] G. Adam and M. Delbrück, in "Structural Chemistry and Molecular Biology" (A. Rich and N. Davidson, eds.), p. 199, Freeman, San Francisco, California, 1968.
[82] C. R. Hackenbrock, *Trends Biochem. Sci.*, p. 152 (1981).
[83] D. E. Koppel, *Tech. Life Sci.* **B4/11,** 1 (1982).
[84] L. D. Frye and M. Edidin, *J. Cell Sci.* **7,** 319 (1970).
[85] P. Devaux and H. M. McConnell, *J. Am. Chem. Soc.* **94,** 4475 (1972).
[86] H. Träuble and E. Sackmann, *J. Am. Chem. Soc.* **94,** 4499 (1972).
[87] A. G. Lee, N. J. M. Birdsall, and J. C. Metcalfe, *Biochemistry* **12,** 1650 (1973).
[88] P. R. Cullis, *FEBS Lett.* **70,** 223 (1976).
[89] P. A. Kroon, M. Kainosho, and S. I. Chan, *Biochim. Biophys. Acta* **433,** 282 (1976).
[90] G. Lindblom and H. Wennerström, *Biophys. Chem.* **6,** 167 (1977).
[91] R. W. Fisher and T. L. James, *Biochemistry* **17,** 1177 (1978).
[92] A. L. Kuo and C. G. Wade, *Biochemistry* **18,** 2300 (1979).
[93] A. E. Sowers and C. R. Hackenbrock, *Proc. Natl. Acad. Sci. U.S.A.* **78,** 6246 (1981).
[94] M. M. Poo, *Annu. Rev. Biophys. Bioeng.* **10,** 245 (1981).

tions in the membrane (excimer formation, excited state quenching, and some magnetic resonance techniques). For longer distances, i.e., several orders of magnitude larger than molecular dimensions, some nuclear magnetic resonance methods, photodepletion and redistribution of chromophores, cell fusion, electrophoresis, and electron microscopy are more appropriate.

There are a number of optical methods for the study of translational diffusion in the plane of a membrane (Table III). The interested reader is referred to the original literature cited for a more detailed description of the methods and for ways to analyze the data obtained.

In this chapter, we restrict our attention to the so-called "fluorescence recovery after photobleaching" (FRAP) method,[95-98] with which most of the present information regarding translational diffusion in lipid bilayers and biological membranes has been obtained. The technique has also been called "fluorescence photobleaching recovery" (FPR),[96] and "fluorescence microphotolysis" (FM).[99] The principle of the method is shown in Fig. 8. The test molecule, the translational diffusion of which we are interested in examining, must be a fluorescent species, e.g., a fluorescently labeled protein or a fluorescent lipid derivative or analog. A small area of a large planar membrane containing the fluorescent test particles is illuminated with a focused laser beam (monitoring beam) with an intensity that does not result in a significant photobleaching over the time course of the experiment. The baseline fluorescence intensity is measured; it should be proportional to the concentration of fluorescing entities in the illuminated area. It is convenient that the latter constitute a small fraction ($\leq 1\%$) of the total continuous membrane surface and that it be—usually but not necessarily—circular in shape. The experiment begins when a fraction of these fluorophores is eliminated by photobleaching simply by increasing the intensity of the illuminating beam by a factor of $\sim 10^4$ (bleaching beam) for a short period (usually a few milliseconds). The bleaching probably involves the participation of oxygen since, in many cases, it is at least partly inhibited by reducing the oxygen concentration in the sample. Immediately after photobleaching, the laser beam intensity is reduced to its prebleaching level and the intensity of fluorescence from the illuminated area is followed as a function of time. Diffusion of un-

---

[95] R. Peters, J. Peters, K. H. Tews, and W. Bähr, *Biochim. Biophys. Acta* **367**, 282 (1974).
[96] D. Axelrod, D. E. Koppel, J. Schlessinger, E. Elson, and W. W. Webb, *Biophys. J.* **16**, 1055 (1976).
[97] K. Jacobson, Z. Derzko, E. S. Wu, Y. Hou, and G. Poste, *J. Supramol. Struct.* **5**, 565 (1976).
[98] M. Edidin, Y. Zagyansky, and T. J. Lardner, *Science* **191**, 466 (1976).
[99] R. Peters, A. Brünger, and K. Schulten, *Proc. Natl. Acad. Sci. U.S.A.* **78**, 962 (1981).

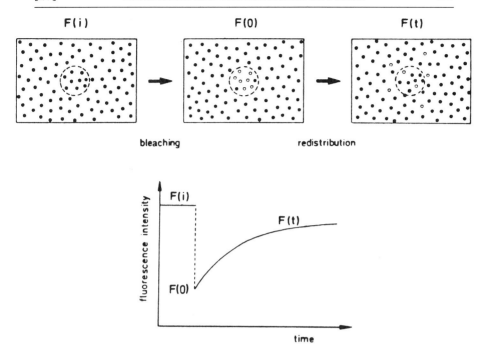

FIG. 8. Principle of a FRAP experiment. The fluorescence intensity, $F(i)$, arising from the chromophores in a small area of the membrane, indicated by the broken line measured (upper left) by illumination with the monitoring beam. The experiment begins by irreversibly photobleaching a fraction of the fluorophores in the designated area by illumination with the bleaching beam (upper center). The recovery of fluorescence due to diffusion of bleached particles out of and unbleached particles from the rest of the membrane into the illuminated area (upper left) is followed with time using excitation with the monitoring beam $[F(t)]$. The temporal behavior of the fluorescence intensity from the illuminated area is shown in the lower part of the figure. Data reproduced from Vaz et al.[15]

bleached molecules into the illuminated area from the surrounding membrane surface results in the repopulation of fluorescent species. The time course of the fluorescence recovery is a function of the characteristics of the illuminated spot as will be discussed below, and provides a direct measure of the diffusion rate.

Instrumentation

The equipment required to do a typical FRAP experiment is shown schematically in Fig. 9 and describes the essential elements of the instrument in use in the authors' laboratory. An argon ion laser is used as the source of excitation. For most dyes (NBD, fluorescein and its derivatives,

TABLE III
OPTICAL METHODS FOR STUDY OF TRANSLATIONAL DIFFUSION IN MEMBRANES AND LIPID BILAYERS

Method	Instrumentation required	Advantages	Disadvantages	General comments
Short-range translational diffusion				
Excimer and exciplex formation[a-c]	Steady-state or time-resolved fluorimetry	Commonly available equipment; very short range if excimer is formed from a singlet excited state	High probe concentration unless intramolecular; local concentration artifacts and perturbation of membrane structure	May result in an overestimation of the diffusion constants
Excited state quenching through collisional or resonance energy transfer[d-f]	Steady-state or time-resolved fluorimetry or phosphorimetry	Very short range to short range; characteristic length depends on excited state lifetime	As above	As above
Long-range translational diffusion				
Fluorescence recovery or redistribution after photobleaching (FRAP, FPR, FM, CFM)[g-m]	Laser-illuminated fluorescence microscope; stable and stationary cells and membranes	Long ranges (≥ 1 μm); data reduction is direct; wide dynamic range ($D_t \sim 10^{-6}$ to $\sim 10^{-11}$ cm^2 sec^{-1}); anisotropic transport	Expensive instrumentation; analysis depends on membrane topography; possible protein cross-linking; no short-range diffusion information	Most widely used technique; also useful for diffusion in three dimensions
Fluorescence photoactivation and dissipation (FPD)[n]	Fluorescence microscope	As above; long-term monitoring possible	Need photoactivatable fluorophores	New technique

Fluorescence redistribution after fusion (FRAF)[o-q]	Fluorescence microscope	Long range (diameter of a cell); wide dynamic range ($D_t \sim 10^{-8}$ to $\sim 10^{-12}$ cm^2 sec^{-1}); slow diffusion	Requires fusion of cells; not an exact method	Useful technique for study of slow diffusion on cell surfaces
Fluorescence correlation spectroscopy (FCS)[r,s]	Fluorescence microscope	Temporal or spatial correlation is possible	Long measuring times; possible photobleaching; slow diffusion cannot be studied	Experimentally difficult method

[a] H. J. Galla and E. Sackmann, *Biochim. Biophys. Acta* **339**, 103 (1974).
[b] J. M. Vanderkooi, S. Fischkoff, M. Andrich, F. Podo, and C. S. Owens, *J. Chem. Phys.* **63**, 3661 (1975).
[c] K. A. Zachariasse, W. Kühnle, and A. Weller, *Chem. Phys. Lett.* **73**, 6 (1980).
[d] K. Razi-Naqvi, J. P. Behr, and D. Chapman, *Chem. Phys. Lett.* **26**, 440 (1974).
[e] S. M. Fernandez and R. D. Berlin, *Nature (London)* **264**, 411 (1976).
[f] D. D. Thomas, W. F. Carlsen, and L. Stryer, *Proc. Natl. Acad. Sci. U.S.A.* **75**, 5746 (1978).
[g] M. M. Poo and R. A. Cone, *Nature (London)* **247**, 438 (1974).
[h] R. Peters, J. Peters, K. H. Tews, and W. Bähr, *Biochim. Biophys. Acta* **367**, 282 (1974).
[i] D. Axelrod, D. E. Koppel, J. Schlessinger, E. Elson, and W. W. Webb, *Biophys. J.* **16**, 1055 (1976).
[j] K. Jacobson, Z. Derzko, E. S. Wu, Y. Hou, and G. Poste, *J. Supramol. Struct.* **5**, 565 (1976).
[k] B. A. Smith and H. M. McConnell, *Proc. Natl. Acad. Sci. U.S.A.* **75**, 2759 (1978).
[l] R. Peters, A. Brünger, and K. Schulten, *Proc. Natl. Acad. Sci. U.S.A.* **78**, 962 (1981).
[m] Fringe pattern fluorescence photobleaching; J. Davoust, C. M. B. Butor, G. van der Meer, R. Stricker, R. W. van Resandt, submitted for publication.
[n] B. R. Ware, L. J. Brvenik, R. T. Cummings, R. H. Furukawa, and G. A. Krafft, in "Applications of Fluorescence in the Biomedical Sciences" (D. L. Taylor, A. S. Waggoner, R. F. Murphy, F. Lanni, and R. R. Birge, eds.), p. 141. Liss, New York, 1986.
[o] L. D. Frye and M. Edidin, *J. Cell Sci.* **7**, 319 (1970).
[p] M. Edidin and D. Fambrough, *J. Cell Biol.* **57**, 27 (1973).
[q] M. Schindler, D. E. Koppel, and M. P. Sheetz, *Proc. Natl. Acad. Sci. U.S.A.* **77**, 1457 (1980).
[r] E. L. Elson and D. Magde, *Biopolymers* **13**, 1 (1974).
[s] N. O. Petersen, *Biophys. J.* **49**, 809 (1986).

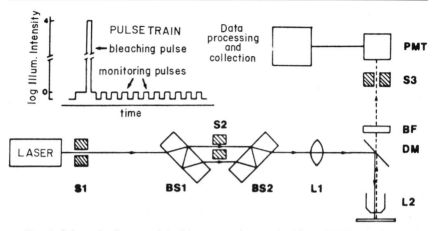

Fig. 9. Schematic diagram of the instrumentation required for a FRAP experiment. S1, S2, and S3 are fast electromechanical shutters, BS1 and BS2 are beam-splitting plates, L1 is a laser-focusing lens and L2 is a microscope objective, DM is a dichroic mirror, BF is a cut-off filter, and PMT is a photomultiplier. The upper left part of the figure shows the intensity of the illuminating beam as a function of time in the FRAP experiment. The functioning of the instrument is described in the text.

rhodamine, carbocyanines, and oxacyanines) an argon ion laser with a 500 mW output in the 488 and 514 nm lines is adequate. The laser beam is split into a monitoring beam and a bleaching beam by a set of beam-splitting plates, BS1 and BS2 (Fig. 8).[100] Care must be taken that the two beams are coaxial on emergence from the beam splitters. This is conveniently done by mounting the latter on rotatable tilt tables which allow a very sensitive manipulation of their positions. The bleaching beam between the two beam splitters can be obstructed for a programmable interval by a fast electromechancial shutter, S2. The laser beams are then focused using a lens, L1, onto an image plane of a microscope objective. Thus, the focused beam has a Gaussian profile at the object plane, the quality of which can be improved by the use of a spatial filter placed before the laser focusing lens. The illuminating beam is reflected onto the object through the microscope objective by a dichroic mirror, DM, which reflects at the illuminating wavelength and transmits at the wavelength of the light emitted by the fluorophores in the sample. Another cut-off filter, BF, can be placed above the dichroic mirror for added security. The fluorescence emission is detected and quantified by a photomultiplier, PMT, which is further protected by a shutter, S3. S3 is closed when S2 is open to avoid damage to the photomultiplier by reflections of the bleach-

[100] D. E. Koppel, *Biophys. J.* **28**, 281 (1979).

ing beam, which can be quite strong in spite of the dichroic mirror and cut-off filters. For slowly diffusing samples it often becomes necessary to measure for relatively long times (often >30 min). Bleaching may be a problem at almost any illuminating intensity, depending on the fluorophore's stability to photobleaching. It is usually negligible for short measurement times but has to be considered for long periods at the monitoring beam intensity usually used. For particularly extended measurements, it is convenient to measure the fluorescence intensity noncontinuously at regular intervals after the photobleaching event. This is done by placing a shutter, S1, in front of the laser and activating it so as to briefly illuminate the sample at desired intervals.

Several modifications of the above scheme are possible. Splitting of the laser beam into the weak monitoring and intense bleaching beams may be achieved with the use of an acoustooptic modulator or a Pockels cell. This allows considerably shorter bleaching pulses than are possible with the system described above. Such fast bleaching pulses are required when fluorescence depletion[42,101] experiments for studying rotational diffusion are performed on the microscope. For translational diffusion experiments on membranes the electromechanical shutters are fast enough. In addition to the usual Gaussian intensity profile for the illuminating beam, other distributions such as a uniform circular profile or a uniform pattern can be used. The former has the advantage over the Gaussian beam that the fluorescence recovery rate is not a function of the degree of photobleaching. With uniform pattern illumination and photobleaching, such as in the case of a striped pattern, the analysis of the recovery curve is significantly simpler.

A uniform circular profile can be achieved by using a slightly defocused beam to illuminate a diaphragm placed at the image plane of the objective. If the radius of the defocused beam is several times larger than that of the diaphragm, only the central portion of the Gaussian laser beam is selected and an essentially uniform circular profile is obtained at the object plane.[102] The latter can also be achieved by illuminating the polished end of an optical fiber, the other polished end of which is placed at the image plane of the objective. This method has an advantage over the previous one in that there is less loss of intensity inasmuch as the whole output of the laser is used. Speckle can be removed almost entirely by vibrating the fiber so that the illuminated spot at the object plane has a uniform intensity.[103] Placing a Ronchi ruling at the image plane of the objective results in a uniform striped pattern illumination at the object

[101] P. B. Garland, *Biophys. J.* **33**, 481 (1981).
[102] H. G. Kapitza and E. Sackmann, *Biochim. Biophys. Acta* **595**, 56 (1980).
[103] H. Gaub, E. Sackmann, R. Büschl, and H. Ringsdorf, *Biophys. J.* **45**, 725 (1984).

plane.[104] The use of a two-dimensional pattern has been used to study anisotropic diffusion on cell surfaces.[105] Other configurations are also possible.[106]

All of the above methods observe only the bleached spot after the photobleaching event, which gives a mean value for the rates of all lateral transport processes (diffusion and flow) occurring in the vicinity of the observation area. In order to get a better idea of the lateral transport process, it is desirable to observe the fluorescence intensity over the entire membrane simultaneous with the recovery of fluorescence in the test spot. The first attempt in this direction involved the introduction of a scanning mirror before the focusing lens, which permitted sequential focusing of the laser beam at several positions in and around the bleached area, thereby providing information about anisotropic transport in the sample.[100] A recent and potentially very useful extension of this concept has been the use of a video camera instead of the photomultiplier described above.[107] This method allows one simultaneously to follow the recovery kinetics in the photobleached area and quantitate the fluorescence intensity throughout the sample.

A further modification, the so-called "continuous fluorescence microphotolysis" (CFM)[99] method, increases the sensitivity of the FRAP technique. A continuous illumination of the sample with a relatively high intensity beam (about 10- to 100-fold more intense than a typical monitoring beam) monitors the emission from the exposed area at the same time as it effects a photobleaching of the fluorophores in it. The resultant curve, representing a decrease in fluorescence intensity with time, is a composite of the bleaching rate and the rate at which the bleached particles are replaced from the nonexposed pool. The method has the advantage of an enhanced sensitivity in comparison with the classical technique, but suffers from the drawback that the continuous exposure to high light levels may provoke photodamage,[108] particularly in biological samples. The curves are also difficult to interpret since analysis involves obtaining a best fit to the experimental data after having made assumptions regarding the rate of photobleaching, which is often dependent on sample preparation and components.

In the actual performance of a FRAP experiment several aspects have

---

[104] B. A. Smith and H. M. McConnell, *Proc. Natl. Acad. Sci. U.S.A.* **75**, 2759 (1978).

[105] B. A. Smith, W. R. Clark, and H. M. McConnell, *Proc. Natl. Acad. Sci. U.S.A.* **76**, 5641 (1979).

[106] D. Axelrod, in "Spectroscopy and the Dynamics of Molecular Biological Systems" (P. M. Bayley and R. E. Dale, eds.), p. 163. Academic Press, New York, 1985.

[107] H. G. Kapitza, G. McGregor, and K. A. Jacobson, *Proc. Natl. Acad. Sci. U.S.A.* **82**, 4122 (1985).

[108] M. P. Sheetz and D. E. Koppel, *Proc. Natl. Acad. Sci. U.S.A.* **76**, 314 (1979).

to be considered. Many dyes (such as the fluorescein derivatives) are quite photolabile. Conditions have to be found which do not result in a significant degree of photobleaching of the sample during monitoring but at the same time give reasonable signal-to-noise ratios. An important aspect is that the bleaching and monitoring beams have to be entirely coaxial. Noncoaxiality can lead to significant errors in the experimental results.[109] Heating of the sample is usually not a problem[110] but could result in serious membrane and cellular damage if the sample contains large concentrations of nonfluorescent substances that absorb the exciting light, as is the case of intact erythrocytes. Photoinduced cross-linking of membrane components (especially of proteins in cellular membranes) can be a serious problem and has been studied in some detail.[108] The degree of membrane protein cross-linking has been shown to be a function of the density of photobleachable groups in the membrane, above a certain threshold, and of the irradiation time. For a given dose, a smaller intensity of the irradiating beam applied for a longer time causes more photodamage. It was also found that the presence of reducing agents such as cysteamine, 2-mercaptoethanol, and reduced glutathione in the samples provided some protection against photoinduced protein cross-linking. Arguments against some potential artifacts in FRAP results have been presented.[111] Potential problems related to the specific configuration and to possible probe motion have also to be considered.[112]

The topography of the membrane under study is an important consideration. Most of the treatments of FRAP data analysis assume that the membrane surface being studied is planar. A theoretical analysis of FRAP results obtained on nonplanar microvillous membrane surfaces[113] shows that the diffusion coefficient on such a surface can be underestimated by conventional analysis. Similar considerations also apply to other nonplanar forms, though there may be differences, depending on the type of surface irregularities. Thus, quantitative comparisons between different nonplanar membranes may be made only if their topography is known and the FRAP results are corrected for it.

Data Analysis

Analysis of the FRAP curves depends on the intensity profile of the bleaching beam at the object. For a Gaussian profile and a fluorescence

[109] B. G. Barisas, *Biophys. J.* **29,** 545 (1980).
[110] D. Axelrod, *Biophys. J.* **18,** 129 (1977).
[111] D. E. Wolf, M. Edidin, and P. R. Dragsten, *Proc. Natl. Acad. Sci. U.S.A.* **77,** 2043 (1980).
[112] R. E. Dale, *Eur. Biophys. J.* **14,** 179 (1987).
[113] B. M. Aizenbud and N. D. Gershon, *Biophys. J.* **38,** 287 (1982).

recovery which is due only to the diffusion or uniform flow of a single fluorescing species, the fluorescence intensity as a function of time, $F(t)$, is given by the following series solutions,[96] valid for all degrees of bleaching and all times:

for diffusion only

$$F(t) = F(i) \sum_{n=0}^{\infty} [(-K)^n/n!][1 + n(1 + 2t/\tau_D)]^{-1} \tag{10a}$$

for uniform flow

$$F(t) = F(i) \sum_{n=0}^{\infty} [(-K)^n/(n + 1)!] \exp\{[-2n/(n + 1)](t/\tau_F)^2\} \tag{10b}$$

where $F(i)$ is the fluorescence intensity before the photobleaching event; K is a "bleaching parameter" given by $F(0)/F(i) = [1 - \exp(-K)]/K$ where $F(0)$ is $F(t)$ at $t = 0$; τ_D is the characteristic diffusional recovery time given by $\tau_D = \omega^2/4\mathbf{D}_t$; and τ_F is the characteristic flow recovery time given by $\tau_F = \omega/V$ where V is the flow velocity and ω is the radius of the focused laser beam at the e^{-2} intensity. Expressions for more complicated situations, such as multiple diffusing species or combinations of diffusion and uniform flow on the sample,[96] and linearized forms of these expressions[114,115] are available.

In case of an uniform circular profile, the fractional fluorescence recovery, $f(t) = [F(t) - F(0)]/[F(\infty) - F(0)]$, is given by a series solution for translational diffusion [Eq. (14) in Ref. 96, corrected by inclusion of the brackets in Eq. 11a below],

$$f(t) = 1 - \left\{(\tau_D/t) \exp(-2\tau_D/t)[I_0(2\tau_D/t) + I_2(2\tau_D/t)] \right. \\ \left. + 2 \sum_{k=0}^{\infty} [(-1)^k(2k + 2)! (k + 1)! (\tau_D/t)^{k+2}](k!)^{-2}[(k + 2)!]^{-2}\right\} \tag{11a}$$

and, by an analytical solution for uniform flow,

$$f(t) = (2/\pi) \sin^{-1}(t/2\tau_F) + (t/\tau_F)(1 - t^2/4\tau_F^2)^{1/2} \tag{11b}$$

where I_0 and I_2 are modified Bessel functions. The series solution for translational diffusion given above is inconvenient for numerical work in the short time regime ($t < 0.1\tau_D$) for which an alternative solution has been presented.[116,117] Nevertheless, fitting experimental recovery curves that arise from more than one diffusing component is not simple.

[114] J. Yguerabide, J. A. Schmidt, and E. E. Yguerabide, *Biophys. J.* **39**, 69 (1982).
[115] E. J. J. van Zoelen, L. G. J. Tertoolen, and S. W. de Laat, *Biophys. J.* **42**, 103 (1983).
[116] D. M. Soumpasis, *Biophys. J.* **41**, 95 (1983).
[117] T. J. Lardner and N. Solomon, *J. Theor. Biol.* **60**, 433 (1976).

In cases where it is known that the membrane is planar and recovery of fluorescence is due to only one diffusing species, the translational diffusion constant or the flow velocity may be simply evaluated from the time, $t_{1/2}$, required for the fractional fluorescence recovery, $f(t)$, to attain a value of 0.5, using the expressions[96]

$$\mathbf{D}_t = \omega^2 \gamma_D / 4 t_{1/2} \tag{12a}$$

and

$$\mathbf{V} = \omega \gamma_F / t_{1/2} \tag{12b}$$

where $\gamma_D = t_{1/2}/\tau_D$ and $\gamma_F = t_{1/2}/\tau_F$, and are dependent upon the beam profile and the bleaching parameter K. For an uniform circular profile $\gamma_D = 0.88$ and $\gamma_F = 0.81$, both independent of K.

The third commonly used beam profile is the periodic striped pattern. In this case, the concentration, C, along the direction, x, perpendicular to the direction of the stripes, as a function of time, t, after photobleaching is given by[104]

$$C(x, t) = A + B \exp[-\mathbf{D}_t a^2 t] \sin ax + E \exp[-9\mathbf{D}_t a^2 t] \sin 3ax + F \exp[-25\mathbf{D}_t a^2 t] \sin 5ax + \ldots \tag{13}$$

where $A, B, E, F, \ldots$ are related to the concentration of probe before photobleaching, the duration and intensity of the photobleaching burst, and the contrast and resolution of the stripe image in the sample; $a = 2\pi/P$ is the spatial frequency of the pattern and P its period. After a time $t > 0.1/\mathbf{D}_t a^2$, $C(x, t)$ can be described by only the first two terms of the above expression so that the recovery of fluorescence is monoexponential for a single diffusing component. The measured time constant τ_D is given by $\mathbf{D}_t = 1/[a^2 \tau_D]$. The analysis of FRAP experiments using two-dimensional periodic pattern illumination has been discussed elsewhere.[105] However, this method probably will be supplanted by video analysis combined with photobleaching,[107] as mentioned above.

Sample Preparation

The preparation of samples for FRAP experiments has been described in several of the earlier works in considerable detail. For studies on lipid bilayers, so-called multibilayers, which are planar and oriented multi-lamellar lipid–water systems that are a few microns thick, are the most commonly used. Their preparation has been described by different laboratories.[118,119] Essentially, a solution of the desired lipid, usually in a mixture of chloroform and methanol, is evaporated on a small (~1 cm²) area

---

[118] L. Powers and N. A. Clark, *Proc. Natl. Acad. Sci. U.S.A.* **72**, 840 (1975).
[119] E. S. Wu, K. Jacobson, and D. Papahadjopoulos, *Biochemistry* **16**, 3936 (1977).

of a clean glass slide. The residue is then heated at ~80° for 5–10 min in an oven, in a nitrogen atmosphere if unsaturated lipids are involved, to remove all traces of the solvent. The amount of lipid is typically between 1 and 3 mg. The residue is hydrated with 50–100 μl of water (or buffer) added as a hanging drop on the underside of a microscope coverslip. The entire process is carried out at a temperature above the phase transition temperature of the lipid. After ~10 min the slide may be pressed between optical flats, using a 2-kg weight and avoiding shear, though this is not always necessary. The slide is then sealed and stored at a temperature that is higher than the phase transition temperature of the lipid. Storage in a chamber with 100% humidity such as a cell incubator is recommended in order to avoid possible loss of water from the slide. In any case, it is advisable to check the sample for possible leaks in the sealant. A continuous spreading of the lipid smear is an indication of inadequate sealing. The choice of sealant is dependent on the temperature at which the sample will be studied. In the authors' experience a mixture of paraffin wax and a viscous fraction of liquid paraffin or petroleum jelly is best suited for the purpose. Mixtures with lower melting temperatures (higher liquid paraffin content) are better for samples that are studied at low temperatures since cracks in the sealant lead to evaporation of the water on the slide. The use of sealants containing organic solvents is not recommended since these eventually find their way into the lipid multilayers. We find that the use of siliconized glass slides and coverslips or of optical flats instead of slides leads to the formation of multilayers with a considerably higher degree of orientation. The orientation of a multilayer domain is conveniently observed under the microscope using crossed polarizers where it appears as a dark zone with highly birefringent edges.[118,119]

FRAP experiments have also been performed on isolated bilayers deposited on a grid[120] or on isolated monolayers[121] and bilayers deposited on siliconized glass or surface-oxidized silicon wafers.[122] The preparation of isolated bilayers ("black films") on grids involves the use of organic solvents and suffers from the drawback that considerable amounts of the solvent may be included in the bilayers. In the monolayers and bilayers adsorbed onto glass or silicon wafers[121,122] it is not quite clear how the proximity of the layer being examined to an immobile surface affects the diffusion results. Monolayers at an air–water interface have also been studied.[123,124] Special chambers have to be constructed for such studies

[120] P. F. Fahey and W. W. Webb, *Biochemistry* **17**, 3046 (1978).
[121] V. von Tscharner and H. M. McConnell, *Biophys. J.* **36**, 421 (1981).
[122] L. K. Tamm and H. M. McConnell, *Biophys. J.* **47**, 105 (1985).
[123] J. Teissie, J. F. Tocanne, and A. Baudras, *Eur. J. Biochem.* **83**, 77 (1978).
[124] R. Peters and K. Beck, *Proc. Natl. Acad. Sci. U.S.A.* **80**, 7183 (1983).

to avoid convective flow in the lipid monolayer during the FRAP measurement.

The preparation of protein-containing lipid bilayer vesicles or multibilayers depends on the nature of the interaction between the protein and the bilayer. Surface-attached or extrinsic proteins may be added in the hydrating medium to a dry lipid residue.[125] Care must be taken to allow enough time for all of the protein in the aqueous phase to adsorb onto the bilayers and distribute itself homogeneously over them. A 24-hr period is usually sufficient for this purpose. Reconstituted vesicles and multilayers containing intrinsic proteins can be prepared by fusion of reconstituted small vesicles.[126] In this method an integral membrane protein is reconstituted into liposomes using any one of several classical techniques. The reconstituted vesicles are sedimented by centrifugation from the suspension, a thick suspension of the pellet is deposited on a clean glass slide, and the excess water is removed by placing the slide in a chamber with a relative humidity of close to 100% or by placing it in a dessicator over 1 M NaCl or another suitable salt solution.[127] This form of dehydration is a slow process; faster desiccation is possible over anhydrous silica or calcium chloride but care must be exercised to avoid complete dehydration of the deposited material. Under these conditions, the proteoliposome suspension is brought to a state which is not fully dehydrated but has no *excess* water; as a consequence, a fusion of the liposomes results. When sufficient proteoliposomes have been deposited and fused, the slide is hydrated under excess water conditions using the method described earlier for the preparation of lipid multilayers. In this way, it is possible to obtain multilayers with spontaneous formation of pauci- and multilamellar liposomes with diameters of 30–100 μm on the edge of the preparations.[126] It is not certain that all membrane proteins withstand this treatment without denaturation. The acetylcholine receptor protein[128] retains its ability to bind bungarotoxin and the sarcoplasmic reticulum Ca^{2+}-activated adenosinetriphosphatase[129] retains its ability to hydrolyze ATP when reconstituted in this way.

The labeling of cell surfaces can be done by any one of several meth-

---

[125] W. L. C. Vaz, K. Jacobson, E. S. Wu, and Z. Derzko, *Proc. Natl. Acad. Sci. U.S.A.* **76**, 5645 (1979).

[126] W. L. C. Vaz, H. G. Kapitza, J. Stümpel, E. Sackmann, and T. M. Jovin, *Biochemistry* **20**, 1392 (1981).

[127] E. W. Washburn (ed.), "International Critical Tables of Numerical Data," Vol. 1, pp. 67–68; Vol. 3, pp. 366–374 and p. 385. McGraw-Hill, New York, 1926.

[128] M. Criado, W. L. C. Vaz, F. J. Barrantes, and T. M. Jovin, *Biochemistry* **21**, 5750 (1982).

[129] W. L. C. Vaz, M. Criado, V. M. C. Madeira, G. Schoellmann, and T. M. Jovin, *Biochemistry* **21**, 5608 (1982).

ods. Tagging a fluorescent-labeled antibody or ligand to the protein or lipid whose diffusion on a cell surface is being studied, followed by extensive washing of the cells or centrifugation through a gradient to remove free ligand, probably results in the cleanest preparations. The direct conjugation of membrane components with reactive fluorophores is not recommended in general since there is no selectivity in the labeling and it is difficult to remove unreacted dye that is adsorbed to or even dissolved within the cell. To study lipid diffusion in cellular membranes, an extrinsic lipid-like amphipathic fluorescent probe such as the alkyloxacyanines or indocyanines or a fluorescent lipid derivative such as NBD-PE is usually incorporated into the cell membrane. These substances may be added directly, as aqueous suspensions or as adsorbates on an inert support, to a suspension of cells or may be added to the cellular suspension as an alcoholic solution. The exact method of addition probably has to be empirically established separately for each case. It is essential to confirm whether the added amphipathic probes are really incorporated within the membranes or just adsorbed to their surfaces.[130]

Representative Data and Concluding Remarks

Figure 10 shows typical, representative FRAP curves obtained on various membranes. The expected theoretical recovery curves are also shown where these were reported. A few selected results for translational diffusion in membranes using FRAP are presented in Tables IV and V. No attempt was made to review the literature on the subject since such an undertaking is beyond the scope of this chapter; the reader is directed to the reviews of diffusion in model membranes[15-17] and cell membranes[8-13] cited earlier. The criterion for selection of the results presented in Table IV was the availability of translational diffusion data for model membranes and various cellular systems, in some cases using different methods, and/or the possibility of a comparison with rotational diffusion results presented in Table II.

A detailed discussion of comparisons between experimental results and the theoretical expectations of models for diffusion in membranes is available elsewhere.[16,17] Typically, values of D_t for lipids are in the range of 10^{-8}–10^{-7} cm^2 sec^{-1} in pure lipid bilayers and in the range of 10^{-9}–10^{-8} cm^2 sec^{-1} in the plasma membrane of cells. Several reasons have been proposed for the observed differences: the high protein concentration in cellular membranes,[129,131] the existence of effectively immobile protein

[130] F. Szoka, K. Jacobson, Z. Derzko, and D. Papahadjopoulos, *Biochim. Biophys. Acta* **600**, 1 (1980).
[131] K. Jacobson, Y. Hou, Z. Derzko, J. Wojcieszyn, and D. Organisciak, *Biochemistry* **20**, 5268 (1981).

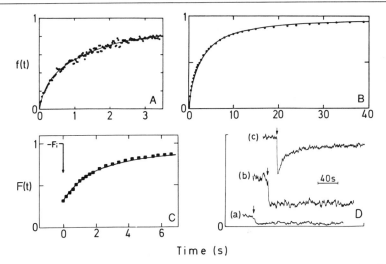

FIG. 10. Typical FRAP curves. (A) NBD-POPC in POPC bilayers. Data reproduced from W. L. C. Vaz, D. Hallmann, R. M. Clegg, A. Gambacorta, and M. De Rosa, *Eur. Biophys. J.* **12**, 19 (1985). (B) FITC-acetylcholine receptor monomer in soybean lipid bilayers. Data reproduced from Criado *et al.*[128] (C) FITC-Apo-C III in egg-PC bilayers (W. L. C. Vaz, unpublished data). (D) Rh-EGF attached to its receptors in A431 cell plasma membranes. Data reproduced from A. R. Rees, M. Georgiou, P. Johnson, and P. B. Garland, *EMBO J.* **3**, 1843 (1984). Trace a is a control for autofluorescence, trace b corresponds to high-affinity sites, and trace c corresponds predominantly to low-affinity sites. Fluorescence is in arbitrary units, the gain for traces a and b being 5 times greater than for trace c. In A–C, the points are experimental and the lines are theoretical fits for one diffusing component. The data in A and B were obtained using a uniform circular beam and in C and D using a Gaussian beam.

domains in the cellular membranes causing the diffusion of lipids to be similar to percolation in an archipelago,[132] possible interactions between effectively immobile proteins and lipids,[133,134] and very high viscosities at the membrane surface.[135,136] It is probable that a combination of some or all of these factors play a role.

A more complicated situation presents itself with regard to the translational diffusion results for proteins. Here we see differences of about 10^3-fold in the values of $\mathbf{D}_t$ in model systems as compared with cellular mem-

[132] M. J. Saxton, *Biophys. J.* **9**, 165 (1982).
[133] E. L. Elson and J. A. Reidler, *J. Supramol. Struct.* **12**, 481 (1979).
[134] M. Schindler, M. J. Osborn, and D. E. Koppel, *Nature (London)* **283**, 346 (1980).
[135] B. D. Hughes, B. A. Pailthorpe, L. R. White, and W. H. Sawyer, *Biophys. J.* **37**, 673 (1982).
[136] W. L. C. Vaz, J. Stümpel, D. Hallmann, A. Gambacorta, and M. De Rosa, *Eur. Biophys. J.* **15**, 111 (1987).

TABLE IV
REPRESENTATIVE FRAP RESULTS ON THE TRANSLATIONAL DIFFUSION OF LIPID PROBES IN DIFFERENT MEMBRANES

Diffusant	Membrane	Temperature (°C)	D_t (cm^2 sec^{-1})	Mobile fraction	Ref.
NBD-PE[a]	DMPC	25	4.4×10^{-8}	1.0	b
		15	6×10^{-11}	1.0	c
	DMPC–cholesterol (1:1)	30	7×10^{-9}	1.0	d
	POPC	25	4.2×10^{-8}	1.0	b
diI[e]	DMPC	24	8.4×10^{-8}	1.0	f
NBD-MSPE[g]	POPC	24	2.8×10^{-8}	1.0	h
diI	Fibroblast plasma membrane lipids	20	2.5×10^{-8}	—	i
	Fibroblast plasma membrane	20	8×10^{-9}	—	i
	Murine spleen lymphocytes, plasma membranes	—	1.5×10^{-8}	0.9	j
	Murine spleen lymphocytes, plasma membranes, blebs	—	3.2×10^{-8}	0.9	j
	Erythrocyte plasma membranes	25	8×10^{-9}	1.0	k
		25	2×10^{-9}	1.0	l
	Inner mitochondrial membrane	25	5×10^{-9}	—	m

[a] NBD-PE, N(7-Nitro-2,1,3-benzoxadiazol-4-yl)-phosphatidylethanolamine, used as the dimyristoylphosphatidylethanolamine analog in DMPC multibilayers and as the 1-palmitoyl-2-oleoylphosphatidylethanolamine analog in POPC multibilayers in Footnote b, and as the egg phosphatidylethanolamine analog in Footnote c.
[b] W. L. C. Vaz, R. M. Clegg, and D. Hallmann, *Biochemistry* **24**, 781 (1985).
[c] B. A. Smith and H. M. McConnell, *Proc. Natl. Acad. Sci. U.S.A.* **75**, 2759 (1978).
[d] C. H. Chang, H. Takeuchi, T. Ito, K. Machida, and S. Ohnishi, *J. Biochem. (Tokyo)* **90**, 997 (1981).
[e] diI, 3,3'-Dioctadecyl- or 3,3'-dihexadecylindocarbocyanine iodide.
[f] Z. I. Derzko and K. Jacobson, *Biochemistry* **19**, 6050 (1980).
[g] NBD-MSPE, N-(7-Nitro-2,1,3-benzoxadiazol-4-yl) derivative of a membrane-spanning glyceroldialkylglycerol tetraether diphosphatidylethanolamine.
[h] W. L. C. Vaz, D. Hallmann, R. M. Clegg, A. Gambacorta, and M. De Rosa, *Eur. Biophys. J.* **12**, 19 (1985).
[i] K. Jacobson, Y. Hou, Z. Derzko, J. Wojcieszyn, and D. Organiesciak, *Biochemistry* **20**, 5268 (1981).
[j] E. S. Wu, D. W. Tank, and W. W. Webb, *Proc. Natl. Acad. Sci. U.S.A.* **79**, 4962 (1982).
[k] J. A. Bloom and W. W. Webb, *Biophys. J.* **42**, 295 (1983).
[l] H. G. Kapitza and E. Sackmann, *Biochim. Biophys. Acta* **595**, 56 (1980).
[m] B. Chazotte, E. S. Wu, and C. R. Hackenbrock, *Biochem. Soc. Trans.* **12**, 464 (1983).

TABLE V
REPRESENTATIVE FRAP RESULTS ON THE TRANSLATIONAL DIFFUSION
OF PROTEINS IN MEMBRANES

Diffusant	Membrane	Temperature (°C)	D_t (cm^2 sec^{-1})	Mobile fraction	Ref.
Extrinsic proteins					
Apo-C III	Egg-PC	25	4×10^{-8}	1.0	a
	Egg-PC cholesterol (1:1)	25	2×10^{-9}	1.0	a
Plastocyanine	Egg-PC	23	$5 \times 10^{-8\,b}$	0.7^b	c
			$8 \times 10^{-10\,b}$	0.3^b	
Cytochrome c	Mitochondrial membrane		$3.5-7 \times 10^{-10}$		d
Integral proteins					
Glycophorin	DMPC	25	1.5×10^{-8}	1.0	e
Bovine rhodopsin	DMPC	25	1.8×10^{-8}	1.0	f
	Rod outer segment membranes	20	4×10^{-9}	—	g
		22	3×10^{-9}	0.4	h
Ca^{2+}-ATPase	Sarcoplasmic reticulum lipids	25	1.8×10^{-8}	1.0	f
Erythrocyte band 3 anion transporter	DMPC	30	1.6×10^{-8}	—	i
	Normal erythrocyte plasma membranes	24	4.5×10^{-11}	—	j
	Spherocytic erythrocyte plasma membranes	24	2.5×10^{-9}	—	j
	Erythrocyte plasma membranes (46 mM NaPO$_4$)	21	4×10^{-11}	0.1	k
	Erythrocyte plasma membranes (13 mM NaPO$_4$)	37	2×10^{-9}	0.9	k
Acetylcholine receptor					
Monomer	Soybean lipid	25	2.2×10^{-8}	1.0	l
Dimer	Soybean lipid	25	2.2×10^{-8}	1.0	l
Tetramer	Soybean lipid	25	2.1×10^{-8}	1.0	m
	Single muscle fiber plasma membrane	—	$\leq 10^{-12}$	—	n
	Single muscle fiber plasma membrane, blebs	—	2×10^{-9}	1.0	o
EGF receptor	A431 cells (sparse culture) plasma membrane	3–37	$2.6 \times 10^{-10\,p}$	0.8	q
			$\leq 10^{-13\,p}$		q
	A431 cells (confluent cultures) plasma membrane	25	6×10^{-10}	0.6	r
Con A receptor	Murine spleen lymphocyte plasma membranes	—	$\leq 3 \times 10^{-11}$	≤ 0.2	s
	Murine spleen lymphocyte plasma membranes, blebs	—	8.7×10^{-9}	1.0	s

(*continued*)

TABLE V (continued)

Diffusant	Membrane	Temperature (°C)	D_t (cm² sec⁻¹)	Mobile fraction	Ref.
Sendai virus glycoproteins (HN, F)	Human erythrocyte (adsorbed, not fused)	4	$\leq 5 \times 10^{-12}$	≈ 0.05	t
	(fused)	37	3×10^{-10}	0.5–0.6	t

[a] W. L. C. Vaz, K. A. Jacobson, E. S. Wu, and Z. Derzko, *Proc. Natl. Acad. Sci. U.S.A.* **76,** 5645 (1979).
[b] The FRAP recovery curve was accounted for by two diffusing components having the diffusion coefficients shown and the amplitudes shown as mobile fractions.
[c] M. Fragata, S. Ohnishi, K. Asada, T. Ito, and M. Takahashi, *Biochemistry* **23,** 4044 (1984).
[d] J. Hochman, S. Ferguson-Miller, and M. Schindler, *Biochemistry* **24,** 2509 (1985).
[e] W. L. C. Vaz, H. G. Kapitza, J. Stümpel, E. Sackmann, and T. M. Jovin, *Biochemistry* **20,** 1392 (1981).
[f] W. L. C. Vaz, M. Criado, V. M. C. Madeira, G. Schoellmann, and T. M. Jovin, *Biochemistry* **21,** 5608 (1982).
[g] M. Poo and R. A. Cone, *Nature (London)* **247,** 438 (1974).
[h] C. L. Wey, R. A. Cone, and M. A. Edidin, *Biophys. J.* **33,** 225 (1981).
[i] C. H. Chang, H. Takeuchi, T. Ito, K. Machida, and S. Ohnishi, *J. Biochem. (Tokyo)* **90,** 997 (1981).
[j] M. P. Sheetz, M. Schindler, and D. E. Koppel, *Nature (London)* **285,** 510 (1980).
[k] D. E. Golan and W. Veatch, *Proc. Natl. Acad. Sci. U.S.A.* **77,** 2537 (1980).
[l] M. Criado, W. L. C. Vaz, F. J. Barrantes, and T. M. Jovin, *Biochemistry* **21,** 5750 (1982).
[m] W. L. C. Vaz and M. Criado, *Biochim. Biophys. Acta* **819,** 18 (1985).
[n] D. Axelrod, P. Ravdin, D. E. Koppel, J. Schlessinger, W. W. Webb, E. L. Elson, and T. R. Podleski, *Proc. Natl. Acad. Sci. U.S.A.* **73,** 4594 (1976).
[o] D. W. Tank, E. S. Wu, and W. W. Webb, *J. Cell Biol.* **92,** 207 (1982).
[p] The higher value of D_t is for low-affinity receptors and the lower value is for high-affinity receptors.
[q] A. R. Rees, M. Gregoriou, P. Johnson, and P. B. Garland, *EMBO J.* **3,** 1843 (1984).
[r] G. M. Hillman and J. Schlessinger, *Biochemistry* **21,** 1167 (1982).
[s] E. S. Wu, D. W. Tank, and W. W. Webb, *Proc. Natl. Acad. Sci. U.S.A.* **79,** 4962 (1982).
[t] Y. I. Henis, O. Gutman, and A. Loyter, *Exp. Cell Res.* **160,** 514 (1985).

branes. Technical problems, such as photoinduced protein cross-linking during photobleaching may effect FRAP measurements of protein diffusion in cellular membranes since, as seen in Table IV for the case of the acetylcholine receptor protein, other techniques[94,137] seem to yield considerably higher values for D_t. However, apart from the one case in question, no other proteins have been examined by more than one tech-

[137] M. M. Poo, *Nature (London)* **295,** 332 (1982).

nique. It has been shown[138-140] that interactions between integral membrane proteins in the plasma membrane and cytoskeletal structures are probably responsible for the apparent slow diffusion. A cytoskeletal network just below and in contact with or close to the plasma membrane of a cell will certainly form a barrier to diffusion of proteins which protrude out into the cytoplasmic side of the membrane. From this point of view, comparisons between diffusion in model systems where one has effectively "infinite" planar sheets and natural membranes may not be theoretically justified.[17] Besides interactions with the cytoskeleton, interactions of membrane-bound proteins in natural membranes with the glycocalyx and the relatively viscous cytoplasm, as well as homologous and heterogeneous associations[133] (see also previous discussion on section or rotational diffusions) or the effects of high protein concentration upon membrane viscosity,[141] may also have to be considered.

A theoretical treatment of the diffusion of membrane components (proteins and lipids) within or on a membrane poses some problems. Several models have been proposed.[17,69-71,142-144] It appears, from studies on diffusion of integral proteins in model membranes[129,141,144] that some of the predictions of theoretical models based on continuum fluid hydrodynamics,[69-71,142] regarding the weak dependence of D_t on protein radius and the translational diffusion behavior of "porous" protein aggregates, have been experimentally confirmed, whereas other predictions of these models, regarding the relationship between the rotational and translational frictional drags and the influence of bounding fluid viscosity on the translational diffusion, are still untested due to insufficient experimental results. The measured rates of lipid diffusion in bilayers have been compared with theoretical predictions from models based on continuing fluid hydrodynamics and "free-volume" theory for liquids.[16,17,136,143,144] The conclusion is that further theoretical refinements are required.

Acknowledgments

We acknowledge the contributions of numerous colleagues in the development and utilization of the instrumentation described in this chapter. We also thank Drs. M. Edidin and Y. Henis for critical reading of the manuscript.

[138] D. E. Golan and W. Veatch, *Proc. Natl. Acad. Sci. U.S.A.* **77**, 2637 (1980).
[139] M. P. Sheetz, M. Schindler, and D. E. Koppel, *Nature (London)* **285**, 510 (1980).
[140] D. W. Tank, E. S. Wu, and W. W. Webb, *J. Cell Biol.* **92**, 207 (1982).
[141] R. Peters and R. J. Cherry, *Proc. Natl. Acad. Sci. U.S.A.* **79**, 4317 (1982).
[142] F. W. Wiegel, *Lect. Notes Phys.* **121**, 1 (1980).
[143] H. J. Galla, W. Hartmann, U. Theilen, and E. Sackmann, *J. Membr. Biol.* **48**, 215 (1979).
[144] W. L. C. Vaz, R. M. Clegg, and D. Hallmann, *Biochemistry* **24**, 781 (1985).

[28] Membrane Protein Molecular Weight Determined by Low-Angle Laser Light-Scattering Photometry Coupled with High-Performance Gel Chromatography

By Yutaro Hayashi, Hideo Matsui, and Toshio Takagi

Introduction

Intrinsic membrane proteins are converted into complexes composed of protein, lipid, and surfactant by solubilization of the membrane with a surfactant.[1] To elucidate the molecular structure of this kind of protein, the molecular weights of the total protein moiety and the constituent polypeptides must be determined by means of reliable physicochemical methods such as sedimentation equilibrium and light scattering.

The intensity of the light scattered from a solution of a protein depends both on the solute concentration and on the scattering angle. Therefore, when the conventional technique is applied to obtain the molecular weight of a solubilized membrane protein, the scattering at each protein concentration must be extrapolated to zero angle, and that at each angle to zero concentration[2-4]; such a plot is known as the Zimm plot.[5] Thus, the technique requires not only a large amount of sample but also a long time. A special light-scattering instrument, a low-angle laser light-scattering photometer, with unique optics and employing a laser beam as the light source, was developed by Kaye *et al.*[6,7] This new type of light-scattering photometer allows measurement at a scattering angle of as low as 6° and at very low concentrations of protein. By making measurements at such low angle and low concentration of solute, the molecular weight can be determined from data based on a single measurement without tedious extrapolation.[6-8]

With the low-angle laser light-scattering (LALLS)[9] photometer

---

[1] C. Tanford and J. A. Reynolds, *Biochim. Biophys. Acta* **457**, 133 (1976).
[2] E. P. Pittz, J. C. Lee, B. Bablouzian, R. Townend, and S. N. Timasheff, this series, Vol. 27, p. 209.
[3] C. Tanford, "Physical Chemistry of Macromolecules," p. 275. Wiley, New York, 1961.
[4] M. Bier, this series, Vol. 4, p. 147.
[5] B. H. Zimm, *J. Chem. Phys.* **16**, 1099 (1948).
[6] W. Kaye, A. J. Havlik, and J. B. McDaniel, *Polym. Lett.* **9**, 695 (1971).
[7] W. Kaye and A. J. Havlik, *Appl. Opt.* **12**, 541 (1973).
[8] T. Takagi, J. Miyake, and T. Nashima, *Biochim. Biophys. Acta* **626**, 5 (1980).
[9] Abbreviations used are: LALLS, low-angle laser light scattering; HPGC, high-performance gel chromatography; HPGC/LALLS, monitoring of elution from a HPGC column with a LALLS photometer and supplementary equipment.

equipped with a flow cell, Fukutomi et al.[10] and Takagi and collaborators[11,12] have measured protein molecular weight by monitoring the elution of protein from a high-performance gel chromatography (HPGC)[9] column, which can be used with aqueous solvents. The combined use[13–16] of the two methods has opened a new era in the estimation of the molecular weights of solubilized membrane proteins which bind appreciable amounts of surfactant and lipid. In the HPGC/LALLS[9] method,[15–17] the sole parameter necessary for estimation of the molecular weight of the protein moiety is the extinction coefficient at 280 nm of the objective protein itself. A single run can be completed within a few hours. On the other hand, the ultracentrifuge technique requires several days to attain sedimentation equilibrium, as well as the estimation of several parameters of each of the substances bound to the protein.[1,18]

The purpose of this chapter is to present the basic principle of measurement of the molecular weight of a solubilized membrane protein and to describe the practical techniques involved in the measurement. The reader is referred to a review by Takagi[19] for a detailed description of the LALLS photometer and application of the technique to water-soluble proteins.

Theoretical Background

The fundamental equation for light scattering by a protein or its complex with any ligands that are sufficiently small compared with the wavelength of light is as follows[3]:

$$\frac{i_\theta}{I_0} = \frac{2\pi^2 n_0^2 (dn/dc)^2 (1 + \cos^2\theta) c}{N\lambda^4 r^2 (1/M + 2Bc)} \tag{1}$$

[10] M. Fukutomi, M. Fukuda, and T. Hashimoto, *Toyo Soda Kenkyu Hokoku* **24**, 33 (1980).
[11] T. Takagi, *J. Biochem. (Tokyo)* **89**, 363 (1981).
[12] T. Takagi, in "Protides of the Biological Fluids" (H. Peeters, ed.), p. 701. Pergamon, Oxford, England, 1982.
[13] K. Kameyama, T. Nakae, and T. Takagi, *Biochim. Biophys. Acta* **706**, 19 (1982).
[14] S. Maezawa, Y. Hayashi, T. Nakae, J. Ishii, K. Kameyama, and T. Takagi, *Biochim. Biophys. Acta* **747**, 291 (1983).
[15] S. Maezawa and T. Takagi, *J. Chromatogr.* **280**, 124 (1983).
[16] Y. Hayashi, T. Takagi, S. Maezawa, and H. Matsui, *Biochim. Biophys. Acta* **748**, 153 (1983).
[17] Y. Hayashi, H. Matsui, S. Maezawa, and T. Takagi, in "The Sodium Pump" (I. M. Glynn and J. C. Ellory, eds.), p. 51. Company of Biologists, Cambridge, England, 1985.
[18] J. C. H. Steele, Jr., C. Tanford, and J. A. Reynolds, this series, Vol. 48, p. 11.
[19] T. Takagi, in "Progress in HPLC" (H. Parvez, Y. Kato, and S. Parvez, eds.), Vol. 1, p. 27. VNU Science Press, Utrecht, The Netherlands, 1985.

where n_0 is the refractive index of the solvent; (dn/dc) and c, the specific refractive index increment and the weight concentration of the protein, respectively; θ, the scattering angle; r, the distance between the scattering sample and the photodetector; I_0, the intensity of the incident beam; i_θ, the intensity of the scattered light; M, the molecular weight of the protein; B, the second virial coefficient; N, Avogadro's number; λ, the wavelength. When the intensity of the scattered light is measured for solutes dissolved in a given solvent at a given scattering angle with a scattering photometer, most terms in Eq. (1) can be replaced by a single constant, k'. Thus, the output of the detector, $(\text{Output})_{\text{LS}}$, proportional to the right-hand term of Eq. (1), can be expressed as follows:

$$(\text{Output})_{\text{LS}} = \frac{k'(dn/dc)^2 c}{1/M + 2Bc} \tag{2}$$

When the differential refractive index of the solution (difference in refractive index between the solution and the solvent) and the light scattering are measured with a single aliquot, the output of the refractometer, $(\text{Output})_{\text{RI}}$, can be expressed as follows:

$$(\text{Output})_{\text{RI}} = k''(dn/dc)c \tag{3}$$

where k'' is a proportional constant. Then, Eq. (1) can be converted to the following equation, where k''' is a constant:

$$\frac{k'''(\text{Output})_{\text{RI}}(dn/dc)}{(\text{Output})_{\text{LS}}} = 1/M + 2Bc \tag{4}$$

Solubilized membrane protein generally exists as a complex composed of protein, surfactant, lipid, and carbohydrate. The molecular weight of the protein moiety of the complex, M_p, and the weight concentration of the protein moiety, c_p, are interrelated with those of the complex as follows:

$$M = M_p(1 + \delta) \tag{5}$$
$$c = c_p(1 + \delta) \tag{6}$$

where δ is the amount (g) of all the substances bound per gram of the protein moiety. When M and c in Eq. (4) are replaced by M_p and c_p, respectively, the following equation is obtained:

$$\frac{k_1(\text{Output})_{\text{RI}}(dn/dc_p)}{(\text{Output})_{\text{LS}}} = \frac{1}{M_p} + 2B(1 + \delta)^2 c_p \tag{7}$$

where (dn/dc_p) is the specific refractive index increment of the complex expressed in terms of c_p, and k_1 is a constant. When a laser is used as the light source and the scattered light is collected through an annular slit set at a scattering angle as low as 5°, protein solution at a concentration as low as 0.3 mg/ml can produce scattered light with adequate intensity at the detector even when the scattering volume is as little as 0.1 μl, as in a LALLS photometer. The second term of the right-hand side of Eq. (7) can be neglected for a protein solution at such a low concentration as above to give the following equation:

$$M_p = \frac{k_1(\text{Output})_{\text{LS}}}{(dn/dc_p)(\text{Output})_{\text{RI}}} \quad (8)$$

The specific refractive index increment can be assumed to be independent of particle concentration. Therefore, it can be obtained by dividing the differential refractive index of the solution by c_p, both of which values are obtained for an aliquot of the solution at any protein concentration. If the substances bound to the protein moiety have no absorbance at 280 nm, c_p can be determined by measuring the absorbance of the solution at 280 nm with a UV spectrophotometer according to the following equation:

$$c_p = \frac{k''''(\text{Output})_{\text{UV}}}{A} \quad (9)$$

where k'''' is a constant; A, the extinction coefficient expressed in terms of weight concentration of c_p; $(\text{Output})_{\text{UV}}$, the output of the spectrophotometer. Thus, the equation for (dn/dc_p) can be obtained from both Eqs. (3) and (9) as follows:

$$(dn/dc_p) = \frac{k_2 A(\text{Output})_{\text{RI}}}{(\text{Output})_{\text{UV}}} \quad (10)$$

where k_2 is a constant. From Eqs. (8) and (10) the molecular weight can be expressed as

$$M_p = \frac{k(\text{Output})_{\text{LS}}(\text{Output})_{\text{UV}}}{A(\text{Output})_{\text{RI}}^2} \quad (11)$$

where k is a constant. Equation (11) shows that this procedure enables one to obtain M_p without estimation of the amounts of substances bound to the protein moiety. The value of (dn/dc_p), however, which reflects the amount of the substances, is a very important parameter for molecular weight determination of solubilized membrane proteins. Thus, (dn/dc_p) as well as M_p can usually be determined by using Eqs. (10) and (8).

Instrumentation

As a typical example, the whole measuring system being operated daily in the laboratory of one of the authors (TT) is schematically illustrated in Fig. 1.

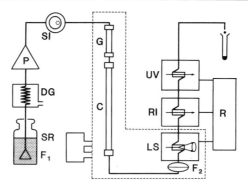

FIG. 1. Schematic diagram of instrumentation for measurement of the molecular weight of membrane proteins by low-angle laser light-scattering photometry coupled with high-performance gel chromatography (HPGC/LALLS system). The dotted line shows the portions that are subjected to temperature control by circulating medium. See the text for details.

Each of the components in Fig. 1 will be discussed briefly, in order of occurrence along the flow path. The solvent reservoir (SR) is filled with a buffer solution containing a surfactant, which has been filtered through an ultrafilter with a pore size of 0.8 μm (Millipore, type AA). The buffer solution is sucked in via a filter (F_1) made of sintered stainless steel. Air dissolved in the buffer solution is eliminated with a degasser (DG; Elma Optical Works, Model ERC-3310). The buffer solution is supplied at a constant speed in the range from 0.3 to 0.7 ml/min with a HPLC pump (P; Toyo Soda Manufacturing Co. Ltd., Model CCPD). An aliquot of solubilized membrane protein less than 500 μl in volume is applied to the columns through a sample injector (SI; Rheodyne, Model 7125). The flow goes through a guard column (G; TSK guard column SW, 7.5 × 75 mm) to the main column (C; TSK-gel G3000SW, 7.5 × 600 mm). Elution from the column is monitored successively with the following three kinds of detectors; a low-angle laser light-scattering photometer (LS; Toyo Soda Manufacturing Co. Ltd, TSK Model LS-8000), a precision differential refractometer (RI; TSK Model RI-8011), and a UV spectrophotometer (UV; TSK Model UV-8000). The eluate is filtered through an ultrafilter with a pore size of 0.5 μm (F_2; Millipore, Type FHLP 01300) immediately before entering the light-scattering photometer. Each column is provided with a jacket. The flow-through cell of the scattering photometer is provided with a metallic holder through which water or a water–ethanol mixture at constant temperature is circulated. The holder was manufactured to the design of the authors by Toyo Soda Manufacturing Co. Ltd.[20] Cell win-

---

[20] This product is now available commercially.

dows were purged with nitrogen gas to avoid fogging when measurements were made near 0°. The lines connecting the columns and the cell are wrapped with a pair of synthetic rubber tubes with semicircular section and good heat conductivity. The temperature of the columns, the cell, and the connecting tubes was controlled with an accuracy of ±0.1° by circulation of water or water–ethanol mixture through the jacket. The outputs of the three kinds of detectors were registered by a three-pen recorder (R).

For the estimation of the molecular weight of a membrane protein, the system has the following advantages: (1) The LALLS apparatus equipped with a flow cell makes it possible to measure automatically and immediately the light scattering after elution of a fractionated protein; (2) the protein solution eluted is in "dialysis" equilibrium with the elution buffer containing micelles as well as monomeric surfactant.

Methods

Solubilization of Membrane Proteins with Nonionic Surfactant

Nonionic surfactants do not usually denature proteins.[1,21] Thus, to solubilize a membrane protein without denaturation, a nonionic surfactant with polyoxyethylene,[1,21] sugar,[21–23] or other[24,25] head group is used. It is essential for elucidation of the structure–function relationship of a membrane protein to ascertain experimentally that the solubilized form retains biological activity. For membrane proteins with measurable biochemical activity such as enzymatic activity or a partial reaction involving substrate or related ligand binding, retention of the activity indicates that no disruption severely affecting the primary function has occurred. General precautions in carrying out the solubilization of membrane proteins have been described elsewhere in this series.[26]

A nonionic surfactant must fulfill the following requirements to be used for molecular weight determination by the HPGC/LALLS method. First, the surfactant should not bind to the water-soluble proteins used as standards, so that the values of their specific refractive indices in the presence of the surfactant can be assumed to be the same as those in its absence. Second, a micelle of the surfactant (often incorporating lipids derived from the membrane) should be smaller in size than any of the

---

[21] A. Helenius, D. R. McCaslin, E. Fries, and C. Tanford, this series, Vol. 56, p. 734.
[22] T. Shimamoto, S. Saito, and T. Tsuchiya, *J. Biochem. (Tokyo)* **97,** 1807 (1985).
[23] C. Baron and T. E. Thompson, *Biochim. Biophys. Acta* **382,** 276 (1975).
[24] J. E. K. Hildreth, *Biochem. J.* **207,** 363 (1982).
[25] M. Hanatani, K. Nishifuji, M. Futai, and T. Tsuchiya, *J. Biochem. (Tokyo)* **95,** 1349 (1984).
[26] L. M. Hjelmeland and A. Chrambach, this series, Vol. 104, p. 305.

proteins to be measured, otherwise the elution peak of the protein may be obscured by the overlap with that of the micellar surfactant. Third, the surfactant should be transparent enough to allow monitoring of the elution of the proteins at 280 nm.

Membrane-bound Na^+,K^+-ATPase could be successfully solubilized with a nonionic surfactant, octaethylene glycol n-dodecyl ether ($C_{12}E_8$). The ATPase was purified from dog kidney by the method of Jørgensen with minor modifications.[16] An aliquot (60 to 120 μl) of the membrane protein was solubilized in buffer solution containing $C_{12}E_8$ to give a final volume of 300 μl and a final composition of 2 mg/ml protein, 6 mg/ml $C_{12}E_8$, 0.2 M KCl, 2 mM dithioerythritol, 10% (w/v) glycerol, and 13 mM imidazole and 8 mM HEPES [4-(2-hydroxyethyl)-1-piperazineethanesulfonic acid] at pH 7.1 and 0°. Centrifugation was started after incubation of the mixture for 5 min and was continued for 15 min after reaching 140,000 g at 0°. The supernatant was used as solubilized Na^+,K^+-ATPase. This solubilization gave the maximum protein recovery of 80 to 90% for the solubilized protein. The solubilized enzyme thus obtained showed ATPase activity equivalent to the membrane-bound activity when the assay was done with 0.03 mg/ml protein of the enzyme at 20° in reaction medium containing 0.1 mg/ml $C_{12}E_8$. Furthermore, the solubilized enzyme could bind ligands such as ATP, Rb, and Tl.[17] These data exclude the possibility of irreversible denaturation of the solubilized ATPase.

Use of SDS for Molecular Weight Determination of Constituent Polypeptides

Sodium dodecyl sulfate (SDS) is a typical anionic surfactant used to analyze the individual polypeptide chains of membrane proteins as well as water-soluble proteins. The surfactant usually dissociates oligomeric proteins into their constituent polypeptide chains at the concentration and temperature that have to be adopted to solubilize membrane protein. In contrast, a major matrix protein of *Escherichia coli* outer membrane, porin, retains its trimeric structure even after solubilization by SDS at room temperature, and dissociates into the monomeric form only on heat treatment.[27] Therefore, heat treatment of proteins after solubilization with SDS is very effective in some cases to confirm complete monomerization.

Porin was isolated by gel chromatography on a Sephacryl S-300 column (1.6 × 80 cm) which had been equilibrated and eluted with 2 mM SDS, 2 mM sodium azide, and 0.1 M sodium phosphate buffer at pH 6.9 and room temperature. After concentration and dialysis against the elu-

[27] T. Nakae, J. Ishii, and M. Tokunaga, *J. Biol. Chem.* **254**, 1457 (1979).

tion buffer, porin was subjected to heat treatment at 100° for 5 min at a protein concentration of 1.6 mg/ml to dissociate it into the constituent polypeptides. The molecular weights of porin with and without the heat treatment were found to be 34,000 ± 2,000 and 103,000 ± 5,000, respectively, by the HPGC/LALLS method.[13]

Columns for High-Performance Gel Chromatography

Columns of TSK-gel SW type (Toyo Soda Manufacturing Co. Ltd.) packed with porous modified silica gel are suitable for molecular weight estimation by the HPGC/LALLS method. There are three kinds of columns of different pore size, i.e., G2000SW, G3000SW, and G4000SW, which have effective separation ranges of 5,000–100,000, 10,000–500,000 and 20,000–7,000,000, respectively.[28] The ranges were determined with the use of globular proteins. In choosing a column for membrane proteins, it must be taken into consideration that a membrane protein binds large amounts of nonionic surfactant in the solubilized form, whereas a water-soluble globular protein binds little.[29–31] The protein component of Na^+,K^+-ATPase, $\alpha\beta$-protomer with a molecular weight of 154,000 (see Table II) was eluted earlier than yeast glutamate dehydrogenase, a globular protein with a molecular weight of 297,354, when the two proteins were chromatographed on a TSK-gel G3000SW column in the presence of 0.2 mg/ml $C_{12}E_8$ (compare Fig. 2 with Fig. 3). The pH of the elution buffer must be between 2 and 7.5 so as not to destroy the silica gel. Superose 6 and 12 columns (Pharmacia AB)[32] were also found to be adequate for HPGC/LALLS; they are stable between pH 1 and 13.[32]

Elution Buffer

The elution buffer must contain an appropriate surfactant at a concentration higher than the critical micelle concentration to keep membrane proteins adequately dispersed. The ionic strength of the buffer should be higher than about 0.1 to eliminate the ion-exclusion effect[33] in gel chromatography. The presence of a substance having high absorbance at 280 nm, such as a high concentration of ATP, interferes with the UV spectrophotometry. The pH must be within the range where the gel is stable.

[28] Y. Kato, K. Komiya, H. Sasaki, and T. Hashimoto, *J. Chromatogr.* **190**, 297 (1980).
[29] A. Helenius and K. Simons, *J. Biol. Chem.* **247**, 3656 (1972).
[30] A. Helenius and K. Simons, *Biochim. Biophys. Acta* **415**, 29 (1975).
[31] S. Makino, J. A. Reynolds, and C. Tanford, *J. Biol. Chem.* **248**, 4926 (1973).
[32] T. Andersson, M. Carlsson, L. Hagel, and P. Pernemalm, *J. Chromatogr.* **326**, 33 (1985).
[33] B. Stenlund, *Adv. Chromatogr.* **14**, 37 (1976).

Elution buffer with the following composition was used as the standard buffer for HPGC of solubilized Na^+,K^+-ATPase: 0.37 mM $C_{12}E_8$ (Nikko Chem. Co., Tokyo), 0.1 M KCl or NaCl, 1 mM EDTA, 10 mM imidazole, and 16 mM HEPES at pH 7.0.[17] The cmc of $C_{12}E_8$ was reported to be 0.087 mM.[34] To estimate the molecular weight of porin trimer or monomer in the presence of SDS, elution buffer consisting of 2 mM SDS, 2 mM sodium azide, and 0.1 M sodium phosphate at pH 6.9 was used.[13] The azide is present to prevent bacterial growth in the buffer.

Elution Temperature

Na^+,K^+-ATPase solubilized with $C_{12}E_8$ behaved very differently in gel chromatography depending on the temperature of the column. The $\alpha\beta$-protomeric and $(\alpha\beta)_2$-diprotomeric components can be separated at about 0°, but not at higher temperature in the presence of 0.1 M NaCl.[35] This is the result of the temperature dependence of the association–dissociation equilibrium. Characterization of the molecular species eluted from the column with respect to molecular weight is, therefore, only possible when the flow cell of the LALLS photometer is thermostatted. Phospholipids essential for Na^+,K^+-ATPase activity are removed during the chromatography at 20°. They are, however, partially retained at temperatures near 0°. Extreme care is required in experiments at such low temperature to protect SW type TSK-gel from freezing, which destroys it.

Flow Rate

Elution is recommended to be carried out at a flow rate of less than 0.35 ml/min at 0° and less than 0.7 ml/min at 20° to ensure long life of the column.

Calibration of the System with Standard Proteins

The constants k_1 and k_2 in Eqs. (8) and (10) can be estimated by using a standard protein with known molecular weight, extinction coefficient, and specific refractive index increment. To ensure accuracy, a mixture of three or four proteins chosen from among the standard proteins listed in Table I may be applied to the HPGC/LALLS system. Nonionic surfactants do not bind to most of the typical water-soluble globular proteins.[29–31] Thus the values of (dn/dc_p) of the standard proteins in the presence of a nonionic surfactant can be assumed to be 0.187 ml/g (see Table I), virtually the same as the values for the proteins in the absence of

---

[34] W. L. Dean and C. Tanford, *Biochemistry* **17**, 1683 (1978).
[35] Y. Hayashi and T. Takagi, unpublished data (1985).

TABLE I
PARAMETERS OF STANDARD PROTEINS USED TO CALIBRATE THE HPGC/LALLS SYSTEM

Protein[a]	Source	Molecular weight	(dn/dc)[b] (ml/g)	$A_{280\ nm}$ (ml mg^{-1}cm^{-1})
Glutamate dehydrogenase	Yeast	297,354[c]	0.187	1.27[d]
Lactate dehydrogenase	Pig heart	142,000[e]	0.187	1.38[f]
Enolase[g]	Yeast	93,020[h]	0.187	0.895[h]
Albumin	Bovine serum	66,267[i]	0.187	0.665[j]
Adenylate kinase	Yeast	24,077[k]	0.187	—

[a] The proteins were purchased from Oriental Yeast Co., Ltd. (Tokyo) except the albumin (Armour Pharmaceutical Co., USA).
[b] Nine values of specific refractive index increment for the following proteins, measured at a wavelength of 546 nm in aqueous solvent without any surfactant, were chosen from among those listed in the "Handbook of Biochemistry and Molecular Biology" [G. D. Fasman (ed.), Vol. 2, 3rd Ed., p. 372. CRC Press, Cleveland, OH, 1976], and the average with standard deviation of 0.005 ml/g was adopted: bovine serum albumin, S-carboxymethylkeratin, $\alpha_{s1,2}$-caseins, chymotrypsinogen, γ-globulin (human), hemoglobin (human), β-lactoglobulin (bovine), lysozyme, and ovalbumin. The increments show almost no dependence on the wavelength of the light source adopted [G. E. Perlmann and L. G. Longsworth, J. Am. Chem. Soc. 70, 2719 (1948)].
[c] W. S. Moye, N. Amuro, J. K. M. Rao, and H. Zalkin, J. Biol. Chem. 260, 8502 (1985).
[d] Unpublished data (1985).
[e] R. Jaenicke and S. Knof, Eur. J. Biochem. 4, 157 (1968).
[f] Average of four values (SD 0.07) listed in the "Handbook of Biochemistry and Molecular Biology" [G. D. Fasman (ed.), Vol. 2, 3rd Ed., p. 462. CRC Press, Cleveland, Ohio, 1976].
[g] This protein cannot be used as a standard in the absence of MgCl$_2$ because of dissociation into two components [J. M. Brewer, CRC Crit. Rev. Biochem. 11, 209 (1981)].
[h] C. C. Q. Chin, J. M. Brewer, and F. Wold, J. Biol. Chem. 256, 1377 (1981).
[i] R. G. Reed, F. W. Putnam, and T. Peters, Jr., Biochem. J. 191, 867 (1980).
[j] Average of four values (SD 0.010) listed in the "Handbook of Biochemistry and Molecular Biology" [G. D. Fasman (ed.), Vol. 2, 3rd Ed., p. 386. CRC Press, Cleveland, Ohio, 1976].
[k] A. G. Tomasselli, E. Mast, W. Jares, and E. Schiltz, Eur. J. Biochem. 155, 111 (1986).

surfactant. A set of standard proteins was dissolved in elution buffer, and then applied to the HPGC/LALLS system. Figure 2 shows a typical elution curve thus obtained. Values of k_1 or k_2 obtained from two or more kinds of protein agreed to within 3% of the mean.

When SDS is used as a surfactant, the refractive index increments of standard proteins are not constant, but depend on the conditions used due to variations in the binding of SDS to the proteins. Therefore, another method can be adopted to obtain M_p of the constituent polypeptides of a membrane protein. First, the constant k in Eq. (11), in place of k_1 and k_2 as described above, is determined by application of standard proteins with known M_p and A values to the HPGC/LALLS system. Second, the mem-

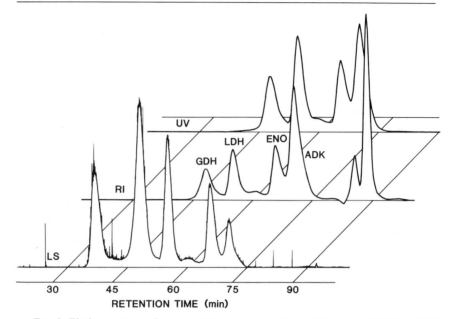

FIG. 2. Elution patterns of water-soluble proteins used to calibrate the HPGC/LALLS system. A mixture of glutamate dehydrogenase (GDH, 250 µg), lactate dehydrogenase (LDH, 360 µg), enolase (ENO, 390 µg), and adenylate kinase (ADK, 520 µg) was dissolved in 280 µl of elution buffer consisting of 0.37 mM $C_{12}E_8$, 0.1 M KCl, 1 mM EDTA, 10 mM imidazole, and 16 mM HEPES, pH 7.0. An aliquot of 220 µl of the protein solution was applied to the HPGC/LALLS system illustrated in Fig. 1, and chromatographed with the same elution buffer. The sources of the standards and the parameters which are required to calibrate the system are listed in Table I. Flow rate, 0.30 ml/min; temperature, 1.6°; gain setting for the detectors, 8 for LS, 16 for RI, and 0.64 absorbance unit (full scale) for UV.

brane protein is applied to the same system. The M_p of the sample protein can be calculated by using Eq. (11), without estimation of (dn/dc_p). An example is the estimation of M_p of human α_1-acid glycoprotein.[15] A special method[13] is required to evaluate the refractive index increments of the standard proteins in the presence of SDS: the protein solution in "dialysis" equilibrium is obtained by a conventional technique of gel chromatography, and the differential refractive index of the solution is determined in a batchwise operation with a precision differential refractometer.

Determination of Extinction Coefficient at 280 nm of Solubilized Protein

Equation (11) shows that the accuracy of M_p depends on that of the extinction coefficient of the sample protein. The protein concentration must be determined absolutely to obtain a reliable extinction coefficient. In the case of Na^+,K^+-ATPase, the protein concentration of the intact membrane-bound protein after purification was preliminarily determined by the Bio-Rad protein assay method described in Bio-Rad Technical Bulletin 1051 (April, 1977), which was originally reported by Bradford.[36] A solution of crystalline bovine serum albumin (Armour Pharmaceutical Co., Chicago, IL) was used as a standard. The protein concentration was calibrated by quantitative amino acid analysis.[37] For every sample at least seven kinds of amino acid were used to estimate the amount of protein in the aliquot subjected to the analysis, and the relative amounts were internally consistent with the composition of $\alpha\beta$-complex calculated from the published compositions[38] for α and β chains of Na^+,K^+-ATPase from sheep kidney. The yield of norleucine, which was added to each sample as an internal standard, was $88 \pm 14\%$ ($n = 6$). The ratio of the concentration determined by the Bio-Rad method mentioned above to that obtained by the amino acid analysis was 0.874 ± 0.055 ($n = 6$).[37] The extinction coefficient of solubilized Na^+,K^+-ATPase at 280 nm was determined by dissolving the purified membrane-bound protein in 0.2 M sodium phosphate buffer containing 1% SDS. The absorbance was corrected for light scattering in the same manner as described by Reddi.[39] The extinction coefficient at 280 nm thus obtained was 1.22 ± 0.05 mg^{-1}ml cm^{-1} ($n = 15$).[16] The same value was obtained with the use of a buffer solution containing 4 mg/ml $C_{12}E_8$ instead of the SDS.

Calculation of (dn/dc_p) and M_p

When a solubilized membrane protein is eluted as a single peak or substantially separated from the other proteins by HPGC, the outputs of the three kinds of detector can be obtained instantaneously as the heights of the elution peaks. From the values of outputs expressed in unit of length, as well as A, k_1 and k_2, (dn/dc_p) and M_p of the solubilized membrane protein can be calculated by using Eqs. (10) and (8), respectively. Figure 3 shows a set of elution patterns obtained when Na^+,K^+-ATPase solubilized with $C_{12}E_8$ was applied to the HPGC/LALLS system at 1.6°. The two main protein components, $\alpha\beta$-protomer and $(\alpha\beta)_2$-diprotomer, could be substantially separated. The heights of the two peaks gave the values of M_p and (dn/dc_p) listed in Table II.

When two or more protein components of the solubilized membrane

---

[36] M. M. Bradford, *Anal. Biochem.* **72**, 248 (1976).
[37] Y. Hayashi and R. L. Post, unpublished data (1980).
[38] L. K. Lane, J. D. Potter, and J. H. Collins, *Prep. Biochem.* **9**, 157 (1979).
[39] K. K. Reddi, *Biochim. Biophys. Acta* **24**, 238 (1957).

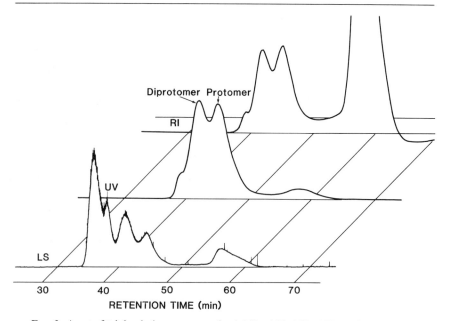

FIG. 3. A set of triple elution patterns of solubilized Na^+,K^+-ATPase for estimation of the molecular weight. The membrane-bound ATPase purified from dog kidney was solubilized with $C_{12}E_8$ as described in the text. An aliquot of 220 μl containing 0.37 mg of protein was injected into the HPGC/LALLS system under the same conditions as described in the legend to Fig. 2. Gain setting for the detectors: 2 for LS, 8 for RI, and 0.32 absorbance unit (full scale) for UV.

protein cannot be separated completely from one another, the resultant overlapped peak can be simulated by a linear summation of a set of Gaussian distribution curves to evaluate the peak height of each of the components.[16] A Gaussian curve is specified by three parameters, peak position (X), peak height (H), and half-width (W). In the simulation, the value of X must be common among the three kinds of elution pattern after correction for the displacement in the patterns due to the differences in positions of the detector cells and recorder pens. The value of W for the elution pattern seen by the detectors located behind the LALLS photometer must be wider than that adopted for the pattern of the LALLS photometer. The degree of widening is determined by comparing the values of W of peaks of standard proteins in the elution patterns observed with the different detectors.

Evaluation of Amount of Detergent Bound to Protein from (dn/dc_p)

The value of (dn/dc_p) of a solubilized membrane protein allows one to estimate the amount of substances bound to it, if the protein per se in the

TABLE II
CHARACTERISTICS OF $C_{12}E_8$-SOLUBILIZED Na^+,K^+-ATPASE DETERMINED BY THE HPGC/LALLS METHOD[a]

Parameter	Unit	$(\alpha\beta)_2$-Diprotomer	$\alpha\beta$-Protomer
M_p[b]		304,000 ± 14,000	154,000 ± 7,000
(dn/dc_p)[b]	(ml/g)	0.279 ± 0.001	0.304 ± 0.002
Phospholipid bound[c]	(g/g protein)	0.072 ± 0.005	0.077 ± 0.031
Carbohydrate[d]	(g/g protein)	0.073	0.073
$C_{12}E_8$ bound[e]	(g/g protein)	0.542	0.723

[a] The conditions for solubilization and for chromatography were as described in the text and in the legend to Fig. 2, respectively. From Y. Hayashi, H. Matsui, and T. Takagi, unpublished data (1985).
[b] Average of three determinations.
[c] Determined by measuring organic phosphorus content according to the ultramicro method of G. R. Bartlett [*J. Biol. Chem.* **234**, 466 (1959)] and by assuming that the molecular weight of phospholipid is 770.
[d] The content for pig kidney enzyme [J. W. Freytag and J. A. Reynolds, *Biochemistry* **20**, 7211 (1981)] was adopted.
[e] Calculated according to Eq. (12), assuming that no lipids other than phospholipid are retained in the solubilized membrane protein and that (dn/dc) of carbohydrate is 0.134 ml/g [P. Becher, *J. Colloid Sci.* **16**, 49 (1961)].

complex is assumed to show the value of specific refractive index increment of 0.187 ml/g, which is appropriate for a simple protein. The increment for the solubilized protein component, (dn/dc_p), can be expressed in terms of parameters of bound substances as follows:

$$(dn/dc_p) = 0.187 + \sum_i \delta_i (dn/dc)_i \quad (12)$$

where δ_i is the amount (grams) of i-substance bound to 1 g of protein, and $(dn/dc)_i$ is the specific refractive index increment of i-substance. If the complex is composed of protein, phospholipid, and $C_{12}E_8$, Eq. (12) can be rewritten as follows, since the increment of phospholipid[40] is almost the same as that of $C_{12}E_8$, that is, 0.134 ml/g[41]:

$$(dn/dc_p) = 0.187 + 0.134(\delta_{CE} + \delta_{PL}) \quad (13)$$

where δ_{CE} and δ_{PL} are the amounts of bound $C_{12}E_8$ and phospholipid expressed in the unit of grams per gram of protein, respectively. The δ_{PL} of the isolated fraction of solubilized membrane protein is determined by measuring organic phosphorus content according to the ultramicromethod of Bartlett.[42] Thus, δ_{CE} can be evaluated by using Eq. (13).

[40] K. Kameyama and T. Takagi, unpublished data (1985).
[41] P. Becher, *J. Colloid Sci.* **16**, 49 (1961).
[42] G. R. Bartlett, *J. Biol. Chem.* **234**, 466 (1959).

Comments

The HPGC/LALLS system as described here allows one to estimate the molecular weight of a membrane protein immediately after its elution from a high-performance gel column in the presence of $C_{12}E_8$, provided that the extinction coefficient at 280 nm is known for the protein. This method is valuable for investigation of the structure–function relationship of intrinsic membrane proteins.

Acknowledgment

The methodology described here is the outcome of a project on the application of the low-angle laser light-scattering technique to molecular weight determination of solubilized membrane proteins that has been carried out since 1979 at the Institute for Protein Research. The authors wish to thank to the many research associates and graduate students who contributed to the initial phase of the project. The project has been supported by Grants-in-Aid for Scientific Research from the Ministry of Education, Science and Culture of Japan [Nos. 59570115 (YH), 59108003 (HM), and 56580110 and 60470157 (TT)] and by a Naito Foundation Research Grant for 1983 (TT).

[29] Critical Micellar Concentrations of Detergents

By M. Zulauf, U. Fürstenberger, M. Grabo, P. Jäggi, M. Regenass, and J. P. Rosenbusch

Introduction

Membrane-spanning proteins are amphiphilic macromolecules that are incorporated vectorially in the lipid membrane in a quasisolid state. Part of their surface that is in contact with the lipid bilayer core is hydrophobic in character, while other parts that are exposed to the aqueous environment on either side of the membrane are more hydrophilic. On disruption of the membrane and removal of the lipids, such proteins tend to aggregate unspecifically and to precipitate in bulk, thus avoiding an entropically unfavorable exposure of their hydrophobic parts to water. Most conventional biochemical and biophysical methods depend on the proteins being in dilute, nearly ideal aqueous solutions. Although some properties of membrane proteins may be studied *in situ,* it is usually a prerequisite for the investigation of their functional and structural properties at a molecular level to solubilize them in aqueous solutions. The easiest and least disruptive method to achieve this is the replacement of membrane lipids by detergents[1] which, at the concentration used in biochemistry, do not form membranelike structures.

Both phospholipids and detergents are amphiphilic, flexible molecules, the most significant difference being that lipids aggregate in water into extended lamellar structures,[2] usually bilayers forming vesicles. Detergents, on the other hand, form small colloidal aggregates called micelles. The concentration of free phospholipid monomers (not participating in the bilayer structure) is always vanishingly small ($<10^{-12}$ M). In contrast, micelles can only exist above a substantial detergent concentration (10^{-1}–10^{-6} M) which is called the critical concentration of micellization (cmc). Its value depends on the specific detergent used. Above the cmc, micelles always coexist with unaggregated monomers, present at the concentration given by the cmc. These colloidal particles are highly dynamic, with continuous exchange of monomers between micelles and the nonaggregated molecules.[3]

In detergent micelles, the apolar tail of the constituent molecules (most often alkyl chains) are stowed away in the interior of the structure to reduce energetically unfavorable interactions with the surrounding water. Micelle formation is thus a manifestation of the hydrophobic effect.[4] The polar head groups (ionic or nonionic) are exposed toward the aqueous phase. Particle diameters are around 5 nm, forming solutions which are transparent, isotropic and of low viscosity. At high concentration of detergents (>30%), aggregation properties change and extended periodic structures form.[5,6] Lipid vesicles are considerably larger, with diameters between 50 and several hundred nanometers, so that the corresponding solutions look turbid with the intensity of scattered light depending on the lipid concentration. Solubilized membrane proteins form so-called mixed micelles with the detergent. Under appropriate conditions, their hydrophobic surface parts are covered by a (curved) monolayer film of detergent.[7] As this film only forms at adequate detergent concentration, the final solution after membrane solubilization contains mixed protein–detergent micelles, mixed lipid–detergent micelles, pure detergent micelles, and detergent monomers, the latter at the cmc. However, just below or at

[1] A. Helenius, R. Darrell, R. McCaslin, E. Fries, and C. Tanford, this series, Vol. 56, p. 734.
[2] V. Luzzati, in "Biological Membranes" (D. Chapman, ed.), pp. 71–123. Academic Press, London, 1968.
[3] M. Zulauf and J. P. Rosenbusch, *J. Phys. Chem.* **87**, 856 (1983).
[4] C. Tanford, "The Hydrophobic Effect." Wiley, New York, 1980.
[5] H. Wennerström and B. Lindman, *Phys. Rep.* **52**, 2 (1979).
[6] D. J. Mitchell, G. J. T. Tiddy, L. Waring, T. Bostock, and M. P. McDonald, *J. Chem. Soc., Faraday Trans. 1* **79**, 975 (1983).
[7] M. Zulauf, M. Grabo, and J. P. Rosenbusch, submitted for publication.

the cmc, mixed micelles may contain more than one protein molecule in one and the same complex.[8,9] To avoid this situation of incomplete and evasive solubilization, experiments should be performed at a detergent concentration at which the free-to-mixed micelle ratio[8] is at least 10. Hence it is appropriate in membrane protein biochemistry to have a knowledge and understanding of the significance of the cmc, which is particularly critical for membrane protein reconstitution[10] and crystallization.[11] In the following, the determination of the cmc is described.

Critical Phenomena Associated with Detergent Solutions

Micelles should not be perceived as rigid particles, because their aggregation properties depend in general on both detergent concentration (c) and temperature (T). There are upper and lower boundaries in c and T where micellar solutions form, and these boundaries are associated with phase transitions to detergent aggregation which may be different from micelle formation.[5,6] A convenient way to visualize these boundaries is to draw phase diagrams exhibiting the boundaries as lines in plots in which the temperature is shown as a function of concentration. An example of a phase diagram of a hypothetical detergent is shown in Fig. 1A. With the exception of octylhydroxyethane sulfoxide,[11] not all boundaries are actually observed with any one of the detergents most often used in biochemistry. The schematic diagram nevertheless allows some of the most common boundaries to be discussed by referring to a single figure.

When crossing the cmc boundary, shown at the left as nearly vertical line in Fig. 1A, monomer and micelle concentrations vary according to the diagram shown in Fig. 1B. This is again an idealized representation. Particularly for short detergents containing less than 10 carbon-atoms in the alkyl chain, a sharp boundary, as it is suggested by the figure, is not observed, but the transition is more gradual. This is due to the dynamical nature of the micelles and possibly also to premicellar aggregates. Thus, independent of the accuracy of measurements or of detergent purity, the cmc is not measurable as a precise concentration, but may rather be described as a "gray zone" of some 5–10% width, as suggested by the hatched area in Fig. 1B. Corresponding provisos apply also to some of the other boundaries in the phase diagram (Fig. 1A). As indicated in A, the

[8] J. A. Reynolds, in "Physics of Amphiphiles: Micelles, Vesicles and Microemulsions" (V. Degiorgio, ed.), pp. 555–562. North-Holland, Amsterdam, 1985.
[9] M. Grabo, Ph.D. thesis. University of Basel, Basel, Switzerland, 1982.
[10] E. Racker, "Mechanisms in Bioenergetics." Academic Press, New York, 1965.
[11] R. M. Garavito and J. P. Rosenbusch, this series, Vol. 125, p. 309.

cmc does not depend strongly on temperature. For detergents with alkyl tail groups, the value of the cmc decreases by roughly a factor of 3 if the chain length is increased by one carbon-atom, or by an order of magnitude for additional C_2 units. *Salts* drastically decrease the cmc of *ionic* detergents such as sodium dodecyl sulfate (SDS),[12] whereas the effect on the cmc of *nonionic* detergents is rather small.[13]

At *low temperature,* detergents may form hydrated crystals in the micellar solution, thus precipitating in bulk. This reflects a critical temperature of crystallization boundary (ccb), also known as the Krafft point, cmt (critical micelle temperature), or T_c (critical temperature of crystallization). This is well known for SDS,[14] and has also been described for octyl-α-glucoside[15] and octylhydroxyethane sulfoxide.[11] At *high temperature,* spontaneous demixing of the single iostropic micellar phase into two phases, differing substantially in detergent concentration, may occur, as has been described for alkyl polyoxyethylenes.[16] The corresponding lower consolute boundary (lcb), or the critical temperature of demixing (T_d), is shown as the curved upper line in Fig. 1A. The phase separation is preceded by critical opalescence, and thus the phase boundary may be evaluated from the "cloud points" or "clouding temperatures" as a function of concentration. Upon clouding, demixing into two clear, isotropic, immiscible phases occurs. At high detergent concentrations (>30% w/w), complex mesophases are often observed; instead of aggregation into micelles, detergents form *viscous* phases consisting of extended structures such as cylinders or stacked sheets.[2,15] At elevated temperature, they melt (Fig. 1A). These temperature-phase transitions are highly sensitive to the presence of salts, polymers, or impurities, as discussed elsewhere.[11]

The nature of these transitions may be ignored for the practical aspects considered in the present context. Before starting work with an unknown detergent, it may nevertheless be advantageous to be aware of their existence and to test qualitatively for such boundaries by heating and cooling between $-5°$ and $100°$ at various detergent concentrations. Thus, it is possible to remain within a "safe" zone with regard to both temperature and concentration as it is indicated by the hatched area in Fig. 1A.

---

[12] J. A. Reynolds and C. Tanford, *Proc. Natl. Acad. Sci. U.S.A.* **66,** 1002 (1970).
[13] P. Becher, in "Nonionic Detergents" (M. J. Schick, ed.), pp. 478–515. Marcel Dekker, New York, 1966.
[14] N. A. Mazer, G. B. Benedek, and M. C. Carey, *J. Phys. Chem.* **80,** 1075 (1976).
[15] D. L. Dorset and J. P. Rosenbusch, *Chem. Phys. Lipids* **29,** 299 (1981).
[16] M. Zulauf, K. Weckström, J. B. Hayter, V. Degiorgio, and M. Corti, *J. Phys. Chem.* **89,** 3411 (1985).

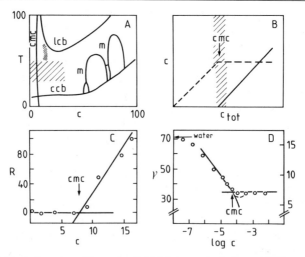

FIG. 1. Critical phase transitions of colloidal detergent solutions. (A) Phase diagram of a hypothetical detergent, as mentioned in the text. Temperature (T) is given in °C, the concentration (c) as % wt. ccb, critical crystallization boundary; m, various mesophase boundaries; lcb, lower consolute boundary. Demixing into two isotropic phases may also occur at low temperature for β-octylglucoside in the presence of polyethylene glycol (see Ref. 11). The sequence of mesophases is well understood,[6] but is not relevant in the present context. If an appropriate detergent selection has been made, and conditions carefully chosen, methods of solution biochemistry may be applicable to membrane proteins within the shaded area. For their crystallization, it has been shown that the neighborhood of a phase boundary[11] may be exploited as indicated schematically as darkly shaded area in this figure. (B) On the ordinate, the concentrations (c) of monomers (broken line) and of micelles (solid line) are presented as a function of the total detergent concentration (c_{tot}). The cmc appears where the monomer concentration reaches a plateau value and the first micelles appear. Due to the possible formation of premicellar aggregates, but particularly because of micelle dynamics,[3] this transition corresponds to a "gray zone" (shaded). (C) Shown is the scattered light expressed as Rayleigh ratio ($R10^{-6}$ cm^{-1}) as a function of detergent concentration c (mg ml^{-1}) for β-decylglucoside. The intersection of the lines yields a cmc of 0.79 mg ml^{-1}, corresponding to 2.5 mM. (D) The cmc is obtained from the break in the surface tension γ (mN m^{-1}) as a function of the concentration c (plotted logarithmically in molar concentration M) for dodecyloctaoxyethylene. Counting the number of turns in the winding (right ordinate) is sufficient to obtain the cmc value in the simple set-up described in Fig. 2. Conversion to surface tension follows the equations given in Appendix B. The hatched line indicates an "undershoot" that a contaminant may cause.

Determination of cmc

Strategy

In 1963, Shinoda[17] described 22 different methods to determine the cmc. It is not the aim of this chapter to evaluate or extend this list, but

---

[17] K. Shinoda, in "Colloidal Surfactants" (K. Shinoda, B.-I. Tamamuski, T. Nakagawa, and T. Isemura, eds.), pp. 1–96. Academic Press, New York, 1963.

rather to mention the fundamental concepts and to give one qualitative and one quantitative method. These are both based on a single approach and appear to be particularly easy to perform and well-suited to the biochemist's needs. In general terms, there are two basically different approaches that may be used to determine the cmc. In the first, the *formation of micellar aggregates* is directly observed. The appropriate techniques include light scattering (intensity versus concentration; an example is shown in Fig. 1C), nuclear magnetic resonance (self diffusion measurement), and electron spin resonance, the latter requiring appropriately labeled molecules. Electrical conductivity and dye solubilization (spectral shifts upon incorporation of dye molecules into micelles) may also be used. The first three techniques require sophisticated instrumentation, the other two either cannot be applied to nonionic detergents or may themselves affect the boundaries of the phase diagram. The second approach consists of measuring *surface tension* as a function of detergent concentration: in a solution containing amphiphilic molecules, these molecules partition into the water–air interface, with their hydrophobic moieties exposed to the air. Thereby, these surface-active substances lower the surface tension (hence the name "surfactants" as a synonym for detergents). The typical variation of surface tension as a function of detergent concentration is shown in Fig. 1D. There is a break point at the concentration above which further addition of detergent does not affect the surface tension anymore. Rather, it participates in the formation of micelles. The break point thus indicates the cmc of the detergent. An exception to this clear-cut behavior is observed only with bile salts[17a] and saponins, where a plateau in surface tension is not reached. The survey of extensive empirical work[18] established surface tension as a convenient method to measure cmc values.

The most accurate method to measure surface tension is that introduced by Du Nouy.[19] Excellent commercial equipment is available, but the measurements are both time consuming and material-intensive.[9] A much simpler, though somewhat less accurate, method is based on the observation that drops forming at the orifice of a tube cannot exceed a maximum size before they fall; this volume depends on surface tension. A quantitative method to obtain the surface tension from drop volume measurements has been described by Tornberg,[20] who has also developed a

---

[17a] J. P. Kratohuil, W. P. Hsu, M. A. Jacobs, T. M. Aminabhaui, and Y. Mukunoki, *Colloid Polymer Sci.* **261**, 781 (1983).

[18] P. Mukerjee and K. J. Mysels, "Critical Micelle Concentration of Aqueous Surfactant Systems," U.S. National Bureau of Standards NSRDS–NBS36. U.S. Government Printing Office, Washington, D.C., 1971.

[19] P. L. Du Nouy, *J. Gen. Physiol.* **1**, 521 (1919).

[20] E. Tornberg, *J. Colloid Interface Sci.* **60**, 50 (1977).

commercially available instrument. Here we discuss first a qualitative assessment of the cmc by drop formation. We then describe a quantitative determination of the cmc using a rather simple instrument with an accuracy that is adequate for biochemical applications.

Qualitative cmc Determination

As a drop of water falls from some height onto a hydrophobic surface, it either conserves the drop form or splashes into shallow puddles. This behavior is governed by surface tension. Additional drops aimed at sites near the first drop tend to coalesce with other drops below the cmc, whereas above it, little coalescence is observed. The onset of splashing and drop deformation gives the cmc value within a factor of 2–5 within a matter of minutes with very small amounts of material. The detailed recipe is given in the Appendix (A).

Quantitative Assessments

Precise values of the cmc may be obtained by measuring the volume of successive drops falling from a pipe of appropriate diameter. A convenient way to measure this is to advance the piston of a syringe by a screw. The number of turns necessary for successive drops to detach can be monitored as a function of the logarithm of detergent concentration, yielding the plot shown in Fig. 1D. The cmc is obtained from the intersection of two extrapolating lines, as indicated. In the Appendix (B) and (C), the realization of such measurements is described in some detail, together with the conversion of the number of screw turns into SI units of the surface tension (γ). The accuracy of this method is 5–10%, which compares well with the reproducibility reported in the literature,[18] and is in keeping with the notion of the critical concentration of micellization as a gray zone (see above). The temperature is not easily controlled by performing measurements in the air at ambient temperature. Also, water may evaporate if determinations proceed too slowly, but these restrictions are rather relaxed and can be evaluated by comparing experimental values with well-studied detergents such as SDS[1,18] or nonionic detergents.[1,11,13] A particular benefit of surface tension measurements is that they allow the detergent purity to be assessed also: If contaminants are present, such as other surface-active compounds or remnants of organic solvents, an undershoot at the break point is often seen.

Perspectives

Knowledge of the phase transition phenomena has been exploited recently for membrane protein crystallization.[11] The significance of the critical concentration of micellization (cmc) for solubilization, characterization, and particularly for reconstitution into lipid membranes[10] and crystallization[11] is now also well established. For many membrane proteins, it will be necessary to search for new detergents that extract "functional units," containing a protein monomer or oligomer (possibly complexed with lipids or other components) that will allow the reconstitution of a functional state. The process of selection will include considerations of several guidelines such as the cmc and the choice of optimal concentrations of salt and organic molecules (such as polyethylene glycol), a task which is still empirical to a very large degree. The requirement to determine cmc values is therefore likely to be frequent in extensive searches for appropriate conditions. Due to an abundant and often bewildering literature, cmc values have been mystified and have often been thought of as considerable problems in membrane protein biochemistry. With the qualitative and quantitative methods described in this chapter, it is hoped that this will no longer be a limiting factor so that new detergents as well as synthetic phospholipids with short alkyl chains[20a] may become applicable and broaden the approaches to be taken in the future.

Appendix: Determination of cmc Values

A. Qualitative Determination

Prepare solutions with increasing concentrations (0, 0.01, 0.1, 1% of the detergent) in separate test tubes. Place a fresh sheet of Parafilm on the bench and remove the paper sheet from the top side. Draw 0.1 ml into a Pasteur pipette. With your elbow steadying the hand, drip one drop of the solution onto the Parafilm. Below the cmc, the drop will retain a nearly spherical shape. If additional drops are aimed near the site of the first one, drops will coalesce spontaneously. Above the cmc, the first drop will splash and subsequent drops will show little tendency to coalesce. The cmc is taken to be reflected by that concentration at which drops become distorted and no longer coalesce spontaneously.

B. Quantitative Measurement

Prepare tubes with 2–3 ml of an ascending concentration series starting from a concentration approximately one-fiftieth of the qualitatively

[20a] J.-L. Eisele and J. P. Rosenbusch, *J. Mol. Biol.*, in press (1989).

estimated cmc. Once the cmc is established approximately, prepare solutions of concentrations allowing a precise evaluation as indicated in Fig. 1D. Before starting a set of measurements, soak the Teflon block with the brass pipe (Fig. 2) for some minutes in acetone and carefully avoid subsequent contact with the skin, etc. Proceed to determine the drop volume of pure water by drawing a solution into the prepared syringe. Before measuring drop size, make sure that the drops forming at the orifice of the brass pipe wet its surface completely without soiling the Teflon block. If that happens, wipe the orifice with a fresh Kleenex wetted with acetone. Then count turns of the piston directly, or use an electronic turn counter (Fig. 2). Count number of turns between drop detachments until the syringe is empty. Note the average number of turns for at least a dozen drops. Repeat the procedure for each detergent solution with increasing concentration by rinsing the syringe at least five times with the new solution before starting to determine the number of drops for the next concentration.

C. Calculation of the Surface Tension

Just before detachment, the weight of a drop hanging at the mouth of a tube is balanced by a force given by the surface tension multiplied by the length of the circumference of the tube to which it is attached. For a round pipe of radius r,

$$V_{tot} \rho g = 2\pi r \gamma$$

where V_{tot} is the hanging drop volume, ρ is the solvent density, g the earth's gravitational acceleration, and γ the surface tension (in mN m^{-1}). After detachment, some liquid remains at the orifice. Thus, the falling drop with a volume V is smaller than V_{tot}:

$$V = V_{tot} f$$

The surface tension is consequently

$$\gamma = \frac{V \rho g}{2\pi r f}$$

The falling drop volume can be measured easily by averaging the readings obtained with several successive drops. The hanging drop volume requires wiping of the liquid remaining at the orifice from a previous drop. Alternatively, the ratio of the two volumes, f, can be calculated from measurements of V, the falling drop alone, using the approximation given by Wilkinson.[21] For these measurements, the radius of the pipe should be

[21] M. C. Wilkinson, *J. Colloid Interface Sci.* **40**, 14 (1972).

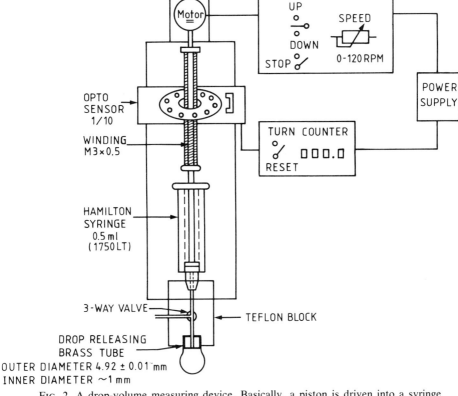

FIG. 2. A drop-volume measuring device. Basically, a piston is driven into a syringe (Hamilton 0.5 ml, 1750 LT) at whose tip a brass pipe is mounted from which the extruded solution drips. The drop volume is measured by counting the number of turns of the screw in the winding by an Optosensor that is directly wired to a counter indicating one-tenth of a turn. The speed of the motor is regulated by a motor-control unit; before drop detachment, motor speed may be reduced. The critical part of the device is a brass pipe which is machined on a precision lathe with a sharp outer edge. Its diameter must be known accurately while the inner diameter of the pipe is not critical. The position of the Teflon block enclosing the brass pipe prevents the solution from creeping along the outer cylindrical surface of the slightly protruding brass tube (0.1 mm). Also within the Teflon block is a three-way valve that allows rinsing of the syringe and the inside of the brass pipe, as well as filling the pipette. For the description of the procedure, see the Appendix.

chosen such that the drop surface continues along the pipe mantle (see the drop drawn in Fig. 2). For aqueous detergent solutions, a value of $r \approx 2.5$ mm meets these requirements. The density used is, strictly speaking, a difference density between the solution and air, or between two solutions.

For the practical purposes discussed here, the solution density can be used as such.

D. Calculation of the Area Covered by Detergent Monomers at the Interphase

This area, A, can be calculated using the Gibbs adsorption isotherm according to Barry and El Eini[22] from the slope of the descending surface tension by the following formula.

$$A = -2.303kT(d\gamma/d \log c)^{-1}$$

where k is the Boltzmann constant and T the absolute temperature.

Acknowledgments

This work was supported by grants from the Swiss National Science Foundation (3.656.80 and 3.294.85). We thank P. Henz for his help with the electronic hardware, R. Niederhauser for secretarial assistance, and M. Jäggi for artistic help.

[22] B. W. Barry and D. I. D. El Eini, *J. Colloid Interface Sci.* **54**, 339 (1976).

[30] High-Performance Liquid Chromatography of Membrane Lipids: Glycosphingolipids and Phospholipids

By R. H. McCluer, M. D. Ullman, and F. B. Jungalwala

The isolation and measurement of glycosphingolipids and phospholipids is essential for many studies on complex biological systems. The preparative and analytical separations of these lipids by conventional column or thin-layer (TLC) techniques are tedious and time-consuming processes. Because of the difficulties inherent in separations by older chromatographic techniques, the use of high-performance liquid chromatography (HPLC) in the purification and analysis of these lipids is a welcome addition to the research tools for membrane biology.

Older methods for the separation and analysis of complex glycosphingolipids and phospholipids have primarily depended on class separation. The degree of molecular heterogeneity has generally been established by analysis of the products of hydrolysis. HPLC, however, provides the means to separate lipid classes and their molecular species.

Modern HPLC technology implies the use of injection ports for sample application, reusable columns packed with microparticulate (down to 3 μm) stationary phases, pumps for uniform solvent flow, and on-line sample detection. It provides the potential for rapid and highly sensitive methods for the separation and analysis of lipids. Numerous books,[1-4] reviews, and courses are now available which present the theoretical and practical aspects of HPLC.

Currently, individual classes of glycosphingolipids and/or phospholipids, obtained by solvent partition and chromatography, are separated by normal-phase HPLC according to their polarity. Separation of their molecular species can subsequently be obtained by reversed-phase HPLC according to their nonpolar moieties. Argentation chromatography can further resolve components by their degree of unsaturation.

General HPLC Techniques

Most chromatography of glycosphingolipids and phospholipids involves either adsorption chromatography on silica for class separations, or reversed-phase chromatography on bonded phases for molecular species separations. For adsorption chromatography, two types of silica are predominantly used: totally porous and pellicular.

Totally porous silicas have mean particle diameters of 3, 5, or 10 μm and have either spherical or irregular shapes. This silica has large linear (loading) capacities, high efficiencies, and some selectivity for fatty acid chain length. The high efficiencies provide excellent resolution but the selectivity for the fatty acid chain length can cause multiple-peak configurations (e.g., "doublets") or peak broadening, based on the population of medium- and long-chain fatty acids.

Pellicular packings have a thin porous layer (pellicle) of stationary phase coated on solid (approximately 25 μm mean particle diameter) glass beads. They have lower linear capacities than totally porous packings and their efficiencies approach those of 5-μm-particle-size totally porous silica. Further, two features make pellicular columns especially attractive for the analysis of glycosphingolipids: selectivity and economy.

---

[1] L. R. Snyder and J. J. Kirkland, "Introduction to Modern Liquid Chromatography," 2nd Ed. Wiley (Interscience), New York, 1979.

[2] E. L. Johnson and R. Stevenson, "Basic Liquid Chromatography." Varian, Palo Alto, California, 1978.

[3] J. Q. Walker, M. T. Jackson, Jr., and J. B. Maynard, "Chromatographic Systems: Maintenance and Troubleshooting." Academic Press, New York, 1977.

[4] R. W. Yost, L. S. Ettre, and R. D. Conlon, "Practical Liquid Chromatography: An Introduction." Perkin-Elmer, Norwalk, Connecticut, 1980.

The pellicular material has less selectivity for the fatty acid portion of the complex lipids compared to that observed for the totally porous silica. Less selectivity avoids peak splitting or broadening and thus actually makes it more advantageous for the quantitative analysis of neutral glycosphingolipids.

Pellicular columns can be inexpensively dry-packed in less than 20 min and they routinely provide very high resolution for long periods of time (many months of extensive usage). The column can be easily packed in about 15 min by the "tap–fill" method.[1] Briefly, the bottom of a 500 mm × 2.1 mm i.d. stainless-steel tube is fitted with a male end-fitting which contains a stainless-steel frit. The top of the tube is fitted with a male end-fitting which has a 4-cm piece of tubing and no frit. The tubing is filled with pellicular silica (Zipax, E.I. Dupont de Nemours, Wilmington, DE) to about 1–2 cm high. The bottom of the column is then tapped on a bench top about 80 times. During this procedure the column is also rotated. The silica is added and packed repetitively until the column is filled. The top end-fitting is replaced with one which contains a stainless-steel frit so that no dead space remains and the column is ready to be equilibrated and used.

The mobile phases for the HPLC of glycosphingolipids and phospholipids are mixtures of relatively apolar solvents, such as hexane, with reasonably polar solvents, such as dioxane, 2-propanol or 2-propanol and water. The mobile phase is commonly presented in a gradient of increasing or decreasing concentration of the polar solvent during adsorption or reversed-phase HPLC, respectively. The choice of mobile phases is limited by the mixing compatibility of the different solvents and by the method of detection.

Poor mixing characteristics of the solvents which comprise the mobile phase create baseline instabilities. A dual-chamber dynamic (mechanical) mixer or two single-chambered mixers in series are required for baselines sufficiently stable to allow detection at high sensitivity such as 0.04 absorbance units full scale (AUFS).

Some of the polar solvents have residual UV absorption at the absorption maximum of the compounds of interest (e.g., 230 nm for perbenzoylated glycosphingolipids). This causes a positive baseline drift during gradient elutions. The drift can be so great that it either precludes the use of a given mobile phase, or decreases the sensitivity of the procedure. The drifting baseline is eliminated by directing the mobile phase through a preinjector flow-through reference cell (Fig. 1). This generates a horizontal baseline (because the *rate* of change in the polar solvent concentration is the same in the reference and sample cells when a linear gradient is generated) during the major part of the gradient run, with a negative and

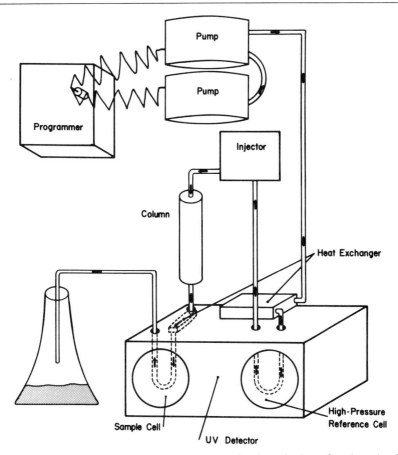

FIG. 1. Diagram of chromatographic system showing flow of solvent first through reference cell followed by that through injector, column, and sample cell, respectively. Reproduced from Ref. 5.

positive deflection at the beginning and end of the gradient, respectively. To minimize the magnitude of these deflections, the volume of the injector sample loop should be 100 μl or less and the volume of tubing between the detector reference cell outlet and sample cell inlet should be minimal. Such a flow path can also be useful for systems which have only a minor amount of absorption (e.g., 2-propanol/hexane), but where high sensitivity is sought.

At high sensitivities UV detectors are vulnerable to ambient temperature fluctuations, particularly at the tubing which leads from the column outlet to the detector inlet. Inadequate insulation around this tubing will lead to unstable baselines. This piece of tubing should be well insulated. We use a piece of rubber tubing slit from end to end and slipped over the tubing to be insulated. It is also helpful if there is a heat exchanger in the detector to stabilize the mobile phase temperature.[5] Some manufacturers have incorporated such an exchanger into their detectors. An easy method to detect ambient temperature problems is to simply place your finger on the tubing in question for a few seconds. If there is a baseline disturbance it is an indication that there is an ambient temperature problem.

Glycosphingolipids

HPLC Analysis of Perbenzoylated Derivatives

General Comments on Perbenzoylation. Glycolipids do not possess a characteristic chromophore that permits their quantitative detection where high sensitivity is required. However, they can be derivatized with benzoyl chloride or benzoic anhydride to form stable O-, N-, or O-benzoylated products, respectively. These products can be quantitatively measured with high sensitivity by their absorption of UV light.

Nonhydroxy fatty acid (NFA)- and hydroxy fatty acid (HFA)-glycosphingolipids can be perbenzoylated with either benzoyl chloride or benzoic anhydride. Benzoyl chloride derivatizes the hydroxyl groups and amide nitrogen of NFA cerebrosides. When alkaline methanolysis is performed on these diacylamine derivatives, a mixture of parent cerebrosides and *N*-benzoylpsychosine is obtained (Fig. 2).[6,7] When HFA cerebrosides are perbenzoylated the substitution of the amide hydrogen is stearically blocked so that only O-benzoylation occurs. Thus, mild methanolysis of these HFA derivatives yields only the parent compounds.

Both NFA- and HFA-glycosphingolipids derivatized with benzoic anhydride yield only the *O*-benzoyl derivatives.[8] Therefore, native glycolipids are recovered after mild alkaline hydrolysis. Because sphingolipids which contain only HFA as *N*-acyl substituents form the same derivative with either benzoyl chloride or the anhydride reaction, they can easily be distinguished from NFA-containing sphingolipids which form different derivatives that are separated by HPLC. The benzoyl chloride reaction is most useful for analytical purposes because resolution of components

---

[5] M. D. Ullman and R. H. McCluer, *J. Lipid Res.* **19**, 910 (1978).
[6] R. H. McCluer and J. E. Evans, *J. Lipid Res.* **14**, 611 (1973).
[7] R. H. McCluer and J. E. Evans, *J. Lipid Res.* **17**, 412 (1976).
[8] S. K. Gross and R. H. McCluer, *Anal. Biochem.* **102**, 429 (1980).

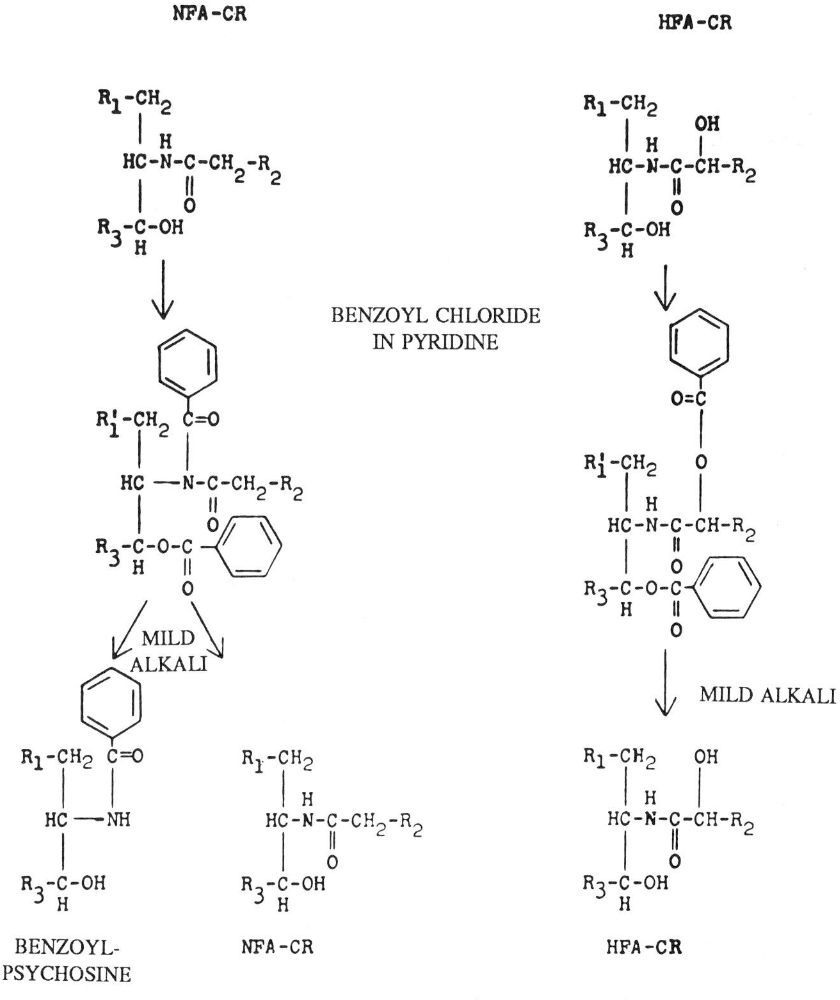

FIG. 2. The formation of benzoylated cerebroside derivatives by reaction with benzoyl chloride and their degradation with mild alkali.

R_1 = GALACTOSYL-O-
R_1' = TETRABENZOYLGALACTOSYL-O-
R_2 = $-(CH_2)_{20}-CH_3$
R_3 = $-CH=CH-(CH_2)_{12}-CH_3$

containing HFA and/or phytosphingosine (PHYTO) is superior to that obtained with the O-benzoates formed with benzoic anhydride.

Several factors influence the benzoyl chloride derivatization. For example, it is important that isolated neutral glycosphingolipids and/or ganglioside samples to be perbenzoylated are relatively free of silica or silicic acid particles. These particles arise from either dissolved silica after extraction of the glycosphingolipids from TLC or from "fines" that avoid filtration. Reversed-phase chromatography on sample cartridges can eliminate silica gel from gangliosides after TLC.[9]

Benzoyl chloride is susceptible to the presence of moisture. It reacts instantaneously with water (e.g., atmospheric moisture) to form benzoic acid and hydrochloric acid. Thus, it is important to store benzoyl chloride in dry surroundings. A 500-ml bottle of benzoyl chloride can usually be used for 3 to 6 months.

The pyridine and toluene (for ganglioside derivatization) must be dried by storage over 4 Å molecular sieves. Some batches of pyridine may have a very high moisture content and the molecular sieves may not trap all of the water. This situation is evident upon mixing the pyridine with the benzoyl chloride because a crystalline precipitate forms almost immediately. There are other indications that moisture has been introduced into the reaction medium. For example, the "solvent front" of the chromatogram may be broader than usual or there may be a broad peak that elutes just after the "solvent front." These anomalies are due to benzoic acid. It is also important to dry the sample *in vacuo* over phosphorus pentoxide at least 1 hr before the perbenzoylation is performed.

Procedure for Mono-, Di-, Tri-, and Tetraglycosylceramides: Per-O,N-benzoylation with Benzoyl Chloride. Samples or standards (at least 200 ng of each glycolipid) and if desired, a known amount of N-acetylpsychosine, as internal standard[5] are placed in a 13 mm × 100 mm screw-capped culture tube with a Teflon-lined cap. The glycosphingolipids are per-O,N-benzoylated by adding 0.5 ml of 10% benzoyl chloride in pyridine (v/v) directly to the bottom of the tube and incubating the sample at 37° for 16 hr.[5] The pyridine is evaporated with a stream nitrogen at room temperature until the residue appears as an oil-covered solid. Then 3 ml of hexane is added to the reaction tube. The hexane is washed three times with 1.8 ml of methanol–water 80:20 (v/v) which is saturated with sodium carbonate (1.2 g sodium carbonate in 300 ml of methanol–water 80:20). In removing the lower aqueous methanol layer, a slight positive pressure must be exerted on the pipette bulb as the pipette tip passes through the hexane layer. The hexane layer is then washed with 1.8 ml of

---

[9] H. Kubo and M. Hoshi, *J. Lipid Res.* **26**, 638 (1985).

methanol–water 80:20 (v/v). The lower phase is withdrawn and discarded and the hexane is evaporated with a stream of nitrogen. The derivatives are dissolved in carbon tetrachloride and an aliquot is injected into the HPLC column.

Alternatively, the derivatives can be isolated from the reagents and reaction by-products with a C_{18} reversed-phase cartridge (C_{18} Bond Elut, Analytichem International, Harbor City, CA) by the procedure that was developed for per-O,N-benzoylated gangliosides (see section below). The procedure is fast and the recoveries are comparable to the partition workup.[5] However, N-acetylpsychosine cannot be used as an internal standard with this method because the partition characteristics of its per-O,N-benzoylated derivative on the reversed-phase cartridge are substantially different from those for the per-O,N-benzoylated derivatives of neutral glycosphingolipids. The per-O,N-benzoylated products are stabile for months providing they are completely free of alkali.

The derivatives are separated on a pellicular (Zipax, E.I. DuPont de Nemours, Inc., Wilmington, DE) column (2.1 mm i.d. × 500 mm) with a 10-min linear gradient of 2–17% water-saturated ethyl acetate in hexane and a flow rate of 2 ml/min.[10] They are detected by their UV absorption at 280 nm. The minimum level of detection (twice baseline noise) by this procedure is approximately 70 pmol of each neutral glycosphingolipid. The sensitivity of the procedure can be increased by the use of a mobile phase that is transparent at 230 nm, the absorption maximum of the derivatives.

The sensitivity increase with 230 nm detection is accomplished with a 13-min linear gradient of 1–20% dioxane in hexane and a flow rate of 2 ml/min (Fig. 3). The mobile phase is directed through a preinjector flow-through reference cell to cancel the residual absorption of the dioxane (Fig. 1). Several UV detectors have flow-through reference cells that are pressure rated high enough to be utilized in this system. The maximum pressure rating required is about 1000 psi. A modification of this procedure (15-min linear gradient of 0.23–20% dioxane in hexane) can be used to separate and quantitate per-O,N-benzoylated ceramides (derivatized by the same procedure) and neutral glycosphingolipids in a single chromatographic run.[11]

A consistent separation of per-O,N-benzoylated glucocerebroside from per-O,N-benzoylated galactocerebroside, at all ratios, is obtained by HPLC of the per-O,N-benzoylated derivatives by the pellicular silica column maintained at 60° and a 10-min linear gradient of 1–20% dioxane

---

[10] M. D. Ullman and R. H. McCluer, *J. Lipid Res.* **18**, 371 (1977).
[11] K.-H. Chou and F. B. Jungalwala, *J. Neurochem.* **36**, 394 (1981).

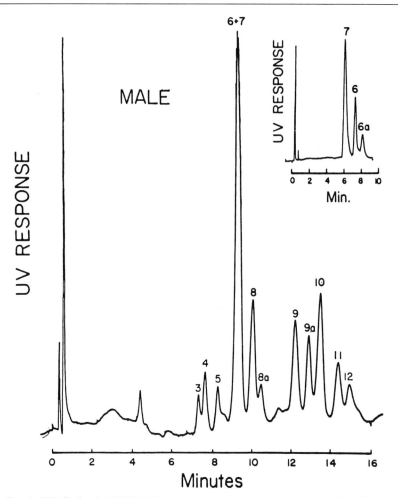

FIG. 3. HPLC of male (C57BL/6J) mouse kidney perbenzoylated glycosphingolipids on a Zipax column with detection at 230 nm. A 13-min linear gradient of 1–20% dioxane in hexane at a flow rate of 2 ml/min was used as the mobile phase. Inset shows the separation of the collected peak 6 and 7 when reinjected on the same column but with a 14-min linear gradient of 0.25–1% 2-propanol in hexane as the solvent. Peak 3, Glc-Sph-Nfa; 4, Gal-Sph-Nfa; 5, Glc-Phyto-Nfa; 6, Glc-Sph-Hfa + Glc-Phyto-Hfa; 6a, Gal-Sph-Hfa + Gal-Phyto-Hfa; 7, GaOSe2-Sph-Nfa; 8, GaOSe2-Sph-Hfa + GaOSe2-Phyto-Nfa; 8a, Lac-Sph-Nfa; 9, GbOSe3-Sph-Nfa; 9a, GbOSe3-Phyto-Nfa, 10, GbOSe3-Sph-Hfa; 11, GbOSe4-Sph-Nfa; and 12, GbOSe4-Phyto-Nfa + GbOSe4-Sph-Nfa. Sph refers to C_{18}-sphingosine; Phyto, C_{18}-phytosphingosine; Nfa, nonhydroxy fatty acid; Hfa, hydroxy fatty acid. Reproduced from Ref. 15.

in hexane at a flow rate of 2 ml/min.[12] The system is particularly useful for the analysis of brain cerebrosides since the ratio of galactocerebroside to glucocerebroside is usually very high.

Per-O-benzoylation with Benzoic Anhydride. The per-*O*-benzoyl derivatives of neutral glycosphingolipid are produced via their reaction with 10% benzoic anhydride in pyridine with 5% 4-*N*-dimethylaminopyridine (DMAP) as a catalyst. The reaction is run at 37° for 4 hr.[8] Excess reagents are removed by solvent partition between hexane and aqueous alkaline methanol as described previously. The derivatives are then separated by HPLC on either pellicular or totally porous silica columns as described for the *O,N*-benzoyl derivatives (Fig. 4). The per-*O*-benzoyl derivatives of HFA monoglycosylceramides (formed by reaction with benzoic anhydride and benzoyl chloride) elute earlier than the per-*O*-benzoyl derivatives NFA-monoglycosylceramides (formed with benzoic anhydride) but later than the per-*O,N*-benzoyl derivatives of NFA-monoglycosylceramides (formed with benzoyl chloride).

Comments. In our laboratories, the formation of *O,N*-benzoyl derivatives is used most frequently for the quantitative analysis of neutral glycosphingolipids. This procedure provides excellent resolution of the derivatives, including the HFA- and NFA-glycosphingolipids. However, the *O*-benzoyl derivatives are also used analytically, especially as an adjunct procedure for the tentative identification of HFA-glycosphingolipids. The per-*O*-benzoyl derivatives can be used to distinguish between HFA- and NFA-glycosylceramides by comparing their elution times to the per-O,N-benzoylated derivatives. The *O*-benzoyl derivatives have also been used for preparative purposes since native neutral glycosphingolipids are recovered by mild alkaline methanolysis of the isolated derivatives.[13]

A variety of chromatographic systems have been used to separate perbenzoylated derivatives. Mobile phases of 2-propanol in hexane have been used which do not have to be directed through a preinjector flow-through reference cell since they are reasonably UV transparent at 230 nm.[14,15] These systems are used with a totally porous 5- or 10-μm silica column, but they can also be used with the pellicular silica columns.

Both systems, dioxane/hexane and 2-propanol/hexane, have specific advantages and disadvantages. The dioxane/hexane system provides consistent resolution of the major glycosphingolipids and, in our laboratories,

---

[12] E. M. Kaye and M. D. Ullman, *Anal. Biochem.* **138**, 380 (1984).

[13] R. H. McCluer and M. D. Ullman, in "Cell Surface Glycolipids" (C. C. Sweeley, ed.), p. 1. Am. Chem. Soc., Washington, D.C., 1980.

[14] P. M. Strasberg, I. Warren, M. A. Skomorowski, and J. A. Lowden, *Clin. Chim. Acta* **132**, 29 (1983).

[15] R. H. McCluer and S. K. Gross, *J. Lipid Res.* **26**, 593 (1985).

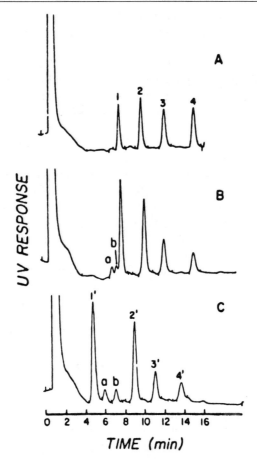

FIG. 4. HPLC of per-O- and per-O,N-benzoylated standard and plasma glycosphingolipids. The derivatized glycosphingolipids were injected onto a Zipax column (2.1 mm × 50 cm) and eluted with a 13-min linear gradient of 2.5–25% dioxane in hexane with detection at 230 nm. (A) Standard glycosphingolipids (GSL) per-O-benzoylated with benzoic anhydride and 4-dimethylaminopyridine (DMAP). (B) Plasma GSL per-O-benzoylated with benzoic anhydride and DMAP. (C) Plasma GSL perbenzoylated with benzoyl chloride. Glycosphingolipid peaks are identified as (1) glucosylceramide, (2) lactosylceramide, (3) galactosyllactosylceramide, (4) N-acetylgalactosaminylgalactosyllactosylceramide. Peak a is unidentified, and peak b is hydroxy fatty acid containing galactosylceramide. Reproduced from Ref. 8.

provides a better resolution of the minor plasma and tissue neutral glycosphingolipids. Further, the selectivity and efficiency of this system is such that the neutral glycosphingolipid derivatives yield only one peak for each neutral glycosphingolipid rather than further partial separation of each neutral glycosphingolipid on the basis of its fatty acid composition. This provides more consistent automatic integration of the peaks and yields a less complex chromatogram. The disadvantage of the system is that the mobile phase, in a gradient run but not in an isocratic run, must pass through a preinjector flow-through reference cell. Many detectors will accommodate the rather modest pressure requirements and the hexane/dioxane system is the one of choice in our laboratories.

The 2-propanol/hexane mobile phase used with a porous silica column produces adequate resolution of the major plasma glycosphingolipids. It also shows some selectivity for the fatty acid portion of derivatized neutral glycosphingolipid and the chromatographic peaks are often broadened or split into doublets based on the population of medium- and long-chain fatty acids.

The 2-propanol/hexane mobile phase can be used with a pellicular silica column but the resolution of the neutral glycosphingolipid derivatives is not as high as that obtained with either the hexane/dioxane with a pellicular column or the 2-propanol/hexane system with a totally porous silica column. The 2-propanol/hexane mobile phase with a pellicular column will, however, provide resolution of some HFA- or phytosphingosine-containing neutral glycosphingolipids which are not resolved in the dioxane system. Sometimes combinations of systems can be used. For example, if one system is not adequately resolving peaks of interest, the peaks can be collected and separated in the other system (Fig. 3, inset). Obviously, the system of choice in any laboratory is dependent on the specific objectives.

Finally, for analytical procedures, it is important to know the absolute recoveries of the glycosphingolipids. There are several examples in the literature in which losses of neutral glycosphingolipid that occur during the isolation procedures are not taken into consideration. It is documented that neutral glycosphingolipid recoveries from silicic acid columns decrease with increasing length of neutral glycosphingolipid carbohydrate chains.[16] It is important to add, if possible, a high-specific-activity radiolabeled (or other) standard that can be perbenzoylated and used to determine the total recovery, and, consequently, the absolute lipid concentrations of the tissue source. N-Acetylpsychosine has been used as an internal standard[5] for neutral glycosphingolipid determinations. However, difficulty is encountered because the short-chain N-acetyl group imparts characteristics that are slightly different from those of long-chain fatty acid glycolipids.

---

[16] R. E. Vance and C. C. Sweeley, *J. Lipid Res.* **8**, 621 (1967).

Procedure for Gangliosides

Mono- and polysialogangliosides are quantitated by HPLC of their per-O,N-benzoylated derivatives.[17] Gangliosides (at least up to 20 nmol) are per-O,N-benzoylated in a 1-ml reaction vial with 100 μl of 5% benzoyl chloride in 25% toluene–pyridine (v/v). The mixture is heated at 45° for 16 hr. It is terminated by drying the mixture under nitrogen and adding 0.8 ml of methanol to the residue. The methanolic extract is then transferred to a prewashed (3 ml of methanol) reversed-phase cartridge (Bond Elut, Analytichem International, Inc., Harbor City, CA). The solvent is suctioned through the cartridge with a slight vacuum and saved. The reaction vial is rinsed with an additional 0.8 ml of methanol. The methanol rinse is added to the cartridge, suctioned through as before, and combined with the initial methanol eluate. The sample of combined methanol fractions is applied to the cartridge. The cartridge is then washed with 4 ml of methanol and the per-O,N-benzoylated derivatives are eluted with 3 ml of methanol–benzene (8:2). The methanol–benzene is removed with a stream of nitrogen and the derivatives are dissolved in carbon tetrachloride and an aliquot is injected into the HPLC column.

The derivatives are separated on a silica (3-μm mean particle size) column (4.6 mm i.d. × 100 mm) which is maintained at 90°. The derivatives are eluted from the HPLC column with a 15-min linear gradient of 1.8–12% 2-propanol in hexane and a flow rate of 2 ml/min. They are detected by their UV absorption at 230 nm and are quantitated by comparing the response with an external standard (Fig. 5).

External standards are prepared from pure individual gangliosides. The pure gangliosides are dissolved in chloroform–methanol–water 10:10:0.25 (v/v) (approximately 100 μg ganglioside/ml). The absolute concentration of each ganglioside standard is determined by the resorcinol reaction.[18]

If only monosialogangliosides are to be analyzed, they can be perbenzoylated with 10% benzoyl chloride in pyridine at 60° for 1 hr.[19] These derivatives are separated by the above described HPLC system or on a custom-packed silica Si-4000 column with an 18-min linear gradient of 7–23% dioxane in hexane.

Comments. These HPLC procedures for gangliosides are convenient, highly sensitive, and reproducible. It is worthwhile to note the following aspects of these procedures. (1) Toluene is included in the reaction medium for the derivatization of polysialogangliosides because it may act as

---

[17] M. D. Ullman and R. H. McCluer, *J. Lipid Res.* **26**, 501 (1985).
[18] R. W. Ledeen and R. K. Yu, this series, Vol. 83, p. 139.
[19] E. G. Bremer, S. K. Gross, and R. H. McCluer, *J. Lipid Res.* **20**, 1028 (1979).

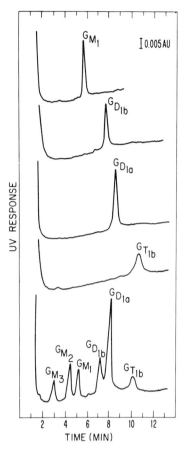

FIG. 5. HPLC separation of brain gangliosides as their perbenzoylated derivatives on silica (3 μm) column (4.6 mm × 15 mm) maintained at 90°. A 15-min linear gradient of 1.8–12% 2-propanol in hexane with a flow rate of 2 ml/min was used as the mobile phase. Detection was at 230 nm. Reproduced from Ref. 17.

a *pi* base and suppress the activity of the HCl generated during the reaction. This virtually eliminates the formation of multiple peaks from G_{Tlb}. Further, it slows the reaction rate so that a convenient (16 hr) reaction time is provided. Finally, it also facilitates the evaporation of the HCl which is generated during the reaction. (2) The isolation of the derivatives from reversed-phase cartridges minimizes losses of time and sample. (3) The carboxyl groups of gangliosides are not derivatized under these perbenzoylation conditions. When the derivatives are chromatographed at ambient temperature, the protonated and unprotonated forms of the polysialoganglioside carboxyl groups cause the chromatographic peaks to broaden and/or form doublets. Chromatography on a column that is maintained at 90° eliminates this broadening and doublet formation. (4) These analytical, nondestructive procedures allow the derivatives to be collected for further analysis (e.g., separation and quantification of their individual ceramide molecular species on a reversed-phase HPLC column).

Molecular Species Separation by Reversed-Phase HPLC

Native Gangliosides. The analysis of ganglioside fatty acids and long-chain bases has customarily been performed after acid hydrolysis. However, these types of data do not reveal the exact combination of the fatty acid with the sphingoid base in an individual ganglioside molecular species. Reversed-phase HPLC of underivatized gangliosides has been used to isolate and provide information about individual monosialoganglioside molecular species.[20] For such analyses, it is necessary first to isolate highly purified individual gangliosides.

Isolation of Individual Monosialogangliosides. Total brain ganglioside fractions are isolated by extraction of fresh or frozen tissue twice with 20 volumes of chloroform–methanol–water (4:8:3) and partitioned by adding water to give a final chloroform–methanol–water ratio of 4:8:5.6 (v/v) as described by Svennerholm and Fredman.[21] The lower phase is partitioned twice more with water and the lipids in the combined upper phases recovered by passing over a column of C_{18} reversed-phase packing (Lichroprep, 0.3 g/g of wet wt. brain tissue extracted) by a modification of the procedure of Williams and McCluer.[22] After salt and water-soluble materials are removed by washing with 50 column volumes of water, gangliosides are eluted from the column by elution with 20 column volumes of methanol. The dried residue is treated with 0.25 N sodium hydroxide in methanol at 37° for 2 hr to destroy ester lipids and, after neutralization with acetic acid, the gangliosides are recovered with Lichroprep as described above. The dried ganglioside fraction is dissolved in chloroform–methanol–water (30:60:8, v/v) and fractionated on a DEAE-Sephadex (acetate) column as described by Ledeen and Yu,[18] except that Lichroprep reversed-phase chromatography is used instead of dialysis. Briefly, the gangliosides are dissolved in chloroform–methanol (6:1) and applied to a Unisil column. The column is successively washed with 10 column volumes of chloroform–methanol (6:1), chloroform–methanol (5:1), 30 column volumes of chloroform–methanol–water

---

[20] H. Kadowaki, J. E. Evans, and R. H. McCluer, *J. Lipid Res.* **25**, 1132 (1984).
[21] L. Svennerholm and P. Fredman, *Biochim. Biophys. Acta* **617**, 97 (1980).
[22] M. A. Williams and R. H. McCluer, *J. Neurochem.* **35**, 266 (1980).

(10:10:3), and 15 volumes of methanol. The later two solvent fractions which contain gangliosides are combined and the solvent evaporated to dryness. The ganglioside fraction is then dissolved in methanol and applied to a DEAE-Sephadex (acetate) column to separate mono-, di- and trisialogangliosides as described.[18] The monosialoganglioside fraction which is eluted with 0.02 M ammonium acetate in methanol is further resolved into individual monosialogangliosides by silicic acid chromatography. The monosialoganglioside fraction (~100 mg) is dissolved in approximately 1 ml of chloroform–methanol–water (65:35:3) and applied to a 50 g Iatrobead 6RS-8060 column (1.2 × 100 cm). The column is eluted with chloroform–methanol–water (65:35:3), 10-ml fractions collected, and gangliosides in each fraction examined by HPTLC on Merck plates with chloroform–methanol–0.25% $CaCl_2$ (55:45:10) as the developing solvent; gangliosides are visualized with resorcinol spray reagent.[18] G_{M4} elutes after a bed volume of solvent and then G_{M3}, G_{M2}, and G_{M1} elute in succession. Fractions containing the overlapping end of the G_{M3} peak and the beginning of the G_{M2} peak are reapplied to the Iatrobead column and eluted with the same solvent system. The fractions which contain the end of G_{M4} and beginning of G_{M3} are reapplied to the column and eluted with chloroform–methanol–water (70:30:3). Recoveries of each pure monosialoganglioside following the Iatrobead chromatography have been estimated to be over 95% for G_{M4} and G_{M3} and over 90% for G_{M2} and G_{M1}.

Separation of Molecular Species by Reversed-Phase HPLC. A 25–400 μg sample of each brain monosialoganglioside, G_{M4}, G_{M3}, G_{M2}, or G_{M1} is dissolved in a minimum volume of methanol and separated into molecular species by reversed-phase HPLC on 5-μm-mean-particle-size ODS column (4.6 × 250 mm, Altex).[20] The ganglioside molecular species are eluted isocratically with methanol–water mixtures of approximately 85:15 at 28° and with detection at 205 nm (Fig. 6). Effects of varying solvent composition and temperature are also shown.

The ceramide composition of the peaks has been determined by direct ammonia CI–MS of the intact gangliosides, and by GLC analysis of the fatty acids with HPLC analysis of the long-chain bases after acetonitrile–HCl hydrolysis.[20] The amount of ganglioside in each peak was determined by normal-phase HPLC analysis of the per-O,N-benzoylated derivatives.[17] Approximately half of the peaks shown were found to be homogeneous. Separation is not achieved with ceramides[22a] that have the same effective methylene numbers such as d18:1-C20:0 versus d20:1-C18:0 or d18:1-C22:0 versus d18:1-C24:1. The percentages of each component in the heterogeneous peaks must be calculated from the ratio of fatty acids and long-chain bases in each peak. The relative retention times

---

[22a] Nomenclature of Breimer *et al.*, *Biochim. Biophys. Acta* **348**, 232 (1974).

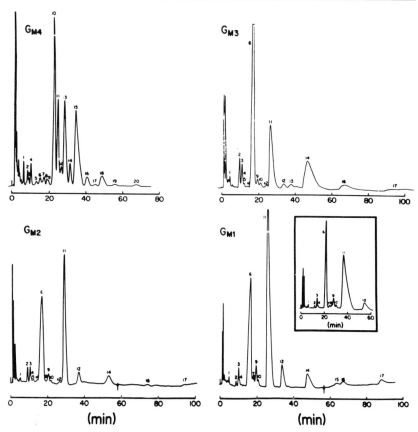

FIG. 6. HPLC separation of the molecular species of human brain G_{M4}, G_{M3}, G_{M2}, and G_{M1} on a C_{18} reversed-phase column. The chromatographic conditions: G_{M4} (58 μg) was chromatographed with methanol–water (89:11, v/v) at a flow rate of 1.3 ml/min, at the column temperature of 26°; G_{M3} (51 μg) with methanol–water (87:13, v/v) and 1.5 ml/min at 25°; G_{M2} (46 mg) with methanol–water (84.7:15.3, v/v) and 1.5 ml/min at 28°; G_{M1} (81 μg) with methanol–water (85.2:14.8, v/v) and 1.5 ml/min at 28°; G_{M1} (inset) with methanol–water (83.2:16.8, v/v) and 1.3 ml/min at 28°. Detection was by absorption at 205 nm (0.1 absorbance units full scale and then changed to 0.05 as shown by the arrow of G_{M2} and G_{M1}). Peaks are numbered in sequence of elution. Reproduced from Ref. 20.

(RRT) of the major molecular species are shown in Table I. The RRT of any ganglioside with a particular ceramide composition is the same for all of the monosialogangliosides. Thus, with any monosialoganglioside, the order of elution of molecular species is constant and independent of the composition of the component sugars. The separation of the molecular species of monosialogangliosides from normal adult human brain allowed

TABLE I
DISTRIBUTION OF MOLECULAR SPECIES FROM G_{M3}, G_{M2}, AND G_{M1}[a]

Peak number[b]	Molecular species	G_{M3}	G_{M2}	G_{M1}
1	d18:1–C14:0	0.32 ± 0.08	0.28 ± 0.06	0.26 ± 0.12
2	d16:1–C18:0	0.51 ± 0.13	0.33 ± 0.07	0.20 ± 0.09
	d18:1–C16:0	2.05 ± 0.52	1.32 ± 0.30	0.15 ± 0.07
3	d18:2–C18:0	1.20 ± 0.98	1.42 ± 0.47	0.77 ± 0.23
4	d16:0–C18:0	0.46 ± 0.12	0.56 ± 0.31	0.18 ± 0.01
5	d18:1–C17:0	0.31 ± 0.08	0.21 ± 0.05	ND
X1	Unidentified	0.22 ± 0.07	0.33 ± 0.07	ND
6	d18:1–C18:0	39.94 ± 4.55	35.93 ± 6.09	25.50 ± 1.58
7	d20:2–C18:0	ND	ND	0.54 ± 0.39
8	d16:0–C20:0	0.37 ± 0.22	0.81 ± 0.43	0.60 ± 0.28
9	d18:0–C18:0	1.39 ± 0.43	2.60 ± 0.65	1.94 ± 0.37
10	d18:1–C19:0	0.75 ± 0.41	0.67 ± 0.47	0.29 ± 0.07
X2	Unidentified	0.53 ± 0.16	0.54 ± 0.12	ND
11	d18:1–C20:0	3.33 ± 0.50	6.37 ± 1.08	7.39 ± 0.98
	d20:1–C18:0	15.15 ± 2.26	33.45 ± 5.67	45.37 ± 6.01
12	d18:0–C20:0	0.10 ± 0.04	0.51 ± 0.07	0.40 ± 0.06
	d20:0–C18:0	0.79 ± 0.32	2.89 ± 0.39	3.60 ± 0.51
13	d18:1–C21:0	0.28 ± 0.10	ND	ND
	d18:1–C23:1	1.70 ± 0.59	ND	ND
14	d18:1–C22:0	2.20 ± 0.60	0.47 ± 0.17	0.52 ± 0.05
	d18:1–C24:1	11.89 ± 3.21	1.26 ± 0.45	0.75 ± 0.07
	d20:1–C20:0	0.59 ± 0.16	1.41 ± 0.51	1.98 ± 0.20
15	d20:0–C20:0	ND	ND	0.30 ± 0.05
16	d18:1–C23:0	0.91 ± 0.13	0.36 ± 0.10	0.26 ± 0.20
	d18:1–C25:1	1.31 ± 0.18	0.43 ± 0.12	0.27 ± 0.21
17	d18:1–C24:0	1.41 ± 1.11	0.49 ± 0.11	0.14 ± 0.10
	d18:1–C26:1	0.53 ± 0.42	0.25 ± 0.05	0.11 ± 0.07
	d20:1–C22:0	0.07 ± 0.06	0.17 ± 0.04	0.15 ± 0.10
	d20:1–C24:1	0.41 ± 0.33	0.32 ± 0.07	0.23 ± 0.16
Remaining peaks[c]		0.40 ± 0.10	0.61 ± 0.44	0.67 ± 0.35
Recovery		89.09 ± 2.91	93.12 ± 15.48	92.54 ± 8.69

[a] Results are shown as the mean ± SD from triplicate HPLC separation of the G_{M3}, G_{M2}, and G_{M1} (see Fig. 6). Each peak collected from a C_{18} reversed-phase HPLC column was perbenzoylated and then quantitated by normal-phase HPLC on a LiChrosphere SI 4000 column. The amounts of individual molecular species in heterogeneous peaks were calculated from the results of long-chain base and fatty acid analyses. ND, Not detectable. Reproduced from Ref. 20.
[b] From chromatograms shown in Fig. 6.
[c] Remaining peaks = gangliosides between major peaks.

the detection of minor species not previously reported and the observation that the percentage composition of d20 long-chain bases increases in the order $G_{M3} < G_{M2} < G_{M1}$.

Comments. The method permits good separation of as much as 400 μg or as little as 25 μg of monosialogangliosides when quantitative analysis of the per-O,N-benzoylated derivatives is to be performed. If only qualitative information about molecular species present is desired, as little as 5 μg of ganglioside can be used. Improved resolution with increased retention times is obtained with slightly elevated temperature and less methanol in the mobile phase (inset, Fig. 6). Regulation of column and solvent temperature is critical for reproducible separations. It may be possible to further resolve the components of heterogeneous peaks by silica-bonded silver columns or by addition of a small percentage of acetonitrile to the mobile phase.[23] This methodology should be useful not only for the analysis of actual molecular species of gangliosides but also for the preparation of ganglioside species for physical, chemical, and radioisotope tracer studies. A preparative reversed-phase HPLC procedure for the isolation of molecular species of G_{M1} and G_{D1a} gangliosides containing a single long-chain base has also been described.[24,25]

Derivatized Glycosphingolipids: Neutral Glycosphingolipids and Ceramides. The molecular species of glucosyl- and galactosylceramides, lactosylceramides, or free ceramides can be separated and quantitated by the reversed-phase HPLC method. Before quantitative molecular species analysis, it is first necessary to benzoylate the lipids and separate them into individual classes of sphingolipids, including the hydroxy and nonhydroxy fatty acid types, by normal-phase HPLC as described above. Briefly, the glycolipid mixture containing each of about 10–100 μg NFA- and HFA-ceramides, galactocerebrosides, and ceramide lactosides is perbenzoylated with benzoyl chloride. These derivatives are then resolved on the pellicular (50 cm × 2.1 mm) column with the hexane–dioxane gradient previously described. Individual perbenzoylated lipid fractions are collected and the solvent is evaporated under nitrogen. The lipids are then redissolved in a minimal volume of methanol–dichloromethane (85 : 15) and suitable aliquots (about 2–5 μg) are injected into a C_{18}-reversed phase (3 μm) column (100 × 4.6 mm, Rainin's Microsorb-C_{18}-short one). The molecular species of individual perbenzoylated glycolipids are

---

[23] J. P. Roggero and S. V. Coen, *J. Liq. Chromatogr.* **4**, 1817 (1981).

[24] S. Sonnino, R. Ghidoni, G. Gazzotti, G. Kirschner, G. Galli, and G. Tettamanti, *J. Lipid Res.* **25**, 620 (1984).

[25] S. Sonnino, G. Kirschner, R. Ghidoni, D. Acquotti, and G. Tettamanti, *J. Lipid Res.* **26**, 248 (1985).

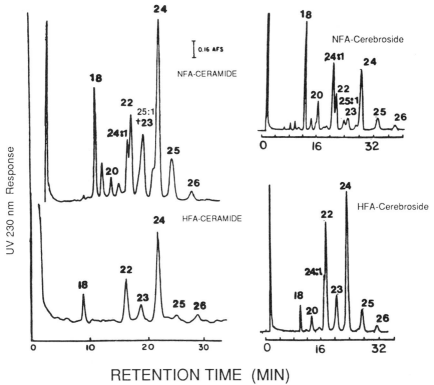

FIG. 7. Reversed-phase chromatography of per-O,N-benzoylated cerebrosides and ceramides. The perbenzoylated derivatives (2–5 µg) of cerebrosides or ceramides were dissolved in methanol–dichloromethane and aliquots (2–5 µl) injected onto a C_{18}-reversed-phase (3 µm particle) column (100 × 4.6 mm, Rainin's Microsorb-C_{18}-short one). The column was eluted with methanol–dichloromethane (85:15, v/v) at a flow rate of 1 ml/min and detection was at 230 nm.

resolved with an isocratic mobile phase of methanol–dichloromethane (85:15) and a flow rate of 1 ml/min. The detection is at 230 nm.

Representative chromatograms of bovine brain NFA- and HFA-ceramides and cerebrosides are shown in Fig. 7. The fatty acid components are listed near each peak. The separation is essentially based on the relative hydrophobic interaction of the lipid's fatty acid and sphingosine base moieties to the C_{18} arms with the reversed-phase material. Under the same chromatographic conditions, the more polar benzoylated cerebrosides are eluted earlier than the corresponding ceramides and more polar HFA species are eluted earlier than the corresponding NFA species. The percentage distribution of each molecular species can be calculated directly from the area under each peak. The absolute amounts of each molecular species also can be determined by comparing the peak areas of the test sample with those from a standard sample. Usually, we have used bovine brain NFA- and HFA-ceramides and cerebrosides with known fatty acid and sphingosine base composition as the standards. From the known total amount of benzoylated lipid standard injected, one can calculate the absolute amount in individual molecular species and these are compared with the unknown test material. Peak identification is reliable if the sphingosine base composition of a sample is homogeneous, i.e., mostly C_{18}-sphingenine as in the case of bovine brain cerebrosides and ceramides. If the sphingosine base composition is heterogeneous, i.e., a mixture of C_{18}, C_{20}-sphingenine, sphinganine, and phytosphingosine, a complex pattern with more unresolved molecular species peaks is obtained. In such cases, reliable identification becomes difficult and other methods such as GLC of the fatty acids and sphingosine bases of individual peaks or HPLC-MS methods as discussed in later sections should be used.

Neutral Glycosphingolipids. Molecular species of neutral glycosphingolipid have also been obtained by chromatography of their O-acetyl-N-p-nitrobenzoyl derivatives[26–28] by reversed-phase chromatography.

Gangliosides. The molecular species of per-O,N-benzoylated monosialogangliosides can be separated by reversed-phase HPLC. The separation of molecular species of per-O,N-benzoylated G_{M1}, L_{M1}, and G_{M3} on a C_{18} reversed-phase, Accupak (3 μm) column with methanol–dichloromethane (4 : 1) as the mobile phase has recently been reported.[29]

Analysis of Long-Chain Bases

Gas chromatographic methods for the analysis of long-chain bases of neutral glycosphingolipids and gangliosides, produced by aqueous acid methanolysis, have been very useful.[30–32] However, they have not provided resolution of all sphingoid isomers and hydrolysis products. Long-chain bases can be conveniently analyzed by HPLC as their stable N-biphenylcarbonyl derivatives which have an extinction coefficient of

---

[26] A. Suzuki, S. Handa, and T. Yamakawa, *J. Biochem. (Tokyo)* **82,** 1185 (1977).
[27] T. Yamazaki, A. Suzuki, S. Handa, and T. Yamakawa, *J. Biochem. (Tokyo)* **86,** 803 (1979).
[28] A. Suzuki, S. Handa, and T. Yamakawa, *J. Biochem. (Tokyo)* **80,** 1181 (1976).
[29] K. H. Chou, C. E. Nolan, and F. B. Jungalwala, *J. Neurochem.* **44,** 1898 (1985).
[30] C. C. Sweeley and E. A. Moscatelli, *J. Lipid Res.* **1,** 40 (1959).
[31] R. C. Gaver and C. C. Sweeley, *J. Am. Oil Chem. Soc.* **42,** 294 (1965).
[32] H. E. Carter and R. C. Gaver, *J. Lipid Res.* **8,** 391 (1967).

26,000 at 280 nm. These derivatives can be separated by normal-phase HPLC, which elutes all sphingoid bases together. This provides a measure of total long-chain bases in a sphingolipid sample. They can also be analyzed by reversed-phase HPLC which separates the most commonly occurring individual bases and side products formed during aqueous acid methanolysis. Gangliosides require more vigorous hydrolysis conditions for quantitative liberation of sphingoid bases than do other sphingolipids. These more vigorous conditions result in the production of O-methyl ethers that are not easily separated by HPLC techniques. To eliminate the production of O-methyl ethers, a procedure for hydrolysis of gangliosides in an aprotic solvent, acetonitrile, was developed.[33,34] The HPLC analysis of the long-chain bases as their biphenylcarbonyl derivatives is nondestructive, which allows the collection of separated components for further analysis or determination of radioactivity. As little as 0.1 μg (0.33 nmol) can be measured by the technique.

Procedures

Liberation and Extraction of Long-Chain Bases. Unique hydrolysis conditions are required to provide maximum yields of sphingoid bases from different sphingolipids. For maximum yields, sphingomyelins (500 μg or less) are hydrolyzed with 1 ml of 1 N HCl in water–methanol (67:33, v/v) at 70° for 16 hr, and cerebrosides (500 μg or less) with 0.5 ml of 3 N HCl in water–methanol (1:1) at 60° for 1.5 hr, while gangliosides require more vigorous conditions. Gangliosides (10–200 μg) are hydrolyzed with 0.3 ml of aqueous acetonitrile–HCl (0.5 N HCl and 4 M H$_2$O in acetonitrile) at 72° for 2 hr.[33,34] After hydrolysis, the sample is evaporated with nitrogen and the long-chain bases are partitioned between 1 ml of alkaline upper-phase, methanol–saline–chloroform (48:47:3) containing 0.05 N NaOH and 5 ml of a lower phase, consisting of chloroform–methanol–water (86:14:1). The lower phase is washed once with alkaline upper phase and then three times with neutral upper phase. The lower phase is evaporated to dryness and the residue is used for derivatization.

Derivatization and Reversed-Phase HPLC. Standard long-chain bases and bases liberated from sphingolipids are reacted with 50 μl of 1% biphenylcarbonyl chloride in tetrahydrofuran and 100 μl of 50% aqueous sodium acetate. The biphasic mixture is agitated for 2 hr at room temperature. Five milliliters of upper-phase solvent is then added and the mixture is shaken, centrifuged, and the upper phase is removed. The lower phase

---

[33] F. B. Jungalwala, J. E. Evans, E. Bremer, and R. H. McCluer, *J. Lipid Res.* **24**, 1380 (1983).

[34] H. Kadowaki, E. G. Bremer, J. E. Evans, F. B. Jungalwala, and R. H. McCluer, *J. Lipid Res.* **24**, 1389 (1983).

is washed twice with 1 ml of methanol–0.9% saline–chloroform–15 M ammonia (96:92:2:3) followed by three washes with upper phase without alkali. The lower phase is evaporated to dryness and dissolved in methanol for HPLC analysis.

The biphenylcarbonyl derivatives are separated and quantitated by reversed-phase chromatography on a column (2.5 cm × 4.6 mm) packed with 5-μm ODS particles (Ultrasphere-ODS, Altex). Two solvent systems, tetrahydrofuran–methanol–water (25:40:20) or methanol–water (94:6), have been used to separate ganglioside long-chain bases.[34] Chromatograms illustrating the HPLC separation of long-chain-base biphenylcarbonyl derivatives are given in the following section on HPLC mass spectrometry (see Fig. 10).

Comments. The use of tetrahydrofuran–methanol–water as a mobile phase for HPLC analysis has some advantages over methanol–water. For example, d20:0 is clearly resolved from the minor side products. Methanol–water, however, provides better resolution of the peaks with longer retention times. The derivatization of long-chain bases with biphenylcarbonyl chloride and HPLC analysis provides highly sensitive detection. Theoretically, 1-μg samples of brain gangliosides can be quantitatively analyzed if an internal standard is used to correct for sample loss during extraction and derivatization. C_{18}-Psychosine can be used as an internal standard since it elutes before the long-chain-base derivatives with minimal interference and since it contains galactose, it can be easily calibrated with standard sugar methods. C_{18}-Psychosine should be added to the sample obtained after hydrolysis and prior to derivatization. The amount added should be about equivalent to the expected amount of long-chain bases in the sample. If only relative percentage composition is desired, however, addition of psychosine is not necessary.

Phospholipids

Class Separations of Phospholipids by Normal-Phase HPLC

Native Phospholipids. The phospholipid classes can be separated on a silica HPLC column with online UV light detection at about 200 nm.[35,36] The absorption of low UV light for phospholipids is mainly due to double bonds in the fatty acid side-chains as well as functional groups such as carbonyl, carboxyl, phosphate, and amino groups. The response of the detector varies with the number of functional groups, primarily the num-

[35] F. B. Jungalwala, J. E. Evans, and R. H. McCluer, *Biochem. J.* **155**, 55 (1976).
[36] F. B. Jungalwala, S. Sanyal, and F. LeBaron, in "Phospholipids in the Nervous System" (L. Horrocks, G. B. Ansell, and G. Porcellati, eds.), Vol. 1, pp. 91–103. Raven, New York, 1982.

ber of double bonds. Thus, direct quantitation of phospholipids with an unknown degree of unsaturation is not possible with the ultraviolet detection mode. However, separated phospholipids can be collected and quantitated by independent micromethods.

Procedure. Phospholipids in the lower phase of tissue lipid extracts obtained by the procedure of Folch *et al.*[37] can be analyzed directly by HPLC if phospholipids are the major components. Otherwise, the phospholipids are separated from the nonpolar lipids and glycolipids on a Unisil column as described by Vance and Sweeley.[16] The phospholipid mixture (about 10–200 μg) is dissolved in 10–20 μl of acetonitrile–methanol–15 M ammonia (1:1:0.005, v/v) and injected onto a Micropak-Si-5 column (30 cm × 2.1 mm) (Varian). The elution is with a 20-min linear gradient of acetonitrile–methanol–water–15 M ammonia from 95:3:2:0.05 to 65:21:14:0.35, pumped at 2 ml/min. The detection is at 205 nm. After the phospholipids are eluted, the solvent gradient is reversed in 3 min linearly and, after the baseline is stabilized, the next sample can be injected.[35,36] HPLC separation of commonly occurring phospholipids is shown in Fig. 8.

Comments. Approximately 1 nmol of a phospholipid containing at least one double bond per molecule can be detected. This method has been used for the separation of phospholipids from a variety of tissues and cells. Almost baseline resolution is obtained for all phospholipids in natural mixtures. However, the separation of phosphatidylethanolamine and phosphatidylserine is at times variable if the amount of phosphatidylserine is relatively large. The individual phospholipids are collected from several chromatographic runs and the amount of each is quantitated by phosphorus analysis. The injected samples can be entirely recovered after chromatography for further analysis such as radioactivity, phosphate or fatty acid species determinations. However, there are certain drawbacks to this mode of detection. Direct quantitation of phospholipids is not possible, as discussed previously. Sensitivity of detection for the same molar amount of phospholipid mainly varies with the number of double bonds. Thus, disaturated fatty acid-containing phospholipids such as dipalmitoylglycerylphosphorylcholine, are detected with rather low sensitivity. Small amounts of UV-absorbing impurities can interfere with the analysis. Several variations of the HPLC method for the analysis of native phospholipids have been published.[38–42]

[37] J. Folch, M. Lees, and G. H. Sloan Stanley, *J. Biol. Chem.* **226**, 497 (1957).
[38] R. W. Gross and B. E. Sobel, *J. Chromatogr.* **197**, 79 (1980).
[39] V. L. Hanson, J. Y. Park, T. W. Osborn, and R. M. Kiral, *J. Chromatogr.* **205**, 393 (1981).
[40] J. R. Yandrasitz, G. Berry, and S. Segal, *J. Chromatogr.* **225**, 319 (1981).
[41] S. S. Chen and A. Y. Kou, *J. Chromatogr.* **227**, 25 (1982).
[42] T. L. Kaduce, K. C. Norton, and A. A. Spector, *J. Lipid Res.* **24**, 1398 (1983).

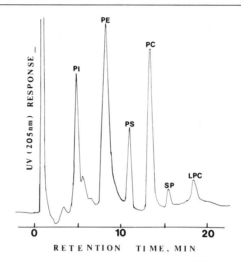

FIG. 8. HPLC separation of pig liver phosphatidylinositol (PI), bovine brain phosphatidylethanolamine (PE), bovine brain phosphatidylserine (PS), bovine brain sphingomyelin (SP), and lysophosphatidylcholine (LPC) on a Micropak-SI-5 μm column. The elution was with a 20-min linear gradient of acetonitrile–methanol–water–ammonium hydroxide (15 M) from 95:3:2:0.05 to 65:21:14:0.35 (v/v), pumped at 2 ml/min. The rise in the baseline at 205 nm was corrected by using a Schoffel memory module with the monitor. Reproduced from Ref. 36.

Phospholipid Derivatives. An alternate approach to quantitative analysis of phospholipids is to form UV-absorbing or fluorescent derivatives and analyze these by HPLC. Most of the phospholipids have functional groups available for such derivatization except for phosphatidylcholine. The inability to derivatize phosphatidylcholine precludes simultaneous analysis of all the phospholipids after derivatization. However, many phospholipids have been derivatized and quantitatively analyzed in the picomole range by HPLC.

Amino Phospholipids. HPLC of biphenylcarbonyl derivatives of amino group-containing phospholipids, such as phosphatidylethanolamine, phosphatidylserine, lysophosphatidylethanolamine, and lysophosphatidylserine with UV detection at 280 nm has been reported.[43,44] The amino phospholipids containing vinyl ether bonds (plasmalogens) can be determined separately from diacyl and alkylacyl phospholipids by

[43] R. H. McCluer and F. B. Jungalwala, in "Current Trends of Sphingolipidoses and Allied Disorders" (B. W. Volk and L. Schneck, eds.), p. 533. Plenum, New York, 1976.
[44] F. B. Jungalwala, R. J. Turel, J. E. Evans, and R. H. McCluer, *Biochem. J.* **145,** 517 (1975).

HPLC before and after treatment of the samples with HCl as done with the TLC method.[45] The lower limit of detection by HPLC of these lipids is about 10–15 pmol or 0.3–0.4 ng of phospholipid phosphorus.

Procedure. The procedure for the preparation of the biphenylcarbonyl derivatives of the amino phospholipids is the same as that for the long-chain bases described in the section on the analysis of long-chain bases. The phospholipid derivatives (1–100 ng) are separated and quantitated on a MicroPak SI-10 (50 cm × 2.5 mm) column (Varian) with a 10-min linear gradient of dichloromethane–methanol–15 M ammonia from 94:7:0.75 to 80:15:3 at 2 ml/min. Detection is at 280 nm.

Comments. A similar method for the analysis of amino group-containing phospholipids after formation of their fluorescent dimethylaminonaphthalene-5-sulfonyl derivatives has been described.[46] The sensitivity of the fluorescent method is similar to that reported for the biphenylcarbonyl group UV absorbance method.

Sphingomyelin, phosphatidylglycerol, and phosphatidylinositol can be derivatized with 10% benzoic anhydride in tetrahydrofuran with dimethylaminopyridine as catalyst to form benzoyl derivatives.[43,47] The derivatized phospholipids are then quantitatively analyzed by HPLC. Thus, derivatization of phospholipids permits sensitive quantitative HPLC analysis of some phospholipid classes; however, a universal method using this approach has not emerged.

Separation of Molecular Species of Phospholipids by Reversed-Phase HPLC

Recently, great interest in the analysis of the molecular species of individual phospholipids has been evident. Reversed-phase HPLC has been generally used to achieve molecular species separation. The important feature of the reversed-phase chromatographic processs is the magnitude of the hydrophobic interaction which is determined by the nonpolar contact area between the solute and the ligand of the stationary phase.[48]

Sphingomyelin

Native Sphingomyelins. The molecular species of sphingomyelin from bovine brain, sheep, and pig erythrocytes can be resolved into 10–12 separate peaks on a μBondaPak C_{18} (Waters) or Nucleosil 5 μm C_{18} (Merck) column with methanol–5 mM potassium phosphate buffer, pH

---

[45] L. A. Horrocks, *J. Lipid Res.* **9**, 469 (1968).
[46] S. S. Chen, A. Y. Kou, and H. Y. Chen, *J. Chromatogr.* **208**, 339 (1981).
[47] M. Smith, P. Monchamp, and F. B. Jungalwala, *J. Lipid Res.* **22**, 714 (1981).
[48] C. Horvath and W. Melander, *Chromatographia* **11**, 262 (1978).

7.4 (97:3, v/v), as the isocratic solvent with a flow rate of 1 ml/min and detection at 203–205 nm.[49] The separation is based on the number of carbon atoms and degree of unsaturation in the fatty acid and sphingoid side-chains of the various sphingomyelin species.

Benzoylated Sphingomyelins. The HPLC analysis of sphingomyelin species is made quantitative and more sensitive by separating the benzoylated sphingomyelins on a "fatty acid analysis" column from Waters Associates.[47] About 5 μg of benzoylated sphingomyelin is separated into various molecular species in 30 min with detection at 230 nm.

Procedure. Sphingomyelin (10–500 μg) is benzoylated with 0.05 ml of 10% benzoic anhydride in tetrahydrofuran at 70° for 2 hr. After the reaction, the solvent is evaporated under N_2 and the product is dissolved in 1 ml of chloroform. The sample is chromatographed on a silicic acid (200–500 mg) column with 10 ml each of chloroform, chloroform–methanol (8:1), and methanol, respectively. The methanol fraction contains the 3-O-benzoylated sphingomyelin. A suitable aliquot of the benzoylated sphingomyelin is injected onto a "fatty acid analysis" column (25 cm × 4.6 mm) from Water's Associates or on a Nucleosil 5 μm C_{18} (Merck) column (30 cm × 4.6 mm) with methanol–15 M ammonia (99:1) pumped at a flow rate of 2 ml/min.[47,50] Detection is at 230 nm.

Comments. Although good separation is achieved on the reversed-phase columns, sphingomyelins having a double bond in the acyl side-chain (e.g., 24:1) are not resolved from the lipid having a saturated fatty acid with two less carbon atoms (e.g., 22:0). This difficulty can be overcome by the combined use of argentation HPLC and reversed-phase HPLC.[47] First, separation based on the degree of unsaturation is achieved with a commercially prepared silica-bonded silver column on which 3-O-benzoylated sphingomyelin is resolved into two peaks. One contains the lipid with only saturated fatty acids whereas the other contains sphingomyelin with only the monounsaturated fatty acids. The two peaks separated on the silver column are collected and individually reinjected on the reversed-phase column to resolve all the possible molecular species of sphingomyelin.

Other Phospholipids

Comments. Separation of molecular species of other phospholipids has been also achieved by reversed-phase and argentation–HPLC. Smith and Jungalwala[51] reported the reversed-phase HPLC separation of phos-

---

[49] F. B. Jungalwala, V. Hayssen, J. M. Pasquini, and R. H. McCluer, *J. Lipid Res.* **20,** 579 (1979).
[50] F. N. LeBaron, S. Sanyal, and F. B. Jungalwala, *Neurochem. Res.* **6,** 1081 (1991).
[51] M. Smith and F. B. Jungalwala, *J. Lipid Res.* **22,** 697 (1981).

phatidylcholine from various sources. It was shown that, from the chromatographic behavior of each phosphatidylcholine species, one can determine the relative hydrophobicity of various molecular species. All molecular species are not resolved on the reversed-phase column. However, separation based on the degree of unsaturation can be achieved on a silver-coated silica gel HPLC column. Patton et al.[52] have also described a similar separation of molecular species of phosphatidylcholine, phosphatidylethanolamine, phosphatidylinositol, and phosphatidylserine by reversed-phase HPLC with detection at 205 nm. Hsieh et al.[53] reported preparation of phosphatidic acid dimethyl esters from phosphatidylcholine after enzymatic hydrolysis of phosphatidylcholine followed by esterification of the phosphatidic acid with diazomethane. The dimethyl esters of phosphatidic acid were then chromatographed on a reversed-phase column. Improved resolution using two Partisil-10 ODS columns in tandem was later reported for these derivatives.[54] Batley et al.[55] reported separation of diacylglycerol p-nitrobenzoates by reversed-phase HPLC. Phospholipids were first degraded by the action of phospholipase C to diacylglycerols which were then derivatized and quantitatively analyzed. The latter method is based on partial degradation of phospholipids and may not be suitable if phospholipids labeled with radioisotopes in the base or phosphorus moiety are to be analyzed. HPLC methods for the separation of alkenylacyl-, alkylacyl-, and diacylacetylglycerols derived from ethanolamine glycerophospholipids and for separation of the individual molecular species from each of the separated classes have been described by Nakagawa and Horrocks.[56]

Liquid Chromatography–Mass Spectrometric Methods (LC–MS)

The on-line combination of liquid chromatography with mass spectrometry (LC–MS) has attracted attention because of the success of the related technique of GC–MS. It is recognized that the combination of the versatility of liquid chromatography with the specificity and sensitivity of mass spectrometry will provide a procedure of enormous power. We have utilized the Finnigan polyimide moving belt interface for the LC–MS analysis of long-chain bases, glycosphingolipids, and phospholipids.[57,58]

[52] G. M. Patton, J. M. Fasulo, and S. J. Robins, *J. Chromatogr.* **23**, 190 (1982).
[53] J. Y. K. Hsieh, D. K. Welch, and J. G. Turcotte, *J. Chromatogr.* **208**, 398 (1981).
[54] J. Y. K. Hsieh, D. K. Welch, and J. G. Turcotte, *Lipids* **16**, 761 (1981).
[55] M. Batley, N. H. Packer, and J. W. Redmond, *J. Chromatogr.* **198**, 520 (1980).
[56] Y. Nakagawa and L. A. Horrocks, *J. Lipid Res.* **24**, 1268 (1983).
[57] F. B. Jungalwala, J. E. Evans, and R. H. McCluer, *J. Lipid Res.* **25**, 738 (1984).
[58] F. B. Jungalwala, J. E. Evans, H. Kadowaki, and R. H. McCluer, *J. Lipid Res.* **25**, 209 (1984).

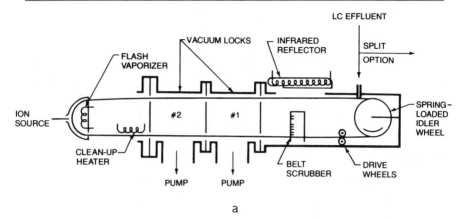

Fig. 9. Simplified drawing of LC–MS interface. Reproduced from the Finnigan catalog.

With this interface (see Fig. 9), the HPLC eluate is applied directly onto a moving polyimide belt which proceeds through vacuum chambers where most of the chromatographic solvent is removed. The residue on the belt is transported into an evaporator chamber adjacent to the ion source where it is rapidly thermally desorbed.

General LC–MS Techniques

A common problem for all LC–MS methods which utilize a moving belt is the maintenance of chromatographic resolution during transport of the sample into the ion chamber of the mass spectrometer. Karger and co-workers[59] have developed a hot-gas nebulizer which provides uniform application of the solvent on the belt and maintains good chromatographic resolution. We have utilized a fine liquid stream, generated with a small orifice, to apply the chromatographic eluate as an even coating and to maintain resolution. This form of application performs well until the water composition of the mobile phase exceeds about 50%. At this water concentration, beads begin to form on the imide belt and resolution is degraded. The chromatographic effluent is formed in the stream by inserting a 10-cm section of polyimide-coated fused silica capillary GC column (0.25 mm i.d.), pulled to a fine orifice, into the bore of the 0.009 inch i.d. × 1/16 inch Teflon tube from the HPLC column. A 1/16-inch Swagelok fitting is then tightened over the sleeved portion of the tubing to provide a compression seal. A flow rate of 0.2 ml/min usually results in streaming,

[59] E. P. Lankmayer, M. J. Hayes, B. L. Karger, P. Vourus, and J. M. McGuire, *Int. J. Mass. Spectrom. Ion Phys.* **46**, 177 (1983).

depending on solvent composition and exact size of the fine orifice. The capillary orifice is positioned about 1–2 cm above the moving belt so that the stream impacts the middle of the belt.

Long-Chain Bases. A reversed-phase HPLC method for the analysis of long-chain bases as their biphenylcarbonyl derivatives has been devised[33,34] as described above. During the hydrolysis of sphingolipids, several unidentified side products are formed, particularly from gangliosides which require more vigorous conditions for quantitative liberation of the long-chain bases. These side products are not completely resolved after HPLC analysis and lead to uncertainty of peak identification. The LC–MS method for the analysis of long-chain bases which utilizes chemical ionization mass spectrometry provides on-line analysis of the components.[58]

Procedure. Sphingolipids are hydrolyzed and the long-chain bases derivatized with biphenylcarbonyl chloride as described above in the section on HPLC analysis of long-chain bases. Five to 20 μg of the biphenylcarbonyl derivatives is injected on an Accupak ODS (3 μm) column with methanol–water (94:6, v/v) as the mobile phase pumped at 0.4 ml/min. The eluant is monitored with an UV monitor set at 280 nm and then streamed onto the moving belt of the LC–MS interface (described above) with a Finnigan 4000 mass spectrometer in the chemical ionization (CI) mode. The flash evaporizer temperature is set at 270° and the clean-up heater at 240°. The drive chamber heater is used at 90% of the maximum to aid evaporation of the solvent. Ammonia (0.3 Torr) is used as the reagent gas with an ion source temperature of 220°. Mass spectra are acquired under the control of a Teknivent model 56K data system. The multiple products observed by UV detection are identified directly by their CI-mass spectra obtained as they are eluted from the HPLC column. Examples of mass spectra and specific ion plots are shown in Fig. 10. The characteristic major ions in the ammonia CI–MS of the biphenylcarbonyl derivatives are given in Table II.

Interpretation of Data. In the ammonia CI mass spectra of the long-chain bases, the base ion is a m/z 198 or 215 corresponding to the adduct formation of the biphenylcarbonyl group with ammonia. The other major ions found in the spectra are $[M + H]^+$, $[M - OH]^+$, and $[M - CH_2OH]^+$. Other ions common to all the sphingoids are seen at m/z 222, 224, and 240. The m/z 240 ion is apparently produced by cleavage between carbons 2 and 3 of the long-chain base and the 224 ion is produced by the loss of an oxygen from the m/z 240 ion. The m/z 222 may result from the loss of two protons from the m/z 224 ion. Ions characteristic of specific bases are $[M - 197]^+$, $[M - 213]^+$, $[M - 240 + 17]^+$, and $[M - 227]^+$. The $[M - 187]^+$ is produced by the loss of biphenyl-$CONH_2$, $[M - 213]^+$ is produced by

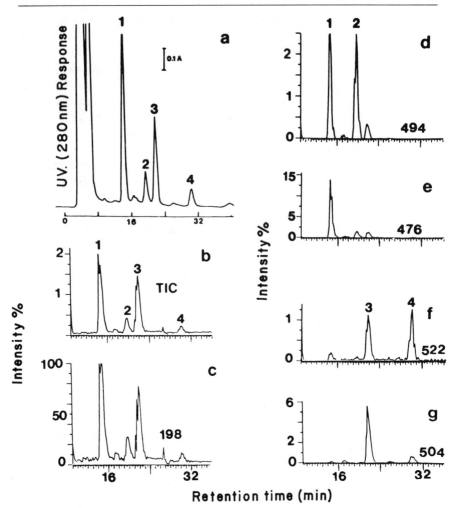

FIG. 10. HPLC–MS of biphenylcarbonyl derivatives of an O-methylsphingoid fraction obtained from brain gangliosides. Ten micrograms of the sphingoid derivatives was analyzed by HPLC–MS in ammonia CI mode. a, UV chromatogram; b, reconstructed plot of total ion current (TIC). Reconstructed plot of specific ions monitored at c, m/z 198; d, m/z 494; e, m/z 476; f, m/z 522; and g, m/z 504. In a, the biphenylcarbonyl derivatives are peak 1, 5-O-methyl-C_{18}-sphingenine; peak 2, 3-O-methyl-C_{18}-sphingenine; peak 3, 5-O-methyl-C_{20}-sphingenine, and peak 4, 3-O-methyl-C_{20}-sphingenine. Other minor peaks were unidentified. Reproduced from Ref. 58.

the further loss of an oxygen from $[M - 197]^+$, and $[M - 227]^+$ is probably produced by loss of CH_2OH from $[M - 196]^+$. These ions are useful in the determination of the structures of the long-chain bases and their products.

Comments. The solvent system, 6% aqueous methanol, used for the

TABLE II
CHARACTERISTIC MAJOR IONS AND THEIR RELATIVE INTENSITY IN AMMONIA CI–MS OF BIPHENYLCARBONYL DERIVATIVES OF SPHINGOID BASES[a]

Ion	C_{18}-Sphingenine	C_{18}-Sphinganine	C_{18}-Hydroxy-sphinganine[b]	C_{20}-Sphingenine	C_{18}-3-OCH$_3$-Sphingenine	C_{18}-5-OCH$_3$-Sphingenine	C_{20}-3-OCH$_3$-Sphingenine	C_{20}-5-OCH$_3$-Sphingenine
M$^+$ + 1	480 (75)	482 (40)	498 (15)	508 (10)	494 (26)	494 (6)	522 (9)	522 (5)
M$^+$ − 17	462 (78)	464 (36)	480 (23)	490 (23)	476 (3)	476 (15)	504 (2)	504 (11)
M$^+$ − 31	448 (8)	450 (5)	466 (2)	476 (1)	462 (8)	462 (58)	490 (3)	490 (16)
M$^+$ − 35	444 (18)	446 (8)	462 (11)	472 (3)	444[c] (22)	444[c] (10)	472[c] (3)	472[c] (3)
M$^+$ − 197	282 (12)	284 (10)	300 (10)	310 (9)	296 (4)	296 (4)	324 (4)	324 (8)
M$^+$ − 213	266 (8)	268 (8)	284 (3)	294 (7)	280 (6)	280 (5)	308 (7)	308 (5)
M$^+$ − 240 + 17	256 (22)	258 (2)	274 (10)	284 (45)	270 (4)		298 (2)	298 (2)
M$^+$ − 240					253 (12)	253 (3)	281 (12)	281 (3)
M$^+$ − 227	252 (14)	254 (5)	270 (5)	280 (14)	266 (10)	266 (8)	294 (6)	294 (6)
240	(32)	(22)	(61)	(54)	(22)	(4)	(11)	(10)
198	(100)	(100)	(43)	(85)	(68)	(100)	(100)	(100)
198 + 17	(25)	(18)	(100)	(100)	(100)	(17)	(79)	(85)

[a] Numbers in parentheses represent percent relative intensities to the base ion m/z 198 or 215 which are produced as due to biphenylcarbonyl-ammonia adducts (181 + 17). Reproduced from Ref. 58.
[b] Phytosphingosine.
[c] m/z 444 and 472 are M$^+$ − 17-CH$_3$OH.

HPLC separation flows evenly onto the moving belt with the streaming orifice, and the evaporation is complete at a flow rate of up to about 0.6 ml/min so that HPLC resolution is not lost during application, and mass spectra of good quality can be obtained under conditions of CI–MS. However, at a higher flow rate or with an eluant with higher water content, the evaporation is not complete and this interferes with the mass spectrometer function. If higher flow rates and/or use of more aqueous solvents are necessary, then one may consider using the solvent nebulizer device reported by Hayes et al.,[60] or employ a stream splitter to limit the flow of the eluant onto the belt. The specific ion monitoring and analysis of mass spectra of the HPLC resolved peaks allow direct characterization of the complex mixtures of long-chain base components and their products formed during hydrolysis.

Phospholipids. As discussed previously, although significant improvement in the separation of individual phospholipids and molecular species of phospholipids have been achieved by HPLC, facile quantitative detection of these lipids is not quite satisfactory.

An LC–MS procedure for separation of standard phospholipid mixtures was developed after testing a variety of solvents and columns.[57] The MS monitoring is most satisfactory with ammonia as a reagent gas in the positive-ion mode.

Procedure. The separation of phospholipids (5–10 μg of each phospholipid) is achieved on a Brownlee silica gel (5 μm) cartridge column (2 × 60 mm). The column is developed with a linear gradient of solvent A, dichloromethane–methanol–water (93 : 6.5 : 0.5, v/v), and B, dichloromethane–methanol–water–15 M ammonia (65 : 31 : 4 : 0.2, v/v), starting at 88% A + 12% B to 55% A + 45% B in 10 min and programming to 100% B in 2 min at a flow rate of 0.8 ml/min. The eluate is applied in a fine stream to the moving belt of a Finnigan HPLC–MS interface. The solvent is removed by heating the belt at 330° under vacuum and the phospholipid is introduced into the ion source (150°) of the mass spectrometer. Positive-ion mass spectra are continuously collected in the CI mode with ammonia as the reagent gas from m/z 100–900, every 7 sec, under the control of Teknivent model 56K MS data system. Under these conditions fairly good reproducibility is achieved. All the phospholipids are well resolved except phosphatidylserine, which tails to some extent into the lysophosphatidylinositol peak. A reconstructed plot of the total ion current after HPLC–MS analysis of rat brain phospholipids (80 μg) is shown in Fig. 11a.

Interpretation of Data. Major characteristic fragment ions in the am-

---

[60] M. J. Hayes, E. P. Lankmayer, P. Vouros, B. L. Karger, and J. M. McGiure, *Anal. Chem.* **55**, 1745 (1983).

TABLE III
Major Characteristic Fragment Ions in Positive Ammonia Chemical Ionization–Mass Spectra of Various Phospholipids[a]

Phospholipid	Ion	Phospholipid	Ion
PC	142 (from phosphocholine) $[M + 1]^+$ $M^+ - 41$ $M^+ - 182$ $[M + 35]^+ - 182$	SP	142 (from phosphocholine) $M^+ - 182$ $M^+ - 182 - 18$
		Lyso-PE	141 (ethanolamine phosphate) $M^+ - 140$ $[M + 35]^+ - 140$ $M^+ - 43$ $M^+ - 17$@@
PE	141 (ethanolamine phosphate) $M^+ - 140$ $[M + 35]^+ - 140$ $[M + 17]^+ - 140$		
PS	105 (serine) $M^+ - 184$ $[M + 35]^+ - 184$	Lyso-PC	142 (from phosphocholine) $M^+ - 182$ $[M + 35]^+ - 182$ $M^+ - 31$ $M^+ - 41$
PI	180 (inositol) 198 (inositol + 18) $M^+ - 259$ $[M + 35]^+ - 259$	PG	172 (glycerol phosphate) $M^+ - 171$ $[M + 35]^+ - 171$

[a] Reproduced from Ref. 57.

monia CI–positive-ion mass spectra of various phospholipids are given in Table III. The specific ion plots for m/z 105, 141, and 142 are given in Fig. 11b–d. These ions are fairly specific for the individual phospholipid bases. Thus m/z 105, specific for phosphatidylserine, is not found to be associated with other phospholipids. Similarly, m/z 141 is relatively specific for ethanolamine-containing phospholipids, whereas m/z 142 is specific for choline-containing phospholipids and m/z 198 is found to be associated only with inositol-containing phospholipids (not shown).

The data also show that individual phospholipids are chromatographically resolved to some extent based on molecular species. The ethanolamine-containing phospholipids from rat brain are resolved into two separate peaks (Fig. 11c). The mass spectral analysis of the earlier peak showed that it contained mostly alkenylacyl-GPE, m/z 561 (Fig. 11e), 587 (Fig. 11g), and 589 corresponding to plasmalogens with 16:1–18:1 (33%), 18:1–18:2 (36%), and 18:2–18:2 (31%), respectively. The front peak,

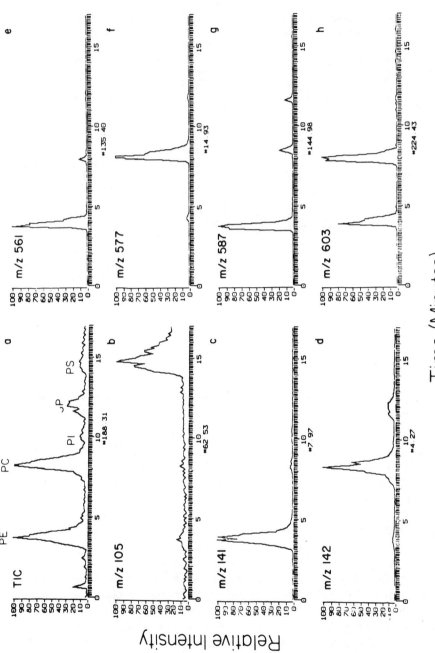

however, also contained some diacyl-GPE, m/z 623, 627, and 651 corresponding to 16:0–22:6 (20%), 18:0–20:4 (28%), and 18:0–22:6 (19%), respectively. The later eluting peak contained mostly diacyl-GPE with 18:0–18:1 (m/z 605, 15%) and 18:0–18:2 (m/z 603, 19%, Fig. 11h). The ratio of alkenylacyl GPE to diacyl GPE was 58:42. These results agree well with previously published analyses.[44,61]

The rat brain phosphatidylcholine peak also split into two major peaks, the earlier peak contained mostly long-chain fatty acid-containing species, whereas the latter contained short-chain fatty acid species. From the relative intensity of the diglyceride ions of phosphatidylcholine it was calculated that of the total rat brain phosphatidylcholine, 5.4% was ether-containing phosphatidylcholine, mostly 16:1–18:0 (m/z 563, 27%) and 18:0–20:4 (m/z 613, 25%) containing species. The major diacyl-containing phosphatidylcholines were 16:0–18:1 (m/z 577, 43%, Fig. 11f), 16:0–16:0 (m/z 551, 19%), 18:0–18:1 (m/z 605, 11%), 16:0–18:0 (m/z 579, 9%), 16:0–16:1 (m/z 549, 6%), and 14:0–16:0 (m/z 523, 6%). Eight other minor species were also recognized in the mass spectrum. It is surprising to note that rat brain phosphatidylcholine contained several bis-saturated fatty acid-containing species in large amounts. Saturated species represented 35% of the total diacylphosphatidylcholine. The percentage composition of individual fatty acids in phosphatidylcholine was calculated and found to be in good agreement with the previously reported amounts.[62]

The amount of phosphatidylinositol in rat brain phospholipids is very small (about 4%). The specific ion m/z 198, however, clearly identified the peak as phosphatidylinositol. The relative intensity of the $[M + 35]^+$ − 259 ions indicated that the major species of rat brain phosphatidylinositol are 18:0–20:4 (m/z 662 and 627, 90%), 18:0–20:3 (m/z 664, 5%), and 18:0–22:5 (m/z 688, 5%).

The rat brain sphingomyelin is resolved into two major peaks. The

[61] W. T. Norton and S. E. Poduslo, *J. Neurochem.* **21**, 759 (1973).
[62] D. A. White, in "Form and Function of Phospholipids" (G. B. Ansell, J. N. Hawthorne, and R. M. C. Dawson, eds.), p. 441. Elsevier, Amsterdam, 1973.

FIG. 11. Reconstructed plots of total ion current (a) and various specific ions (b–h) monitored after HPLC–CI–MS of rat brain phospholipids (80 μg) on a Brownlee silica gel (5 μm) cartridge HPLC column and eluted as described in the text. The eluate was applied to the moving belt of a Finnigan HPLC–MS interface. The solvent was removed by heating the belt at 330° under vacuum and the phospholipid was introduced into the ion source (150°) of the mass spectrometer. Positive-ion mass spectra were continuously collected in the CI mode with ammonia as the reagent gas from m/z 100–900, every 7 sec, under the control of Teknivent model 56K MS data system. Reproduced from Ref. 57.

front peak is mostly sphingomyelin with C_{18}-sphingenine and 22:0 (m/z 586 and 604, 4%), 24:1 (m/z 612 and 630, 12%), and 24:0 (m/z 614 and 632, 6%) fatty acids. The later peak contained sphingomyelin with C_{18}-sphingenine and 16:0 (m/z 502 and 520, 5%), 18:0 (m/z 530 and 548, 54%), 20:0 (m/z 558 and 576, 5%) fatty acids, and also C_{18}-sphinganine with 16:0 (m/z 522, 1%) and 18:0 (m/z 532 and 550, 10%) fatty acids. Small amounts (<3%) of other minor species of sphingomyelin can also be identified.

The last peak in the chromatogram is phosphatidylserine (m/z 105). The major molecular species of rat brain phosphatidylserine was identified as 18:0–18:1 fatty acid-containing phosphatidylserine.

Comments. Coupling of MS with HPLC offers an important advance in analytical methodology. The mass spectrometer not only functions as a highly sensitive universal detector but also serves to provide valuable structural information not obtainable otherwise on small quantities of compounds. A transport type of interface has been found quite suitable for removal of HPLC solvent and introduction of the residue on the belt into the mass spectrometer. Previously it has been demonstrated that moving belt-type interface works well for such a purpose and that peak broadening is minimal.[57,60] Although, theoretically, any solvent could be used for HPLC analysis in this system, practical considerations require that solvents with high volatility be utilized to facilitate evaporation from the belt. Naturally occurring phospholipids are a complex mixture of compounds having similar polarity. Thus separation of phospholipids on a chromatographic column is difficult. Separation is also hampered by the existence of various salt forms of the same phospholipids or a mixture of salts and free bases. In our experience the HPLC of phospholipids on silica columns is best achieved by the use of solvents containing ammonia. The same separation is consistently obtained providing the solvent is well equilibrated with the column. Most of the commonly occurring phospholipids are well-resolved; however, the separation could be improved, especially for phosphatidylserine, which tails in this system, and to accommodate the resolution of other minor phospholipids in natural mixtures, such as phosphatidic acid, phosphatidylglycerol, and lysophospholipids.

Quantitative analysis of individual phospholipids by HPLC–MS technique is yet to be developed. However, this should not be difficult since each phospholipid produces specific ions in the low mass range. We have also not explored the limits of sensitivity of this system; however, we routinely inject about 5 µg of individual phospholipid for complete HPLC–MS analysis. The sensitivity can easily be increased 5-fold to obtain reliable information of the molecular species of individual phospholipids. If one is just interested in knowing the amount of phospholipid in a mixture, specific ion monitoring in the low mass range for individual phospholipid should provide detection capability at subnanogram levels.

In summary, HPLC–CI–MS overcomes the limitations of detection method encountered in HPLC. Simultaneously, the method provides extensive information on the molecular structure of each phospholipid and on the relative abundance of each molecular species of an individual class of phospholipids. Such information is obtained by this method in a few minutes. By conventional methods, such analysis would require separation of large quantities of individual phospholipids by TLC or another chromatographic method, enzymatic and chemical degradation of the individual phospholipids, and gas chromatography.

Acknowledgments

This work was supported by USPHS grants MS 16447, NS 24405, CA 16853, NS 10437, and HD 05515 and by the Veterans Administration. HPLC–MS results were obtained in collaboration with Mr. J. E. Evans.

[31] Circular Dichroism for Determining Secondary Structure and State of Aggregation of Membrane Proteins

By MAARTEN P. HEYN

Secondary Structure

The determination of the secondary structure of water-soluble proteins by means of far-UV circular dichroism (CD) spectra is a widely used method. These spectra, in the range from 185 to 240 nm, are mainly due to the identical peptide group chromophores of the protein backbone and are quite sensitive to the backbone conformation. The measured ellipticities are converted into mean residue ellipticities and a least-squares fit is made to the spectrum with a linear combination of reference spectra for the α-helix, β-sheet, β-turn, and random coil states. The coefficients directly provide the percentages of the various secondary structures present. Some of the possible errors and ambiguities inherent in this approach are: (1) the results depend on the choice of the basis set of reference spectra; (2) the results are sensitive to the wavelength range over which the fit is carried out, and in particular to the value of the lower wavelength limit; (3) the results depend on the proper calibration of the spectrometer; and (4) significant errors are caused by incorrect determinations of the protein concentration.

Of all the secondary structures, the percentage of α-helix can be determined with the greatest accuracy. A reliable value for the percentage of β-sheet requires data down to at least 190 nm.

When these methods were first applied to membrane proteins, three additional problems became apparent: (1) The standard reference sets for water-soluble proteins may be inappropriate for intrinsic membrane proteins. (2) Light scattering may lead to distortions of the spectra. (3) Membranes have very high local concentrations of chromophores, giving rise to an inhomogeneous distribution of absorbers and an apparent reduction in absorption (Duysens absorption flattening). The second and third problems have recently been discussed in detail.[1,2] Some of the experimental manifestations of these problems are abnormally high or low ellipticities, CD signals that vary with the acceptor angle of the detector, and CD at wavelengths where no absorption occurs. Commercial CD spectrometers determine the extinction coefficient by measuring the intensity of the transmitted light. Since both absorption and scattering lead to a reduction of the incident light intensity, both contribute to the apparent extinction. Light scattering will contribute in this way to the CD spectrum only if it differs for left and right circularly polarized light. This is often the case for large chiral molecules or assemblies of molecules. The circular differential light scattering depends on the angle between the directions of the incident and scattered light. Therefore when only some of the scattered light is collected by moving the photomultiplier closer to the sample cell, the contribution of differential light scattering to the apparent CD signal may either increase or decrease. Whether this procedure will lead to a reduction or an increase in the light-scattering artifacts in a CD spectrum will in fact depend on the detailed angular dependence of the circular differential scattering of the system under study. However, when all of the scattered light is collected so that all the light that is not absorbed reaches the detector, the true CD spectrum will be recorded. This can be achieved by using fluorescence-detected circular dichroism (FDCD): a nonchiral fluorescent molecule is added directly to the sample solution, providing a 4π steradian detector that collects all of the scattered and transmitted light.[1,3] The problems associated with circular differential scattering can thus be solved experimentally. For optically active particles, circular differential flattening will occur and the reduction of the CD

---

[1] I. Tinoco, Jr., M. F. Maestre, and C. Bustamente, *Trends Biochem. Sci.* **8**, 41 (1983).
[2] B. A. Wallace and D. Mao, *Anal. Biochem.* **142**, 317 (1984).
[3] C. Reich, M. F. Maestre, S. Edmonson, and D. M. Gray, *Biochemistry* **19**, 5208 (1980).

signal will be largest at wavelengths corresponding to the absorption maxima.[2] For proteins this means that the CD extrema around 195 and 208 nm will be reduced more than the peak around 222 nm. It seems that these spectral distortions may be largely abolished by eliminating the close packing of the chromophores and dispersing them uniformly. This may be achieved by solubilizing the membrane proteins to the state of monomers or by reconstituting them in small lipid–protein vesicles at very high lipid-to-protein ratios,[4] provided that the protein secondary structure is unaltered by these manipulations. This last condition is very difficult to verify and can at best be made plausible by indirect arguments.

Bacteriorhodopsin (bR) has been a favorite test case for these methods, since secondary structure in the purple membrane is believed to be 70–80% α-helical, based on diffraction results showing seven α-helices spanning the membrane, and since methods for its solubilization to the monomer state and for reconstitution into vesicles are available. For bR in native purple membranes, FDCD measurements have shown that there are virtually no differences between the normal CD spectrum and the FDCD spectrum.[5] The conclusion is that circular differential light scattering effects are negligible for this membrane system. Purple membranes have average diameters of 0.5 μm. A substantial number of far-UV CD measurements have been performed with bR in purple membranes,[4-8] in sonicated purple membranes,[4,7,8] in detergent micelles,[4,5,9-11] and with bR reconstituted in DMPC vesicles.[4,8] The values of the percentage of α-helix vary over a wide range from a low of 38% to a high of 83%. From an analysis of the causes for these variations, it is apparent that all of the four sources of errors mentioned for water-soluble proteins contribute. In addition, there are problems concerning the question whether the sum of the coefficients should be constrained to be one and the associated renormalization.[4,12] For purple membrane suspensions, the percentage of α-helix is lowest and the fit is worst. For bR incorporated in small unilamellar vesicles the percentage is highest and the quality of the fit is quite good.

---

[4] D. Mao and B. A. Wallace, *Biochemistry* **23**, 2667 (1984).
[5] B. K. Jap, M. F. Maestre, S. B. Hayward, and R. M. Glaeser, *Biophys. J.* **43**, 81 (1983).
[6] B. Becher and J. Y. Cassim, *Biophys. J.* **16**, 1183 (1976).
[7] M. M. Long, D. W. Urry, and W. Stoeckenius, *Biochem. Biophys. Res. Commun.* **75**, 725 (1977).
[8] E. Nabedryk, A. M. Bardin, and J. Breton, *Biophys. J.* **48**, 873 (1985).
[9] J. A. Reynolds and W. Stoeckenius, *Proc. Natl. Acad. Sci. U.S.A.* **74**, 2803 (1977).
[10] K.-S. Huang, H. Bayley, M.-J. Liao, E. London, and H. G. Khorana, *J. Biol. Chem.* **256**, 3802 (1981).
[11] N. A. Dencher and M. P. Heyn, *FEBS Lett.* **96**, 322 (1978).
[12] R. M. Glaeser and B. K. Jap, *Biochemistry* **24**, 6398 (1985).

Going from purple membrane to detergent-solubilized micelles to vesicles a trend in spectral shape is discernible with the peak at 208 nm gaining relative to the peak at 222 nm. Using these spectral data and a theoretical analysis,[13] it has recently been argued that differential flattening effects can be ruled out for bR in purple membranes.[12] The question whether the secondary structure of bR is unaltered in detergent micelles and vesicles remains unanswered. Small conformational changes are hard to rule out in going from the aggregated state in the purple membrane to the monomeric state. In fact, for both detergent-solubilized bR and bR in DMPC vesicles above T_c, a reduction has been observed in the extinction coefficient and λ_{max} in the visible absorption spectrum and there is a reduction in the extent of light–dark adaptation; these effects appear to be a result of the elimination of the protein–protein interactions and are characteristic for the monomeric state.[11,14,15] The studies finding a low percentage of α-helix in the purple membrane conclude that there must be a substantial amount of β-sheet conformation.[5,12] Infrared measurements in the amide I band also indicate the presence of some β-sheet component.[5,8,16,17]

In conclusion, it appears that substantial progress has been made in understanding the sources of artifacts in the far-UV CD spectra of membrane proteins. At present a combination of FDCD and incorporation in small vesicles offers the best way to obtain reliable secondary structure estimates for membrane proteins.

State of Aggregation

Chromophore–chromophore interactions are well known in the absorption and CD spectra of proteins and nucleic acids. When the identical peptide chromophores of the protein backbone are regularly arranged such as in an α-helix, exciton coupling effects are observed in the far-UV which are particularly evident in the CD spectra and which are used for secondary structure determination. Likewise, when the nearly identical nucleic acid bases stack in the regular structure of the double helix, pronounced exciton coupling effects are observed in the CD spectra which depend on the detailed stacking geometry and on the helix sense. Similar effects may be expected in membranes provided the membrane proteins

---

[13] D. J. Gordon and G. Holzwarth, *Arch. Biochem. Biophys.* **142**, 481 (1971).
[14] N. A. Dencher, K.-D. Kohl, and M. P. Heyn, *Biochemistry* **22**, 1323 (1983).
[15] R. J. Cherry, U. Müller, R. Henderson, and M. P. Heyn, *J. Mol. Biol.* **121**, 283 (1978).
[16] N. W. Downer, T. J. Bruchman, and J. H. Hazzard, *J. Biol. Chem.* **261**, 3640 (1986).
[17] D. C. Lee, J. A. Hayward, C. J. Restall, and D. Chapman, *Biochemistry* **24**, 4364 (1985).

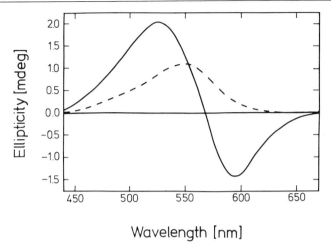

FIG. 1. CD spectra of a suspension of bR–DMPC vesicles at 0.6° (——) when bR is in the hexagonally aggregated state, and at 35° (----) when bR is monomeric. Molar phospholipid-to-bR ratio is 1:117. The vesicles were suspended in 0.1 M sodium acetate (pH 5.0).

have appropriate chromophores and are arranged in well-defined aggregates of fixed geometry. The first example was discovered[18] in the visible CD spectra of the purple membrane of *Halobacterium halobium,* which consists of a two-dimensional lattice of bR molecules, each containing as the chromophore one molecule of retinal. A typical CD spectrum of hexagonally arranged bR is shown in Fig. 1. Whereas the absorption spectrum of bR appears to consist of a single broad band centered at 568 nm in the light-adapted state, the corresponding CD spectrum has a positive and a negative lobe with zero crossover around 574 nm. The pair of bands is also asymmetric, the positive peak being larger than the negative one (nonconservative spectrum). The initial interpretation of this CD spectrum was[18] that it represents a superposition of a conservative exciton couplet due to retinal–retinal interactions, with equal amplitudes of positive and negative lobes and with a crossover near 568 nm and a positive monomer CD band with the shape of the absorption spectrum which is due to the interaction of the chromophore with the opsin to which it is bound. In this way, both the amplitude asymmetry and the red-shifted position of the crossover wavelength could be explained.[18] A substantial

[18] M. P. Heyn, P.-J. Bauer, and N. A. Dencher, *Biochem. Biophys. Res. Commun.* **67,** 897 (1975).

amount of evidence has accumulated in favor of this explanation:

1. Taking the lattice apart by solubilizing to the state of monomers should make the coupling effects disappear and only the monomer band should remain. For bR solubilized in Triton X-100 and octylglucoside this is indeed what is observed.[11,18,19] In both detergents it could be shown by analytical ultracentrifugation[9] and gel filtration[11] that the bR–lipid–detergent micelles contain only one bR. The CD spectra are routinely used to monitor the solubilization process of bR.

2. A second way to unmask the monomer band is to use apomembranes.[19,20] It is possible to reconstitute the chromophore-free apomembrane by the successive addition of aliquots of all-*trans*-retinal. In the initial stages of such a reconstitution experiment, only single isolated retinal binding sites will be occupied and one expects to observe only the positive band. The coupling bands, which are due to the interaction between pairs of neighboring chromophores, will make their appearance at a later stage in the titration and are expected to increase in amplitude quadratically with the percentage reconstitution. The observed spectra are in good quantitative agreement with these considerations.

3. A third way to expose the underlying monomer band is by using oriented films of purple membrane. When the direction of propagation of the incident light is perpendicular to the planes of the membranes and thus to the vectors connecting interacting chromophores, the exciton coupling bands are expected to vanish and this is indeed observed.[21] A positive CD band centered around the absorption maximum remained.[21] This monomer band is very sensitive to the local environment of the chromophore and to external factors such as pH and solvent.

4. At low pH a blue form develops with absorption maximum around 605 nm. In a coordinated fashion both CD peaks shift to the red.[22] The negative lobe of the exciton couplet now has the largest amplitude. This could be explained if the monomer band at low pH were negative. This was indeed observed using oriented low pH films.[22]

5. Addition of glycerol and polyhydric alcohols leaves the absorption spectrum unchanged, but leads to a blue shift of the crossover wavelength and an ellipticity ratio of the positive and negative lobes below one.[23] This effect would be in perfect agreement with the superposition explanation if the monomer band in glycerol were to be negative. Using oriented films in

---

[19] B. Becher and T. G. Ebrey, *Biochem. Biophys. Res. Commun.* **69**, 1 (1976).
[20] P.-J. Bauer, N. A. Dencher, and M. P. Heyn, *Biophys. Struct. Mech.* **2**, 79 (1976).
[21] D. D. Muccio and J. Y. Cassim, *Biophys. J.* **26**, 427 (1979).
[22] D. D. Muccio and J. Y. Cassim, *J. Mol. Biol.* **135**, 595 (1979).
[23] J. E. Draheim and J. Y. Cassim, *Biophys. J.* **41**, 331a (1983).

glycerol, this was shown to be the case.[23] Previous reports that the asymmetry of the positive and negative lobes was largely due to light-scattering artifacts[24] (since index of refraction matching with glycerol reduced the asymmetry[24]) and that the true CD spectrum was in fact conservative with a very small monomer amplitude,[25] appear to be incorrect. Rather, glycerol affects the sign of the monomer band by changing the retinal–opsin interaction.

6. The exciton splitting is expected to be quite small since no wavelength dependence of the linear dichroism in the 568-nm band was observed[26] and since the absorption maxima of oriented film and solution are almost the same.[21] With a splitting of the order of 5 nm, the exciton coupling spectrum is expected to be proportional to the derivative of the absorption spectrum. The wavelength position of the two peaks is thus determined by the bandwidth of the absorption spectrum rather than by the much smaller exciton splitting. The predicted apparent splitting of the couplet is in reasonable agreement with the experimental value.[18,20]

7. Organic solvents, such as DMSO and diethyl ether, and volatile anesthetics cause a reversible coordinated reduction of both CD peaks,[18,27] which is accompanied by the appearance of rotational mobility of bR around axes perpendicular to the membrane.[28] This randomization of chromophore directions is indeed expected to lead to a loss of exciton coupling.[28]

8. The two enantiomers of 5,6-epoxyretinal (+ and −) both bind to apomembranes.[29] After regeneration the (+)-enantiomer absorbs at 485 nm, has a bilobed exciton CD spectrum, and leads to a normal hexagonal lattice. The (−)-enantiomer on the other hand absorbs at 445 nm, but has a positive CD band and does not induce the hexagonal protein lattice. These experiments show clearly that exciton CD bands are observed if an ordered chromophore–chromophore arrangement exists and are absent in the disordered state in which the coupling effects average out.

Not all of the experimental evidence is readily explainable using an exciton coupling explanation. The CD spectrum of the photocycle inter-

---

[24] M. Brith-Lindner and K. Rosenheck, *FEBS Lett.* **76**, 41 (1977).
[25] T. G. Ebrey, B. Becher, B. Mao, P. Kilbride, and B. Honig, *J. Mol. Biol.* **112**, 377 (1977).
[26] M. P. Heyn, R. J. Cherry, and U. Müller, *J. Mol. Biol.* **117**, 607 (1977).
[27] S. Nishimura, T. Mashimo, K. Hiraki, T. Hamanaka, Y. Kito, and I. Yoshiya, *Biochim. Biophys. Acta* **818**, 421 (1985).
[28] M. P. Heyn, P.-J. Bauer, and N. A. Dencher, in "Biochemistry of Membrane Transport" (G. Semenza and E. Carafoli, eds.), p. 96. Springer-Verlag, Berlin, Federal Republic of Germany, 1977.
[29] K. Hiraki, T. Hamanaka, K. Yoshihara, and Y. Kito, *Biochim. Biophys. Acta* **891**, 177 (1987).

mediate M also shows a positive and a negative peak in the M absorption band (412 nm).[18,30] The positive CD lobe is about five times larger than the negative lobe[30] and an exciton interpretation of this spectrum would require a very large monomer band. The lattice is preserved in this state at $-100°$[31] and the structure of M is not much different from that in the ground state.[31] The transition dipole moment at 412 nm makes the same angle with the membrane as in the ground state.[30]

One alternative explanation for the CD spectrum is that the broad absorption band consists of two nearly degenerate bands, one of which is weak in absorbance but has a large rotational strength. Its rotational strength should be about equal in amplitude and opposite in sign to that of the band that is strong in absorbance. It is very difficult to explain the experimental evidence presented above on this basis and there is at present no evidence in favor of it.

The range of the exciton interaction with intertrimer retinal distances of the order of 30 Å is considerably larger than that encountered in proteins or nucleic acids. The prerequisites for the observation of this effect are: (1) large extinction coefficients. The amplitude of the exciton CD spectrum is proportional to the square of the maximum extinction coefficient.[32] With a molar extinction coefficient of 62,700 M^{-1} cm^{-1} this condition is satisfied.[33] (2) The chromophore should be immobilized within bR. This was shown to be the case.[34] (3) The transition dipole moment geometry should be favorable. The unfavorable geometry with the transition dipole moment in the plane of the membrane, leading to aggregates with a plane of symmetry and zero rotational strength, does not occur since linear dichroism measurements show that the transition dipole moments make an angle of about 20° with the plane of the membrane.[26] Attempts based on standard CD theory[35] to gain structural information from the CD spectra have met with limited success. So far, only trimer calculations have been performed which take into account only the intertrimer interactions.[20,25,36,37,38] Recent neutron diffraction experiments on the location of the polyene chain of retinal show, however, that this simplification is

---

[30] J. E. Draheim and J. Y. Cassim, *Biophys. J.* **47**, 497 (1985).
[31] R. M. Glaeser, *Biophys. J.* **47**, 322a (1985).
[32] M. P. Heyn, *J. Phys. Chem.* **79**, 2424 (1975).
[33] M. Rehorek and M. P. Heyn, *Biochemistry* **18**, 4977 (1979).
[34] T. Kouyama, Y. Kimura, K. Kinosita, Jr., and A. Ikegami, *FEBS Lett.* **124**, 100 (1981).
[35] I. Tinoco, Jr., *Adv. Chem. Phys.* **4**, 113 (1962).
[36] T. Kouyama, Y. Kimura, K. Kinosito, Jr., and A. Ikegami, *J. Mol. Biol.* **153**, 337 (1981).
[37] A. N. Kriebel and A. C. Albrecht, *J. Chem. Phys.* **65**, 4575 (1976).
[38] C. A. Hasselbacher, J. L. Spudich, and T. G. Dewey, *Biochemistry* **27**, 2540 (1988).

incorrect.[39,40] Now that the in-plane position and orientation of the chromophore are known,[39,40] it is clear that the chromophore–chromophore distances within a trimer are only 12% smaller than between adjacent trimers. The angular part of the transition dipole geometry is also very similar for these pairs. It is thus necessary to redo the calculations taking the whole lattice into account. Measurements on contracted purple membrane lattices, in which the trimers move closer together after partial delipidation, show that the exciton CD amplitudes increase by 40% and provide experimental evidence for intertrimer contributions.[41] In these shrunken lattices the hexagonal unit cell dimension is reduced to 57–59 Å and all interchromophore distances are about equal. Calculations including all nearest neighbors of each chromophore can explain the observed increase.[42]

The bilobe CD spectrum is observed not only in the purple membrane lattice but is also expected for small specific aggregates of bR, starting with the dimer. The exciton CD signal thus provides a rapid way to determine whether the state of aggregation of bR is monomeric or not. When bR is reconstituted as a monomer in DMPC vesicles it remains monomeric above T_c (provided the molar lipid-to-protein ratio is larger than 100) and aggregates in the same hexagonal lattice as *in vivo* below T_c.[15] This reversible self-assembly was monitored using X-ray diffraction, freeze-fracture electron microscopy, CD, and measurements of the rotational diffusion of bR.[15] Figure 1 shows the CD spectrum of such reconstituted vesicles above and below T_c. Above T_c bR rotates rapidly around an axis perpendicular to the membrane and there is no evidence for a lattice from freeze fracturing and X-ray diffraction. The CD spectrum is indicative of a monomer. Below T_c the rotational motion is frozen and both freeze fracture and X-ray diffraction show evidence for a hexagonal lattice. The CD spectrum is typical for the aggregated state. The CD spectra can therefore be used as a routine check on the state of aggregation of bR in various systems. This has been particularly useful in studies on the properties of bR monomers.[14,43] It could be shown that the aggregated state is not required for bR's function as a proton pump.[43] The extent of

---

[39] F. Seiff, I. Wallat, P. Ermann, and M. P. Heyn, *Proc. Natl. Acad. Sci. U.S.A.* **82**, 3227 (1985).
[40] M. P. Heyn, J. Westerhausen, I. Wallat, and F. Seiff, *Proc. Natl. Acad. Sci. U.S.A.* **85**, 2146 (1988).
[41] S. C. Hartsel and J. Y. Cassim, *Biochemistry* **27**, 3720 (1988).
[42] S. Grzesiek and M. P. Heyn, submitted.
[43] N. A. Dencher and M. P. Heyn, *FEBS Lett.* **108**, 307 (1979).
[44] M. P. Heyn, R. J. Cherry, and N. A. Dencher, *Biochemistry* **20**, 840 (1981).

light–dark adaptation is, however, quite different in the monomeric state.[14] For investigations of lipid–protein interactions with bR the exciton signal was also quite valuable.[44,45] We expect further applications in studies on the assembly of the bR lattice. Whether this CD method to monitor protein aggregation will be applicable to other self-assembling membrane protein systems remains to be seen. The list of requirements is rather stringent and relatively few membrane proteins have chromophores in the visible spectrum.

[45] M. P. Heyn, A. Blume, M. Rehorek, and N. A. Dencher, *Biochemistry* **20**, 7109 (1981).

[32] Cross-Linking Techniques

By H. G. Bäumert and H. Fasold

Various techniques have been employed to introduce covalent linkages between amino acid side chains in proteins. Most of these investigations make use of bifunctional compounds, designed to react with one or only a few of the side-chain families.

Apart from this procedure, direct cross-linking of protein molecule neighbors is possible by oxidative reactions, as in the reaction of *o*-phenanthroline and Cu^{2+} with biomembranes. It seems to involve more than just the cysteine side chains.[1,2] The presence of glycoproteins gives rise to the rather complex reaction with periodate, followed by lysine and paraformaldehyde treatment.[3] More clear-cut are cross-links introduced by glutamyltransferase, which have probably not been explored to their fullest.[4,5] Ozone is another oxidant[6] used occasionally for the fusion of lipids to proteins in membranes. However, we will not deal with these direct reactions here as reviews on the topic of cross-links are available,[6,7] rather we will describe a few characteristic applications of bifunctional reagents.

[1] H. G. Bäumert, L. Mainka, and G. Zimmer, *FEBS Lett.* **132**, 308 (1981).
[2] Y. Hiroshige and J. Tohuku, *J. Exp. Med.* **130**, 385 (1980).
[3] D. C. Hixson, J. M. Yep, J. R. Glenney, Jr., T. Hayes, and E. F. Walborg, Jr. *J. Histochem. Cytochem.* **29**, 561 (1981).
[4] D. F. Mosher, *Ann. N.Y. Acad. Sci.* **312**, 38 (1978).
[5] H. Verwei, K. Christianse, and J. Van Steveninck, *Biochim. Biophys. Acta* **701**, 180 (1982).
[6] F. Wold, this series, Vol. 11, p. 617.
[7] K. Peters and F. M. Richards, *Annu. Rev. Biochem.* **46**, 523 (1977).

Originally, cross-linkers were devised for the determination of intramolecular distances in protein tertiary or quaternary structures.[8,9] The reagents were envisaged as extended yardsticks, and some of the measurements were in good agreement with the final analyses of some proteins by X-ray crystallography.[10] It became evident, however, that most amenable reagents contain several highly flexible bonds, and bind to long protein side chains, e.g., lysines. This leads to evaluations of distances within a certain range, usually varying at least over ±10 Å. For longer cross-links, a very small number of really rigid compounds have been synthesized on a polyproline basis.[11]

However, bifunctional reagents have proved very useful in the elucidation of protein neighborhoods in enzyme complexes and biomembranes, in the determination of the complexity of protein oligomers, and even for the modification of biological properties of single membrane-bound protein molecules (see [33] in this volume). In a few cases, it has been possible to change rheological properties of membranes, and several enzymes have been successfully stabilized by internal cross-linking.[11,12] Some allosteric proteins, such as hemoglobin[10] or phosphorylase b,[13] were cross-linked between well-defined points on their subunits with a concomitant loss of cooperativity, furnishing some insight on the mechanism of subunit interaction.

Hydrophobicity and Length

Most of the reagents listed in Table I contain a distinctly hydrophobic moiety in their structure. Only a few compounds in use so far carry a sufficient number of strongly cationic or anionic functional groups to confine them to one side of the biological membrane. Even sulfonic acids such as 2-nitro-4-azidophenyltaurine[14] slowly penetrate, e.g., the red cell membrane. Only the introduction of a second ionic site, as in stilbenedisulfonic acid derivatives (see chapter [33] in this volume), effectively precludes passage through the membrane.

[8] H. Zahn, *Angew. Chem.* **67**, 56 (1955).
[9] H. Fasold, *Biochem. Z.* **339**, 482 (1964).
[10] H. Fasold, J. Klappenberger, C. Meyer, and R. Remold, *Angew. Chem.* **83**, 875 (1971).
[11] H. Fasold, H. G. Bäumert, and G. Fink, *Adv. Exp. Med. Biol.* **86A**, 207 (1977).
[12] N. Maeda, K. Kon, K. Imaizumi, M. Sekiya, and T. Shiga, *Biochim. Biophys. Acta* **735**, 104 (1983).
[13] H. Fasold, in "Interconvertible Enzymes," pp. 60–64. Springer-Verlag, Berlin, Federal Republic of Germany, 1976.
[14] J. V. Staros and F. M. Richards, *Biochemistry* **13**, 2720 (1974).

TABLE I
FREQUENTLY USED CROSS-LINKING REAGENTS FOR PROTEINS AND COMPLEXES OF PROTEINS
WITH OTHER NATURALLY OCCURRING COMPOUNDS

Structure and name	Specificity	Cleavability	Ref.
Dimethyl diimidates ($n = 1$–12)	Lys, R-NH$_2$	+	F. C. Hartman and F. Wold, *J. Am. Chem. Soc.* **88**, 3890 (1966)
Tartryldiazide	Lys, R-NH$_2$	+	L. C. Lutter *et al.*, *FEBS Lett.* **48**, 288 (1974)
Tartryldi(glycylazide)	Lys, R-NH$_2$	+	L. C. Lutter *et al.*, *FEBS Lett.* **48**, 288 (1974)
Tartryldi(ε-aminocaproylazide)	Lys, R-NH$_2$	+	L. C. Lutter *et al.*, *FEBS Lett.* **48**, 288 (1974)

Structure	Name	Reactivity	+	Reference
(disuccinimidyl tartrate structure)	Disuccinimidyl tartrate	Lys, R-NH$_2$	+	J. R. Coggins et al., *Biochemistry* **15**, 2527 (1976); R. J. Smith et al., *Biochemistry* **17**, 3719 (1978)
H$_2$C—CH—CH—CH$_2$ (with epoxide rings)	Diepoxybutane	Lys, R-NH$_2$, Cys, nucleic acids	+	H. G. Bäumert et al., *Eur. J. Biochem.* **89**, 353 (1978)
H$_2$N$^+$\\C—(CH$_2$)$_2$—S—symm. / H$_3$CO	Dimethyl-3,3'-dithiobispropionimidate	Lys, R-NH$_2$	+	K. Wang and F. M. Richards, *J. Biol. Chem.* **250**, 6622 (1975)
(iminothiolane ring structure)	2-Iminothiolane	1. Lys, R-NH$_2$ 2. Cys	+	R. R. Traut et al., *Biochemistry* **12**, 3266 (1973)

(continued)

TABLE I (*continued*)

Structure and name	Specificity	Cleavability	Ref.
⟨sulfosuccinimidyl⟩—O—C(=O)—(CH$_2$)$_2$—S—┊—symm.	Lys, R-NH$_2$	+	J. V. Staros, *Biochemistry* **21**, 3950 (1982)
3,3'-Dithiobis(sulfosuccinimidyl propionate)			
H$_2$N$^+$=C(OCH$_3$)—(CH$_2$)$_2$—S(=O)$_2$—CH$_2$—CHOH—┊—symm.	Lys, R-NH$_2$	+	H. Fasold *et al.*, *Mod. Methods Protein Chem.* **2**, 261 (1985)
DDDD			
⟨succinimidyl⟩—O—C(=O)—(CH$_2$)$_2$—S—┊—symm.	Lys, R-NH$_2$	+	A. J. Lomant and G. Fairbanks, *J. Mol. Biol.* **104**, 243 (1976)
Dithiobis(succinimidyl propionate)			
⟨succinimidyl⟩—O—C(=O)—(CH$_2$)$_3$—┊—symm.	Lys, R-NH$_2$	−	M. Hill *et al.*, *FEBS Lett.* **102**, 282 (1979)
Disuccinimidyl suberate			

Structure	Reactive toward	Cleavable	Reference
Bis[2-(succinimidoxycarbonyloxy)ethyl] sulfone	Lys, R-NH$_2$	+	D. A. Zarling et al., J. Immunol. **124**, 913 (1980)
Ethylene glycol bis(succinimidyl succinate)	Lys, R-NH$_2$	+	P. M. Abdella et al., Biochem. Biophys. Res. Commun. **87**, 734 (1979)
Epichlorohydrin H$_2$C—CH—CH$_2$—Cl	Cys, Lys, R-NH$_2$, nucleic acids	− unless reacted with Lys	J. L. Hoffman, Anal. Biochem. **33**, 209 (1970)

(*continued*)

TABLE I (continued)

Structure and name	Specificity	Cleavability	Ref.
N-(3-Fluoro-4,6-dinitrophenyl)cystamine	Cys, Lys, R-NH$_2$ (Tyr, His)	+	K. Peters and F. M. Richards, *Annu. Rev. Biochem.* **46**, 523 (1977)
Difluorodinitrobenzene	Cys, Lys, R-NH$_2$ (Tyr, His)	−	H. Zahn, *Angew. Chem.* **67**, 56 (1955)
Succinimidyl-4-(*p*-maleimidophenyl)butyrate	Cys, Lys, R-NH$_2$	−	T. Kitagawa and T. Aikawa, *J. Biochem. (Tokyo)* **79**, 233 (1976)
	Cys, Lys, R-NH$_2$	+	J. Carlsson *et al.*, *Biochem. J.* **173**, 723 (1978)

Structure	Target		Reference
2-Bromoethylmaleimide	Cys	−	S. Simon and W. Königsberg, *Proc. Natl. Acad. Sci. U.S.A.* **56**, 749 (1966)
Methylenebismaleimide	Cys	−	S. Simon and W. Königsberg, *Proc. Natl. Acad. Sci. U.S.A.* **56**, 749 (1966)
N,N'-p-Phenylenedimaleimide	Cys	−	E. Reisler *et al.*, *Biochemistry* **13**, 3887 (1974)
2,2′-Dicarboxy-4,4′diisothiocyanatoazobenzene	Lys, R-NH_2	+	H. Fasold, *Biochem. Z.* **339**, 482 (1964)

(*continued*)

TABLE I (*continued*)

Structure and name	Specificity	Cleavability	Ref.
![structure] I—CH$_2$—C(O)—N(H)—[benzene with COOH]—N=N—symm.	Cys, Lys	+	H. Fasold *et al.*, *Angew. Chem.* **83**, 875 (1971)
2,2'-Dicarboxy-4,4'-iodoacetamidoazobenzene			
S=C=N—[benzene with SO$_3^-$]—CH=CH—symm.	Lys, R-NH$_2$	−	L. Zaki *et al.*, *J. Cell. Physiol.* **86**, 471 (1975)
DIDS			
Br—CH$_2$—[benzene with -O$_3$S]—CH$_2$—Br	Cys, (His, Lys)	−	C. B. Hiremath and R. A. Day, *J. Am. Chem. Soc.* **86**, 5027 (1964)
2,5-Dibromomethylbenzenesulfonic acid			
O$_2$N—[benzene]—C(=CH$_2$)—CH$_2$—[benzene with COOH and NO$_2$]—S—	Cys	+	*J. Am. Chem. Soc.* **101**, 3097 (1979)
2-(4-Nitrophenyl)allyl-4-nitro-3-carboxy-			

Structure	Specificity	Reactivity	Reference

4-Maleimidobenzyl-*N*-hydroxysuccinimide ester — Cys, Lys, R-NH$_2$ — − — T. Kitagawa and T. Aikawa, *J. Biochem. (Tokyo)* **79**, 233 (1976)

Ethylene glycol bis[3-(2-ketobutyraldehyde) ether] — Lys, Arg, nucleic acids — + — L. A. Brewer et al., *Biochemistry* **22**, 4303 (1983)

Glutardialdehyde — Lys, R-NH$_2$ — − — E. Engvall and P. Perlmann, *J. Immunol.* **109**, 129 (1972)

N-Methylbis(2-chloroethylamine) — Cys, Lys, nucleic acids — − — V. C. E. Burnop et al., *Biochem. J.* **66**, 504 (1957); P. D. Lawley and P. Brookes, *Annu. Rep. Cancer Res. Campaign* **37**, 68 (1959); P. Brookes and P. D. Lawley, *J. Chem. Soc.* p. 539 (1960)

(*continued*)

TABLE I (continued)

Structure and name	Specificity	Cleavability	Ref.
Cl—(CH$_2$)$_2$—N(H)—C(=O)—N(NO)—(CH$_2$)$_2$—Cl 1,3,-Bis(2-chloroethyl)-1-nitrosourea	Cys, nucleic acids		Z. M. Banjar et al., Biochem. Biophys. Res. Commun. **114**, 767 (1983)
(4-N$_3$-C$_6$H$_4$)—C(=O)—CH$_2$—Br 4-Azidophenacyl bromide	Cys, nucleic acids	−	S. H. Hixson and S. S. Hixson, Biochemistry **14**, 4251 (1975)
(4-N$_3$-C$_6$H$_4$)—C(=O)—C(H)=O 4-Azidophenylglyoxal	Arg, Lys, nucleic acids	− +	S. M. Politz et al., Biochemistry **20**, 372 (1981)
4-N$_3$-2-[O—(CH$_2$)$_3$—C(=NH$_2^+$)—OCH$_3$]-C$_6$H$_3$—CHO 4-(6-Formyl-3-azidophenoxy)butyrimidate	Lys, R-NH$_2$	−	J. A. Maassen, Biochemistry **18**, 1288 (1979)

Structure	Name	Reactive with	+/−	Reference
3-nitro-5-azidobenzoyl-Se-(CH₂)₂-COOH	3-(4-Azido-2-nitrobenzoylseleno)propionic acid	Cys, nucleic acids	+	K. Friebel et al., *Hoppe-Seyler's Z. Physiol. Chem.* **362**, 421 (1981)
Br—CH₂—C(O)—(CH₂)₂—SO₂—(CH₂)₂—C(O)—O—N-hydroxysuccinimide	3-(4-Bromo-3-oxobutane-1-sulfonyl)propionic-*N*-hydroxysuccinimide ester	Cys, Lys, nucleic acids	−	G. Fink et al., *Anal. Biochem.* **108**, 394 (1980)
4-fluoro-3-nitrophenylazide structure	4-Fluoro-3-nitrophenylazide	Cys, Lys, nucleic acids	−	R. Bisson et al., *J. Biol. Chem.* **253**, 1874 (1978)

(continued)

TABLE I (continued)

Structure and name	Specificity	Cleavability	Ref.
4-Azidophenyl isothiocyanate	Lys, R-NH$_2$, nucleic acids	–	H. Sigrist and P. Zahler, *FEBS Lett.* **113**, 307 (1980)
N-(Maleimidomethyl)-2-(p-nitrophenoxy)carboxamidoethane	Lys, R-NH$_2$, Cys	+	P. C. Jelenc et al., *Proc. Natl. Acad. Sci. U.S.A.* **75**, 3564 (1978)
N-Hydroxysuccinimidyl-4-azidobenzoate	Lys, R-NH$_2$, nucleic acids	–	C. C. Yip et al., *J. Biol. Chem.* **253**, 1743 (1978)

Structure	Target	Reference
N-Succinimidyl-6-(4'-azido-2'-nitrophenylamino)hexanoate	Lys, R-NH$_2$, nucleic acids	J. U. Baenziger and D. Fiete, *J. Biol. Chem.* **257**, 4421 (1982)
p-Azidophenylacetimidate	Lys, R-NH$_2$, nucleic acids	G. Fink *et al.*, *Anal. Biochem.* **108**, 394 (1980); J. Rinke *et al.*, *J. Mol. Biol.* **137**, 301 (1980)
Methyl-4-azidobenzoimidate	Lys, R-NH$_2$, nucleic acids	T. H. Ji, *J. Biol. Chem.* **252**, 1566 (1977)

(*continued*)

TABLE I (continued)

Structure and name	Specificity	Cleavability	Ref.
N-Succinimidyl-(4-azidophenyldithio)-propionate	Lys, R-NH$_2$, nucleic acids	+	J. U. Baenziger and D. Fiete, *J. Biol. Chem.* **257**, 4421 (1982)
N-Hydroxysuccinimidyl-4-azidosalicylic acid	Lys, R-NH$_2$, nucleic acids	−	T. H. Ji and I. Ji, *Anal. Biochem.* **121**, 286 (1982)
N-Hydroxysuccinimidyl-4-azidosalicylic acid	Lys, R-NH$_2$, nucleic acids	+	T. H. Ji and I. Ji, *Anal. Biochem.* **121**, 286 (1982)
Sulfosuccinimidyl-2-(*p*-azidosalicylamido)-			

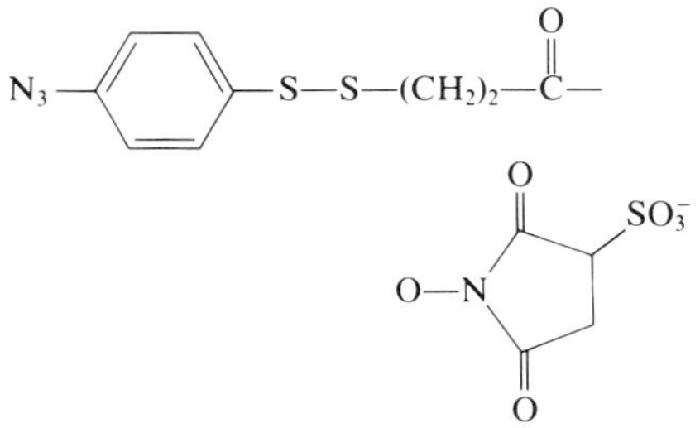

Sulfosuccinimidyl-(4-azidophenyldithio)-propionate

Lys, R-NH$_2$, nucleic acids

J. U. Baenziger and D. Fiete, *J. Biol. Chem.* **257**, 4421 (1982)

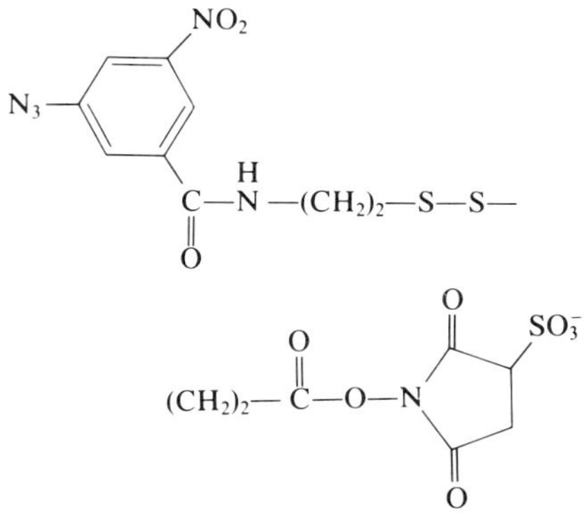

Sulfosuccinimidyl-2-(*m*-azido-*o*-nitrobenz-amido)ethyl-1,3'-dithiopropionate

Lys, R-NH$_2$, nucleic acids

(*continued*)

TABLE I (continued)

Structure and name	Specificity	Cleavability	Ref.
N-Bromoacetyl-N-methylaminodiethoxy-ethanemethylaminomethyltrimethylpsoralen	Cys, nucleic acids	—	H. Fasold et al., *Mod. Methods Protein Chem.* **2**, 261 (1985)
Ethidium bromide diazide	Nucleic acids	—	R. Bastos, *J. Biol. Chem.* **250**, 7739 (1975)

Structure	Reactive toward	Cross-linking	Reference
Bis-*N*-(2-nitro-4-azidophenyl)cystamine *S,S*-dioxide	Cys, nucleic acids	+	C.-K. Huang and F. M. Richards, *Fed. Proc., Fed. Am. Soc. Exp. Biol.* **35**, 1378 (Abstr.) (1976)
2,2′-Disulfonyl-4,4′-diazidoazobenzene	Nucleic acids (Trp, Tyr, His)	+	H. G. Bäumert, manuscript in preparation
Different psoralenes	Nucleic acids	−	C. J. Shen and J. E. Hearst, *Proc. Natl. Acad. Sci. U.S.A.* **73**, 2649 (1976) S. T. Isaacs *et al.*, *Biochemistry* **16**, 1058 (1977)

A few reagents having a very distinct and uniformly hydrophobic nature, such as diisothiocyanates of stilbene, anthracene derivatives, or azo dyes,[10] have been proposed for cross-links aimed at intrinsic membrane proteins. Usually, however, the side reactions with reactive sites of the membrane lipids are prohibitive.

Thus, most of the frequently applied compounds, such as the partially hydrophobic diimidates, will attach to proteins at or near the border between the hydrophobic interior of membranes and the aqueous milieu, and will also tend to penetrate into the hydrophobic cores of the protein molecules. In some cases, this can be observed in the distinctly reduced reactivity of the molecule. Thus, the azo dye (**1**) binds to the anion carrier

$$S=C=N-\underset{\underset{SO_3H}{|}}{\overset{\overset{SO_3H}{|}}{\bigcirc}}-N=N-\bigcirc-N=C=S$$

(1)

of the red cell membrane with the same affinity as the corresponding stilbene derivative. In aqueous solution, the azo bond is easily cleaved by low concentration of dithionite ions after a few minutes at room temperature and neutral pH. Within the membrane, however, even the prolonged reaction at temperatures up to 50° leaves the cross-links entirely intact.

These interactions between hydrophobic portions of the protein and reagent molecules pose a drawback when the aim of the experiment is the linkage of neighboring protein molecules, as in topographical studies. In experiments on proteins in aqueous solution, as in the cross-linking of hemoglobin[10] or F-actin, it was found that the ratio of intramolecular to intermolecular or intersubunit linkages was not clearly related to the length of the cross-linker. When a series of reagents with an approximately 10–18 Å span between the lysine side-chain attachment points was tested, the yield of intermolecular and intersubunit cross-links increased distinctly with the content of hydrophilic moieties in the compounds.[11] In the case of F-actin, which provides a number of lysines within suitable distance for intermolecular cross-linking, only 5% of the molecules of *p*-diisothiocyanatoazobenzenedicarboxylic acid bound produced such linkages. The extremely hydrophilic tartaric acid derivatives (see Table I) reached values of up to 45%. In both cases, the reaction was limited to the attachment of a maximum of five reagent molecules per protein monomer.

This effect of internal protein hydrophobicity may be partly counteracted by the properties of the lipid bilayer of the biological membrane, causing the hydrophobic part of the reagent to run out from the surface of the protein within the membrane. In our experience, however, the intermolecular cross-linking reaction is also favored by the hydrophilic nature of the reagent under these circumstances.

Two-Step Reagents

Two different functional reactive groups are sometimes introduced into the reagent. This is advantageous when the first step of the reaction is carried out with a large excess of reagent without denaturation of the proteins, as is frequently the case with fairly hydrophilic diimido esters. The relatively large number of amidino groups thus introduced does not change the overall charge pattern of the protein. After removal of the excess reagent, the second step, e.g., an acylating or alkylating reaction, may be carried out with reduced denaturation effects. These reagents are especially suitable as labels for various counterparts, when proteins, like ferritin, avidin, or immunoglobulins, are to be coated by the reagent in the first step and then activated in the presence of the macromolecule to be labeled in the linking reaction. The technique has been used for the binding of peroxidase or alkaline phosphatase to immunoglobulins for the development of ELISA tests, or for the linkage of hormones to their receptor (see below).

Frequently, the functional group for the second step is an aromatic azide or diazoketone group which permits photolytic labeling of the protein. Small changes in the structure have a strong influence on the yield of the reaction. When p-azidophenylacetic acid imido ester (**2**)[15] was com-

$$N_3-\left\langle\underset{}{}\right\rangle-CH_2-C\underset{OCH_3}{\overset{NH}{\diagup\!\!\!\diagdown}}$$

(2)

pared to p-azidobenzoic acid imido ester in an exhaustive reaction with alkaline phosphatase by the imidate group (first step) and immunoglobulin G (second step) under identical experimental conditions, the benzoic acid derivative produced a linkage between the enzyme and 35% of the IgG, while the imido ester reached a yield of 61%.

[15] G. Fink, H. Fasold, W. Rommel, and P. Brimacombe, *Anal. Biochem.* **108,** 394 (1980).

Cleavable Reagents

This family of bifunctional protein-reactive compounds contains a covalent bond that can be cleaved under mild conditions without denaturation of the proteins involved. Disulfides (which may be reduced by various methods or subjected to disulfide exchange), azo bonds, (cleaved by dithionite), or vicinal diols (oxidized by periodate) have been found to be particularly useful (see Table I). In topographical studies, the technique can be used for fast and reliable "diagonal" electrophoresis (see below).

Cleavage of cross-links can also help to clarify the effect of structural restrictions on the function of proteins. Moreover, it serves as conclusive evidence that the changed properties of the cross-linked protein are not due to partial denaturation, but to the inhibition of conformational changes. Thus, artificial actin dimers linked by tartrylbis-ε-aminocapryl linkages are not able to polymerize to F-actin filaments, but serve as polymerization nuclei for G-actin. After cleavage of the cross-link, normal polymerization of the restored G-actin monomers is observed.

Protein–Nucleic Acid Cross-Linking

During the life cycle and biological function of several membrane systems, a close interaction between nucleic acids and proteins becomes essential. Some examples are nuclear envelopes as well as mitochondrial and endoplasmic reticulum membranes. Therefore, a number of reagents have been adapted in the effort to identify the protein partners of nucleic acids in such complexes, usually on the basis of photoactivated functional groups (see Table I). A principal difficulty lies in the fact that all functional groups of the aromatic bases in nucleic acids suitable for substitution reactions are also present in amino acid side chains. Moreover, in all reactions that can be carried out under physiological conditions, protein side chains are more reactive than nucleic acids. Nevertheless, several reagents have been devised that produce a fair yield of nucleic acid–protein cross-links, e.g., ethylene glyco bis[3-(2-ketobutyraldehyde) ether].[16] Among the two-step reagents, α-bromoketones (**3**) preferentially

$$CH_3-\overset{O}{\underset{}{C}}-\underset{Br}{CH}-CH_2-\overset{O}{\underset{O}{S}}-CH_2-CH_2-\overset{O}{\underset{}{C}}-O-\underset{}{\langle\rangle}-NO_2$$

(3)

---

[16] C. A. Brewer, S. Goelz, and H. F. Noller, *Biochemistry* **22**, 4303 (1983).

react with adenine and cytosine at pH 5–6, while a second functional group can to a large part be kept intact for a protein linkage step at pH 7–8. A similar procedure can be employed for psoralene or ethidium bromide derivatives that intercalate into nucleic acids and can be covalently linked to them by irradiation.[17]

Sterically Enforced Reactions

Most of the functional groups employed in direct or two-step bifunctional reagents form covalent bonds with several amino acid side chains. A typical example is furnished by the bromoacetyl group which can alkylate, under physiological conditions, cysteine, histidine, or lysine side chains. At pH 7.4, the reaction rates with these three functional groups form a ratio of 1 : 0.1 : 0.001 in the sequence named above. Thus, a bifunctional reagent bearing bromoacetyl residues should primarily couple to sulfhydryl groups, even if the two partner reactants are not available within the span of the reagent. However, once the cross-linking reagent has been attached to the protein at the first site, the second functional group shows reaction rates with a sterically favored group that are enhanced by several orders of magnitude over the isolated reaction. Thus, in the cross-linking of hemoglobin by two reagents alkylating a cysteine in position 93 of the β chain, the second functional group reacts with lysine side chains.[10]

Examples

Hormone–Receptor Coupling

Hormone–receptor coupling has been frequently used to cross-link proteohormones to their membrane-bound receptors. The reason for these successful experiments is the small dissociation constant of the hormone–receptor complex. Thus, under saturation conditions, the bifunctional reagent will always find a second partner during the reaction time, and the low concentration of the receptors in the reaction mixture is sufficient. Therefore, the designation "affinity cross-linking" is sometimes employed.

Although the experimental conditions vary among the various proteohormones so far studied, a general working procedure is feasible.[18–25]

[17] H. Fasold, H. G. Bäumert, and C. Meyer, *Mod. Methods Protein Chem.* **2**, 261 (1985).
[18] P. F. Pilch and M. P. Czech, *J. Biol. Chem.* **255**, 1722 (1980).
[19] M. Kasuga, E. Van Olberghen, S. P. Nissley, and M. M. Rechler, *J. Biol. Chem.* **256**, 5305 (1981).

In the first stage, the proteohormone is labeled, usually with ^{125}I by conventional methods.[20,22,24,26] In the second stage, the purified membranes, pelleted in Krebs–Ringer phosphate buffer, pH 7.4, are suspended in the same buffer, with or without the addition of protective serum albumin (0.1–1%). The labeled hormone is added to the membrane preparation in the nanomolar range (1–20 nM), and incubated at room temperature for 1–2 hr. The protein concentration during the binding incubation ranges between 0.2 and 2 mg/ml. The reaction is terminated by 3- to 10-fold dilution with cold buffer, and the excess of hormone is removed by centrifugation.

In the third stage, the cross-linking reagent is added after resuspension of the membrane pellet. Bisimidate reagents, such as suberimidate, are frequently employed at 0.05–0.1 mM final concentration, and photolabile reagents sometimes used at concentrations of up to 1 mM. The reaction is terminated by addition of a large excess of amino groups, for example, in the form of Tris buffer. The pelleted membrane fraction is then subjected to SDS-gel electrophoresis and autoradiography to detect the bands conjugated to the labeled hormone.

Cross-Linking of the Rabbit Muscle Sarcoplasmic Reticulum Membrane Ca^{2+}-ATPase

The reagent bispyridoxal phosphate (BPP) was chosen because it has several advantages for investigations on complicated protein mixtures in biological membranes. The substance is hydrophilic, and permeates through membranes extremely slowly. It is cleavable, and can be labeled by reduction with boron [^{3}H]hydrides after the cross-linking reaction, thus avoiding a tedious radioactive synthesis. The reactive aldehyde groups are stable.

---

[20] S. D. Chernansek, S. Jacobs, and J. J. Van Wyk, *Biochemistry* **20**, 7345 (1981).

[21] T. Tsushima, Y. Ohmori, H. Muzukami, K. Shizume, and Y. Hirata, *Diabetologia* **24**, 387 (1983).

[22] G. L. Johnson, V. J. MacAndrew, Jr., and P. F. Pilch, *Proc. Natl. Acad. Sci. U.S.A.* **78**, 875 (1981).

[23] P. M. Grob, C. H. Berlot, and M. A. Bothwell, *Proc. Natl. Acad. Sci. U.S.A.* **80**, 6819 (1983).

[24] S. Paglin and J. D. Jamieson, *Proc. Natl. Acad. Sci. U.S.A.* **79**, 3739 (1982).

[25] R. V. Rebois, F. Omedo-Sale, R. O. Brady, and P. H. Fishman, *Proc. Natl. Acad. Sci. U.S.A.* **78**, 2086 (1981).

[26] M. Kasuga, E. Van Olberghen, S. P. Nissley, and M. M. Rechler, *Proc. Natl. Acad. Sci. U.S.A.* **79**, 1864 (1982).

$$\text{(4)}$$

Structure (4): Pyridoxal phosphate–azo–phenyl–CH$_2$— unit with O=P(O$^-$Na$^+$)$_2$—O—CH$_2$— on pyridine ring bearing HO—, H$_3$C—, O=CH—, and —N=N—C$_6$H$_4$—CH$_2$— (SYMM.)

Bispyridoxal phosphate can be prepared by slowly adding a solution of 1 mmol diazotized 1,2-bis(4-aminophenyl)ethane (prepared with 2 mmol NaNO$_2$ in 30 ml 3 N HCl at 0°) to a solution of 5 mmol pyridoxal phosphate in water, pH 8.5, at 0–5°. During the reaction, the pH is kept constant with 2 N NaOH. After completion, the product is precipitated by the addition of several volumes of acetone. The product can be further purified by redissolving the precipitate in 0.1 N NaOH and repeated acetone precipitations. As the final purification step, the reagent is precipitated by acidifying a neutral solution with 2 N HCl to pH 2. The purity can be checked by thin-layer chromatography on silica gel with 0.2% NH$_4$OH.

Vesicles from rabbit muscle sarcoplasmic reticulum are prepared according to Hasselbach et al.[27,28] At 0°–25° cross-links are introduced at protein concentrations of between 0.1 and 2 mg/ml. Stock solutions of BPP in 100 mM KCl, 0.1 M triethanolamine HCl, pH 7 and pH 7.5, are prepared and the vesicles treated under MgATP (2 mM) and Ca^{2+} (50 μM) protection with final concentrations of 0.15 to 0.8 mM BPP. After completion of the reaction (usually about 2 min), it is terminated by reducing protein–BPP–protein conjugates and free BPP with 0.05 to 0.3 mM NaBH$_4$, making protein–BPP products irreversible and at the same time inactivating excess free reagent. After dissolving in 2% SDS solution, samples are analyzed for their protein–protein cross-linking extent by SDS–PAGE.[29] For the identification of cross-linked protein pairs the diagonal technique is used,[30] cleaving cross-linked complexes after the first

---

[27] W. Hasselbach and M. Makinose, *Biochem. Z.* **333**, 518 (1961).
[28] L. de Meis and W. Hasselbach, *J. Biol. Chem.* **246**, 4759 (1971).
[29] U. K. Laemmli, *Nature (London)* **227**, 680 (1970).
[30] L. C. Lutter, F. Ortanderl, and H. Fasold, *FEBS Lett.* **48**, 288 (1974).

dimension with 1 mM Na$_2$S$_2$O$_4$. Other samples are freed from excess reagent and tested for their capability to perform ATP hydrolysis.

At 0.15 and 0.3 mM BPP the Ca^{2+}-ATPase is mainly cross-linked to dimers after 2 min. This gel pattern does not change even after a 20-min incubation time, leaving the amount of monomer (about 70%) constant. Only increasing the pH to 9 and the BPP concentration to 0.8 mM yields trimers and a few tetramers of the Ca^{2+}-ATPase. The diagonal technique shows that cleavage of ATPase complexes yielded only monomers of this protein, indicating that no other proteins were involved. Samples which had been cross-linked with 0.15 mM BPP showed a 50% inhibition of ATPase activity which could be reversed by treatment with Na$_2$S$_2$O$_4$.

From consideration of the symmetry of the ATPase complex and statistical calculations concerning the cross-linking reaction, we conclude that the ATPase dimer is the most probable unit for the Ca^{2+}-pumping protein. A monomer–dimer–tetramer equilibrium in comparison with a monomer–dimer equilibrium should bring about a considerable amount of tetramer in this cross-linking reaction at an almost equal rate to the formation of trimers. Here, trimers are formed, but only after the yield of dimer production has reached 30% of the total protein, while only traces of artificial tetramers can be detected at very high reagent concentrations.

Topographical Investigations

Cross-linking reagents have frequently been used to elucidate neighborhood properties of proteins, e.g., in organelles, such as ribosomes,[30] or in biological membranes. In a few cases, the reagent employed yielded noncleavable cross-links.[31,32] Then one-dimensional PAGE should be used and the molecular weight of the cross-linked proteins estimated. Usually, however, either photoactivated[33,34] or primarily protein-reactive[35,36] cleavable reagents are used. The analytical detection of the cross-linked protein partners is then very frequently carried out by diagonal electrophoresis.[30,37]

After the cross-linking experiment, and sometimes after preliminary separations, e.g., by gel filtration, a symmetrical two-dimensional SDS–PAGE is applied. After the first dimension in cylindrical gels (0.5–5 mm,

[31] L. I. Takemoto and J. S. Hansen, *Exp. Eye Res.* **33**, 267 (1981).
[32] K. S. Lam and C. B. Kasper, *J. Biol. Chem.* **254**, 11713 (1979).
[33] C. R. Middaugh and T. H. Ji, *Eur. J. Biochem.* **110**, 587 (1980).
[34] E. T. Palva, *Biochim. Biophys. Acta* **596**, 235 (1980).
[35] J. Peters and G. Drews, *Eur. J. Cell Biol.* **29**, 115 (1983).
[36] N. Koch and D. Haustein, *Hoppe-Seyler's Z. Physiol. Chem.* **361**, 885 (1980).
[37] H. G. Bäumert, S. E. Sköld, and C. G. Kurland, *Eur. J. Biochem.* **89**, 353 (1978).

5–15% acrylamide), the gel is soaked (20 min for 0.5 mm, 5 hr for 5 mm gels) in 0.1% buffered SDS, if the original buffer reacts with the cleavage reagent. The gel is then bathed for the same time in a buffered solution of the cleavage reagent (sodium periodate for vicinal diols, mercaptoethanol and other sulfhydryl compounds for disulfide bonds, and sodium dithionite for azo bonds). The gel is equilibrated for the same time again in the buffer of the second electrophoresis, placed on top of the second-dimension gel slab, and polymerized into place with spacer gel. A marker lane of the same protein mixture is positioned at one side of the second dimension.

The free proteins run as a diagonal series from the upper left-hand corner of the slab, while the cross-linked pairs appear as single spots, in vertical arrangement, on the left-hand lower half of the gel. Artifact spots may arise in the upper right-hand half of the gel by reassociation of proteins, or overloading in one region of the gel, due to heavy cross-linking and formation of protein aggregates.

[33] Membrane-Impermeant Cross-Linking Reagents

By JAMES V. STAROS and P. S. R. ANJANEYULU

Introduction

The decade beginning with the work of Henderson and Unwin[1] on bacteriorhodopsin, through the elucidation of the high-resolution structure of the photoreaction center of *Rhodopseudomonas viridus* by Deisenhofer *et al.*[2] witnessed great advances in the application of diffraction techniques to solving the three-dimensional structures of membrane proteins. However, for most membrane proteins, the determination of a high-resolution structure is a distant hope at best. Therefore, information concerning the tertiary and quaternary structure of most membrane proteins cannot profitably await the application of diffraction methods.

Protein cross-linking is a method that has proved useful in the analysis of the tertiary and quaternary structure of proteins, both soluble and in membranes. (For general reviews on cross-linking, see Refs. 3–6; for

[1] R. Henderson and P. N. T. Unwin, *Nature (London)* **257**, 28 (1975).
[2] J. Deisenhofer, O. Epp, K. Miki, R. Huber, and H. Michel, *Nature (London)* **318**, 618 (1985).
[3] H. Fasold, J. Klappenberger, C. Meyer, and H. Remold, *Angew. Chem., Int. Ed. Engl.* **10**, 795 (1971).

reviews which treat aspects of cross-linking as applied to membrane proteins, see Refs. 7–10.) One of the earliest membrane protein cross-linking studies to yield useful structural information was that of Steck,[11] who applied several reagents to the study of erythrocyte membrane proteins. A particularly fruitful part of that study was the use of a Cu^{2+}/o-phenanthroline complex to oxidize endogenous thiols to disulfides.[12] In these studies, cross-linking was followed by observing the loss of particular gel bands on SDS–PAGE[13] and the appearance of new bands at higher molecular weight. Further, as a control, the disulfide cross-links could be cleaved with thiols to yield the original gel bands. The limitation of this method was that it could only be applied to proteins with appropriately positioned thiol groups.

Wang and Richards[14] addressed two important problems in cross-linking of membrane proteins. They synthesized dimethyl-3,3′-dithiobispropionimidate, a cleavable analog of dimethyl suberimidate, in which the two central methylene residues in the suberate backbone were replaced by a disulfide group. They also developed a two-dimensional SDS-polyacrylamide gel electrophoresis procedure in which the first dimension is run under nonreducing conditions, so as to separate cross-linked species by the molecular weights of the *oligomers,* and the second dimension is run under reducing conditions, so as to separate the protein species by the molecular weights of the *subunits.* Using this procedure, one could analyze directly not only for the formation of cross-linked complexes, but also for the constituents of such complexes. This reagent and this analytical system have since been applied to many cross-linking studies.

One important structural feature of membrane proteins is their asymmetric orientation in the membrane. However, in cross-linking experi-

---

[4] F. Wold, this series, Vol. 25, p. 623.
[5] M. Das and C. F. Fox, *Annu. Rev. Biophys. Bioeng.* **8,** 165 (1979).
[6] T. H. Ji, this series, Vol. 91, p. 580.
[7] K. Peters and F. M. Richards, *Annu. Rev. Biochem.* **46,** 523 (1977).
[8] T. H. Ji, *Biochim. Biophys. Acta* **559,** 39 (1979).
[9] B. J. Gaffney, *Biochim. Biophys. Acta* **882,** 289 (1985).
[10] H. G. Bäumert and H. Fasold, this volume [32].
[11] T. L. Steck, *J. Mol. Biol.* **66,** 295 (1972).
[12] K. Kobashi and B. L. Horecker, *Arch. Biochem. Biophys.* **121,** 178 (1967).
[13] Abbreviations: SDS–PAGE, sodium dodecyl sulfate-polyacrylamide gel electrophoresis; BSSDP, bis(sulfosuccinimidyl)-4-doxyl pimelate; $HOSu(SO_3)$, N-hydroxysulfosuccinimide; DIDIT, diisethionyl-3,3′-dithiobispropionimidate; DTBP, dimethyl-3,3′-dithiobispropionimidate; BS^3, bis(sulfosuccinimidyl) suberate; DTSSP, 3,3′-dithiobis(sulfosuccinimidyl propionate); H_2DIDS, 4,4′-diisothiocyanodihydrostilbene-2,2′-disulfonic acid.
[14] K. Wang and F. M. Richards, *J. Biol. Chem.* **249,** 8005 (1974).

ments using conventional reagents, information concerning asymmetry is generally lost, because membranes are permeable to such reagents. In order to extract information concerning the topological orientation of the cross-linked protein, we[15-17] introduced a series of membrane-impermeant cross-linking reagents, utilizing both alkyl imidate[15] and active ester[17] functional groups. Using such reagents, one can ascertain not only the proximity of two functional groups, or of two subunits of an oligomer, but also the topological orientation—cytoplasmic or extracytoplasmic—of the groups involved in the cross-link.

Principles and Practice of Cross-Linking

The Cross-Linking Reaction

Cross-linking reactions differ from monofunctional modification in several important ways. Here we will consider the case of homobifunctional reagents, i.e., reagents that have the same reactive group at both ends, though the general principles can be extended to heterobifunctional reagents, as well. (For a discussion of some of the specific features of heterobifunctional reagents, in particular that subclass in which one end is photoreactive, see Ref. 18.)

Two of the most important aspects in which cross-linking reactions differ from monofunctional modifications are in their reaction kinetics and in the functional group specificity of the reactive functionalities. Consider a cross-linking reagent that is added to the cross-linking reaction with an initial concentration in the micromolar–millimolar (μM–mM) range. The first step of the cross-linking reaction is, to a first approximation, second order, first order with respect to the reagent concentration and first order with respect to the effective concentration of the target nucleophile. However, the reaction of the second end of the reagent occurs somewhat differently. Once the reagent has reacted at one end, it is tethered. The second end of the reagent has a very high concentration within the span of the tether and is excluded from the rest of the solution. If, for example the span is 10 Å and one makes the simplifying assumptions that the reagent is tethered to a planar surface and that the second end is allowed all degrees of freedom within the span of the reagent, the effective local concentra-

---

[15] J. V. Staros, D. G. Morgan, and D. R. Appling, *J. Biol. Chem.* **256**, 5890 (1981).
[16] J. V. Staros, *Biophys. J.* **37**, 21 (1982).
[17] J. V. Staros, *Biochemistry* **21**, 3950 (1982).
[18] H. Bayley and J. V. Staros, in "Azides and Nitrenes: Reactivity and Utility" (E. F. V. Scriven, ed.), pp. 433–490. Academic Press, Orlando, Florida, 1984.

tion of the second end is 1 molecule/$\frac{1}{2} \times \frac{4}{3} \pi r^3$, where r is 10 Å, or

$$\frac{(6 \times 10^{23})^{-1} \text{ moles}}{\frac{1}{2} \times \frac{4}{3} \pi \times 10^3 \text{ Å}^3} \times \frac{10^{24} \text{ Å}^3}{\text{cm}^3} \times \frac{10^3 \text{ cm}^3}{1} = 0.8 \; M$$

This 10^3- to 10^6-fold enhancement relative to the initial concentration will obviously have a profound effect on the rate of the second reaction, assuming that a second target nucleophile resides within that hemisphere.

However, what is the fate of the second end of the cross-linker if a second target nucleophile is not within the span of the reagent? The second end could hydrolyze, so that that cross-linker molecule would not participate in a cross-link. This is a possible fate for any cross-linking reagent, as all reactive groups yet devised are subject to hydrolysis at finite rates. For rapidly hydrolyzed functional groups, this may account for a significant proportion of the reagent molecules which initially react. Another possible fate of the second end is that, due to its high local concentration, it reacts with a secondary nucleophile, i.e., a nucleophilic group that reacts with the functional group of the reagent very much more slowly than does the target nucleophile.

A classic example of this type of reaction is the cross-linking of filamentous actin with p-phenylene-N,N'-dimaleimide, which results in cross-links between a Cys on one subunit and a Lys on an adjacent subunit.[19] Maleimides are usually considered highly specific for thiols,[20] and with good reason. At pH 7, reaction with a primary thiol is $\sim 10^3$ faster than with the analogous primary amine.[21] However, at higher pH, the reaction with amines becomes appreciable.[21] When filamentous actin was treated with p-phenylene-N,N'-dimaleimide at pH 9.5, the reagent first reacted with the Cys, then more slowly formed the cross-link with the Lys on the neighboring subunit.[19] The high degree of participation of Lys in the cross-link formed is most likely caused by the high local concentration due to the tethering of the reagent to the Cys, abetted by the high pH which would enhance the rate of the second reaction with the Lys.

Protein Concentration in Studies of Quaternary Structure

Most cross-linking studies are carried out to address some aspect of the quaternary structure of the system of interest, i.e., which protomers in the system are associated with other protomers in a functional oligo-

---

[19] P. Knight and G. Offer, *Biochem. J.* **175**, 1023 (1978).
[20] L. A. Cohen, *Annu. Rev. Biochem.* **37**, 695 (1968).
[21] G. E. Means and R. E. Feeney, "Chemical Modification of Proteins," pp. 110–117. Holden-Day, San Francisco, California, 1971.

mer. Davies and Stark[22] demonstrated that, for oligomeric proteins in dilute solution, intramolecular, i.e., intraoligomer, cross-linking is very much favored over intermolecular cross-linking. They treated several oligomeric proteins with dimethyl suberimidate and analyzed the reaction mixtures by SDS–PAGE. For proteins known to be tetrameric, bands corresponding to monomers, dimers, trimers, and tetramers were clearly visible, with very little material running as higher-order polymers. Likewise, for a dimeric protein, bands corresponding to monomer and dimer were visible, with little material appearing at higher apparent molecular weights. The appearance of material running as higher-order polymers could be further reduced by carrying out the reaction at a lower concentration of protein. Lowering the protein concentration in this way would reduce the probability of random collisions between oligomers without affecting the spatial arrangement of subunits within an oligomer. Conversely, raising the protein concentrations would increase the likelihood of collisions between oligomers and would therefore increase the probability of formation of higher-order cross-linked polymers. At the upper limit of protein concentration, the distance between subunits on adjacent oligomers would be comparable to the distance between subunits within an oligomer. Such a situation is approached in the cytoplasm of erythrocytes in which the concentration of hemoglobin is approximately 34% (w/v).[23] Wang and Richards[24] treated intact erythrocytes with the membrane-permeant reagent dimethyl-3,3'-dithiobispropionimidate, lysed the cells, and analyzed the hemoglobin for cross-linking by SDS–PAGE. With increasing concentrations of reagent, hemoglobin oligomers up to octamers were observed.

For most native biological membranes, it is difficult to vary significantly the concentrations of the constituent proteins. Further, the overall concentration of proteins is quite high. Lowering the temperature at which a cross-linking reaction is carried out would have the effect of increasing the effective viscosity of the membrane and therefore of reducing the probability of collisions among the constituent proteins. However, lowering the temperature would also have the effect of lowering the rates of the chemical reactions between the target nucleophiles of the proteins and the cross-linking reagent. Despite these problems, many successful cross-linking experiments on membranes have been reported. Careful

---

[22] G. E. Davies and G. R. Stark, *Proc. Natl. Acad. Sci. U.S.A.* **66,** 651 (1970).
[23] J. W. Harris and R. W. Kellermeyer, "The Red Cell. Production, Metabolism, Destruction: Normal and Abnormal," 2nd Ed., p. 282. Harvard Univ. Press, Cambridge, Massachusetts, 1970.
[24] K. Wang and F. M. Richards, *J. Biol. Chem.* **250,** 6622 (1975).

optimization of reagent concentration is important. In addition, there is a compensating factor in the properties of membranes, in particular of plasma membranes of intact cells. That is, that many plasma membrane proteins appear to have hindered mobility, presumably due to interaction of these proteins with the cytoskeleton. This type of hindered mobility can be useful in a cross-linking experiment in helping to minimize random cross-linking due to collisions between molecules.

Intrasubunit Cross-Linking

Whether the objective of a cross-linking experiment is to obtain information about the tertiary or quaternary structure of a protein, intrasubunit cross-linking almost invariably occurs. In experiments designed to probe quaternary structure, it might seem that intrasubunit cross-linking would simply account for a nonproductive loss of reagent and would otherwise have no bearing on the results of the experiment. However, intrasubunit cross-linking can have an effect on the results of such an experiment, especially in those experiments in which the reaction mixture is analyzed by SDS–PAGE. This is due to the steric constraints placed on the protein when it is denatured in the presence of SDS in preparation for analysis by SDS–PAGE. Often a protomer containing an internal cross-link will form a complex with SDS that has a smaller Stoke's radius than does the native protomer–SDS complex. This cross-linked complex will then be observed on SDS–PAGE as a band migrating with *lower* apparent molecular weight than does the uncross-linked protomer. This effect can also be observed as multiple species of dimer, trimer, etc., on SDS–PAGE, reflecting the multiple conformations that these species may have due to different degrees and sites of intrasubunit cross-linking. The observation of this effect on SDS–PAGE separations of cross-linked species is usually more pronounced in separation systems that are more sensitive to subtle differences in conformation, such as the discontinuous system of Laemmli.[25] In continuous gel systems, such as that of Fairbanks *et al.*,[26] intrasubunit cross-linking usually results in a broadening of the bands corresponding to monomer, dimer, etc., without the appearance of discrete subspecies. However, one does often observe a decline in apparent molecular weight of each species as the degree of intrasubunit cross-linking becomes greater.

Intrasubunit cross-linking can reveal useful information concerning the tertiary structure of proteins, including membrane proteins. For ex-

---

[25] U. K. Laemmli, *Nature (London)* **227**, 680 (1970).
[26] G. Fairbanks, T. L. Steck, and D. F. H. Wallach, *Biochemistry* **10**, 2606 (1971).

ample, Jennings and Passow[27] used the bifunctional anion channel inhibitor H_2DIDS to probe the structure of the anion-exchange channel (band 3) in human erythrocytes. It had been previously shown that, when intact erythrocyes are treated with chymotrypsin, band 3 subunits are cleaved at a single site in the extracytoplasmic domain to yield two fragments with relative molecular weights of approximately 58,000 and 38,000 on SDS–PAGE.[28,29] Further, if cells were treated sequentially in isotonic buffer at physiological pH with $^3H_2DIDS$ and chymotrypsin in either order, SDS–PAGE analysis of the membrane proteins revealed that $^3H_2DIDS$ migrated primarily with the larger chymotryptic fragment of band 3.[28] If, after the initial incubation with H_2DIDS, the pH was adjusted to 9.5, the second isothiocyano group of H_2DIDS would react with a group (presumably a Lys) on the smaller fragment, thereby demonstrating the spatial proximity of residues of both of the chymotryptic fragments to the inhibitor binding site.[27] Careful kinetic studies of the reaction of H_2DIDS with the anion channel have revealed that over a wide range of pH, reaction with the larger fragment always occurs more rapidly than with the smaller fragment.[30] Since the isocyano group on H_2DIDS rarely reacts first with the smaller fragment, the intrasubunit cross-linking of band 3 by H_2DIDS appears to depend on the tethering effect, discussed above.

The advent of highly sensitive protein sequencing techniques,[31] together with the commercial availability of radiolabeled cross-linking reagents, should lead to more precise cross-linking studies of tertiary structure.

Development of Membrane-Impermeant Cross-Linking Reagents

Reagent Design

Four major characteristics had to be considered in the design of these reagents: membrane impermeance, reactive functionalities, span between reactive ends, and cleavability. In order to be membrane-impermeant, a reagent must be strongly hydrophilic, i.e., the oil:water partition coefficient must be essentially zero, so as to exclude the possibility of passive

---

[27] M. L. Jennings and H. Passow, *Biochim. Biophys. Acta* **554**, 498 (1979).
[28] Z. I. Cabantchik and A. Rothstein, *J. Membr. Biol.* **15**, 227 (1974).
[29] T. L. Steck, B. Ramos, and E. Strapazon, *Biochemistry* **16**, 2966 (1976).
[30] L. Kampmann, S. Lepke, H. Fasold, G. Gritzsch, and H. Passow, *J. Membr. Biol.* **70**, 199 (1982).
[31] R. M. Hewick, M. W. Hunkapillar, L. E. Hood, and W. J. Dreyer, *J. Biol. Chem.* **256**, 7990 (1981).

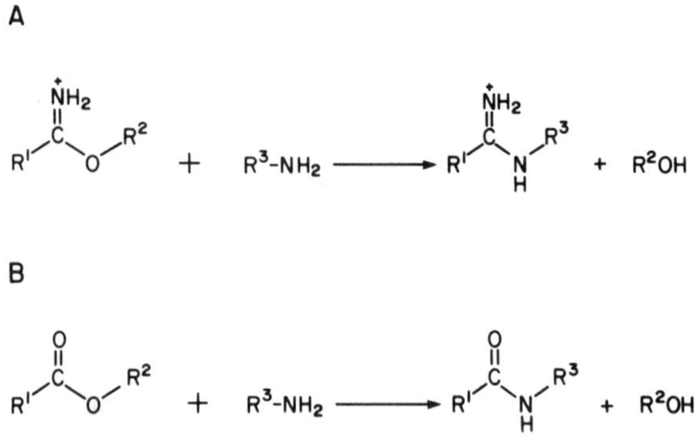

FIG. 1. Reactions of an alkyl imidate and active ester with amines. (A) An alkyl imidate can react with a primary amine to yield an amidine and an alcohol as a leaving group. (B) An active ester can react with an amine to yield an amide and an alcohol as a leaving group.

diffusion across the membrane. Also, the reagent should not bear too close a resemblance to a natural hydrophilic ligand, lest it be recognized by a specific permease and be transported across a membrane. Inclusion of ionic groups is a practical way of meeting these criteria. The ionic groups utilized must have pK_as far from neutrality. For example, alkyl imidates are ionizable, yet bifunctional alkyl imidates have been shown to freely permeate membranes.[24] Since an imidate has a $pK_a < 8$,[32] a significant fraction of the imidate will be uncharged, and the uncharged species apparently partition into membranes and thereby readily diffuse across them. One is, therefore, restricted to "harder" charges, e.g., phosphates, sulfates, sulfonates, and quaternary ammonium groups.

One potential problem with the inclusion of "hard" charges in the reagent is that modification of a protein with such a reagent might grossly alter the physical properties of the protein, rendering subsequent identification and separation difficult. A solution to this problem is to incorporate the ionic groups into a leaving group of the reactive functionality of the reagent, so that, on reaction with a target nucleophile in a protein, the ionic groups would be released. Alkyl imidates (Fig. 1A) and active esters (Fig. 1B) are two classes of reactive functionalities having leaving groups into which the introduction of ionic species is relatively straightforward. They have the further advantage of reacting readily with amino groups, which are the most common nucleophilic groups on the surface of proteins[33,34] and which are therefore excellent target nucleophiles for general-purpose cross-linking reagents.

[32] N. M. Whiteley and H. C. Berg, *J. Mol. Biol.* **87**, 541 (1974).
[33] M. H. Klapper, *Biochem. Biophys. Res. Commun.* **78**, 1018 (1977).
[34] C. Chothia, *J. Mol. Biol.* **105**, 1 (1976).

A question that must be addressed in the design of any cross-linking reagent is the distance that can be spanned by the reagent between the two protein nucleophilic groups that react with it. In a rigid molecule, such as 1,5-difluoro-2,5-dinitrobenzene,[35] the geometry and therefore the distance between the two groups that add to the reagent is completely defined. In a flexible cross-linker, the distance between addends is variable, in the limiting case, ranging from van der Waals contact to separation by the maximum span of the reagent. Our reagents have a flexible backbone, as this has the advantage in principle that, once reacted, a tethered reagent with a flexible backbone will have a higher probability of encountering a suitable nucleophile for reaction at the second end than would a rigid reagent. In choosing a maximum span for our reagents, we have borne in mind that the structural questions that we wish to address are intramolecular, whether it is the neighboring relationship of subunits in an oligomer or of residues within a single polypeptide chain. But what is a useful intramolecular distance to probe? One that might be useful is the distance of closest approach of neighboring units of secondary structure, e.g., α-helices, in a protein. For myoglobin, which has a largely α-helical structure, the center-to-center distance of closest approach for the most closely packed α-helices is 7.3–10.3 Å, with some interhelical contacts observed between helices up to 12 Å apart.[36] The reagents described below have spans in the upper part of this range.

Finally, there is the question of cleavability of the cross-link. For some applications, a noncleavable reagent is preferable; but for experiments in which it is important to be able to cleave the cross-link, how should it be done? Three commonly used cleavable bonds are disulfides, *vic*-hydroxyls, and azo groups. Disulfides can be cleaved under gentle reducing conditions; however, when disulfide-containing cross-linkers are used, reducing agents such as 2-mercaptoethanol or dithiothreitol cannot be used in the experiment between the cross-linking step and in the step in which the cross-link is meant to be cleaved. The use of *vic*-hydroxyls obviates the problems with reducing agents, as they are cleaved by oxidation with periodate; however, periodate also cleaves sugars in the carbohydrate moieties of glycoproteins, and the conditions for quantitative cleavage of *vic*-hydroxyls are somewhat more vigorous than are the conditions for disulfide cleavage. Azo bonds, like disulfides, are cleaved by reduction, though under substantially more vigorous conditions. For the cleavable reagents discussed here, we have incorporated disulfide linkages; for our purposes, the ease and gentleness of the cleavage step outweigh their limitations.

---

[35] H. Zahn, *Angew. Chem.* **67**, 561 (1955).
[36] T. J. Richmond and F. M. Richards, *J. Mol. Biol.* **119**, 537 (1978).

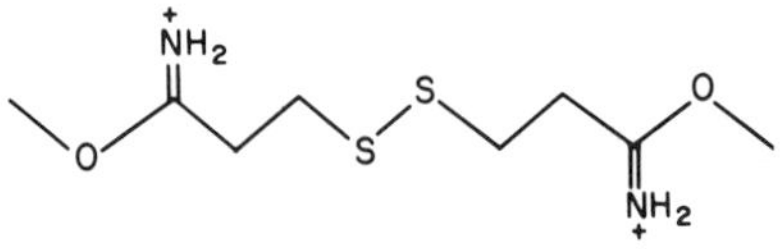

FIG. 2. Structures of DTBP and DIDIT.

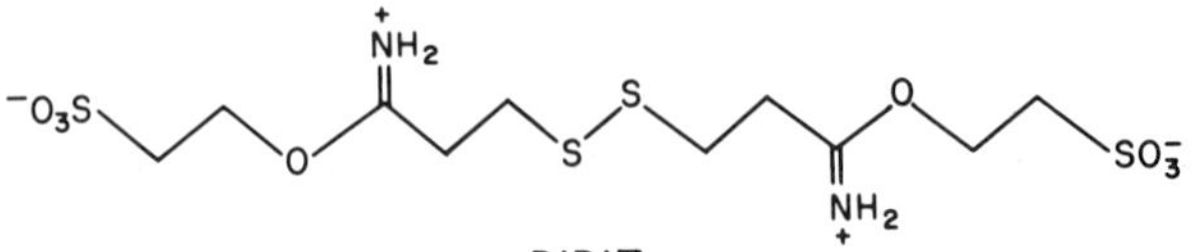

DIDIT

The first membrane-impermeant cross-linking reagent that we devised was the bis(alkyl imidate) DIDIT (Fig. 2).[15] This is an impermeant homolog of the Wang and Richards reagent DTBP,[14] in which the methoxy leaving groups of DTBP are replaced by 2-oxyethane sulfonate. This leaving group had previously been employed by Whitely and Berg in their monofunctional reagent isethionylacetimidate.[32]

Alkyl imidates, which are quite specific for primary amino groups, have the distinct advantage that, at neutral pH, the product amidines are charged like the primary amines from which they were derived (Fig. 1A).[37] Even extensive amidination does not markedly change the physical properties of a number of proteins.[38] Therefore, one can use them with some confidence that initial reaction with the reagent will not result in major structural changes that might be sampled by subsequent reaction.

In contrast, alkyl imidates have the disadvantage that the pH maximum for their reaction with primary amines is far above physiological, e.g., pH 9–10.5 for the reaction of methylacetimidate with α- or ε-amino groups.[37] Below this maximum, alkyl imidates hydrolyze very rapidly. Even when the reaction is carried out at high pH, hydrolysis can severely reduce the cross-linking yield. For example, Niehaus and Wold[38a] showed that treatment of erythrocyte membranes with dimethyl adipimidate at pH 9.6 resulted in extensive modification of lysyl residues; however, only 20% of the modified lysyl residues were cross-linked, while the remainder were modified with reagent that had hydrolyzed at the second end. In

---

[37] M. J. Hunter and M. L. Ludwig, this series, Vol. 25, p. 585.
[38] L. Wofsy and S. J. Singer, *Biochemistry* **2**, 104 (1963).
[38a] W. G. Niehaus, Jr. and F. Wold, *Biochim. Biophys. Acta* **196**, 170 (1970).

addition to the problems of low yields due to hydrolysis, below pH 8 a side reaction becomes significant in which N,N'-disubstituted amidines are formed by reaction of two amino groups with a single alkyl imidate.[39] The noncleavable cross-links thus formed can be a problem in experiments in which cleavability of the cross-link is important. (For a more in-depth discussion of the reaction mechanisms of alkyl imidates, the reader is referred to the review by Peters and Richards.[7])

Despite the limitations of alkyl imidates, DIDIT demonstrated that membrane-impermeant cross-linking reagents are feasible and that the strategy of incorporating a highly charged moiety in the leaving group is viable. DIDIT continues to be a useful reagent for experiments in which retention of charge on the reacted amino groups is essential.

BS^3 and DTSSP

For membrane-impermeant cross-linking reagents to have really broad utility, it was clear that substantial enhancements in yield would be necessary. An approach that had led to high-yield conventional cross-linking reagents was the use of active esters of N-hydroxysuccinimide,[40,41] which hydrolyze very slowly compared with their reaction with amino groups.[40] However, these reagents are typically quite hydrophobic, requiring dissolution in a water-miscible solvent for their introduction into aqueous solution.[40,41] Motivated by the successful application of a strongly charged leaving group in DIDIT, we synthesized a sulfonated analog of N-hydroxysuccinimide, N-hydroxysulfosuccinimide (Fig. 3),[17] and we and others used it to prepare a number of monofunctional and bifunctional active esters.[17,42–49]

The reactions of sulfosuccinimidyl active esters with water and with the various nucleophilic groups found in proteins have been studied with monofunctional model compounds.[45] Like succinimidyl active esters,[40] sulfosuccinimidyl active esters hydrolyze very slowly as compared with

---

[39] D. T. Browne and S. B. H. Kent, *Biochem. Biophys. Res. Commun.* **67**, 126 (1975).
[40] A. J. Lomant and G. Fairbanks, *J. Mol. Biol.* **105**, 243 (1976).
[41] P. F. Pilch and M. P. Czech, *J. Biol. Chem.* **254**, 3375 (1979).
[42] J. V. Staros and B. P. Kakkad, *J. Membr. Biol.* **74**, 247 (1983).
[43] N. J. Kotite, J. V. Staros, and L. W. Cunningham, *Biochemistry* **23**, 3099 (1984).
[44] A. H. Beth, T. E. Conturo, S. D. Venkataramu, and J. V. Staros, *Biochemistry* **25**, 3824 (1986).
[45] P. S. R. Anjaneyulu and J. V. Staros, *Int. J. Peptide Protein Res.* **30**, 117 (1987).
[46] J. B. Denny and G. Blobel, *Proc. Natl. Acad. Sci. U.S.A.* **81**, 5286 (1984).
[47] M. L. Jennings and J. S. Nicknish, *J. Biol. Chem.* **260**, 5472 (1985).
[48] F. R. Ludwig and F. A. Jay, *Eur. J. Biochem.* **151**, 83 (1985).
[49] J. A. Donovan and M. L. Jennings, *Biochemistry* **25**, 1538 (1986).

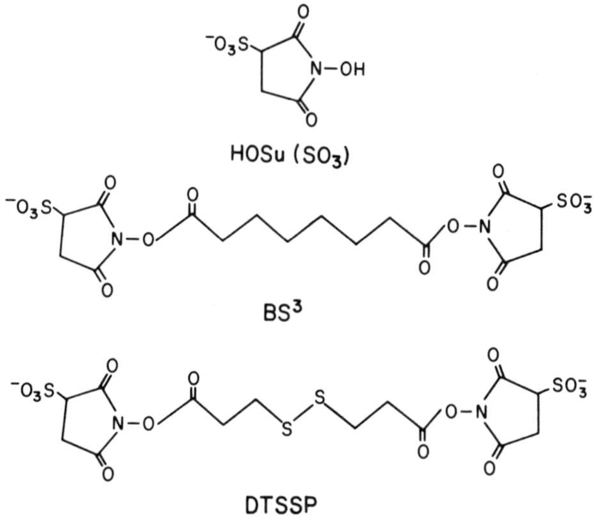

FIG. 3. Structures of N-hydroxysulfosuccinimide [HOSu(SO$_3$)], BS^3, and DTSSP.

their rates of reaction with nucleophiles, particularly nitrogen nucleophiles in proteins.[45] For example, in the absence of nucleophiles, the half-life at 25° of sulfosuccinimidyl propionate in 20 mM HEPES, pH 7.4, is 1.9 hr. When this compound was incubated with a variety of amino acid derivatives, the order of reactivity of nucleophilic functional groups was found to be imidazole > ε-amino ≈ α-amino ≫ thiolate ≈ phenolate. However, the acylimidazole was found to be a transient product, subject to hydrolysis or to reaction with another nucleophile.[45] These model studies suggest that the predominant stable adducts formed by the reaction of sulfosuccinimidyl esters with proteins are amides from the acylation of ε- and α-amino groups. However, as noted above, the tethering of one end of a cross-linker can have a substantial effect on the reaction specificity of the second reactive end. Thus, the high cross-linking yields observed with these reagents may in part be due to their ability to react with secondary nucleophiles once they are tethered.

The most commonly used cross-linking reagents employing sulfosuccinimidyl active esters are BS^3 and DTSSP (Fig. 3). A number of examples of the use of these reagents to problems of membrane protein structure and function are summarized in Table I. (There have also been many reported applications of these reagents in cross-linking studies of soluble protein complexes, but such studies are not the focus of this chapter.) The remainder of this article will address the preparation, handling, and application of these reagents.

TABLE I
APPLICATIONS OF MEMBRANE-IMPERMEANT CROSS-LINKING REAGENTS

Application	Reagents	Ref.
Tertiary and quaternary structure of erythrocyte anion channel	BS^3, DTSSP bis(sulfosuccinimidyl)adipate	a–d
Rotational diffusion of erythrocyte anion channel by EPR spectroscopy	Bis(sulfosuccinimidyl)-4-doxyl pimelate	e
Platelet membrane glycoproteins and their interaction with collagen	BS^3, DTSSP	f–h
Quaternary structure of receptor for IgE on mast cells	BS^3, DTSSP	i, j
Quaternary structure of photoreaction center of *Rhodopseudomonas viridis*	DTSSP, 4,4'dithiobis(sulfo-succinimidyl)butyrate	k
Quaternary structure of receptor for transforming growth factor beta	BS^3	l
Identification of atrial natriuretic factor receptors in adrenal membranes	BS^3	m
Identification of receptors for trophoblast-derived growth factor in placental membranes	BS^3, DTSSP	n
Association of anaphylatoxin C3a with chymase on surface of mast cells	BS^3	o

[a] J. V. Staros, *Biochemistry* **21,** 3950 (1982).
[b] J. V. Staros and B. P. Kakkad, *J. Membr. Biol.* **74,** 247 (1983).
[c] M. L. Jennings and J. S. Nicknish, *J. Biol. Chem.* **260,** 5472 (1985).
[d] M. L. Jennings, R. Monaghan, S. M. Douglas, and J. S. Nicknish, *J. Gen. Physiol.* **86,** 653 (1985).
[e] A. H. Beth, T. E. Conturo, S. D. Venkataramu, and J. V. Staros, *Biochemistry* **25,** 3824 (1986).
[f] S. M. Jung and M. Moroi, *Biochim. Biophys. Acta* **761,** 152 (1983).
[g] N. J. Kotite, J. V. Staros, and L. W. Cunningham, *Biochemistry* **23,** 3099 (1984).
[h] J. V. Staros, N. J. Kotite, and L. W. Cunningham, this series, in press.
[i] W. T. Lee and D. H. Conrad, *J. Immunol.* **234,** 518 (1985).
[j] J. V. Staros, W. T. Lee, and D. H. Conrad, this series, Vol. 150, p. 503.
[k] F. R. Ludwig and F. A. Jay, *Eur. J. Biochem.* **151,** 83 (1985).
[l] B. O. Fanger, L. M. Wakefield, and M. B. Sporn, *Biochemistry* **25,** 3083 (1986).
[m] S. Meloche, H. Ong, M. Cantin, and A. DeLean, *J. Biol. Chem.* **261,** 1525 (1986).
[n] A. Sen-Majumdar, U. Murthy, and M. Das, *Biochemistry* **25,** 627 (1986).
[o] J. E. Gervasoni, D. H. Conrad, T. E. Hugli, L. B. Schwartz, and S. Ruddy, *J. Immunol.* **136,** 285 (1986).

Methods

Synthesis of N-Hydroxysulfosuccinimide [HOSu(SO₃)] Sodium Salt[17]

HOSu(SO$_3$)Na is prepared by sulfitolysis of N-hydroxymaleimide in an inert atmosphere. [HOSu(SO$_3$)Na has recently become commercially available from at least two sources, Pierce Chemical Co. and Fluka Chemical Corp.] It is very important to keep the starting material, N-hydroxymaleimide, in an inert atmosphere. Therefore, solutions of this compound are prepared in a glove bag filled with N$_2$ and are transferred in a syringe to a reaction vessel that has been purged and refilled with N$_2$. Thus, N-hydroxymaleimide (Fluka Chemical Corp.) (1.45 g, 12.8 mmol) is dissolved in absolute ethanol (15 ml) and is transferred to the reaction vessel. An aqueous solution (10 ml) of Na$_2$S$_2$O$_5$ (1.22 g, 6.4 mmol) is then slowly added, with stirring. After 2 hr of stirring under N$_2$ at room temperature, the reaction vessel is opened to the atmosphere and the solvent is removed by rotary evaporation (bath temperature 40°). The product, a thick yellow oil, is dissolved in H$_2$O (50 ml), filtered, and lyophilized. The resulting light yellow solid is then triturated overnight with anhydrous ether, and the product, an off-white powder, is recovered by filtration. Typical yields are ~95%. The product gives a single spot when subjected to thin-layer chromatography on silica gel plates (0.20 mm, with fluorescent indicator on Al backing, from EM laboratories) developed in 1-butanol : acetic acid : water (5 : 2 : 3, v/v). The product gives a single peak when subjected to ion-pair reversed-phase HPLC using a C$_{18}$ column (Alltech, cat. no. 600RP) and a mobile phase of 60 : 40, aqueous 10 mM tetrabutylammonium formate, pH 4.0 : methanol, with monitoring at 254 nm. If necessary, the product may be recrystallized from 95% ethanol. For confirmation of structure, the product can be characterized by elemental analysis,[17] ^{1}H and ^{13}C NMR,[17] and by fast atom bombardment mass spectroscopy.[44,45]

Preparation of Cross-Linking Reagents: Synthesis of DTSSP[17]

Homobifunctional cross-linking reagents that are bis(sulfosuccinimidyl) esters of dicarboxylic acids can be prepared by coupling HOSu(SO$_3$) to the diacid with dicyclohexylcarbodiimide under anhydrous conditions. (A number of these reagents are now commercially available. BS^3 and DTSSP in nonradioactive form are available from Pierce Chemical Co. and [^{14}C]BS^3 is available from Amersham Corp.) The synthesis of DTSSP is presented as a specific example of the general method used to prepare such reagents.[17]

HOSu(SO$_3$)Na (0.44 g, 2.0 mmol), 3,3'-dithiodipropionic acid (Aldrich Chemical Co.) (0.21 g, 1.0 mmol), and N,N'-dicyclohexylcarbodiimide (Aldrich) (0.46 g, 2.2 mmol) are dissolved in 5.0 ml of anhydrous dimethylformamide. A magnetic stir bar is added, the vessel is capped, and the reaction mixture is stirred. Within a few minutes, N,N'-dicyclohexylurea, a side product of the reaction, can be seen precipitating from the reaction mixture. The reaction mixture is stirred overnight at room temperature, then the temperature is lowered to 3°, and stirring is continued for 2–3 hr to maximize the precipitation of the dicyclohexylurea. The dicyclohexylurea is removed by filtration and is washed with a small quantity of dry dimethylformamide. The product is then precipitated from the pooled filtrate by addition of 20 volumes of ethyl acetate. Sometimes it is helpful to store the product in suspension in ethyl acetate overnight at 3° to maximize precipitation. If this is done, the suspension should be allowed to come to room temperature before the product is recovered. The product is recovered by filtration and stored in a vacuum desiccator. Typical yields for DTSSP are ~65%, assuming a pure anhydrous product. Some of these reagents, e.g., BSSDP, are extremely hygroscopic, and it is very helpful to carry out the recovery of product in a dry atmosphere, e.g., in a N$_2$-filled glove bag. This can also be useful in the preparation of the less hygroscopic compounds when the humidity is high.

A useful method for characterizing these reagents is to dissolve a small quantity in water (or buffer) and analyze by ion-pair reversed-phase HPLC,[17,44] e.g., with the system described above. If analyzed immediately after dissolution, the product appears as a single major peak, though a small amount of free HOSu(SO$_3$) is usually seen as well. Subsequent injections, with increasing time of incubation in aqueous solution, yield relatively decreasing amounts of the cross-linker and increasing amounts of free HOSu(SO$_3$) and a transient peak corresponding to the monosulfosuccinimidyl ester of the diacid. After exhaustive hydrolysis, only HOSu(SO$_3$) and the starting diacid are observed. Fast atom bombardment mass spectroscopy is also useful for characterizing these reagents.[44]

Testing Reagents and Reaction Conditions: Cross-Linking of Rabbit Muscle Aldolase[17]

It is important to test the cross-linking activity of each new reagent or each new batch of reagent in a well-defined system. It is also important to compare the cross-linking activity of a reagent in new reaction conditions, e.g., a change of buffer, pH, or temperature, with its cross-linking activity in a well-defined system. For these purposes, a simple assay has been devised in which the reagent is tested for its ability to cross-link rabbit

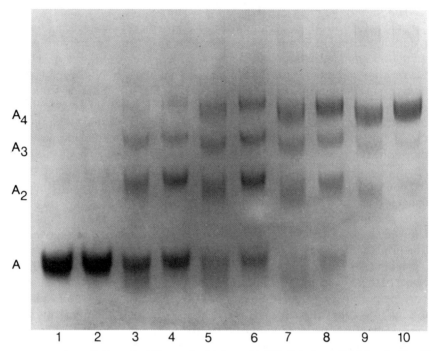

FIG. 4. Cross-linking of rabbit muscle aldolase with BS³. Solutions of aldolase in 50 mM sodium phosphate, pH 7.4 (even-numbered lanes), and 50 mM HEPES, pH 7.4 (odd-numbered lanes), were cross-linked with the following concentrations of BS³: lanes 1 and 2, no cross-linker; lanes 3 and 4, 50 μM; lanes 5 and 6, 0.20 mM; lanes 7 and 8, 0.50 mM; lanes 9 and 10, 1.0 mM. Samples were subjected to SDS–PAGE.[26] The Coomassie Blue-stained gel is shown. Positions of migration of aldolase monomers (A), dimers (A_2), trimers (A_3), and tetramers (A_4) are indicated to the left of the gel. See the text for experimental details.

muscle aldolase, a tetrameric protein,[50] under standard conditions.[17] In Fig. 4 are shown the results of such an assay in which two buffers are compared as media for the cross-linking of aldolase with BS³.

A suspension of crystalline rabbit muscle aldolase in 2.5 M $(NH_4)_2SO_4$ (type IV, Sigma Chemical Co.) is exhaustively dialyzed against 50 mM sodium phosphate, pH 7.4 (the standard buffer). If, as in this example, an alternative buffer is to be tested, a second sample of aldolase is dialyzed against that buffer. The final concentration of the dialyzed aldolase is determined by absorbance at 280 nm ($\varepsilon_{280}^{1\%} = 9.38$).[51] Aliquots of the pro-

[50] B. L. Horecker, O. Tsolas, and C. Y. Lai, in "The Enzymes" (P. D. Boyer, ed.), Vol. 7, 3rd Ed., p. 213. Academic Press, New York, 1972.
[51] J. W. Donovan, Biochemistry 3, 67 (1964).

tein solution are diluted with the appropriate buffer so that the final protein concentration, after addition of the cross-linker, is 1 mg/ml. The cross-linker is prepared as a fresh 10 mM stock solution in the appropriate buffer. Each stock solution is prepared *immediately* before use. After addition of the cross-linker, the reaction mixtures are incubated at room temperature for 30 min, and then are quenched by addition of one-sixth of a volume of 50 mM ethanolamine in the appropriate buffer, readjusted to pH 7.4. In experiments employing a cleavable (disulfide) cross-linker such as DTSSP, 20 mM N-ethylmaleimide is also added to the quench buffer so as to alkylate endogenous thiol groups and thereby block thiol–disulfide exchange. Other primary amines can replace ethanolamine in the quench buffer. Two-thirds of a volume of a solubilizing solution containing 3% (w/v) SDS, 25% (w/v) sucrose, 25 mM Tris-HCl, pH 8.0, 2.5 mM EDTA, 25 μg/ml of pyronin Y, with or without 0.10 M dithiothreitol, is added to each reaction mixture. In experiments employing noncleavable cross-linkers such as BS^3, the dithiothreitol is usually included. For experiments employing cleavable cross-linkers such as DTSSP, the sample is usually split, and half is solubilized in the absence and half in the presence of dithiothreitol. The samples are then heated at 50° for 3 min and then subjected to SDS–PAGE.[26]

In the example shown in Fig. 4, HEPES is compared with sodium phosphate as an appropriate buffer for cross-linking with BS^3. Though the products of cross-linking in these two buffers appear to migrate somewhat differently, the yields of cross-linked aldolase subunits obtained in the two buffers are essentially identical. This result is in concert with model studies that show that hydrolysis of sulfosuccinimidyl esters, the most important side reaction that can limit cross-linking yields, occurs at similar rates with these two buffers.[45]

Cross-Linking of the Anion-Exchange Channel in Intact Erythrocytes with DTSSP[17]

The anion-exchange channel of the human erythrocyte has been the subject of a number of studies employing bifunctional sulfosuccinimidyl ester cross-linking reagents.[17,42,44,47,52,53] At low concentrations, these reagents act as affinity labels for the anion-binding site, producing an intrasubunit cross-link.[44,47,52,53] This affinity for the anion-binding site has been

---

[52] M. L. Jennings, R. Monaghan, S. M. Douglas, and J. S. Nicknish, *J. Gen. Physiol.* **86**, 653 (1985).
[53] A. H. Beth, T. E. Conturo, N. A. Abumrad, and J. V. Staros, in "Membrane Proteins: Proceedings of the Membrane Protein Symposium" (S. C. Goheen, ed.), pp. 371–382. Bio-Rad Laboratories, Richmond, CA, 1987.

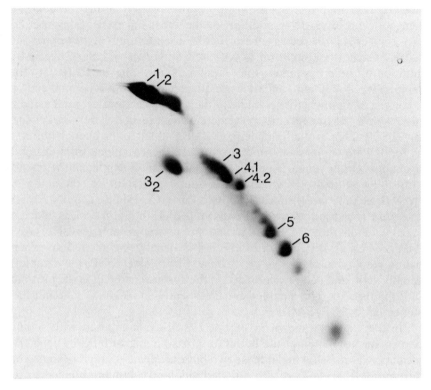

FIG. 5. Cross-linking of the anion channel in intact human erythrocytes with DTSSP. Washed erythrocytes that had been treated with 5.0 mM DTSSP were lysed, and membranes were prepared as described in the text. The membranes were solubilized and subjected to the two-dimensional SDS–PAGE technique devised by Wang and Richards,[14] in which the sample is run under nonreducing conditions in the first dimension, then under reducing conditions to cleave the cross-link in the second. Proteins that were not cross-linked appear as a diagonal pattern in the gel. Proteins which had been cross-linked to form higher-molecular-weight species run more slowly in the first dimension (when they are cross-linked) than they do in the second dimension (when the cross-link is cleaved) and so appear as spots to the left of the diagonal. In the pattern shown, most of the membrane-associated proteins, including spectrin (labeled 1 and 2), actin (labeled 5), and glyceraldehyde-3-phosphate dehydrogenase (labeled 6) fall entirely on the diagonal, indicating that they had not been cross-linked. However, a spot labeled 3_2 appears to the left of the position on the diagonal at which subunits of the anion channel migrate (labeled 3). The mobility in the first dimension of the complex that gave rise to 3_2 was approximately that of two anion channel subunits. Further, there are no other off-diagonal spots in vertical register with 3_2, suggesting that 3_2 arose from a dimer of anion channel subunits cross-linked by DTSSP in the extracytoplasmic domain.

exploited in studies in which BSSDP, a bifunctional sulfosuccinimidyl cross-linker containing a nitroxide free radical, has been employed to specifically label the anion channel for studies of the rotational dynamics of the ion channel in intact erythrocytes.[44,53] At higher concentrations of reagent, cross-links are formed between the two subunits of the anion channel.[17,42,44] The results of an experiment of the latter type are shown in Fig. 5.

For such an experiment, erythrocytes are prepared from a fresh sample of human blood drawn into acid–citrate–dextrose or heparin.[54] After several washes in phosphate-buffered saline, the cells are washed once in 106 mM sodium phosphate, pH 7.4, and are resuspended in the same buffer at a hematocrit of approximately 50%. The cross-linker is then added from a freshly prepared 10 mM stock solution in the same buffer and the cell suspension is gently agitated for 30 min at room temperature. The reaction is then quenched by addition of 10 volumes of 0.15 M NaCl, 25 mM Tris-HCl, pH 7.4. The cells are pelleted by centrifugation, and the A_{550} of the supernatant is measured to monitor lysis during the reaction. By this measure, lysis is typically <0.5%, using the supernatant of a control lysate for comparison. After one additional wash in the quench buffer, the cells are lysed at a hemolytic ratio of 80:1 in 0.1 mM Na$_2$EDTA, 20 mM N-ethylmaleimide, adjusted to pH 7.4 with Tris. The N-ethylmaleimide is included to alkylate free thiols, as mentioned above in the section on cross-linking of aldolase. The resulting ghosts are washed 2–3 times in 15 mM NaCl, 1.7 mM Tris-HCl, 0.1 mM Na$_2$EDTA, pH 7.4, resuspended in the same buffer, and solubilized in the SDS-gel solubilizing solution described above, *without* dithiothreitol.

Conclusions

Membrane-impermeant cross-linking reagents based on sulfosuccinimidyl ester chemistry have made accessible several new approaches to the study of membrane protein structure and function. The impermeant character of these reagents allows the experimenter to address not only the question of which subunits of an oligomeric membrane protein are proximate to which others, but also at which face of the membrane such intersubunit contacts occur. The high yields allow the experimenter to design cross-linking experiments involving proteins which comprise a very small fraction of the total membrane protein, especially when such cross-linking studies utilize immunochemical methods of detection, such

---

[54] J. V. Staros and F. M. Richards, *Biochemistry* **13**, 2720 (1974).

as immunoblotting.[55] Spectroscopic ligands can be incorporated into the reagents, and these can report dynamic parameters of membrane proteins of interest.[44,53] The commercial availability of radiolabeled cross-linkers, together with highly sensitive methods of protein sequencing,[31] should allow more precise tertiary structural studies in which the spatial proximity of specific residues in the folded protein can be directly ascertained.

Acknowledgments

Research grant support from the National Institutes of Health (R01 DK31880) is gratefully acknowledged. The authors thank J. Stephens for excellent technical assistance and S. Heaver for assistance in the preparation of the manuscript.

[55] K. Wang, B. O. Fanger, C. A. Guyer, and J. V. Staros, this volume [35].

[34] Photochemical Labeling of Apolar Phase of Membranes

By JOSEF BRUNNER

Introduction

The purpose of this chapter is to review the method of photolabeling of the apolar core of membranes. Introduced in the mid-1970s,[1,2] this method has since evolved into an important tool in studies of a variety of problems related to membrane structure and dynamics.

A number of earlier review articles have appeared on hydrophobic photolabeling of membranes.[3-8] For this reason the present article focuses

[1] A. Klip and C. Gitler, *Biochem. Biophys. Res. Commun.* **60,** 1155 (1974).
[2] P. Chakrabarti and H. G. Khorana, *Biochemistry* **14,** 5021 (1975).
[3] J. Brunner, *Trends Biochem. Sci.* **6,** 44 (1981).
[4] H. Bayley, in "Membranes and Transport" (A. Martonosi, ed.), Vol. 1, p. 185. Plenum, New York, 1982.
[5] H. Bayley, "Photogenerated Reagents in Biochemistry and Molecular Biology." Elsevier/North-Holland, Amsterdam, 1983.
[6] R. Bisson and C. Montecucco, in "Progress in Protein–Lipid Interactions" (A. Watts and J. J. H. H. M. de Pont, eds.), p. 259. Elsevier, Amsterdam, 1985.
[7] R. Bisson and C. Montecucco, in "Techniques for the Analysis of Membrane Proteins" (R. Cherry and C. I. Ragan, eds.), p. 153. Chapman & Hall, London, 1986.
[8] R. J. Robson, R. Radhakrishnan, A. H. Ross, Y. Takagaki, and H. G. Khorana, in "Lipid–Protein Interactions" (P. C. Jost and O. H. Griffiths, eds.), Vol. 2, p. 149. Wiley, New York, 1982.

on more recent ideas and developments. Because of the scope of this series, emphasis is laid on the methodological aspects of labeling and, equally important, the analysis of labeled proteins. This review is divided into three parts. The first is concerned with the general principles of hydrophobic photolabeling and includes paragraphs describing the various areas of usefulness, the requirements with respect to the reagents, the properties of some photogenerated intermediates and their precursors, the design and synthesis of reagents, and, finally, the main elements of a typical labeling experiment. In the second part, a number of applications are reviewed. Following a discussion of the results obtained from labeling the human erythrocyte membrane with various types of reagents, the remaining sections are devoted to studies with reagents containing the 3-trifluoromethyl)-3-phenyl-3H-diazirine group, a particularly attractive photoactivatable unit. A principal aim is to outline and discuss the recent developments in what appears to become increasingly important, the identification of individual labeled amino acids within a polypeptide chain. Finally, the Appendix contains a list of most of the reagents thus far developed for the purpose of hydrophobic photolabeling of membranes with references to their syntheses and applications.

Areas of Application

The main goal of this chapter is to give an overview of the various areas in which hydrophobic photolabeling of membranes has been successfully applied, or is of potential use. Initially, however, it seems worthwhile to describe briefly the main features of the reagents thus far developed for this purpose.

From a physical point of view, the reagents can be divided into two major categories, hydrophobic and amphipathic ones. In the presence of membranes, reagents of the former type can be assumed to dissolve and diffuse freely within the entire hydrocarbon core of the lipid bilayer. In this respect, they differ from amphipathic compounds which, when inserted into membranes, show restricted transverse mobility (flip-flop).[9–13] This has important implications since it provides a basis for labeling mem-

---

[9] L. W. Johnson, M. E. Hughes, and D. B. Zilversmit, *Biochim. Biophys. Acta* **375**, 176 (1975).
[10] B. Bloj and D. B. Zilversmit, *Biochemistry* **15**, 1277 (1976).
[11] A. Rousselet, C. Guthmann, J. Matricon, A. Bienvenue, and P. F. Devaux, *Biochim. Biophys. Acta* **426**, 357 (1976).
[12] J. Brunner, M. Spiess, R. Aggeler, P. Huber, and G. Semenza, *Biochemistry* **22**, 3812 (1983).
[13] R. G. Sleight and R. E. Pagano, *J. Biol. Chem.* **260**, 1146 (1985).

branes in an asymmetrical fashion, either from within a single leaflet, or from within different depths of the lipid bilayer.

The vast majority of the photoactivatable reagents is based on nitrene or carbene chemistry. The corresponding light-sensitive precursors are azides, diazo compounds, and diazirines. In spite of the poor characterization of most of the transient reactive species, it is nonetheless clear that they encompass a very wide range regarding reactivity and chemical selectivity. While the most reactive ones efficiently insert into C—H bonds of alkyl residues, others exhibit pronounced chemical selectivity and form covalent adducts with nucleophiles only (e.g., with NH_2 or SH groups). Undoubtedly, these differences in reactivity and chemical selectivity are the main reasons why, in several cases, labeling of the same membrane with different types of reagents has led to conflicting results and conclusions.

Due to the high reactivity of photogenerated intermediates, they react predominantly with the "solvent" molecules, the phospholipids. Labeling of integral proteins is, in general, relatively inefficient and, particularly in cases in which the protein of interest is present at low abundancy (receptors, carriers, etc.), or has a small membrane-embedded portion, analysis of label distribution patterns may require extremely sensitive methods. In general, reagents containing an isotope of high specific radioactivity are employed. It should be pointed out that quite often the introduction of a suitable isotope, of high activity, in a stably bound form, is the most difficult step in the design and synthesis of a reagent.

Labeling of Integral Proteins and of Their Membrane-Embedded Domains

The most common goal of hydrophobic photolabeling is the identification of integral proteins or of their domains which make contact with the lipid bilayer. Presumably, it is the most direct method for that purpose, and in several cases this technique has rendered possible or considerably facilitated the identification and subsequent isolation of such polypeptides.[14–20]

[14] M. Spiess, J. Brunner, and G. Semenza, *J. Biol. Chem.* **257,** 2370 (1982).
[15] T. Frielle, J. Brunner, and N. P. Curthoys, *J. Biol. Chem.* **257,** 14979 (1982).
[16] S. Stieger, U. Brodbeck, B. Reber, and J. Brunner, *FEBS Lett.* **168,** 231 (1984).
[17] R. A. Nicholas, *Biochemistry* **23,** 888 (1984).
[18] T. A. Dutta-Choudhury and T. L. Rosenberry, *J. Biol. Chem.* **259,** 5653 (1984).
[19] I. Kahan and M. A. Moscarello, *Biochemistry* **24,** 538 (1985).
[20] J. Giraudat, C. Montecucco, R. Bisson, and J.-P. Changeux, *Biochemistry* **24,** 3121 (1985).

In order to be able to make unambiguous assignments, it is necessary to achieve a high selectivity in labeling integral versus peripheral proteins. Furthermore, it is essential that all integral proteins are labeled and not only those containing particularly reactive amino acid side chains. The selective or preferential labeling of integral proteins is generally assumed to be the result of high lipid–water partition coefficient of the reagent and the short lifespan of the reactive intermediate(s). Although these are, no doubt, important factors, one should also take into consideration that within a lipid bilayer a (singlet) carbene or nitrene can add much more easily to a biomolecule than within an aqueous environment where such species are scavenged essentially instantaneously.

Mapping the Protein–Lipid Interface

If, as suggested in the previous section, labeling is confined to the hydrophobic core of a membrane, it should be possible to identify those polypeptide segments that make contact with the bilayer. Increased topological information [for example, the orientation of a polypeptide across a bilayer, or the relative position (depth) of (an) individual amino acid residue(s)] may then be obtained by restriction of the labeling to spatially definable subcompartments of the membrane.

With the current (amphipathic) reagents it is possible to confine labeling to a single leaflet (the outer or inner of a closed vesicle) or to relatively thin layers near the lipid–water interface. The selective labeling of a (narrow) central layer appears to be a more difficult task. The difficulties seem to originate mainly from the fluidity of a membrane and from what can be described as "looping back" of fatty acyl chains. It appears as if a photoreactive group that is attached to a (flexible) fatty acyl chain would sample an increasingly broad distribution across a lipid bilayer the further away from the polar (and relatively fixed) end of the chain this group is positioned. However, as suggested by Bayley,[4,5] labeling of membranes at a defined depth within the bilayers might be possible with reagents more rigid than the ones so far used.

The true problem in "mapping" the lipid exposed surface of integral proteins, as outlined above, originates from the lack of efficient methods for determining individual labeled amino acid residues within a polypeptide chain. In spite of recent progress made in the isolation, purification, and characterization of photolabeled hydrophobic polypeptides, this kind of work is not straightforward and, in general, requires efforts of substantial proportions. However, in those few cases where the covalently bound label has been traced back to individual amino acid residues, interesting details concerning the folding patterns of these segments have been unraveled. Naturally, it is one of the main objectives of what follows to discuss the main reasons for these difficulties and to carefully evaluate the various approaches so far pursued to overcome these problems.

Studying Lipid–Lipid and Lipid–Protein Interactions

The (relative) extent to which an integral protein is labeled depends, among other factors, upon the amount (fraction) of reagent that, at the time of photoactivation, is present within a certain neighborhood of this protein. While it can be assumed that this fraction is similar for simple hydrophobic molecules (containing the same type of photoreactive group), marked differences would probably result when labeling is carried out with photoactivatable phospholipids, due to a nonuniform distribution of lipids in the plane of the bilayer and because of specific interactions of at least some proteins with certain types of lipids.[21–28]

Photoactivatable lipids which cross-link to structures in their immediate vicinity are potentially powerful probes to study lipid–lipid (lateral distribution, patching) and lipid–protein interactions. Chakrabarti and Khorana,[2] who first gave a detailed account of this general approach, outlined a scheme for the synthesis of fatty acids and phospholipids containing a photoactivatable group. As shown in the Appendix, in the meantime quite a respectable number of such analogs have been synthesized. Most of the phospholipids are of the lecithin type (glycerophosphorylcholines) but, especially over the last few years, several other lipids (e.g., sphingomyelins, phosphatidylinositol) have been made as well. A most remarkable and encouraging finding has been that fatty acids containing a photoreactive group are incorporated into membranes of *Escherichia coli* (an auxotroph strain which requires unsaturated fatty acids for growth).[29,30] More recently, biosynthetic utilization of photoreactive fatty

[21] W. Kundig and S. Roseman, *J. Biol. Chem.* **246,** 1407 (1971).
[22] L. S. Milner and H. R. Kaback, *Proc. Natl. Acad. Sci. U.S.A.* **65,** 683 (1970).
[23] P. Ronner, P. Gazzotti, and E. Carafoli, *Arch. Biochem. Biophys.* **179,** 578 (1977).
[24] B. Roelofsen and H. J. Schatzmann, *Biochim. Biophys. Acta* **464,** 17 (1977).
[25] H. Sandermann, *Biochim. Biophys. Acta* **515,** 209 (1978).
[26] P. Gazzotti, H.-G. Bock, and S. Fleischer, *Biochem. Biophys. Res. Commun.* **58,** 309 (1974).
[27] S. B. Vik, G. Georgevich, and R. A. Capaldi, *Proc. Natl. Acad. Sci. U.S.A.* **78,** 1456 (1981).
[28] Y. Nishizuka, *Trends Biochem. Sci.* **8,** 13 (1983).
[29] G. R. Greenberg, P. Chakrabarti, and H. G. Khorana, *Proc. Natl. Acad. Sci. U.S.A.* **73,** 86 (1976).
[30] S. C. Quay, R. Radhakrishnan, and H. G. Khorana, *J. Biol. Chem.* **256,** 4444 (1981).

acids into lipids has also been reported for mammalian cells lines in tissue culture.[31-35] As noted above, upon activation these lipids can form covalent adducts with neighboring molecules and it is obvious that corresponding cross-linking patterns can provide useful information concerning the structural basis of both lipid–lipid and lipid–protein interactions.

Studying Dynamic Phenomena

Lipid phase photolabeling also exhibits large potential for the investigation of various aspects of time-dependent changes in membranes. A main prerequisite is that the reagent can be activated within a period of time that is short when compared with the duration of the process under investigation. Furthermore, in order to define the state of the system at the time point of reagent activation and labeling, the process to be studied must somehow be synchronized. With the development of some diazirine-based reagents which are both highly photolabile and give rise to a single, short-lived reactive intermediate, the former requirement can surely be satisfied. Time-resolved photolabeling was first used by Wisniesky and Bramhall[36,37] in exploring the penetration of cholera toxin into the membrane of Newcastle disease virus. These experiments nicely illustrate the potential of photolabeling to the study of dynamic phenomena. Evidently, a wide area of interesting problems is now open to experimental attack.

Requirements for Reagents Designed to Label the Lipid Core of Membranes

In order to perform successfully the types of experiments outlined in the previous section, the reagents have to meet several requirements. First, and most important, the reactive species must be capable of inserting into C—H bonds of paraffinic chains. This is because of the fact that membranous polypeptide segments are extremely rich in amino acids containing aliphatic side chains; some may consist exclusively of such amino acids. Although it was implicated by earlier work, and notably that

---

[31] W. Stoffel, K. Salm, and U. Körkemeier, *Hoppe-Seyler's Z. Physiol. Chem.* **357**, 917 (1976).
[32] A. Franchi and G. Ailhaud, *Biochimie* **59**, 813 (1977).
[33] P. Leblanc, J. Capone, and G. E. Gerber, *J. Biol. Chem.* **257**, 14586 (1982).
[34] J. Capone, P. Leblanc, G. E. Gerber, and H. P. Ghosh, *J. Biol. Chem.* **258**, 1395 (1983).
[35] P. Leblanc and G. E. Gerber, *Can. J. Biochem. Cell. Biol.* **62**, 375 (1984).
[36] B. J. Wisnieski and J. S. Bramhall, *Biochem. Biophys. Res. Commun.* **87**, 308 (1979).
[37] B. J. Wisnieski and J. S. Bramhall, *Nature (London)* **289**, 319 (1981).

by Westheimer and colleagues on photoaffinity labeling,[38–41] that such species can be generated *in situ* by photodissociation of a suitable precursor, it has become clear lately that this is not a trivial reaction and, as will be evident from the subsequent discussion, much of the progress made in this field can be directly ascribed to the development of molecules with this ability. Photogenerated intermediates which, under normal conditions, are very poor reagents for insertion into aliphatic residues are arylnitrenes (or the products derived therefrom by intramolecular rearrangements), the group of reagents most often used in the past.[4,42,43] Species which do insert are nitrenes generated from other sources and, above all, carbenes which may be obtained by photolysis of two main groups of precursors: diazo compounds and diazirines.[44–49] As discussed below, several of these precursors suffer, however, from other disadvantages.

A second requirement is that irradiation of the sample does not per se cause damage to the biological system. Since both proteins and nucleic acids absorb below 300 nm, the reagent should be activatable with light above this wavelength and photolysis should occur more rapidly than the rate at which biological samples are damaged. Many aliphatic diazo compounds and azides, which otherwise would combine favorable properties, cannot be tolerated for most applications because of the harsh conditions they need for activation.

Third, photoactivation of the reagent should yield a single reactive (short-lived) intermediate that does not undergo rearrangement to other, less-reactive species. Unfortunately, photoinduced isomerizations and intramolecular rearrangements are quite common in carbene and nitrene chemistry and often lead to long-lived intermediates which themselves can react and interfere with photolabeling. A well-known case is the

---

[38] A. Singh, E. R. Thornton, and F. H. Westheimer, *J. Biol. Chem.* **237**, PC3006 (1962).

[39] J. Shafer, P. Baronowsky, R. Laursen, F. Finn, and F. H. Westheimer, *J. Biol. Chem.* **241**, 421 (1966).

[40] H. Chaimovich, R. J. Vaughan, and F. H. Westheimer, *J. Am. Chem. Soc.* **90**, 4088 (1968).

[41] R. J. Vaughan and F. H. Westheimer, *J. Am. Chem. Soc.* **91**, 217 (1969).

[42] H. Bayley and J. R. Knowles, this series, Vol. 46, p. 69.

[43] H. Bayley and J. V. Staros, *in* "Azides and Nitrenes" (E. F. V. Scriven, ed.), p. 433. Academic Press, New York, 1984.

[44] W. Kirmse, "Carbene Chemistry," Academic Press, New York, 1964.

[45] W. Lwowski, "Nitrenes." Wiley, New York, 1970.

[46] W. Lwoswski, *in* "Reactive Intermediates" (M. Jones and R. A. Moss, eds.), Vol. 1, p. 197. Wiley, New York, 1978.

[47] R. A. Moss and M. Jones, "Carbenes," Vol. 1. Wiley, New York, 1973.

[48] R. A. Moss and M. Jones, "Carbenes," Vol. 2. Wiley, New York, 1975.

[49] R. A. Moss and M. Jones, *in* "Reactive Intermediates" (M. Jones and R. A. Moss, eds.), Vol. 1, p. 69. Wiley, New York, 1978.

photolysis of diazoacetic acid esters which leads in large measure to the ketene, the product of the photochemical analog of the Wolff rearrangement.[40,41] Other examples will be discussed in the following sections.

The (synthetic) availability of the photolabeling group, its chemical versatility (introduction of additional functional groups), and chemical stability in the dark are further important points to be considered. In general, it is also essential that the reagent can be prepared with a high specific radioactivity. Since most analyses of labeled membranes rely solely on distributions of radiolabel among individual proteins or protein fragments, it is crucial that this isotope is firmly connected with the photoreactive moiety of the reagent so that it is not lost during work-up of the labeled components. Finally, it should be reiterated that the reagent must partition to a large extent into the lipid bilayer of a membrane and that it should be moderate in size in order to minimize steric perturbation of the membrane.

Properties of Some Photogenerated Intermediates and Their Precursors

The chemical modification of membranous polypeptides that do not contain reactive amino acid side chains represents a central problem in hydrophobic photolabeling. In fact, the method relies to a considerable extent on the availability of reagents which display low chemical selectivity and are capable of attacking even the most inert amino acid side chains. It is not surprising, therefore, that the search for photoactivatable groups endowed with this capability has played a decisive role in the development of this method. For this reason, the main features of nitrene and carbene chemistry shall be outlined and, where appropriate, discussed in detail.

Some General Properties of Carbenes and Nitrenes

The chemistry of nitrenes and carbenes is complex and in detail poorly understood.[50-54] For example, little is known concerning absolute or even relative rates of those reactions that lead to labeling of biological targets.

---

[50] A. Reiser and H. M. Wagner, in "The Chemistry of the Azido Group" (S. Patai, ed.), p. 441. Wiley (Interscience), New York, 1971.
[51] C. Wentrup, "Reaktive Zwischenstufen I," Thieme, Stuttgart, Federal Republic of Germany, 1979.
[52] N. Turro, "Modern Molecular Photochemistry." Benjamin/Cummings, Menlo Park, California, 1979.
[53] E. F. V. Scriven, in "Reactive Intermediates" (R. A. Abramovitch, ed.), p. 1. Plenum, New York, 1982.
[54] P. A. S. Smith, in "Azides and Nitrenes" (E. F. V. Scriven, ed.), p. 95. Academic Press, New York, 1984.

Moreover, many of the properties of carbenes have been derived from experiments carried out in pure organic solvents, in the gas phase or in matrix-isolated systems at low temperatures. These are conditions that are vastly different from those that exist within the two-phase system of a membrane. This must be considered in any attempt to interpret photochemical labeling results at the molecular level.

The most simple members of the family of nitrenes and carbenes are methylene and nitrene: CH_2 and NH. Formally, they correspond to the products of the abstraction of two hydrogen atoms each from methane and ammonia, respectively. Nitrenes and carbenes are isoelectronic species with only six valence electrons. A consequence of this electron deficiency is that they normally behave as strong electrophiles. Although carbenes are, in general, more reactive than related nitrenes, their chemistry is at least qualitatively similar.

Nitrenes and carbenes can exist in two energetically proximate states, referred to as the singlet and triplet states. It is this electronic configuration which determines the types of reactions a nitrene or carbene will undergo. Typical reactions of a singlet are (1) electrophilic attack of π-electron pairs (e.g., addition to C=C double bonds), (2) electrophilic attack of σ-electron pairs (insertion into C—H bonds), (3) electrophilic attack of a nonbonding electron pair, and (4) rearrangements to fully bonded molecules (e.g., Wolff rearrangement). Typical reactions of triplet intermediates are (1) radical attack on σ-bonds (e.g., hydrogen abstraction to form a triplet radical pair), (2) radical attack on π-bonds (e.g., addition to a C=C bond to form a triplet diradical), and (3) radical attack on a nonbonding electron pair. Triplets are much less likely to undergo rearrangements than singlet species because they can do so only after intersystem crossing.

The direct ultraviolet irradiation of azides, diazo compounds, and diazirines yields singlet nitrenes and carbenes. This is because these precursors normally exist in the singlet state and due to a principle requiring that the spin state must be preserved in the ensuing reaction (Wigner's spin rule). A triplet state nitrene or carbene can be obtained, however, indirectly by the decomposition of an azide, diazo compound, or diazirine that has initially been converted to the triplet state with a special photosensitizing agent. For common photolabeling purposes, sensitized photolysis has not been used so far. However, triplet states derived from irradiation of α,β-unsaturated ketones, aryl and diaryl ketones, are in several respects promising photolabeling groups.[55-57] For certain applications,

---

[55] R. E. Galardy, L. C. Craig, and M. P. Printz, *Nature (London) New Biol.* **242**, 127 (1973).
[56] R. J. Martyr and W. F. Benisek, *Biochemistry* **12**, 2172 (1973).
[57] P. Campell and T. L. Gioannini, *Photochem. Photobiol.* **29**, 883 (1979).

such reagents even have an advantage over nitrenes and carbenes in that they are inert toward water and, hence, are not wasted by the solvent. Triplet-mediated cross-linking of surfactants containing a keto group have been used, for example, as probes of the microenvironment of micelles and model membranes.[58-60] The triplet excited state of ketones and the hydrogen atom abstraction reaction have been well studied from a photophysical as well as chemical standpoint and do not require further treatment here.

Nitrenes and Their Precursors

Nitrenes Generated from Arylazides. Arylazides are, in general, easy to synthesize, are relatively stable over a wide range of conditions, and exhibit satisfactory or excellent spectral properties (they absorb around 300 nm or above, depending on additional ring substituents). These are the main reasons for their frequent use in photolabeling reagents. Photochemical experiments in solution with simple lipid bilayers and with biological membranes have shown, however, that the generated intermediates are not very reactive by photochemical standards and are relatively specific for certain functional groups.

The principal elements of arylazide photochemistry, and especially those relevant to labeling, have been reviewed recently by Bayley and Staros.[43] Using laser flash photolysis, Schrock and Schuster[61] were able to provide further details with respect to the various transients generated in both inert and reactive (nucleophilic) solvents at room temperature. There is little doubt now that the bicyclic aziridine and 1,2,4,6-azacyloheptatetraene are the intermediates primarily responsible for reactions with nucleophiles in the photolysis of phenylazide at ambient temperature in organic solution.[61] It is also clear that substituents on phenylazide may dramatically change the position and rate of attaining equilibrium of the singlet intermediates. For example, pentafluorophenylazide does not give the above mentioned species and the initial singlet nitrene rapidly and irreversibly intersystem crosses to the triplet.[62] A related situation appears to exist also for *p*-nitrophenylazide which does not give nucleophile incorporation products.[63]

Simple nonsubstituted arylnitrenes are inefficient reagents for labeling

---

[58] R. Breslow, S. Kitabatake, and J. Rothbard, *J. Am. Chem. Soc.* **100**, 8156 (1978).
[59] M. F. Czarniecki and R. Breslow, *J. Am. Chem. Soc.* **101**, 3675 (1979).
[60] J. R. Winkle, P. R. Worsham, K. S. Schanze, and D. G. Whitten, *J. Am. Chem. Soc.* **105**, 3951 (1983).
[61] A. K. Schrock and G. B. Schuster, *J. Am. Chem. Soc.* **106**, 5228 (1984).
[62] I. R. Dunkin and P. C. P. Thomson, *J. Chem. Soc., Chem. Commun.* 1192 (1982).
[63] H. Nakayama, M. Nozawa, and Y. Kanaoka, *Chem. Pharm. Bull.* **27**, 2775 (1979).

alkyl residues in a lipid bilayer. This was shown first by Bayley and Knowles[64] in an experiment in which dimyristoyllecithin liposomes were equilibrated with phenylazide, irradiated, and subsequently analyzed for C—H insertion products. They found traces only (approximately 0.25%) of the radiolabeled reagent to become covalently associated with the acyl chains. In subsequent experiments with phenylazide-containing phospholipids, cross-linking yields of 1 and 2.8%, respectively, were obtained.[65,66] This increased labeling efficiency could have been due to a decreased rate of reactions that lead to quenching of the reactive species.[66]

One of the most widely used groups in photolabeling is the *m*-nitrophenylazide. Does the electron-withdrawing substituent increase the efficiency of insertion reactions into aliphatic C—H bonds? Regrettably, corresponding experiments did not lead to a consistent answer to this obviously important question. For example, Hu and Wisnieski,[67] using the amphipathic probe 12-(4-azido-2-nitrophenoxy)stearoyl[1-^{14}C]glucosamine, obtained considerable labeling of dimyristoyllecithin present as bilayer vesicles. In contrast, we measured barely detectable labeling (less than 1%) of the same saturated lipid with a reagent (1-palmitoyl-2-[11-(4-azido-2-nitroanilino)undecanoyl]-*sn*-glycero-3-phosphorylcholine) predicted to exhibit photochemical properties very similar to those of the former compound.[68] Although the reasons for this discrepancy are not known, it should be pointed out that different experimental approaches were used in the two studies to measure cross-linking yields. While we separated lipid monomers and dimers by Sephadex LH-20 gel filtration, Hu and Wisnieski[67] used thin-layer chromatography for this purpose which, in our view, bears more risks of misinterpreting experimental data. Clearly, additional work is needed to characterize nitroarylazides as photolabeling reagents, but we tend to believe that *m*-nitroarylazides are hardly more effective than nonsubstituted arylazides for the photolabeling of aliphatic residues. This view is in agreement with recent conclusions drawn by Nielsen and Buchardt.[69]

Nitrenes Generated from Other Precursors: Acylazides and Nitrenes. Following the suggestion of Lwowski,[70] these may be defined as those

---

[64] H. Bayley and J. R. Knowles, *Biochemistry* **17**, 2414 (1978).
[65] C. M. Gupta, R. Radhakrishnan, G. E. Gerber, W. L. Olsen, S. C. Quay, and H. G. Khorana, *Proc. Natl. Acad. Sci. U.S.A.* **76**, 2595 (1979).
[66] J. Brunner and F. M. Richards, *J. Biol. Chem.* **255**, 3319 (1980).
[67] V. W. Hu and B. J. Wisnieski, *Proc. Natl. Acad. Sci. U.S.A.* **76**, 5460 (1979).
[68] J. Brunner, unpublished results.
[69] P. E. Nielsen and O. Buchardt, *Photochem. Photobiol.* **35**, 317 (1982).
[70] W. Lwowski, in "Azides and Nitrenes" (E. F. V. Scriven, ed.), p. 205. Academic Press, New York, 1984.

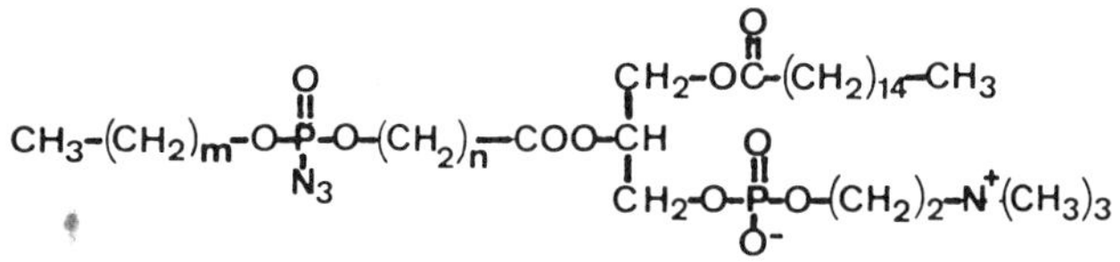

$$\text{CH}_3\text{-(CH}_2)_m\text{-O-}\overset{\overset{\displaystyle O}{\|}}{\underset{\underset{\displaystyle N_3}{|}}{P}}\text{-O-(CH}_2)_n\text{-COO-CH} \quad \begin{array}{l} \text{CH}_2\text{-O}\overset{\overset{\displaystyle O}{\|}}{C}\text{-(CH}_2)_{14}\text{-CH}_3 \\ \text{CH} \\ \text{CH}_2\text{-O-}\overset{\overset{\displaystyle O}{\|}}{\underset{\underset{\displaystyle O^-}{|}}{P}}\text{-O-(CH}_2)_2\text{-N}^+\text{(CH}_3)_3 \end{array}$$

FIG. 1. General structure of analogs of lecithin containing the phosphorylazido group at variable positions along the sn-2 fatty acyl chain. From Aggeler.[72]

that have their nitrogen function attached to electron-withdrawing groups that do not possess a heteroatom bearing an unshared electron pair next to the nitrogen function. Acylazides are most often made by nucleophilic substitution on acyl halides by azide ions. For the photolabeling of biological targets, acylazides have been found to have a very limited application. This is mainly because of the limited stability of acylazides (all of them are more or less rapidly hydrolyzed in water) and the fact that they must be irradiated with UV light below 300 nm. Furthermore, many acylnitrenes are prone to intramolecular rearrangements; the photochemical analog of the Curtius rearrangement being perhaps best known.

Singlet nitrenes are generated by direct photolysis of acylazides, triplet nitrenes by spontaneous intersystem crossing of singlet nitrenes, or by dissociation of triplet photoexcited precursors (triplet sensitized photolysis). Acylnitrenes have a triplet ground state but intersystem crossing is slow enough that, upon direct photolysis of azides, the chemistry observed is that of singlet nitrenes.[70]

The most reactive acylnitrenes are phosphorylnitrenes which have been shown to give substantial C—H insertion even if generated in *tert*-butanol.[71] Recently, Aggeler has synthesized phosphorylazide-containing phospholipids of the general structure shown in Fig. 1.[72] When liposomes, prepared from a mixture of such a lipid ($m = 7$; $n = 7$), in a ^{14}C-labeled form, and [^{3}H]dipalmitoyllecithin (molar ratio approximately 1:5) were irradiated in a quartz vessel with the unfiltered light of a high-pressure mercury lamp, approximately 25% of the generated nitrene reacted via C—H insertion with the saturated lipid as assessed by ^{31}P NMR spectroscopy and chemical degradation studies. Unfortunately, phosphorylazides have such unfavorable spectral properties (their molar extinction coefficient is only 20 M^{-1} cm^{-1} at 250 nm[73]) that photoactivation of the reagent would presumably severely damage biological molecules.

[71] R. Breslow, A. Feiring, and F. Herman, *J. Am. Chem. Soc.* **96**, 5937 (1974).
[72] R. Aggeler, Ph.D. dissertation. Eidegenössische Technische Hockschule, Zurich, Switzerland, 1984.
[73] B. Lakatos, Á. Hesz, Z. Vetéssy, and G. Horváth, *Acta Chim. Acad. Sci. Hung.* **60**, 308 (1969).

Alkylazides. The long-wave UV absorption maximum of alkylazides is around 285 nm with a molar extinction coefficient of approximately 25 M^{-1} cm^{-1}.[74] Although it is believed that photolysis leads to the corresponding nitrene, intermolecular C—H insertion is not observed in organic solution because of a very rapid intramolecular 1,2-hydrogen migration which the nitrene undergoes.[75,76]

In spite of this rearrangement reaction, alkylazides have been evaluated and successfully applied by Stoffel and colleagues in labeling membranes.[77–82] From a photochemical point of view, the most remarkable finding was that, in lipid bilayers, upon photoactivation, alkylazidophospholipids can form covalent adducts with neighboring saturated fatty acyl chains.[79,81] This finding was unexpected and contradictory to results obtained by Khorana and co-workers who noted barely detectable lipid–lipid cross-linking in similar experiments.[65] However, a possibly important difference between the two studies was the rigorous exclusion of oxygen during photolysis in Stoffel's experiments which, indeed, has been claimed to be an essential requirement.[78]

In an attempt to define more precisely the conditions that lead to intermolecular reactions of alkylazides, Aggeler measured photoinduced lipid–lipid cross-linking in liposomes prepared from a 1 : 6 molar mixture of 1-palmitoyl-2-(12-azidostearoyl)-*sn*-glycero-3-phosphorylcholine and [³H]dipalmitoyllecithin.[72] As suggested by Stoffel, irradiation was performed under strict exclusion of oxygen. Following separation of lipid monomers and dimers, Aggeler calculated that up to 25% of the nitrene generated reacted with the saturated lipid, a result which at least in a qualitative sense confirmed Stoffel's finding. On the other hand, by using a Pyrex glass filter of 4 mm thickness to screen out irradiation below about 300 nm but otherwise identical conditions, Aggeler was unable to photolyze a significant fraction of the azide within a reasonable period of

---

[74] W. D. Clossen and H. B. Gray, *J. Am. Chem. Soc.* **85**, 290 (1963).
[75] J. March, "Advanced Organic Chemistry," 2nd Ed., p. 178. McGraw-Hill, New York, 1977.
[76] F. A. Carey and R. J. Sundberg, "Advanced Organic Chemistry," Parts B, p. 301. Plenum, New York, 1977.
[77] W. Stoffel, C. Schreiber, and H. Scheefers, *Hoppe-Seyler's Z. Physiol. Chem.* **359**, 923 (1978).
[78] W. Stoffel, K.-P. Salm, and M. Müller, *Hoppe-Seyler's Z. Physiol. Chem.* **363**, 1 (1982).
[79] W. Stoffel, W. Därr, and K.-P. Salm, *Hoppe-Seyler's Z. Physiol. Chem.* **358**, 453 (1977).
[80] R. A. Heller, R. Klotzbücher, and W. Stoffel, *Proc. Natl. Acad. Sci. U.S.A.* **76**, 1721 (1979).
[81] E. R. Podack, W. Stoffel, A. F. Esser, and H. J. Müller-Eberhard, *Proc. Natl. Acad. Sci. U.S.A.* **78**, 4544 (1981).
[82] W. Stoffel and P. Metz, *Hoppe-Seyler's Z. Physiol. Chem.* **363**, 19 (1982).

time (2 hr), implying that the Pyrex filters employed by Stoffel had not been thick enough to completely remove light below 300 nm. This finding also points to the possibility that the failure of Gupta et al.[65] to obtain a significant amount of lipid–lipid heterodimers may simply have been due to inefficient conversion of the azide into the nitrene. The chemical filters used in those experiments may have effectively removed short-wave UV which evidently is required for photoactivation of the azide.

It remains a challenging goal to elucidate the mechanism of the intermolecular C—H insertion reaction. In discussing their results, Stoffel and colleagues pointed to the fact that a membrane is a two-phase system with a high degree of order that may affect the chemistry of an alkylnitrene.[78] Bayley[4] suggests that, within this ordered environment, hydrogen or alkyl group migration may be under certain conformational constraints which may permit the formation of a triplet nitrene. If the rigorous exclusion of oxygen turns out indeed to be a requirement, then a mechanism involving a reactive triplet is very likely. Much additional work is necessary to characterize alkylazides as photolabeling reagents and to specify the degree of damage to proteins and membranes that, at least with sensitive systems, is likely to occur in the course of reagent activation.

Carbenes and Their Precursors

Diazo Compounds. Esters of diazoacetic acid, the reagents first used for photolabeling purposes, suffer from the two main disadvantages that they have rather limited stability in the dark and that, upon activation, they undergo Wolff rearrangement to a considerable extent.[40,41] Furthermore, they exhibit only weak absorption at long wavelengths, with an ε of about 15 M^{-1} cm^{-1} at 340 nm. As far as the stability and rearrangement is concerned, a major improvement has been achieved with the introduction of the 2-diazo-3,3,3-trifluoropropionic acid.[83] Corresponding esters undergo photolysis with substantially less rearrangement and, in addition, have very much increased stability in acid. These esters, integrated into phospholipids, have been used extensively by Khorana's group in their studies with membranes.[65,84–89]

---

[83] V. Chowdhry, R. Vaughan, and F. H. Westheimer, *Proc. Natl. Acad. Sci. U.S.A.* **73**, 1406 (1976).
[84] C. H. Gupta, C. E. Costello, and H. G. Khorana, *Proc. Natl. Acad. Sci. U.S.A.* **76**, 3139 (1979).
[85] W. Curatolo, R. Radhakrishnan, C. M. Gupta, and H. G. Khorana, *Biochemistry* **20**, 1374 (1981).
[86] G. E. Gerber, R. Radhakrishnan, C. M. Gupta, and H. G. Khorana, *Biochim. Biophys. Acta* **640**, 646 (1981).

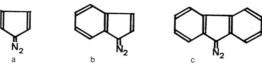

FIG. 2. Structures of (a) diazocyclopentadiene, (b) diazoindene, and (c) 9-diazofluorene.

Diazo compounds more photolabile than those mentioned above are p-toluene sulfonyldiazoacetates and (dansyldiazomethyl)phosphinates, reagents which were also developed by Westheimer and colleagues.[90,91] However, these compounds may be too polar and too bulky to be suitable for labeling the apolar phase of membranes.

A very promising class of photoactivatable reagents is formed by diazocyclopentadiene, diazoindene, and 9-diazofluorene (Fig. 2). All three absorb strongly at around 300 nm or above and are photolyzed very rapidly at these wavelengths. The photochemistry of these compounds has been the subject of several systematic investigations and led to the discovery of the existence of a rapid thermal equilibrium between the triplet and singlet states.[92-100] Cyclopentadienylidene is one of the most reactive carbenes known.[92] When generated in 2,3-dimethylbutane, its tertiary/primary C—H insertion selectivity is approximately 7.3. The extreme reactivity is also documented by the relatively low addition/insertion ratio (20.9) observed in tetramethylethylene.[92]

---

[87] A. H. Ross, R. Radhakrishnan, R. J. Robson, and H. G. Khorana, *J. Biol. Chem.* **257**, 4152 (1982).
[88] Y. Takagaki, R. Radhakrishnan, C. M. Gupta, and H. G. Khorana, *J. Biol. Chem.* **258**, 9128 (1983).
[89] Y. Takagaki, R. Radhakrishnan, K. W. A. Wirtz, and H. G. Khorana, *J. Biol. Chem.* **258**, 9136 (1982).
[90] V. Chowdhry and F. H. Westheimer, *J. Am. Chem. Soc.* **100**, 309 (1978).
[91] J. Stackhouse and F. H. Westheimer, *J. Org. Chem.* **46**, 1891 (1981).
[92] R. A. Moss, *J. Org. Chem.* **31**, 3296 (1966).
[93] W. Ando, Y. Saiki, and T. Migita, *Tetrahedron* **29**, 3511 (1973).
[94] R. A. Moss and M. A. Joyce, *J. Am. Chem. Soc.* **100**, 4475 (1978).
[95] M. S. Baird, I. R. Dunkin, N. Hacker, M. Poliakoff, and J. J. Turner, *J. Am. Chem. Soc.* **103**, 5190 (1981).
[96] D. Griller, C. R. Montgomery, J. C. Scaiano, M. S. Platz, and L. Hadel, *J. Am. Chem. Soc.* **104**, 6813 (1982).
[97] B.-E. Brauer, P. B. Grasse, K. J. Kaufmann, and G. B. Schuster, *J. Am. Chem. Soc.* **104**, 6814 (1982).
[98] R. A. Moss and C. M. Young, *J. Am. Chem. Soc.* **105**, 5859 (1983).
[99] P. B. Grasse, B.-E. Brauer, J. J. Zupancic, K. J. Kaufmann, and G. B. Schuster, *J. Am. Chem. Soc.* **105**, 6833 (1983).
[100] D. Griller, A. S. Nazran, and J. C. Scaiano, *Acc. Chem. Res.* **17**, 283 (1984).

Bayley and Knowles[101] were the first to suggest that diazocyclopentadiene might be a useful reagent for labeling the lipid core of membranes. More recently, Anjaneyulu and Lala proposed diazofluorene for the same purpose[102]; the insertion yields into phosphatidylcholine vesicles were reported, however, to be rather low. In an attempt to evaluate diazocyclopentadiene in more detail, a tritriated form was made in our laboratory and some initial experiments have already been carried out.[103] Upon equilibration of [^{3}H]diazocyclopentadiene with liposomes prepared from dimyristolyllecithin and activation of the reagent [a single pulse from a xenon flash tube (Power supply type PS4302 from Noblelight Limited, Cambridge, England; 1 kilojoule at total pulse duration of 300 μsec) was sufficient to photolyze nearly completely the diazo compound], as much as 60% of the radiolabel was found to be covalently associated with the lipid and, as suggested from chemical degradation experiments, labeling most probably occurred through C—H insertion reactions. This is the highest efficiency that has so far been measured for photogenerated species reaction with a saturated lipid bilayer.

Diazirines. This important class of compounds was discovered around 1960.[104,105] In the early 1970s, Smith and Knowles[106,107] described the synthesis of 3-aryl-3*H*-diazirines. Meanwhile, this class of reagents has proved to be the most promising one for photolabeling of the lipid core of membranes.

GENERAL PROPERTIES OF DIAZIRINES. Diazirines are remarkably stable compounds. They generally resist treatment with strong acid or base and are not reduced by mild reducing agents such as NaBH$_4$ or thiols. Most diazirines show a characteristic absorption maximum at around 360 nm with a molar extinction coefficient of around 50 M^{-1} cm^{-1} (alkyldiazirines) and 200–300 M^{-1} cm^{-1} [3*H*(alkyl),3-aryldiazirines].[107,108] Absorption maxima at around 300 nm have been reported for 3,3-diazirinomalonic acid derivatives.[109] Upon irradiation, diazirines undergo both fragmentation to carbenes as well as photoisomerization to linear diazo compounds (Fig. 3). In all cases reported, the diazo isomer is formed to

[101] H. Bayley and J. R. Knowles, *Biochemistry* **17**, 2420 (1978).
[102] P. S. R. Anjaneyulu and A. K. Lala, *FEBS Lett.* **146**, 165 (1982).
[103] R. Aggeler, R. Conforti, G. Semenza, and J. Brunner, unpublished.
[104] S. R. Paulsen, *Angew. Chem.* **72**, 781 (1960).
[105] E. Schmitz and R. Ohme, *Angew. Chem.* **73**, 115 (1961).
[106] R. A. G. Smith and J. R. Knowles, *J. Am. Chem. Soc.* **95**, 5072 (1973).
[107] R. A. G. Smith and J. R. Knowles, *J. Chem. Soc., Perkin Trans. 2*, p. 686 (1975).
[108] R. F. R. Church and M. J. Weiss, *J. Org. Chem.* **35**, 2465 (1970).
[109] G. V. Shustov, N. B. Tavakalyan, and R. G. Kostyanovsky, *Angew. Chem.* **93**, 206 (1981).

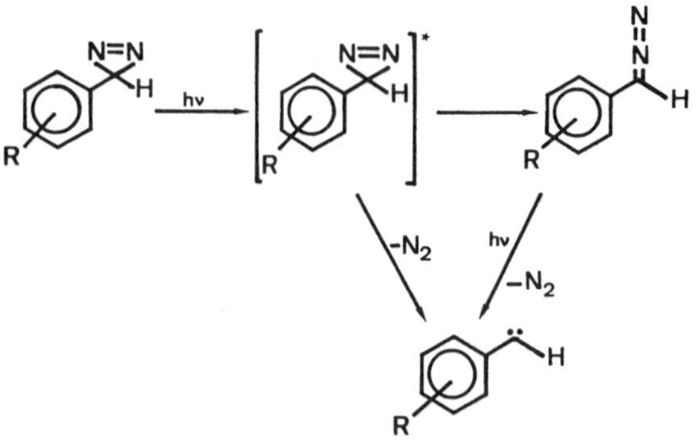

FIG. 3. General scheme of the photolysis of 3-aryl-3H-diazirine. This photolytic pathway also applies, in principle, to other diazirines.

the extent of 30 to 70%. Although the diazo compounds can themselves be photolyzed at suitable wavelengths to yield the carbenes, their transient accumulation during a photolabeling experiment is undesirable since it could lead to protein labeling via alkylation reactions (see below). Such unintended reactions do not occur, however, if diazirines are used which generate diazoisomers that are unreactive under normal conditions of labeling. On the basis of these considerations 3-aryl-3-(trifluoromethyl)-3H-diazirine was introduced.[110,111] Since the reagents derived from this carbene generator are particularly valuable, both preparation and properties shall be considered in more detail.

SYNTHESIS OF 3-(TRIFLUOROMETHYL)-3-ARYL-3H-DIAZIRINES. The synthesis of this class of diazirines starts from the corresponding trifluoromethyl ketones and follows the route depicted in Fig. 4. It seems worthwhile to mention the following modifications and improvements of the original procedure. (1) It was shown by Nassal that the ketone can also be obtained upon replacing the halide by lithium (with n-butyllithium) and subsequent conversion of the organometallic compound with N-trifluoroacetylpiperidine.[112] The yields of ketone by this procedure were reported to be considerably higher than those obtained by the original Grignard procedure.[111] (2) Several oxime intermediates could be obtained in nearly quantitative yields by carrying out the condensation reaction under anhydrous conditions (pyridine or pyridine–ethanol rather than in aqueous ethanol.[112–114] (3) Mesylation of the oxime appears to be superior to tosyla-

[110] J. Brunner, H. Senn, and F. M. Richards, *J. Biol. Chem.* **255,** 3313 (1980).
[111] F. M. Richards and J. Brunner, *Ann. N.Y. Acad. Sci.* **346,** 144 (1980).
[112] M. Nassal, *Liebigs Ann. Chem.* p. 1510 (1983).
[113] M. Nassal, *J. Am. Chem. Soc.* **106,** 7540 (1984).
[114] L. B. Shih and H. Bayley, *Anal. Biochem.* **144,** 132 (1985).

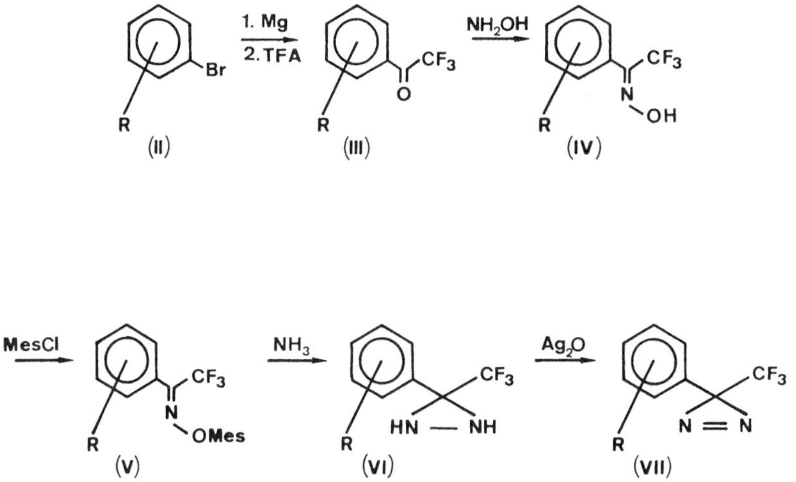

FIG. 4. Synthesis of 3-(trifluoromethyl)-3-aryl-3H-diazirines. For further details, see Brunner et al.,[12,110] Nassal,[112,113] and Shih and Bayley.[114]

tion as fewer by-products are formed and work-up of the reaction mixture is facilitated.[12] (4) Oxidation of the diaziridine with silver oxide turns out to be another critical step in the synthesis. As a matter of fact, this reaction is occasionally very sluggish and, in some instances, several weeks have been necessary to achieve substantial conversion. The rate of oxidation depends strongly on the "activity" of the silver oxide (for example, it is crucial that freshly precipitated oxidant is used), and appears also to be influenced by additional substituents on the phenyl ring. Recently, a modification of the original procedure was brought to our attention that should give higher yields of the diazirine within a much shorter period of time.[115] The procedure is that originally reported by Church et al.[116]

CHEMICAL PROPERTIES OF 3-(TRIFLUOROMETHYL)-3-ARYL-3-H-DIAZIRINES. In general, these are exceptionally stable. For example, it was possible to build up the entire amino acid skeleton in 4'-(1-azi-2,2,2-trifluoroethyl)phenylalanine, a carbene-generating analog of phenylalanine, in the presence of the intact diazirine group.[113] This involved reactions such as side-chain bromination with N-bromosuccinimide (in boiling tetrachloromethane), an alkylation of diethyl acetamidomalonate (Na-salt) in ethanol (heating to 65° for 2 hr), base-catalyzed ester hydrolysis, and a decarboxylation reaction (80°; 60 min). Quite surprisingly, however,

---

[115] H.-P. Michel, personal communications.
[116] R. F. R. Church, A. S. Kende, and M. J. Weiss, *J. Am. Chem. Soc.* **87**, 2665 (1965).

p-alkoxy and *p*-benzoxy substituents were found to destabilize strongly the diazirine. Thus, 3-(4'-benzoxyphenyl)-3-(trifluoromethyl)-3*H*-diazirine is decomposed slowly even at $-20°$.[117,118]

PHOTOCHEMICAL REACTIONS IN ORGANIC SOLUTIONS. Irradiation of dilute (approximately 1 mM) solutions of 3-(trifluoromethyl)-3-phenyl-3*H*-phenyldiazirine in dioxane or methanol and of 4'(1-azi-2,2,2-trifluoroethyl)phenylalanine in ethanol showed a time-dependent decrease in the diazirine concentrations which is consistent with the expected first-order kinetics.[110,113] In Table I are listed the principal products that were obtained upon photolysis of the above mentioned compounds in various solvents. The high yields with which ethers were formed in alcohols is indicative of singlet carbene chemistry. However, Nassal[113] identified three minor components (**IV, V,** and **VI**) which are believed to have arisen from the triplet carbene. Another remarkable result was the formation of product **VIII**, which presumably resulted from insertion of the carbene into the alcoholic O—H bond, accompanied by elimination of one HF molecule. This parallels an observation made by Gupta *et al.*[84] with a phospholipid containing the 3,3,3-trifluoro-2-diazopropionyl group photolyzed in multilamellar liposomes. The lack of formation of C—H insertion products in alcohol was interpreted by Nassal to be due to the presence of oxygen which is known to be a powerful scavenger of triplets.[113]

PHOTOCHEMICAL REACTIONS IN LIPID BILAYERS. Evidence for the occurrence of C—H insertion reaction comes from experiments with liposomes which have been prepared from mixtures of [^{3}H]dipalmitoyllecithin and a lecithin analog carrying the 3-(trifluoromethyl)-3-aryl-3*H*-diazirine group.[66] Using gel filtration on Sephadex LH-20 for separation of lipid monomers and dimers, it could be shown that up to 20% of the photogenerated carbene reacted with the saturated lipid to form lipid heterodimers. Cross-linking occurred through C—H insertion reactions as assessed by ^{19}F NMR spectroscopy. It is interesting to note that the labeling yields (lipid–lipid cross-linking) obtained with trifluoromethylarylcarbene were only about half of those measured by Khorana and associates with phospholipids containing either the 3*H*,3-aryldiazirine or the 3,3,3-trifluoromethyl-2-diazopropionyl group.[65] Presumably, this is due to a somewhat lower reactivity of trifluoromethylarylcarbene when compared with the carbenes generated from the latter precursors.

[117] A. Azzi, personal communication.
[118] H. Acha-Orbea, Diplomarbeit, Master's thesis, Eidgenössiche Technische Hochschule, Zurich, Switzerland, 1980.

TABLE I
DISTRIBUTIONS OF PRODUCTS OBTAINED ON PHOTOLYIS OF TWO
3-(TRIFLUOROMETHYL)-3-ARYL-3H-DIAZIRINES IN ORGANIC SOLUTION[a]

Compound photolyzed	Conditions of photolysis (solvent)	Product		Yield (%)
R_1 = 4-methylphenyl (diazirine with CF₃)	3 mM in methanol	R_1—CH(OCH$_3$)CF$_3$	(I)	95
	3 mM in cyclohexane	R_1—CH(C$_6$H$_{11}$)CF$_3$	(II)	55
R_2 = 4-(NHCHO-CH(COOCH$_3$)-CH$_2$-)phenyl (diazirine with CF₃)	2 mM in ethanol	R_2—CH(OC$_2$H$_5$)CF$_3$	(III)	73
		R_2—CO—CF$_3$	(IV)	17
		R_2—CH$_2$(OH)—CF$_3$	(V)	6
		R_2—CH$_2$—CF$_3$	(VI)	5
	2 mM in *tert*-butanol	R_2—CH(OC$_4$H$_9$)—CF$_3$	(VII)	59
		R_2—CO—CF$_3$	(IV)	9
		R_2—CH$_2$(OH)—CF$_3$	(V)	10
		R_2—C(OC$_4$H$_9$)=CF$_2$	(VIII)	22
	2 mM in cyclohexane	R_2—CH(C$_6$H$_{11}$)CF$_3$	(IX)	47 (73)[b]
		R_2—CO—CF$_3$	(IV)	34 (18)[b]
		R_2—CH$_2$(OH)—CF$_3$	(V)	18 (9)[b]

[a] Summarized from Brunner *et al.*[110] and Nassal.[113]
[b] Solution of the diazirine purged with nitrogen prior to photolysis.

The Design and Synthesis of Reagents

The high reactivity and relatively indiscriminate nature of carbenes and some nitrenes is not a blessing in every respect. Unlike the modification of proteins with conventional reagents which usually give rise to (near) stoichiometric modification of one or a few residues, photolabeling efficiencies are, in general, low. The reason for this is that carbenes and nitrenes are rapidly quenched by reactions with the fatty acyl chains of the lipid bilayer. Therefore, only those reactive species generated in the immediate vicinity of an integral protein have a chance at all to lead to its successful modification. To provide a reasonable basis for the analysis of minor, weakly labeled proteins, it is often necessary to start with considerable quantities of the radiolabeled reagent. Its proper design, which permits synthesis with a high specific radioactivity or equips it with other properties that allow detection with high sensitivity, is, therefore, an important prerequisite. In this section, some of the considerations in the design of two promising reagents shall be discussed. The compounds referred to are 3-(trifluoromethyl)-3-(m-[^{125}I]iodophenyl)-3H-diazirine ([^{125}I]TID),[119] one of the well-characterized lipid-soluble reagents, and an analog of a phospholipid, 1-palmitoyl-2-[10-[4-(trifluoromethyl)diazirinyl]-phenyl]-8-[9-^{3}H]oxadecanoyl-sn-glycero-3-phosphorylcholine[12] for which the abbreviation [^{3}H]PTPC will be used (Fig. 5).

3-(Trifluoromethyl)-3-(m-[^{125}I]iodophenyl)-3H-diazirine ([^{125}I]TID)

Having demonstrated the favorable photochemical properties of 3-(trifluoromethyl)-3-phenyl-3H-diazirine, the next obvious task was to introduce a radioactive isotope into the molecule for further evaluation. Due to the very high specific radioactivity (up to 2000 Ci/mmol) with with iodine-125 is available, we have chosen this isotope, rather than ^{3}H or ^{14}C. It could be expected that iodination would markedly increase the lipid-to-water partitioning and, at the same time, reduce the volatility of the compound. This latter point is crucial if the substance has to be prepared at the microgram scale and has to be free of residual solvent. The placement of iodine into the meta-position to the diazirinyl group was largely the result of considerations related to synthesis. However, as discussed above, there are now reasons to believe that iodine in the para-position might have affected the stability of the diazirine. The synthetic route, finally worked out for the preparation of [^{125}I]TID, involves radioiodination in a simple step at the end of the synthesis. Although [^{125}I]TID is now commercially available at a specific radioactivity of approximately 10 Ci/

---

[119] J. Brunner and G. Semenza, *Biochemistry* **20**, 7174 (1981).

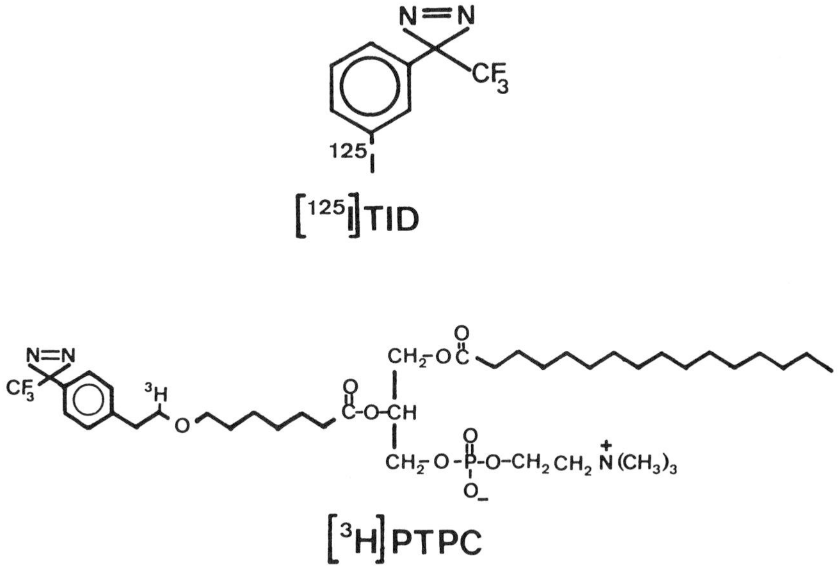

FIG. 5. Chemical structures of [^{125}I]TID[119] and [^{3}H]PTPC.[12]

mmol (from Amersham), the ease of introducing iodine in high yields enables its preparation at much higher specific radioactivity. For the purpose of autoradiography, the use of iodine-131 has distinct advantages[120] and the inconvenience of a short life-time of this isotope (6 days) should not represent a severe problem in the preparation of [^{131}I]TID.

1-Palmitoyl-2-[10-[4-(trifluoromethyl)diazirinyl]phenyl]-8-
[9-^{3}H]oxadecanoyl]-sn-glycero-3-phosphorylcholine ([^{3}H]PTPC)[12]

The rationale behind making this lipid analog was to have a reagent which shares with [^{125}I]TID the photochemical properties but is distinctly different in the way it interacts with membranes. As illustrated in the Appendix, essentially all of the photoreactive phospholipids are of the lecithin type. To introduce ^{3}H or ^{14}C, Bisson and Montecucco used a strategy involving exchange with phospholipase D of the choline group by ethanolamine followed by methylation of the primary amine with [^{14}C]- or [^{3}H]methyliodide.[121] Although this procedure is relatively simple and, in the case of ^{3}H, yields phosphatidylcholine derivatives with specific radioactivities of more than 100 Ci/mmol, these lipids could suffer from the

[120] N. H. H. Teng and L. B. Chen, *Nature (London)* **259**, 578 (1976).
[121] R. Bisson and C. Montecucco, *Biochem. J.* **193**, 757 (1981).

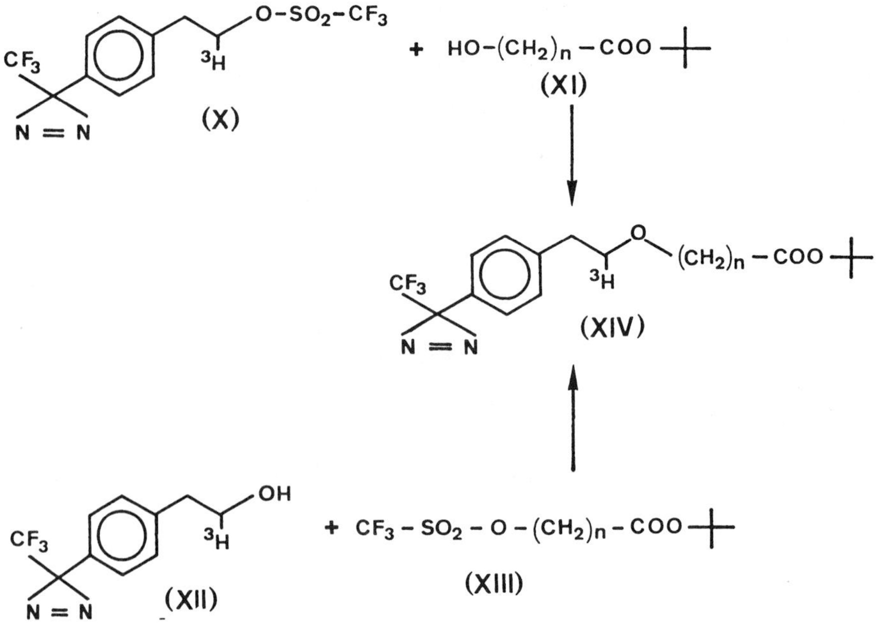

FIG. 6. Formation of the ether linkage between the photoactivatable group and the ω-position of the fatty acyl chain in the course of the synthesis of [^{3}H]PTPC. For further details, see text.

limited stability of the ester bond, which connects the photosensitive fatty acyl chain with that of the molecule that carries the radioisotope. However, for most kinds of analyses (including Edman degradation of labeled proteins), the ester appears to be sufficiently stable.[122,123] In the majority of the lipid synthesized, the radiolabel is within the fatty acyl chain that also contains the photoreactive group. Loss of label is less likely to occur but, in general, greater efforts are needed to synthesize such lipids. In addition, the overall yields with respect to the radioactivity are usually low.

The following aspects in the design and synthesis of [^{3}H]PTPC seem worthwhile noting (Fig. 6): (1) Attempts made in our laboratory to avoid preparation of the tritiated and extremely moisture-sensitive intermediate **X** by forming the trifluoromethane sulfonate **XIII** and subsequent alkylation of alcohol **XII** have not so far been successful. Presumably, compound **XIII** undergoes β-elimination. β-Elimination of trifluoromethane

---

[122] J. Hoppe, C. Montecucco, and P. Friedl, *J. Biol. Chem.* **258**, 2882 (1983).
[123] R. Bisson, G. C. M. Steffens, and G. Buse, *J. Biol. Chem.* **257**, 6716 (1982).

sulfonic acid from **X** is not the dominant reaction, which is surprising in view of the finding by Nassal[112] that the corresponding tosylate reacted predominantly in this way. (2) Acylation of the lysophospholipid is now generally carried out in the presence of 4-dimethylaminopyridine (or 4-pyrrolidinopyridine), a very powerful acylation catalyst,[124] which was first used in phospholipid synthesis by Stoffel *et al.*[125] and was more systematically examined by Khorana's group.[126,127] In our hands, highest yields (with regard to the photoreactive, radioactive fatty acids) have been obtained by using the fatty acyl chlorides (rather than anhydrides),[128] but other derivatives (e.g., mixed anhydrides with trifluoroacetic acid) might be equally efficient.

Perspectives

All phospholipids so far designed for lipid phase labeling of membranes contain either ^{3}H or ^{14}C (see Appendix). However, the use of these isotopes has the disadvantage that labeled membrane samples cannot be readily recovered after scintillation counting and reutilized for other purposes. For this reason, it would be a distinct advantage to have phospholipids which contain radioiodine in place of the weak β-emitters. The availability of both ^{125}I and ^{131}I would further enable the design of double-label experiments at very high specific radioactivity. Due to the nonvolatility of phospholipids, such a reagent would also be less inconvenient to handle than many of the hydrophobic, slightly volatile reagents.

One approach, currently pursued in our laboratory to increase both the scope and sensitivity of hydrophobic membrane labeling, is based on an "indirect" detection method. Toward this goal, we have synthesized the photoactivatable lipid shown in Fig. 7. This phospholipid does not contain a radioisotope but, instead, has a biotinyl residue linked to the polar head group via a hydrophilic spacer. The enormous affinity of biotin for avidin and streptavidin (K_d approximately 10^{-15} M^{-1}) enables its detection at an unprecedented level of sensitivity.[129–133] Following incorporation of such a phospholipid into a membrane and photocross-linking to

---

[124] W. Steglich and G. Hoefle, *Angew. Chem., Int. Ed. Engl.* **8**, 981 (1969).
[125] W. Stoffel, O. Zierenberg, and B. D. Tunggal, *Hoppe-Seyler's Z. Physiol. Chem.* **353**, 1962 (1972).
[126] C. M. Gupta, R. Radhakrishnan, and H. G. Khorana, *Proc. Natl. Acad. Sci. U.S.A.* **74**, 4315 (1977).
[127] R. Radhakrishnan, R. J. Robson, Y. Takagaki, and H. G. Khorana, this series, Vol. 72, p. 408.
[128] H. Meister, R. Bachofen, G. Semenza, and J. Brunner, *J. Biol. Chem.* **260**, 16326 (1985).
[129] N. M. Green, *Biochem. J.* **89**, 585 (1963).
[130] L. Chaiet and F. J. Wolf, *Arch. Biochem. Biophys.* **106**, 1 (1964).

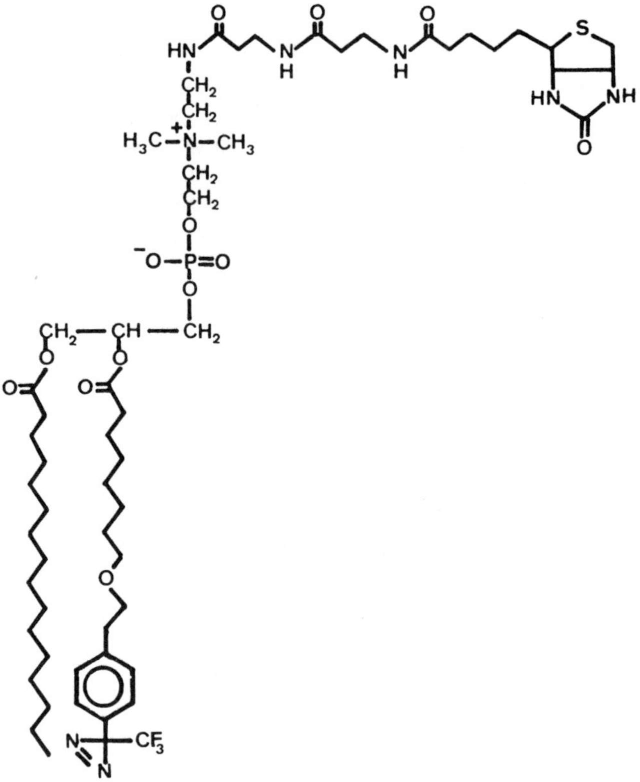

FIG. 7. Structure of an analog of a phospholipid containing the photoactivatable 3-(trifluoromethyl)-3-aryldiazirine, linked to the ω-position of the sn-2 fatty acyl chain, and the biotinyl residue, connected via a hydrophilic "spacer" to the polar head group of the phospholipid. From Aggeler.[72]

phospholipids and proteins, the component of interest is then isolated (e.g., by immunological techniques) and separated from the bulk of the biotinylated probe originally present in the sample. The following steps would then assist in the detection and quantitation of the minute amount of biotin bound to the isolated component. To aid this goal, avidin and streptavidin of very high specific radioactivity, and as conjugates with highly active enzymes, are available.

---

[131] K. Hofmann, F. M. Finn, H.-J. Friesen, C. Diaconescu, and H. Zahn, *Proc. Natl. Acad. Sci. U.S.A.* **74**, 2697 (1977).
[132] F. M. Finn, G. Titus, J. A. Montibeller, and K. Hofmann, *J. Biol. Chem.* **255**, 5742 (1980).
[133] J. J. Leary, D. J. Brigati, and D. C. Ward, *Proc. Natl. Acad. Sci. U.S.A.* **74**, 4045 (1983).

Experimental Aspects of Hydrophobic Photolabeling of Membranes

In a typical photolabeling experiment, three steps may be distinguished: (1) reagent delivery and incorporation into the membranes, (2) irradiation of the sample, and (3) analysis of (the) photolabeled membrane constituent(s). Each step will be discussed here.

Incorporation of the Reagent into the Membrane

With hydrophobic reagents, this is usually done simply by injecting a solution of the probe in a water-miscible organic solvent (ethanol, dimethyl sulfoxide, or dimethylformamide) into the suspensin of the membranes and equilibrating for a suitable period.[134–140] The amount of reagent which is needed for an experiment depends on its specific radioactivity, the efficiency with which the component(s) of interest is (are) labeled, and on the particular requirements for its (their) analyses. To avoid perturbation of the bilayer, the concentration of the reagent within the lipid core should not exceed a certain value. Presumably, the millimolar range can be tolerated in most experiments. However, it should be ascertained that neither reagent partitioning into the lipid bilayer nor, following activation of the probe, its distribution among the various membrane components is dependent on the reagent concentration chosen. The principal idea behind performing such measurements is to rule out the existence of specific and saturable binding sites for the reagent in a protein. When injected into the aqueous medium, it is possible that the reagent initially precipitates (which may be visible by a transient turbidity). In the case of reagents which are relatively bulky and extremely insoluble in water, equilibration with the membrane may be relatively slow and incubation periods of several minutes may be necessary. Smaller reagents such as [^{3}H]adamantanediazirine or [^{125}I]TID can be assumed, however, to equilibrate almost instantaneously with the membrane.

A point requiring special consideration and care is that hydrophobic labeling reagents are readily adsorbed by all kinds of plastics (polyethylene, polypropylene and, to a lesser extent, by Teflon). Hence, it is

---

[134] T. Bercovici and C. Gitler, *Biochemistry* **17**, 1484 (1978).
[135] V. Šator, J. M. Gonzalez-Ros, P. Calvo-Fernandez, and M. Martinez-Carrion, *Biochemistry* **18**, 1200 (1979).
[136] D. W. Goldman, J. S. Pober, J. White, and H. Bayley, *Nature (London)* **280**, 841 (1979).
[137] N. Cerletti and G. Schatz, *J. Biol. Chem.* **254**, 7746 (1979).
[138] H. Bayley and J. R. Knowles, *Biochemistry* **17**, 3883 (1980).
[139] M. J. Owen, J. C. A. Knott, and M. J. Crumpton, *Biochemistry* **19**, 3092 (1980).
[140] D. P. Smith, M. R. Kilbourn, J. H. McDowell, and P. A. Hargrave, *Biochemistry* **20**, 2417 (1981).

crucial that these reagents are never allowed to come into contact with these materials. Furthermore, the vapor pressure of [^{125}I]TID, and presumably most other hydrophobic reagents as well, is sufficient to permit fairly rapid "transfer" through the gas phase. Vials or photolysis cells containing these reagents should, therefore, not be sealed with Parafilm or plastic caps. Glass stoppers or aluminum foil-coated seals should be used instead.

Purging membranes with nitrogen, or argon, prior to irradiation has been shown to reduce photooxidative damage to proteins and may enhance labeling efficiencies through C—H insertion reactions. It should be clear that this must be done before the (volatile) reagent is added!

When experimenting with amphipathic reagents, loss of label by evaporation can be ruled out and adsorption to hydrophobic surfaces does not occur to the same extent as for hydrophobic molecules. Amphipathic reagents can be delivered as micellar solutions in buffer or, in the case of phospholipids, as sonicated liposomes. The fact that some may have detergent-like properties must also be considered in outlining an experiment. Although analog of phospholipids are the most interesting class of labeling reagents, their incorporation into a (preformed) membrane is less straightforward and special techniques, such as membrane fusion or lipid transfer catalyzed by phospholipid exchange protein, may be required.[12,66,89,141] In several studies, photoactivatable lipids have been used in reconstituted membranes of defined lipid and protein composition.[87-89,122,142] [^{3}H]PTPC, a phospholipid with a critical micelle concentration of approximately 6×10^{-9} M, has been found to be transferred spontaneously at a reasonable rate from sonicated liposomes (prepared from this lipid) to erythrocyte membranes.[12] Spontaneous exchange of phospholipids between membranes appears, however, to be a more general phenomenon.[143-148] Using preparations of sealed membranes, it can be assumed that the insertion is initially restricted to the outer leaflet of the bilayer. Incidently, from the distribution of [^{3}H]PTPC label among the

---

[141] P. Moonen, H. P. Haagsman, L. L. M. Van Deenen, and K. W. A. Wirtz, *Eur. J. Biochem.* **99**, 439 (1979).

[142] C. Montecucco, R. Bisson, C. Gache, and A. Johannsson, *FEBS Lett.* **128**, 17 (1981).

[143] F. J. Martin and R. C. MacDonald, *Biochemistry* **15**, 321 (1976).

[144] G. Duckwitz-Peterlein, G. Eilenberger, and P. Overath, *Biochim. Biophys. Acta* **469**, 311 (1977).

[145] M. A. Roseman and T. E. Thompson, *Biochemistry* **19**, 439 (1980).

[146] Y. Lange, A. L. Molinaro, T. R. Chauncey, and T. L. Steck, *J. Biol. Chem.* **258**, 6920 (1983).

[147] M. DeCuyper, M. Joniau, and H. Dangreau, *Biochemistry* **22**, 415 (1983).

[148] J. E. Ferrel, Jr., K.-J. Lee and W. H. Huestis, *Biochemistry* **24**, 2857 (1985).

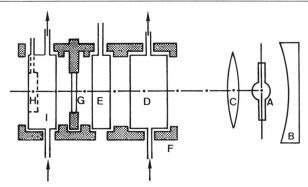

FIG. 8. Scheme of a photolysis apparatus: (A) light source, (B) reflector, (C) lens, (D) infrared filter (circulating cold water, 50 mm), (E, G) filter combination consisting of a chemical (e.g., a saturated solution of copper sulfate, 20 mm) and an optical filter, (H) sample cuvette, (I) thermostatted cell, (F) tube.

two halves of the transmembrane stretch of glycophorin A labeled with [^{3}H]PTPC, this has been shown to be the case.[12]

Photoactivtion of the Reagent

Photoactivation is carried out by illuminating the cuvette containing the membranes equilibrated with the reagent. The type of light source and combination of filters most appropriate for an experiment depend primarily on the spectral properties of reagent. High-intensity light beams will normally only be required for time-resolved labeling experiments. The main goal is to prevent damage to the membrane which, for example, can be assessed by measuring functional integrity. Fortunately, the long-wave absorption maximum of diazirines at around 360 nm allows activation of the reagent under rather mild conditions. If particularly sensitive membranes are to be labeled, one might consider activating only a fraction of the reagent. This, of course, is at the cost of labeling efficiency. Incomplete reagent activation may also have to be tolerated in time-resolved labeling studies in which activation must be conducted within a limited period of time.

The apparatus that we have used for most studies is schematically shown in Fig. 8. It consists of a high-pressure mercury arc lamp (type 350-1008 from Illumination Industries, Inc.) and a filter of circulating cold water (50 mm) and one of a saturated solution of copper sulfate (20 mm) to cut off radiation shorter than 315 nm.[149] With this system, the half-time of

---

[149] J. A. Katzenellenbogen, H. J. Johnson, K. E. Carlson, and H. N. Myers, *Biochemistry* **13**, 2986 (1974).

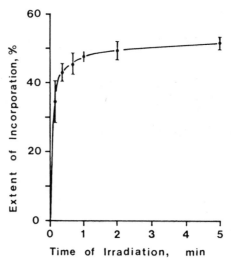

FIG. 9. Time course of the covalent incorporation of [^{125}I]TID into erythrocyte membranes. The curve represents the percentage of radioactivity originally present that could not be removed from aliquots of membranes upon irradiation for increasing periods of time. For reagent activation an apparatus as shown in Fig. 8 was used. From Brunner and Semenza.[119]

decay of [^{125}I]TID in the presence of erythrocyte membranes (ghosts) is approximately 7 sec.[119] Detailed experimental protocols for determining photolysis times have been given earlier.[119,150] The result of a typical experiment is depicted in Fig. 9. Very useful descriptions of most relevant aspects of experimental photochemistry, including a survey of the types of lamps, the various chemical and optical filters, and the design of photolysis cells are found in the recent monograph by Bayley.[5]

Label Distribution Analysis

An experiment is usually completed by determining the distribution of label among lipids and proteins, individual segments of integral proteins, or even single amino acids. The first step normally consists of the removal of residual unphotolyzed and photolyzed reagent not covalently bound to membrane constituents. In the case of hydrophobic reagents, this can be done by repeated sedimentation of the membranes in a buffer containing

[150] J. Brunner, R. Aggeler, and B. Reber, in "Enzymes, Receptors and Carriers of Biological Membranes" (A. Azzi, U. Brodbeck, and P. Zahler, eds.), p. 55. Springer-Verlag, Berlin, Federal Republic of Germany, 1984.

0.5 to 2% bovine serum albumin until no further radioactivity is released into the supernatant.

All major techniques of protein purification and analyses have been used to determine label distribution patterns. SDS–polyacrylamide gel electrophoresis, in conjunction with fluorography or autoradiography, is the method most frequently used. In order to identify those segments which, in the intact protein, are in contact with the bilayer, the labeled protein has to be degraded and the resulting fragments separated.

A first point of concern is the anticipated chemical instability of some of the adducts formed between reagent and certain amino acid residues. Such labile bonds may be detected by exposing the labeled peptides to mild base or acid and either monitoring the radioactivity released or that which remains bound. Liberation of label from a polypeptide chain may also be due to dissociation from protein of (labeled) lipids which, in some cases, are difficult to remove completely. A more systematic approach is by chemically characterizing the adducts obtained with individual amino acids which may be isolated upon degradation or obtained by chemical synthesis. Using these approaches, Ross et al.[87] were able to show that their carbene-generating phospholipids reacted in glycophorin A with glutamic acid 70 by forming an ester bond and that this bond was stable under the conditions employed for protein purification. In a [^{125}I]TID-labeling study of bacteriorhodopsin, it was of particular interest to clarify whether or not the absence of label in two particular aspartic acid residues was due to true nonlabeling of those residues or whether it simply reflected a limited stability of the ester, the putative covalent adduct formed with the carboxyl group. On the basis of stability measurements with a synthetic model compound, the latter possibility could be ruled out.[151]

The second problem is related to the purification and identification of photolabeled polypeptides. Although, on average, less than one reagent molecule can be assumed to be incorporated under normal conditions per polypeptide chain, those copies which receive label may be difficult to characterize. The reason is that the additional group(s) may alter the physical properties to such an extent that they no longer copurify with the nonlabeled counterparts. Such unequal behavior may particularly manifest itself when high-resolution separation techniques (SDS–polyacrylamide gel electrophoresis and reversed-phase HPLC) are employed, and is presumably more distinct for bulkier reagents, on the one hand, and for smaller peptides on the other. For example, in the cases of the polypeptides from succinate–cytochrome-*c* reductase and complex **III** labeled

---

[151] J. Brunner, A. J. Franzusoff, B. Lüscher, C. Zugliani, and G. Semenza, *Biochemistry* **24**, 5422 (1985).

with an arylazide-containing phospholipid, Gutweniger et al.[152] observed a shift in the radioactivity peaks with respect to the Coomassie blue bands. As expected, this effect was larger for the low-molecular-weight peptides. A more complex situation (discussed in a subsequent section) was experienced in the course of the purification of a 50-residue cyanogen bromide fragment of bacteriorhodopsin labeled with [^{125}I]TID.[151]

The unequivocal assignment of photolabeled peptides generated by chemical or proteolytic fragmentation of integral proteins is a difficult problem to which no general solution can yet be offered. This is mainly because of the very minute quantities of actually labeled peptides available which are almost always only traced by their radioactivity and may be contaminated by unrelated unlabeled peptides. At the time being, the most promising approach to tackle this problem appears to consist of the further degradation of these peptides by solid-phase Edman sequencing and the subsequent analysis of the individual photolabeled PTH amino acid derivatives. Both the rationale behind this method and the main results obtained with bacteriorhodopsin and other proteins are discussed in the following section.

Some Applications of Hydrophobic Membrane Labeling

Considering the wide range of reactivity of photogenerated intermediates, it is not surprising that the various reagents respond in different ways when applied to the same membrane system. In several cases, this has led to conflicting pictures as to which proteins are or are not inserted in the bilayer. An example, already referred to in previous articles, is the Na$^+$,K$^+$-ATPase, a protein complex consisting of two subunits. Whereas in all studies the α-subunit was strongly labeled, modification of the β-subunit was hardly detectable with some reagents.[142,153–157] Inconsistent results were also obtained with the mouse lyt-2/3 antigen complex,[158,159] the acetylcholine receptor,[160,161] and the H$^+$-ATP-synthase from E.

[152] H. Gutweniger, R. Bisson, and C. Montecucco, *J. Biol. Chem.* **256**, 11132 (1981).
[153] S. J. D. Karlish, P. L. Jørgensen, and C. Gitler, *Nature (London)* **269**, 715 (1977).
[154] R. A. Farley, D. W. Goldman, and H. Bayley, *J. Biol. Chem.* **255**, 860 (1980).
[155] P. L. Jørgensen, S. J. D. Karlish, and C. Gitler, *J. Biol. Chem.* **257**, 7435 (1982).
[156] P. L. Jørgensen and J. Brunner, *Biochim. Biophys. Acta* **735**, 291 (1983).
[157] M. Girardet, K. Geering, B. C. Rossier, J. P. Kraehenbuhl, and C. Bron, *Biochemistry* **22**, 2296 (1983).
[158] B. Luescher, H. Y. Naim, H. R. MacDonald, and C. Bron, *Mol. Immunol.* **21**, 329 (1984).
[159] H. Y. Naim, B. Luescher, G. Corradin, and C. Bron, *Mol. Immunol.* **21**, 337 (1984).
[160] R. Tarrab-Hazdai, T. Bercovici, V. Goldfarb, and C. Gitler, *J. Biol. Chem.* **255**, 1204 (1980).
[161] D. S. Middlemas and M. A. Raftery, *Biochem. Biophys. Res. Commun.* **115**, 1075 (1983).

coli.[122,162,163] On the other hand, remarkably similar labeling patterns were noted for other multisubunit protein complexes. For example, when the complex of the complement proteins C5b, C6, C7, C8, and C9 was labeled with reagents based on arylazides, alkylazides, or diazirines, the majority of the label was associated with C-8–C-9 in all cases.[81,164–166] Of course, this does not mean, or imply, that the same sets of amino acid residues had been modified. On the contrary, it is almost certain that major differences would be found in the label distribution patterns upon increasing the resolution in the analysis to the level of small fragments or single amino acids.

This section is mainly concerned with the application of hydrophobic photolabeling reagents to biological membranes and with the interpretation of the results obtained thereby. In the first part, experiments with the human erythrocyte membrane will be discussed. Due to its moderately complex and yet relatively well-characterized composition, this membrane has served as a model in most evaluations of new reagents. Subsequent subsections are then devoted to recent studies in which Edman sequencing was applied in order to determine the individual residues labeled within a polypeptide chain. The application of this analytical tool has not only greatly extended our knowledge concerning chemical reactivities and selectivities of the various reagents but has unraveled interesting details concerning the folding and topological arrangement of membrane-embedded polypeptides.

Labeling of the Lipid Core of the Human Erythrocyte Membrane

Composition of the Membrane. The protein-to-lipid ratio of the red cell membrane is approximately 1 : 1 (w/w). Its main integral proteins are band 3, present at approximately 0.9×10^6 copies per cell, and glycophorin A, the major sialoglycoprotein, with approximately 0.45×10^6 monomers per cell.[167] Additional integral proteins are those known as band 4.5, which include the carriers of D-glucose and of nucleosides.[168–171]

---

[162] J. Hoppe, P. Friedl, and B. B. Jørgensen, *FEBS Lett.* **160**, 239 (1983).
[163] J. Hoppe, J. Brunner, and B. B. Jørgensen, *Biochemistry* **23**, 5610 (1984).
[164] V. W. Hu, A. F. Esser, E. R. Podack, and B. J. Wisnieski, *J. Immunol.* **127**, 380 (1981).
[165] E. W. Steckel, B. E. Welbaum, and J. M. Sodetz, *J. Biol. Chem.* **258**, 4318 (1983).
[166] P. Amiguet, J. Brunner, and J. Tschopp, *Biochemistry* **24**, 7328 (1985).
[167] T. L. Steck, *J. Cell Biol.* **62**, 1 (1974).
[168] S. Lepke, H. Fasold, M. Pring, and H. Passow, *J. Membr. Biol.* **29**, 147 (1976).
[169] M. Kasahara and P. C. Hinkle, *J. Biol. Chem.* **252**, 7384 (1977).
[170] D. C. Sogin and P. C. Hinkle, *Biochemistry* **19**, 5417 (1980).
[171] J. D. Young, S. M. Jarvis, M. J. Robins, and A. R. P. Paterson, *J. Biol. Chem.* **258**, 2202 (1983).

TABLE II
DISTRIBUTION OF RADIOLABEL AMONG COMPONENTS OF THE HUMAN ERYTHROCYTE MEMBRANE

Component	Radioactivity (%) when labeled in the presence of		
	[5-^{125}I]Idonaphthyl-1-azide^a	[^{3}H]Adamantane-diazirine^b	[^{125}I]TID^c
Total radioactivity in photolysis sample	100	100 (0.3–3 mg protein/ml)	100 (1.1 mg protein/ml)
Membranes after extraction of non-bound label	60	40	52
Label associated with lipid	40	32	50
Label associated with protein	18	8	2.5
Label in band 3	Minor labeling	3	1.2–1.3
Glycophorin			
Isolated by wheat germ agglutinin	8	1.8 (−GSH)^d 0.9–1.2 (+GSH)^d	
Isolated by LIS–phenol extraction^d	10		0.35
Spectrin (bands 1 and 2)	Minor labeling	0.4	0.01

^{a} From Kahane and Gitler[179] and Gitler and Bercovici.[181]
^{b} From Bayley and Knowles.[138]
^{c} From Brunner and Semenza.[119]
^{d} Abbreviations: LIS, lithium diiodosalicilate; GSH, reduced glutathione.

The two major peripheral proteins of the red cell membrane are spectrin 1 and 2, accounting for as much as 30% of the total protein.[168]

Band 3, the anion transporter, is assumed to span the membrane 12 times,[172] whereas glycophorin A traverses it only once.[173] This implies that the membrane-embedded domain of band 3 is more than 20 times larger in mass than that of glycophorin A and, consequently, is predicted to be much more heavily labeled. However, as we shall see, in most cases opposite results were obtained.

Covalent Incorporation of Label in Lipids and Proteins. Table II shows the results from labeling of the red cell membrane with three well-characterized reagents, 5-[^{125}I]iodonaphthyl-1-azide, [^{3}H]adamantane-diazirine, and [^{125}I]TID. Although the various experiments were not carried out under strictly identical conditions, the data, nonetheless, make clear that these reagents behaved in vastly different ways both with regard to the lipid-to-protein labeling ratios and the distributions of label among individual proteins.

[172] R. P. Kopito and H. F. Lodish, *Nature* (*London*) **316**, 234 (1985).
[173] M. Tomita and V. T. Marchesi, *Proc. Natl. Acad. Sci. U.S.A.* **72**, 2964 (1975).

The extent of total incorporation of radiolabel was similar for all reagents and varied only within the range from approximately 40 to 60%. This level seems to be typical for most classes of hydrophobic labeling reagents when applied to biological membranes. Marked differences were found, however, in the lipid-to-protein labeling ratio. While this was about 1 to 2 (depending on the basis of calculation) for 5-[^{125}I]iodonaphthyl-1-azide, approximately 4 was measured for [^{3}H]adamantanediazirine and as much as 16 for [^{125}I]TID. It is almost certain now that the relatively high protein labeling efficiency observed with the first two reagents is due to the occurrence of long-lived (electrophilic) intermediates that can react with proteins but not with lipids. In the case of [^{3}H]adamantanediazirine, it is the diazo isomer which most likely accounts for this modification, whereas an aziridine is thought to be the species primarily responsible for protein labeling with 5-[^{125}I]idodonaphthyl-1-azide.[174]

Distribution of Label among Proteins. Not all reagents compared in Table II were equally able to distinguish between integral and peripheral proteins of the red cell membrane. The highest degree of discrimination was achieved with [^{125}I]TID which, at most, gave barely detectable labeling of any of the peripheral proteins (bands 1, 2, 4.1, 4.2, 5, and 6). For example, spectrin, the major peripheral component, was labeled to an extent corresponding to approximately 0.01% of the label that had been present originally (or about 0.2% of that associated with protein). In contrast, these components were significantly labeled with both 5-[^{125}I]iodonaphthyl-1-azide and [^{3}H]adamantane diazirine, in spite of the fact that the lipid-to-water partition coefficient of the former reagent is even higher than that of [^{125}I]TID. Again, relatively long-lived reactive intermediates which may diffuse out of the lipid bilayer are presumably responsible for the observed modification of peripheral components. Unlike carbenes or nitrenes, which are immediately quenched upon contact with water, species like ketenes, azirines, or diazo compounds have lifetimes long enough to enable them to diffuse to and react with functional groups exposed on peripheral proteins.

In contrast to the "background labeling" of peripheral proteins just mentioned, several soluble proteins have been found during the past years which are labeled strongly with hydrophobic reagents (including [^{125}I]TID). One such protein is serum albumin, whose physiological role is to bind and carry fatty acids. Apparently, the hydrophobic pocket in this protein can also accommodate (some of) these reagents and upon photoactivation becomes labeled. A second example is calmodulin in its

[174] A. K. Schrock and G. B. Schuster, *J. Am. Chem. Soc.* **106**, 5234 (1984).

Ca^{2+}-bound form, which is also presumed to expose a hydrophobic domain.[175,176] Substantial labeling with both a hydrophobic and a phospholipid analog has also been noted for the β-subunit of the F_1 moiety of F_1F_0-ATP synthase from *E. coli,* a peripheral membrane protein complex.[122,163] There are also proteins, for example, certain bacterial toxins, which, depending on the conditions, exist in either a water-soluble or a membrane-bound form. Hydrophobic photolabeling is currently used in several laboratories to investigate binding of such proteins to membranes.

Distribution of Label among Integral Proteins. The protein by far most heavily labeled in erythrocyte membranes with 5-[^{125}I]iodonaphthyl-1-azide is glycophorin A, while band 3 was modified only to an extent comparable to that of spectrin. Evidently, a negative result has little meaning and, in particular, does not imply a peripheral disposition of the respective protein. With [^{3}H]adamatanediazirine, a somewhat "better" result was achieved but data closest to those predicted were obtained with [^{125}I]TID.[119] The approximately four times heavier labeling of band 3 than that of glycophorin A was the first clear indication that modification with this reagent is not relying on the presence of particularly reactive amino acids. In fact, even proteins such as the small intestinal sucrase–isomaltase, whose membranous portion appears to contain not a single reactive amino acid,[177] is substantially more strongly labeled with [^{125}I]TID than are extrinsic proteins.[14] In the following, we shall examine in more detail the labeling of a particular protein, glycophorin A.

Labeling of Glycophorin A. A question addressed in several studies was whether or not labeling of this protein was restricted to that peptide traversing the membrane. Following proteolysis of glycophorin A with trypsin, in all investigations, the majority of the label was recovered within a fragment, termed T-6,[178] that contains the membrane-spanning region.[87,136,179–181] Of course, this finding does not rule out that some of the label was bound just outside the bilayer (but within T-6). Surprisingly, labeling with [^{3}H]adamantanediazirine was found to be reduced substantially when glutathione (50 mM) had been present in the incubation buffer.[138] Although this thiol can be assumed to be restricted to the aque-

---

[175] J. Krebs, J. Buerkler, D. Guerini, J. Brunner, and E. Carafoli, *Biochemistry* **23,** 400 (1984).
[176] J. Buerkler and J. Krebs, *FEBS Lett.* **182,** 167 (1985).
[177] W. Hunziker, M. Spiess, G. Semenza, and H. F. Lodish, *Cell* **46,** 227 (1986).
[178] H. Furthmayr, R. E. Galardy, M. Tomita, and V. T. Marchesi, *Arch. Biochem. Biophys.* **185,** 21 (1978).
[179] I. Kahane and C. Gitler, *Science* **201,** 351 (1978).
[180] E. Wells and J. B. C. Findlay, *Biochem. J.* **179,** 265 (1979).
[181] C. Gitler and T. Bercovici, *Ann. N.Y. Acad. Sci.* **346,** 199 (1980).

ous phase, it may, nonetheless, have affected labeling within the bilayer by virtue of a rapid exchange of the (a) reactive intermediate into the aqueous medium and efficient scavenging within this phase.[138] Since it is known from another study that glutathione does not reduce labeling of the lipid core of model membranes by carbenes,[101] it is very likely that diazoadamantane is the species that was scavenged by the thiol and, hence, that it was partly responsible also for the observed modification of glycophorin A.

Glycophorin A contains two methionines, Met-8 and Met-81, of which, according to current ideas, the latter is located in the middle of the transmembrane segment.[87] Therefore, cyanogen bromide treatment (cleavage at methionines) of glycophorin A yields fragments which either contain that part of the transmembrane segment which traverses the outer leaflet of the lipid bilayer (fragment 8–81) or the inner half (fragment 82–131). Since the two fragments can be separated by SDS–polyacrylamide gel electrophoresis, it is possible to determine the distribution of label among the two segments, each crossing half of the lipid bilayer. Upon labeling with [^{3}H]adamantanediazirine, Goldman and colleagues[136] were able to demonstrate that both fragments were labeled to nearly the same extent, a result which excluded the possibility that labeling was confined to a single, particularly reactive amino acid (e.g., Glu-70; see below).

Using Edman degradation of the membrane-spanning segment, Ross *et al.*[87] have determined the major sites of modification obtained with various phospholipids containing the $3H$,3-phenyldiazirine group. Like adamantanediazirine, upon photolysis these give rise to a reactive diazo isomer. Two results of this study are of interest here: First, measurements of the kinetics of cross-linking of the photoreactive phospholipid to glycophorin A revealed a "slow component," accounting for approximately 50–75% of the labeling, that was ascribed to be due to the diazo isomer formed.[87] Second, upon cyanogen bromide cleavage, essentially all of the label was found within fragment 9–81 (and 1–81) whereas little, if any, was recovered in fragment 82–131. Using Edman sequencing, the label in fragment 9–81 could be traced back to essentially a single residue (Glu-70). Very likely, this modification arose mostly through the diazo isomer (not the carbene). This demonstrates that modification of proteins may be far more efficient with long-lived species than with a short-lived carbene, provided a suitable, reactive residue(s) is (are) present and accessible.

Proton-Translocating Portion F_0 of F_1F_0-ATP Synthase

ATP synthase from *E. coli* is composed of a hydrophobic component F_1 and a membrane-integrated part F_0 which catalyzes H^+-conduction across the membrane.[182] The primary structure of the five F_1 and three F_0 polypeptides has been derived from corresponding nucleotide sequences.[183–185] Two of the F_0 subunits, a and c, are very hydrophobic while subunit b is an amphipathic protein thought to interact with the membrane via its apolar N-terminal region.[186]

Three reagents were used by Hoppe and colleagues[122] to investigate the membrane topology of the F_0 subunits. The first one was 1-pamitoyl-2-(2-azido-4-nitro)benzoyl-*sn*-glycero-3-[^{3}H]phosphorylcholine, which was found to label predominantly subunit b whereas only little, if any, of the label was recovered in the two other F_0 subunits. Solid-phase Edman degradation of subunit b showed that most of the label was bound to Cys-21 and that a minor fraction was associated with Asn-2 and Trp-26. A very similar label distribution pattern was also obtained with 5-[^{125}I]iodonaphthyl-1-azide which, according to Edman sequencing data, labeled exclusively Cys-21 in subunit b.[162] The curious failure of both reagents to label the major membranous polypeptides a and c is very likely due to the limited reactivity of the intermediates generated from both arylazides, and presumably reflects the absence of sufficiently reactive and accessible amino acids in these subunits. The two studies illustrate once more that a negative result has little meaning, at most perhaps that no reactive residues are available. On the other hand, they clearly reveal the quite limited applicability of arylazides as probes of integral proteins.

The third reagent applied was [^{125}I]TID.[163] Labeling was performed in different conformational (structural) states of the proteins and, again, Edman sequencing was used to determine the individual residues modified. In contrast to the previous studies, all F_0 subunits were labeled. Due to difficulties in purifying and sequencing polypeptide a, the analysis was restricted to subunits b and c. In agreement with the other labeling studies and consistent with the model proposed for the interaction of subunit b with the membrane,[186] labeling was confined to the N-terminal, hydrophobic portion. Again, Cys-21 was the predominant site of labeling but there were additional residues which were clearly labeled as well (Leu-8, Ile-12, Phe-14, Phe-17, and Trp-26). Interestingly, the label distribution patterns were nearly the same when labeling was carried out with pure F_1F_0 complex, F_1-depleted membranes, or SDS-solubilized protein. This was inter-

[182] A. E. Senior and J. G. Wise, *J. Membr. Biol.* **73**, 105 (1983).
[183] J. N. Gay and J. E. Walker, *Nucleic Acids Res.* **9**, 3919 (1981).
[184] J. Nielsen, F. G. Hansen, J. Hoppe, P. Friedl, and K. von Mayenburg, *Mol. Gen. Genet.* **184**, 33 (1981).
[185] H. Kanazawa, K. Mabuchi, T. Kayano, T. Noumi, T. Sekiya, and M. Futai, *Biochem. Biophys. Res. Commun.* **103**, 613 (1981).
[186] J. E. Walker, M. Saraste, and N. J. Gay, *Nature (London)* **298**, 867 (1982).

preted to suggest that the membranous segment of subunit b must be equally accessible in all states compared and, since in SDS there are no interaction with other F_0 subunits, it was inferred that also in the intact system this segment does not make extensive contact with other polypeptides.

Labeling of subunit c resulted in patterns which were completely different upon labeling either in intact F_1F_0 and F_1-depleted membranes or in the SDS-dissociated form. In the intact subunit, labeling was confined to a few sites that lie within two regions located near the N- and C-termini of the polypeptide chain whereas, in the SDS-denatured state, labeling was spread over the entire chain, resulting in a large number of modified residues. The sites of dominant labeling were methionyl residues which evidently have a high intrinsic reactivity toward the carbene. However, in spite of the obvious preference of the carbene for labeling sulfur-containing centers (Cys in subunit b and Met in subunit c), it was an important result that different sets of methionines were labeled in intact F_0 and in SDS-dissociated subunit c. This suggested that labeling patterns are primarily determined by reagent accessibility and not by the intrinsic reactivity of the individual amino acid residues.

As mentioned, only a few discrete residues were labeled in the native structure of subunit c. This may reflect extensive intra- and intermolecular contact among membrane-embedded polypeptides. Most interestingly, those residues within the N-terminal portion labeled in the intact F_0 were distributed in such a manner that they would all be located on the same side of the cylindrical envelope circumscribing an α-helix formed by this segment. From this, it was postulated that radioactivity distribution patterns with an average periodicity of three to four residues are indicative of (1) a helical structure of the respective segment and (2) of tight interactions with other helical rods so that only one face of the surface is exposed and accessible to the reagent. The absence of label in the middle region of subunit c polypeptide chain (when labeled in intact F_0) provided supporting evidence to a previous study proposing that this part of the chain extends from the lipid phase into the cytoplasm.

Occasionally, it is not possible to decide on the basis of radioactivity distribution profiles, as obtained from Edman sequencing, whether a particular residue is labeled or not. To make assignments less ambiguous, Hoppe introduced a thin-layer chromatographic method to analyze the radioactive components released during Edman sequencing. There is increasing evidence that the radioactivity patterns, as visualized by autoradiography, are quite characteristic for a given amino acid. Therefore, this method allows one to differentiate between a truly labeled amino acid and artifactually released radioactivity, such as "carry-over" from a preceding heavily labeled site or release of peptide from glass. More details concerning this valuable analytical tool will be given below.

Bacteriorhodopsin

Both the finding that labeling of integral proteins with [^{125}I]TID is strongly affected by interactions among membrane-embedded polypeptide segments, and the occurrence of a three- to four-residue periodicity in the distribution of labeled residues along the polypeptide segment suggested that labeling with this reagent is confined to amino acid residues which are in contact with the lipid bilayer. To provide further support of this view, an analysis of the labeling pattern was performed which was obtained with a protein of rather well-defined structural features: bacteriorhodopsin. This light-driven proton pump of *Halobacterium halobium* has an M_r of 27,788 and its polypeptide chain has been proposed to cross the membrane seven times via predominantly α-helical rods.[187-189] Since all of these helices show extensive contact among themselves and most or all are also exposed to the lipid bilayer, a label distribution pattern was expected showing features similar to those described above, i.e., showing periodicities that reflect the number of amino acid residues per turn of the helix. For several reasons (discussed elsewhere[151]), the analysis was focused to a particular transmembrane helical segment, the third in the bacteriorhodopsin sequence.

First, it was shown that the partitioning of [^{125}I]TID into purple membrane was not dependent on the reagent concentration, indicating that [^{125}I]TID did not bind to specific (and saturable) sites or pockets and that the (relative) extent of labeling of bacteriorhodopsin was not a function of the reagent concentration chosen. At 0.12 mg of protein/ml and total [^{125}I]TID concentrations ranging from 2×10^{-8} M to 10^{-4} M, the fraction of reagent that partitioned into purple membrane was between 76 and 82% and the fraction which, upon reagent activation, became covalently bound was between 5.7 and 6.3%. Second, following delipidation of the [^{125}I]TID-labeled bacteriorhodopsin and degradation of the protein by cyanogen bromide, the fragment from residue 69 to 118 (CNBr-9a according to the nomenclature of Gerber and Khorana[190]) was isolated using gel permeation chromatography on Sephadex LH-60 and reversed-phase

---

[187] R. Henderson and P. N. T. Unwin, *Nature (London)* **257**, 28 (1975).
[188] H. G. Khorana, G. E. Gerber, W. C. Herlihy, C. P. Gray, R. J. Anderegg, K. Nihei, and K. Biemann, *Proc. Natl. Acad. Sci. U.S.A.* **76**, 5046 (1979).
[189] D. M. Engelman, R. Henderson, A. D. McLachlan, and B. A. Wallace, *Proc. Natl. Acad. Sci. U.S.A.* **77**, 2023 (1980).
[190] G. E. Gerber and H. G. Khorana, this series, Vol. 88, p. 56.

HPLC as devised by Khorana's group.[191,192] Purification and quantitative recovery of [^{125}I]TID-labeled CNBr-9a turned out to be difficult. This was mainly because the modified species no longer coeluted with the unlabeled peptide. The critical requirement then became the unambiguous identification of the trace amounts of actually labeled CNBr-9a not coeluting with unlabeled peptide. In the case of CNBr-9a, the labeled and unlabeled peptides behaved almost identically upon electrophoresis on urea-SDS polyacrylamide gels and this method (in conjunction with autoradiography) may be the most suitable one for screening individual fractions, as for example obtained from HPLC, for a particular labeled peptide. However, as mentioned already, final verification of the identity of a labeled peptide had to include Edman sequencing of the respective peptide and an analysis of the individual [^{125}I]TID-modified PTH amino acids by thin-layer chromatography.

A 17-residue long stretch within the [^{125}I]TID-labeled CNBr-9a was then analyzed for modified amino acids. A total of eight residues were found to be derivatized by the carbene. Their distribution along the chain was consistent with the idea that, in the native protein, this segment formed a helix whose cylindrical envelope was mainly accessible from a single face. In fact, as shown by the axial projection of that segment in an α-helical conformation, seven out of the eight labeled residues are located on one-half of the helix wheel (Fig. 10). Hence, this result suggests a meaningful correlation between lipid contact of individual residues and labeling by [^{125}I]TID. That labeling occurred at the less polar half of the helix surface (assuming 3.6 residues per turn of the coil), was further consistent with the view that bacteriorhodopsin is an "inside-out" protein with the majority of polar and charged residues facing the interior of the molecule and nonpolar residues exposed to the lipid bilayer.[193]

Light-Harvesting Polypeptide B870-α from Rhodospirillum rubrum

The last example of a labeling study considered in more detail in this article is that of the light-harvesting protein B870-α polypeptide chain of *R. rubrum*.[128] This is a rather short peptide, consisting only of 52 amino acid residues, whose function is the collection and transfer of light energy to the reaction center, where it is converted into an electrochemical gradi-

---

[191] Y. Takagaki, G. E. Gerber, K. Nihei, and H. G. Khorana, *J. Biol. Chem.* **255**, 1536 (1980).
[192] G. E. Gerber, R. J. Anderegg, W. C. Herlihy, C. P. Gray, K. Biemann, and H. G. Khorana, *Proc. Natl. Acad. Sci. U.S.A.* **76**, 227 (1979).
[193] D. M. Engelman and G. Zaccai, *Proc. Natl. Acad. Sci. U.S.A.* **77**, 5894 (1980).

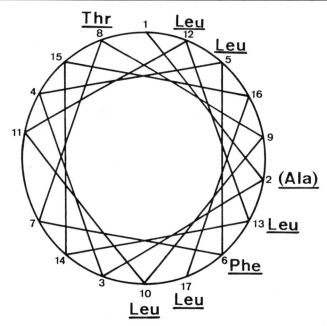

Fig. 10. Axial projection of the putative third transmembrane helix of bacteriorhodopsin depicting those amino acid residues (marked by the three-letter code) which, in intact purple membrane, were labeled with [^{125}I]TID. From Brunner et al.[151]

ent.[194,195] The sequence shows a typical tripartite structure: the polar N- and C-terminal regions being connected by a sequence of approximately 20 predominantly hydrophobic amino acids which, as shown by other methods, crosses the lipid bilayer.[196] From a methodological point of view, this study is of interest because labeling patterns have been compared which were derived from three reagents differing in molecular structure and size but exhibiting (nearly) identical photochemical properties. The reagents referred to are [^{125}I]TID, 1-[4-[(trifluoromethyl)diazirinyl]phenyl]-9-[10-^{3}H]oxaundecanoic acid ([^{3}H]TOA) (an analog of a fatty acid), and [^{3}H]PTPC.

Using the general methodology outlined above (solid-phase Edman sequencing in conjunction with TLC analysis of the labeled PTH amino acids), the radioactivity originating from [^{125}I]TID-labeling could be

[194] R. K. Clayton and W. R. Sistrom, "The Photosynthetic Bacteria." Plenum, New York, 1978.
[195] A. N. Glazer, Annu. Rev. Biochem. **52**, 125 (1983).
[196] R. A. Brunisholz, F. Suter, and H. Zuber, Hoppe-Seyler's Z. Physiol. Chem. **365**, 675 (1984).

traced back to two regions of the polypeptide, the transmembrane stretch (as was expected) and the N-terminal portion. Figure 11A shows the corresponding profile of the [^{125}I]TID-radioactivity distribution among individual amino acid residues along with the TLC patterns of the [^{125}I]TID-modified PTH amino acids (Fig. 11B). In addition to those residues which were clearly labeled (Met-1, Ile-4, Leu-7, Phe-8, Leu-14, Leu-17, Phe-20, Val-23, Leu-26, Leu-27, and His-29), some contained only traces of radioactivity (Arg-3, Ile-21, Phe-22, and Phe-30). From the TLC patterns of the radioactive components, visualized by autoradiography (Fig. 11B), evidence for minor labeling of Phe-22 and Phe-30 (same radioactivity pattern as Phe-20) was obtained, whereas labeling of Ile-21 could be excluded because the pattern seen at this cycle is identical with that of Phe-20, suggesting that it was due to some "carry over" from the strongly labeled Phe-20. Again, as observed previously and noted above for the distributions of label among segments of the c subunit of F_1F_0-ATP synthase[163] and CNBr-9a from bacteriorhodopsin,[151] there was a striking regularity in the spacing of labeled sites along these stretches which is indicative of a helical structure of those segments that are accessible from one side only. Further, the spatial distribution within the N-terminal portion of both labeled and polar amino acids was suggestive of a helix that is amphipathic perpendicular to its axis and extends along the water–lipid interface of the membrane.

As far as the transmembrane part is concerned, similar overall label distribution patterns were also obtained with the two other amphipathic reagents. However, marked differences were noted in the labeling and, hence, apparent access of the various reagents to the N-terminal portion of the polypeptide, the putative amphipathic helix. The meaning of this finding is not yet clear, but can be interpreted either to reflect different distribution profiles of the three reagents within the lipid bilayer or, in a more speculative manner, a change in the rotational orientation of that helical segment (possibly induced by the introduction of label into the chromatophore membrane).[128]

A point of particular interest that was raised in the course of this study concerns the possible damage to the membrane by the high-intensity irradiation required for activation of the reagents. This required particular consideration since chromatophores absorb light over much of the ultraviolet and visible range. Although the primary event(s) of photoinduced, irreversible alterations (e.g., cleavage of covalent bonds) may be very rapid, larger, multiresidue, conformational changes, as may be detectable by the current photolabeling method, are presumably slower. To reduce the possibility that the labeling results reflected such extensive structural (conformational) rearrangements within the light-harvesting polypeptide,

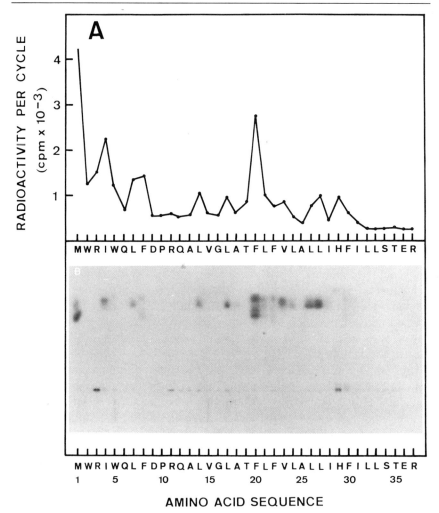

AMINO ACID SEQUENCE

FIG. 11. Distribution of radioactivity among individual amino acid residues of the B870-α polypeptide chain labeled with [^{125}I]TID in intact chromatophores of *Rhodospirillum rubrum* G-9$^+$. Labeled peptide was subjected to solid-phase Edman degradation and the radioactivity eluted with each cycle was plotted (A). Following conversion into phenylthiohydantoin derivatives, aliquots were applied onto silica gel thin-layer plates and the chromatogram was developed in chloroform–ethanol (98:2, v/v). The [^{125}I]TID-labeled derivatives of the individual phenylthiohydantoin amino acids were visualized by autoradiography (B). From Meister et al.[128]

additional experiments were carried out in which the reagent was activated with a powerful flash from an air-cooled xenon flash tube which was fired at 2.25 kV and discharged within approximately 200 μs. Under these conditions, approximately 20 to 40% of the reagent could be converted into the carbene with a single flash. Edman sequencing of thus [^{125}I]TID-labeled B870-α gave a radioactivity distribution pattern which was indistinguishable from that obtained upon long-term irradiation. Since the total inputs of light were different in both types of experiments (activation by the flash and continuous light, respectively), as were the spectral properties of the two light sources (and, hence, contained different proportions of potentially damaging irradiation), the absence of any significant difference in the label distribution patterns virtually rules out major structural rearrangements of the B780-α polypeptide as a result of the illumination of the chromatophores.

Concluding Remarks

All reagents discussed in the present chapter, including those listed in the Appendix, are monofunctional, which means that they contain a single center designed to react with a biomolecule. It should be pointed out, however, that a number of bifunctional reagents have been developed which at one end contain a "group-specific" function (intended to react first with the target) and a photoactivatable group on the other end. Although some of these reagents are very hydrophobic and, presumably, partition almost quantitatively into the lipid bilayer, they are usually not designed to react (exclusively) within the hydrophobic core but within the aqueous phase as well. What has been said in this chapter about monofunctional photoactivatable reagents is, in principle, applicable to bifunctional reagents of the type mentioned above. However, since the questions asked and goals envisaged are different from those of hydrophobic photolabeling, additional aspects must also be considered. For example, the fact that (singlet) carbenes are very efficiently scavenged by water does imply that cross-linking of proteins, within an aqueous phase, with such reagents would probably be very inefficient, if detectable at all. For this reason, arylazides or nitroarylazides, though generating less reactive species, might make the more suitable reagents. This and other features of cross-linking reagents have been discussed in other reviews.[197–200]

[197] K. Peters and F. M. Richards, *Annu. Rev. Biochem.* **46**, 523 (1977).
[198] S. H. Hixson, T. F. Brownie, C. C. Chua, B. B. Crapster, L. M. Satlin, S. S. Hixson, C. O. L. Boyce, M. Ehrich, and E. K. Novak, *Ann. N.Y. Acad. Sci.* **346**, 104 (1980).
[199] T. H. Ji, this series, Vol. 91, p. 580.
[200] H. Sigrist and P. Zahler, in "The Enzymes of Biological Membranes" (A. Martonosi, ed.), Vol. 1, p. 333. Plenum, New York, 1985.

Appendix: Application of Hydrophobic and Amphipathic Photolabeling Reagents

Structure of reagent	Subject of study/labeling	Ref.[a]
I. Lipid-soluble reagents		
	Egg lecithin/myelin membrane vesicles	201
	Model membranes	64
	Human erythrocyte membranes	202
	Rabbit skeletal muscle sarcoplasmic reticulum	1
	Human erythrocyte membrane	202
	Human erythrocyte membrane/glycophorin A	180
	Pig kidney microvillar membrane	203
	Human erythrocyte membrane/band 3	204
	Rabbit skeletal muscle sarcoplasmic reticulum	1
	Na^+,K^+-ATPase from pig kidney	153, 155
	Brush border membrane (small intestine)	205
	Membrane proteins and ionophoric channels	206
	Rabbit skeletal muscle sarcoplasmic reticulum/Ca^{2+}-ATPas	134
	Human erythrocyte membrane/glycophorin A	179, 181
	Cytochrome oxidase from baker's yeast	137

Acetylcholine receptor (*Torpedo californica*)	160
Pig microvillus aminopeptidase	207
Chromatophores from *Rhodospirillum rubrum*	208
Membrane-associated proteins of stomatitis virus	209, 210
Membrane insertion of prothrombin	211
F_0-subunits of F_1F_0-ATPase from *E. coli*	162
Uncoupling of response of adenylate cyclase to gonadotropins	212
Cytochrome oxidase from bakers' yeast	137
Retinal rod outer disk membrane/rhodopsin	140
Labeling of lymphocyte surface antigens	139
Interactions of complement proteins with membranes	165
Na^+,K^+-ATPase from *Bufo marinus* toad kidney	157
Mouse Lyt-2/3 antigen complex	158

(*continued*)

Appendix (*continued*)

Structure of reagent	Subject of study/labeling	Ref.[a]
[pyrene-SO₂N₃ structure]	Acetylcholine receptor subunits (*Torpedo californica*)	135
[adamantyl diazo structure]	Model membranes	101
	Hydrophobic segments of intrinsic proteins	136
	Human LA-DR-antigen polypeptides	213
	Human erythrocyte membrane	138
	Na⁺,K⁺-ATPase from canine kidney	17, 154
	Acetylcholine receptor (*Torpedo californica*)	161
[phenyl diazo structure]	Synthesis	106, 107, 214
	Model membranes	101

[diazirine-CF₃-phenyl structure]	Synthesis	110, 111
[diazirine-CF₃-iodophenyl structure]	Synthesis and properties	119, 150
	Membrane domains of γ-glutamyltransferase	115
	Interactions of enkephalins and dynorphin with liposomes	215
	Interactions of adrenocorticotropin peptides with liposomes	216, 217
	Na⁺,K⁺-ATPase from pig kidney	156
	Calmodulin	175, 176
	F₀ part of F₁F₀-ATP synthase from *E. coli*	163
	Mouse Lyt-2/3 antigen complex	158
	Ca²⁺-ATPase from human erythrocyte	218
	F₀ part of F₁F₀-ATP synthase from *Neurospora crassa*	219
	Acetylcholinesterase from human erythrocyte membrane	18
	Acetylcholinesterase from *Torpedo marmorata*	16
	Bacteriorhodopsin	151
	Light-harvesting protein B870- from *Rhodospirillum rubrum*	128
	Lipophilin from human myelin	19
	Membrane attack complex of complement	166

(*continued*)

Appendix (*continued*)

Structure of reagent		Subject of study/labeling	Ref.[a]
II. Amphipathic reagents			
A. Fatty acid derivatives			
[sugar-NH—CO—(CH$_2$)$_m$—CH(O-aryl-NO$_2$,N$_3$)—(CH$_2$)$_n$—CH$_3$ structure]	$m = 10; n = 5$	M13 coat protein/liposomes	67
		Cholera toxin	36, 37
		Newcastle disease virus proteins	220
		Ninth component of human complement	221
		Ricin toxin insertion into membranes	222
		Phospholipid vesicles and human erythrocyte membrane	223
		Diphtheria toxin	224
	$m = 0; n = 13$	Phospholipid vesicles and human erythrocyte membrane	223
N$_3$—(aryl-NO$_2$)—NH—(CH$_2$)$_{11}$—CONH—CH$_2$—COOH		Na$^+$,K$^+$-ATPase from *Electrophorus electricus*	142

Structure	Application	Ref.
[Sugar derivative with NH—(CH$_2$)$_{11}$—CO—NH linked to 2-nitro-4-azidophenyl group]	Cholera toxin insertion into membranes Na$^+$,K$^+$-ATPase from *Electrophorus electricus*	225 142
[3-azidophenyl—O—(CH$_2$)$_{10}$—COOH]	Biosynthetic incorporation into phospholipids of *E. coli*	30
CH$_3$—(CH$_2$)$_5$—CH(—O-2-nitro-4-azidophenyl)—(CH$_2$)$_{10}$—COOH	Biosynthetic incorporation in *E. coli* Biosynthetic utilization in cultured cells	29 32

(*continued*)

Appendix (*continued*)

Structure of reagent		Subject of study/labeling	Ref.[a]
$N_3-(CH_2)_{15}-COOH$		Vesicular stomatitis virus	77
$CH_3-(CH_2)_5-CH-(CH_2)_{10}-COOH$ $|$ N_3		Biosynthetic incorporation in tissue culture	31
$CH_3-(CH_2)_5-CH-CH_2-CH=CH-(CH_2)_7-COOH$ $|$ N_3			
$N_3-(CH_2)_5-CH=CH-CH_2-CH=CH-(CH_2)_7-COOH$			
(cyanophenyl)$-O-(CH_2)_n-COOH$	$n = 15$	Biosynthetic incorporation into phospholipids of *E. coli*	30
	$n = 8, 10, 12, 14$	Synthesis and biosynthetic utilization	33
	$n = 8$	*In vivo* incorporation into vesicular stomatitis virus	34
$CH_3-(CH_2)_5-CH-(CH_2)_{10}-CO-NH-R$ (2-NO$_2$, 4-N$_3$ phenoxy)		Synthesis	226

R = CH$_2$—CO—NH—CH$_2$—COOH
(plus four additional polar residues)

B. Phospholipids
  1. Phosphatidylcholines

$$\begin{array}{l} \text{R}_2\text{—C—O—CH}_2 \\ \quad\ \|\quad\quad\ | \\ \quad\ \text{O}\quad\quad\ \text{CH—O—C—R}_1 \\ \quad\quad\quad\quad\ |\quad\quad\ \| \\ \quad\quad\quad\quad\ \text{CH}_2\text{—O—PO}_2^-\text{—O—(CH}_2)_2\text{—}^+\text{N—(CH}_3)_3 \end{array}$$

Structure	Application	Ref.
R$_1$ = CH$_3$—(CH$_2$)$_{14}$—	Phospholipid synthesis	126
R$_2$ = —O—(CH$_2$)$_{10}$—〔phenyl-N$_3$〕	Cross-linking to phospholipids	65
R$_1$ = CH$_3$—(CH$_2$)$_{14}$— (n = 10)	Cross-linking to phospholipids	126
R$_2$ = —O—(CH$_2$)$_n$—〔phenyl-NO$_2$, N$_3$〕 (n = 6)	Phosphatidylcholine exchange protein	141

(*continued*)

Appendix (continued)

Structure of reagent	Subject of study/labeling	Ref.[a]
$R_1 = CH_3-(CH_2)_{12}-$ $R_2 =$ (2-azido-4-nitrophenyl with methyl)	D-β-Hydroxybutyrate apodehydrogenase	227
$R_1 = CH_3-(CH_2)_{14(16)}-$ $R_2 =$ (2-azido-4-nitrophenyl with methyl)	Calf liver cytochrome b_5	228
	Beef heart cytochrome-c oxidase subunits	229
	Mitochondrial ATPase complex	230
	Na^+,K^+-ATPase subunits	142
	Lipid insertion of cholera toxin	225
	Beef heart ubiquinone–cytochrome-c reductase (complex III)	152
	Subunit II of beef heart cytochrome-c oxidase	123
	Sarcoplasmic reticulum ATPase	121
	ATP synthase from bovine heart mitochondria and $E.\ coli$	231
$R_1 = CH_3-(CH_2)_{12}-$ $R_2 =$ NH-$(CH_2)_{11}$- (2-nitro-4-azidophenyl)	Subunit b of F_1F_0-ATP synthase from $E.\ coli$	122
	Rabbit and rat sarcoplasmic reticulum	232
	Acetylcholine receptor from *Torpedo californica*	20
	Tetanus toxin	233

$R_1 = R_2 = CH_3-(CH_2)_{10}-\overset{N_3}{\overset{|}{CH}}-(CH_2)_3-$	Synthesis and cross-linking in HDL[b]	78, 82
$R_1 = R_2 = N_3-(CH_2)_{15}-$	Synthesis and cross-linking in HDL[b]	78, 82
$R_1 = R_2 = CH_3-(CH_2)_5-\overset{N_3}{\overset{|}{CH}}-CH_2-CH=CH-(CH_2)_7-$	Synthesis and cross-linking in HDL[b]	78, 82
$R_1 = R_2 = N_3-(CH_2)_5-CH=CH-CH_2-CH=CH-(CH_2)_7-$	Synthesis and cross-linking in HDL[b]	78, 80
$R_1 = CH_3-(CH_2)_{14(16)}-$ $R_2 = CH_3-(CH_2)_5-\overset{|}{CH}-(CH_2)_{10}-$ $\overset{|}{N_3}$	Synthesis Synthesis and cross-linking in HDL[b]	2 78, 80
$R_1 = CH_3-(CH_2)_{14(16)}-$ $R_2 = CH_3-(CH_2)_5-\overset{N_3}{\overset{|}{CH}}-CH_2-CH=CH-(CH_2)_7-$	Synthesis Synthesis and cross-linking in HDL[b]	2 78, 80
$R_1 = CH_3-(CH_2)_{14(16)}-$ $R_2 = CH_3-(CH_2)_{10}-\overset{|}{CH}-(CH_2)_3-$ $\phantom{R_2 = CH_3-(CH_2)_{10}-}\overset{|}{N_3}$	Synthesis Synthesis and cross-linking in HDL[b]	78 80
$R_1 = CH_3-(CH_2)_{14(16)}-$ $R_2 = N_3-(CH_2)_5-CH=CH-CH_2-CH=CH-(CH_2)_7-$	Cross-linking to high-density apoprotein A-I Interaction of complement proteins with membranes Synthesis and cross-linking in HDL[b]	79 81 78, 80
$R_1 = CH_3-(CH_2)_{14}-$ $R_2 = CH_3-(CH_2)_5-\overset{|}{CH}-(CH_2)_8-$ $\overset{|}{N_3}$	Cross-linking to phospholipids	65

(continued)

Appendix (continued)

Structure of reagent		Subject of study/labeling	Ref.[a]
R_1 = Alkyl R_2 = $CH_3—(CH_2)_5—CH—CH_2—CH=CH—(CH_2)_7—$ $\quad\quad\quad\quad\quad\quad\quad\;\; O—C(=O)—N_3$		Cross-linking to phospholipids	234
$R_1 = CH_3—(CH_2)_m—$	$m = 14, 16$	Synthesis	126
$R_2 = CF_3—C(N_2)—COO—(CH_2)_n—$	$n = 10, 11, 15$	Intermolecular cross-linking in model membranes	65, 84
	$m = 14; n = 11$	Membranous segment of cytochrome b_5	88
	$m = 16; n = 11$	Cross-linking to cholesterol	86
$R_1 = CH_3—(CH_2)_m—$	$m = 14; n = 10$	Phospholipid synthesis	126
$R_2 = $ (3-(diazirinyl)phenoxy-$(CH_2)_n$-)		Cross-linking to phospholipids	65
		Phase separation in model membranes	85
		Cross-linking to cholesterol	86
		Transmembrane domain of glycophorin A	87
		Sites of intermolecular cross-linking to phospholipids	235
		Membrane-embedded segment of cytochrome b_5	88
	$m = 16; n = 10$	Cross-linking to phospholipids	65
	$m = 14; n = 5$	Transmembrane domain of glycophorin A	87
		Sites of intermolecular cross-linking to phospholipids	235

$R_1 = CH_3-(CH_2)_{14}-$	$n = 6$	Synthesis/labeling of the outer leaflet of the human erythrocyte membrane	12
$R_2 =$ (CF$_3$)(N=N)-C$_6$H$_4$-(CH$_2$)$_2$-O-(CH$_2$)$_n$-	$n = 7$	Synthesis/labeling of chromatophores from *R. rubrum* Tetanus toxin Interaction of vinculin with membranes	128 233 236
$R_1 = CH_3-(CH_2)_{14}-$	$n = 3$	Cross-linking to phospholipids	66
$R_2 =$ 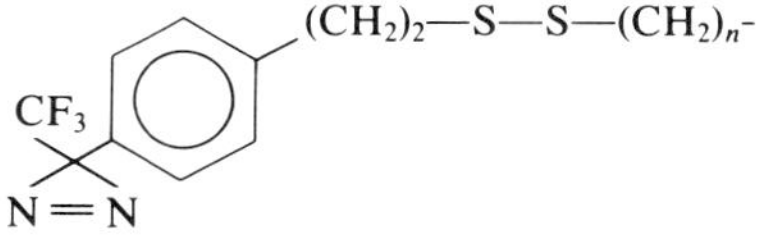 (CF$_3$)(N=N)-C$_6$H$_4$-(CH$_2$)$_2$-S-S-(CH$_2$)$_n$-	$n = 7$	Cross-linking to phospholipids/gramicidin A	66

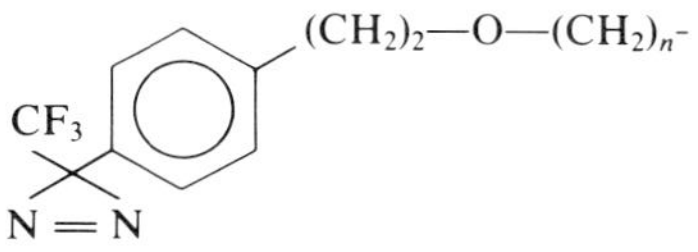

2. Sphingomyelins

$$C_{13}H_{29}-CH=CH-CH(OH)-CH(NH-C(=O)-R)-CH_2-O-P(=O)(O^-)-O-(CH_2)_2-{}^+N(CH_3)_3$$

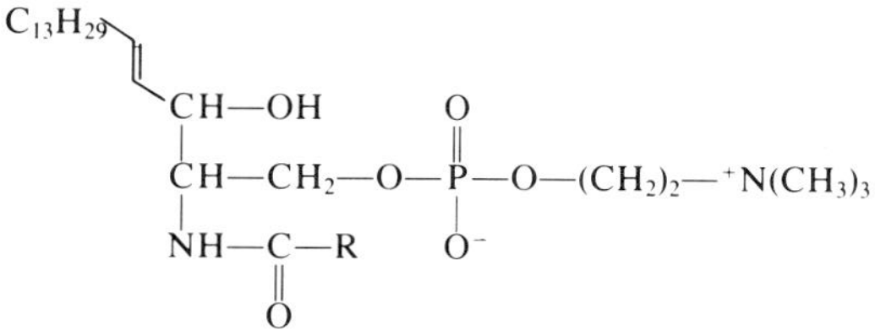

$R = N_3-(CH_2)_{15}-$ ⎫
$R = CH_3-(CH_2)_5-CH(N_3)-CH_2-CH=CH-(CH_2)_7-$ ⎬ Synthesis 78
$R = N_3-(CH_2)_5-CH=CH-CH_2-CH=CH-(CH_2)_7-$ ⎭

(*continued*)

Appendix (*continued*)

Structure of reagent	Subject of study/labeling	Ref.[a]
3. Phosphatidylinositol	Synthesis	237
C. Glycolipids	Synthesis and photochemical properties	238

Structure	Description	Ref.
$R_1 = $ —OH; $R_2 = $ —N_3 (25-Azidonorcholesten-3β—ol)	Synthesis Inhibition of cholesterol biosynthesis	78 239
$R_1 = $ —O—CO—$(CH_2)_7$—CH=CH—$(CH_2)_7$—CH_3 $R_2 = $ —N_3	Synthesis	78
$R_1 = $ —N_3; $R_2 = $ —CH_3 (3-azido-5-cholestene)	Synthesis	239
$R_1 = N_3$—$(CH_2)_{15}$—C(=O)—O—; $R_2 = $ —CH_3	Synthesis	78
$R_1 = CH_3$—$(CH_2)_5$—CH(—N_3)—CH_2—CH=CH—$(CH_2)_7$—COO— $R_2 = $ —CH_3	Synthesis	78
$R_1 = $ —O—C(=O)—CH=N_2; $R_2 = $ —CH_3	Synthesis Photolysis in model membranes	240 241

[a] Numbers refer to text footnotes.
[b] HDL; High-density lipoprotein.

Acknowledgments

This work was supported by the Swiss National Science Foundation, Bern. The author wishes to thank Prof. G. Semenza in whose laboratory much of the work described has been carried out. Valuable discussions over the years with Hagan Bayley, Fred Richards, Cesare Montecucco, Robert Aggeler, Bernhard Reber, and many other colleagues are also cordially acknowledged. Thanks are also due to Frances Jay, Martin Spiess, and Carman Zugliani for their criticism and help in preparing this manuscript.

---

[201] K. M. Abu-Salah and J. B. C. Findlay, *Biochem. J.* **161**, 223 (1977).
[202] E. Wells and J. B. C. Findlay, *Biochem. J.* **179**, 257 (1979).
[203] A. G. Booth, L. M. L. Hubbard, and A. J. Kenny, *Biochem. J.* **179**, 397 (1979).
[204] E. Wells and J. B. C. Findlay, *Biochem. J.* **187**, 719 (1980).
[205] K. Sigrist-Nelson, H. Sigrist, T. Bercovici, and C. Gitler, *Biochim. Biophys. Acta* **468**, 163 (1977).
[206] C. Gitler, *Dev. Bioenerg. Biomembr.* **1**, 11 (1977).
[207] O. Norén and H. Sjöström, *Eur. J. Biochem.* **104**, 25 (1980).
[208] E. Odermatt, M. Snozzi, and R. Bachofen, *Biochim. Biophys. Acta* **591**, 372 (1980).
[209] J. J. Zakowski and R. R. Wagner, *J. Virol.* **36**, 93 (1980).
[210] D. A. Mancarella and J. Lenard, *Biochemistry* **20**, 6872 (1981).
[211] M. F. Lecompte, I. Rosenberg, and C. Gitler, *Biochem. Biophys. Res. Commun.* **125**, 381 (1984).
[212] Y. Raviv, T. Bercovici, C. Gitler, and Y. Salomon, *Biochemistry* **23**, 503 (1984).
[213] J. F. Kaufman and J. L. Strominger, *Proc. Natl. Acad. Sci. U.S.A.* **76**, 6304 (1979).
[214] P. Leblanc and G. E. Gerber, *Can. J. Chem.* **62**, 1767 (1984).
[215] B. Gysin and R. Schwyzer, *Arch. Biochem. Biophys.* **225**, 467 (1983).
[216] B. Gysin and R. Schwyzer, *FEBS Lett.* **158**, 12 (1983).
[217] B. Gysin and R. Schwyzer, *Biochemistry* **23**, 1811 (1984).
[218] M. Zurini, J. Krebs, J. T. Penniston, and E. Carafoli, *J. Biol. Chem.* **259**, 618 (1984).
[219] J. Hoppe and W. Sebald, in "H^+-ATPsynthases" (S. Papa, K. H. Altendorf, L. Ernster, and L. Packer, eds.), p. 173. JCSU Press, Bari, Italy, 1984.
[220] J. S. Bramhall, M. A. Shiflett, and B. J. Wisnieski, *Biochem. J.* **177**, 765 (1979).
[221] B. Ishida, B. J. Wisnieski, C. H. Lavine, and A. F. Esser, *J. Biol. Chem.* **257**, 10551 (1982).
[222] B. Ishida, D. B. Cawley, K. Reue, and B. J. Wisnieski, *J. Biol. Chem.* **258**, 5933 (1983).
[223] T. A. Berkhout, A. van Amerongen, and K. W. A. Wirtz, *Eur. J. Biochem.* **142**, 91 (1984).
[224] L. S. Zahlman and B. J. Wisnieski, *Proc. Natl. Acad. Sci. U.S.A.* **81**, 3341 (1984).
[225] M. Tomasi and C. Montecucco, *J. Biol. Chem.* **256**, 11177 (1981).
[226] J. Bramhall, B. Ishida, and B. Wisnieski, *J. Supramol. Struct.* **9**, 399 (1978).
[227] M. S. El Kebbaj, J. M. Berrez, T. Lakhlifi, C. Morpain, and N. Latruffe, *FEBS Lett.* **182**, 176 (1985).
[228] R. Bisson, C. Montecucco, and R. A. Capaldi, *FEBS Lett.* **106**, 317 (1979).
[229] R. Bisson, C. Montecucco, H. Gutweniger, and A. Azzi, *J. Biol. Chem.* **254**, 9962 (1979).
[230] C. Montecucco, R. Bisson, F. Dabbeni-Sala, A. Piotti, and H. Gutweniger, *J. Biol. Chem.* **255**, 10040 (1980).
[231] C. Montecucco, F. Dabbeni-Sala, P. Friedl, and Y. M. Galante, *Eur. J. Biochem.* **132**, 189 (1983).
[232] H. E. Gutweniger and C. Montecucco, *Biochem. J.* **220**, 613 (1984).

[233] C. Montecucco, G. Schiavo, J. Brunner, E. Duflot, P. Boquet, and M. Roa, submitted for publication.
[234] V. A. Vaver, A. N. Ushakov, and M. L. Tsirenina, *Bioorg. Khim.* **5,** 1520 (1979).
[235] R. Radhakrishnan, C. E. Costello, and H. G. Khorana, *J. Am. Chem. Soc.* **104,** 3990 (1982).
[236] V. Niggli, D. P. Dimitrov, J. Brunner, and M. M. Burger, submitted for publication.
[237] J. Brunner and C. Zugliani, manuscript in preparation.
[238] J. Wydila and E. R. Thornton, *J. Org. Chem.* **49,** 244 (1984).
[239] W. Stoffel and R. Klotzbücher, *Hoppe-Seyler's Z. Physiol. Chem.* **359,** 199 (1978).
[240] S. A. Keilbaugh and E. R. Thornton, *J. Am. Chem. Soc.* **105,** 3283 (1983).
[241] S. A. Keilbaugh and E. R. Thornton, *Biochemistry* **22,** 5063 (1983).

[35] Electrophoretic Transfer of High-Molecular-Weight Proteins for Immunostaining

By KUAN WANG, BRADFORD O. FANGER, CHERYL A. GUYER, and JAMES V. STAROS

Introduction

Since its introduction by Towbin *et al.*,[1] the technique of electrophoretic transfer of proteins from SDS[2]–polyacrylamide gels to solid supports for identification of specific proteins by immunological (or other) methods has become widely applied to many biochemical problems (reviewed in Refs. 3–8). A serious limitation of the original method is that it results in little or no transfer of high-molecular-weight proteins ($M_r > 100,000$). A number of modifications in the original technique have been introduced which facilitate the transfer of high-molecular-weight proteins.[8] The purpose of this brief chapter is to update previous reviews in this series[3,4] by describing a set of methods that we have successfully applied to the efficient electrophoretic transfer of high-molecular-weight (M_r to

[1] H. Towbin, T. Staehelin, and J. Gordon, *Proc. Natl. Acad. Sci. U.S.A.* **76,** 4350 (1979).
[2] Abbreviations: SDS, sodium dodecyl sulfate; PAGE, polyacrylamide gel electrophoresis; Tris, tris(hydroxymethyl)aminomethane; EGF, epidermal growth factor; BS³, bis(sulfosuccinimidyl) suberate.
[3] A. Haid and M. Suissa, this series, Vol. 96, p. 192.
[4] J. Reiser and G. R. Stark, this series, Vol. 96, p. 205.
[5] J. M. Gershoni and G. E. Palade, *Anal. Biochem.* **131,** 1 (1983).
[6] H. Towbin and J. Gordon, *J. Immunol. Methods* **72,** 313 (1984).
[7] G. Bers and D. Garfin, *BioTechniques* **3,** 276 (1985).
[8] J. M. Gershoni, *Trends Biochem. Sci.* **10,** 103 (1985).

~500,000) or very-high-molecular-weight (M_r to ~2,000,000) membrane and cytoskeletal proteins.

Methods

Materials and Reagents

Electrophoretic transfer apparatus (Hoefer TE42 for slab gels up to 20 × 20 cm, TE22 for minislab gels up to 9 × 9 cm, or Bio-Rad Trans-Blot Cell)

Power supply (Hoefer Model PS 250/2.5, Bio-Rad Model 250/2.5 or other medium-voltage, high-current power supply)

Refrigerated circulator (Neslab HX50, Lauda K-2/R or Haake FK10)

Nitrocellulose sheets (Schleicher and Schuell BA85, 0.45 μm pore size, or ≤0.45 μm pore size *pure* nitrocellulose membrane from this or another vendor)

Cellulose acetate sheet (Schleicher and Schuell ST69, 1.2 μm pore size, or equivalent *pure* cellulose acetate from other vendors)

Nylon mesh (monofilament nylon mesh: CMN-105 from Small Parts Inc., P.O. Box 381736, Miami, FL 33238-1736)

Filter paper (Whatman 3MM or Whatman 17Chr)

Sodium dodecyl sulfate (SDS) (>99% C_{12}, BDH 44244 or equivalent)

Buffers

Stock buffer: 125 mM Tris base, 960 mM glycine (a pH of 8.3 is obtained on dilution without adjustment)

Soaking buffer: 10× dilution of stock buffer, 0.1% (w/v) SDS

Transfer buffer A: 10× dilution of stock buffer, 20% (v/v) methanol, 0.1% (w/v) SDS

Transfer buffer B: 5× dilution of stock buffer, 20% (v/v) methanol, 0.1% (w/v) SDS

Procedures for the Transfer of Very-High-Molecular-Weight Proteins (M_r to ~2,000,000)

SDS–PAGE. Proteins with subunits ranging from M_r ~10,000 to 2,000,000 can be resolved on two gradient gel systems: a 2 to 12% polyacrylamide gradient gel[9] carried out in the buffer system of Fairbanks *et*

---

[9] L. L. Somerville and K. Wang, *Biochem. Biophys. Res. Commun.* **102**, 53 (1981).

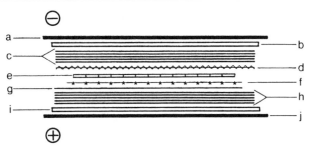

FIG. 1. Diagrammatic representation of the transfer sandwich. Each component of the sandwich is applied individually and assembled in a tray containing soaking buffer: a and j, plastic grid; b and i, sponge; c and h, filter paper; d, nylon mesh; e, polyacrylamide gel; f, cellulose acetate sheet; g, nitrocellulose membrane.

al.[10] and a 3 to 22% polyacrylamide gradient gel[11] carried out in the buffer system of Laemmli.[12]

Electrophoretic Transfer. Gloves should be worn during all manipulations. The nitrocellulose sheet is cut to a size slightly larger than the SDS–polyacrylamide gel that is to undergo transfer and is soaked in water along with 6–10 sheets of filter paper. Following electrophoresis, the polyacrylamide gel is briefly (15–20 min) placed into soaking buffer. These soaking procedures are designed to help minimize any change in size of the nitrocellulose paper or of the gel during the transfer.

Assembly of the transfer "sandwich" (Fig. 1) is accomplished in a tray containing soaking buffer. Each component is individually applied to avoid trapping air bubbles, which would block the flow of current and thus distort or completely prevent protein transfer into the nitrocellulose sheet. It is particularly important to avoid air bubbles between the gel and the nitrocellulose sheet. The perforated plastic grid (a) in Fig. 1 is successively overlayered with (b) a sponge pad, (c) 3–5 sheets of presoaked Whatman filter paper, (d) a sheet of nylon mesh, (e) the polyacrylamide gel, (f) a sheet of cellulose acetate that is large enough to completely cover the low percentage region of the gel, (g) the nitrocellulose sheet, which has been wetted first in water, then in soaking buffer just prior to use, (h) 3–5 additional pieces of presoaked Whatman filter paper, (i) a second sponge pad, and (j) a second plastic grid used to lock the sandwich together. The nylon mesh prevents the very sticky, low-concentration gel

---

[10] G. Fairbanks, T. L. Steck, and D. F. H. Wallach, *Biochemistry* **10**, 2606 (1971).
[11] K. Wang, manuscript in preparation.
[12] U. K. Laemmli, *Nature (London)* **227**, 680 (1970).

region from adhering to the Whatman paper. The cellulose acetate membrane, being nonabsorptive and porous, prevents the gel from sticking to the nitrocellulose sheet, while allowing eluted proteins to pass through freely with little loss. For most transfers, three layers of nitrocellulose paper are used, and eluted protein bands are replicated with varying efficiency on all sheets.

The sandwich is positioned in the transfer apparatus with the low-concentration end of the gel pointing downward and the nitrocellulose sheets facing the anode. Only one sandwich is used per transfer and the movable electrodes are placed as closely as possible (~3 cm) to provide higher field strength for rapid elution. Electrophoresis is performed for 2 hr at 900 mA in transfer buffer A. The temperature is kept low and constant (7–10°) by circulating refrigerated polyethylene glycol coolant through the cooling coil attachment.

Following transfer, the polyacrylamide gel is stained in Coomassie Blue (0.05% in 25% 2-propanol, 10% acetic acid) for 1–2 hr and destained in 25% 2-propanol, 10% acetic acid to determine the efficiency of transfer. The third nitrocellulose blot is routinely stained with India Ink[13] (0.1% (v/v) Pelikan India Ink, 0.9% (w/v) NaCl, 0.4% (w/v) Tween 20, 10 mM Tris-Cl, pH 7.4) for at least 1 hr at room temperature and destained with several changes of distilled water to assess fidelity of transfer. Regions that show distortion or circular clear zones (caused by trapped air bubbles) are marked; these marked zones are useful in interpreting immunostaining of other blots from the same transfer.

The remaining blots are rinsed with appropriate buffers to remove residual SDS and used immediately for immunostaining. Alternatively, the blots can be rinsed with several changes of distilled water to remove SDS, air-dried between layers of Whatman No. 3MM filter paper, and stored in zip-lock plastic bags at room temperature or in a refrigerator for subsequent immunostaining.

Procedures for the Transfer of High-Molecular-Weight Proteins (M_r to 500,000)

SDS–PAGE. High-molecular-weight proteins are resolved in polyacrylamide gels with an acrylamide gradient of 3–10% in the buffer system of Laemmli.[12]

Electrophoretic Transfer. The transfer "sandwich" (Fig. 1) is assembled in a tray containing transfer buffer B as described above, except that the nylon mesh (d), the cellulose acetate (f), and the upper sponge pad (b)

---

[13] K. Hancock and V. C. W. Tsang, *Anal. Biochem.* **133**, 157 (1983).

are not needed. The filter paper is sufficient to hold the gel in place. We have found that the 3–5 sheets of Whatman 3MM paper can be replaced by one (h) or two (c) sheets of the thicker Whatman 17Chr paper, and that assembly in reverse order from that described in the section above allows air bubbles between the gel and the nitrocellulose to be removed by rolling a Pasteur pipette over the gel before addition of the last sheets of Whatman paper. Electrophoresis is performed in transfer buffer B for 2 hr at 750 mA with cooling, with the electrodes 10 cm apart.

Applications of the Methods

Very-High-Molecular-Weight-Proteins

Myofibrillar proteins from rabbit psoas muscle or from the fibrillar flight muscle of *Lethocerus* are separated by SDS–PAGE and transferred to nitrocellulose as described. As shown in Figs. 2 and 3, the highest-molecular-weight subunits ($M_r >200,000$) are generally the most difficult ones to elute quantitatively from the gel. Indeed, we have rarely observed the complete elution of titin[14] (M_r 1.2–1.4 × 10^6 subunit) from acrylamide gels. In contrast, the lowest-molecular-weight subunits ($M_r < 40,000$) are easily eluted but rarely retained to a significant extent by the several layers of nitrocellulose under the same transfer conditions.

The inclusion of SDS in the transfer buffer is essential for the increased efficiency of transfer of very large proteins. Without SDS, less than 10% of titin can be eluted from the acrylamide gel. Although methanol is not necessary for elution, it is critical for the absorption of proteins to nitrocellulose in the presence of SDS. The high field strength and short transfer time are important to minimize distortion caused by the significant shrinkage of the low-concentration gel *during* prolonged transfer (despite the step of presoaking the gel *prior* to transfer).

The sticky and soft low-concentration acrylamide gel causes mechanical problems during handling of the sandwich. With too much pressure, the gel is easily smashed, and with too little pressure, the soft gel deforms by its own weight; either creates distorted patterns of transfer. The use of nylon mesh and cellulose acetate sheets alleviates these problems.

High-Molecular-Weight Proteins

Receptors for EGF and covalently cross-linked[15] complexes containing EGF receptors are separated by SDS–PAGE, transferred to nitrocel-

[14] K. Wang, this series, Vol. 85, p. 264.
[15] J. V. Staros and P. S. R. Anjaneyulu, this volume [33].

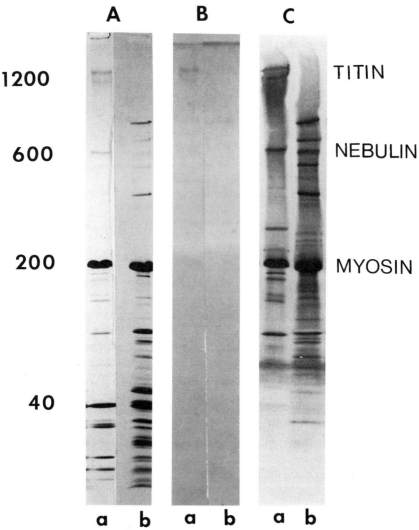

FIG. 2. Electrophoretic transfer of myofibrillar proteins of rabbit skeletal muscle (a) and water bug flight muscle (b) separated in 3–22% gradient Laemmli gels (15 cm × 17 cm × 2 mm). The positions of migration of protein standards ($M_r \times 10^{-3}$) are indicated. (A) Coomassie Blue-stained gel prior to transfer; (B) Coomassie Blue-stained gel after transfer; (C) India Ink-stained nitrocellulose blot. Note the slight distortion near the top due to the deformation of the very soft gel.

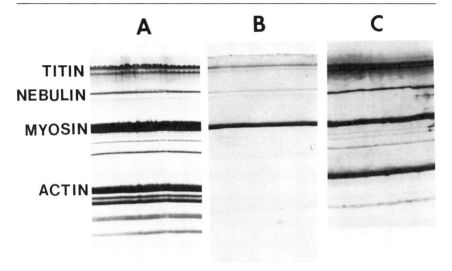

FIG. 3. Electrophoretic transfer of myofibrillar proteins of rabbit skeletal muscles separated in 2–12% gradient Fairbanks gels (6 cm × 8 cm × 2 mm). (A) Coomassie Blue-stained gel prior to transfer; (B) Coomassie Blue-stained gel after transfer; (C) India Ink-stained nitrocellulose blot. Note the nonuniform transfer on the upper right region.

lulose as described above, and immunostained with anti-EGF receptor antiserum (Fig. 4). The M_r for the cross-linked EGF receptor complexes is estimated to be ~350,000, based on extrapolation from the positions of migration of molecular weight markers to 400,000. Formation of the M_r ~350,000 complexes is EGF-dependent, and yield of the covalent complexes is a function of cross-linker concentration.[16]

EGF receptors possess an intrinsic kinase activity and are subject to autophosphorylation (reviewed in Ref. 18). Cross-linked EGF receptors retain their kinase activity and thus can be labeled with [γ-^{32}P]ATP, allowing the efficiency of transfer to be determined. Very little of either the cross-linked receptor complexes or the native receptors is detected in the gel by autoradiography after transfer. Transfer of the M_r ~200,000 and 400,000 subunits of laminin has also been examined by staining; a small amount of the larger subunit and a trace of the smaller subunit are left in the gel. None of the above proteins is detected on a second sheet of 0.45

---

[16] B. O. Fanger, J. E. Stephens, and J. V. Staros, *FASEB J.* **3**, 71 (1989).
[17] C. M. Stoscheck and G. Carpenter, *Arch. Biochem. Biophys.* **227**, 457 (1983).
[18] J. V. Staros, S. Cohen, and M. W. Russo, in "Molecular Mechanisms of Transmembrane Signalling" (P. Cohen and M. D. Houslay, eds.), p. 253. Elsevier, Amsterdam, 1985.

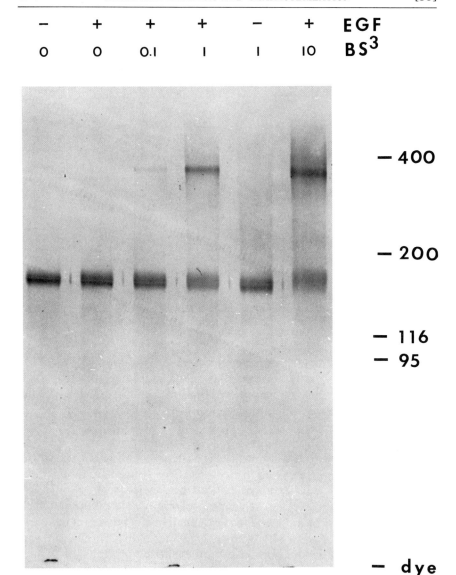

FIG. 4. Transfer of native and cross-linked EGF receptor.[16] Solubilized A431 cell extracts were incubated with or without 4.4 µg/ml EGF and cross-linked for 10 min with 0–10 mM BS^3 as indicated. Samples were separated in 3–10% gradient Laemmli gels, transferred as described, and immunostained with anti-EGF receptor antiserum.[17] The positions of migration of protein standards ($M_r \times 10^{-3}$) are indicated.

μm nitrocellulose, although proteins of lower M_r are transferred through to some degree. The use of 0.05 μm nitrocellulose reduces this loss significantly[19] and results in a lower background when immunostained with alkaline phosphatase-linked antibodies. We have noted a difference in the efficiency of transfer to the primary nitrocellulose layer of myosin and laminin (both M_r ~200,000), suggesting that factors other than size (perhaps glycosylation) affect this process.

General Considerations

The two procedures for the transfer of high- and very-high-molecular-weight proteins can best be understood as two points in a methodological continuum. While these procedures yield excellent results for the examples described here, their application to other proteins may require some modification. In applying one of these procedures, the polyacrylamide gel should be stained after the transfer step (or dried and subjected to autoradiography, if the proteins are radiolabeled), in order to ensure that proteins in the molecular weight range of interest have been transferred out of the gel. It is also worthwhile, especially in initial trials of the procedure, to use more than one layer of nitrocellulose membrane and to subject all the layers used to the chosen staining procedure. In this way, it can be assessed whether proteins of interest are being electrophoretically driven *through* the primary nitrocellulose layer. Inspection for staining on the back side of the nitrocellulose sheet can also give an indication of whether protein is being driven through the sheet. Transfer of proteins through the primary nitrocellulose layer can be caused either by overly vigorous conditions during the electrophoretic transfer or by saturation of the binding capacity of the nitrocellulose. Purposeful overloading can be useful in providing two or more images of the same gel. Adjustments to undertransfer, i.e., protein remaining in the SDS gel, can include increasing the time of transfer, lowering the ionic strength of the transfer buffer, etc. Overtransfer can be compensated by the converse strategy, or by use of nitrocellulose with a smaller pore size.[19]

Conclusions

Electrophoretic transfer from SDS–PAGE gels to nitrocellulose or other matrices is being applied to a growing catalog of proteins. The expansion of this technique to high-molecular-weight proteins has opened

---

[19] W. Lin and H. Kasamatsu, *Anal. Biochem.* **128**, 302 (1983).

this technique to many proteins of the membrane, cytoskeleton, and contractile apparatus.

Acknowledgments

Studies in the authors' laboratories were supported by research and training grants from the National Institutes of Health: R01 DK20270, R01 HL31491, R01 DK25489, R01 DK31880, T32 DK07061. The authors thank Dr. G. Carpenter for anti-EGF receptor antiserum; G. Gutierrez, M. Gutierrez, and J. Stephens for excellent technical assistance; and S. Heaver for assistance in the preparation of the manuscript.

[36] Size and Shape of Membrane Protein–Detergent Complexes: Hydrodynamic Studies

By STEVEN CLARKE and MURRAY D. SMIGEL

Mild detergents such as Triton X-100 are effective in solubilizing membrane proteins in largely native conformation.[1] The mechanism of this solubilization appears to be the substitution of bound detergent molecules for previously bound phospholipid/cholesterol molecules and the result of the process is generally a lipid-free protein–detergent complex. Using sufficient quantities of pure preparations of membrane proteins, traditional hydrodynamic or other physical methods can be used to characterize the size and shape of these complexes. However, when a membrane protein can only be obtained in limited amounts or as a mixture with other proteins, these methods are not applicable. We describe here a general method which has been successfully used over the last few years to obtain estimates of the hydrodynamic size and shape of a variety of membrane protein–detergent complexes. These methods can provide information on proteins in crude mixtures if it is possible to identify the species of interest by a biological activity (enzyme activity, ligand, or antibody binding), or if the protein can be specifically radiolabeled.

Theoretical Considerations

The basis of the methods described here is measuring three hydrodynamic properties of the detergent–protein complex: the sedimentation coefficient in water at 20°C ($s_{20,w}$), the partial specific volume ($\bar{v}$), and the

[1] A. Helenius and K. Simons, *Biochim. Biophys Acta* **415**, 29 (1975).

diffusion coefficient in water at 20° ($D_{20,w}$). The molecular weight of the complex can then be directly determined from the Svedberg equation:

$$MW = (s_{20,w}RT)/[D_{20,w}(1 - \bar{v}\rho_{20,w})]$$

where R is the gas constant (8.31×10^7 g cm^2/sec^2 mol °K), T is the absolute temperature (293°K), and $\rho_{20,w}$ is the density of water at 20°C (0.998). When the sedimentation coefficient is expressed in units of sec [one Svedberg (S) unit $= 10^{-13}$ sec] and the diffusion coefficient in units of cm^2 sec^{-1}, the molecular weight is obtained in units of g/mol.

Some information on the shape of the complex can then be obtained by using these same data to calculate a frictional ratio:

$$f/f_0 = \frac{kT}{6\pi\eta_{20,w}D_{20,w}}\left(\frac{4\pi N}{3MW\bar{v}}\right)^{1/3}$$

where k is Boltzmann's constant (1.38×10^{-16} g cm^2 sec^{-2} °K^{-1}), N is Avogadro's number (6.02×10^{23}), and $\eta_{20,w}$ is the viscosity of water at 20°C (0.01002 g cm^{-1} sec^{-1}).

The frictional ratio reflects both the hydration and the asymmetry of the complex. A value of 1.00 indicates a nonhydrated sphere. The portion of the frictional ratio that is due to the asymmetry of the particle itself can be calculated as

$$f/f_0 \text{ asymmetry} = \frac{kT}{6\pi\eta_{20,w}D_{20,w}}\left(\frac{4\pi N}{3MW(\bar{v} + \delta/\rho_{20,w})}\right)^{1/3}$$

where δ is the amount of water bound per gram of complex. An average value of about 0.4 g H$_2$O/g protein can be taken,[2] although the hydration of Triton X-100 micelles can be higher (up to 1.4 g/g).[3] This corrected frictional ratio can then provide an estimate of the asymmetry of the protein–detergent complex.[2]

These calculations provide information on the structure of the complex of protein and detergent. Often, it is desirable to know the molecular weight of the protein portion of the complex. When it is possible to measure directly the amount of bound detergent to the protein,[4] the molecular weight of the protein portion of the protein–detergent complex can be found by:

MW (protein portion) = MW (complex)/(1 + g detergent/g protein)

---

[2] C. R. Cantor and P. R. Schimmel, "Biophysical Chemistry," pp. 549–565. Freeman, San Francisco, California, 1980.
[3] S. Yedgar, Y. Barenholz, and V. G. Cooper, *Biochim. Biophys. Acta* **363**, 98 (1978).
[4] S. Clarke, *J. Biol. Chem.* **250**, 5459 (1975).

In the absence of such data, it is possible to estimate the detergent binding from the difference of the measured partial specific volume of the complex and the partial specific volume of the protein portion itself:

g detergent/g protein =
$$(\bar{v} \text{ complex} - \bar{v} \text{ protein})/(\bar{v} \text{ detergent} - \bar{v} \text{ complex})$$

This calculation is subject to rather large errors because it depends on an estimate of the protein $\bar{v}$ (a value in the range of 0.71 to 0.75 cm^3/g is generally taken). This value is not only dependent on the composition of the protein itself, but on the presence of non-amino acid components, such as bound carbohydrate or lipid. Nevertheless, for proteins of "average" composition that are not associated with other components, one can arrive at a good estimate of the number of detergent-binding sites on the surface of a membrane protein. It is then possible to calculate a "hydrophobic binding area" on the protein[4,5] which gives an estimate of the fraction of the protein surface which may be in direct contact with the lipid bilayer.

Finally, the subunit structure of the protein can be arrived at if the polypeptide composition and molecular weight are known from other techniques. For pure proteins, this can be readily established by gel electrophoresis in the presence of sodium dodecyl sulfate. For impure proteins, this procedure can also be effective if the denatured polypeptide can be identified after electrophoresis. Such identification can be accomplished by immunoblotting techniques, determination of enzyme activity after renaturation, or by following radioactivity from affinity-labeled preparations.

Experimental Determination of Diffusion Coefficient for Membrane Protein–Detergent Complexes

This value can be readily determined by comparing the elution position of the protein–detergent complex with that of marker proteins of known Stokes radius (or $D_{20,w}$) in gel filtration chromatography.[6] The choice of the column material and solvent is generally dependent on the approximate size of the protein–detergent complex—examples can be found in Refs. 4, 5, and 7. It should be pointed out that the diffusion coefficient is mathematically related to the Stokes radius by the equation $a = kT/6\pi\eta D$ where a is the Stokes radius, and η is the viscosity of the

---

[5] M. Smigel and S. Fleischer, *J. Biol. Chem.* **252**, 3689 (1977).
[6] L. M. Siegel and K. J. Monty, *Biochim. Biophys. Acta* **112**, 346 (1966).
[7] J. E. Sadler, J. I. Rearick, J. C. Paulson, and R. L. Hill, *J. Biol. Chem.* **254**, 4434 (1979).

medium. Since one measures a value for the diffusion coefficient in relation to the $D_{20,w}$ values of marker proteins, the value obtained for the protein–detergent complex represents the $D_{20,w}$ even if the gel filtration chromatography is not performed in water at 20°C.

Experimental Determination of Sedimentation Coefficient and Partial Specific Volume for Membrane Protein–Detergent Complexes

The sedimentation coefficient of the protein–detergent complex is measured by observing the migration of the complex relative to that of marker proteins when samples containing a mixture of these proteins are centrifuged through sucrose density gradients. Fractions can be collected after centrifugation and assayed for the marker proteins of known sedimentation properties and for the protein–detergent complex of interest. When the unknown protein has the same value of the partial specific volume as the marker proteins, one can directly obtain a value of the $s_{20,w}$ of the unknown from comparison with the $s_{20,w}$ values of the marker proteins.[8] However, this is generally not the case for protein–detergent complexes where the detergent often has a lower density than the protein and thus a higher partial specific volume. For these proteins, techniques have been specifically developed that result in obtaining reliable estimates of both $s_{20,w}$ and $\bar{v}$.[4,5] These techniques involve performing *two* centrifugation experiments: one in a H$_2$O-based sucrose gradient and one in a D$_2$O-based sucrose gradient. The difference in the density of these media affects the sedimentation rate of proteins according to their partial specific volume, and comparisons of both unknown and marker protein migration can give the $s_{20,w}$ and the $\bar{v}$ of the protein–detergent complex.

We describe here two analytical treatments of these centrifugation data to obtain $s_{20,w}$ and $\bar{v}$ of protein–detergent complexes. Both treatments require knowledge of the migration distance of the unknown protein and several marker proteins as well as knowledge of the density and viscosity profile of the sucrose gradient. Although the migration positions can be readily determined, and the density can be measured or estimated with confidence, determination of the viscosity of the gradient is much more difficult and represents the largest problem in these methods. The viscosity is not a linear function of sucrose concentration and is extremely dependent on the temperature. In the method described by Clarke,[4] the viscosity is not directly determined but is back-calculated as the factor which predicts the correct migration of the marker proteins of known sedimentation properties. The method of Smigel and Fleischer[5] ap-

[8] R. G. Martin and B. N. Ames, *J. Biol. Chem.* **236**, 1372 (1961).

proaches the same problem differently, using viscosity values calculated from measured sucrose densities. Because the actual temperature of the sucrose gradient during centrifugation cannot be controlled exactly,[5] this method also makes use of a correction factor that predicts the observed migration of the marker proteins. Examples of calculations using each method are given below.

Example Using the Method of Clarke[4] to Determine $s_{20,w}$ and $\bar{v}$ of the Erythrocyte Band 3 Anion Transporter–Triton X-100 Complex

Theoretical Background. A sedimentation coefficient is defined as the ratio of the instantaneous velocity of a macromolecule to the instantaneous centrifugal force it is experiencing.

$$s_{T,m} = (dr/dt)/\omega^2 r \qquad (1)$$

Here, r is the distance of the macromolecule from the axis of rotation, ω is the angular velocity of the rotor, and $s_{T,m}$ refers to the coefficient at a given temperature in a given medium. For a single protein component in the analytical ultracentrifuge, it is possible to measure r at various time intervals during the sedimentation. However, when complex mixtures are analyzed on sucrose gradients in preparative ultracentrifuges, it is only possible to obtain two values of r—the initial distance of the applied sample from the axis of rotation and the final migration position after centrifugation during a given time interval.

However, it is possible to simplify Eq. (1) above to make it useful in this latter case. In appropriately prepared sucrose density gradients, the velocity of travel in the centrifugal field is approximately constant and independent of the angular acceleration.[8] This is so because the increase in the field can be approximately balanced by the increase in the density and viscosity of the solvent. The value of $s_{T,m}$ is then proportional to $1/r$:

$$s_{T,m} = v/\omega^2 r \qquad (2)$$

where the velocity, v (dr/dt), is a constant. The value of v can be obtained by measuring the distance traveled by the macromolecule in the gradient during the time of centrifugation:

$$v = (r - r_o)/t \qquad (3)$$

where r_o is the distance of the applied band from the center of rotation, and r is the distance of the band peak at time t. The value of the sedimentation coefficient at the position r_{avg}, defined as the half-distance of travel by the relationship $r_{avg} = (r_o + r)/2$, is calculated from

$$s_{T,m_{(r_{avg})}} = [(r - r_o)/t]/\omega^2 r_{avg} \qquad (4)$$

Sedimentation coefficients can be calculated in this way for both marker proteins of known $s_{20,w}$ and $\bar{v}$ and for the protein under investigation. These sedimentation coefficients can be corrected to standard conditions (water at 20°C) by the following equation:

$$s_{20,w} = s_{T,m}(\eta_{T,m}/\eta_{20,w})[(1 - \bar{v}\rho_{20,w})/(1 - \bar{v}\rho_{T,m})] \tag{5}$$

This correction requires a knowledge of the viscosity and the density of the medium as well as the partial specific volume of the protein. The density of the medium is a linear function of the sucrose concentration and can be calculated from tables for each point in the gradient. For the purpose of this correction, the density at r_{avg} for each protein is utilized. It is more difficult to estimate the viscosity, which is not a linear function of the gradient and is highly temperature dependent. It is possible, however, to calculate the viscosity in the gradient from the calculated average densities, the partial specific volumes and corrected sedimentation coefficients of the marker proteins (from the literature), and the experimentally determined apparent sedimentation coefficients $s_{T,m}$.

$$\frac{\eta_{T,m_{avg}}}{\eta_{20,w}} = \left(\frac{s_{20,w}}{s_{T,m}}\right)\left(\frac{1 - \bar{v}\rho_{T,m_{avg}}}{1 - \bar{v}\rho_{20,w}}\right) \tag{6}$$

The average viscosity experienced by the unknown protein is determined by interpolation of a plot of η_{avg} for the marker proteins against gradient position.

Equation (5) can be solved for the unknown protein if the partial specific volume is known. This can be obtained by repeating the experiments outlined above in a medium containing D_2O. Equation (5) is a general relation between these parameters in any two media, and one can write

$$s_H = s_D\left(\frac{\eta_{D_{avg}}}{\eta_{H_{avg}}}\right)\left(\frac{1 - \bar{v}\rho_{H_{avg}}}{1 - \bar{v}\rho_{D_{avg}}}\right) \tag{7}$$

where D subscripts refer to values measured in sucrose gradients in D_2O and H subscripts to those values measured similarly in H_2O. Equation (7) can be solved for $\bar{v}$ explicitly for an unknown protein.

$$\bar{v} = \frac{s_D\eta_{D_{avg}}/s_H\eta_{H_{avg}} - 1}{\rho_{H_{avg}}(s_D\eta_{D_{avg}}/s_H\eta_{H_{avg}}) - \rho_{D_{avg}}} \tag{8}$$

It is now possible to use Eq. (5) to calculate $s_{20,w}$ for the unknown protein. This calculation can be done independently for both the gradient run in H_2O and that run in D_2O.

Experimental Procedures. An extract of human erythrocyte membranes in 2% Triton X-100 was mixed with four marker proteins of known

TABLE I
SEDIMENTATION PROPERTIES OF "MARKER" PROTEINS FOR SUCROSE GRADIENT CENTRIFUGATION

Protein	Source	$s_{20,w}$ ($\times 10^{13}$ sec)	$\bar{v}$ (cm^3/g)	Ref.
Malate dehydrogenase	Pig heart, cytoplasmic	4.32	0.734	Thorne[9]
Aldolase	Rabbit muscle	7.35	0.742	Taylor and Lowry[10]
Fumarase	Pig heart	9.09	0.738	Hill and Teipel[11]
Catalase	Bovine liver	11.30	0.730	Sumner and Gralen[12]
Cytochrome c	Horse heart	1.9[a]		Nozaki[13]
Ovalbumin	Hen egg	3.6[b]		Svedberg and Pedersen[14] Kegeles and Gutter[15]
Serum albumin	Bovine	4.5		Putnam[16]
γ-Globulins	Equine	7.0		Edsall[17]

[a] Assumed to have the same value as bovine heart.
[b] Average of the two literature values.

sedimentation coefficients and partial specific volumes. These proteins included pig heart malate dehydrogenase, rabbit muscle aldolase, pig heart fumarase, and bovine liver catalase (see Table I). Centrifugation of a 0.25-ml sample was performed at 42,000 rpm for 9.5 hr ($\omega^2 t = 6.62 \times 10^{11}$ sec^{-1}) in a Beckman SW 50L rotor at 4°C through a 4.75-ml 5–20% (wt/wt) sucrose gradient containing 0.01 M Tris–H$_2$SO$_4$, pH 7.5, 0.1 M Na$_2$SO$_4$, and 0.05% Triton X-100. Thirty-seven fractions were collected from the bottom of the tube and the positions of the four marker proteins were determined by enzymatic assays of each of the fractions.[4] The position of the band 3 anion transporter was determined by quantitative densitometry of Coomassie-stained sodium dodecyl sulfate gels of each fraction. With a knowledge of the dimensions of the centrifuge tube and the rotor, one can calculate that the top of the 5% sucrose is 5.20 cm from the

[9] C. J. R. Thorne, *Biochim. Biophys. Acta* **59**, 624 (1962).
[10] J. F. Taylor and C. Lowry, *Biochim. Biophys. Acta* **20**, 109 (1956).
[11] R. L. Hill and J. W. Teipel, in "The Enzymes" (P. D. Boyer, ed.), Vol. 5, 3rd Ed., p. 539. Academic Press, New York, 1971.
[12] J. B. Sumner and N. Gralen, *J. Biol. Chem.* **125**, 33 (1938).
[13] M. Nozaki, *J. Biochem. (Tokyo)* **47**, 592 (1960).
[14] T. Svedberg and K. O. Pedersen, in "The Ultracentrifuge." Oxford Univ. Press, London, 1940.
[15] G. Kegeles and F. J. Gutter, *J. Am. Chem. Soc.* **73**, 3770 (1951).
[16] F. W. Putnam, in "The Proteins" (H. Neurath, ed.), Vol. 3, 2nd Ed., p. 174. Academic Press, New York, 1965.
[17] J. T. Edsall, in "The Proteins" (H. Neurath and K. Bailey, eds.), Vol. 1, 1st Ed., p. 549. Academic Press, New York, 1953.

axis of rotation and that the midpoint of the applied sample was 5.10 cm from this axis. The inner diameter of the centrifuge tube was 1.25 cm, and the total volume of sample and gradient was 5.00 ml. The change in radial position per fraction can then be calculated as 0.110 cm. The origin was determined to be at fraction 35.8 by extrapolating a plot of fraction number versus $s_{20,w}$ of the marker proteins to zero sedimentation coefficient. The density at r_{avg} was calculated from densities of the 5 and 20% sucrose solutions of 1.0327 and 1.0921, respectively. Assuming that the 5% sucrose started at fraction 34.6, the density at each protein's average position was determined assuming a linear gradient of density with fraction number. The viscosity at r_{avg} was calculated using Eq. (6) for each of the standard proteins. This value for the band 3 anion transporter was obtained by interpolation of a plot of the fraction number versus viscosity for these marker proteins. Table II summarizes data obtained from this experiment, as well as data from a similar experiment in which the sucrose gradient was prepared in D_2O and centrifuged at 48,000 rpm for 18 hr. In this latter gradient, 40.3 fractions were collected (0.101 cm/fraction), and the $\omega^2 t$ term was 16.37×10^{11} sec^{-1}. The density of the 5% sucrose in this case was 1.122 and 1.168 for the 20% sucrose.

From these data, it is possible to calculate that the partial specific volume of the anion transporter is 0.824 cm^3/g [Eq. (8)], and that the $s_{20,w}$ is 7.68×10^{-13} sec [or 7.68 S; Eq. (5)]. The latter value is an average of the value obtained using the H_2O and the D_2O data.

Alternative Calculations. Explicit corrections can be made for differences in the binding of detergent to membrane proteins in H_2O and D_2O by the method described by Clarke.[4] Since these binding data are usually not available for impure membrane proteins, the approach described above is based on the assumption that equal amounts of detergent are bound in H_2O and D_2O solution. For a series of membrane protein–detergent complexes (including rhodopsin, the Na^+,K^+-ATPase, as well as the erythrocyte anion transporter), this assumption results in less than a 10% error of the calculated molecular weights of the complex.[4]

Sadler et al.[7] have described a modification of the method described above where the point of reference is the top of the gradient instead of the r_{avg} position.

Example Using the Method of Smigel and Fleischer[5]
to Determine $s_{20,w}$ and $\bar{v}$ of the Rat Liver Prostaglandin
E_1 Binding Protein–Triton X-100 Complex

Theoretical Background. The integrated form of the equation which defines the sedimentation coefficient [Eq. (1)], corrected for the density

TABLE II
CALCULATION OF $s_{20,w}$ AND $\bar{v}$ FOR BAND 3 BY THE METHOD OF CLARKE[a]

Material	Fraction of peak	r (cm)	r_{avg} (cm)	$s = \dfrac{(r-r_0)/t}{\omega^2 r_{avg}}$ ($\times 10^{13}$ sec) [Eq. (4)]	ρ_{avg} (g/cm^3)	η_{avg} ($\times 10^2$ g cm^{-1} sec^{-1}) (cp) [Eq. (6)]	$\bar{v}$ (cm^3/g) [Eq. (8)]	$s_{20,w}$ ($\times 10^{13}$ sec) (S) [Eq. (5)]
A. Sucrose velocity gradient in H$_2$O				s_H				
Origin	35.8	5.10						
Malate dehydrogenase	28.5	5.90	5.50	2.21	1.0369	1.751		
Aldolase	23.4	6.46	5.78	3.57	1.0413	1.808		
Fumarase	20.6	6.77	5.93	4.26	1.0437	1.864		
Catalase	16.6	7.21	6.15	5.19	1.0471	1.893		
Band 3	24.0	6.40	5.75	3.42	1.0408	(1.802)[b]	0.824	7.68
B. Sucrose velocity gradient in D$_2$O				s_D				
Origin	39.0	5.10						
Malate dehydrogenase	24.5	6.57	5.83	1.54	1.1258	1.831		
Aldolase	14.8	7.55	6.32	2.36	1.1322	1.920		
Fumarase	11.5	7.88	6.49	2.62	1.1344	2.151		
Catalase	6.2	8.42	6.76	3.00	1.1380	2.356		
Band 3	23.0	6.72	5.91	1.67	1.1268	(1.849)[a]		7.68

[a] See Ref. 4.
[b] Value determined from interpolation of a plot of r_{avg} versus η_{avg} for the marker proteins.

and viscosity of the solvent [Eq. (5)], is given by

$$s_{20,w} = \frac{1 - \bar{v}\rho_{20,w}}{\omega^2 t \eta_{20,w}} \int_{r(0)}^{r(t)} \frac{\eta(r)\,dr}{r[1 - \bar{v}\rho(r)]} \quad (9)$$

where $r(0)$ indicates the initial distance of the particle from the axis of rotation, $r(t)$ indicates this distance at time t, $\eta(r)$ indicates the viscosity of the solvent at a distance r from the axis of rotation, and $\rho(r)$ indicates the density of the solvent at this distance. Although it is a simple matter to integrate this equation under conditions in which the solvent density and viscosity are not radial functions (such as in the determination of $s_{20,w}$ in the analytical ultracentrifuge), this is not the case when a sucrose gradient is used to stabilize sample bands in a preparative ultracentrifuge.

However, if the values of the density and viscosity along the gradient can be determined accurately, it is possible to utilize a single density gradient experiment to define a relationship of $s_{20,w}$ and $\bar{v}$ for an unknown particle [Eq. (9)]. Another sedimentation experiment through a different gradient, for example where the density is shifted by D_2O, will define a second curve of $s_{20,w}$ versus $\bar{v}$. Where these curves intersect defines the actual value of $s_{20,w}$ and $\bar{v}$ of the particle of interest.

The density and viscosity of the gradient are calculated from measurements of the refractive index of gradient fractions (see below). The integration of Eq. (9) is then, utilizing a computer, performed numerically using a Simpson's rule routine with adaptive step size for various values of $\bar{v}$, thus determining the relationship between $s_{20,w}$ and $\bar{v}$. This is done for sedimentation experiments performed in H_2O- and D_2O-containing sucrose gradients and, from the intersection, the $s_{20,w}$ and $\bar{v}$ of the protein–detergent complex are found.

The foremost practical problem in this method is maintaining constant temperature in the gradient during the experiment. In the Beckman L2-75B centrifuge, the refrigeration is switched on and off ("bang-bang regulation") in response to changes in temperature felt at a radiometer monitoring the relatively massive rotor body. The buckets that hold the gradient tubes are much less massive and much closer to the cooled wall of the evacuated chamber; consequently, they will experience considerable thermal cycling as the control system regulates to achieve some average temperature. The viscosity of sucrose solutions is a nonlinear function of temperature. The viscosity will increase more on the downswing of the cycle than it decreases on the upswing. Thus, the average viscosity will be higher than the viscosity found at the average temperature. In this method, an empirical correction for this is made by assuming an actual rotor temperature that makes the protein standard running nearest the unknown peak assume its correct s value. This correction amounted to 1 to 6°C.

The calculation of the density and viscosity of the sucrose solutions from refractive index data was done as described by Smigel.[18] The densities of mixtures were derived from simple weighted sums of the partial specific volumes of sucrose and water. The temperature dependence of the partial specific volume of sucrose was taken from Barber,[19] as were the data for the viscosity of H_2O sucrose mixtures as a function of temperature and composition. The densities of H_2O and D_2O as a function of temperature were taken from the work of Kravchenko.[20] The ratio of the viscosities of pure H_2O and D_2O was fitted to a second-order polynomial in the temperature from data given by Kirschenbaum.[21] Equilibrium of the proteins with D_2O was assumed to increase the density of the proteins by a factor of 1.0155, as given by Edelstein and Schachman.[22] A FORTRAN IV program for performing the data reduction is available upon request from M. Smigel.

Experimental Procedures. Rat liver plasma membrane (190 mg of protein) was suspended in 40 ml of 0.15 M NaCl containing 10% (w/w) sucrose and 10 mM PIPES, pH 6.5 (SSP), and was incubated with 1.3 nM [5,6-^{3}H]prostaglandin E_1 (about 71,000 cpm/ml) for 80 min at 37°C. The suspension was cooled to 0°C and centrifuged at 20,000 rpm (31,000 g at r_{avg}) for 30 min in a Beckman JA-20 rotor. The pellet was resuspended in 45 ml of 0.25 M sucrose and was recentrifuged at 20,000 rpm for 30 min as before. The pellet was resuspended in a total volume of 20 ml with SSP and 3.4 ml of 10% (w/v) Triton X-100 (pH 7.0) was added. After 10 min at 0°C, the mixture was centrifuged at 40,000 rpm (123,000 g at r_{avg}) for 90 min in a Beckman 42.1 rotor at 2°C. The supernatant was decanted and used in the centrifugation experiments.

Sets of three linear sucrose density gradients (5 ml) were formed in $\frac{1}{2} \times 2$ inch cellulose nitrate tubes with the use of a six-syringe Beckman Density Gradient Former. The low- and high-density solutions were, respectively, 10% (w/v) and 34% (w/v) sucrose, each containing 0.15 M NaCl, 10 mM HEPES (pH 7.5), and 2 mg/ml of Triton X-100. In sucrose/D_2O gradients, D_2O replaced greater than 94% of the H_2O.

Standard proteins used for calibration included horse heart cytochrome c, hen egg ovalbumin, bovine serum albumin, and equine γ-globu-

---

[18] M. Smigel, Ph.D. thesis. Vanderbilt University, Nashville, Tennessee, 1976.
[19] E. J. Barber, *Natl. Cancer Inst. Monogr.* **21**, 219 (1966).
[20] V. S. Kravchenko, *At. Energ.* **20**, 168 (1966).
[21] I. Kirschenbaum, in "Physical Properties of Heavy Water," p. 33. McGraw-Hill, New York, 1951.
[22] S. J. Edelstein and H. K. Schachman, this series, Vol. 27, p. 83.

lins (Table I). Standard protein, 0.2 ml of a 10 mg/ml solution in H_2O, or 0.2 ml of a Triton X-100-solubilized plasma membrane preparation (see above) was layered on top of each gradient. Each set of six tubes was centrifuged at 45,000 rpm (189,000 g at r_{avg}) for either 20 hr (H_2O gradients) or 48 hr (D_2O gradients) in a Beckman SW 50.1 rotor in a Beckman L2-75B centrifuge. The temperature was set at 2–8° for various runs. After deceleration with the brake off, the tubes were pierced, 55% (w/w) sucrose was pumped into the bottom, and fractions of eight drops each were collected from the top. Alternate fractions were taken for measurement of refractive index. The remaining tubes of the standards were analyzed for protein, while those containing solubilized plasma membranes were taken for scintillation counting.

The refractive index data were converted to a series of 25 points of $\rho(r)$ and $\eta(r)$, which were equally spaced along the tube. Linear interpolation was used to arrive at values between these points. These data were then used with W. Quire's numerical integration routine (SIMPER)[23] to integrate Eq. (9) for a series of values of $\bar{v}$, for each experimental value of the peak position at time t, $r(t)$, to generate the required curve of $s_{20,w}$ versus $\bar{v}$. From plots of the H_2O and D_2O data, the point of intersection defined the actual value of the $s_{20,w}$ (5.65 S) and the $\bar{v}$ (0.805) of the prostaglandin E_1 binding protein (Fig. 1).

Alternative Calculations. Newman *et al.*[24] have described a similar method for analyzing sedimentation data based on the integration of the sedimentation equation. In this method, longer centrifuge tubes (and shallower gradients) were used so that the viscosity could be estimated as a linear function of the radial distance.

Comparison of Methods of Analyzing Data from Sucrose Gradient Centrifugation Experiments

The methods of Clarke[4] and Smigel and Fleischer[5] both utilize the sedimentation of proteins of known properties to compensate for the lack of knowledge of the precise value of the solution viscosity during the sedimentation run. In the method of Clarke,[4] the average viscosity experienced by the unknown protein is determined from the variation of calculated average viscosity (determined from the $s_{20,w}$ and $\bar{v}$ values of the proteins of known hydrodynamic properties) with the distance sedimented. This method is based on the assumption that proteins migrate with constant velocity in the gradient. In the method of Smigel and

---

[23] W. Squire, "Integration for Engineers and Scientists," pp. 282–283. Elsevier, New York, 1970.
[24] S. A. Newman, G. Rossi, and H. Metzger, *Proc. Natl. Acad. Sci. U.S.A.* **74**, 869 (1977).

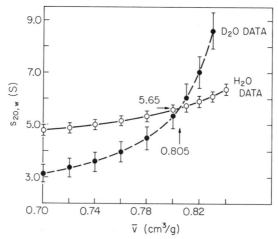

FIG. 1. Determination of $s_{20,w}$ and $\bar{v}$ for the prostaglandin E_1 binding protein of rat liver by the method of Smigel and Fleischer.[5] Data were calculated as described in the text for sucrose gradients containing H_2O and D_2O. Error bars are ±1 SD for calculation of H_2O data ($n = 4$) and D_2O data ($n = 5$). Horizontal and vertical arrows indicate, respectively, the values of $s_{20,w}$ which are consistent with the sedimentation in both sets of gradients.

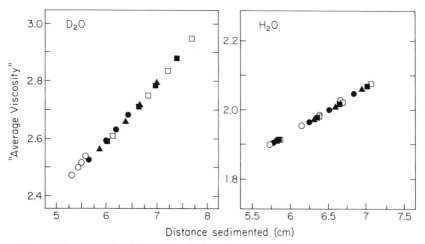

FIG. 2. Computer simulations of calculated "average viscosity" as a function of the distance sedimented and the partial specific volume of four marker proteins. The conditions were set up to match the experiment described in Table II, except that the partial specific volumes of the marker proteins (malate dehydrogenase, aldolase, fumarase, and catalase) were varied from 0.73 cm³/g (□), 0.77 cm³/g (■), 0.81 cm³/g (▲), 0.85 cm³/g (●), to 0.89 cm³/g (○). The observation that the calculated viscosity is independent of the partial specific volume of the marker proteins confirms the validity of this method.

Fleischer,[5] an explicit description of the sedimentation process is employed which, in theory at least, takes exact account of the differences in retardation due to both buoyancy and viscosity felt by proteins of arbitrary partial specific volumes.

In order to test whether the calculation of "average viscosity" in the method of Clarke[4] gives rise to significant errors, especially in the general case where the "marker" proteins are usually soluble proteins of lower partial specific volume than the detergent–membrane complex of interest, a series of numerical simulations of the sedimentation process were performed. In these simulations, which closely mimic the conditions of sedimentation described for these proteins (cf. Table II), calibration plots of average viscosity versus distance sedimented were constructed for proteins of various partial specific volumes and $s_{20,w}$ values. As can be seen in Fig. 2, the data from proteins with $\bar{v}$ values ranging from 0.73 to 0.89 cm^3/g could be described by a single curve. The only exception that arises is in the case of proteins very near to neutral buoyancy which fail to move an appreciable distance into the D$_2$O gradient. Since there is no experimental need to work under such conditions, the approximation of this method appears to be well validated.

Acknowledgments

SC wishes to thank Prof. Guido Guidotti (Harvard University) for suggesting the mathematical treatment of the sedimentation data used in the first method described here.

Author Index

Numbers in parentheses are footnote reference numbers and indicate that an author's work is referred to although the name is not cited in the text.

A

Abarbanel, R. M., 451, 454(42)
Abbing, A., 66
Abdella, P. M., 589
Abdulaev, N. G., 440
Abu-Salah, K. M., 686
Abumrad, N. A., 625, 627(53)
Acha-Orbea, H., 646
Acquotti, D., 556
Acuto, O., 349
Adorante, J. S., 121
Affolter, H., 151, 151(25), 153
Agard, D. A., 400
Aggeler, R., 629, 639, 643, 645(12), 649(12), 652(72), 654(12), 655(12), 656
Ahnen, D. J., 18, 19(6), 24(6), 30(6)
Aikawa, T., 593
Ailhaud, G., 633
Airkawa, T., 590
Aiuchi, T., 112
Aizenbud, B. M., 503
Akerlund, H. E., 13, 24
Akerman, K. E. O., 78, 111, 115, 121(54)
Akiyama, J., 107
Al-Awqati, Q., 12, 50, 51(7), 54, 55, 56(7, 17), 57, 85, 87, 89, 92, 94(1), 156, 306, 307
Alamo, L., 141
Albertsson, P. A., 13
Albertsson, P.-A., 24, 25
Albrecht, A. C., 582
Alder, G. M., 99
Alderton, J., 233
Alexander, J., 241
Alkin, A., 353
Allen, D. G., 164, 165, 168(1), 169(1), 178, 185, 198, 199, 200(77), 202, 203
Allen, J., 439
Allen, T. J. A., 255
Almers, W., 244
Alt, J., 441, 446, 449(31)

Altshuld, R. A., 83
Ambrosini, A., 231
Ameloot, M., 487
Ames, B. M., 699
Amiguet, P., 659
Aminabhaui, T. M., 533
Amman, E., 29
Ammann, D., 136, 144, 148, 149(3), 150
Ammann, E., 353
Amselem, S., 364
Amuro, N., 523
Anataki, P., 429
Anderegg, R. J., 666, 667
Anderson, J. M., 167
Anderson, R. G. W., 93
Anderson, S. L., 444
Anderson, S., 444
Andersson, B., 13, 24
Andersson, T., 521
Andreoli, T. E., 162
Andrich, M., 499
Angelides, K. J., 418, 421(18)
Anglister, L., 111
Anjaneyulu, P. S. R., 619, 629(45), 643, 691
Anner, B. M., 109
Ansager, M., 384, 386(77)
Apell, H. S., 78
Apell, H.-J., 109
Appling, D. R., 611, 618(15)
Aragon, S. R., 370, 380, 386
Argos, P., 440, 441(11)
Arias, I. M., 9, 12(19), 13(19), 16
Armitage, J. P., 110
Arndt-Jovin, D., 476, 481, 482(48), 487(34), 493
Aronson, P. S., 360, 361, 362
Arslan, P., 231, 232, 233(6), 244(13), 252(13), 253(13), 255(6)
Asada, K., 512
Asai, H., 478
Ashwell, G., 419
Auffret, A. D., 448, 449(41)

Aune, K. C., 385
Ausiello, D. A., 418, 428
Austin, R. H., 474, 476, 477(26, 30), 481, 487, 493, 494(48)
Avison, M. S., 84
Avron, M., 81, 82(33)
Axelrod, D., 472, 496, 499, 502, 503, 504(96), 505(96), 508(13), 512
Azzi, A., 76, 646, 686
Azzone, G. F., 63, 111, 121(55)

B

Bablouzian, B., 514
Bachman, L., 207
Bachofen, R., 651, 669(128), 670(128), 686
Baenziger, J. U., 597, 598, 599
Bähr, W., 496, 499
Baird, M. S., 642
Bakeeva, L. E., 121
Baker, P. F., 198, 199, 255
Balaban, A. T., 82
Bamberg, E., 332
Bangham, A. D., 364
Banjar, Z. M., 594
Bankier, A. T., 444
Barac-Nieto, M., 349
Barasch, J., 89, 94
Barber, E. J., 706
Barber, M. J., 429
Bardawill, C. J., 399
Bardin, A. M., 577, 578(8)
Barenholz, Y., 364, 386, 464, 470, 697
Bargeron, C. B., 381
Barhanin, J., 437
Barisas, B. G., 489, 503
Barnard, E. A., 421, 423, 487, 493
Barnett, A. J., 29
Baron, C., 519
Baronowsky, P., 634
Barrantes, F. J., 487, 493, 507
Barrell, B. C., 444
Barrett, P. Q., 361
Barry, B. W., 538
Bartholdi, M., 474, 476(26), 477(26), 481, 482(48), 487, 493
Bartlett, G. R., 527
Barzilai, A., 35, 36(12)

Bashford, C. L., 63, 72(2, 6), 76(6), 78(2), 98, 99, 102, 110, 112(3)
Bass, D. A., 109
Bastos, R., 600
Batley, M., 565
Batzri, S., 364
Baudras, A., 506
Bauer, P.-J., 579, 580, 581, 582(18, 20)
Baughman, G. D., 213
Baumann, K., 12, 17
Baümert, H. G., 584, 585, 587, 601, 605, 608
Baumgarten, C. M., 142
Bayley, H., 577, 611, 628, 631(4, 5), 634, 637(43), 638, 643, 644, 645(114), 653, 656(5), 658, 660(138), 662(136)
Bayley, P. M., 472
Baylor, S. M., 102, 110
Beauregard, G., 419, 423, 428, 429, 437, 438
Beaven, M. A., 246
Becher, B., 577, 580, 581
Becher, P., 527, 531, 534(13)
Beck, K., 506
Beeler, T. J., 115
Behr, J. P., 499
Belanger, M., 383
Belfrage, P., 393, 397(1), 398(1)
Belisle, M., 438
Beliveau, R., 347, 357(13)
Belkin, M., 423
Beltrame, M., 232, 244(13), 252(13), 253(13)
Ben-Zeev, O., 438
Benedek, G. B., 531
Benet, L. Z., 10
Bengtsson-Olivecrona, G., 428
Benisek, W. F., 636
Bennett, M. V. L., 254
Bennette, V., 17
Bentley, P. J., 304
Bentz, J., 126
Benz, R., 332
Bercovici, T., 653, 658, 662, 686
Berenski, C. J., 420, 426, 427, 429(53)
Berg, H. C., 616, 618(32)
Bergara, J., 262
Berger, S. J., 349
Bergström, J., 448
Berkhout, T. A., 686
Berlin, R. D., 499
Berliner, L., 344
Berlot, C. H., 606

Bernal, S. D., 112
Berne, B., 366, 367(25)
Bernhardt, J., 325
Bernoit, V., 356
Berrez, J. M., 686
Berry, E. A., 80, 81(31)
Berry, G., 561
Bers, D. M., 43, 196, 256
Bers, G., 687
Bersch, B., 78
Berthon, B., 237, 241
Beth, A. H., 619, 621, 622(44), 623(44), 625, 627
Bevins, T. L., 50
Bezanilla, F., 111
Bibb, M. J., 444
Biber, J., 346, 348(3), 349, 353, 357(29)
Bieber, L. L., 393
Biemann, K., 666, 667
Biempica, L., 9, 12(19), 13(19)
Bienvenue, A., 629
Bifano, E. M., 119
Bildstein, C. L., 352
Binder, H. J., 363
Binet, A., 237, 241(27)
Birdsall, N. J. M., 495
Bishop, L. J., 457
Bisson, R., 595, 628, 630, 649, 650, 654, 658, 686
Black, M. T., 448
Blackmore, P. F., 241
Blagrove, R. J., 378
Blair, O. C., 109
Blasie, J. F., 448
Blasie, J. K., 289, 400, 403
Blatt, E., 473, 476, 477(23, 32), 480(32), 485(32), 489(23, 32)
Blaustein, M. P., 311
Blinks, J. R., 164, 165, 166, 168, 169(1), 170(5), 178, 183, 184(23), 197, 198, 199, 200(77), 202(4, 33, 77), 245
Blobel, G., 619
Bloj, B., 629
Bloom, J. A., 510
Bloomfield, V., 487
Blume, A., 464, 584
Blumenthal, R., 127
Bock, H.-G., 632
Bockus, B. J., 111
Bode, F., 12, 17

Bolger, G., 437
Bond, M., 207, 217(4), 289
Bone, J. V., 312
Bonner, W. D., Jr., 111
Bonting, S. L., 399, 437
Boos, K. S., 135
Booth, A. G., 6, 7(10), 11(10), 12(10), 349, 357(25), 686
Bootman, M., 107, 108(25)
Boquet, P., 687
Borle, A. B., 180, 181, 193(47)
Boron, W. F., 63, 79(9)
Borsotto, M., 437
Bostock, T., 529, 532(6)
Bothwell, M. A., 606
Bott, S., 379, 386(68), 387(59)
Boucek, R. J., Jr., 428
Boulnois, G. J., 454, 455(63)
Boumendil-Podevin, E. F., 352
Bourne, P. K., 177
Bowman, J., 427
Boxer, D. H., 487, 493
Boxer, D., 476, 493
Boyce, C. O. L., 671
Boyer, D. E., 349
Boyle, M. B., 110
Bradford, M. M., 525
Brady, R. O., 606
Brahm, J., 121
Bramhall, J. S., 82, 633, 686
Brand, M. D., 106, 110(17), 121(17), 437, 438
Brandl, C. J., 453
Brandolin, G., 135
Brandt, P. W., 311
Brandt, W. R., 424
Branham, S., 352
Brauer, B.-E., 642
Braun, V., 461
Bräuner, T., 110
Braw, R., 162
Bremer, E. G., 550, 559, 567(34)
Breslow, R., 637, 639
Breton, J., 577, 578(8)
Brewer, C. A., 604
Brewer, J. M., 523
Brewer, L. A., 593
Bridge, M., 126
Brierley, G. P., 83
Brigati, D. J., 652
Bright, G. R., 186

Brimacombe, P., 603
Brinley, F. J., Jr., 199
Brith-Lidner, M., 581
Brock, W., 332
Brodbeck, U., 109, 630
Brodsky, W. A., 6
Bron, C., 658
Brookes, P., 593
Brown, J. C., 384
Brown, J. E., 232
Brown, K. T., 138, 140
Brown, M. L., 388
Brown, M. S., 415
Brown, T. R., 84
Browne, D. T., 619
Brownie, T. F., 671
Bruchman, T. J., 578
Brunden, K. R., 453, 454, 458(54)
Brunette, M. G., 347, 357(13)
Brünger, A., 496, 499
Brunisholz, R. A., 668
Brunner, J., 364, 628, 629, 630, 638, 643, 644, 645(12, 110), 646(66, 110), 647(110), 648, 649(12, 119), 651, 654(12, 66), 655(12), 656, 657, 658, 659, 660(119), 662, 664(163), 668(151), 669(128, 151), 670(128), 687
Brvenik, L. J., 499
Buchardt, O., 638
Buchsbaum, R. N., 83
Buchsbaum, R. S., 85
Buerkler, J., 662
Burckhardt, G., 8, 12(18), 346
Burger, M. M., 687
Burnham, C., 10, 158, 162
Burnop, V. C. E., 593
Büschl, R., 501
Buse, B., 446
Buse, G., 650
Bustamente, C., 576
Butor, C. M. B., 499
Buyon, J. P., 106

C

Cabantchik, Z. I., 6, 123, 124(6, 7), 125(6, 7), 126, 127(6, 8), 130(6, 7, 11), 131(6), 133(8), 134(11, 12), 615
Cafiso, D. S., 66, 71(17), 72(17), 333, 334, 338, 339(14), 341, 342(8, 18), 343(8), 344

Cafiso, D., 332
Cain, J. E., 400
Callahan, P. J., 463
Callingham, B. A., 420, 438
Calvo-Fernandez, P., 653
Cambier, J. C., 109
Camerini-Otero, R. D., 380
Campbell, A. K., 180, 181
Campell, P., 636
Campos de Carbalho, A., 254
Cannell, M. B., 185
Cannon, C., 50, 51(7), 54, 56(7, 17), 87, 92
Cantin, M., 621
Cantley, L., 106, 109(16)
Cantor, C. R., 697
Capaldi, R. A., 632, 686
Caplan, S. R., 317
Capone, J., 633
Caputo, C., 141, 262
Carafoli, E., 154, 155, 352, 632, 662, 686
Carey, F. A., 640
Carey, M. C., 531
Carlsen, W. F., 499
Carlson, F. D., 366, 367(31), 386
Carlson, K. E., 655
Carlsson, J., 590
Carlsson, M., 521
Caroni, P., 155
Carpenter, G., 693
Carter, H. E., 558
Casadio, R., 81
Casey, M. L., 415
Caspar, D. L. D., 400
Cassano, G., 362
Cassim, J. Y., 577, 580, 582, 583
Castillo, C. L., 84
Castle, J. D., 335
Castranova, V., 114
Caswell, A. H., 424
Cavieres, J. D., 437
Cawley, D. B., 686
Cerda, J. J., 349
Cerletti, N., 653
Cha, C., 437
Chabre, M., 400
Chaiet, L., 651
Chaimovich, H., 634, 635(40)
Chakrabarti, P., 628, 632
Chamberlain, B. K., 420, 426(22), 427(22)
Chan, M., 347, 357(13)

Chan, P. T., 454, 455(61)
Chan, S. I., 495
Chan, S. L., 476, 477(30), 493
Chan, S. Y., 35, 36(7, 9), 37(7), 38(7)
Chandler, W. K., 110
Chang, C. H., 510, 512
Chang, C.-F., 227
Chang, E. L., 386
Chang, K. J., 17
Chang, L.-R., 421
Change, C. H., 439
Changeux, J.-P. 630
Chany, S. H., 441
Chapman, C. E., 114
Chapman, D., 472, 487, 493, 499, 578
Charbonneau, H., 170
Charcosset, J.-Y., 109
Charest, R., 241
Chartier, M., 454, 455(64)
Chase, H., 156
Chase, H., Jr., 306, 307, 308
Chase, M., 12
Chauncey, T. R., 654
Chazotte, B., 510
Chen, H. Y., 563
Chen, L. B., 111, 112, 649
Chen, S. S., 561, 563
Chen, S.-H., 366
Cheng, K., 121
Cheng, L., 348
Chernansek, S. D., 606
Cherry, R. J., 469, 471, 472, 474, 476(25), 478(4), 487(3), 578, 581, 583
Cherry, R., 482
Cheung, H. C., 487
Chi, S. V., 127
Chiesi, M., 43
Chin, C. C. Q., 523
Chin, G., 34, 35, 39(5)
Chothia, C., 440, 441(9), 616
Chou, K. H., 558
Chou, K.-H., 545
Chovaniec, M. E., 114
Chowdhry, V., 641, 642
Chrambach, A., 519
Christianse, K., 584
Chu, B., 366, 367(24, 28), 376(24), 378, 381
Chua, C. C., 671
Church, R. F. R., 643, 645
Churchill, P., 424, 426, 429(53)

Chused, T. M., 106, 109
Civan, E. D., 157, 159(9)
Civan, M. M., 157, 159(9)
Claret, M., 237, 241(27)
Clark, N. A., 403, 405(9), 505, 506(118)
Clark, R. A., 114
Clark, W. R., 502
Clarke, S., 697, 699(4), 700(4), 703(4), 704(4), 707(4), 709(4)
Clausen, C., 52
Clayton, D. A., 444
Clayton, R. K., 668
Clegg, R. M., 472, 474, 489(28), 495(17), 508(17), 510
Clevden, P. B., 180
Cleveland, B. M., 453
Clossen, W. D., 640
Coachman, D., 52
Cobbold, P. H., 172, 175, 177, 233, 241
Coen, S. V., 556
Coggins, J. R., 587
Cogoli, A., 474, 476(25)
Cohen, 95
Cohen, C. J., 147
Cohen, F. S., 453, 454(42), 457, 458(54)
Cohen, H. J., 114
Cohen, L. A., 612
Cohen, L. B., 102, 110, 111
Cohen, R. E., 451
Cohen, S. M., 84
Cohen, S., 693
Cohn, M., 301
Coke, M., 493
Coleman, R., 353
Coles, J. A., 138, 142, 143(5)
Coll, K., 241
Collins, J. H., 525
Colson, A.-M., 446
Cone, R. A., 499, 512
Conforti, R., 643
Conlon, R. D., 539
Conover, T. E., 107
Conrad, D. H., 621
Conteas, C. N., 18, 24(2, 4), 30, 31(4)
Conturo, T. E., 619, 621, 622(44), 623(44), 625, 627(44, 53)
Cooper, V. G., 697
Corin, A. F., 476, 477, 480(32), 481(40), 482(40), 485(32), 489(32), 493
Cork, R. J., 232

Cormier, M. J., 165, 168, 170
Cornell, B. A., 386
Corradin, G., 658
Corti, M., 531
Coruzzi, G., 446
Costa, T., 421
Costello, C. E., 641, 646(84), 687
Coulson, A. R., 444
Courtney, K. R., 110
Cox, R. N., 311
Craig, L. C., 636
Cramer, W. A., 441, 444, 445(16), 446, 448, 453, 454, 457, 458(53, 54), 461
Crane, R. K., 349, 357(27)
Cranney, M., 463
Crapster, B. B., 671
Criado, M., 493, 507, 512
Crofts, A. R., 446
Crofts, R. A., 81, 82(35)
Crozel, V., 461
Crumpton, M. J., 653
Cullis, P. R., 495
Cummings, R. T., 499
Cummins, H. Z., 366, 374
Cundall, R. B., 463, 464(1), 466(1), 472
Cunningham, L. W., 619, 621
Cuppoletti, B. K., 420
Cuppoletti, J., 427
Curatolo, W., 641
Curthoys, N. P., 630
Cushman, S. W., 50
Cuthbertson, K. S. R., 241
Cuthbertson, S. R., 172, 177
Czarniecki, M. F., 637
Czech, M. P., 605, 619

D

D'Souza, M. P., 14
Dabbeni-Sala, F., 686
Dahneke, B. E., 366
Daimatsu, T., 112
Daldal, F., 444, 446
Dale, R. E., 463, 464(1), 466(1), 472, 487, 503
Damjanovich, S., 476, 487(34), 493
Dangreau, H., 654
Daniele, R. P., 121
Danisi, G., 356
Dankert, J. R., 453, 458(53)

Danovich, G. R., 383
Dantzig, J. A., 289
Darmon, A., 123, 124(6), 125(6), 126, 127(6), 130(6), 131(6)
Darr, W., 640
Darrell, R., 529, 534(1)
Darszon, A., 156, 482
Das, M., 610, 621
David, M. M., 399
Davidson, E., 444
Davidson, N., 6
Davidson, V. L., 453, 454, 457, 458(54)
Davies, G. E., 613
Davies, R. W., 444
Davis, C. G., 415
Davis, D. G., 118
Davis, M. H., 83
Davoust, J., 499
Daw, R. A., 181
Dawson, A. J., 441
Dawson, G., 438
Dawson, R. M. C., 45, 349
Day, R. A., 592
de Bruijn, M. H. L., 444
de Duve, C., 3, 4, 15
de Laat, S. W., 87, 504
de Meis, L., 607
de Moura, J. L., 50, 52(6)
de Nouy, P. L., 533
de Pont, J. J. H. H. M., 399, 437
de Rago, J. P., 446
De Rosa, M., 509, 510
Deamer, D. W., 81, 82(35)
Deamer, D., 364
Dean, W. L., 522
Debye, P., 370
DeCuyper, M., 654
DeFeo, T. T., 179
Degiorgio, V., 378, 531
deGroot, E. J., 135
Deisenhofer, J., 439, 451(3), 609
Del Castillo, J. R., 352
deLaat, S. W., 240
Delbrück, M., 487
DeLean, A., 621
Delves, P. J., 109, 110
DeMaeyer, L., 382
Demarest, T. R., 83
Demou, P. C., 298
Demura, M., 66

Dencher, N. A., 464, 466(8), 468(8), 469, 470, 577, 578, 579, 580, 581, 582(18, 20), 583, 584
Denny, J. B., 619
Derzko, I., 472, 497(15), 508(15)
Derzko, Z., 496, 499, 507, 508, 510, 512
Deutsch, C. J., 84, 121
Deutsch, C., 84, 102
Devaux, P., 495, 629
Dever, C. M., 453
DeWeer, P., 120
Dewey, T. G., 582
Deyhimi, F., 138, 142, 143(5)
Diaconescu, C., 652
Dibona, D. R., 158
Dietz, R., 384
Dilger, T. P., 65
Dillon, S., 110
Dimitrov, D. P., 687
DiNapoli, A., 381
Ding, J. L., 438
DiPolo, R., 141, 199, 262
Dise, C. A., 121
DiVirgilio, F., 231, 232, 244(13), 252(13), 253(13), 255(7)
Dixon, T. E., 52
Dobbins, J. W., 362
Doell, G., 423
Dolly, J. O., 423
Dolphin, A. C., 289
Donelson, J. E., 444
Donovan, J. A., 619
Donovan, J. W., 624
Doolittle, R. F., 440, 441(8), 451(8)
Dorberska, C. A., 438
Dormer, R. L., 180
Dormier, K., 493
Dorset, D. L., 531
Douglas, A. P., 29, 353
Douglas, S. M., 621, 625
Dousa, T. P., 347, 357(14)
Douste-Blazy, L., 438
Downer, N. W., 578
Dragsten, P. R., 503
Dragutan, I., 82
Draheim, J. E., 580, 582
Dratz, E. A., 400
Drews, G., 608
Dreyer, W. J., 615
Dronin, J., 444

Dubinsky, W. P., 315
Duckwitz-Peterlein, G., 654
Dudak, V., 79
Dudley, S., 338, 339(14)
Duflot, E., 687
Duggan, P. F., 424
Dunger, I., 451
Dunkin, I. R., 637, 642
Dunn, R., 441
Dupont, Y., 135
Durand, P., 438
Dutta-Choudhury, T. A., 630
Dvorak, A. M., 180
Dzhandzhugazyan, K. N., 10

E

Eads, T. M., 478
Earnshaw, J. C., 366, 381
Eaton, W. A., 300
Ebina, Y., 454, 455(60)
Ebrey, T. G., 580, 581
Eddy, A. A., 97, 98, 99(4), 100(4), 101(4), 106
Edelman, I. S., 10, 53, 406
Edelstein, S. J., 706
Edidin, M., 472, 493, 495, 496, 499, 503, 508(8), 512
Edmonson, S., 576
Edsall, J. T., 702
Edwards, P. A., 425, 427(50)
Ehrenberg, B., 111
Ehrenberg, M., 485
Ehrenspeck, G., 6
Ehrich, M., 671
Eichberg, J., 393
Eichholz, A., 349
Eidelman, O., 123, 124(6, 7), 125(6, 7), 126, 127(6, 8), 130(6, 7, 11), 131(6), 134(11, 12)
Eilam, Y., 122
Eilenberger, G., 654
Einstein, A., 369
Eisenberg, D., 439, 440(1), 441(1), 445, 457, 460(67, 68), 461(68)
El Eini, D. I. D., 538
El Kebbaj, M. S., 686
Elgersma, O., 170
Ellens, H., 126
Elliot, D. C., 45

Elliot, W. H., 45
Ellory, C., 437
Ellory, J. C., 420, 421, 422, 428, 437, 438
Elmer, L. W., 418, 421(18)
Elson, E. L., 472, 499, 508(12), 509, 512
Elson, E., 496, 499, 504(96), 505(96)
Emaus, R. K., 79
Endoh, M., 198
Engelman, D. M., 440, 666, 667
Engels, J., 290
Engvall, E., 593
Eperon, I. C., 444
Eperson, I. E., 444
Epp, O., 439, 451(3), 609
Epstein, F. H., 16
Erecinska, M., 448
Erickson, S. K., 425, 427(50)
Ermann, P., 583
Erne, D., 144, 148(13)
Esser, A. F., 640, 659, 686
Ettre, L. S., 539
Evans, J. E., 542, 552, 553(20), 554(20), 559, 560, 561(35), 562, 565, 567(34, 58), 568(58), 569(58), 570(57), 571(57), 573(44, 57), 574(57)
Evans, M. C. W., 110
Evers, C., 349, 357(28)
Evers, J., 16
Exton, J. H., 241
Ezaki, O., 50

F

Fabiato, A., 186, 199, 256
Fabiato, F., 256
Fahey, P. F., 506
Fahr, A., 332
Faini, G. J., 167
Fairbanks, G., 588, 614, 619, 625(26), 689
Fairclough, P., 349, 357(27)
Falkenberg, F., 12
Fambrough, D., 499
Fanger, B. O., 621, 628, 693
Farber, I. C., 111
Farley, R. A., 658
Farmen, R. H., 115

Farquhar, M. G., 17
Farr, A. L., 393, 397(9), 398(9), 399(9)
Fasold, H., 585, 588, 591, 592, 600, 602(10), 603, 605, 607, 608(30), 609, 610, 615, 659, 660(168)
Fasulo, J. M., 565
Faust, R. G., 349
Fay, F. S., 233, 241(19), 245
Feeney, R. E., 612
Feery, D. R., 437
Feher, G., 439
Feinstein, M. B., 118
Feiring, A., 639
Felber, S. M., 106, 110(17), 121(17)
Felsenfeld, H., 118
Fendler, J. H., 476, 487(35), 493
Ferenczi, M. A., 300
Ferguson-Miller, S., 418, 428(16)
Fernandez, J. M., 52
Fernandez, S. M., 499
Ferrel, J. E., Jr., 654
Ferreu, B., 311
Ferry, D. R., 423
Fewtrell, C., 428
Field, J. B., 416, 417(10)
Fiete, D., 597
Finch, S. A. E., 476, 493
Fincham, D. A., 420, 437
Findlay, J. B. C., 662, 686
Finer-Moore, J., 457
Fink, G., 585, 595, 597, 603
Finkelstein, A., 453, 461
Finkelstein, J. N., 109
Finn, F. M., 652
Finn, F., 634
Fischkoff, S., 499
Fisher, R. W., 495
Fishman, P. H., 606
Flamberg, A., 387
Flaming, D. G., 138, 140
Fleischer, S., 403, 420, 426, 427, 428, 429(21, 53), 632, 698, 699(5), 707(5), 708(5), 709(5)
Fletcher, G. C., 386
Fletcher, M. P., 109
Fletterick, R. J., 451, 454(42)
Fleuren-Jakobs, A. M. M., 399
Flewelling, R., 336, 343
Flöge, J., 6, 11(13), 12(13), 16(13)
Fluke, D. J., 423, 424(43), 427

Fogarty, K. E., 233, 241(19)
Folch, J., 561
Foord, R., 378
Forbush, B., III, 288, 290(2), 298(2), 299(2), 300(2)
Ford, N. C., 378
Forgac, M., 34, 35, 39(5)
Forray, C., 180
Forstner, G. G., 349
Forte, J. G., 50, 51(5)
Forte, T. M., 50, 51(5)
Foskett, J. K., 123, 125(5)
Fosset, M., 437
Foster, D. L., 394, 399(13)
Foster, V., 82
Fox, C. F., 610
Fragata, M., 512
Franchi, A., 633
Franzusoff, A. J., 657, 668(151), 669(151)
Frasch, W., 17
Fraser, C. M., 437
Fredman, P., 552
Freedman, J. C., 102, 104, 107, 109(12, 22), 114(22), 117(12), 115(12, 22), 116(12), 118(22), 119, 121(12)
Freeman, H. J., 356
Freudenrich, C. C., 181, 193(47)
Freysz, L., 574
Freytag, J. W., 527
Friebel, K., 595
Friedhoff, L. T., 121
Friedl, P., 650, 659, 662(122), 664, 686
Frielle, T., 630
Fries, E., 519, 529, 534(1)
Friesen, H.-J., 652
Fritzsch, G., 615
Frohnert, P. P., 17
Fromm M., 147
Frömter, E., 143
Frye, L. D., 495, 499
Fryer, M. W., 198
Fujimori, T., 289
Fujita, M., 353
Fukutomi, M., 515
Fuller, S. D., 4, 15(3), 16(3)
Fulpius, B. W., 109
Funder, J., 121
Furthmayr, H., 662
Furukawa, R. H., 499
Futai, M., 519, 664

G

Gabellini, N., 444
Gaber, B. P., 386
Gabriel, N. E., 364
Gache, C., 654, 658(142)
Gaffney, B. J., 610
Gains, N., 364
Galante, Y. M., 686
Galardy, R. E., 636, 662
Galla, H. J., 499
Gallagher, J. G., 385
Galley, W. C., 490
Galli, G., 556
Gallin, J. I., 102, 109, 114, 121(69)
Gallo, R. L., 109
Gambacorta, A., 509, 510
Garavito, R. M., 439, 530, 531(11), 534(11), 535(11)
Garcia, A. M., 309
Garcia-Sancho, J., 254
Gardos, G., 122
Garfin, D., 687
Garfinkel, A. S., 438
Garland, P. B., 246, 474, 476, 478, 485(27), 487, 488, 489(42, 72, 73), 493, 501, 512
Garner, F., 29
Garnier, J., 451, 453(43)
Garty, H., 53, 156, 157, 159(9), 302
Gasko, O. D., 41, 158, 304
Gatti, R., 438
Gaub, H., 501
Gaver, R. C., 558
Gay, J. N., 664
Gay, N. J., 664
Gazzotti, G., 556
Gazzotti, P., 632
Gebler, B., 143
Geddes, L. A., 145
Geering, K., 658
Geersten, J. A., 352
Gelman, N. S., 437
Gennis, R. B., 119
Georgalis, Y., 388, 390(101)
Georgevich, G., 632
Georgolis, Y., 331
Gerber, G. E., 633, 638, 641, 666, 667, 686
Gerlt, J. A., 298
Germann, W. J., 308
Gershon, M. D., 94

Gershon, N. D., 503
Gershoni, J. M., 687
Gervasoni, J. E., 621
Ghidoni, R., 556
Ghosh, B. K., 349
Ghosh, H. P., 633
Gibbes, D., 386
Giedd, K. N., 106
Gietzen, K., 437
Gilbert, C. W., 487
Gilbert, J. C., 167
Gilbert, K., 446
Gilkey, J. C., 207
Gilman, P., Jr., 107
Ginkinger, K., 352
Ginsburg, H., 126, 127(8), 133(8)
Gioannini, T. L., 636
Girardet, M., 658
Giraudat, J., 111, 630
Giroux, S., 418, 423
Girvin, M. E., 448
Giryr, F., 598
Gitler, C., 126, 628, 653, 658, 662, 686
Glaeser, R. M., 577, 578(5), 582
Glatt, E., 493
Glauber, R. J., 376
Glazer, A. N., 668
Glenney, J. R., Jr., 584
Glossman, H., 423, 437
Glossmann, H., 13
Gluck, S., 50, 51(7), 54(7), 56(7), 92
Glucksman, M. J., 406
Glynn, I. M., 428
Glynn, P., 84
Gmaj, P., 346, 347(5), 348(5), 352, 363
Godt, R. E., 198
Goelz, S., 604
Golan, D. E., 512
Goldfarb, V., 658
Goldin, S. M., 34, 35, 36(6, 7, 8, 9), 37(6, 7, 8), 38(7), 39(5, 13)
Goldkorn, T., 425
Goldman, A., 440
Goldman, D. W., 653, 658, 662(136)
Goldman, Y. E., 289, 290, 300(18), 301(18)
Goldstein, J. L., 415
Goldstein, L., 11, 12(24), 16(24)
Goll, A., 437
Goll, J. H., 381
Goll, J., 386

Gomperts, B. D., 52
Gonzalez-Ros, J. M., 653
Gonzalez-Serratos, H., 203, 221(1), 223, 224(25)
Goodenough, D. A., 400
Goodman, D. B. P., 121
Goodsaid-Zalduondo, F., 472, 508(16)
Goody, R. S., 289, 300(5)
Gordon, D. J., 578
Gordon, J., 687
Gorman, A. L., 311
Gornall, A. G., 399
Gotterer, G. S., 353
Gotto, A. M., 385
Goyns, M. H., 172, 177
Grabau, C., 453, 458(53)
Graber, B. P., 386
Grabo, M., 529, 530
Graenek, S., 583
Graf, F., 45
Gralen, N., 702
Gras, W. J., 400
Grasse, P. B., 642
Grassl, S. M., 12
Gratecos, D., 356
Graton, E., 490
Gray, C. P., 666, 667
Gray, D. M., 576
Gray, G. M., 18, 19(6), 24(6), 30
Gray, H. B., 640
Gray, J. C., 448, 449(41)
Gray, M. A., 99
Gray, M. S., 388
Green, F. A., 420
Green, N. M., 651
Greenberg, G. R., 632
Greene, B. I., 300
Greger, R., 162
Gregoriou, M., 512
Greinert, R., 476, 493
Grell, E., 331, 386, 388, 390(101)
Griller, D., 642
Grinius, L. L., 65, 121
Grinvald, A., 110, 111
Grips, H., 13
Grisi, E., 444
Grob, P. M., 606
Gross, R. W., 561
Gross, S. K., 542, 547, 548(8), 550
Gruenberg, H., 451

Grunwald, R., 79
Grynkiewicz, G., 232, 233(10), 242(10), 260(10)
Guerini, D., 662
Guggino, S. E., 360
Gulari, E., 381
Gupta, B., 211, 223
Gupta, C. H., 641, 646(65)
Gupta, C. M., 638, 641, 642, 646(84), 651, 654(88)
Gupta, R. K., 110
Gurney, A. M., 288
Guth, K., 289, 300(5)
Guthmann, C., 629
Gutman, O., 512
Gutter, F. J., 702
Gutweniger, H., 658, 686
Guy, H. R., 453
Guyer, C. A., 628
Gysin, B., 686

H

Haagsman, H. P., 654
Haas, M. N., 437
Haas, S. M., 393
Haase, B. A., 438
Haase, W., 6, 10(11), 14, 349, 353, 357(28, 29)
Hackenbrock, C. R., 495, 510
Hacker, N., 642
Hadel, L., 642
Hagel, L., 521
Hah, J. S., 437
Haid, A., 687
Haigler, H. T., 413, 423, 426, 429(6)
Hall, R. D., 490
Hall, T. A., 211, 223
Hallam, T., 249
Hallett, M. B., 180, 181
Hallmann, D., 509, 510
Hamanaka, T., 581
Hammerman, M. R., 352, 353
Hanatani, M., 519
Hancock, K., 690
Handa, S., 558
Handler, J. S., 17
Hanna, S. D., 19, 29, 353
Hannafin, J. A., 16

Hanning, K., 7, 12(16)
Hansen, F. G., 664
Hansen, J. S., 608
Hanson, V. L., 561
Hardy, W. G., 423
Hare, J. D., 109
Hargrave, P. A., 653
Hargrove, P. A., 440, 441(11)
Harmon, J. T., 417, 420, 423(26), 424(26)
Harper, G., 438
Harrer, G. C., 164, 168(1), 169(1)
Harris, A. L., 254
Harris, J. W., 613
Hart, R. C., 181
Hartline, H. K., 276, 277(2), 280(2)
Hartman, F. C., 586
Hartsel, S. C., 583
Harvey, S. C., 487
Hashimoto, T., 515, 521
Haspei, H. C., 121
Hasse, W., 349
Hasselbach, W., 607
Hasselbracher, C. A., 582
Hassinen, I. E., 111
Hastings, J. W., 167, 197
Haugland, R., 463
Hauser, H., 349, 357(32), 364, 384, 476, 487(35), 493
Hauska, G., 445
Haustein, D., 608
Havlik, A. J., 514
Hay, R. D., 406
Hayashi, Y., 515, 520(17), 522, 525, 527
Haydon, D. A., 332
Hayes, M. J., 566, 570, 573(60), 574(60)
Hayes, T., 584
Hayssen, V., 564
Hayter, J. B., 531
Hayward, J. A., 578
Hayward, S. B., 577, 578(5)
Hazzard, J. H., 578
Hearon, J. Z., 415
Hearst, J. E., 601
Heath, M. F., 29
Hebert, S. C., 162
Hehmke, B., 109
Heinemeyer, W., 441
Heinonen, E., 115
Heinz, E., 12, 316, 318
Heiple, J. M., 83

Heitz, F., 453, 455(52), 458(52)
Helenius, A., 519, 521, 522(29, 30), 529, 534(1), 696
Heller, R. A., 640
Hellingwerf, K. S., 66
Hempling, H. G., 274, 276, 277(3)
Henderson, R., 400, 439, 578, 609, 666
Hendrickx, H., 487
Hendrix, J., 289, 300(5), 382
Henis, Y. I., 512
Henius, G. V., 106
Henkart, P., 127
Henry, J. P., 111
Hensley, C. B., 18, 24(2, 4), 30(4), 31(4)
Herbette, L. G., 403
Herbette, L., 289
Herlihy, J. T., 107
Herlihy, W. C., 666, 667
Herman, F., 639
Hermann, A., 311
Hermodson, M., 453, 458(54)
Herreman, W., 487
Herring, P. J., 181
Herrmann, R. G., 441, 444, 445, 446, 448(25, 32), 449(31)
Hershkowitz, S., 481, 494(48)
Hesketh, T. R., 83, 99, 104, 110(13), 119(13), 126, 240, 246, 249(41), 252, 438
Hess, E. J., 34, 36(7), 37(7), 38(7)
Hess, G. P., 290
Hess, P., 165, 166(5), 170(5), 198(5), 199(5), 245
Hesz, A., 639
Hetherington, H. P., 84
Heuser, J. E., 52
Hewick, R. M., 615
Heyn, M. P., 464, 466(8), 468, 469, 577, 578, 579, 580, 581, 582(18, 20, 25, 26), 583, 584
Heywood-Waddington, D., 25
Hibberd, M. G., 289, 290, 300(18), 301(18)
Hickman, J. A., 109
Hietel, B., 421
Higuchi, H., 423
Hilden, S. A., 349, 357(26)
Hildesheim, R., 111
Hildmann, B., 6, 11(6), 346, 347(4), 348(4), 349, 352, 353(49), 363
Hildreth, J. E. K., 519
Hill, B. C., 110

Hill, G. C., 444
Hill, M., 588
Hill, R. L., 698, 702
Hill, T. L., 495
Hill, W., 145
Hille, B., 309
Hillman, G. M., 512
Hinkle, P. C., 80, 81(31), 304, 659
Hiralú, K., 581
Hirata, Y., 606
Hirayama, B., 33
Hiremath, C. B., 592
Hissin, P. J., 50
Hixson, D. C., 584
Hixson, S. H., 594, 671
Hixson, S. S., 594, 671
Hjelmeland, L. M., 519
Hladky, S. B., 96, 107, 114(23), 119(23), 121(23), 332
Ho, S., 170
Hochachka, P. W., 427
Hochstrasser, R. M., 300
Hockman, J., 512
Hodgkin, A. L., 199
Hoefle, G., 651
Hoffman, E. K., 101
Hoffman, J. F., 99, 102, 104, 107, 108(21), 109(22), 114(22), 115(22), 118(22), 119, 120(21, 22), 122, 288, 290(2), 298(2), 299(2), 300(2)
Hoffman, J. L., 589
Hoffmann, W., 472
Hofmann, K., 652
Hogan, M., 481, 487
Hogeboom, G. H., 7
Holian, A., 121
Holian, S. K., 121
Holtzman, E., 50, 51(1)
Holtzwarth, G., 578
Homsher, E., 300
Hongoh, R., 67
Honig, B., 581
Hood, L. E., 615
Hopfer, U., 16, 29, 315, 316, 318, 323, 324, 330, 346, 347(6, 7, 8), 349, 351, 353, 361, 387
Hopkinson, D. A., 88
Hoppe, J., 650, 654(122), 659, 662(122), 664, 686
Horan, P. K., 109

Horecker, B. L., 610, 624
Hori, R., 352
Horoshige, Y., 584
Horowitz, M. A., 55, 90, 91(13)
Horowitz, P., 145
Horrocks, L. A., 563, 565
Horvath, C., 563
Horváth, G., 639
Hoshi, M., 544
Hotting, E. J., 78
Hou, Y., 496, 499, 508, 510
Houslay, M. D., 438, 471, 495(1)
Housman, D., 106, 109(16)
Howard, P. H., 107
Howell, K., 349
Howell, N., 446
Hruska, K. A., 309
Hsieh, J. Y. K., 565
Hsu, C. H., 454, 455(62)
Hsu, T. L., 437
Hsu, W. P., 533
Hu, V. W., 638, 659
Huang, C., 364, 386
Huang, C.-K., 601
Huang, K.-S., 577
Huang, Y. K., 352, 353(49)
Huang, Y.-K., 6, 11(6)
Hubbard, L. M. L., 686
Hubbel, W. L., 66, 71(17), 72(17)
Hubbell, W. L., 333, 334, 335, 336, 338, 341, 342(8, 18), 343, 344(9)
Huber, P., 629, 645(12), 649(12), 654(12), 655(12)
Huber, R., 439, 451(3), 609
Hudson, B. S., 463
Hudson, R. A., 311
Huestis, W. H., 654
Hughes, B. D., 488, 509
Hughes, M. E., 629
Hughes, S. M., 437, 438
Hugli, T. E., 621
Hugues, M., 437
Hui, D. Y., 415, 418(7)
Hülser, D. F., 110
Hunkapillar, M. W., 615
Hunt, C. A., 364
Hunter, M. J., 618
Hunziker, W., 662
Hutchinson, T. E., 223
Huttner, W. B., 244

Hymel, L., 420, 426(21), 427(21), 429(21), 437

I

Ickowicz, R. K., 311
Ikegami, A., 466, 467, 468, 472, 478, 486(6), 487(6), 582
Ikegami, I., 487
Illingworth, J. A., 194, 262
Im, W. B., 349
Imahori, K., 453
Imaizumi, K., 585
Inesi, G., 43
Innerarity, T. L., 415, 418(7)
Inoue, M., 9, 12(19), 13(19), 16
Inui, K., 352
Isaacs, S. T., 601
Isenberg, I., 486
Ishida, B., 686
Ishigaki, I., 429
Ishii, J., 520
Ishiwata, S., 478
Islam, A., 18, 19(6), 24(6), 30(6)
Isselbacher, K. J., 16, 349
Ito, T., 510, 512
Ito, Y., 25
Itoh, S., 75
Ives, H. E., 13, 18, 24(7), 352

J

Jackson, J. B., 63, 72(8)
Jackson, M. T., Jr., 539
Jacobberger, J. W., 109
Jacobs, M. A., 533
Jacobs, S., 606
Jacobsen, C., 347, 357(12)
Jacobson, K. A., 472, 497(15), 502, 508(15), 512
Jacobson, K., 472, 496, 499, 505, 506(119), 507, 508, 510
Jacquemin-Sablon, A., 109
Jaehnig, F., 493
Jaenicke, R., 523
Jaffee, E. K., 301
Jähnig, F., 468, 486
Jakeman, E., 377, 378

James, S. J., 438
James, T. L., 495
James-Kracke, M. R., 180
Jameson, D. M., 490
Jamieson, J. D., 606
Jansen, P. L. M., 438
Jansonius, J. N., 439
Jap, B. K., 577, 578(5)
Jares, W., 523
Jarori, G. K., 170
Jarvis, S. M., 420, 422, 659
Jarvisalo, J. O., 111
Jasaitis, A. A., 65, 76, 121
Jay, F. A., 619, 621
Jelenc, P. C., 596
Jenkins, M. R., 152
Jenkins, T., 439
Jennings, M. L., 615, 619, 621, 625
Jensen, J., 437
Jenssen, H.-L., 109
Jerguson-Miller, S., 512
Jewell, B. R., 164, 168(1), 169(1)
Jeynolds, J. A., 527
Ji, I., 598
Ji, T. H., 597, 598, 608, 610, 671
Joergensen, K. E., 353
Johannsson, A., 654, 658(142)
Johnson, B. J. B., 444
Johnson, E. L., 539
Johnson, E., 98
Johnson, F. H., 168, 198, 200(69)
Johnson, G. L., 606
Johnson, H. J., 655
Johnson, L. V., 111
Johnson, L. W., 629
Johnson, P. C., 180
Johnson, P., 246, 478, 487(42), 488, 489(42, 72, 73), 493, 501(42), 512
Johnson, R. G., 81, 121
Johnstone, R. M., 98, 107, 108(25)
Jones, D. D., 444
Jones, D. P., 352
Jones, G. R., 463
Jones, K. M., 45
Jones, M., 634
Jones, V. P., 441
Joniau, M., 654
Jordon, P. C., 332
Jorgensen, B. B., 659, 664(163)
Jorgensen, K. E., 347, 357(15)

Jørgensen, P. L., 10, 16, 158, 162(11), 364, 390(8), 658
Jovin, T. M., 473, 474, 476, 477, 478, 480(32), 481, 482, 484, 485(32), 487, 489(23, 24, 28, 32), 493, 494(48, 57), 495, 507, 512
Joyce, M. A., 642
Jung, C. Y., 420, 422, 424(36), 426, 427, 428, 429(21, 53), 437
Jung, D. W., 83
Jung, S. M., 621
Jungalwala, F. B., 545, 558, 559, 560, 561(35, 36), 562, 563, 564, 565, 567(34, 58), 568(58), 569(58), 570(57), 571(57), 573(44, 57), 574(57)
Junge, W., 82

K

Kaback, H. R., 63, 67(4), 111, 394, 399(13), 425, 632
Kadowaki, H., 552, 553(20), 554(20), 565, 567(34, 58), 568(58), 569(58), 571(57), 573(57), 574(57)
Kaduce, T. L., 561
Kadziauskas, T. P., 65
Kaenzig, W., 386
Kagawa, Y., 39, 364, 437
Kahan, I., 630
Kahane, I., 662
Kahn, A. M., 352
Kahn, C. R., 417
Kaila, K., 115
Kainosho, M., 495
Kakkad, B. P., 619, 621, 625(42)
Kaldany, R., 311
Kamat, V. B., 6
Kamentsky, L. A., 109
Kameyama, K., 515, 521(13), 522(13), 523(13), 527
Kamino, K., 110
Kamo, N., 66, 67
Kamp, R. M., 461
Kampmann, L., 615
Kamps, K. M. P., 399
Kanaoka, Y., 637
Kanazawa, H., 664
Kanner, B. I., 323
Kapitza, H. G., 501, 502, 507, 510, 512

Kaplan, J. H., 288, 290(2), 298(2), 299(2), 300(2)
Kaplan, R. S., 393, 396(11), 398(11), 399
Kapreyants, A. S., 437
Karger, B. L., 566, 570, 573(60), 574(60)
Karlin, A., 311
Karlish, S. J. D., 10, 156, 158, 162, 421, 658
Karlish, S., 302
Karniele, E., 50
Karnovsky, M. J., 29
Karnovsky, M. L., 29
Karp, R. D., 208
Kasahara, M., 659
Kasamatsu, H., 695
Kasowaki, H., 559
Kasper, C. B., 608
Kasuga, M., 605, 606
Katchalsky, A., 280, 282(4), 285(4)
Kato, Y., 521
Katzenellenbogen, J. A., 655
Kaufman, J. F., 686
Kaufmann, K. J., 642
Kauppinen, R. A., 111
Kawai, K., 353
Kawamoto, R. W., 424
Kawato, S., 466, 467, 468, 472, 486(5, 6), 487(5, 6)
Kayano, T., 664
Kaye, E. M., 547
Kaye, W., 514
Keana, J. W. F., 336
Kedem, O., 280, 282(4), 285(4)
Kegeles, G., 702
Keilbaugh, S. A., 687
Keith, C. H., 232, 243(11), 246(11)
Keith, J., 493
Keljo, D. F., 16
Kellermeyer, R. W., 613
Kelly, S., 54, 56(17), 87
Kemp, C. M., 493
Kemple, M. D., 170
Kempner, E. S., 413, 414(5), 415, 416, 417, 418, 419, 420, 421, 423, 424(26), 425, 426, 427, 428, 429, 437, 438
Kempner, E., 418, 428, 438
Kendall, D. A., 126
Kende, A. S., 645
Kennedy, E. L., 438
Kenny, A. J., 6, 7(10), 11(10), 12(10), 349, 357(25), 686

Kent, S. B. H., 619
Kenyon, G. L., 298
Kepner, G. R., 412, 413(3)
Kerker, M., 366, 367(19), 370(19)
Kessler, M., 349
Ketterer, B., 332
Khintchine, O., 371
Khorana, H. D., 441
Khorana, H. G., 577, 628, 632, 638, 641, 642, 646(65, 84), 651, 654(87, 88, 89), 657(87), 662(87), 666, 667, 687
Kiehart, D. P., 172
Kihara, M., 66, 115
Kihara, T., 289
Kilbourn, M. R., 653
Kilbride, P., 581
Kim, S., 364
Kimura, Y., 582
Kincaid, R. L., 418
King, M., 316
King, P. A., 11, 12(24), 16(24)
Kinnally, K. W., 107
Kinne, R., 6, 9, 11, 12, 13(19), 14, 15, 16, 17, 317, 346, 348(1), 349, 351, 352, 357(28), 363, 437
Kinne-Saffran, E., 6, 16, 349, 351, 352
Kinosita, K., 466
Kinosita, K., Jr., 467, 468, 472, 478, 486(5, 6), 487, 582
Kippen, I., 33
Kiraki, K., 581
Kiral, R. M., 561
Kirkland, J. J., 539, 540(1)
Kirmse, W., 634
Kirschenbaum, I., 706
Kirschner, G., 556
Kiselev, A. V., 440
Kitabatake, S., 637
Kitagawa, T., 590, 593
Kitazawa, S., 352
Kitazawa, T., 211, 215, 220(13)
Kito, Y., 581
Klahr, S., 352, 353
Klappenberger, J., 585, 602(10), 609
Klapper, M. H., 616
Klein, R. A., 437
Kleinzeller, A., 16
Klempner, M. S., 180
Kleps, R. A., 84
Klip, A., 628

Klotzbücher, R., 640, 687
Klymkowsly, M. W., 400
Knable, N., 374
Knauf, P. A., 134
Knibihler, M., 356
Knickelbein, R., 362
Knight, D. E., 181
Knight, P., 612
Knof, S., 523
Knott, J. C. A., 653
Knott, R., 406
Knowles, A. F., 41, 158, 304
Knowles, J. R., 634, 638, 643, 653, 660(138)
Kobashi, K., 610
Kobatake, Y., 66, 67, 112
Koch, N., 608
Koenig, S. H., 380
Kohl, K.-D., 578, 583(14), 584(14),
Komaromy, M., 457, 460(68), 461(68)
Komiya, K., 521
Komoriya, A., 440, 456(10)
Kon, K., 585
König, B., 16
Königsberg, W., 591
Koning, W. N., 66
Konishi, M., 199
Konisky, J., 454, 455(62)
Kono, T., 50
Kopito, R. P., 660
Koppel, D. E., 379, 380, 384(60), 495, 496, 499, 500, 502, 504(96), 505(96), 509, 512
Korchak, H. M., 106
Korenbrot, J. I., 232
Korenstein, R., 157
Körkemeier, U., 633
Korn, E. D., 364
Koshland, D. E., Jr., 108, 119(30), 121(30)
Kostyanovsky, R. G., 643
Kotite, N. J., 619, 621
Kou, A. Y., 561, 563
Kouyama, T., 582
Kowarski, D., 227
Kozlowski, T. R., 18, 24(2, 4), 30(4), 31(4)
Kraaynhof, R., 78
Krab, K., 78
Kraehenbuhl, J. P., 658
Krafft, G. A., 499
Krag-Hansen, U., 347, 353, 357(12)
Kragh-Hansen, U., 357
Kratohuil, J. P., 533

Kravchenko, V. S., 706
Krebs, J., 662, 686
Kremer, J. M. H., 364
Kriebel, A. N., 582
Krigbaum, W. R., 440, 456(10)
Kroon, P. A., 495
Kruskal, B. A., 232, 243(11), 246(11)
Kubo, H., 544
Kukuda, M., 515
Kuliene, V. V., 121
Kume, T., 429
Kundig, W., 632
Kuntz, I. D., 451, 454(42)
Kuo, A. L., 495
Kurihara, K., 112
Kurihara, S., 185, 199
Kurland, C. G., 608
Kuryatov, A. B., 440
Küthale, W., 499
Kyte, J., 440, 441(8), 451(8)

L

Labotka, R. S., 84
Lad, P. M., 438
Laemmli, U. K., 607, 614, 689, 690(12)
LaFleur, L. E., 336
Lai, C. Y., 624
Lai, F. A., 421
Lakatos, B., 639
Lakhlifi, T., 686
Lakowicz, J. R., 472
Lala, A. K., 643
Lam, K. S., 608
Lambert, I. H., 101
Lampidis, T. J., 112
Lamprecht, J., 487, 493
Lan, S.-F., 425, 427(50)
Lancet, D., 126
Lane, L. K., 525
Lang, I., 135
Lange, Y., 654
Langer, G. A., 43
Langridge-Smith, J. E., 315
Lankmayer, E. P., 566, 570, 573(60), 574(60)
Lanter, F., 144, 148
Lardner, T. J., 496, 504
Laris, P. C., 98, 102, 106, 107, 108(21, 25), 119(22), 120(22)

Larsson, C., 13, 24
Lash, L. H., 352
Lasic, D. D., 384
Lasic, D., 476, 487(35), 493
Latorre, R., 156, 310
Latruffe, N., 686
Läuger, P., 109, 332
Laursen, R., 634
Lavery, A. N., 381
Lavie, E., 119
Lavine, C. H., 686
Laviner, C. S., 29, 353
Lawley, P. D., 593
Lazdunski, C., 453, 454, 455(52, 64), 458(52)
Lazdunski, M., 437
Leaf, A., 158
Leary, J. J., 652
Leaver, C. J., 441
LeBaron, F., 560, 561(36), 564
Leblanc, P., 633, 686
Lechene, C. P., 203
Lecompte, M. F., 686
Ledeen, R. W., 550
Lee, A. G., 495
Lee, C. P., 81, 82(34)
Lee, D. C., 578
Lee, J. C., 514
Lee, K.-J., 654
Lee, M. Y., 121
Lee, S., 444
Lee, W. T., 621
Lees, M. B., 393
Lees, M., 561
Leeuwen, M., 197
LeGrange, J., 487
Leibowitz, J., 454, 455(61)
Lemasters, S. T., 79
Lenard, J., 686
Lentz, B. R., 463
LePecq, J.-B., 109
Lepke, S., 615, 659, 660(168)
Lesher, S., 110
Lessing, H. E., 477
Lester, H. A., 288, 290
Letellier, L., 83
Levenson, R., 106, 109(16)
Lever, J. E., 6
Levine, M., 111
Levinson, S. R., 437
Levinthal, C., 453, 461

Levinthal, F., 461
Levitsky, D. O., 121
Levitt, M., 451
Lew, D. P., 231, 244, 255(7)
Lew, V. L., 254
Lewis, B. A., 353
Lewis, S. A., 50, 52(6)
Liang, C. T., 348, 352
Liao, M.-J., 577
Liberman, E. A., 65
Lichtenberg, D., 364
Liedtke, C. M., 315
Likhtenshtein, G. I., 344
Lin, J.-T., 437
Lin, S. H., 35, 36(10)
Lin, W., 695
Lindblom, G., 495
Lindemann, B., 156
Lindman, B., 529
Lindsay, J. G., 14
Ling, K. J., 349
Lipari, G., 467, 468(12), 486
Lipson, L. G., 30
Litman, B. J., 464
Litman, G. B., 386
Liu, D., 352
Liu, Q.-R., 461
Liverman, E. A., 121
Lloubes, R., 454, 455(64)
Lo, M. M. S., 423
Lo, M. M., 487, 493
Lodish, H. F., 660, 662
Loew, L. M., 111, 126
Loewenstein, W. R., 186
Loh, Y. P., 111
Lolkema, T. S., 66
Lomant, A. J., 588, 619
Lombet, A., 437
London, E., 577
Long, M. M., 577
Longsworth, L. G., 523
Lopez, J. R., 141
Lopez, R., 262
Lottspeich, F., 451
Lowden, J. A., 547
Lowe, A., 85
Lowe, M. E., 413, 414(5)
Lowry, C., 702
Lowry, O. H., 393, 397(9), 398(9), 399(9)
Loyter, A., 512

Lu, C. C., 18, 19(3), 24(2, 3), 30, 31(3), 32(3)
Lübbecke, 423
Luché, B., 276, 277(2), 280(2)
Luciane, S., 35, 36(11)
Lücke, H., 317, 356
Lüdi, H., 109
Ludwig, F. R., 619, 621
Ludwig, M. L., 618
Luescher, B., 658
Luippold, D. A., 403, 405(9)
Lüscher, B., 657, 668(151), 669(151)
Lutter, L. C., 586, 607, 608(30)
Luzzati, V., 529, 531(2)
Lwowski, W., 634, 638

M

Maag, U., 347, 357(13)
Maassen, J. A., 594
Mabuchi, K., 664
MacAndrew, V. J., Jr., 606
Macara, I. G., 106, 109(16)
MacDonald, H. R., 658
MacDonald, R. C., 126, 654
Macey, R. I., 121, 412, 413(3)
Machen, T. E., 50, 51(5), 83, 87
Machida, K., 510, 512
Macknight, A. D. C., 158
MacNab, R. M., 115
Macrae, D. R., 10
Macy, R., 82
Madeira, V. M. C., 507
Madeira, V., 152
Madsen, K. M., 50, 51(9)
Maeda, N., 585
Maeda, Y., 289, 300(5)
Maestracci, D., 349
Maestre, M. F., 576, 577, 578(5)
Maezawa, S., 515, 520(17), 523(15)
Magde, D., 499
Mahley, R. W., 415, 418(7)
Maim, H. Y., 658
Mainka, L., 584
Majumdar, A., 441
Makino, S., 521, 522(31)
Makinose, M., 607
Makowski, L., 400, 406
Malathi, P., 427

Mallett, P., 349, 357(27)
Mameli, L., 438
Mamelok, R. D., 10, 352
Mancarella, D. A., 686
Mandel, L., 376
Manganiello, V. C., 418
Mankovich, J., 454, 455(62)
Mantle, T. J., 438
Mao, B., 581
Mao, D., 576, 577
Marban, E., 148, 149(23)
March, J., 640
Marchesi, V. T., 660, 662
Marcus, M. M., 109
Maret, A., 438
Marin, R., 352
Markwell, M. A. K., 393
Marshall, M. W., 110
Martell, A. E., 249
Martin, B. R., 438
Martin, F. J., 654
Martin, G. J., 360
Martin, G. M., 364
Martin, R. G., 699
Martinez, C., 453, 455(52), 458(52)
Martinez-Carrion, M., 653
Martinosi, A., 424
Martonosi, A. N., 115
Martyr, R. J., 636
Masamoto, K., 75
Mashimo, T., 581
Mason, J. T., 386
Masotti, L., 472
Mast, E., 523
Masur, S. K., 50, 51(1)
Masuzawa, T., 15
Matayoshi, E. D., 477, 481(40), 482, 493
Matayoshi, E. M., 493
Matkó, J., 473, 489(24)
Matricon, J., 629
Matsui, H., 353, 515, 520(17), 527
Matsumoto, K., 423
Matsunaga, M., 112
Matsura, K., 75
Mattingly, P. H., 164, 168(1), 169(1), 197
Maunsbach, A. B., 364, 390(8)
Maurer, A., 420, 426, 427(21), 429(21, 53)
Maxfield, F. R., 55, 90, 91(13), 232, 243(11), 246(11)

Maynard, J. B., 539
Mazer, N. A., 531
Mazer, N., 386
McCally, R. L., 381
McCann, R. O., 168, 170
McCaslin, D. R., 519
McCaslin, R., 529, 534(1)
McClellan, G., 203, 223, 224(25)
McCluer, R. H., 542, 544(5), 545, 547, 548(8), 550, 551(18), 552, 553(18, 20), 554(20), 559, 560, 561(35), 562, 563(43), 564, 565, 567(34, 58), 568(58), 569(58), 570(57), 571(57), 573(44, 57), 574(57)
McConnell, H. M., 489, 493, 495, 499, 502, 506, 510
McCoy, T., 441
McCray, J. A., 289, 290, 300(13)
McCreery, M. J., 421, 428, 429, 437
McCutcheon, M., 276, 277(2), 280(2)
McDaniel, J. B., 514
McDonald, M. P., 529, 532(6)
McDonough, A. A., 18, 24(2, 4), 30(4), 31(4)
McDowell, J. H., 653
McGiure, J. M., 566, 570, 573(60), 574(60)
McGregor, G., 502
McGrew, S. G., 428
McIntosh, T. S., 65
McIntyre, J. O., 424, 426, 427, 428, 429(53)
McKenna, M. P., 180
McKenney, M., 180
McLachlan, A. D., 666
McLaughlin, A., 332
Mclaughlin, S. G. A., 65
McLaughlin, S., 315, 332
McMillan, D. J., 213
McNab, R. M., 66
McNeil, P. L., 180
McPhail Berks, M., 444
McWhirter, J. G., 379, 382(57)
Means, G. E., 612
Medow, M. S., 352
Meech, R. W., 172
Meier, P. C., 148, 150
Meiri, Z., 111
Meister, H., 651, 669(128), 670(128)
Melander, W., 563
Melandri, B. A., 81, 82

Meldolesi, J., 231, 244
Melhorn, R. J., 82, 342
Meloche, S., 621
Mende, J., 461
Mendocino, J., 29
Metcalfe, J. C., 83, 110, 126, 240, 246, 249(41), 252, 438, 495
Metz, P., 640
Metzger, H., 428, 707
Meyer, C., 585, 602(10), 605, 609
Michel, H., 439, 451, 609
Michel, H.-P., 645
Michell, R. H., 29, 353
Micklem, K. J., 99
Middaugh, C. R., 608
Middlehurst, J., 386
Middlemas, D. S., 658
Mie, G., 370
Mihashi, K., 487
Mikes, V., 79
Miki, K., 439, 451(3), 609
Miller, C. J., 156
Miller, C., 156, 309, 310
Miller, D. J., 196, 256
Miller, D. L., 232
Miller, J. B., 108, 119(30), 121(30)
Miller, J. H., 415, 427, 437
Miller, P., 349
Mills, S., 309
Milner, L. S., 632
Milon, A., 386
Milsmann, M. H. W., 386
Mimms, L. T., 364
Minocherhomjee, A. M., 437
Mircheff, A. K., 13, 18, 19, 24(2, 3, 4, 6, 7), 29, 30, 31(3, 4), 32(3), 33, 352, 353
Mitchell, D. J., 529, 532(6)
Mitchell, G. W., 167
Mitchell, G., 197
Mitchell, P., 6
Miyagishima, T., 199
Miyake, J., 514
Miyata, T., 454, 455(60)
Moczydlowski, E. G., 34
Moisescu, D. G., 196, 257
Mokrasch, L. C., 393
Molinaro, A. L., 654
Møller, O. J., 15
Monaghan, R., 621, 625

Monchamp, P., 563
Monitto, C. L., 481, 494(48)
Monroe, J. G., 109
Montal, M., 156, 482
Montecucco, C., 99, 104, 107, 110, 119(13), 628, 630, 649, 650, 654, 658, 662(122), 686, 687
Montgomery, C. R., 642
Montibeller, J. A., 652
Monty, K. J., 698
Moolenaar, W. H., 240
Moolenar, W. H., 87
Moon, R. B., 84
Moonen, P., 654
Moore, A. L., 111
Moore, C. H., 474, 476(27), 485(27), 487, 493
Moore, C., 476, 493
Moore, E. D. W., 166, 196(8), 197(8)
Moore, H. P. H., 126
Moore, H.-P. H., 309
Moore, J. P., 126, 246, 249(41)
Mooris, J. D. R., 240
Morad, M., 110
Moran, A., 5, 29, 323, 347, 357(9, 10, 11), 358(10), 359(10)
Moreno, G., 112
Moreton, R. B., 223
Morf, W. E., 136, 150
Morgan, D. G., 611, 618(15)
Morgan, J. P., 179
Morgan, K. G., 179
Moriyama, Y., 111
Morlon, J., 454, 455(64)
Moroi, M., 621
Morpain, C., 686
Morris, W. J., 25
Morriset, J. D., 385
Moscarello, M. A., 630
Moscatelli, E. A., 558
Mosher, D. F., 584
Moss, R. A., 634, 642
Moye, W. S., 523
Moyle, J., 6
Muccio, D. D., 580
Muir, W. A., 316
Muirhead, K. A., 134
Mukerjee, P., 533, 534(18)
Mukunoki, Y., 533
Müller, M., 349, 640, 641(78)

Müller, U., 578, 581
Müller-Eberhard, H. J., 640, 659(81)
Mullins, L. J., 199
Munoz, J.-L., 142
Murakami, K., 112
Muratsugu, M., 67
Murer, B., 351
Murer, H., 6, 10(11), 11(6), 14, 16, 29, 317, 318, 346, 347(4, 5), 348(1, 3, 4), 349, 352, 353, 356, 357(28, 29), 361, 362, 363
Murov, S. L., 479
Murphy, E., 241
Murray, E. K., 472, 487
Murthy, U., 621
Murvanidze, G. B., 112
Muzukami, H., 606
Myers, H. N., 655
Mysels, K. J., 533, 534(18)

N

Nabedryk, E., 577, 578(8)
Naftalin, L., 29
Nakae, T., 515, 520, 521(13), 522(13), 523(13)
Nakagawa, Y., 565
Nakamura, Y., 112
Nakatani, Y., 386
Nakaya, K., 112
Nakayama, H., 637
Nakazawa, A., 454, 455(60)
Nakazawa, T., 454, 455(60)
Nargeot, J., 290
Narlock, R., 418, 428(16)
Naser, R., 207
Nashima, T., 514
Nassal, M., 644, 645(112, 113), 647(113)
Natale, P. J., 109
Nauta, H., 437, 438
Navon, G., 84
Nazran, A. S., 642
Neering, I. R., 198
Neher, E., 52, 145, 156, 171, 244
Nelson, K., 16, 349
Nerbonne, J. M., 290
Nerbonne, J., 254
Nerl, C., 109
Ness, G. C., 421

Neumann, E., 325
Neumcke, B., 332
Neville, D. M., Jr., 349
Newcomb, K., 352
Newman, E. L., 421
Newman, M. J., 394, 399(13)
Newman, S. A., 707
Nguyen Hong, V., 438
Nguyen, H. V., 438
Nichimura, S., 581
Nicholas, R. A., 630
Nicholls, D. B., 63, 72(8)
Nicknish, J. S., 619, 621, 625
Nielsen, J., 664
Nielsen, M., 437
Nielsen, P. E., 638
Nielsen, T. B., 416, 417(10), 420, 423(26), 424(26), 438
Nierlich, D. P., 444
Nieva-Gomez, D., 119
Niggli, V., 687
Nihei, K., 666, 667
Nikazy, J., 438
Nishifuji, K., 519
Nishimura, M., 75
Nishio, T., 487
Nishizuka, Y., 632
Nissley, S. P., 605, 606
Niswender, G. D., 477
Nitschke, W., 445
Njus, D., 111
Nobrega, F. G., 444
Noël, H., 438
Nolan, C. E., 558
Noller, H. F., 604
Norén, O., 686
Norman, R. I., 437
Norman, R., 437
Norris, J., 439
Norton, K. C., 561
Norton, W. T., 573
Nossal, R., 366
Notter, R. H., 109
Noumi, T., 664
Novak, E. K., 671
Novak, T. S., 104, 109(12), 115(12), 116(12), 117(12), 119, 121(12)
Nozaki, M., 702
Nozaki, Y., 364
Nozawa, M., 637
Nunez, E. A., 94
Nutter, T. J., 418, 421(18)

O

O'Brien, T. A., 119
Obaid, A. L., 111
Ochs, D. L., 232
Odermatt, E., 686
Oesterhelt, D., 451
Oetliker, H., 109
Offer, G., 612
Ogawa, S., 84
Ohkuma, S., 90, 111
Ohme, R., 643
Ohmori, H., 454, 455(61)
Ohmori, Y., 606
Ohnishi, S., 510, 512
Ohno-Iwashita, Y., 453
Ohta, H., 353
Okano, T., 352
Okimasu, E., 107
Oku, N., 126
Olivecrona, T., 428
Oliver, C. J., 378
Olsen, W. L., 638, 641(65), 646(65)
Olson, F., 364
Omedo-Sale, F., 606
Ong, H., 621
Oppenheimer, N. J., 298
Oppliger, M., 474, 476(25)
Orchard, C. H., 202
Organiesciak, D., 510
Organisciak, D., 508
Orme, F. W., 121
Ortanderl, F., 607, 608(30)
Osborn, M. J., 509
Osborn, T. W., 561
Osborne, J. C., Jr., 418, 428
Osguthorpe, D. J., 451, 453(43)
Osotimehin, B., 121
Ostrovskii, D. N., 437
Ostrowsky, N., 382, 386
Ott, P., 109
Ott, S., 421
Ottolenghi, P., 421, 437
Ourisson, G., 386
Ovchinnikiv, Y. A., 440
Overath, P., 654

Owen, M. J., 653
Owens, C. S., 499

P

Pachence, J. M., 406
Packer, L., 82, 342
Packer, N. H., 565
Pagano, P., 364
Pagano, R. E., 629
Paglin, S., 606
Pailthorpe, B. A., 488, 509
Palade, G. E., 7, 687
Palmer, L. G., 308
Palmer, L., 52
Palva, E. T., 608
Papahadjopoulos, D., 364, 505, 506(119), 508
Papazian, D. M., 34, 35, 36(6, 8, 9), 37(6, 8)
Paradiso, A. M., 83, 87
Parce, J. W., 109
Parente, R. A., 463
Parilvssen, N., 382
Park, J. Y., 561
Parker, P., 382
Parkinson, D., 420, 438
Pasquini, J. M., 564
Passow, H., 122, 134, 615, 659, 660(168)
Pasternak, C. A., 98, 99, 110
Paterson, A. R. P., 420, 437, 659
Pathak, R. K., 93
Pathmamanoharan, C., 364
Patton, G. M., 565
Pattus, F., 453, 455(52), 458(52)
Paul, S. M., 421
Paulsen, S. R., 643
Paulson, J. C., 698
Paxman, S., 393
Peacocke, A. R., 378
Pecht, I., 126
Pecora, R., 366, 367(25, 29), 370, 371, 380, 386, 387
Pedersen, K. O., 702
Pedersen, P. L., 393, 396(11), 398(11), 399
Peers, E., 421
Peiser, H., 349, 357(27)
Pendleton, L. C., 421
Pennington, R. J., 29

Penniston, J. T., 45, 686
Pereiza-da-Silva, L., 78
Perevucnic, G., 384
Perkins, W. R., 341
Perlmann, G. E., 523
Perlmann, P., 593
Pernemalm, P., 521
Perotto, J., 16
Perrin, D. D., 256
Perrotto, I., 349
Pershadsingh, H. A., 107, 108(25)
Peters, J., 496, 499, 608
Peters, K., 584, 590, 610, 619(7), 671
Peters, R., 123, 125(4), 134, 135, 472, 496, 499, 506, 508(11)
Peters, T. J., 25, 438
Peters, T., Jr., 523
Peters, W. H. M., 399, 438
Petersen, N. O., 499
Peterson, A. A., 453, 461
Peterson, G. L., 393, 397(5, 6), 398(5, 6), 399(6)
Phillips, D. L., 379, 382(54)
Phillips, W. C., 400
Philo, R. D., 97, 98(4), 99(4), 100(4), 101(4), 106
Pierce, D. H., 289
Pietrobon, D., 63
Pike, E. R., 366, 377, 378, 379, 382
Pilch, P. F., 605, 606, 619
Piotti, A., 686
Pittz, E. P., 514
Plattner, H., 207
Platz, M. S., 642
Pober, J. S., 653, 662(136)
Podack, E. R., 640, 659
Podevin, R. A., 352
Podleski, T. R., 512
Podo, F., 499
Poduslo, S. E., 573
Poenie, M., 232, 233, 242(10), 260(10)
Poliakoff, M., 642
Politz, S. M., 594
Pollard, E. C., 412
Pollet, R. J., 438
Poo, M. M., 495, 499, 512
Poo, M., 512
Poole, B., 90
Poole, K. J. V., 289, 300(5)

Post, R. L., 525
Poste, G., 496, 499
Potier, M., 418, 419, 423, 428, 429, 437, 438
Pottel, H., 487
Potter, J. D., 525
Poujeol, P., 361
Powers, L., 505, 506(118)
Poy, G., 428
Pozzan, T., 83, 87, 107, 110, 230, 231, 232, 233, 237(4), 239(3), 241, 242(4), 243(4, 35), 244, 255(4, 6, 7)
Prasher, D., 168, 170(17)
Preiser, H., 349, 427
Preiss, J. W., 423
Prendergast, F. G., 165, 166(5), 169, 170, 198, 199(5), 245, 463
Preston, M. S., 438
Prince, R. C., 81, 82(35)
Pring, M., 659, 660(168)
Printz, M. P., 636
Proberbio, F., 352
Provencher, S. W., 330, 331(21), 379, 382, 383, 453, 455(52), 458(52)
Proverbio, T., 352
Prusiner, S. B., 10
Purves, R. D., 136, 141(1), 145(1)
Pusch, H., 196, 257
Pusey, P. N., 366, 367(30), 380, 384
Putnam, F. W., 523, 702

Q

Quamme, G. A., 356
Quatrecasas, P., 17
Quay, S. C., 632, 638, 641(65), 646(65)
Quigley, J. D., 353
Quintaniha, A. T., 342
Quinton, P. M., 15

R

Rabon, E., 29, 353
Racker, E., 39, 41, 83, 85, 158, 304, 364, 530, 535(10)
Radhakrishnan, R., 628, 632, 641, 642, 646(65), 651, 654(87, 88, 89), 657(87), 662(87), 687

Radian, R., 323
Raftery, M. A., 126, 309, 658
Rahamimoff, H., 35, 36(6, 7, 8, 9, 12), 37(6, 7, 8), 38(7)
RajBhandary, U. L., 441
Ramos, B., 615
Ramos, S., 63, 67(4)
Randall, R. J., 393, 397(9), 398(9), 399(9)
Rao, A. K., 29
Rao, B. D. N., 170
Rao, J. K. M., 440, 441(11), 523
Rapp, G., 289, 300(5)
Rauth, A. M., 412
Ravdin, P., 512
Raviv, Y., 686
Ray, B. D., 170
Rayleigh, Lord, 370, 375
Raymond, L., 453, 461
Razi-Naqvi, K., 499
Razin, M., 126, 127(8), 133(8)
Rearick, J. I., 698
Reber, B., 630, 656
Rebois, R. V., 606
Rechler, M. M., 605, 606
Reddi, K. K., 525
Redmond, J. W., 565
Reed, P. W., 65
Reed, R. G., 523
Reed, W., 476, 487(35), 493
Rees, A. R., 512
Rees, D. C., 439
Reese, T. S., 52
Regateiro, F., 476, 487(34)
Regaterok, R., 493
Rehorek, M., 464, 466(8), 468(8), 470, 582, 584
Reich, C., 576
Reid, G. P., 14, 290, 300(13)
Reidler, J. A., 509
Reinlib, L., 155
Reiser, A., 635
Reiser, J., 687
Reisler, E., 591
Remold, H., 609
Remold, R., 585, 602(10)
Requena, F., 199
Restall, C. J., 472, 487, 493, 578
Reue, K., 686
Revzin, A., 418, 428(16)

Reynolds, G. T., 186
Reynolds, J. A., 364, 514, 515(1), 519(1), 521, 522(31), 530, 531, 577, 580(9)
Reynolds, M. A., 298
Rhoden, V., 34
Ricapito, S., 16
Rice, V., 172, 177(27)
Rich, A. M., 106
Rich, T. L., 241
Richard, F. M., 617
Richard, S., 290
Richards, D. E., 428
Richards, F. M., 584, 585, 587, 590, 601, 610, 613, 619(7), 625(14), 627, 638, 644, 646(66,110), 647(110), 654(66), 671
Richards, J. T., 463
Richards, T. H., 84
Richelson, E., 180
Richie, R. S., 66
Richmond, T. J., 617
Ricka, J., 386
Ridgway, E. B., 199
Riete, D., 599
Rigaud, J. L., 83
Rigler, R., 485
Rimon, G., 425
Ringsdorf, H., 501
Rink, T. J., 87, 96, 99, 102, 104, 107, 110, 114(23), 119(13, 23), 121(23), 143, 144, 148, 149(23), 230, 231, 233, 237(4), 241, 242(4), 243(4, 35), 244(4), 249, 255(4, 6, 8)
Rink, T. S., 83
Rink, T., 107
Rinke, J., 597
Ripoche, P., 361
Ritherford, L. E., 106
Roa, M., 687
Roberts, K. D., 429
Roberts, M. F., 364
Robertson, D. E., 73, 75(21), 76(21)
Robins, M. J., 659
Robins, S. J., 565
Robinson, F. W., 50
Robinson, S. C., 104, 107
Robson, B., 451, 453(43)
Robson, R. J., 628, 642, 651, 654(87), 657(87), 662(87)
Rodbell, M., 428, 438
Rodman, J. S., 17
Rodriquez, H. J., 10
Roe, B. A., 444
Roelofsen, B., 632
Roess, D. A., 477
Rogers, J., 246
Rogers, T., 83
Roggero, J. P., 556
Roigaard-Petersen, H., 347, 353, 357
Roisin, M. P., 111
Rojas, H., 262
Rommel, W., 603
Ronner, P., 632
Roos, A., 63, 79(9)
Roos, D., 106
Rose, B., 186
Rose, G. D., 454
Rosebrough, N. J., 393, 397(9), 398(9), 399(9)
Roseman, M. A., 654
Roseman, S., 632
Rosenberg, I., 126, 686
Rosenberry, T. L., 630
Rosenbloom, I. L., 348
Rosenbusch, J. P., 439, 529, 530, 531(11), 532(3), 534(11), 535(11)
Rosenheck, K., 581
Rosenthal, K. S., 109
Ross, A. H., 628, 642, 654(87, 88, 89), 657(87), 662(87)
Ross, M. J., 400
Rossi, G., 707
Rossier, B. C., 658
Roth, K. S., 352
Rothbard, J., 637
Rothschild, K. J., 403, 405(9)
Rothstein, A., 615
Rottenberg, H., 63, 66, 69(1, 18), 70(18), 72(18), 73, 75(21), 76(20, 21), 79(1), 81, 82(33, 34), 84, 85
Roufogalis, B. D., 437
Rousselet, A., 629
Roux, M., 298
Ruddy, S., 621
Rudy, B., 156, 302
Ruf, H., 331, 386, 388, 390(101)
Russell, J. T., 111
Russo, M. W., 693
Rusznak, I., 254
Rutherford, L. E., 106
Ryall, M. E. T., 181

S

Sabesin, S. M., 349
Sabolic, I., 8, 12(18)
Saccomani, G., 427
Sachs, G., 15, 16, 29, 352, 353, 427
Sackmann, E., 495, 499, 501, 507, 510, 512
Sacktor, B., 348, 349, 352, 357(26)
Sadler, J. E., 698
Saffman, G., 487
Saffman, P. G., 487
Saiga, Y., 198, 200(69)
Saito, A., 403
Saito, S., 519
Saito, T., 15
Sakai, T., 199
Sakmann, B., 156, 171
Salama, G., 81, 121
Salans, L. B., 50
Salet, C., 112
Salm, K., 633
Salm, K.-P., 640, 641(78)
Saloman, Y., 686
Salovey, R., 423
Salvayre, R., 438
Salzberg, B. M., 102, 110, 111
Salzman, E. W., 180
Sample, C. E., 421
Samuel, D., 157
Sandermann, H., 632
Sanger, F., 444
Santiago, N. A., 18, 19(6), 24(6), 30
Sanyal, S., 560, 561(36), 564
Saraste, M., 448, 664
Sariban-Sohraby, S., 17
Sartorelli, A. C., 109
Sasaki, H., 521
Satlin, L. M., 671
Sato, F., 15
Šator, V., 653
Sattler, E.-L., 423
Sawyer, H. R., 477
Sawyer, W. H., 477, 482, 493, 509
Saxton, M. J., 509
Sayce, I. G., 256
Scaiano, J. C., 642
Scalera, V., 6, 10(11), 11(6), 352, 353
Scarpa, A., 61, 121, 199, 289, 403
Scazzocchio, C., 444
Schaber, J. S., 437

Schachman, H. K., 706
Schaefer, A., 14
Schaefer, D. W., 380
Schaffner, W., 394, 397(12), 398(12)
Schamber, F. H., 213
Schanze, K. S., 637
Schatz, G., 653
Schatzmann, H. J., 632
Scheefers, H., 640
Schendel, D. J., 109
Scherberich, J. E., 12
Scherman, D., 111
Schiabo, G., 687
Schiffer, M., 439
Schimmel, P. R., 697
Schindler, H. G., 156
Schindler, H., 482, 499, 509, 512
Schlaeger, E. J., 290
Schlatter, E., 162
Schlatz, L. J., 6
Schlegel, W., 415, 417, 421, 428, 429, 438
Schlessinger, J., 472, 478, 481(45), 487, 493, 494(57), 495(45, 57), 496, 499, 504(96), 505(96), 508(12), 512
Schlimme, E., 135
Schlitz, E., 523
Schmid, A., 437
Schmid-Antomarchi, H., 437
Schmidt, A., 363
Schmidt, J. A., 504
Schmitz, E., 643
Schmitz, J., 349
Schneider, G., 474, 476(25)
Schneider, R. F., 107
Schneider, W. C., 7
Schneider, W. J., 415
Schober, M., 437
Schoellmann, G., 493, 507
Schoenborn, B. P., 406
Schölermann, B., 16
Scholts, M. S. G., 78
Schonknecht, G., 82
Schotz, M. C., 438
Schramm, E., 461
Schreiber, C., 640
Schrier, P. H., 444
Schrijen, J. J., 437
Schrock, A. K., 637, 661
Schröder, H., 487
Schron, C. M., 362

Schuldiner, S., 63, 67(4), 81, 82(33), 111
Schulten, K., 496, 499
Schultz, S. G., 147, 308
Schulz, H., 109
Schulz-Dubois, E. O., 366, 367(26), 377
Schurtenberger, P., 384, 386
Schuster, G. B., 637, 642, 661
Schütz, H., 16
Schwab, S. J., 353
Schwartz, A. L., 438
Schwartz, G. J., 54, 57
Schwartz, G., 89
Schwartz, I. L., 6
Schwartz, L. B., 621
Schwartz, R. D., 421
Schwartz, S., 400
Schwarz, E., 457, 460(68), 461(68)
Schwendener, R. A., 384, 386(77)
Schwendimann, B., 109
Schwyzer, R., 686
Scoble, J. E., 309
Scott, I. G., 115
Scott, R. H., 289
Scriven, E. F. V., 635
Scrutton, M. C., 181
Searle, A., 438
Sebald, W., 686
Seeds, M. C., 109
Segal, S., 352, 561
Seidman, L., 17
Seiff, F., 583
Seifter, J., 362
Seigneuret, M., 83
Sekiya, M., 585
Sekiya, T., 664
Seligmann, B. E., 106, 109(18), 114, 121(69)
Seligmann, B., 102, 109
Semenza, G., 349, 474, 476(25), 629, 630, 643, 645(12), 648, 649(12, 119), 651, 654(12), 655(12), 656(119), 657, 660(119), 662, 668(151), 669(128, 151), 670(128)
Semeriva, M., 356
Sen-Majumdar, A., 621
Senior, A. E., 664
Senn, H., 644, 645(110), 646(110), 647(110)
Separovic, F., 386
Serdyuk, I. N., 383
Setlow, R. B., 412
Seufert, W. D., 364, 383

Sevald, W., 444
Severina, I. I., 112, 121
Seversky, M. C., 134
Sexton, M., 29
Shafer, J., 634
Shanzer, A., 157
Shapira, A. H. V., 198
Shapiro, H. M., 109
Sharrow, S. O., 127
Shechter, E., 83
Sheetz, M. P., 499, 502, 512
Sheikh, M. I., 347, 357
Sheikh, M. J., 347, 353, 357(12)
Shen, C. J., 601
Sheridan, J. P., 386
Shertzer, H. G., 41
Shertzor, H. G., 158
Shetzer, H. G., 304
Shiflett, M. A., 686
Shiga, T., 585
Shih, L. B., 644, 645(114)
Shimamoto, T., 519
Shimomura, O., 166, 168, 170(6), 198, 200(69)
Shimomura, S., 166, 198(7)
Shimume, K., 606
Shinitzky, M., 470, 471, 487(2)
Shinoda, K., 532
Ship, S., 126
Shiraishi, N., 107
Shiver, J. W., 461
Shulman, R. G., 84
Shuman, H., 203, 208, 211, 213(11, 13), 216(11), 218(11), 219, 220(13), 221(1), 223, 226, 227
Shustov, G. V., 643
Siegel, L. M., 698
Siegert, A. J. F., 376
Sigel, E., 151, 154
Sigrist, H., 596, 671, 686
Sigrist-Nelson, K., 361, 686
Silcox, J. C., 208
Silva, P., 16
Silverstein, S. C., 54
Simchowitz, L., 120
Simmons, T. J. B., 98
Simon, B. A., 403, 405(9)
Simon, M., 143
Simon, S., 591
Simon, W., 144, 148, 150

Simons, E. R., 114
Simons, K., 4, 15(3), 16(3), 521, 522(29, 30), 696
Simons, R. E., 83
Simons, S. A., 65
Simons, T. J. B., 104
Simonsen, L. O., 238
Simpson, I. A., 50
Simpson, J. A., 412
Simpson, L. L., 111
Sims, P. J., 99, 104, 119(10)
Simsen, M., 441
Singer, I., 147
Singer, S. J., 618
Singh, A., 634
Singleton, W. S., 388
Sistrom, W. R., 668
Sjöström, H., 686
Sköld, S. E., 608
Skolnick, P., 421
Skomorowski, M. A., 547
Skorecki, K. L., 428
Skorecki, K., 418
Skou, J. C., 16
Skrabal, P., 364
Skriver, E., 364, 390(8)
Skulachev, V. P., 63, 65, 67(3), 76, 112, 121
Slatin, S. L., 453, 461
Slatin, S., 461
Slavik, J., 83
Sleight, R. G., 629
Sloan Stanley, G. H., 561
Sluyterman, L. A. A., 170
Small, E. W., 486
Smigel, M., 698, 699(5), 706, 707(5), 708(5), 709(5)
Smith, A. J. H., 444
Smith, B. A., 499, 502, 510
Smith, D. P., 653
Smith, G. A., 83, 126, 240, 246, 249(41), 252, 438
Smith, G. D., 438
Smith, G. L., 196, 256
Smith, J. C., 63, 72(2), 102, 110(3), 112(3)
Smith, L. M., 489, 493
Smith, L., 127
Smith, M., 180, 421, 563, 564
Smith, P. A. S., 635
Smith, R. A. G., 643
Smith, R. J., 587
Smith, R. L., 106, 109(16)
Smith, R. M., 249
Smith, S. W., 231, 244(8), 255(8)
Smith, T. C., 98, 104, 107
Smith, U., 439
Snider, R. M., 180
Snowdowne, K. W., 180, 181, 193(47)
Snozzi, M., 686
Snyder, L. R., 539, 540(1)
Sobel, B. E., 561
Sodetz, J. M., 659
Sogin, D. C., 659
Solomon, N., 504
Solomonson, L. P., 428, 429
Somerville, L. L., 688
Somlyo, A. P., 203, 204, 208, 211, 213(11, 13), 216, 217(4), 218(11), 219, 220(13), 223, 224(25), 226, 227, 289, 290
Somlyo, A. V., 203, 204, 208, 211, 213(11), 216(11), 217(4), 218(11), 219, 221(1), 223, 224(25), 289, 290
Sommer, J. R., 207
Sonenberg, M., 119, 121
Sonnino, S., 556
Sornette, D., 382, 386
Soulina, E. M., 158
Soumpasis, D. M., 504
Sowers, A. E., 495
Spack, E., Jr., 477
Spanier, R., 35, 36(12)
Spector, A. A., 561
Speez, N., 52
Speirs, A., 487, 493
Spiess, M., 629, 630, 645(12), 649(12), 654(12), 655(12), 662
Spilberg, I., 120
Sporn, M. B., 621
Spray, D. C., 254
Spring, T. A., 17
Spudich, J. L., 582
Spurr, A. R., 218
Squire, W., 707
Stackhouse, J., 642
Staden, R., 444
Staehelin, L. A., 207
Staehelin, T., 687
Staerk, H., 476, 493
Standaert, M. L., 438
Standish, M. M., 364
Stange, G., 356

Stanley, K. K., 471, 495(1)
Stark, G. R., 613, 687
Stark, G., 332
Staros, J. V., 585, 588, 611, 618(15), 619, 621, 622(17, 44), 623(17, 44), 625, 626, 627(44, 53), 628, 634, 637(43), 691, 693
Steck, T. L., 610, 614, 615, 625(26), 654, 659, 689
Steckel, E. W., 659
Steele, J. C. H., Jr., 515
Steer, C. J., 419, 438
Steer, M. W., 366
Steffens, G. C. M., 650
Steffens, G. J., 446
Steglich, W., 651
Steiger, B., 362
Stein, J. M., 438
Stein, W. D., 122
Steiner, R. A., 144, 148
Steinfeld, R. C., 134
Steinhardt, R., 233
Steinmetz, P. R., 50, 51(8)
Steitz, T. A., 440
Stelzer, E., 386, 387
Stenlund, B., 521
Stepanowski, A. L., 109
Stephens, J. E., 693
Stephenson, D. G., 179, 197
Stetson, D., 50, 51(8)
Stevenson, R., 539
Stieger, B., 346, 349, 351, 357(29)
Stieger, S., 630
Stier, A., 476, 493
Stock, G., 381, 386(68)
Stoeckenius, W., 577, 580(9)
Stoffel, W., 633, 640, 641(78), 651, 659(81), 687
Stolte, H., 6, 11(13), 12(13), 16(13)
Storelli, C., 349, 353
Storelli, D., 6, 10(11)
Storelli-Joss, C., 6, 10(11), 353
Stoscheck, C. M., 693
Strambini, G. B., 490
Strapazon, E., 615
Strasberg, P. M., 547
Strasser, R. J., 110
Straub, R. W., 356
Streissnig, J., 437
Stricker, R., 499
Strom, T. B., 109
Strominger, J. L., 686
Stroud, R. M., 400, 457
Stryer, L., 449
Stümpel, J., 507, 509, 512
Styrt, B., 180
Suarez, M. D., 418, 428(16)
Suissa, M., 687
Summerhayes, I. C., 112
Sumner, J. B., 702
Sun, S.-T., 386
Sundberg, R. J., 640
Suolinna, E. M., 41
Suolinna, E.-M., 304
Suter, F., 668
Sutherland, I. A., 25
Sutherland, P. J., 179, 197
Suzuki, A., 558
Svedberg, T., 702
Svennerholm, L., 552
Swanson, J. A., 54
Sweeley, C. C., 549, 558
Sweet, R. M., 445
Sweet, T. R., 45
Swillens, S., 418
Szabo, A., 467, 468(12), 472, 474, 486, 487(21), 489(28)
Szarvas, T., 254
Szejda, P., 109
Szende, B., 254
Szoka, F. C., 364
Szoka, F., 126, 364, 508
Szöllösi, J., 476, 487(34), 493
Szwarc, K., 437

T

Takagaki, Y., 628, 642, 651, 667
Takagi, T., 514, 515, 520(17), 521(13), 522, 523(13, 15), 527
Takahashi, M., 427, 512
Takano, M., 352
Takano, T., 111
Takemoto, L. I., 608
Takeuchi, H., 510, 512
Takeyasu, K., 423
Tam, W. W. H., 111
Tamir, H., 94
Tamm, L. K., 506
Tanabe, H., 112

Tanaka, T., 386
Tanford, C., 364, 514, 515(1, 3), 519, 521, 522, 529, 531, 534(1)
Tank, D. W., 510, 512
Tarrab-Hazdai, R., 658
Tasaki, I., 147
Tatham, P. E. R., 109, 110
Tavakalyan, N. B., 643
Taylor, C. C., 99
Taylor, D. L., 83, 180, 186
Taylor, I., 207
Taylor, J. C., 84
Taylor, J. F., 702
Taylor, J. S., 84
Taylor, M. V., 126, 240, 246, 249(41)
Tedeschi, H., 107
Teipel, J. W., 702
Teissie, J., 506
Teng, N. H. H., 649
Tertoolen, L. G. J., 240, 504
Terwilliger, T. C., 457, 460(67)
Tettamanti, G., 556
Tews, K. H., 496, 499
Thayer, W. S., 304
Thees, M., 16, 352
Thierry, J., 361
Thomas, A. P., 241
Thomas, D. D., 478, 499
Thomas, E. W., 463
Thomas, J. A., 85
Thomas, J. W., 421
Thomas, M. V., 166, 193(10)
Thomas, R. C., 136, 145(2), 147
Thomas, S. A., 83
Thompson, D. A., 418, 428(16)
Thompson, D. S., 381
Thompson, J., 17
Thompson, T. E., 364, 386, 519, 654
Thomson, P. C. P., 637
Thorne, C. J. R., 702
Thornton, E. R., 634, 687
Tiddy, G. J. T., 529, 532(6)
Tiede, D., 439
Tiffert, T., 199, 254
Timasheff, S. N., 514
Tinoco, I., Jr., 576, 582
Tipton, K. F., 438
Tisher, C. C., 50, 51(9)
Titus, G., 652
Tocanne, J. F., 506

Tohuku, J., 584
Tokunaga, M., 520
Tolbert, N. E., 393
Tomasi, M., 686
Tomasselli, A. G., 523
Tomita, M., 660, 662
Tomizawa, J., 454, 455(61)
Topali, V. P., 65
Topp, M. R., 289
Tormey, J. M., 15
Tornberg, E., 533
Tornqvist, H., 393, 397(1), 398(1)
Tosteson, D. C., 118
Totsuka, Y., 416, 417(10)
Towbin, H., 687
Townend, R., 514
Tracey, D., 29
Tran, T., 9, 12(19), 13(19), 16
Tran-Dinh, S., 298
Träuble, H., 495
Traut, R. R., 587
Trebst, A., 441, 444, 445(16), 446, 448(25, 32)
Trentham, D. R., 289, 290, 300, 301(18)
Trezl, L., 254
Triggle, D. J., 437
Trón L., 476, 487(34), 493
Tsang, V. C. W., 690
Tschopp, J., 659
Tse, S. S., 352
Tsfasman, I. M., 437
Tsien, R. Y., 83, 87, 99, 104, 107, 110, 114(23), 119(13, 23), 121(23), 143, 144, 148, 149(23), 230, 231, 232, 233, 237(4), 239(3), 241, 242(4, 10), 243(4, 35), 244, 255(4, 6, 8)
Tsirenina, M. L., 687
Tsolas, O., 624
Tsotine, L. M., 65
Tsuchiya, T., 519
Tsunashima, Y., 381
Tsushima, T., 606
Tunggal, B. D., 651
Turcotte, J. G., 565
Turek, P. J., 99
Turel, R. J., 562, 573(44)
Turner, J. J., 642
Turner, J., 319, 323
Turner, R. J., 5, 29, 347, 357(9, 10, 11), 358(10), 359(10), 360(10, 11), 437

Turner, S. T., 347, 357(14)
Turro, N., 635
Twomey, S., 382
Tyihak, E., 254
Tzagoloff, A., 444, 446

U

Uchida, S., 423
Ullman, M. D., 542, 544(5), 545, 547, 550, 551(18), 553(18)
Unwin, P. N. T., 400, 439, 609, 666
Uratani, Y., 453, 458(53)
Urry, D. W., 577
Ushakov, A. N., 687
Utsumi, K., 107
Uturina, I. Y., 440

V

Vail, W., 364
Valet, G., 109
Vallano, M. L., 121
Valle, V. G. R., 78
van Adelsberg, J., 54, 55, 56(17), 87
van Amerongen, A., 686
Van Chu, L., 76
van de Hulst, H. C., 370
Van Deenen, L. L. M., 654
van den Esker, M. W. J., 364
Van der Bosch, H., 574
van der Meer, W., 487
van der Saag, P. T., 87
Van Dyke, K., 114, 168
Van Etten, R. A., 444
van Groningen-Luyben, W. A. H. M., 437
Van Heeswijk, M. P., 352
van Leeuwen, M., 164, 168(1), 169(1)
van Meer, G., 499
Van Olberghen, E., 605, 606
van Os, C. H., 33, 352
van Resandt, R. W., 499
Van Steveninck, J., 584
Van Walraven, H. S., 78
Van Wyk, J. J., 606
van Zoelen, E. J. J., 504
Vanaman, T. C., 170
Vance, R. E., 549

Vanderkooi, J. M., 499
Varenne, S., 454, 455(64)
Varley, J. M., 454, 455(63)
Varzilay, M., 126
Vasilov, R. G., 440
Vaughan, J. M., 366, 367(30)
Vaughan, M., 418
Vaughan, R. J., 634, 635(40, 41), 641
Vaver, V. A., 687
Vaz, W. L. C., 472, 474, 476, 477(26), 487(26, 34), 493, 495(17), 497(15), 507, 508(15, 16, 17), 509, 510, 512
Veatch, W., 512
Venkataramu, S. D., 619, 622(44), 623(44), 625(44), 627(44)
Venter, J. C., 437
Vercesi, A. E., 78
Vergara, J., 141
Verkman, A. S., 418, 428
Verwei, H., 584
Vetéssy, Z., 639
Vicentini, L. M., 231
Vienne, K., 106
Vier, M., 514
Vignais, P. V., 135
Vigneault, N., 429
Vik, S. B., 632
Vladimirov, M. A., 65
von Jena, A., 477
von Mayenburg, K., 664
von Tscharner, V., 506
Vouros, P., 570, 573(60), 574(60)
Vourus, P., 566

W

Wade, C. B., 495
Wade, J. B., 50, 51(2)
Waggoner, A. S., 63, 72, 95, 99, 102, 104, 107, 110, 119(10)
Wagner, H. M., 635
Wagner, R. R., 686
Wahl, P., 465
Wakefield, L. M., 621
Walbert, M. W., 444
Walborg, E. F., 584
Waldvogel, S., 386
Walker, J. E., 664
Walker, J. Q., 539

Walker, J. W., 289, 290, 300(13)
Wallace, B. A., 406, 576, 577, 666
Wallace, N. R., 207
Wallach, D. F. H., 6, 614, 625(26), 689
Wallat, I., 583
Walling, M. W., 19, 29, 353
Walsh, K. A., 170
Walsh, M. L., 111
Walter, R., 50, 51(1)
Walz, B., 216
Walz, D., 317
Wampler, J. C., 167
Wampler, J. E., 167, 187
Wang, C. H., 99, 104, 110, 119(10)
Wang, C. T., 403
Wang, J., 481, 494(48)
Wang, K., 587, 610, 613, 618(14), 625(14), 628, 688, 689, 691
Wank, R., 109
Ward, D. C., 652
Ware, B. R., 499
Ware, J. A., 180
Waring, L., 529, 532(6)
Waring, R. B., 444
Warner, R. R., 203
Warnock, D. G., 13, 18, 24(7), 352
Warren, G. B., 438
Warren, I., 547
Washburn, E. W., 507
Watkins, J. C., 364
Weaver, A. J., 169
Webb, N. G., 400
Webb, W. W., 472, 496, 499, 504(96), 505(96), 506, 508(9), 510, 512
Weber, G., 472
Weckström, K., 531
Weder, H. G., 364, 384, 386(77)
Weder, H.-G., 386
Wegener, W. A., 487, 489
Weinman, E. J., 352
Weinstein, J. N., 127
Weiser, M. M., 356
Weisman, R. B., 300
Weiss, M. J., 643, 645
Weiss, R. M., 457, 460(67), 489, 493
Weissman, G., 106
Weissmann, C., 394, 397(12), 398(12)
Welbaum, B. E., 659
Welch, D. K., 565
Weller, A., 476, 493, 499

Wells, E., 662, 686
Wendt, I. R., 179
Wennerström, H., 495, 529
Wentrup, C., 635
Wersto, R. P., 109
Westerhauser, J., 583
Westheimer, F. H., 634, 635(40, 41), 641, 642
Wey, C. L., 512
Weyer, K. A., 451
Wheeler, K. P., 437
Whitaker, J., 29
White, J. L., 388
White, J., 653, 662(136)
White, L. R., 488, 509
Whiteley, N. M., 616, 618(32)
Whitin, J. C., 114
Whitten, D. G., 637
Widger, W. R., 441, 444, 445(16), 446, 448, 449(41), 461
Wiener, N., 371
Wier, W. G., 165, 166(5), 170(5), 198(5), 199(5), 245
Wiersma, P. H., 364
Wieth, J. O., 121
Wiff, D. R., 379
Wikstrom, M. F. K., 78
Wilkenfeld, C., 106
Wilkinson, M. C., 536
Williams, D. A., 233, 241(19), 245
Williams, J. A., 232
Williams, M. A., 552
Williamson, J. R., 84, 241
Wilong, R. F., 349
Wilson, D. F., 121, 448
Wilson, H. A., 106, 109(18)
Wilson, S. B., 107, 111
Wilson, T. H., 394, 399(13)
Windus, D. W., 352
Winiski, T., 332
Winkle, J. R., 637
Wirtz, K. W. A., 642, 654, 686
Wise, J. G., 664
Wisnieski, B. J., 633, 638, 659, 686
Wist, A. O., 186
Wofsy, L., 618
Wojcieszyn, J., 472, 508, 510
Wolber, P. K., 463
Wold, F., 523, 584, 586, 610
Wolf, D. E., 503

Wolf, F. J., 651
Woll, R., 457, 460(68), 461(68)
Wollheim, C., 244
Wong, M., 386
Wong, S. M. E., 307
Wood, E., 378
Wood, R., 423
Woodbury, D. J., 426
Woods, N. M., 177, 241
Wootton, J. F., 289
Worsham, P. R., 637
Worthington, C. R., 400
Wright, C. T., 444
Wright, E. M., 15, 19, 29, 33, 102, 353
Wu, E. S., 496, 499, 505, 506(119), 507, 510, 512
Wüster, M., 421
Wydila, J., 687
Wynkoop, E. M., 106

Y

Yamada, Y., 454, 455(60)
Yamamoto, T., 415
Yamane, T., 84
Yamazaki, T., 558
Yan, D. L. S., 438
Yandrasitz, J. R., 561
Yarden, Y., 478, 481(45), 487, 493, 494(57), 495(45, 57)
Yeates, T. O., 439
Yedgar, S., 697
Yee, V. J., 13, 18, 24(7), 352
Yeh, Y., 374
Yep, J. M., 584
Yguerabide, E. E., 504
Yguerabide, J., 504
Yip, C. C., 596
Yirinec, B. D., 54
Yoshida, H., 423
Yoshida, T. M., 489
Yoshihara, K., 581
Yoshikami, S., 127

Yoshimura, H., 478, 487
Yoshiya, I., 581
Yost, R. W., 539
Young, A., 45
Young, C. M., 642
Young, I. G., 444
Young, J. D., 420, 422, 437, 659
Yu, R. K., 550

Z

Zaccai, G., 667
Zachariasse, K. A., 499
Zagyansky, Y., 496
Zahler, P., 596, 671
Zahlman, L. S., 686
Zahn, H., 585, 590, 617, 652
Zaki, L., 592
Zakowski, J. J., 686
Zalkin, H., 523
Zampighi, G., 364
Zangvil, M., 126, 127(8), 133(8)
Zanotti, A., 111, 121(55)
Zaritsky, A., 66, 115
Zarling, D. A., 589
Zarnowski, M. J., 50
Zidovetzki, R., 476, 477, 478, 481, 482(48), 487, 493, 494(48, 57), 495
Zierenberg, O., 651
Zigman, S., 107
Zilversmit, D. B., 629
Zimm, B. H., 514
Zimmer, G., 584
Zimniak, A., 83, 85
Zinov'eva, M. E., 437
Zorati, M., 63
Zuber, H., 668
Zugliani, C., 657, 668(151), 669(151), 687
Zühlke, H., 109
Zulauf, M., 529, 531, 532(3)
Zumbuehl, O., 364
Zupancic, J. J., 642
Zurini, M., 686

Subject Index

A

A23187, 246
Acetic acid, for transmembrane pH difference determination, 80
Acetylcholine-R, studied by phosphorescence anisotropy, 492
Acetylcholine receptor, 109
  hydrophobic labeling of, 658
  in-plane order of protein components, 400–401
  target analysis of, 434
  Torpedo californica
    photochemical labeling of, 673, 674, 680
    subunits, photochemical labeling of, 674
  translational diffusion in membranes, representative FRAP results on, 511
Acetylcholinesterase
  from human erythrocyte membrane, photochemical labeling of, 675
  target analysis of, 434
  Torpedo marmorata, photochemical labeling of, 675
N-Acetylhexosaminidase, target analysis of, 435
N-Acetylpsychosine, as internal standard for neutral glycosphingolipid determinations, 549
Acid phosphatase, as membrane marker, 30
ACMA. *See* 9-Amino-6-chloro-2-methoxyacridine
Acridine orange, 81, 91–92
  to estimate relative pH in cellular compartments, 93–95
  wavelength of maximal fluorescence emission, and concentration, 91–92
Acridines, 91–92
Acrylic plastic light guides, 182
Actin, cross-linking studies of, 612
Action potentials
  cardiac, 110
  nerve, 110

study with optical potentiometric indicators, 104
Active transport, 399
Acylazides, photochemistry of, 638–639
Acylnitrenes, 639
Adenosine diphosphate
  caged
    definition of, 289
    HPLC retention times, 295, 297–298
  HPLC retention times, 295, 297–298
Adenosine triphosphate
  caged, 289
    definition of, 289
    HPLC retention times, 295, 297–298
    nuclear magnetic resonance properties of, 297, 298
    synthesis, 293–294
  HPLC retention times, 295, 297–298
  nuclear magnetic resonance properties of, 297, 298
Adenylate cyclase
  catalytic unit, target analysis of, 435
  for characterization of membrane fractions, 15
  as marker enzyme for basolateral membrane, 347
  response to gonadotropins, uncoupling of, photochemical labeling of, 673
  target inactivation analysis, 428
Adenylate kinase, yeast, calibration of low-angle laser light-scattering photometry plus HPGC system with, 523
ADP. *See* Adenosine diphosphate
Adrenocorticotropin peptides, interactions of, with liposomes, photochemical labeling of, 675
Adsorption chromatography, 539
Aequorin, 165–166, 230–231
  assay of, 168
  calcium concentration–effect curve for, 190, 198
    determination of, 189
    influence of temperature on, 192
  concentration, for microinjections, 177–178

desalting, 170–171
frog muscle cells injected with, method used to estimate fractional luminescence, 200–202
heart muscle injected with
setup for recording light from, 183–185
signal averaging for, 185–186
injection pipettes for, 175
isospecies, 166
loading, into cultured cells, 180
lyophilization, 170–171
rate of luminescent reaction, 198
source, 169–170
L-Alanine transporter, target analysis of, 433
Albumin, bovine serum
calibration of low-angle laser light-scattering photometry plus HPGC system with, 523
and quin2/AM loading, 238
in sucrose gradient centrifugation of protein–detergent complexes, sedimentation properties of, 702
Aldolase
rabbit muscle, cross-linking studies of, 623–625
in sucrose gradient centrifugation of protein–detergent complexes, sedimentation properties of, 702
Alkaline phosphatase
markers for apical (luminal) plasma membranes, 15
as membrane marker, 15, 30, 31, 347–348
in outer cortical and outer medullary brush border vesicle preparations, 358
target analysis of, 435
Alkylazides, intermolecular reactions of, 640–641
Alkyl imidates, 616, 618–619
Amido Black 10B protein assay, 394–399
Amines
distribution, estimation of, 80–82
fluorescent, 81
determination of transmembrane pH difference, 81–82
labeled with spin probes, 82
Amino acid transport, 32–33
pathways, definition of, 33–34

L-Amino acid transporter, target analysis of, 433
9-Aminoacridine, 81, 92
9-Amino-6-chloro-2-methoxyacridine, 81
N-(2-Aminoethyl sulfonate)-7-nitrobenz-2-oxa-3-diazole, as fluorescent substrate of anion transport system, 126
Aminonaphthalenetrisulfonic acid, as impermeant substrate coupler, 126
Aminopeptidase M
as marker enzyme for brush border membrane, 347–348
separation of membranes containing, from bulk of basolateral membranes, 355
Aminopeptidases, markers for apical (luminal) plasma membranes, 15
Analog computer network
A/D interfaces, for delivery of analog signals to microcomputers, 268, 270
to count signals from particle-size analyzer, 274, 275
Analog computer program
for modelling fluxes across three-compartment system
amplifier outputs, 264, 266
potentiometer settings for, 264, 266
of parallel model of three-compartment closed system, 268–269
of series model of three-compartment closed system, 265
Analytical ultracentrifugation, of vesicles, 365
Anaphylatoxin Csa, association with chymase, on surface of mast cells, cross-linking studies of, 621
Anilinonaphthalene sulfonate, 73
fluorescence, calibration
by external diffusion potential, 76–77
by Rb^+ distribution, 76–77
for measurement of transmembrane electrical potential difference, 76
1-Anilino-8-naphthalene sulfonate, complex with valinomycin in presence of potassium, 118
ANS. See Anilinonaphthalene sulfonate
Anthracene derivatives, as cross-linking reagents, 602
Antibodies, in isolating plasma membranes, 13
Antigenic markers, 17

SUBJECT INDEX

Antipyrylazo III, 231
ANTS. See Aminonaphthalenetrisulfonic acid
Apical membrane markers, 15, 31
Apo-C III, translational diffusion in membranes, representative FRAP results on, 511
Aqueous two-phase systems, 24–25
Argentation chromatography, 539
Arsenazo III, 230–231
Artificial kidney, 45–46
  care of, 46–47
  loading, 46–47
Arylazides, photochemistry of, 637–638
3-Aryl-3*H*-diazirine, photolysis of, 644
AS701, 157
ATP. See Adenosine triphosphate
ATP-driven pumps, target analysis of, 430
ATPαS
  nuclear magnetic resonance properties of, 297, 298
  O-caged, nuclear magnetic resonance properties of, 297, 298
  S-caged, nuclear magnetic resonance properties of, 297, 298
ATPβ,γNH
  caged, 289
    definition of, 289
    HPLC retention times, 295, 297–298
    nuclear magnetic resonance properties of, 297, 298
    steady-state photolysis of, 299
  HPLC retention times, 295, 297–298
  nuclear magnetic resonance properties of, 297, 298
ATPγS
  HPLC retention times, 295, 297–298
  O-caged
    definition of, 289
    HPLC retention times, 295, 297–298
  S-caged
    definition of, 289
    HPLC retention times, 295, 297–298
ATP synthase
  from *E. coli*
    proton-translocating portion F_0 of, photochemical labeling of, 663–666, 673, 675
    subunit b, photochemical labeling of, 680
  from *N. crassa*, proton-translocating portion F_0 of, photochemical labeling of, 675
ATP synthetase
  from bovine heart mitochondria, photochemical labeling of, 680
  from *E. coli*, photochemical labeling of, 680
Atrial natriuretic factor receptor, in adrenal membranes, cross-linking studies of, 621
Avidin, cross-linking studies of, 603
Azides, UV irradiation of, products of, 636–637
p-Azidobenzoic acid imido ester, 603
3-(4-Azido-2-nitrobenzoyloylseleno)propionic acid
  cleavability, 595
  specificity, 595
  structure, 595
4-Azidophenacyl bromide
  cleavability, 594
  specificity, 594
  structure, 594
p-Azidophenylacetic acid imido ester, 603
p-Azidophenylacetimidate
  cleavability, 597
  specificity, 597
  structure, 597
4-Azidophenylglyoxal
  cleavability, 594
  specificity, 594
  structure, 594
4-Azidophenyl isothiocyanate
  cleavability, 596
  specificity, 596
  structure, 596
Azo dyes, as cross-linking reagents, 602, 604

B

Bacterial membrane potential, study with optical potentiometric indicators, 103, 111
Bacterial membrane vesicles, optical measurement of membrane potential in, 111
Bacteriorhodopsin. See also DMPC/bacteriorhodopsin vesicles

circular dichroism study of, 583–584
cross-linking studies of, 609
hydropathy function of, cross-correlation between cytochrome b_6 and, 442
hydropathy plot, 441, 443
hydrophobic domains, 442–444
labeling of, with [^{125}I]-TID, 657–658
photochemical labeling of, 666–668, 675
secondary structure, circular dichroism study, 577–578
state of aggregation, circular dichroism study, 579–581
structure, 439
Band 3, photochemical labeling of, 672
Band 3 anion transporter
continuous monitoring of transport by fluorescence, 124–125, 134–135
covalent labeling of, 482
cross-linking studies of, 615
with DTSSP, in intact erythrocytes, 625–627
studied by delayed fluorescence anisotropy, 492
studied by fluorescence recovery anisotropy, 492
studied by phosphorescence anisotropy, 492
translational diffusion in membranes, representative FRAP results on, 511
Band 3 anion transporter–Triton X-100 complex
partial specific volume for, experimental determination of, 700–704
sedimentation coefficient for, experimental determination of, 700–704
Band 3 protein, 659–660
Basolateral membrane
and brush border membranes, separation of, 347–348
contamination, with brush border membranes, 352–355
contamination of brush border membranes, 349
marker enzymes for, 347–352
from rat kidney cortex, Na^+,K^+-ATPase, Na^+/Ca^{2+} exchange, and Ca^{2+}-ATPase in, 363
from rat renal cortex, Ca^{2+} transport in, 363
from rat small intestinal epithelium, Ca^{2+} transport in, 363

separation of membranes containing Na^+,K^+-ATPase from bulk of, 354–355
separation of membranes containing Na^+-P_i cotransport from bulk of, 354–355
Basolateral membrane vesicles
potassium channel in, stopped-flow experiment with, 308–311
transport studies, 346–363
BCECF, 83
to measure cell pH, 87
Behavior, study with optical potentiometric indicators, 104
Benzoic acid, for transmembrane pH difference determination, 80
Benzoic anhydride, 542–544
Benzoyl chloride, 542–544
storage, 544
Berberines, 79
Bialkali cathodes, 182
Bilayers
adsorbed onto glass or silicon wafers, 506
isolated, on grids, preparation of, 506
Bilayer spanning domains, prediction of, 439–461
Binding assay, target inactivation analysis, 425
Bioenergetics, 110
study with optical potentiometric indicators, 103
Bio-Rad protein assay, 525
Biscarboxyethyl-5(6)-carboxyfluorescein. See BCECF
1,3-Bis(2-chloroethyl)-1-nitrosourea
cleavability, 594
specificity, 594
structure, 594
Bis-N-(2-nitro-4-azidophenyl)cystamine S,S-dioxide
cleavability, 601
specificity, 601
structure, 601
Bisoxonol, 110
complex with valinomycin in presence of potassium, 119
Bispyridoxal phosphate, as cross-linking reagent, 606–608
Bis[2-(succinimidoxycarbonyloxy)ethyl]-sulfone

cleavability, 589
specificity, 589
structure, 589
Bis(sulfosuccinimidyl)adipate, applications of, 621
Bis(sulfosuccinimidyl)-4-doxylpimelate, applications of, 621
Boundary potential
  changes, measuring, 341
  external, 332–333
  internal, 332–333
Boundary regions, 332–333
  and bulk aqueous phase, potential difference between, 332, 336–342
BPP. See Bispyridoxal phosphate
Bremsstrahlung, 209–210
N-Bromoacetyl-N-methylaminodiethoxyethanemethylaminomethyltrimethylpsoralene
  cleavability, 600
  specificity, 600
  structure, 600
2-Bromoethylmaleimide
  cleavability, 591
  specificity, 591
  structure, 591
α-Bromoketones, 604
3-(4-Bromo-3-oxobutane-1-sulfonyl)propionic-N-hydroxysuccinimide ester
  cleavability, 595
  specificity, 595
  structure, 595
Brush border membrane
  aminopeptidase M as marker enzyme for, 347–348
  and basolateral membranes, separation of, 347–348
  contamination, with basolateral membranes, 349
  disaccharidase as marker enzyme for, 348
  γ-glutamyltransferase as marker enzyme for, 347–348
  rat ileal, separate Na^+/H^+ and K^+/H^+ exchange in, 363
  small intestinal
    photochemical labeling of, 672
    transport studies in, 346–363
Brush border membrane vesicles
  isolated from renal cortex, D-glucose transport into, 357–360

marker enzymes
  enrichment factors of, 350–351
  specific activities, 350–351
  outer cortical and outer medullary
    alkaline phosphatase in, 358
    maltase in, 358
    Na^+,K^+-ATPase in, 358
  rabbit ileal, coupled electroneutral NaCl absorption in, 362
  sodium-dependent D-glucose transport in, 356
BS^3
  applications of, 621
  cross-linking of rabbit muscle aldolase, 623–625
  development of, 619–620
  structure of, 620
Bufo marinus. See Toad

C

Ca^{2+}. See Calcium
Ca^{2+}-ATPase
  from human erythrocyte, photochemical labeling of, 675
  photochemical labeling of, 672
  studied by phosphorescence anisotropy, 492
  target analysis of, 430
  translational diffusion in membranes, representative FRAP results on, 511
Ca-EGTA buffers, 256–262
Caged compounds, 288–289
Caged nucleotides, 288–301
  application, technical aspects of, 300–301
  characterization, 290, 297–300
  concentration, 298
  nuclear magnetic resonance properties of, 297, 298
  photochemical properties, 298–299
  photolysis, 290–291, 300–301
    on illumination, 298–299
  pulse photolysis, release of nucleotide on, 299–300
  purification, 292–293
  purity of, 297
  structure, 298
  in studies of biological systems, 300–301
  synthesis, 290–297

general considerations, 292–293
UV spectrum, 298
Calcein, leakage from unilamellar vesicles during fusion, 126
Calcium
  cytosolic free, measurement of, with quin2, 230–262
  intracellular, indicators of, 166–167
  intracellular recording of, with microelectrode, 148–150
  membrane permeability and, 12
Calcium buffers, 192–197
  buffering function, 194–195
  calibrating function, 195–196
  effectiveness, factors affecting, 194–196
  solutions, 196–197
Calcium channel, target analysis of, 433
Calcium concentration–effect curves, for photoproteins, determination of, 188–197
Calcium pump
  from mammalian brain synaptosomes, purification of, 35–49
  reconstituted, purification of, 36
  synaptosomal, transport-specific fractionation of, 37
  target analysis of, 429–438
Calcium pump protein, target analysis of, 430
Calcium-regulated photoproteins, use of, 164–203
Calcium transport, 32–33
  assay, by ion-exchange chromatography, 41–45
Calcium transport proteins, transport-specific fractionation of, 35–49
CALCULATING LP, 276–280
Calmodulin, photochemical labeling of, 675
Calquin 2, for measuring intracellular Ca^{2+}, 126
Carbenes, 634, 641–647
  chemistry of, 635–637
  precursors, 641–647
  properties of, 635–637
  singlet and triplet states, 636–637
Carbocyanine dyes
  to assay membrane potential of mouse ascites tumor cells, 95–101
  for monitoring changes in transmembrane electrical potential difference, 76–78
  null point experiments with, using potassium ion membrane potential and valinomycin, 96, 97, 101
Carbonyl cyanide p-trifluoromethoxyphenylhydrazone, complex with valinomycin in presence of potassium, 118–119
Carboxyfluorescein, 83
  leakage of, assessment of, 127
6-Carboxyfluorescein, excitation spectra, 86
Ca^{2+}-selective electrode, 150
Catalase, in sucrose gradient centrifugation of protein–detergent complexes, sedimentation properties of, 702
Cell disruption
  by cavitation, 6–7
  methods of, 5–7
  shearing forces, 5–6
Cell organelles, 110
  in living cells, pH measurement, with fluorescent dyes, 85–95
  membrane potential, optical measurement of, 102–122
  preparations, 3
  proton electrochemical potential gradient in, 63
Cell rupture, extent of, 8
Cells
  anchorage-dependent
    coverslips for, 240
    microcarrier beads, 240
  continuous monitoring of transport by fluorescence assay, 129–130
  membrane potential, optical measurement of, 102–122
  pH gradients in, measurement of, 91–95
  single
    measurement of cell pH in, 89
    study of transport in, 134
Cell surface labeling, 507–508
Cell suspensions
  dense
    autoluminescence, 481
    delayed luminescence anisotropy, 481–482
    measurement of cell pH in, 87–89

Cell volume regulation, study with optical potentiometric indicators, 103
Ceramides, molecular species, separation and quantitation by reversed-phase HPLC, 556–558
Cerebroside
  benzoylated derivatives, formation, by reaction with benzoyl chloride, 542–543
  per-O,N-benzoylated, reversed-phase chromatography, 557–558
Chelating resins, 169
Chelator, apparent association constant for, at particular pH, 194
Chelex, 191–192
Cholera toxin
  insertion into membranes, photochemical labeling of, 667
  lipid insertion of, photochemical labeling of, 680
  photochemical labeling of, 676
Cholesterol
  containing photoactivatable group, applications of, 685
  cross-linking studies of, using photochemical reagents, 682
Cholesterol esters, containing photoactivatable group, applications of, 685
Chromaffin granules, 111
  dye calibration methods with, using pH, 121
Chromatophores, *Rhodospirillum rubrum*, photochemical labeling of, 673, 683
Chromophore–chromophore interactions, circular dichroism study, 579–581
Circadian rhythms, study with optical potentiometric indicators, 103
Circular dichroism
  exciton coupling bands in, 580–581
  study of membrane proteins, 575–584
Colicins
  amino acid sequences of, 454–456
  amphipathic segments, hydrophobic moments for, 458–461
  channel-forming, 453–461
  lethal unit of, 453
  channel-forming COOH-terminal peptides, 453–454
    average cross-correlation coefficients of hydropathy plots of, 454, 456

  candidate regions for amphipathic transmembrane peptides in, 457–458
  hydrophobic region, 454–456
  insertion, pH conditions for, 454
  lethal activity, 453
  transmembrane regions
    hydrophobic moments of, 458–461
    separated by hairpin turn or bend, 454–456
Colloidal detergent solutions, critical phase transitions of, 532
Complement
  human, ninth component of, photochemical labeling of, 676
  membrane attack complex of, photochemical labeling of, 675
Complement proteins
  hydrophobic labeling of, 659
  interaction of, with membranes, photochemical labeling of, 681
  photochemical labeling of, 673
Computers
  analyzing transport kinetics with, 262–288
  desk-top hybrid, rapid calculation of permeability coefficient to water, 274–280
  model for anisotropy decay, 465
  in slow-scan X-ray mapping, 227–229
Con A receptor
  studied by fluorescence recovery anisotropy, 492
  studied by phosphorescence anisotropy, 492
  translational diffusion in membranes, representative FRAP results on, 511
Constant field equation, 119
Constrained equilibrium, 319–320
CONTIN program, 327–331, 383, 387–388, 390
Continuous fluorescence microphotolysis, in study of translational diffusion, 498, 502
Continuous monitoring of transport by fluorescence, 123–135
  applications, 134–135
  data acquisition, 130–131
  data analysis, 131–134

Michaelian mode: ($S_{08}K_m$), 131–134
trace mode: ($S_{08}K_m$), 131–134
experimental methods, 127–130
mirror image approach to, 126
permeant substrate analogs for impermeant quencher couples in, 125–127
setting of spectrofluorometer for, 130
source of material, 127–129
Contraction motility, study with optical potentiometric indicators, 104
Countercurrent distribution, 30–31
with aqueous two-phase system, 24–25
fraction recovery, 28–29
membrane concentration, 28–29
Countertransport, 317–319
Coupled transport, 317–319
stoichiometry of, 319–320
study with optical potentiometric indicators, 103
Critical concentration of micellization, 529–530
Critical micelle concentration
of detergent, 528–538
determination of, strategy, 532–534
qualitative determination, 534, 535
quantitative assessments, 534, 535–536
Critical micelle temperature, 531
Critical temperature of crystallization, 531
Critical temperature of crystallization boundary, 531
Critical temperature of demixing, 531
Cross-linking reagents
bifunctional, 584–585
uses of, 585
cleavability, 617
cleavable, 604
distance spanned by, 617
flexible, 617
hydrophobicity, 585
length, 585
membrane-impermeant, 609–628
design, 615–621
development of, 615–621
reaction conditions, testing, 623–625
testing, 623–625
for proteins and complexes of proteins with other naturally occurring compounds, 586–601
topographical investigations with, 608–609
two-step, 603
Cross-linking techniques, 584–609, 611–615
hormone–receptor coupling, 605–606
intrasubunit, 614–615
principles, 611–615
reaction used for, 611–612
sterically enforced reactions, 605
for study of protein quaternary structure, protein concentration for, 612–614
Cryoultramicrotomy, 207–209
Cu^{2+}/o-phenanthroline complex, oxidative reactions with, 584, 610
Cyanine dyes, 105, 107
chemical nomenclature, 106
measurement of plasma membrane potentials, influence of mitochondrial potentials on, 105–106
Cyanobacteria, 112
Cytochrome b
A. nidulans
cross-correlation between cytochrome b_6 and, 442, 445–446
hydropathy function, cross-correlation between cytochrome b_6 and, 442
bovine
cross-correlation between cytochrome b_6 and, 442, 445–446
hydropathy function, cross-correlation between cytochrome b_6 and, 442
chloroplast, proposed membrane-folding pattern of, 446–448
correlation of hydropathy functions, 442, 444–449
human
cross-correlation between cytochrome b_6 and, 442, 445–446
hydropathy function, cross-correlation between cytochrome b_6 and, 442
hydropathy function, for prediction of folding pattern of heme-binding domain, 446
hydrophobic domains, 442–444
maize
cross-correlation between cytochrome b_6 and, 442, 445–446
hydropathy function, cross-correlation between cytochrome b_6 and, 442
hydropathy plot, 441, 443

mouse
  cross-correlation between cytochrome b_6 and, 442, 445–446
  hydropathy function, cross-correlation between cytochrome b_6 and, 442
photosynthetic bacterium
  cross-correlation between cytochrome b_6 and, 442, 445–446
  hydropathy function, cross-correlation between cytochrome b_6 and, 442
trypanosome, hydropathy function, cross-correlation between cytochrome b_6 and, 442
yeast
  cross-correlation between cytochrome b_6 and, 442, 445–446
  hydropathy function, cross-correlation between cytochrome b_6 and, 442
Cytochrome b_5
  calf liver, photochemical labeling of, 680
  membrane-embedded segment of, photochemical labeling of, 682
  membranous segment of, photochemical labeling of, 682
Cytochrome b_6
  chloroplast, hydropathy plot, 441, 443
    predictive value of, 442
  cross-linking membrane-spanning peptides II and V, transmembrane arrangement of two hemes of, 448–449
  membrane-spanning peptides, turn regions in polar phases linking, 451
Cytochrome b_6–f and b–c_1 complexes, transmembrane electron transport in, structural framework for, 448
Cytochrome c
  in sucrose gradient centrifugation of protein–detergent complexes, sedimentation properties of, 702
  translational diffusion in membranes, representative FRAP results on, 511
Cytochrome f
  hydropathy plot, effect of averaging window on, 448–451
  spinach, hydropathy function of, cross-correlation between cytochrome b_6 and, 442

Cytochrome oxidase
  reconstituted, determination of H$^+$ and charge stoichiometry of, 154–155
  target inactivation analysis, 428, 431
  yeast, photochemical labeling of, 672, 673
Cytochrome-c oxidase, beef heart
  subunit II, photochemical labeling of, 680
  subunits, photochemical labeling of, 680
Cytochrome oxidase II
  bovine, hydropathy function of, cross-correlation between cytochrome b_6 and, 442
  yeast, hydropathy function of, cross-correlation between cytochrome b_6 and, 442
Cytochrome P-450, studied by delayed fluorescence anisotropy, 492

D

DDDD
  cleavability, 588
  specificity, 588
  structure, 588
Delayed fluorescence, 473–488
Delayed luminescence anisotropy, 475–488
  data processing and analysis, 482–488
  electronic gating of photomultiplier, 476, 479
  emission filters, selection of, 479–480
  of fluorescent uranyl glass GG17, 490–491
  instrumentation for, 477–481
  membrane and cellular systems studied by, 491–495
  probes, 476–477
  sample preparation for, 481–482
  temporal sequence of excitation and emission, 473, 476, 479, 480
Density gradient centrifugation, 9–11, 20–23
  media for, 10
  peak on gradient after equilibrium, 29–30
  separation of membranes from gradient media, 22–23
  two-stage, 48–49
  types of, 11

Detergents. *See also* Membrane protein–detergent complexes
  as amphiphilic, flexible molecules, 528–529
  monomers, calculation of area covered by, 538
Detergent solubilization, 16
  of membrane proteins, 696
  with nonionic surfactant, 519–520
Detergent solutions, critical phenomena associated with, 530–532
Dextran, fluorescent, exocytosis, 56–57
Dextran solutions, 13
Di4-ANEPPS, 106, 111
Diazirines, 634
  photochemistry of, 643–647
  for photolabeling of membrane lipid core, 643–647
  properties of, 643–644
  UV irradiation of, products of, 636–637
Diazo compounds, 634
  photochemistry of, 641–643
  UV irradiation of, products of, 636–637
Diazocyclopentadiene, structure of, 642
Diazoethanes, caution for, 292
9-Diazofluorene, structure of, 642
Diazoindene, structure of, 642
DiBa-C_4(3), 106, 110
  structure, 105
2,5-Dibromomethylbenzenesulfonic acid
  cleavability, 592
  specificity, 592
  structure, 592
2,2′-Dicarboxy-4,4′-diisothiocyanatoazobenzene
  cleavability, 591
  specificity, 591
  structure, 591
2,2′-Dicarboxy-4,4′-iodoacetamidoazobenzene
  cleavability, 592
  specificity, 592
  structure, 592
Dichroism, 473. *See also* Circular dichroism
DIDIT
  development of, 618–619
  structure of, 618
DIDS
  cleavability, 592
  specificity, 592
  structure, 592
Diepoxybutane
  cleavability, 587
  specificity, 587
  structure, 587
Differential centrifugation, 9
Differential precipitation, 11–12
Differential rate sedimentation, separation of microsomes and mitochondria by, 23
Different psoralenes
  cleavability, 601
  specificity, 601
Diffraction gratings, 374
Diffusion potentials, study with optical potentiometric indicators, 103
Difluorodinitrobenzene
  cleavability, 590
  specificity, 590
  structure, 590
Digitonin, density perturbation with, 30–31
DiI-C_1(5), 106, 109
DiI-C_3(5), 106, 109
  structure, 105
Diimidates, as cross-linking reagents, 602
Diimido esters, as cross-linking reagents, 603
Diiodofluorescein, 476
Diisothiocyanates, as cross-linking reagents, 602
Dimethyl diimidates
  cleavability, 586
  specificity, 586
  structure, 586
Dimethyl-3,3′-dithiobispropionimidate, 610, 613
  cleavability, 587
  specificity, 587
  structure, 587
Dimyristoylphosphatidylcholine. *See* DMPC
Dinitrophenyl acetate, alkaline hydrolysis of, measurement of, using flow–quench, 312
DiO-C_5(3), 106, 109
DiO-C_6(3), 106, 108–110
  structure, 105
Dipalmitoylphosphatidylcholine, vesicles, 386
1,6-Diphenyl-1,3,5-hexatriene. *See* DPH

Diphtheria toxin, photochemical labeling of, 676
3,3'-Dipropyloxadicarbocyanine, 95
  fluorimetry, 100
3,3'-Dipropyloxadicarbocyanine iodide
  stability, 99
  storage, 99
3,3'-Dipropylthiadicarbocyanine, 95
  stability, 99
  storage, 99
Dirac delta function, 370
Disaccharidase
  as marker enzyme for apical (luminal) plasma membranes, 15
  as marker enzyme for brush border membrane, 348
DiSBa-C_2(3), 106, 110
  structure, 105
DiS-C_3(5), 106
  fluorescence
    calibration, 77-78
    with membrane potential of human red blood cells altered by valinomycin, 113-115
  optical drift, artifactual, 113
  pharmacological effects, 107-108
  with red cells, 107
  structure, 105
  to study ion permeability and ion gradient dissipation, 126-127
DISCRETE, 328-329
Disuccinimidyl suberate
  cleavability, 588
  specificity, 588
  structure, 588
Disuccinimidyl tartrate
  cleavability, 587
  specificity, 587
  structure, 587
Disulfides, as cross-linking reagents, 604
2,2'-Disulfonyl-4,4'-diazidoazobenzene
  cleavability, 601
  specificity, 601
  structure, 601
Dithiobis(succinimidyl propionate)
  cleavability, 588
  specificity, 588
  structure, 588
4,4'-Dithiobis(sulfosuccinimidyl)butyrate, applications of, 621

3,3'-Dithiobis(sulfosuccinimidyl propionate)
  cleavability, 588
  specificity, 588
  structure, 588
Dithiothreitol, 8
DMO, for transmembrane pH difference determination, 80
DMPC/bacteriorhodopsin vesicles
  fluorescence anisotropy data, analysis of, 467
  viscosity of, 469-470
DMPC vesicles, 386
  pure, fluorescence anisotropy data for, 469
DPH
  phospholipid analogs of, 463-464, 470
  surface-anchored analogs, 463-464
DPPC. See Dipalmitoylphosphatidylcholine
Drop volume, measurement of, 536-537
DTBP, 618
  structure of, 618
DTSSP
  applications of, 621
  cross-linking of band 3 anion transporter in intact erythrocytes, 625-627
  development of, 619-620
  structure of, 620
  synthesis of, 622-623
Duysens absorption flattening, 576
Dyes
  calibration
    by measurement of dye binding, 119
    for red cells, 120-121
    using radioactively labeled lipid-permeable cations, 121
  sources, 112
  used in electrophysiological studies, 104-112
Dynamic laser light scattering, to determine size distributions of vesicles, 364-390
Dynorphin, interactions of, with liposomes, photochemical labeling of, 675

E

EGF-receptor
  anisotropy decays for, 493-494
  covalently cross-linked complexes con-

taining, electrophoretic transfer of, for immunostaining, 691–695
electrophoretic transfer of, for immunostaining, 691–695
studied by phosphorescence anisotropy, 492
translational diffusion in membranes, representative FRAP results on, 511
Egg lecithin/myelin membrane vesicles, photochemical labeling of, 672
EGTA
  commercial, purity, 196–197, 256
  dissociation constant
    for Ca^{2+}, 258, 260
    for Mg^{2+}, 258, 260
  pH-dependence, checking, 262
EGTA buffers, for cellular Ca^{2+}, 256–262
Ehrlich ascites tumor cells, 107, 108
  optical measurement of membrane potential in, 111
Electrogenicity, study with optical potentiometric indicators, 103
Electron microscopy, 383
  of vesicles, 365
Electron paramagnetic resonance, 331–345
Electron probe X-ray microanalysis, 203–229
  background X-rays, 209–211
  biological applications, 223–225
  contamination, 223
  count statistics, 212–219
  elemental standards, 212–219
  independent validations of, 220–221
  least-squares fit of experimental spectrum to reference files, 213–219
  mass loss in, 221–223
  principles, 209
  radiation damage, 221–223
  reference file, 213
  sensitivity, 219
  spatial resolution, 219
  spectral processing, 212–219
  of ultrathin biological specimens, 211
  X-ray detectors, 209–212
  X-ray mapping, 225–229
  X-ray photon creation, 209–212
Electron spin resonance, for direct observation of micellar formation, 533
Electron transport machinery, target analysis of, 431

Electrophoretic transfer of high-molecular-weight proteins, for immunostaining, 687–696
  applications of, 691–695
  high-molecular-weight proteins, 690–691, 691–695
  materials, 688
  methods, 688
  reagents, 688
  very-high-molecular-weight proteins, 688–690, 691, 695
Electrophysiological methods, to measure time course of fusion and transport, 53
Endocytosis
  inhibitors, 54
  morphometric analysis, 52
  rate of, effect on rate of transport, 54
  removal of transport proteins, 53–54
Endosomes, in living cells, pH measurement, with fluorescent dyes, 85–95
Energy transduction processes, 399–400
Enkephalins, interactions of, with liposomes, photochemical labeling of, 675
Enolase, yeast, calibration of low-angle laser light-scattering photometry plus HPGC system with, 523
Enrichment factors
  calculation of, 31–32
  definition of, 4
Enzymatic activity, target inactivation analysis, 425
Enzyme recoveries, 4
Enzymes, in particle-free supernatant, 8
Eosin
  delayed fluorescence, 476
  triplet lifetimes, 476
Epichlorohydrin
  cleavability, 589
  specificity, 589
  structure, 589
Epidermal growth factor. See EGF
Epithelial subcellular fractionation, 18
Epithelial transport mechanisms, characterization, 32
5,6-Epoxyretinal, enantiomers, circular dichroism study of, 581
Equilibrium exchange, 323–325
Erythrocyte ghosts, continuous monitoring of transport by fluorescence assay, 129–130

Erythrocytes, 107, 110. *See also* Band 3 anion transporter
  anion channel, cross-linking studies of, 621
  continuous monitoring of transport by fluorescence assay, 124–125, 128–129, 134–135
  diS-C_3(5) with, 107
  dye calibration in
    method for, 120–121
    triphenylmethylphosphonium distribution for, 121–122
  elemental concentrations, determined by EPMA and by other methods, 220
  evaluating new dyes with, 116–119
  fluorescence histograms, determined by flow cytometry, 109–110
  human
    membranes
      composition of, 659–660
      labeling of lipid core of, 659–663
        covalent incorporation of label in lipids and proteins, 660–661
        distribution of label among integral proteins, 662–663
        distribution of label among proteins, 661–662
        photochemical labeling of, 672, 674, 676, 683
      optical measurement of membrane potential in, 111
      surface glycoproteins, covalent labeling of, 482
    intracellular pH, and alterations in dye binding, 96
    membrane proteins, cross-linking studies of, 610
    mouse, optical measurement of membrane potential in, 109
    study with optical potentiometric indicators, 103
Erythrosin, 474
  delayed fluorescence, 476
  triplet lifetimes, 476
Escherichia coli, 110
  biosynthetic incorporation of fatty acids containing photoactivatable groups, 677, 678
  membranes, labeling with fatty acids containing photoactivatable groups, 632
Ethidium bromide diazide
  cleavability, 600
  specificity, 600
  structure, 600
Ethylene glycol bis[3-(2-ketobutyraldehyde) ether], 604
  cleavability, 593
  specificity, 593
  structure, 593
Ethylene glycol bis(succinimidyl succinate)
  cleavability, 589
  specificity, 589
  structure, 589
Excimer and exciplex formation, in study of translational diffusion, 498
Excitation–contraction coupling, in skeletal muscle, 110
Excited state quenching through collisional or resonance energy transfer, in study of translational diffusion, 498
Exocytosis, 52–53
  assays for, 51
  of fluorescent dextran, 56–57
  morphometric analysis, 52
  regulated, 50
  single-cell assay for, 57–59

F

F-actin, cross-linking studies of, 602
Fatty acids, containing photoactivatable group
  applications of, 676–679
  synthesis of, 632
Ferritin, cross-linking studies of, 603
Fetal calf serum, and quin2/AM loading, 238
Fetal death, 64
Fiber-optic probes, 182
Field autocorrelation function, 372–373
Field correlation function, 375–376
FITC-albumin
  fluorescence emission intensity, and concentration, 90
  in measurement of pH of intracellular compartments, 89–91

FITC-dextran, 83
 fluorescence emission intensity, and concentration, 90
 in measurement of pH of intracellular compartments, 89–91
Flow cytometry, dyes used in, 109–110
Flow dialysis
 for determination of lipophilic ion distribution, 67–69
 in measurement of distribution of radiolabeled acids or amines, 81
Flow–quench, 303–305
 definition of, 303–304
 identification of flux, 307–308
 typical experiment, 304–308
Fluorescein, 83
 derivatives, to measure intracellular pH, 86–87
 excitation spectrum, pH-sensitivity, 86
 to measure cell pH of suspended or single cells, 85–89
Fluorescence. *See also* Delayed fluorescence; Polarized fluorescence
 monitoring of transport by, 122–135
 general principles, 123–127
Fluorescence-activated cell sorter, 93–95
Fluorescence anisotropy, 473. *See also* Transient fluorescence anisotropy
Fluorescence correlation spectroscopy, 475
 in study of translational diffusion, 499
Fluorescence depletion, 488
Fluorescence-detected circular dichroism, 576–578
Fluorescence microphotolysis, 496
 of single cells, 134–135
 in study of translational diffusion, 498
Fluorescence photoactivation and dissipation, in study of translational diffusion, 498
Fluorescence photobleaching recovery, 475, 489, 496
 in study of translational diffusion, 498
Fluorescence properties, compartment-independent variations in, 123–124
Fluorescence recovery after photobleaching, 496
 data analysis, 503–505
 experiment
  performance of, 502–503
  principle of, 496–497
 experimental results, comparisons with theoretical expectations for diffusion, 508–509
 instrumentation, 497–502
 modifications of, 501–502
 representative data, 508–512
 sample preparation for, 505–508
 in study of translational diffusion, 498
Fluorescence recovery anisotropy, 488–489
 irreversible bleaching, 475, 489
 membrane and cellular systems studied by, 491–495
 reversible bleaching, 475, 488–489
Fluorescence redistribution after fusion, in study of translational diffusion, 499
Fluorescent indicators, 166
Fluorescent membrane probes, 470–471
N-(3-Fluoro-4,6-dinitrophenyl)cystamine
 cleavability, 590
 specificity, 590
 structure, 590
4-Fluoro-3-nitrophenylazide
 cleavability, 595
 specificity, 595
 structure, 595
Formycin triphosphate, 135
4-(6-Formyl-3-azidophenoxy)butyrimidate
 cleavability, 594
 specificity, 594
 structure, 594
Fraction, definition of, 3
Fractionation, 9–14
Free flow electrophoresis, 12
Freeze-slam method, 51–52
Freeze-thawing, 16
Frog skeletal muscle
 elemental concentrations, determined by EPMA and by other methods, 220, 223–224
 terminal cisternae
  calcium content of, 224
  elemental composition of, 221
Fumarase, in sucrose gradient centrifugation of protein–detergent complexes, sedimentation properties of, 702
Functional markers, 16
Fura-2, 167, 186, 233
Furosemide-sensitive NaCl/KCl cotransporter, 16

G

Galactosyltransferase, as membrane marker, 30
Gamma rays, ionizations caused by, 411–412
Gangliosides
  long-chain bases, analysis of, 558–560
  molecular species of, chromatographic separation of, 558
  native, reversed-phase HPLC of, 552–553
  per-O,N-benzoylated derivatives, HPLC of, 550–552
Gap junction membrane, in-plane order of protein components, 400–401
Gated ion channels, 399
Gated ion transport, 325
GDP. *See* Guanosine diphosphate
Gel chromatography, of vesicles, 365
Gibbs–Donnan equilibrium, 119
  potentials, as primary standard for optical potentiometric indicators, 119–120
  study with optical potentiometric indicators, 103
Glass tubings, 137–138
  shapes of, 137–138
γ-Globulins, in sucrose gradient centrifugation of protein–detergent complexes, sedimentation properties of, 702
D-Glucose, transport
  into brush border membrane vesicles isolated from renal cortex, 357–360
  membrane-specific system, 351–352
  sodium-dependent, in brush border vesicles, 356
Glucose carrier, target analysis of, 432
Glucose carrier–Na$^+$ symport, target analysis of, 432
Glucose-6-phosphate dehydrogenase, from *Leuconostoc mesenteroides*, internal standard for radiation inactivation studies, 424
β-Glucosidase, target analysis of, 435
β-Glucuronidase, in particle-free supernatant, 8
Glutamate dehydrogenase, yeast, calibration of low-angle laser light-scattering photometry plus HPGC system with, 523
γ-Glutamyltransferase
  as marker enzyme for apical (luminal) plasma membranes, 15
  as marker enzyme for brush border membrane, 347–348
  membrane domains of, photochemical labeling of, 675
  as membrane marker, 30, 31
  target analysis of, 435
Glutamyltransferase, cross-links introduced by, 584
Glutardialdehyde
  cleavability, 593
  specificity, 593
  structure, 593
Glycerophospholipids, ethanolamine, molecular species, separation by reversed-phase HPLC, 565
Glycine, as helix breaker or turn promoter, 451–453
Glycolipids, containing photoactivatable group, applications of, 684
Glycophorin
  covalent labeling of, 482
  studied by phosphorescence anisotropy, 492
  translational diffusion in membranes, representative FRAP results on, 511
Glycophorin A, 659–660
  labeling of, in erythrocyte membrane, 662–663
  photochemical labeling of, 672
  transmembrane domain of, photochemical labeling of, 682
Glycoproteins
  in isolating plasma membranes, 13
  target analysis of, 437
Glycosphingolipids
  derivatized, separation and quantitation by reversed-phase HPLC, 556–558
  HPLC, 538–539
  mobile phases for, 540
  hydroxy fatty acid, perbenzoylation, 542–544, 547
  isolation, absolute recoveries in, 549
  liquid chromatography-mass spectrometry, 565
  long-chain bases, analysis of, 558–560

neutral
  molecular species of, chromatographic separation of, 558
  separation and quantitation by reversed-phase HPLC, 556–558
nonhydroxy fatty acid, perbenzoylation, 542–544, 547
perbenzoylated derivatives
  chromatographic separation of, 547–549
  HPLC analysis, 542–549
  on Zipax column, 545–546
per-O- and per-O,N-benzoylated derivatives, HPLC, 547, 548
per-O-benzoylation, with benzoic anhydride, 547–549
separation and analysis of, methods for, 538
Glycosylceramides, per-O,N-benzoylation with benzoyl chloride, 544–547
Gramicidin A, cross-linking studies of, using photochemical reagents, 683
GTP. See Guanosine triphosphate
GTPβ,γNH
  caged, HPLC retention times, 295, 297–298
  HPLC retention times, 295, 297–298
GTPγS
  caged, 289, 295–297
  HPLC retention times, 295, 297–298
  O-caged, HPLC retention times, 295, 297–298
  S-caged, HPLC retention times, 295, 297–298
Guanosine diphosphate
  caged, HPLC retention times, 295, 297–298
  HPLC retention times, 295, 297–298
Guanosine triphosphate
  caged, HPLC retention times, 295, 297–298
  HPLC retention times, 295, 297–298

H

H^+. See Hydrogen ion transport
Halistaurin, rate of luminescent reaction, 198
Halogenated fluoresceins, photophysical properties, 476
H^+-ATPase
  exocytotic insertion of, into luminal membrane, 54
  target analysis of, 430, 431
H^+-ATP-synthase, from $E.\ coli$, hydrophobic labeling of, 658–659
H_2DIDS, 615
Hemoglobin, cross-linking studies of, 602, 613
Hepatocytes, optical measurement of membrane potential in, 111
Heterodyne spectroscopy, 374
High-density lipoproteins, cross-linking studies of, using photochemical reagents, 681–684
High-performance gel chromatography. See also Low-angle laser light-scattering photometry plus HPGC
  columns for, 521
High-performance liquid chromatography
  general techniques, 539–542
  of membrane lipids, 538–575
    dioxane/hexane system, 547–549
    2-propanol/hexane system, 547–549
    preinjector flow-through reference cell, 540–541
  system for, 541
H^+,K^+-ATPase, target analysis of, 430
H^+–K^+ pump, target analysis of, 429–438
HMG-CoA reductase, target analysis of, 434
Homodyne spectroscopy, 374
Homogenization, 4–8
  control of, 8
  definition of, 3
  medium, 7–8
    buffering capacity, 7
    pH of, 7
  starting material, selection of, 5
  temperature, 8
Homogenizer, Teflon–glass, 5–6
Hormone–receptor coupling, cross-linking techniques, 605–606
Human LA-DR-antigen polypeptides, photochemical labeling of, 674
HYBRIP.BA program, 268, 270–272
HYBRIS.BA program, 268
Hydrazides, 476
Hydrogen ion transport
  gastric, by histamine, 50

in urinary epithelia
 by CO_2, 50
 regulation of, by CO_2-induced exocytotic isertion of H^+-ATPases, 54–59
Hydropathy plot, 440–444
 predictive value of, on family of proteins with similar functions and extensive sequence homology, 441–442
Hydrophobic ions
 free-energy profile for, in hydrocarbon membrane, 332–333
 paramagnetic, used to estimate membrane potentials in vesicles, 336
 partitioning, 332–333
 spin-labeled, 336
Hydrophobicity, 11
Hydrophobic moment, 457–458
D-β-Hydroxybutyrate apodehydrogenase, photochemical labeling of, 680
D-β-Hydroxybutyrate dehydrogenase, target analysis of, 429–438
N-Hydroxysuccinimidyl-4-azidobenzoate
 cleavability, 596
 specificity, 596
 structure, 596
N-Hydroxysuccinimidyl-4-azidosalicylic acid
 cleavability, 598
 specificity, 598
 structure, 598
N-Hydroxysulfosuccinimide, 619
 sodium salt, synthesis of, 622
 structure of, 620
H-2K[k] antigen, studied by phosphorescence anisotropy, 492
H-2L[d] antigen, studied by phosphorescence anisotropy, 492

I

IgE-Fc receptor, studied by phosphorescence anisotropy, 492–494
IgE-receptor, on mast cells, cross-linking studies of, 621
2-Iminothiolane
 cleavability, 587
 specificity, 587
 structure, 587
Immunoelectron microscopy, 50

Immunoglobulin E. *See* IgE
Immunoglobulins, cross-linking studies of, 603
Inactivation curve, 413
 complex, 416–418
 multicomponent systems revealed by, kinetic analyses, 418–419
Indo-1, 233
Initial steady-state rate, determination of, 321–323
Intensity autocorrelation function, 374
 related to field autocorrelation function, 375–376
Intensity fluctuation spectroscopy, 367
 for determination of lipid vesicle size distributions in suspensions, 383–390
 of vesicles, 365
Intracellular ion activities, measurement, 145–149
Intrinsic proteins, hydrophobic segments of, photochemical labeling of, 674
Invertase, loss of activity, after exposure to high-energy electrons, 413–414
Iodoacetamides, 476
Ion activities, measurement of, micro- and macroelectrodes for, 136–155
Ion channels
 activity, assay, 156
 in cell membranes, 155–164
  density of, 156
 in membrane vesicles, 302
 target analysis of, 433
Ion fluxes
 channel-mediated, assay, 158
 measurement of, using rapid-reaction methods, 301–313
 rate coefficient, 302
 through vesicle channels, 302
  experimental procedures, 157–158
Ionic pumps, study with optical potentiometric indicators, 103
Ionomycin
 calibration of quin2, 246–247
 source, 246
Ionophoric channels, photochemical labeling of, 672
Ion-selective electrodes
 practical hints, 154
 troubleshooting, 154

Ion-selective membrane solutions, 141–142
Ion-selective microelectrodes, 136–137, 230–231
  fabrication of, 137
Isethionylacetimidate, 618
Isopycnic centrifugation, 11
Isothiocyanates, 476
Isotope fluxes
  channel-mediated, measurement, 156–164
  measurement of, using flow–quench, 303–308
  through cation-specific channels, measurement of, 156–157

K

K^+. *See* Potassium
Kedeem and Katchalsky equations, 280–281
Krafft point, 531
KS62 erythroleukemia cells, anion transport heterogeneity in, 134

L

lac permease, studied by phosphorescence anisotropy, 492
Lactate dehydrogenase
  in particle-free supernatant, 8
  pig heart, calibration of low-angle laser light-scattering photometry plus HPGC system with, 523
D-Lactate dehydrogenase, target analysis of, 431
Lactose carrier
  from *E. coli* membrane vesicles, target analysis of, 437–438
  target analysis of, 432
Large unilamellar vesicles, 365
LDL receptor protein, molecular weight of, target inactivation analysis, 415
Least-squares analysis, of noisy data, 217–218
Lecithin, analogs, containing phosphorylazido group at variable positions on *sn*-2 fatty acyl chain, 639
Lectins, in isolating plasma membranes, 13–14
Lethocerus. *See* Water bug

Lettré cells, 110
Light-beating spectroscopy, 374
Light energy ion gradient transduction, 400
Light scattering, 514. *See also* Low-angle laser light-scattering photometry plus HPGC
  in circular dichroism signal, 576
  for direct observation of micellar formation, 533
  geometry of, 367–368
  measurement of fluctuations of scattered light, 373–378
  as result of local fluctuations of dielectric constant of medium, 368–369
  techniques, for vesicles, 365
  theoretical analysis, 366
  theory, 367–371, 515–517
Light spectra, measurement, 374
Lipid bilayer
  as anisotropic system, order of, 462
  macromolecular diffusion in, 471–472
  preparation, for fluorescence recovery after photobleaching experiment, 505–506
  synthetic, delayed luminescence anisotropy, 482
Lipid bilayer vesicles, protein-containing, preparation of, 507
Lipid probes
  studied by fluorescence recovery anisotropy, 492
  translational diffusion, in different membranes, representative FRAP results for, 510
Lipid–protein interactions, time-resolved fluorescence anisotropy of, 462–471
Lipids
  effect of, on alternative protein assay methods, 395–398
  membrane, HPLC, 538–575
  protein assay in presence of, 393–399
Lipid vesicles, 529
  size distribution
    evaluated by electron microscopy, 384
    evaluated by intensity fluctuation spectroscopy, 384–390
    sensitivity of parameters, 390
Lipophilic cations, 95
  binding and distribution, 66
  determination of transmembrane electri-

cal potential difference in mitochondria using, 69–70
electron paramagnetic resonance (EPR) spectra of, estimation of membrane potential from, 66
Lipophilic ions
  distribution, 64–66
    determination of, methods, 67–71
  EPR spectra, in determination of transmembrane electrical potential difference, 71–72
  membrane diffusion of, 65
  synthetic, 66
Lipophilin, from human myelin, photochemical labeling of, 675
Lipoprotein, target analysis of, 436
Liposomes, optical measurement of membrane potential in, 111
Liquid chromatography–mass spectrometry, general techniques, 566–567
Liquid chromatography–mass spectroscopy
  Finnigan polyimide moving belt interface, 565
  of membrane lipids, 565–575
Liver, elemental concentrations in, determined by EPMA and by other methods, 220
Liver lysosomes, 111
Long-chain bases, liquid chromatography–mass spectrometry, 565, 567–570
Low-angle laser light-scattering photometer, 514–515
Low-angle laser light-scattering photometry plus HPGC
  calculation of (dn/dc_p) and M_p, 525–526
  calibration of system with standard proteins, 522–524
  columns for HPGC, 521
  determination of membrane protein molecular weight by, 514–528
  elution buffer, 521–522
  elution temperature, 522
  evaluation of detergent bound to protein from (dn/dc_p), 526–527
  extinction coefficient at 280 nm of solubilized protein, determination, 524, 525
  flow rate, 522
  instrumentation, 517–519

Lower consolute boundary, 531
Luminescence, photophysica cycle leading to, 473–474
Lymphocytes, 107, 110
  optical measurement of membrane potential in, 109
  study with optical potentiometric indicators, 103
Lymphocyte surface antigens, photochemical labeling of, 673
Lysis method
  for measuring light intensity from photoprotein-containing cells to absolute Ca^{2+} concentrations, 199–203
  for quin2 calibration, 245–246
Lysophosphatidylcholine
  bovine brain, HPLC, 562
  liquid chromatography–mass spectrometry, 571
Lysophosphatidylethanolamine
  HPLC, 562
  liquid chromatography–mass spectrometry, 571
Lysophosphatidylinositol, liquid chromatography–mass spectrometry, 570
Lysophosphatidylserine, HPLC, 562
Lysosomes, in living cells, pH measurement, with fluorescent dyes, 85–95

M

Macroelectrodes, 137
  diameters of, 150
  preparation of, 150–153
  use of, 153–155
Macromolecules, random-walk diffusion of, under influence of Brownian motion, 366, 371
Macrophages, study with optical potentiometric indicators, 103
Magnesium, cytosolic, localization in cells, 243–245
Malate dehydrogenase
  in sucrose gradient centrifugation of protein–detergent complexes, sedimentation properties of, 702
  target analysis of, 432
Maleimides, 476
  as cross-linking reagents, 612

4-Maleimidobenzyl-N-hydroxysuccinimide ester
  cleavability, 593
  specificity, 593
  structure, 593
N-(Maleimidomethyl)-2-(p-nitrophenoxy)-carboxamidoethane
  cleavability, 596
  specificity, 596
  structure, 596
Maltase
  as membrane marker, 30
  in outer cortical and outer medullary brush border vesicle preparations, 358
Mammary tumor cells, rat, 110
Mannitol, 8, 314
Marker antibodies, 14
Marker enzymes, 14–16
  in brush border membrane vesicles
    enrichment factors of, 350–351
    specific activities, 350–351
  characteristic of membrane structures, 348
  definition of, 4
  specific activity of, definition of, 4
Marker functions, 14
Marker proteins, for sucrose gradient centrifugation experiments, with membrane protein–detergent complexes, 702
Markers, 15–17, 30
  for identifying isolated membrane populations, 29, 31
Mast cells, study with optical potentiometric indicators, 103
M13 coat protein, photochemical labeling of, 676
MCV VERSUS TIME, 274–276, 279
Membrane-bound proteins, target analysis of, 434
Membrane dynamics, study, methodology, 344
Membrane fractions
  morphological examination of, 14–15
  purification, assessment of, 14–17
Membrane orientation procedures, 399–410
Membrane populations
  analysis, 29–32
  resolving, 29–32

Membrane potential. *See also* Transmembrane potential difference
  bacterial, study with optical potentiometric indicators, 103, 111
  determination of, optical probes for, 72–79
  EPR probes of, 331–345
  measurement of, 64–65
    optical methods for, 102
  negative, 95
    test ion for, 64
  positive, test ion for, 64
Membrane protein–detergent complexes
  diffusion coefficient for, experimental determination of, 698–699
  hydrodynamic studies of, 696–709
    theoretical considerations, 696–698
  molecular weight of protein portion, determination of, 697–698
  partial specific volume for, experimental determination of, 699–704
  sedimentation coefficient for, experimental determination of, 699–704
  sucrose gradient centrifugation experiments with, 699–707
    data analysis, 707–709
Membrane proteins. *See also* Transmembrane proteins
  asymmetry, cross-linking study of, 610–611
  correlation of changes in multimeric size with functional state, 437–439
  crystallization, phase transition phenomena, 535
  high-resolution structure, determination of, 609
  hydrophobic, folding pattern of, 440–453
  integral, motions of, model for, 482–484
  molecular weight, determination, by low-angle laser light-scattering photometry plus HPGC, 514–528
  photochemical labeling of, 672
  reconstituted, delayed luminescence anisotropy, 482
  secondary structure
    circular dichroism in determination of, 575–578
    determination of, 575
  solubilization, with nonionic surfactant, 519–520

solubilized, 529
  as complex of lipid, protein, surfactant, and carbohydrate, 514, 516
  molecular weight estimation for, 514–528
  state of aggregation, circular dichroism study, 578–584
  structure of, studies of, 440
  target analysis of, 429–438
Membrane resistance, study with optical potentiometric indicators, 103
Membranes
  apolar phase of, photochemical labeling, 628–687
  components, isolation techniques based on, 13–14
  constituents, protein–protein associations of, target inactivation analysis, 411–439
  fluorescence anisotropy data, analysis and interpretation of, 465–468
  hydrophobic labeling of, 628–687
    applications of, 629–630, 658–671
    disadvantages of ^{3}H or ^{14}C in, 651
    experimental aspects of, 653
    incorporation of reagent into membrane, 653–655
    indirect detection method in, 651
    label distribution analysis, 656–658
    photolysis apparatus for, 655
    reagents, 629–630
      adsorption by plastics, 653–654
      amphipathic, 676–685
      lipid-soluble, 672–675
      photoactivation of, 655–656
    study of dynamic phenomena with, 633
  in-plane order of protein components, 409–410
  integral proteins and their membrane-embedded domains, hydrophobic labeling of, 630–631
  isolation, 3–17
    steps, 3
  lipid core, labeling, requirements for reagents, 633–635
  lipid–lipid interactions, studying, with hydrophobic labeling, 632–633
  lipid–protein interactions, studying, with hydrophobic labeling, 632–633
  methods for ordering and orienting, 410
  model
    phase separation in, photochemical labeling of, 682
    photochemical labeling of, 672, 674
  order, analysis, 462–471
  orientation
    centrifugation/complete dehydration/rehydration method, 403
    centrifugation/partial dehydration method, 401–404
    isopotential spin-dry process, 403–406
    Langmuir–Blodgett film technique, 407–409
    magnetic, 406–408
    for time-averaged and time-resolved structure determinations, 399–410
  protein constituents, association of, 410–411
  protein–lipid interface, mapping, with hydrophobic labeling, 631–632
  solute flow across, 321
  subcellular
    differential sedimentation, 19–20
    homogenization of, 19–20
    isolation of, 19–32
  synthetic and native, methods for orienting, 401–409
  third-dimension separations, 23–24
  viscosity, analysis, 462–471
Membrane transport. *See also* Transport
  regulation of, 49–59
Membrane vesicles
  heterogeneity, 346–347
  ion channel-mediated fluxes in, 155–164
  isolated, tracer studies with, 313–331
Merocyanine 540
  complex with valinomycin in presence of potassium, 119
  structure, 105
Metallochromic dyes, 166–167
Methylamine, for transmembrane pH difference determination, 80
Methyl-4-azidobenzoimidate
  cleavability, 597
  specificity, 597
  structure, 597
N-Methylbis(2-chloroethylamine)
  cleavability, 593
  specificity, 593
  structure, 593

Methylene, 636
Methylenebismaleimide
  cleavability, 591
  specificity, 591
  structure, 591
Metrizamide, density gradient, 10–11
Mg^{2+}-ATPase, target analysis of, 430
Micellar aggregates, formation, direct observation of, 533
Micelles
  aggregation properties, 530
  formation of, 529
Microcomputer, modeling solute movement and volume changes with, 280–288
Microelectrodes, 136
  calibration, 145–146
  Ca^{2+}-selective, 166–167
  liquid membrane, 136
    coaxial double-barrelled, 139
    double-barrelled, 139
    ion-exchanger, as reference microelectrodes, 147–148
    ion-selective
      filling of, 141–142
      interference by physiologically relevant ions, 148
      as reference system, 147–148
    single-barrelled, 139
  measurement of transmembrane electrical potential difference, 64
  measuring arrangement, 145–149
  preparation of, 137–144
  reference, 136
    coaxial, 139
    evaluation, 146–147
    filling of, 141
    single-barrelled, 139
  tip diameters, 141
  use of, 145–149
Microinjection, 171–179
  pipette holder for, 172–174
  pressure system for, 172–175
Micropipettes
  bevelling, 140–141
  breaking, 140–141
  loading with photoprotein solution, 175–177
  parts of, nomenclature of, 139
  preparation of, 137–138

  pressure injection with, 172–175, 178–179
  pulling of, 138–140
  silanization of, 142–144
  tip resistance, 178
Microscopic lens systems, 182
Microsomal fraction, definition of, 3–4
Microsomes, definition of, 3–4
Minielectrodes
  calibration, 153
  electrode cell, 150–151
  electronics, 152
  evaluation, 153
  incubation vessel, 152–153
  membrane composition, 150–151
  preparation, 150–153
  setup, 152
Mitochondria
  binding constants for lipophilic cations in, 70
  calcium content, 225
  dye calibration methods with, using pH, 121
Mitochondrial ATPase complex, photochemical labeling of, 680
Mitochondrial membrane potential, 96–101
  study with optical potentiometric indicators, 103
Mitochondrial pellet, calculation of internal volume, 70–71
Mitochondrial respiration, inhibition, by diS-C_3(5), 107–108
Mixed micelles, formation, 529
Molecular weight determination. *See also* Membrane proteins, molecular weight
  from radiation inactivation of biological function, 411
Monoamine oxidase, target analysis of, 435
Monoclonal antibodies, 17
Monolayers, adsorbed onto glass or silicon wafers, 506
Monosialogangliosides
  individual, isolation, 552–553
  molecular species, separation, by reversed-phase HPLC, 553–556
Mouse ascites tumor cells, mitochondrial membrane potential, measurement, with carbocyanine dyes, 97–101
Mouse embryo L cells, 110

Mouse Lyt-2/3 antigen complex
  Bufo marinus, photochemical labeling of, 673
  hydrophobic labeling of, 658
  photochemical labeling of, 675
Multilamellar vesicles, 365
Murine erythroleukemia cells, optical measurement of membrane potential in, 109
Murine leukemia cells, optical measurement of membrane potential in, 109

N

Na^+. *See* Sodium
Na^+/Ca^+ exchange, in heart sarcolemmal vesicles, charge movements during, 155
Na^+/Ca^{2+} exchanger, reconstituted, purification of, 36
N-Acetyl-β-D-glucosaminidase, as membrane marker, 30
NADH dehydrogenase, target analysis of, 431
NADH oxidase, target analysis of, 431
NADPH-cytochrome-c reductase, as membrane marker, 30
Na^+/H^+ antiport, 32–33
Na^+/H^+ exchanger, 54
Na^+,K^+-ATPase
  Bufo marinus, photochemical labeling of, 673
  from canine kidney, photochemical labeling of, 674
  for characterization of membrane fractions, 15
  from *Electrophorus electricus*, photochemical labeling of, 676–677
  hydrophobic labeling of, 658
  as membrane marker, 30, 31, 347–352
  in outer cortical and outer medullary brush border vesicle preparations, 358
  from pig kidney, photochemical labeling of, 672, 675
  reconstitution of
    by chocolate dialysis, 35
    into lipid vesicles, 109
  separation of membranes containing, from bulk of basolateral membranes, 354–355
  solubilization, with nonionic surfactant, 520
  solubilized, characterization, by low-angle laser light-scattering photometry plus HPGC, 527
  studied by phosphorescence anisotropy, 492
  subunits, photochemical labeling of, 680
  target analysis of, 431
Na^+-K^+ pump, target inactivation analysis, 428
Na^+–P_i cotransport, separation of membranes containing, from bulk of basolateral membranes, 354–355
NBD-aminomethane sulfonate, for monitoring anion transport, 126
NBD-aminopropane sulfonate, for monitoring anion transport, 126
Nernst equation, 73, 95, 119
Neuraminidase, target analysis of, 436
Neuroblastoma cells, 111
Neutrophils, 110
  optical measurement of membrane potential in, 109
  study with optical potentiometric indicators, 103
Newcastle disease virus proteins, photochemical labeling of, 676
Nigericin, effect on acridine orange-loaded cells, 94–95
Nitrenes, 634, 637–641
  chemistry of, 635–637
  generated from acylazides, 638–639
  generated from arylazides, 637–638
  generated from other nitrenes, 638–639
  precursors, 637–641
  properties of, 635–637
  singlet and triplet states, 636–637
2-Nitroacetophenone, hydrazone of, characterization, 291–292
2-Nitro-4-azidophenyltaurine, as cross-linking reagent, 585
2-(4-Nitrophenyl)allyl-4-nitro-3-carboxyphenyl sulfide
  cleavability, 592
  specificity, 592
  structure, 592

1-(2-Nitrophenyl)diazoethane
  spectra, 292
  synthesis, 292
2-Nitrosoacetophenone, 290–291, 300
Nitroxide probes
  phase partitioning, 345
  used to quantitate transmembrane pH gradients, 342–343
Nuclear magnetic resonance, 166
  for direct observation of micellar formation, 533
  pH determination using, 84
  of vesicles, 365
Nucleoside transporter, target analysis of, 433
5′-Nucleotidase, target analysis of, 435
Null method, for translating light intensity from photoprotein-containing cells into absolute Ca^{2+} concentrations, 199
Null point method, of dye calibration, 96, 97, 101, 119, 120

O

Obelin, rate of luminescent reaction, 198
Octaethylene glycol n-dodecyl ether, solubilization of Na^+,K^+-ATPase with, 520
Octyl-α-glucoside, critical temperature of crystallization boundary, 531
Octylhydroxyethane sulfoxide, critical temperature of crystallization boundary, 531
Optical-mixing spectroscopy, 374
Optical potentiometric indicators
  abbreviations, 106
  calibration of fluorescence to millivolts, 119–122
  chemical names of, 106
  choice of, 104–112
  interpretation of results with, 122
  problems studied by, 102–104
  procedure for fast dye, 115–116
  procedure for slow dye, 113–115
  screening dyes, 116–119
  structures of, 105
Optical potentiometric signals
  fast, 107
  optimization of, 112–113
  slow, 107

Optical probes
  for determination of membrane potential, 72–79
  fluorescence, calibration
    by imposition of known diffusion potentials, 73–76
    against ion distribution method, 76
  to monitor internal pH, 82–83
  potential sensitive, 64
Organelles. See Cell organelles
Organic acids, for transmembrane pH difference determination, 80
Ovalbumin, in sucrose gradient centrifugation of protein–detergent complexes, sedimentation properties of, 702
Overshooting uptake, observations of, 317–319
Oxonol IV, 78
Oxonols, 106
  chemical nomenclature, 106
OX-V, 106, 110–111
  structure, 105
OX-VI, 106, 110–111
  structure, 105

P

1-Palmitoyl-2-[10[4-(trifluoromethyl)-diazirinyl]phenyl]-8-[9-^{3}H]oxadecanoyl]-sn-glycero-3-phosphorylcholine. See PTPC
Pancreatic islet cells, optical measurement of membrane potential in, 109
Parafollicular cells, fluorescence-activated cell sorter analysis of, 94
Paramagnetic labels, phase partitioning, quantitation of, 334–336
Paramagnetic probes
  dynamics, 344–345
  internal and external binding constants, measurement of, 338–339
  in measurement of ionic currents in vesicle systems, 341–342
  phase partitioning, 342–343
  quantitation of, 332–333
  time-dependent measurements of, 336
  potential or ΔpH-sensitive, binding constants of, 344–345
  sensitivity considerations, 343–344

transmembrane equilibrium, kinetics of, 339–340
Parinaric acids, 463–464
  phospholipid analogs, 463–464, 470
Perbenzoylation, 542–544
Percoll density gradient, 10–11
Permeable ions, equilibrium distribution of, 64–66
PERMEAR2V, 281–284
PERMEA/R2V AND PLOT, 281, 285–288
Perot–Fabry etalon, 374
pH
  control of, in calcium buffer solutions, 193–194
  cytosolic, measurement of, in living cells, 85–89
  dye calibration methods using, 120–121
  endosomal, measurement of, 89–91
  of external suspension, measurement of, 79
  internal, measurement of, 79
  of intracellular compartments, measurement
    with fluorescent dyes, 85–95
    with impermeant dyes, 89–91
  lysosomal, measurement of, 89–91
Phase partitioning, 12–13
  subfractionation by, 23–29
Phase systems, preparation of, 25–27
pH buffers, 193, 195, 262
N,N'-p-Phenylenedimaleimide
  cleavability, 591
  specificity, 591
  structure, 591
p-Phenylene-N,N'-dimaleimide, as cross-linking reagent, for actin, 612
Phenylmethylsulfonyl fluoride, 8
Phialidin, rate of luminescent reaction, 198
pH indicators, 80
  fluorescent dyes, 85–95
  internal, 82–83
    fluorescent, 83
  nuclear magnetic resonance, 84
Phosphatidylcholine
  containing photoactivatable group, applications of, 679–683
  liquid chromatography-mass spectrometry, 571, 573
  reversed-phase HPLC, 564–565

Phosphatidylethanolamine
  bovine brain, HPLC, 562
  HPLC, 562
  liquid chromatography-mass spectrometry, 571
  molecular species, separation by reversed-phase HPLC, 565
Phosphatidylglycerol, benzoyl derivatives, preparation of, 563
Phosphatidylinositol
  benzoyl derivatives, preparation of, 563
  containing photoactivatable group, applications of, 684
  liquid chromatography-mass spectrometry, 571, 573
  molecular species, separation by reversed-phase HPLC, 565
  pig liver, HPLC, 562
Phosphatidylserine
  bovine brain, HPLC, 562
  HPLC, 562
  liquid chromatography-mass spectrometry, 570, 571, 574
  molecular species, separation by reversed-phase HPLC, 565
Phosphodiesterase, target analysis of, 436
Phospholipids
  amino
    biphenylcarbonyl derivatives, preparation of, 563
    class separations, by HPLC, 562–563
  as amphiphilic, flexible molecules, 528–529
  analog, containing photoactivatable 3-(trifluoromethyl)-3-aryldiazirine, 651–652
  class separations, by normal-phase HPLC, 560–563
  containing photoactivatable group
    applications of, 679–684
    synthesis of, 632
  cross-linking studies of, using photochemical reagents, 679–684
  derivatives, class separations, by HPLC, 562
  HPLC, 538–539
    mobile phases for, 540
  liquid chromatography-mass spectrometry, 565, 570–575

molecular species, separation, by reversed-phase HPLC, 563
native, class separations, by HPLC, 560–562
separation and analysis of, methods for, 538
Phospholipid vesicles
ionic currents in, estimation of, 341–342
photochemical labeling of, 676
size, factors affecting, 385
Phosphonium ions
internal and external binding constants, measurement of, 338–339
positive, 336
spin-labeled, EPR spectra of, 334
transmembrane equilibrium, kinetics of, 339–340
Phosphorescence, 473–488
Phosphorylation, of transport protein, 49
Phosphorylnitrenes, 639
Photocathodes, 182–183
Photochemical labeling, of membrane apolar phase, 628–687
Photocount autocorrelation function, related to first-order field autocorrelation function, 377
Photocycle intermediate M, circular dichroism study of, 581–582
Photocytes, 165
Photolabeling reagents
amphipathic, applications of, 676–685
hydrophobic, applications of, 672–675
lipid-soluble, applications of, 672–675
Photometers, 167–168
Ca^{2+} concentration–effect curve and, 190–191
Photomultipliers, 182–183, 373–374, 376
output, processing, 185
Photon correlation spectroscopy, 367
advantages of, 330
data, of polydisperse samples, methods for evaluation of, 378–383
to determine size distribution and mean sizes of spherical vesicle, 330–331
Photon counting, 185, 376
Photopotentials, study with optical potentiometric indicators, 103
Photoproteins, 164–165
advantage of, 166–167
apparatus for recording light intensities from tissues injected with, 182–184

assay of, 167–168
Ca^{2+} contamination, control of, 168–169, 191
calcium-regulated, 165–203
Ca^{2+} concentration–effect curves, 187–197
calcium buffers, 192–197
EGTA-loading technique, 179–180
hypoosmotic shock technique for loading, 180
interpretation of light measurements, 197–199
introduction into cells, 171–181
isospecies, 166
light signals
detection of, 181–187
recorded from living cells, calibration of, 199–203
microinjection, 171–179
source of, 169–170
stock solutions of reagents, 191
Ca^{2+} sensitivity of, factors affecting, 198–199
desalting, 170–171
loading into cells, 171–181
after exposure to hypertonic solutions, 180
lyophilization, 170–171
solution, salt in, 178
storage, 170–171
Photoregenerated intermediates
for hydrophobic labeling of membranes, 634
precursors, properties of, 635
properties of, 635
Pig kidney microvillar membrane, photochemical labeling of, 672
Pig microvillus aminopeptidase, photochemical labeling of, 673
Plant cells, optical measurement of membrane potential in, 111
Plasmalogens, 562
Plasma membranes
apical/basolateral separation, 18–34
density, isolation techniques based on, 9–11
isolation of, 9
from polar cells and tissues, 18–34
shape, isolation techniques based on, 9–11
size, isolation techniques based on, 9–11

surface properties, isolation techniques based on, 11–13
vesicle preparations, osmotic behavior, 314
Plastocyanine, translational diffusion in membranes, representative FRAP results on, 511
Platelet membrane glycoproteins, cross-linking studies of, 621
Platelets, study with optical potentiometric indicators, 103
PMSF. *See* Phenylmethylsulfonyl fluoride
Poisson statistics, 216–217
Polarized fluorescence, 473
Polyethylene glycol, 13
Polypeptide integrity, target inactivation analysis, 425
Porin, molecular weight determination of constituent polypeptides, after solubilization by SDS, 520–521
Potassium channel
  Ba^{2+}-sensitive, in membranes from renal outer medulla, 162
  in basolateral membrane in epithelia, stopped-flow experiment with, 308–311
  Ca-dependent, target analysis of, 434
  identifying flux in, 310–311
Potassium-selective electrode, 150–151
Potentiometry
  with electrodes, 136
  with ion-selective microelectrodes, 136
Prokaryotic cells, proton electrochemical potential gradient in, 63
Proline, as helix breaker or turn promoter, 451–453
Promyelocytic leukemia cells, optical measurement of membrane potential in, 109
Protease inhibitors, 8
Protein
  as membrane marker, 30
  quarternary structure, protein concentration for cross-linking studies of, 612–614
  tertiary and quaternary structure of, 609–610
  cross-linking study of, 585
Protein assay
  BCA, 398
  biuret procedure, 399

effect of lipid, 395–397
general properties of, 395–397
Lowry method, modified, 393, 397–399
in presence of lipid, 393–399
  Amido Black 10B methods, 394–395
  experimental strategies, 393
  method, 394–395
  precautions, 397
Protein–nucleic acid cross-linking, 604–605
Prothrombin, membrane insertion of, photochemical labeling of, 673
Proton electrochemical potential gradient, measurement of, 63–84
Proton-translocating ATPase, of parafollicular granules, 94
C_{18}-Psychosine, as internal standard for analysis of long-chain bases of membrane lipids, 560
PTPC, tritiated, 648–651
  incorporation into membrane, 654–655
  structure of, 649
Purple membrane
  in-plane order of protein components, 400–401
  lattices, circular dichroism study of, 580, 583
Pyranine, 83

Q

Quene-2, 83
Quin2
  advantages and disadvantages, 230–233
  affinity for Ca^{2+}, 241–242
  bleaching, 232
  Ca^{2+} buffering, side effects of, 255–256
  Ca^{2+} dissociation constant
    and ionic strength, 242
    and monovalent cation concentration, 242
    and temperature, 242–243
  in cell monolayers, 240
  excitation and emission spectra, with varying Ca^{2+}, 234–235
  extracellular, 248–251
  fluorescence, 231
    recording, 238–240
  fluorescence signal
    calibration
      with ionophore and Mn^{2+}, 246–247
      by lysis, 245–246

procedure, 240–243
  by ratio of two excitation wavelengths, 246–247
  correction for autofluorescence, 247–248
  from extracellular dye, 248–251
  heavy metal chelation, 253
  intracellular dye loading, measurement of, 252–253
  intracellular quenching, by heavy metals, 251–252
  leakage, 248–249
  loading
    protocol for, 233–238
    side effects of, 253–256
  localization in cells, 243–245
  measurement of cytosolic free Ca^{2+} with, 230–262
  nondisruptive method of loading, 230
  and pH, 243
  quality of, tests for, 234
  sensitivity to Mg^{2+}, 232
  source, 234
  structure of, 231
Quin2/AM, 235–236
  concentration, and cell density, 236–237
  formaldehyde generation, effects on cell function, 253–255
  incubation time, 237–238
  loading
    albumin and serum during, 238
    temperature, 238
  loading efficiency, and cell density, 236–237
  source, 234
  stock solutions, in DMSO, 236
  structure of, 231
  transformed to quin2 tetranion, 237

R

Rabbit ileal brush border vesicles, coupled electroneutral NaCl absorption in, 362
Rabbit skeletal muscle, myofibrillar proteins, electrophoretic transfer of, for immunostaining, 691–693
Radiation, dosimeters, 422–423
Radiation action, physics of, 411–412
Rapid-reaction apparatus, 303
  for photolabeling, 311
  reliability of, 311–312
Rapid-reaction methods, in membrane vesicles, 302–303
Rat brain synaptosomes, 112
Rate zonal centrifugation, 11
Rat heart, optical measurement of membrane potential in, 111
Rat ileal brush border membranes, separate Na^+/H^+ and K^+/H^+ exchange in, 363
Rat liver mitochondria, optical measurement of membrane potential in, 111
Rat liver prostaglandin E binding protein–Triton X-100 complex
  partial specific volume for, 703–708
  sedimentation coefficient for, 703–708
Rat megakaryocytopoietic cells, permeability coefficient to water, in culture, 279–280
Rat papillary muscle, cryosection from, 204–205
Receptor/channel regulatory processes, 400
Receptor functions, 17
Red blood cells. *See* Erythrocytes
Redox enzymes, target analysis of, 431
Renal outer medulla, Ba^{2+}-inhibited K^+ channel, 164
Renal proximal tubules
  transcellular transport, 346–363
  vesicle heterogeneity, 357–360
Resting potential, study with optical potentiometric indicators, 103
Retinal rod outer disk membrane, photochemical labeling of, 673
Reversible ATP synthetase-proton pump, target analysis of, 431
RH-160, 106, 111
  structure, 105
RH-246, 106, 111
  structure, 105
RH-421, 106, 111
  structure, 105
Rhodamines, 111
  123, 79, 111
    structure, 105
  3B, 111
  cationic, 111
  6G, 111
    structure, 105

Rhodopseudomonas sphaeroides, reaction center complex, X-ray structure of, 439
Rhodopseudomonas viridis
  photoreaction center
    cross-linking studies of, 621
    predicted transmembrane spanning regions of, 451–452
    reaction center complex, X-ray structure of, 439
Rhodopsin
  bovine, translational diffusion in membranes, representative FRAP results on, 511
  photochemical labeling of, 673
  studied by delayed fluorescence anisotropy, 492
  studied by phosphorescence anisotropy, 492
  target analysis of, 436
Rhodospirillum rubrum, light-harvesting polypeptide B870-α, photochemical labeling of, 667–671, 675
Ricin toxin, insertion into membranes, photochemical labeling of, 676
Rotational diffusion, 472–495
  in membranes, measurement, by fluorescence and phosphorescence methods, 471–513
  range of, 462
  rate of, 462
  representative data, 490–495
  slow, measurement of, using fluorescent and phosphorescent methods, 475
  studies of
    selectivity, 473
    sensitivity, 473–474
    strategies, 473
    time resolution, 473
Rubidium uptake, Ba^{2+}-sensitive, 163–164

S

Safranine, 78
Safranine O, 111
  structure, 105
Salicylic acid, for transmembrane pH difference determination, 80
Sarcoplasmic reticulum
  ATPase, photochemical labeling of, 680
  membranes, orientation method for, 403, 404
rabbit
  Ca^{2+}-ATPase, cross-linking of, 606–608
  photochemical labeling of, 672, 680
rat, photochemical labeling of, 680
skeletal muscle, quantitative EPMA, 224
Scanning transmission electron microscopy, 226–229
Scattering form factor, 370
Secretory vesicles
  in living cells, pH measurement, with fluorescent dyes, 85–95
  pituitary, 111
Sedimentation equilibrium, 514
Selective ion electrodes, for determination of lipophilic ion distribution, 67
Self-beating spectroscopy, 374
Sendai virus glycoproteins, translational diffusion in membranes, representative FRAP results on, 512
Shot noise, 185–186
Signal averaging, 185–186
Silica
  pellicular, 539–540
  totally porous, 539
Single-cell suspensions, permeability to water, calculation of by computer, 274
SIPP, 383
Small intestines
  brush border, transport studies in, 346–363
  vesicle heterogeneity, 356
Small unilamellar lecithin vesicles, mass-weighted size distributions of
  average radius, 388–389
  effect of ionic strength, 389–390
Small unilamellar vesicles, 365
  translational motion, in suspension, 372
Sodium channel
  amiloride-blockable, 158–162
  amiloride-sensitive, flow–quench experiment with, 304–307
  target analysis of, 433
Sodium dodecyl sulfate
  critical temperature of crystallization boundary, 531
  for molecular weight determination of constituent polypeptides, 520–521

Sodium transport, in urinary bladder, by hydrostatic pressure, 50
Sodium uptake, amiloride-blocked, kinetic properties of, 161–162
Solute movement and cell volume changes, computer modeling of, 280–288
Spectrofluorometers, used with quin2, 239
Sphingomyelin
  benzoylated, molecular species, separation by HPLC, 564
  benzoylation, 564
  benzoyl derivatives, preparation of, 563
  bovine brain, HPLC, 562
  containing photoactivatable group, applications of, 683
  liquid chromatography-mass spectrometry, 571, 573–574
  native, molecular species, separation by HPLC, 563–564
Sphingomyelinase, target analysis of, 435
Squid giant axons, optical measurement of membrane potential in, 111
Static head experiment, 319–320
Steroid sulfatase, target analysis of, 436
Stilbene, as cross-linking reagent, 602
Stilbenedisulfonic acid derivatives, as cross-linking reagents, 585
Stimulus–response coupling, study with optical potentiometric indicators, 103
Stomatitis virus, membrane-associated proteins of, photochemical labeling of, 673
Stopped-flow methodology, 303
  definition of, 308
  typical experiment, 308–310
Styryls
  chemical nomenclature, 106
  structure, 105
Subcellular fractionation, 51
Succinate dehydrogenase
  as membrane marker, 30
  in particle-free supernatant, 8
Succinimidyl active esters, 619
N-Succinimidyl-6-(4'-azido-2'-nitrophenylamino)hexanoate
  cleavability, 597
  specificity, 597
  structure, 597
N-Succinimidyl-(4-azidophenyldithio)-propionate

  cleavability, 598
  specificity, 598
  structure, 598
Succinimidyl-4-(p-maleimidophenyl)-butyrate
  cleavability, 590
  specificity, 590
  structure, 590
N-Succinimidyl-3-(2-pyridyldithio)propionate
  cleavability, 590
  specificity, 590
  structure, 590
Sucrase, as membrane marker, 30
Sucrose, 314
  density gradient, 10–11
  homogenization medium, 7–8
  osmolarity, 10
Sulfonic acids, as cross-linking reagents, 585
Sulfosuccinimidyl active esters, 619–620
Sulfosuccinimidyl-2(m-azido-o-nitrobenzamido)ethyl-1,3'-dithiopropionate
  cleavability, 599
  specificity, 599
  structure, 599
Sulfosuccinimidyl-(4-azidophenyldithio)-propionate
  cleavability, 599
  specificity, 599
  structure, 599
Sulfosuccinimidyl-2-(p-azidosalicylamido)-ethyl-1,3'-dithiopropionate
  cleavability, 598
  specificity, 598
  structure, 598
Surface charge, 11
Surface tension
  calculation of, 536–537
  measurement of, 533–534
Surfactants, 533. See also Detergents
Swinging bucket rotors, 22
Synaptosomes, study with optical potentiometric indicators, 104
Synthetic vesicles, studied by phosphorescence anisotropy, 492
System A, 34
System ASC, 33, 34
System L, 33

T

Target inactivation analysis, 411–439
 analysis of samples after irradiation, 424–425
 basis for, 412
 of complexes, 428
 data
  analysis of, 426–427
  in literature, tabulation of, 429–438
 domains in, 428–429
 dose delivery, 422–423
 dose measurement, 422–423
 functional versus structural analysis of size, 426
 general principles, 425–429
 internal standards, 424
 mathematics of, 412–416
 methodology, 419–425
 nonexponential loss of activity in, 416
 permutation of monomeric versus oligomeric size by, 426–428
 physical conditions for irradiation, 422
 sample preparations, 419–422
 temperature control, 423–424
 temperature corrections, 423–424
Tartaric acid derivatives, as cross-linking reagents, 602
Tartryldi(ε-aminocaproylazide)
 cleavability, 586
 specificity, 586
 structure, 586
Tartryldiazide
 cleavability, 586
 specificity, 586
 structure, 586
Tartryldi(glycylazide)
 cleavability, 586
 specificity, 586
 structure, 586
TEMPO, phase partitioning of, 345
Tetanus toxin, photochemical labeling of, 680, 683
Tetracarboxylate Ca^{2+} indicators, binding of Zn^{2+} inside tumor cells, 252
Tetrahymena, 111–112
Tetraiodofluorescein. See Erythrosin
N,N,N',N'-Tetrakis(2-pyridylmethyl)ethylenediamine. See TPEN

Tetraphenylboron
 anion, dye calibration using, 121
 complex with valinomycin in presence of potassium, 118
Tetraphenylphosphonium
 binding in rat liver mitochondria, 70
 for measurement of transmembrane electrical potential difference, 66
 tritiated, dye calibration using, 121
Thin-layer apparatus, 25–27
 loading, 27–28
Thiosemicarbazides, 476
Third-dimension separations, 23–24
Three-compartment kinetics
 optimizing solutions for, 263–274
 parallel model, 263, 268–269
 series model, 263–267
Three-compartment system, kinetics of isotope distribution among components of, 263
Thyroid membranes, complex radiation inactivation curve of, 416–418
TID, [^{125}I]-, 648–649
 incorporation into erythrocyte membranes, time course of, 656
 labeling of bacteriorhodopsin, 657–658
 structure of, 649
Time-resolved fluorescence anisotropy, measurement of, 463–465
Time-resolved fluorescence depolarization, 462–471
 advantages and disadvantages of, 462
 detection system, sensitivity of, 465
 G-factor, 465
 measurements, with single-photon counting apparatus, 464
 physical principles of, 462–463
Tissue homogenization, 19–20
Toad urinary bladder
 membrane vesicles
  amiloride-blocked Na^+ uptake into, 159–161
  preparation, 157–158
 sodium channel, flow-quench experiment with, 304–308
Toxoplasma, 112
TPEN, testing for dye quenching by endogenous heavy metals with, 252
TPFN, 253
$TPMP^+$. See Triphenylmethylphosphonium

TPP+. *See* Tetraphenylphosphonium
Tracer exchange, at equilibrium, measurements of, 323–325
Tracer studies, with isolated membrane vesicles, 313–331
Transforming growth factor beta, receptor, cross-linking studies of, 621
Transient fluorescence anisotropy, 475, 489
Translational diffusion, 495–513
  comparisons of experimental results with theoretical expectations, 509–513
  long-range, optical methods for study of, 498–499
  in membranes, measurement, by fluorescence and phosphorescence methods, 471–513
  representative FRAP results on, 511–512
  short-range, optical methods for study of, 498
  studies
    methods for, 495–499
    optical methods for, 496–499
Transmembrane channel proteins, 453–461
Transmembrane pH difference, 63
  measurement of, 79–84
Transmembrane pH gradients, quantitation of, 342–343
Transmembrane potential difference, 63, 332–333
  estimation of, 64
  measurement of, 70–71
    from binding of charged lipophilic EPR probes, 71
    centrifugation method with mitochondria, with lipophilic cations, 69, 70
    by flow dialysis, 67–69
    micro- and macroelectrodes for, 136–155
    in mitochondria, by lipophilic cation distribution, 69–70
    with selective electrodes, 67
    vesicle separation technique, 69
  quantitation of, 336–342
Transmembrane proteins. *See also* Membrane proteins
  turn regions, 451–453
Transport
  heterogeneity of, quantitation of, 325–327

  kinetics, analyzing, by computer, 262–288
  mechanisms, fluorescent monitoring methods for, 122–135
  in nonhomogeneous vesicle populations, 325–330
  target inactivation analysis, 425
  of water, in response to vasopressin, in epithelia, 50–52
Transport activity, on surface area basis, 329–330
Transporters
  change in number of, 50–51
  dynamic behavior of, 321
  target analysis of, 430–431
Transport experiments
  design, 314–330
  initial rate kinetics, 321–323
  interpretation, 314–330
  kinetic approaches, 320–325
  with radiolabeled isotopes, 313–331
  thermodynamics approach, 317–320
  uptake versus transport in, 315–317
Transport functions, as markers, 16–17
Transport protein
  endocytic retrieval, 51
  exocytic insertion of, 51
Transport-specific fractionation
  analytical-scale, 37–40
  preparative-scale, 48–49
  for purification of ATP-dependent Ca^{2+} pumps, 34–49
Transport systems, 64–65
  coexistence in same membrane, experimental evidence for, 360–363
  isolation, by use of specific permeability properties, 36–38
  reconstitution of, into vesicles, before purification, 35–36
3-(Trifluoromethyl)-3-aryl-3H-diazirines
  chemical properties of, 645–646
  in organic solution, distribution of products of photolysis of, 646–647
  photochemical reactions in lipid bilayer, 646
  synthesis of, 644–645
3-(Trifluoromethyl)-3-m-[^{125}I]iodophenyl-3H-diazirine. *See* TID, [^{125}I]-
3-Trifluoromethyl-3-phenyl-3H-diazirine group, 629

Trinitrophenol label, negative, 336
Triphenylmethylphosphonium
  binding in rat liver mitochondria, 70
  for measurement of transmembrane electrical potential difference, 66
  tritiated, dye calibration using, 121
Triton X-100
  lysing cells with, in estimating fractional luminescence, in cells loaded with aequorin, 202–203
  solubilization of membrane proteins, 696
Trophoblast-derived growth factor, receptor, in placental membranes, cross-linking studies of, 621
Turn regions
  determination of, 451–453
  with different turn propensities, sequence criteria for, 451–452
Type II pneumocytes, rabbit, optical measurement of membrane potential in, 109

U

Ubiquinone-cytochrome-c reductase, beef heart, photochemical labeling of, 680
UDP-glucuronyltransferase, target analysis of, 435
Ultrarapid freezing, 205–207
  environmental chamber and freezing apparatus, 206
  handling of specimen, 207
Unilamellar vesicles, 365
Uptake data, time dependencies, 321
Uranyl glass GG17, fluorescence decay of, 490–491
UV detectors, 540–542

V

Valinomycin, 157
  complexes with dyes, in presence of potassium, 118–119
  effect on acridine orange fluorescence, 94
  and electrogenic transport of K^+ or Rb^+, 65–66

Vasopressin, 50, 51
Vesicle population, mean transport properties, 327–330
Vesicles
  bacterial membrane, optical measurement of membrane potential in, 111
  continuous monitoring of transport by fluorescence assay, 129–130
  heterogeneity of, 324
    related to tissue heterogeneity, 355–360
  membrane potential, optical measurement of, 102–122
  polarizability of, 368–371
  proton electrochemical potential gradient in, 63
  in suspension, size distribution, 373
Vesicle size
  distribution, 365
    from photon correlation spectroscopy, 330
    in suspension, determination of, by intensity fluctuation spectroscopy, 383–390
  heterogeneity, effects on potential dependent phase partitioning of paramagnetic probes, 340–341
  in transport experiments, 314–315
Vesicular stomatitis virus, photochemical labeling of, 678
Vicinal diols, as cross-linking reagents, 604
Vinculin, interaction of, with membranes, photochemical labeling of, 683

W

Water, transport, in response to vasopressin, in epithelia, 50–52
Water bug flight muscle, electrophoretic transfer of, for immunostaining, 691, 692
Weak acids and bases, distribution of, in measurement of transmembrane pH difference, 79–82
Wiener–Khintchine theorem, 371
Wigner's spin rule, 636
WW781, 106
  complex with valinomycin in presence of potassium, 119

fluorescence
  correlated with Ca^{2+}-activated K^+ conductance, 115–116
  with human red cells hyperpolarized by addition of valinomycin or A23187, 115–116
  fluorescence excitation and emission spectra, before and after addition of valinomycin, 117
  optimization of dye, cell, and VAL concentrations, 118–119
  pharmacological effects, 109–110
  structure, 105

X

X-ray detectors
  calibration, 211–212
  energy-dispersive, 211
X-ray microanalysis
  of Ca^{2+}, Mg^{2+}, and other ions in rapidly frozen cells, 203–229
  rapid freezing for, 205–207
  specimen preparation for, 205–209

Y

Yeast
  adenylate kinase, calibration of low-angle laser light-scattering photometry plus HPGC system with, 523
  cytochrome b
    cross-correlation between cytochrome b_6 and, 442, 445–446
    hydropathy function, cross-correlation between cytochrome b_6 and, 442
  cytochrome oxidase II, hydropathy function of, cross-correlation between cytochrome b_6 and, 442
  enolase, calibration of low-angle laser light-scattering photometry plus HPGC system with, 523
  glutamate dehydrogenase, calibration of low-angle laser light-scattering photometry plus HPGC system with, 523

Z

Zimm plot, 514
Zonal rotors, 20–22

**NO LONGER THE PROPERTY
OF THE
UNIVERSITY OF R.I. LIBRARY**